UNIFIED STRING THEORIES

Workshop on

UNIFIED STRING THEORIES

edited by M Green and D Gross

29 July — 16 August 1985

INSTITUTE FOR THEORETICAL PHYSICS

University of California

Santa Barbara

World Scientific

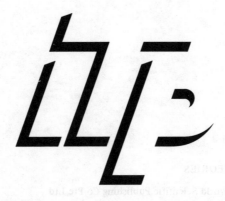

Published by

World Scientific Publishing Co. Pte. Ltd.
P. O. Box 128, Farrer Road, Singapore 9128

Library of Congress Cataloging-in-Publication data is available.

UNIFIED STRING THEORIES

ISBN 9971-50-031-0
9971-50-032-9 pbk

Printed in Singapore by Kyodo-Shing Loong Printing Industries Pte Ltd.

INTRODUCTION

The idea that string theories may provide the framework for a unified description of quantum gravity and the other forces of nature has been the focus of intensive research over the last year. This workshop was designed to provide both a series of lectures of a somewhat pedagogical nature and a series of seminars by leading participants in the field.

The lectures were given by ten speakers, with two or three lectures by each speaker on topics that ranged over most aspects of string theory. C. Thorn discussed the light cone quantization of the bosonic string, a lattice regularization and the proof of the no-ghost theorem; S. Mandelstam developed the functional integral approach, in light cone gauge, to the bosonic and supersymmetric interacting strings; O. Alvarez reviewed the differential geometric approach to string quantization with emphasis on the mathematical issues; the covariant conformal field theory approach to string theory was developed by S. Shenkar, for the bosonic string and by D. Friedan, for the fermionic string; D. Olive surveyed the applications of affine Lie-algebras to string theory; L. Brink discussed the light cone quantization of superstrings; M. Green spoke on anomalies and the light cone gauge string field theory; D. Gross reviewed the structure of the heterotic string; and E. Witten discussed a variety of topological tools useful in analyzing string theories.

The intensive series of seminars was designed to cover most aspects of recent progress in understanding the structure of string theories and their application to nature. We have, somewhat arbitrarily, divided these into three chapters, although in several cases topics covered in different chapters overlap.

As these proceedings illustrate, the workshop consisted of three weeks crammed with lectures and buzzing with excitement over recent developments. The field is rapidly changing and many advances have occurred since the summer. These new ideas have enriched the subject and undoubtedly contribute to future workshops and conferences in this new area of theoretical physics.

Finally, we wish to thank J. R. Schrieffer, R. Sugar, I. Muzinich, and the staff of the Institute for Theoretical Physics at Santa Barbara for their immeasurable aid in the planning and running of the workshop.

Michael B. Green and David J. Gross
November 1985

CONTENTS

OPENING QUESTIONS

David Gross

It is quite appropriate to open this workshop on string theory with a list of questions. In the flurry of excitement generated by recent developments some people have wondered whether we might be witnessing the beginning of the end of physics. For the first time we have a theory which appears, in principle, to be capable of answering all the traditional questions of elementary particle physics. Preliminary investigation of the phenomenology of the $E_8 \times E_8$ heterotic string is quite encouraging. However the game is far from being over, quite the opposite is the case. There are there many unsolved problems and deep mysteries that need to be understood before one can claim success. In addition we have only begun to explore the structure of these new theories. It is not at all clear where these explorations will lead us. If history is a reliable guide to the future then, as our understanding of the theory improves, new domains of physics and new questions will appear. I therefore present below, in the belief that questions are often more important than answers, a list of open questions. Most of these are well known to any worker in this field, are serving as a guide to current research and are addressed in the contributions to this workshop.

1. What is String Theory?

This is a strange question since we clearly know what string theory is to the extent that we can construct the theory and calculate some of its properties. However our construction of the theory has proceeded in an *ad hoc* fashion, often producing, for apparently mysterious reasons, structures that appear miraculous. It is evident that we are far from fully understanding the deep symmetries and physical principles that must underlie these theories. It is hoped that the recent efforts to construct covariant second quantized string field theories will shed light on this crucial question.

2. How Many String Theories are There?

Do there exist more consistent string theories than the known five — the two forms of the closed superstring, the SO(32) open superstring and the two forms of the heterotic string? Do there exist fewer, in the sense that some of the above might be different manifestations (different vacua?) of the same theory? Are some of the known theories actually inconsistent?

3. String Technology

This is not a question but a program of development. Much work remains to be done in developing the calculational techniques of string theory, including control of multiloop perturbation theory and the construction of manifestly supersymmetric and covariant methods of calculation.

4. What is the Nature of String Perturbation Theory?

Our present understanding of string theory has been restricted to perturbative treatments. Does this perturbation theory converge? Most likely it does not. In that case when does it give a reliable asympototic expansion of physical quantities? How can one go beyond perturbation theory and what is the nature of nonperturbative string dynamics? This question is particularly difficult since we currently lack a useful non-perturbative formulation of the theory.

5. String Phenomenology

Here there are many questions that can all be summarized by asking whether one can construct a totally realistic four-dimensional model which is consistent with string theory and agrees with observation?

6. What Picks the Correct Vacuum?

This is one of the great mysteries of the theory which appears, at least when treated perturbatively, to possess an enormous number of acceptable (stable) vacuum states. Why, for example, don't we live in ten dimensions? Does the theory possess a unique vacuum, in which case all dimensionless physical parameters would be calculable or is the vacuum truly degenerate, in which case we would have free parameters? How does the value of the dilaton field get fixed, thereby giving the dilaton a mass? Does the vanishing of the cosmological constant survive the mechanism that lifts the vacuum degeneracy?

7. What is the Nature of High Energy Physics?

By this I mean what does physics look like at energies well above the Planck mass scale? This is a question that is addressable, in principle, for the first time, and might be of more than academic interest for cosmology. In analogy with past theories one might expect physics in this domain to look entirely different. Does the string undergo a phase transition at high temperatures or densities to a new phase, as is perhaps indicated by the existence of a limiting temperature? Can one avoid in string theory the ubiquitous singularities that plague general relativity?

8. Is There a Measurable, Qualitatively Distinctive Prediction of String Theory?

String theories can, in principle, make many "postdictions" (such as the calculation of the mass ratios of quarks and leptons, Higgs masses and couplings, gauge couplings, etc.). They can also make many new predictions (such as the masses of the supersymmetric partners of the observed particles, new gauge interactions, etc.). These would be sufficient to establish the validity of the theory, however in each case one can imagine (although with some difficulty) conventional field theories coming up with similar pre or post dictions. It would be nice to predict a phenomenon, which would be accessible at observable energies and is uniquely characteristic of string theory.

LECTURES

INTRODUCTION TO THE THEORY OF RELATIVISTIC STRINGS*

Charles B. Thorn
Department of Physics, University of Florida
Gainesville, Florida 32611

1. THE CLASSICAL RELATIVISTIC STRING

In analogy to the well-known fact that the dynamics of a free relativistic particle may be derived from the action principle

$$0 = m\delta \int_{\text{world line}} ds = -m\delta \int_0^T d\tau \left\{ - \frac{dx}{d\tau} \cdot \frac{dx}{d\tau} \right\}^{1/2}, \tag{1.1}$$

Nambu[1] proposed that the action for a relativistic string be

$$A = -T_0 \int_{\text{World Sheet}} d^2S = -T_0 \int_0^T d\tau \int_{\sigma_1(\tau)}^{\sigma_2(\tau)} d\sigma \{ (\dot{x}\cdot x')^2 - \dot{x}^2 x'^2 \}^{1/2} \tag{1.2}$$

$$\equiv \int d\tau d\sigma \mathcal{L}(\dot{x}, x').$$

In (1.2) we have made use of the following notation

$x^\mu(\sigma,\tau)$: World Sheet of String, $\mu = 0,1,\ldots,d+1$

$$\dot{x}^\mu \equiv \frac{\partial x^\mu}{\partial \tau}$$

Supported in part by the U.S. Department of Energy under contract No. DE-AS-05-81-ER40008.

$$x^{\mu\prime} = \frac{\partial x^\mu}{\partial\sigma}$$

$$V\cdot W \equiv \underset{\sim}{V}\cdot\underset{\sim}{W} - V^o W^o$$

(σ,τ): arbitrary parametrization of world sheet.

Velocity of light = 1.

1.1 Simple Motions and Physical Interpretation

To interpret T_o, take a static string tied down at $\vec{x} = \vec{0}$ and $\vec{x} = (L,0,0,\ldots)$. Then

$$x^\mu = (t, x^1, 0, 0, ..) \qquad 0 \le x^1 \le L$$

and we may take $\tau = t$, $\sigma = x^1$, $\dot{x} = (1,\vec{0})$, $x' = (0,1,0,\ldots)$,

$$A = -T_o\int_0^T dt\int_0^L dx^1 = -T_o TL = -ET,$$

so T_o is the rest tension in the string.

If one chooses $\tau = t$ in general, (1.2) may be written[2]

$$A = -T_o\int_0^T dt\int_{\sigma_1(t)}^{\sigma_2(t)} d\sigma\,|\vec{x}'|\{1 - \vec{v}_\perp^{\,2}\}^{1/2} \qquad (1.3)$$

where

$$\vec{v}_\perp = \dot{\vec{x}} - \frac{\dot{\vec{x}}\cdot\vec{x}'}{|\vec{x}'|^2}\,\vec{x}'$$

is the velocity perpendicular to the string: there is no kinetic energy for longitudinal vibration. For a string with free ends, the boundary conditions derived from Eq. (1.3) are

$$\vec{v}_{\perp i}^{\,2} = 1$$

$$\dot{\sigma}_i |\vec{x}'| = -\dot{\vec{x}} \cdot \frac{\vec{x}'}{|\vec{x}'|} + O(1-\vec{v}_\perp^{\,2})^{1/2} \quad (\text{i.e. } \vec{v}_i = \frac{d}{dt}\vec{x}(\sigma_i(t),t) = \vec{v}_{\perp i}) \tag{1.4}$$

so a free end moves perpendicular to the string at the speed of light. The equations of motion are (choosing σ so that $\dot{\vec{x}}\cdot\vec{x}' = 0$):

$$\frac{\partial}{\partial t}\vec{\mathcal{P}} = \frac{\partial}{\partial \sigma}\left[T_o\{1-\vec{v}_\perp^{\,2}\}^{1/2}\frac{\partial \vec{x}}{\partial s}\right] \tag{1.5a}$$

$$\vec{\mathcal{P}} = T_o|\vec{x}'|\frac{\vec{v}_\perp}{\{1-\vec{v}_\perp^{\,2}\}^{1/2}} . \tag{1.5b}$$

Thus the effective tension is $T_o\ \{1-\vec{v}_\perp^{\,2}\}^{1/2}$ and the mass per unit length is $T_o|\vec{x}'|\ \{1-\vec{v}_\perp^{\,2}\}^{-1/2}$. The open string boundary conditions assure that $T_{eff} = 0$ at a free end.

A simple motion is a pinwheel rotation of a rigid string:

$$\vec{x} = \sigma\hat{u} \qquad -a \leq \sigma \leq a$$

$$\dot{\hat{u}} = \vec{\omega} \times \hat{u} \qquad \hat{u}\cdot\vec{\omega} = 0$$

where the boundary condition implies $a = 1/\omega$. Energy and angular momentum are

$$E = T_o\int_{-a}^{a} d\sigma \frac{1}{\{1-\omega^2\sigma^2\}^{1/2}} = \pi T_o a = \pi T_o/\omega$$

$$\vec{J} = T_o\int_{-a}^{a} d\sigma \frac{\sigma^2\hat{u} \times (\vec{\omega}\times\hat{u})}{\{1-\omega^2\sigma^2\}^{1/2}} = \frac{\pi T_o\vec{\omega}}{2\omega^3}$$

$$= \hat{\omega}\,(2\pi T_o)^{-1}\,E^2 \qquad \Rightarrow \alpha' = (2\pi T_o)^{-1}.$$

The relativistic string is a model with linear Regge trajectories with slope α'.

1.2 Canonical Dynamics

We return to our original action principle (Eq. 1.2), and work out the momentum conjugate to x^{μ}:

$$\mathcal{P}^{\mu} = \frac{\partial \mathcal{L}}{\partial \dot{x}_{\mu}} = T_o \frac{\dot{x}^{\mu} x'^2 - x'^{\mu}(\dot{x}\cdot x')}{\{(\dot{x}\cdot x')^2 - \dot{x}^2 x'^2\}^{1/2}} \tag{1.6}$$

and notice the primary first class constraints

$$x' \cdot \mathcal{P} = 0 \tag{1.7a}$$

$$\mathcal{P}^2 + T_o^{\,2} x'^2 = 0 \tag{1.7b}$$

as well as the fact that the canonical hamiltonian vanishes. Following Dirac, we can set up a canonical system by taking as hamiltonian

$$H = \int_{\sigma_1}^{\sigma_2} d\sigma \{\frac{\lambda(\sigma,\tau)}{2} (\mathcal{P}^2 + T_o^{\,2} x'^2) + \mu(\sigma,\tau)(x' \cdot \mathcal{P})\} \tag{1.8a}$$

and imposing the constraints <u>after</u> deriving the canonical equations of motion. Note that the phase space action

$$A = \int d\tau \int_{\sigma_1}^{\sigma_2} d\sigma \{\dot{x}\cdot\mathcal{P} - H\} \tag{1.8b}$$

takes Polyakov's form [3] if $\mathcal{P}$ is eliminated:

$$\mathcal{P} = \frac{1}{\lambda} \{\dot{x} - \mu x'\}$$

$$A = \int d\tau \int_{\sigma_1}^{\sigma_2} d\sigma \{\frac{1}{2\lambda} \dot{x}^2 + (\frac{\mu^2}{2\lambda} - \frac{\lambda}{2} T_o^{\,2}) x'^2 + \frac{\mu}{\lambda} \dot{x}\cdot x'\}$$

$$= \frac{T_o}{2} \int d\tau \int_{\sigma_1}^{\sigma_2} d\sigma \sqrt{-g}\;\; g^{ab} \partial_a x \cdot \partial_b x$$

with

$$\sqrt{-g}\ g^{00} = \frac{1}{\lambda T_o}$$

$$\sqrt{-g}\ g^{11} = \left(\frac{\mu^2}{\lambda} - \lambda T_o{}^2\right) \frac{1}{T_o}$$

$$\sqrt{-g}\ g^{01} = \frac{\mu}{\lambda} \frac{1}{T_o} \ .$$

One form of Dirac's procedure, widely used in gauge field theories, is to convert the first class constraints (1.7) to second class ones by specifying a definite parametrization (analogous to a gauge condition), and then to eliminate some of the dynamical variables.

If we want to keep things manifestly covariant, we would avoid eliminating any of the x's or $\mathcal{P}$'s, and might choose, for example,[3,4]

$$\lambda = 1/T_o, \qquad \mu=0 \qquad [g^{00} = -g^{11},\ g^{01} = 0] \tag{1.9}$$

whereupon H becomes

$$H = \frac{1}{2T_o}\int_{\sigma_1}^{\sigma_2} d\sigma\ (\mathcal{P}^2 + T_o{}^2\ x'^2) \tag{1.10}$$

and the constraints (1.7) may be imposed on physical states. (Suitably interpreted they become the Virasoro conditions.)

Alternatively we may sacrifice manifest Lorentz covariance and eliminate some components of x and $\mathcal{P}$. The advantage is that with this choice one can retain a positive definite Hilbert space.

A particularly nice non-covariant choice is the light-cone parametrization of GGRT.[2] For a (d+2) vector V^μ, define

$$V^{\pm} \equiv (x^0 \pm x^3)/\sqrt{2}$$

$$(\underset{\sim}{V})^a \equiv \begin{cases} V^a, & a = 1,2 \\ V^{a+1}, & a = 3,4\ldots,d. \end{cases}$$

Then we choose:

$$x^+(\sigma,\tau) = \tau \tag{1.11a}$$

$$\mathcal{P}^+(\sigma,\tau) = T_0, \qquad \sigma_1(\tau) = 0. \tag{1.11b}$$

Putting (1.11) into (1.8) yields

$$A = \int d\tau \int_0^{\sigma_2} d\sigma \, \{\dot{\underset{\sim}{x}}\cdot\underset{\sim}{\mathcal{P}} - \mathcal{P}^- - T_0\dot{x}^- - \frac{\lambda}{2}(\underset{\sim}{\mathcal{P}}^2 + T_0{}^2 \, \underset{\sim}{x}'^2 - 2T_0\mathcal{P}^-)$$
$$- \mu\,(\underset{\sim}{x}'\cdot\underset{\sim}{\mathcal{P}} - T_0 x^{-\prime})\} \tag{1.12}$$

The equations arising from varying $\mathcal{P}^-$ and λ in (1.12) are the constraints

$$\mathcal{P}^- = \frac{1}{2}\Big(\frac{\underset{\sim}{\mathcal{P}}^2}{T_0} + T_0\,\underset{\sim}{x}'^2\Big) \tag{1.13a}$$

$$\lambda = \frac{1}{T_0} \tag{1.13b}$$

which can be substituted directly in Eq. (1.12) (since they do not involve derivatives):

$$A = \int d\tau \int_0^{\sigma_2} d\sigma \{\dot{\underset{\sim}{x}}\cdot\underset{\sim}{\mathcal{P}} - T_0\dot{x}^- - \frac{1}{2}\Big(\frac{\underset{\sim}{\mathcal{P}}^2}{T_0} + T_0\,\underset{\sim}{x}'^2\Big) - \mu(\underset{\sim}{x}'\cdot\underset{\sim}{\mathcal{P}} - T_0 x^{-\prime})\}. \tag{1.14}$$

The equations arising from varying x^- and μ in (1.14) are constraints involving $x^{-\prime}$ and μ':

$$x^{-\prime} = \frac{1}{T_0}\,\underset{\sim}{x}'\cdot\underset{\sim}{\mathcal{P}} \tag{1.15a}$$

$$\mu' = 0, \tag{1.15b}$$

together with boundary conditions involving $\dot{\sigma}_i$:

$$\mu = \dot{\sigma}_2 = 0 \qquad \text{for open string} \tag{1.16a}$$

$$\dot{\sigma}_2 = 0 \qquad \text{for closed string} \tag{1.16b}$$

where we used (1.15b) to infer $\mu_1 = \mu_2 = \mu$. Note that (1.16b) holds for both cases, and it implies the conservation of the total P^+:

$$P^+ = T_o(\sigma_2 - \sigma_1) = T_o\sigma_2. \tag{1.17}$$

Using (1.15b) and dropping a total time derivative term, we can replace the action with

$$A = \int d\tau\ [\int_0^{\sigma_2} d\sigma\{\dot{\underset{\sim}{x}}\cdot\underset{\sim}{\mathcal{P}} - \tfrac{1}{2}(\underset{\sim}{\mathcal{P}}^2/T_o + T_o\ \underset{\sim}{x}'^2)\} + T_o\dot{\sigma}_2 x^-(\sigma_2,\tau)$$

$$-\mu(\tau)\{\int_0^{\sigma_2} d\sigma\ \underset{\sim}{x}'\cdot\underset{\sim}{\mathcal{P}} - T_o(x^-(\sigma_2,\tau) - x^-(0,\tau))\}] \tag{1.18}$$

In order to use Eqs. (1.16) to simplify (1.18), we must first extract the equation arising from varying σ_2. For the open string we can vary σ_2 holding $\underset{\sim}{x}$, x^- fixed and get

$$\dot{x}^-(\sigma_2,\tau) = \frac{-1}{2T_o}(\underset{\sim}{\mathcal{P}}^2(\sigma_2)/T_o + T_o\ \underset{\sim}{x}'^2(\sigma_2)) + \dot{\underset{\sim}{x}}(\sigma_2)\cdot\underset{\sim}{\mathcal{P}}(\sigma_2)$$

$$= \frac{1}{2T_o}(\underset{\sim}{\mathcal{P}}^2(\sigma_2)/T_o + T_o\ \underset{\sim}{x}'^2(\sigma_2)) + \mu\underset{\sim}{x}'(\sigma_2)\cdot\underset{\sim}{\mathcal{P}}(\sigma_2). \tag{1.19}$$

where we made use of (1.16), the boundary condition $\underset{\sim}{x}'(\sigma_i) = 0$, and the relation $\underset{\sim}{\mathcal{P}} = T_o(\dot{\underset{\sim}{x}} - \mu\underset{\sim}{x}')$ which follows from (1.18). For the closed string one also obtains (1.19), but with the subtlety that the variation of σ_2 must be accompanied by variations in the dynamical variables according to

$$\delta\Omega(\sigma_2) + \delta\sigma_2\Omega'(\sigma_2) = \delta\Omega(0) \tag{1.20}$$

Computing the time derivative of (1.15a)

$$(\dot{x}^-)' = \frac{1}{2}(\boldsymbol{\wp}^2/T_0 + T_0\,\mathbf{x}'^2)' + \mu(\mathbf{x}'\cdot\boldsymbol{\wp})' \tag{1.21}$$

we see that (1.19) and (1.21) together imply

$$\dot{x}^- = \frac{1}{2T_0}(\boldsymbol{\wp}^2/T_0 + T_0\,\mathbf{x}'^2) + \mu\mathbf{x}'\cdot\boldsymbol{\wp} \tag{1.22}$$

The fact that $\mu(\tau)$ is not determined in the closed string case is associated with the fact that (1.18) in that case is invariant under a uniform time dependent translation in σ:

$$\Omega(\sigma,\tau) \to \Omega(\sigma - f(\tau),\tau)$$

$$\dot{\Omega}(\sigma,\tau) \to \dot{\Omega}(\sigma - f(\tau),\tau) - \dot{f}\Omega'(\sigma - f(\tau),\tau)$$

$$\mu(\tau) \to \mu(\tau) - \dot{f}(\tau)$$

where Ω represents $\mathbf{x}$ or $\boldsymbol{\wp}$. Associated with this invariance is the first class constraint obtained from varying μ

$$\mathbb{P} = \int_0^{\sigma_2} d\sigma\, \mathbf{x}'\cdot\boldsymbol{\wp} = 0. \tag{1.23}$$

We may fix this gauge freedom most simply by setting

$$\mu(\tau) = 0 \tag{1.24}$$

keeping in mind that (1.23) must still be imposed. In the open string case this gauge freedom is not present, but $\mu = 0$ is a consequence of the equations of motion (see 1.16a).

Thus in both cases we may simplify Eqs. (1.18) and (1.22) to

$$A = \int d\tau \int_0^{P^+/T_0} d\sigma \{\dot{\mathbf{x}}\cdot\boldsymbol{\wp} - \frac{1}{2}(\boldsymbol{\wp}^2/T_0 + T_0\,\mathbf{x}'^2)\} \tag{1.25a}$$

$$\dot{x}^- = \frac{1}{2T_0}(\boldsymbol{\wp}^2/T_0 + T_0\,\mathbf{x}'^2) \tag{1.25b}$$

$$x^{-\prime} = \frac{1}{T_0} \underset{\sim}{x}' \cdot \underset{\sim}{\mathcal{P}} \;, \qquad (1.25c)$$

where we have used Eqs. 1.16 and 1.17. In (1.25a) P^+ is taken as a constant parameter, P^+/T_0 being the length of parameter space. P^+ is not to be varied since we have used (1.16) to obtain (1.25a), and in any case the information obtained from varying σ_2 is already included in (1.25b). We may at this point also eliminate $\underset{\sim}{\mathcal{P}} = T_0 \dot{\underset{\sim}{x}}$ and write (1.25) as

$$A = \frac{T_0}{2} \int d\tau \int_0^{P^+/T_0} d\sigma (\dot{\underset{\sim}{x}}^2 - \underset{\sim}{x}'^2) \qquad (1.26a)$$

$$\dot{x}^- = \frac{1}{2} (\dot{\underset{\sim}{x}}^2 + \underset{\sim}{x}'^2) \qquad (1.26b)$$

$$x^{-\prime} = \underset{\sim}{x}' \cdot \dot{\underset{\sim}{x}} . \qquad (1.26c)$$

The advantage of the light-cone parametrization should now be apparent. Eqs. (1.25) show that the dynamics of the string is completely described by the dynamics of the transverse variables $\underset{\sim}{x}$ and $\underset{\sim}{\mathcal{P}}$ and the value of the parameter P^+, the total + component of the energy momentum carried by the string. Furthermore the dynamics of $\underset{\sim}{x}$ and $\underset{\sim}{\mathcal{P}}$ are precisely those of an ordinary elastic non-relativistic string free to move about in transverse space. Since the classical and quantum dynamics of such a string is thoroughly understood, we can directly carry over that understanding to the relativistic case. The main novelty here is that the length of parameter space now has a special interpretation as one of the components of energy-momentum. The other components are

$$P^- = \int_0^{P^+/T_0} d\sigma \, \frac{1}{2} \, [\underset{\sim}{\mathcal{P}}^2/T_0 + T_0 \, \underset{\sim}{x}'^2]$$

which is also the hamiltonian for our nonrelativistic analogue system and

$$\underset{\sim}{P} = T_o \int_0^{P^+/T_o} d\sigma\, \dot{\underset{\sim}{x}} .$$

It is remarkable that the dynamics imply that P^+, P^-, $\underset{\sim}{P}$ are components of a Minkowski vector, considering the disparate roles they play in the analogue system.

2. THE QUANTUM RELATIVISTIC STRING ON THE LIGHT-CONE

Once we have chosen the light cone parametrization, quantization is straight forward. A complete set of dynamical variables is $\underset{\sim}{x}$, $\underset{\sim}{\mathcal{P}}$, and the hamiltonian which governs their dynamics can be taken to be

$$H = \int_0^{P^+/T_o} d\sigma\, \frac{1}{2} (\underset{\sim}{\mathcal{P}}^2/T_o + T_o\, \underset{\sim}{x}'^2). \tag{2.1}$$

For the open string (2.1) must be supplemented by the boundary conditions

$$\underset{\sim}{x}'(0,\tau) = \underset{\sim}{x}'(P^+/T_o,\tau) = 0 \tag{2.2}$$

which follow from the action principle. For the closed string, periodic boundary conditions are imposed. In addition, we interpret the constraint (1.23) in the quantum theory as a supplementary condition on the physical closed string state space:

$$\mathbb{P}\,|\text{Phys}\rangle = \int_0^{P^+/T} d\sigma\, \underset{\sim}{x}'\cdot\underset{\sim}{\mathcal{P}}\,|\text{Phys}\rangle = 0. \tag{2.3}$$

Since $\mathbb{P}$ commutes with H, Eq. (2.3) ensures that the choice (2.1) for the closed string hamiltonian is completely equivalent to any of the hamiltonians

$$H_\mu = H + \mu(\tau)\mathbb{P}$$

derived from the action principle (1.18).

The longitudinal variables $\mathcal{P}^+, \mathcal{P}^-$, enter the dynamics only through the identification of P^+/T_0 with the length of parameter space and the identification of P^- with the hamiltonian H. As we shall see, even after we introduce interactions, this is all the knowledge we need of them as long as we are interested only in transition amplitudes between energy (P^-) eigenstates (among which are all S-matrix elements).

One only needs to return to (and interpret) the more detailed longitudinal information encoded in Eqs. (1.25) if he is interested in other quantities. For example, GGRT[2] used these expressions for x^- and $\mathcal{P}^-$ to construct via the correspondence principle the quantum analog of the Lorentz generators

$$M^{\mu\nu} = \int d\sigma(x^{\mu}\mathcal{P}^{\nu} - x^{\nu}\mathcal{P}^{\mu}). \tag{2.4}$$

for which one needs the expressions for x^- and $\mathcal{P}^-$. The generators M^{-i} are then <u>cubic</u> in the transverse variables. Requiring that the cubic terms in the quantum operator occur with the same coefficients as in the classical expression, they find that the Lorentz algebra closes only if <u>d = 24</u> and the <u>first excited state is massless.</u> We shall say more about this result after a discussion of the mass spectrum and wave function of the free quantum string.

2.1 Discrete Free Strings

We choose to interpret the quantum system defined by (2.1) as the continuum limit of a discretized system.[5,6] We divide the interval $0 \leq \sigma \leq P^+/T_0$ into M units: $P^+ = MaT_0$; and we replace

$$\underset{\sim}{x}(\sigma) \rightarrow \underset{\sim}{x}_k \qquad k = 1,2,\ldots,M \tag{2.5a}$$

$$\underset{\sim}{\mathcal{P}}(\sigma) \rightarrow \frac{1}{a}\underset{\sim}{p}_k \qquad k = 1,2,\ldots,M \tag{2.5b}$$

and Eq. (2.1) by

$$H^{closed} = \frac{1}{2aT_o} \sum_{k=1}^{M} p_k^2 + \frac{T_o}{2a} \sum_{k=1}^{M} (\underset{\sim}{x}_{k+1} - \underset{\sim}{x}_k)^2 \tag{2.6}$$

for the closed string with $\underset{\sim}{x}_{M+1} \equiv \underset{\sim}{x}_1$ and by

$$H^{open} = \frac{1}{2aT_o} \sum_{k=1}^{M} \underset{\sim}{p}_k^2 + \frac{T_o}{2a} \sum_{k=1}^{M-1} (\underset{\sim}{x}_{k+1} - \underset{\sim}{x}_k)^2 \tag{2.7}$$

for the open string. Notice that, with discretized σ, the boundary conditions on $\underset{\sim}{x}_k, \underset{\sim}{p}_k$ need not be specified: they are automatically incorporated in the potential energy term of (2.6) or (2.7). However, the periodicity of $x^>$ in the closed string case can be enforced only as a supplementary condition. Since $\int d\sigma\, \underset{\sim}{x}' \cdot \underset{\sim}{\mathcal{P}}$ is the generator of translation in σ, we shall take the condition $\int d\sigma\, \underset{\sim}{x}' \cdot \underset{\sim}{\mathcal{P}} = 0$ to mean, in the discretized theory, that physical states in the closed string sector must be invariant under the transformation which takes $\underset{\sim}{x}_k, \underset{\sim}{p}_k$ into $\underset{\sim}{x}_{k+1}, \underset{\sim}{p}_{k+1}$ and which leaves invariant the ground state of H. With this understanding, the quantum theory is now completely specified by the canonical commutation relations

$$[x_k^a, p_\ell^b] = i\delta_{ab}\delta_{k\ell}, \tag{2.8}$$

all others vanishing, where (a,b) denote the components of the vectors $\underset{\sim}{x}_k, \underset{\sim}{p}_\ell$.

2.1.1 The free closed string

The Hamiltonian (2.6) is that of a set of coupled oscillators. Because of the nearest neighbor interactions and the cyclic symmetry, the transformation to normal modes is simply a Fourier transform:

$$x_k^a = \frac{1}{\sqrt{MT_o}} \sum_{m=0}^{M-1} A_m^a\, e^{\frac{2\pi i km}{M}} \tag{2.9a}$$

$$p_k^a = \left(\frac{T_o}{M}\right)^{1/2} \sum_{m=0}^{M-1} B_m^a \, e^{\frac{2\pi i k m}{M}}, \tag{2.9b}$$

with $\underset{\sim}{A}_m^\dagger = \underset{\sim}{A}_{M-m}$, $\underset{\sim}{B}_m^\dagger = \underset{\sim}{B}_{M-m}$ for $m \neq 0$ and $\underset{\sim}{A}_0^\dagger = \underset{\sim}{A}_0$, $\underset{\sim}{B}_0^\dagger = \underset{\sim}{B}_0$. aH becomes:

$$aH = \frac{\underset{\sim}{B}_0^2}{2} + \frac{1}{2} \sum_{m=1}^{M-1} \left[\underset{\sim}{B}_m \cdot \underset{\sim}{B}_{M-m} + \omega_m^2 \, \underset{\sim}{A}_m \cdot \underset{\sim}{A}_{M-m}\right] \tag{2.10}$$

where

$$\omega_m = 2 \sin \frac{\pi m}{M}. \tag{2.11}$$

To complete the solution, introduce for $m \neq 0$ raising and lowering operators $\underset{\sim}{a}_m^\dagger$, $\underset{\sim}{a}_m$:

$$\underset{\sim}{a}_m = \left(\frac{\omega_m}{2}\right)^{1/2} \left(\underset{\sim}{A}_m + i \frac{\underset{\sim}{B}_m}{\omega_m}\right) \tag{2.12}$$

with commutation relations

$$[a_m^a, a_n^{\dagger b}] = \delta_{ab} \delta_{mn},$$

all others zero, so that aH becomes

$$aH = \frac{\underset{\sim}{B}_0^2}{2} + \sum_{m=1}^{M-1} \omega_m \, \underset{\sim}{a}_m^\dagger \cdot \underset{\sim}{a}_m + \frac{d}{2} \sum_{m=1}^{M-1} \omega_m \tag{2.13}$$

where d is the number of transverse dimensions.

In the limit $M \to \infty$ with $aMT_o = P^+$ fixed, H as given in Eq. (2.13) should serve as the P^- operator for the closed string. According to Lorentz invariance, the spectrum of P^- should be of the form

$$P_r^- = \frac{\underset{\sim}{P}^2 + M_r^2}{2P^+} \tag{2.14}$$

where M_r^2 are the $(\text{mass})^2$ eigenvalues for the energy momentum eigenstates of the string, with $\underset{\sim}{P}$ and P^+ the transverse and plus components of P^μ. The part of the excitation spectrum of (2.13) which is finite as $M \to \infty$ is compatible with (2.14). This is because $\underset{\sim}{B}_0 = \underset{\sim}{P}(MT_o)^{-1/2}$ and ω_m/a is infinite in the continuum limit except for m such that $\omega_m = 0\,(\frac{1}{M})$, i.e. for either $\frac{m}{M} = 0(a)$ or $(1 - \frac{m}{M}) = 0(a)$. Also, the zero point energy becomes in the continuum limit,

$$\frac{d}{2a}\sum_{m=1}^{M-1}\omega_m \underset{MaT_o=P^+}{\underset{M\to\infty}{\sim}} \frac{2d}{a\pi}M - \frac{\pi d T_o}{6P^+} + 0\,(\frac{1}{M}) \tag{2.15}$$

Calling $\underset{\sim}{a}_{M-m} \equiv \tilde{\underset{\sim}{a}}_m$, we may conclude that in the continuum limit, the finite part of the excitation spectrum of H is identical to that of

$$\bar{H} = \frac{2d}{a\pi}M + \frac{1}{2P^+}\{\underset{\sim}{P}^2 + 4\pi T_o\sum_{m=1}^{\infty} m(\underset{\sim}{a}_m^\dagger\cdot\underset{\sim}{a}_m + \tilde{\underset{\sim}{a}}_m^\dagger\cdot\tilde{\underset{\sim}{a}}_m) - \frac{2\pi d T_o}{6}\} \tag{2.16}$$

Next we record the continuum form of the mode expansions (2.9)

$$\underset{\sim}{x}(\sigma) = \underset{\sim}{Q} + \frac{1}{\sqrt{4\pi T_o}}\sum_{m=1}^{\infty}\frac{1}{\sqrt{m}}[(\underset{\sim}{a}_m + \tilde{\underset{\sim}{a}}_m^\dagger)e^{2\pi i m\sigma T_o/P^+} + (\tilde{\underset{\sim}{a}}_m + \underset{\sim}{a}_m^\dagger)e^{-2\pi i m\sigma T_o/P^+}]$$

$$\underset{\sim}{\mathcal{P}}(\sigma) = \frac{\underset{\sim}{P}T_o}{P^+} - \frac{iT_o\sqrt{\pi T_o}}{P^+}\sum_{m=1}^{\infty}\sqrt{m}[(\underset{\sim}{a}_m - \tilde{\underset{\sim}{a}}_m^\dagger)e^{2\pi i m\sigma T_o/P^+} + (\tilde{\underset{\sim}{a}}_m - \underset{\sim}{a}_m^\dagger)e^{-2\pi i m\sigma T_o/P^+}]$$

These can be written more compactly by putting

$$\sqrt{m}\,\underset{\sim}{a}_m = i\underset{\sim}{\alpha}_m \quad m>0$$

$$\underset{\sim}{\alpha}_{-m} \equiv \underset{\sim}{\alpha}_m^\dagger,$$

$$\sqrt{m}\,\tilde{\underset{\sim}{a}}_m = i\tilde{\underset{\sim}{\alpha}}_m \quad m>0$$

$$\tilde{\underset{\sim}{\alpha}}_{-m} \equiv \tilde{\underset{\sim}{\alpha}}_m^\dagger$$

so that

$$\underset{\sim}{\mathcal{P}} \pm T_o\underset{\sim}{x}' = \left(\frac{2\pi T_o}{P^+}\right)\left(\frac{T_o}{\pi}\right)^{1/2}\sum_{m=-\infty}^{\infty}\binom{\tilde{\underset{\sim}{\alpha}}_m}{\underset{\sim}{\alpha}_m} e^{\mp i m\sigma(2\pi T_o/P^+)} \tag{2.17}$$

where we have put

$$\underset{\sim}{\alpha}_o = \underset{\sim}{\tilde{\alpha}}_o = \underset{\sim}{P}/2\sqrt{\pi T_o}.$$

The divergent first term in $\overline{H}$ is exactly proportional to the number of lattice sites. Thus all multi-string states with the same total P^+ will have identical values for this term. Since P^+ is exactly conserved in any transition, this term just contributes a constant overall phase to every S-matrix element between states of the same total P^+. Thus, it has no dynamical significance and may be dropped. Hence, we may take, without altering the theory,

$$\begin{aligned} P^- &\equiv \lim_{\substack{M \to \infty \\ MaT_o = P^+}} \left[H - \frac{2d}{a\pi} M\right] \\ &= \frac{1}{2P^+} \left\{\underset{\sim}{P}^2 + 4\pi T_o \sum_m m\left(\underset{\sim}{a}_m^\dagger \cdot \underset{\sim}{a}_m + \underset{\sim}{\tilde{a}}_m^\dagger \cdot \underset{\sim}{\tilde{a}}_m\right) - \frac{2\pi d T_o}{6}\right\}. \end{aligned} \tag{2.18}$$

Comparison of (2.18) with (2.15) shows that the levels of P^- are consistent with Lorentz invariance, with the $(\mathrm{Mass})^2$ operator for the closed string given by

$$\mathcal{M}^2_{\text{closed string}} = 4\pi T_o \sum_{m=1}^{\infty} m\left(\underset{\sim}{a}_m^\dagger \cdot \underset{\sim}{a}_m + \underset{\sim}{\tilde{a}}_m^\dagger \cdot \underset{\sim}{\tilde{a}}_m\right) - \frac{2\pi d T_o}{6}. \tag{2.19}$$

To analyze the continuum limit of the supplementary condition we note that the operator, which by conjugation sends $\underset{\sim}{x}_k, \underset{\sim}{p}_k \to \underset{\sim}{x}_{k+1}, \underset{\sim}{p}_{k+1}$, is just

$$T = e^{i \frac{2\pi}{M} \sum_{m=1}^{M-1} m\, \underset{\sim}{a}_m^\dagger \cdot \underset{\sim}{a}_m} \tag{2.20}$$

where we have insisted that the lowest energy eigenstate of aH be invariant under T. Thus in the discretized theory physical states must satisfy

$$(T - 1)\,|\text{Phys}> = 0. \tag{2.21}$$

Eigenstates of H which lie in the finite excitation spectrum in the continuum limit satisfy

$$\underset{\sim}{a}_m\,|\text{finite Exc}> = 0$$

unless $\frac{m}{M} = 0(a)$ or $1 - \frac{m}{M} = 0(a)$. Thus, in the continuum limit, Eq. (2.21) may be written for these states:

$$[e^{i\frac{2\pi}{M}\sum_{m=1}^{\infty} m(\underset{\sim}{a}_m^{\dagger}\cdot\underset{\sim}{a}_m - \tilde{\underset{\sim}{a}}_m^{\dagger}\cdot\tilde{\underset{\sim}{a}}_m)} - 1]\,|\text{Phys; Finite Exc}> = 0$$

Since for such states, the eigenvalues of $\sum_{m=1}^{\infty} m(\underset{\sim}{a}_m^{\dagger}\cdot\underset{\sim}{a}_m - \tilde{\underset{\sim}{a}}_m^{\dagger}\cdot\tilde{\underset{\sim}{a}}_m)$ are finite, this becomes simply,

$$\sum_{m=1}^{\infty} m(\underset{\sim}{a}_m^{\dagger}\cdot\underset{\sim}{a}_m - \tilde{\underset{\sim}{a}}_m^{\dagger}\cdot\tilde{\underset{\sim}{a}}_m)\,|\text{Phys; Finite Exc}> = 0. \tag{2.22}$$

Eq. (2.22) is of course just the familiar statement, in the continuum theory, that the mode number operator for right moving modes equal that for left moving modes on closed string physical states.

The lowest few energy momentum eigenstates for the closed string are:

$$\mathcal{M}^2 = -\frac{2\pi d T_o}{6} \qquad |0>$$

$$\mathcal{M}^2 = 8\pi T_o - \frac{2\pi d T_o}{6} \qquad a_1^{\dagger a}\,\tilde{a}_1^{\dagger b}\,|0>$$

$$\mathcal{M}^2 = 16\pi T_o - \frac{2\pi d T_o}{6} \qquad a_1^{\dagger a}\,a_1^{\dagger b}\,\tilde{a}_1^{\dagger c}\,\tilde{a}_1^{\dagger d}\,|0>$$

$$a_1^{\dagger a}\,a_1^{\dagger b}\,\tilde{a}_2^{\dagger c}\,|0>$$

$$a_2^{\dagger a}\,\tilde{a}_1^{\dagger b}\,\tilde{a}_1^{\dagger c}\,|0>$$

$$a_2^{\dagger a}\,\tilde{a}_2^{\dagger b}\,|0>$$

where $|0\rangle$ is the ground state of all the oscillators:

$$\underset{\sim}{a}_m |0\rangle = 0 \qquad m=1,2,\ldots, M-1.$$

The lowest state is the familiar spin 0 tachyon. The first excited states transform as an O(d) tensor, whereas Lorentz covariance would require it to transform as an O(d+1) tensor for $\mathcal{M}^2>0$, or an O(d,1) tensor for $\mathcal{M}^2<0$. Only for $\mathcal{M}^2=0$ is the tensor structure of this state compatible with Lorentz invariance: thus the requirement that d=24 (space-time dimension = 26). The state, symmetric and traceless in a, b, has the kinematic properties of the graviton. The states at the second excited level fit into a four index O(d+1) tensor symmetric and traceless in the first two and second two indices:

$$T^{\alpha\beta\gamma\delta} \quad \text{with} \quad \begin{aligned} T^{\alpha\alpha\gamma\delta} &= T^{\alpha\beta\gamma\gamma} = 0, \\ T^{\alpha\beta\gamma\delta} &= T^{\beta\alpha\gamma\delta} \\ T^{\alpha\beta\gamma\delta} &= T^{\alpha\beta\delta\gamma} \end{aligned}$$

So their transformation properties are compatible with Lorentz invariance. It can be shown that this feature persists for all higher excited states regardless of the value of d.

2.1.2 The free open string

In this case there is no term in the potential energy coupling $\underset{\sim}{x}_M$ to $\underset{\sim}{x}_1$. The decomposition of $\underset{\sim}{x}_k$ and $\underset{\sim}{p}_k$ into normal modes is therefore:

$$\underset{\sim}{x}_k = \frac{\underset{\sim}{q}_0}{\sqrt{MT_o}} + \left(\frac{2}{MT_o}\right)^{1/2} \sum_{m=1}^{M-1} \underset{\sim}{q}_m \cos\frac{m\pi}{M}(k-1/2) \qquad (2.23a)$$

$$\underset{\sim}{p}_k = \tilde{\underset{\sim}{p}}_0\left(\frac{T_o}{M}\right)^{1/2} + \left(\frac{2T_o}{M}\right)^{1/2} \sum_{m=1}^{M-1} \tilde{\underset{\sim}{p}}_m \cos\frac{m\pi}{M}(k-1/2). \qquad (2.23b)$$

The canonical commutation relations for $\underset{\sim}{x}_k$, $\underset{\sim}{p}_k$ imply

$$[q_m^a, \tilde{p}_n^b] = i\,\delta_{m,n}\delta_{ab} \quad \text{for } m,n=0,1,\ldots,M-1 \tag{2.24}$$

and aH becomes

$$aH = \frac{\tilde{\underset{\sim}{p}}_0^2}{2} + \frac{1}{2}\sum_{m=1}^{M-1}[\tilde{\underset{\sim}{p}}_m^2 + \kappa_m^2 \underset{\sim}{q}_m^2] \tag{2.25}$$

with

$$\kappa_m = 2\sin\frac{m\pi}{2M}\,. \tag{2.26}$$

Raising and lowering operators are defined as usual for $m \neq 0$,

$$q_m = \frac{1}{\sqrt{2\kappa_m}}[\underset{\sim}{b}_m + \underset{\sim}{b}_m^\dagger]; \qquad \tilde{\underset{\sim}{p}}_m = \frac{1}{i}\left(\frac{\kappa_m}{2}\right)^{1/2}[\underset{\sim}{b}_m - \underset{\sim}{b}_m^\dagger], \tag{2.27}$$

in terms of which aH becomes

$$aH = \frac{\tilde{\underset{\sim}{p}}_0^2}{2} + \frac{d}{2}\sum_{m=1}^{M-1}\kappa_m + \sum_{m=1}^{M-1}\kappa_m \underset{\sim}{b}_m^\dagger \cdot \underset{\sim}{b}_m\,. \tag{2.28}$$

There is, of course, no supplementary condition for the open string case.

The zero point energy term in Eq. (2.28) becomes, in the continuum limit,

$$\frac{d}{2}\sum_{m=1}^{M-1}\kappa_m \underset{M\to\infty}{\sim} \frac{2d}{\pi}M - \frac{d}{2} - \frac{d\pi}{24M} + O\left(\frac{1}{M^2}\right). \tag{2.29}$$

As noted in the discussion of the closed string the divergent first term on the RHS of Eq. (2.29), being exactly proportional to the number of lattice sites, is of no dynamical significance and may be dropped. The same cannot be said of the second term, $-\frac{d}{2}$, since states with different numbers of free open strings but the same total P^+ will have different values for this term. The only way to get rid of this term is to start with a different lattice hamiltonian for the open string, namely:

$$aH^{open} \equiv \frac{d}{2} + \frac{1}{2T_o} \sum_{Sites} \underset{\sim}{p}_k^2 + \frac{T_o}{2} \sum_{Links} (\Delta \underset{\sim}{x})_k^2 \ . \tag{2.30}$$

The extra first term in Eq. (2.30) may be interpreted locally in σ: it is just $\frac{d}{4}$ x (number of free ends). Thus, one can write the total free hamiltonian for a system of any number of strings locally on the lattice

$$aH_0^{Total} = \frac{d}{4} \text{ (no. of free ends)} + \frac{1}{2T_o} \sum_{Sites} \underset{\sim}{p}_k^2 + \frac{T_o}{2} \sum_{Links} (\Delta \underset{\sim}{x})_k^2 . \tag{2.31}$$

The levels of H_0^{Total} are such that the levels of P_0^{-Total} defined by

$$P_0^{-Total} = \lim_{\substack{M \to \infty \\ MaT_o = P^{+Total}}} \left(H_0^{Total} - \frac{2dM}{\pi a}\right)$$

are consistent with Lorentz invariance. Also, the term in the zero point energy proportional to $\frac{1}{M}$ cannot be altered by such a locally prescribed counter-term.

The continuum limits of the expansions (2.23) read

$$\underset{\sim}{x}(\sigma) = \underset{\sim}{Q} + \frac{1}{\sqrt{\pi T_o}} \sum_{m=1}^{\infty} \frac{1}{\sqrt{m}} [\underset{\sim}{b}_m + \underset{\sim}{b}_m^\dagger] \cos m\sigma\pi T_o/P^+$$

$$\underset{\sim}{\mathcal{P}}(\sigma) = \frac{\underset{\sim}{P} T_o}{P^+} - \frac{iT_o\sqrt{\pi T_o}}{P^+} \sum_{m=1}^{\infty} \sqrt{m}\, [\underset{\sim}{b}_m - \underset{\sim}{b}_m^\dagger] \cos m\sigma\pi T_o/P^+ .$$

We also record these equations in terms of $\underset{\sim}{\beta}_m = -i\sqrt{m}\, \underset{\sim}{b}_m$, $m>0$, $\underset{\sim}{\beta}_{-m} \equiv \underset{\sim}{\beta}_m^\dagger$:

$$\underset{\sim}{x}(\sigma) = \underset{\sim}{Q} + \frac{i}{\sqrt{\pi T_o}} \sum_{m\neq 0} (\underset{\sim}{\beta}_m/m) \cos m\sigma\pi T_o/P^+$$

$$\underset{\sim}{\mathcal{P}}(\sigma) = \frac{\pi T_o}{P^+}\left(\frac{T_o}{\pi}\right)^{1/2} \sum_{m=-\infty}^{\infty} \underset{\sim}{\beta}_m \cos m\sigma\pi T_o/P^+ \tag{2.32}$$

where we have put

$$\beta_0 = P/\sqrt{\pi T_o}$$

The lowest few energy momentum eigenstates for the open string are:

$$\mathcal{M}^2 = -\frac{2\pi d T_o}{24} \qquad |0\rangle$$

$$\mathcal{M}^2 = 2\pi T_o - \frac{2\pi d T_o}{24} \qquad b_1^\dagger |0\rangle$$

$$\mathcal{M}^2 = 4\pi T_o - \frac{2\pi d T_o}{24} \qquad b_1^{\dagger a} b_1^{\dagger b} |0\rangle$$

$$b_2^{\dagger c} |0\rangle$$

where $|0\rangle$ is the ground state of all the oscillators:

$$b_m |0\rangle = 0 \qquad m=1,2,\ldots, M-1.$$

The lowest state is a spin zero tachyon. The first excited states transform as an O(d) vector, whereas Lorentz invariance requires an O(d+1) or O(d,1) transformation law unless $\mathcal{M}^2=0$. Thus Lorentz invariance requires d=24 which fortunately coincides with the requirement for closed strings. The states at the second excited level have O(d) transformation properties consistent with the components of a second-rank, symmetric, and traceless O(d+1) tensor, as required by Lorentz invariance for $\mathcal{M}^2>0$.

2.2 Representations of the Lorentz Group

So far, our discussion of Lorentz covariance has been limited to verifying the covariance of the spectrum, and confirming that the representations of O(d) which occur are sufficient to fill out O(d+1) representations for all massive energy eigenstates. The only states that fail this latter test are the first excited states in the closed and open sectors: compatibility with Lorentz covariance requires that these states be massless. If we take the zero point energy calculation seriously, this gives the critical dimension ($d = 24$).

But a skeptic might wonder whether this is really forced on us. Why not ignore the zero point energy calculation and simply declare that these states must be massless and add a constant to the $(\text{mass})^2$ operator to achieve this? Then, as we have said, all of the remaining levels support a representation of $O(d+1)$ (and hence of $O(d+1,1)$), and we don't need to restrict d. No one could quarrel with this for the free string. To see why this might lead to trouble in the interacting case, we return to the GGRT calculation[2] of the Lorentz algebra. For simplicity we take the open string. Also, we only consider the $O(d+1)$ subalgebra of the Lorentz group and consider only massive states at rest, so $\underset{\sim}{P} = 0$ and $P^+ = P^- = \mathcal{M}/\sqrt{2}$. The GGRT generators for the $O(d+1)$ algebra are then, effectively,

$$M^{ab} = -i \sum_{n=1}^{\infty} \frac{\beta^a_{-n}\beta^b_n - \beta^b_{-n}\beta^a_n}{n} \tag{2.33a}$$

$$M^{a3} = -\frac{i}{\mathcal{M}\sqrt{2}} \sum_{n=1}^{\infty} \frac{\mathcal{L}_{-n}\beta^a_n - \beta^a_{-n}\mathcal{L}_n}{n} \tag{2.33b}$$

where a, b are transverse indices and

$$[\beta^a_n, \beta^b_m] = n\delta_{n,-m}, \qquad \underset{\sim}{\beta}^{\dagger}_n = \underset{\sim}{\beta}_{-n} \tag{2.34}$$

$$\mathcal{L}_n = \frac{1}{2} \sum_{k=-\infty}^{\infty} \underset{\sim}{\beta}_{-k} \cdot \underset{\sim}{\beta}_{k+n} \; . \tag{2.35}$$

GGRT then find for this special situation:

$$[M^{a3}, M^{b3}] = iM^{ab} - \frac{1}{\mathcal{M}^2} \sum_{m=1}^{\infty} \left[m\left(1 - \frac{d}{24}\right) + \frac{1}{m}\left(\frac{d}{24} - \alpha_0\right)\right](\beta^a_{-m}\beta^b_m - \beta^b_{-m}\beta^a_m) \tag{2.36}$$

where α_0 is inserted in the $(\text{mass})^2$ operator (in units with $T_o = 1/2\pi$),

$$\mathcal{M}^2 = \sum_{n=1}^{\infty} \underset{\sim}{\beta}_{-n}\cdot\underset{\sim}{\beta}_{n} - \alpha_0 .$$

The second term should vanish giving d = 24, $\alpha_0 = 1$.

On the other hand, we know our state space admits a representation of O(d+1); let's find it for the second excited level, where the states are:

$$\beta^a_{-1}\beta^b_{-1}\,|0\rangle$$

$$\beta^a_{-2}\,|0\rangle .$$

The O(d) transformation properties are consistent with the traceless symmetric tensor representation of O(d+1), we identify

$$T^{a3}_{-2}\,|0\rangle = \beta^a_{-2}\,|0\rangle 2^{-1/2}$$

$$T^{ab}_{-2}\,|0\rangle = (c_1\beta^a_{-1}\beta^b_{-1} + c_2\delta_{ab}\underset{\sim}{\beta}_{-1}\cdot\underset{\sim}{\beta}_{-1})|0\rangle$$

We then require:

$$M^{a3}T^{b3}_{-2}\,|0\rangle = -i(\delta_{ab}T^{cc}_{-2} + T^{ab}_{-2})|0\rangle$$

$$M^{a3}T^{bc}_{-2}\,|0\rangle = i(T^{3b}_{-2}\,\delta_{ac} + T^{3c}_{-2}\delta_{ab})|0\rangle$$

and find the solution

$$M^{a3} = i[\beta^a_{-1}\underset{\sim}{\beta}_{-1}\cdot\underset{\sim}{\beta}_2 - \underset{\sim}{\beta}_{-2}\cdot\underset{\sim}{\beta}_1\beta^a_1]/\sqrt{2} + \frac{1+\sqrt{d+1}}{d\sqrt{2}}\,[\beta^a_{-2}\underset{\sim}{\beta}^2_1 - \underset{\sim}{\beta}^2_{-1}\beta^a_2]$$

compared to

$$M^{a3}_{GGRT} = \frac{i}{\sqrt{2(2-\alpha_0)}}\,\{\beta^a_{-1}\underset{\sim}{\beta}_{-1}\cdot\underset{\sim}{\beta}_2 - \underset{\sim}{\beta}_{-2}\cdot\underset{\sim}{\beta}_1\beta^a_1 + \tfrac{1}{4}\,[\beta^a_{-2}\underset{\sim}{\beta}^2_1 - \underset{\sim}{\beta}^2_{-1}\beta^a_2]\}.$$

In other words, we would have to modify the relative coefficients of different modes appearing in (2.33). This would destroy the locality of these expression in σ. Since the interactions are local in σ, we should expect[12] that they would not respect Lorentz invariance unless $d = 24$, $\alpha_0 = 1$. We shall confirm this later.

3. COVARIANT QUANTIZATION AND THE NO-GHOST THEOREM

We should now like to discuss an alternative approach to quantization which retains manifest Lorentz covariance.[2,3,4] Starting with the Dirac form of the hamiltonian (Eq. 1.8a), we make the choice Eq. (1.9) for the Lagrange multipliers. With this choice, the momentum conjugate to $x^\mu(\sigma,\tau)$ is

$$\mathcal{P}^\mu = T_0 \dot{x}^\mu \tag{3.1}$$

and the constraints (1.7) may be written

$$\dot{x}\cdot x' = 0 \tag{3.2a}$$

$$\dot{x}^2 + x'^2 = 0, \tag{3.2b}$$

i.e. we have chosen <u>Gaussian</u> coordinates for the two dimensional world sheet. We note in passing that the light-cone parametrization of the preceding section is the special case of Eq. 3.2 with $x^+ = \tau$. The boundary conditions following from the action principle (1.8b) are

$$\dot{\sigma}_i \dot{x}^\mu(\sigma_i,\tau) + x'^\mu(\sigma_i,\tau) = 0 \qquad \text{for open string} \tag{3.3a}$$

$$\dot{\sigma}_1 = \dot{\sigma}_2 \qquad \text{for closed string.} \tag{3.3b}$$

Because of (3.2a) both terms in (3.3a) must vanish i.e.

$$\dot{\sigma}_i = 0, \quad x'^\mu(\sigma_i) = 0 \qquad \text{for open string.}$$

Thus we may take $\sigma_1 = 0$ and σ_2 = constant. We shall choose $\sigma_2 = \pi$ for the open string and $\sigma_2 = 2\pi$ for the closed string.

If we compare (1.10) to the light cone hamiltonian (2.1) we see that $x^\mu, \mathcal{P}^\mu$ obey the same dynamics as $\underset{\sim}{x}, \underset{\sim}{\mathcal{P}}$ if we put $P^+/T_o = \pi,\ 2\pi$ for the open and closed string respectively. Thus we can immediately adapt Eqs. (2.17) and (2.32) to this case:

$$\mathcal{P}^\mu \pm T_o x^{\mu\prime} = (T_o/\pi)^{1/2} \sum_{m=-\infty}^{\infty} \binom{\tilde{\alpha}^\mu_m}{\alpha^\mu_m} e^{\mp im\sigma} \tag{3.4a}$$

$$\alpha^\mu_0 = \tilde{\alpha}^\mu_0 = P^\mu/2\sqrt{\pi T_o} \tag{3.4b}$$

$$[\alpha^\mu_m, \alpha^\nu_n] = m\delta_{m+n}\, g^{\mu\nu} \tag{3.5a}$$

$$[\tilde{\alpha}^\mu_m, \tilde{\alpha}^\nu_n] = m\delta_{m+n}\, g^{\mu\nu} \tag{3.5b}$$

$$[\tilde{\alpha}^\mu_m, \alpha^\nu_n] = 0 \tag{3.5c}$$

for the closed string, and

$$x^\mu = Q^\mu + i(\pi T_o)^{-1/2} \sum_{m\neq 0} \frac{\beta^\mu_m}{m} \cos m\sigma \tag{3.6a}$$

$$\mathcal{P}^\mu = \left(\frac{T_o}{\pi}\right)^{1/2} \sum_{m=-\infty}^{\infty} \beta^\mu_m \cos m\sigma \tag{3.6b}$$

$$\beta^\mu_0 = P^\mu/\sqrt{\pi T_o} \tag{3.6c}$$

$$[\beta^\mu_m, \beta^\nu_n] = m\delta_{m+n} g^{\mu\nu} \tag{3.7}$$

for the open string.

It remains to interpret the constraints (1.7). We first expand the (LHS) of Eqs. (1.7) in normal modes. For the closed string we have

$$(\mathcal{P} \pm T_o x')^2 = \frac{2T_o}{\pi} \sum_{n=-\infty}^{\infty} L^\pm_n\, e^{\mp in\sigma} + \text{c-number} \tag{3.8a}$$

with

$$L_n^{\pm} = \frac{1}{2} \sum_{m=-\infty}^{\infty} \begin{pmatrix} :\tilde{\alpha}_{-m} \cdot \tilde{\alpha}_{m+n}: \\ :\alpha_{-m} \cdot \alpha_{m+n}: \end{pmatrix} \qquad (3.8b)$$

whereas for the open string we have (3.8a) with $L_n^{\pm}$ replaced by L_n

$$L_n = \frac{1}{2} \sum_{m=-\infty}^{\infty} :\beta_{-m} \cdot \beta_{m+n}: \ . \qquad (3.8c)$$

We have chosen to define the various L_n's in their normal ordered forms (i.e. positive indexed operators on the right), and have compensated for this by allowing a c-number in (3.8a): we are, however, restricting the interpretation of the LHS of (3.8a) somewhat by insisting that the ambiguity be a c-number. Furthermore, since only $L_0^{\pm}$ and L_0 are changed by reordering, we shall assume the c-number is independent of σ.

Each set of L's satisfies the Virasoro algebra:

$$[L_m^{\pm}, L_n^{\pm}] = (m-n)L_{m+n}^{\pm} + \frac{d+2}{12}(m^3-m)\delta_{m+n} \qquad (3.9a)$$

$$[L_m, L_n] = (m-n)\, L_{m+n} + \frac{d+2}{12}(m^3-m)\delta_{m+n} \qquad (3.9b)$$

and for the closed string L_m^{+} commutes with L_n^{-}:

$$[L_m^{+}, L_n^{-}] = 0. \qquad (3.9c)$$

We now come to the interpretation of Eqs. (1.7), which state, for the classical case, that all $L_n^{\pm}$, L_n are zero. In the quantum theory, these can't be taken as operator statements but must be interpreted as putting restrictions on the physical state space. The c-number in (3.8a) comes only into the zero mode constraint, which we write as

$$(L_0^{\pm} - \alpha_0^{\pm})|\text{Phys}\rangle = 0 \qquad (3.10)$$

for the closed string. In a left-right symmetric theory we should take $\alpha_0^+ = \alpha_0^-$, but not in, for example, the heterotic string. Writing out (3.10) in modes we have

$$\left(\sum_{n=1}^{\infty} \tilde{\alpha}_{-n}\cdot\tilde{\alpha}_n + \frac{P^2}{8\pi T_o} - \alpha_0^+\right)|\text{Phys}\rangle$$

$$= \left(\sum_{n=1}^{\infty} \alpha_{-n}\cdot\alpha_n + \frac{P^2}{8\pi T_o} - \alpha_0^-\right)|\text{Phys}\rangle = 0$$

which we can rearrange as the pair

$$\mathcal{M}^2|\text{Phys}\rangle = -P^2|\text{Phys}\rangle$$

$$= 4\pi T_0\left[\sum_{n=1}^{\infty}(\alpha_{-n}\cdot\alpha_n + \tilde{\alpha}_{-n}\cdot\tilde{\alpha}_n) - \alpha_0^+ - \alpha_0^-\right]|\text{Phys}\rangle$$

$$\left[\sum_{n=1}^{\infty}(\tilde{\alpha}_{-n}\cdot\tilde{\alpha}_n - \alpha_{-n}\cdot\alpha_n) - \alpha_0^+ + \alpha_0^-\right]|\text{Phys}\rangle = 0. \quad (3.11)$$

For the open string, we have only one zero mode constraint

$$(L_0 - \alpha_0)|\text{Phys}\rangle = 0 \quad (3.12)$$

which in modes becomes

$$\mathcal{M}^2|\text{Phys}\rangle = -P^2|\text{Phys}\rangle = 2\pi T_0\left[\sum_{n=1}^{\infty} \beta_{-n}\cdot\beta_n - \alpha_0\right]|\text{Phys}\rangle . \quad (3.13)$$

The Virasoro algebra precludes imposing all of the non-zero mode constraints. However, since $L_n^\dagger = L_{-n}$, $L_n^{\pm\dagger} = L_{-n}^{\pm}$, we need only impose the constraints for positive n on the kets and the negative n constraints will then follow for the bras. This is sufficient for the correspondence principle and consistent with the Virasoro algebra:

$$L_n^{\pm}|Phys\rangle = 0 \qquad n > 0 \tag{3.14}$$

for the closed string and

$$L_n|Phys\rangle = 0 \qquad n > 0 \tag{3.15}$$

for the open string.

The constraints (3.14) and (3.15) restrict the physical state space to a subspace of the Fock space generated by α_n, $\tilde{\alpha}_n$ for the closed string or β_n for the open string. Since $g^{00} = -1$, Eq. 3.7 shows that this Fock space is full of ghosts (negative norm states). The No-ghost theorem[7,8] states that the physical subspace defined by (3.14) or (3.15) is ghost free if $D = d + 2 = 26$ and $\alpha_0^+ = \alpha_0^- = 1$ or $\alpha_0 = 1$. A corollary shows that the physical subspace is also ghost free if $D \leq 25$ and α_0^+, $\alpha_0^- \leq 1$ or $\alpha_0 \leq 1$. There are always ghosts[9] (1) if $D > 26$, or any of the α_0's are > 1, or (2) if $D = 26$ and any of the α_0's are $\neq 1$. Sometimes the effective value of D in some models can be fractional; in such a case, if

$$D = 25 + \frac{6}{n(n+1)} \qquad n = 3,4,\ldots$$

the results of Freidan, Qiu, and Shenker[10] imply a ghost free physical subspace for certain discrete values of α_0[9].

3.1 Proof of the No-ghost Theorem[7]

Since the physical subspace of the closed string Fock space is isomorphic to the tensor product of two open string physical subspaces it is enough to prove the open string case. The proof proceeds in two steps. We first select a light-like vector k^μ such that $\beta_0 \cdot k \neq 0$ and define

$$K_n \equiv k \cdot \beta_n . \tag{3.16}$$

Then, it is a simple matter to confirm that

$$[K_n, L_m] = n\, K_{n+m} \tag{3.17a}$$

and

$$[K_n, K_m] = 0. \tag{3.17b}$$

Next, define the subspace $\mathcal{J}$ of Fock space by

$$|\psi\rangle \varepsilon \mathcal{J} \quad \Leftrightarrow \quad K_n|\psi\rangle = L_n|\psi\rangle = 0 \qquad n > 0. \tag{3.18}$$

$\mathcal{J}$ is the subspace of "highest weight vectors" associated with the Lie algebra (3.17), (3.9b). Because k is light-like, there are no ghosts in $\mathcal{J}$:

$$|\psi\rangle \varepsilon \mathcal{J} \quad \Rightarrow \quad \langle\psi|\psi\rangle \geq 0. \tag{3.19}$$

Now for any non-null element $|\tau\rangle \varepsilon \mathcal{J}$, which is also an eigenstate of L_0: $L_0|\tau\rangle = h|\tau\rangle$, define the matrix

$$\mathcal{M}^n_{\{\lambda,\mu\};\{\lambda',\mu'\}}(h, \beta\cdot k) = \langle\tau|K_n^{\mu_n}\cdots K_1^{\mu_1} L_n^{\lambda_n}\cdots L_1^{\lambda_1} L_{-1}^{\lambda_1'}\cdots L_{-n}^{\lambda_n'} K_{-1}^{\mu_1'}\cdots K_{-n}^{\mu_n'}|\tau\rangle/\langle\tau|\tau\rangle$$

with $\Sigma k\lambda_k + \Sigma k\mu_k = \Sigma k\lambda_k' + \Sigma k\mu_k' = n$, and which depends only on h and $\beta\cdot k$. Then we have the

<u>Lemma 1</u>. $\det \mathcal{M}^n \neq 0$ for $\beta\cdot k \neq 0$. (3.20)

The proof is by straight-forward inspection.

e.g. $$\mathcal{M}^1 = \begin{pmatrix} 2L_0 & \beta\cdot k \\ \beta\cdot k & 0 \end{pmatrix}, \quad \det \mathcal{M}^1 = -(\beta\cdot k)^2 \neq 0.$$

$\mathcal{M}^n$ is always triangular up to row rearrangements with non-zero diagonal elements. Because of (3.20) and (3.19) we can conclude that is a positive definite subspace isomorphic to the transverse Fock space generated by d dimensional $\underset{\sim}{\beta}_n$'s. Also, the full Fock Space is spanned by the set

$$L_{-1}^{\lambda_1} \cdots L_{-n}^{\lambda_n} K_{-1}^{\mu_1} \cdots K_{-n}^{\mu_n} |\tau) \tag{3.21}$$

where $|\tau>$ is a basis of $\mathcal{J}$. The span (3.21) is linearly independent.

To prove the No-ghost theorem, we need

<u>Lemma 2</u>. If D = 26, L_1 and $\tilde{L}_2 \equiv L_2 + \frac{3}{2} L_1^2$ map spurious states with L_0 eigenvalue 1 into spurious states with L_0 eigenvalues 0, -1 respectively.

Spurious states are by definition of the form

$$|s> = \sum_{n=1}^{\infty} L_{-n}|\phi_n> = L_{-1}|\psi_1> + L_{-2}|\psi_2>. \tag{3.22}$$

To prove Lemma 2 we apply L_1 and $\tilde{L}_2$ to $|s>$:

$$L_1|s> = L_{-1}L_1|\psi_1> + 2L_0|\psi_1> + L_{-2}L_1|\psi_2>$$
$$+ 3L_{-1}|\psi_1> = |s'> + 2L_0|\psi_1> .$$

If $L_0|s> = |s>$, $L_0|\psi_1> = 0$ so L_1 has the announced property. Next

$$(L_2 + \frac{3}{2} L_1^2)|s> = L_{-1}\tilde{L}_2|s> + L_{-2}\tilde{L}_2|s>$$
$$+ 3L_1|\psi_1> + 3L_1L_0|\psi_1> + 3L_0L_1|\psi_1>$$
$$+ (4L_0 + \frac{D}{2})|\psi_2> + 9L_0|\psi_2> + \frac{9}{2} L_{-1}L_1^2|s>$$

$$(L_2 + \frac{3}{2} L_1^2)|s\rangle = |s''\rangle + 3L_1(2L_0)|\psi_1\rangle + (13L_0 + \frac{D}{2})|\psi_2\rangle$$

$$= |s''\rangle$$

for $L_0|s\rangle = |s\rangle$, $D = 26$.

The No-ghost theorem now follows very simply. Using the span (3.21) we can write any state $|\psi\rangle$ as

$$|\psi\rangle = |s\rangle + |\phi\rangle \tag{3.23}$$

where $|s\rangle$ is spurious and $|\phi\rangle$ is in the subspace spanned by $K_{-1}^{\mu_1} \cdots K_{-n}^{\mu_n}|\tau\rangle$; i.e. $K_n|\phi\rangle = 0 \quad n > 0$. For $\alpha_0 = 1$, the physical state conditions are

$$L_0|\psi\rangle = |\psi\rangle$$

$$L_n|\psi\rangle = 0 \qquad n > 0 .$$

So if $|\psi\rangle$ is physical, $\alpha_0 = 1$, and $D = 26$, Lemma 2 implies

$$L_1|s\rangle = L_1|\phi\rangle = \tilde{L}_2|s\rangle = \tilde{L}_2|\phi\rangle = 0 \tag{3.24}$$

which in turn imply

$$L_n|s\rangle = L_n|\phi\rangle = 0 \qquad n > 0.$$

Thus $|s\rangle$ is a null state and $|\phi\rangle \varepsilon \mathcal{J}$. Thus $\langle\psi|\psi\rangle = \langle\phi|\phi\rangle \geq 0$.

For $D \leq 25$, $\alpha_0 \leq 1$, we can consider the physical subspace as a subspace of the $D = 26$, $\alpha_0 = 1$ physical subspace characterized by

$$\beta_n^{\alpha}|\psi\rangle = 0 \qquad n>0,\ \alpha = D+1,\ldots,26$$

and $(\beta_0^{26})^2/2 = 1 - \alpha_0$. So the theorem also applies to this case. An alternative proof applicable to continuous $D \leq 25$ and $\alpha_0 \leq 1$ is also available.[11] This latter proof makes use of the celebrated Kac formula for the determinant of the contravariant form associated with the Virasoro algebra.

4. INTERACTING DISCRETIZED STRINGS AND THE CONTINUUM LIMIT

As discussed by Mandelstam[12], interactions may be introduced into string theories by an ansatz which allows two types of alterations of portions of string to instantaneously occur. First, any piece of string may break at a point in its interior, or inversely, two ends of string may join. Secondly, if a point in one portion of some string coincides with a point in another portion of not necessarily the same string, there may be an instantaneous exchange of parts of each portion.

It is important that these alterations occur with the same probability amplitude at each point of the string, and that they do not depend on the global configuration of the string, in particular on whether it is open or closed. Thus, the first type of alteration can change a closed string into an open string (or vice versa), and it can change an open string into two open strings (or vice versa). The second type of alteration can change a closed string into two closed strings (or vice versa), it can change an open string into an open and a closed string (or vice versa), and it can change two open strings into two altered open strings. Technically, Mandelstam employs path integrals over the transverse string coordinates $\underset{\sim}{x}(\sigma,\tau)$ in the light-cone parametrization of GGRT to compute the transition amplitudes for these processes.

For the discretized string described in section 2, these ideas may be implemented through the sudden approximation of ordinary quantum mechanics. In the discretized theory a multi-string state, with total $P^+=MaT_0$, is a system of M particles of mass aT_0 moving in d dimensions. Each particle interacts with at most two others via the harmonic potential

$$V(\Delta \underset{\sim}{x}) = \frac{T_0}{2a} (\Delta \underset{\sim}{x})^2 . \tag{4.1}$$

The number of strings, their character (open or closed), and their P^+ values are determined completely by the specification of the total potential energy term, i.e. by the pattern in which the particles interact.

The alterations allowed in Mandelstam's interacting string formalism may be adapted to our discretized string theory by allowing certain instantaneous alterations of this pattern of interactions. The "breaking" interaction is just the instantaneous disappearance (or appearance) of an interaction between a pair of these particles. We shall sometimes refer to this as the destruction or creation of a bond. The "interchange" interaction is implemented by interchanging the bond structure of two pairs of bonded particles.

Since the bond structure of our system of particles is changing as a function of τ; we must introduce a new dynamical variable, a matrix $\eta_{k\ell}$, $k,\ell=1,\ldots,M$. If the chains possess an orientation, we employ the concept of a directed bond so that

$$\eta_{k\ell} = \begin{cases} 1 & \text{if there is a bond directed from particle } k \text{ to particle } \ell \\ 0 & \text{if there is no bond directed from particle } k \text{ to particle } \ell \text{ or if } k=\ell \end{cases} \tag{4.2a}$$

With this concept it is easy to specify the restriction on $\eta_{k\ell}$ so that only linear nearest neighbor bond structures are allowed: each row and column of the MxM matrix $\eta_{k\ell}$ has at most one nonvanishing entry. The potential energy for the system with bond structure η is just

$$\frac{T_0}{2a} \sum_{k,\ell} \eta_{k\ell} (\underset{\sim}{x}_k - \underset{\sim}{x}_\ell)^2 .$$

Clearly the Hamiltonian is unchanged if one reverses the direction of all the bonds in any of the chains, implying the degeneracy appropriate to orientation.

For non-oriented chains we cannot use the concept of a directed bond. In this case, we require the matrix η to be symmetric: $\eta_{k\ell} = \eta_{\ell k}$ and

$$\eta_{k\ell} = \begin{cases} 1/2 \text{ if there is a bond between particles } k \text{ and } \ell \\ 0 \text{ if there is no bond between particles } k \text{ and } \ell \text{ or if } k=\ell \end{cases} \tag{4.2b}$$

Now the restriction on η so that only linear nearest neighbor bond structures are allowed is: each row and column of η has at most two non-zero entries.

The complete dynamics of the interacting discretized string may be given by the following prescription. Let $P^+ = aMT_0$ be the total P^+ for the system of strings. Let the system's initial state be given in Schrodinger picture by the wave function

$$\Psi_\eta[\underset{\sim}{x}_1, \underset{\sim}{x}_2, \ldots, \underset{\sim}{x}_M; 0] \tag{4.3}$$

where the matrix η specifies the initial bond structure. The wave function of the system at some later x^+, $\tau=T$ is computed by a two step procedure. First, for each sequence of alterations, specified by the type, x^+, and location of each alteration, solve the Schrodinger equation

$$i \frac{\partial}{\partial\tau} \psi[\{\underset{\sim}{x}\},\tau] = \Big\{-\frac{1}{2aT_0}\sum_k \frac{\partial^2}{\partial \underset{\sim}{x}_k^2} + \frac{T_0}{2a}\sum_{k,\ell}(\underset{\sim}{x}_k - \underset{\sim}{x}_\ell)^2\,\eta_{k,\ell}(\tau) + \frac{d}{2}\Big(M - \sum_{k,\ell}\eta_{k,\ell}\Big)\Big\}\psi \tag{4.4}$$

The matrix $\eta_{k,\ell}(\tau)$ characterizes the pattern of alterations of the potential energy term as a function of τ. Call the solution of Eq. (4.4) which coincides at $\tau=0$ with (4.3) where $\eta \equiv \eta(0)$,

$$\psi_{[\eta_{k\ell}]}[\{\underset{\sim}{x}\},\tau].$$

Then the second step of our procedure is to define

$$\Psi_{\eta'}[\{\underset{\sim}{x}\},T] = \sum_{\substack{\{\eta_{k\ell}(\tau)\} \\ \eta(T)=\eta'}} \int_0^T \prod_{\substack{i=1 \\ \tau_1 \geq \tau_2 \geq \ldots \geq \tau_{N_1+N_2}}}^{N_1+N_2} \left(\frac{d\tau_i}{a}\right) \psi_{[\eta_{k\ell}(\tau)]}[\{\underset{\sim}{x}\},\tau] g^{N_1} G^{N_2} \tag{4.5}$$

where N_1 is the total number of "breaking" alterations and N_2 is the total number of "interchange" alterations occurring between $\tau=0$ and T. The alterations occur at a discrete set of times, τ_i, and Eq. (3.5) prescribes that these times are to be integrated from 0 to T, with weight $\prod_i d\tau_i/a$.

Eq. (4.5) is written displaying two apparently independent coupling constants. However, it is well known that the continuum limit of scattering amplitudes will not possess crossing symmetry unless G and g are related. To see the relationship, consider a process in which the interchange alteration has the same net effect as two successive breaking alterations, for example, the decay of an open string to a closed one and an open one. This can occur through an interchange of bonds at any location on the initial string. The sum over all such locations will, in the continuum limit, become an integral:

$$\sum_{\ell oc} \rightarrow \frac{1}{a}\int d\sigma.$$

Alternatively, the initial open string can break and, at a later time τ, the ends of one of the fragments can rejoin to form a closed string. According to Eq. (3.5) this τ will be integrated with weight $\frac{1}{a}$:

$$\frac{1}{a}\int d\tau.$$

A moment's thought reveals that the $\tau \to 0$ limit of the integrand of the τ integral has precisely the effect of a particular interchange alteration. Continuity of the integrand of the σ integral with that of the τ integral at this common point demands

$$G = g^2. \tag{4.6}$$

This continuity and hence Eq. (4.6) are necessary for the ultimate consistency of the interacting string theory. Thus, after the continuum limit, which eliminates the parameter a, the interacting string theory will have two free parameters, the coupling g and the rest tension T_0. Of course, the condition (4.6) does not arise in the version of the theory with only closed strings, because it comes only from processes which involve interacting open strings.

The dynamics encoded in Eqs. (4.4) and (4.5) may be transcribed into the language of second quantization, which will facilitate our subsequent discussion. For the unregulated continuum theory this transcription has been given by Kaku and Kikkawa,[13] and for our discretized theory it can be found in Ref. 6. Here we summarize the results.

Two non-hermitian quantum field operators are introduced:

$$\psi_M(\underset{\sim}{x}_1,\ldots,\underset{\sim}{x}_M,\tau) = \psi_M(\underset{\sim}{x}_2,\ldots,\underset{\sim}{x}_M,\underset{\sim}{x}_1,\tau) \tag{4.7a}$$

for the discretized closed string with M units of P^+ and

$$\chi_M(\underset{\sim}{x}_1\ldots\underset{\sim}{x}_M,\tau) \tag{4.7b}$$

for the open string with M units of P^+. The operator ψ and χ are postulated to satisfy the equal time commutation relations

$$[\psi_M(\{\underset{\sim}{x}\},\tau),\ \psi_N^\dagger(\{\underset{\sim}{y}\},\tau)] = \delta_{MN} \sum_{i=1}^{M} \prod_{k=1}^{M} \delta(\underset{\sim}{x}_k - \underset{\sim}{y}_{k+i-1}) \tag{4.8a}$$

$$[\chi_M(\{\underset{\sim}{x}\},\tau),\ \chi_N^\dagger(\{y\},\tau)] = \delta_{MN} \prod_{k=1}^{M} \delta(\underset{\sim}{x}_k - \underset{\sim}{y}_k) \tag{4.8b}$$

with all other commutators vanishing. Then the Heisenberg equations for ψ and χ with

$$P_0^- = H_0 = \frac{1}{a} \sum_{M=2}^{\infty} \int d^M\underset{\sim}{x}\ \{\frac{1}{2MT_o} \sum_{k=1}^{M} \frac{\partial\psi_M^\dagger}{\partial\underset{\sim}{x}_k}\cdot\frac{\partial\psi_M}{\partial\underset{\sim}{x}_k} + \frac{T_o}{2M}\sum_{k=1}^{M}(\underset{\sim}{x}_{k+1}-\underset{\sim}{x}_k)^2\psi_M^\dagger\psi_M$$

$$+ \frac{1}{2T_o}\sum_{k=1}^{M}\frac{\partial\chi_M^\dagger}{\partial\underset{\sim}{x}_k}\cdot\frac{\partial\chi_M}{\partial\underset{\sim}{x}_k} + \frac{T_o}{2}\sum_{k=1}^{M-1}(\underset{\sim}{x}_{k+1}-\underset{\sim}{x}_k)^2\chi_M^\dagger\chi_M + \frac{d}{2}\chi_M^\dagger\chi_M\} \tag{4.9}$$

taken as the Hamiltonian are just the free closed string and free open string Schrodinger equations. This is just the standard second quantization procedure. If we define the empty state $|0\rangle$ by

$$\psi_M|0\rangle = \chi_M|0\rangle = 0 \qquad \text{at } \tau=0, \tag{4.10}$$

then $\langle 0|\psi_M(\{\underset{\sim}{x}\},\tau)|\Psi\rangle$ and $\langle 0|\chi_M(\{\underset{\sim}{x}\},\tau)|\Psi\rangle$ are proportional to the free closed string and free open string wave functions respectively, if the dynamics is given by Eq. (4.9).

The crucial step is to identify P_1^- such that the dynamics given by the Hamiltonian

$$P^- = H = P_0^- + P_1^- \tag{4.11}$$

reproduce that given by Eqs. (4.4) and (4.5). To identify P_1^-, rewrite Eq. (4.5) as

$$\psi_{\eta'}[\{\underset{\sim}{x}\},T] = \int d^M\underset{\sim}{y} \sum_{\eta} \mathcal{G}_{\eta'\eta}(\{\underset{\sim}{x}\},T;\{\underset{\sim}{y}\},0)\psi_\eta[\{\underset{\sim}{y}\},0]. \tag{4.12}$$

Then applying time dependent perturbation theory about P_0^- shows that

$$\langle 0|T\psi_M(\{\underset{\sim}{x}\},\tau)\psi_M^\dagger(\{\underset{\sim}{y}\},0)|0\rangle = \sum_{i=1}^{M}\mathcal{G}_{\eta\eta}(\{\underset{\sim}{x}_k\},\tau;\{\underset{\sim}{y}_{k-i-1}\},0) \tag{4.13}$$

where η is that for a single closed string and provided:

$$P_1^- = \frac{1}{a}\{H_1 + H_2 + H_3 + H_4 + H_5\} \tag{4.14}$$

where

$$H_1 = g^2 \sum_{M=4}^{\infty} \int d^M \underset{\sim}{x} \sum_{K=2}^{M-1} [\psi_M^\dagger(\{\underset{\sim}{x}\})\psi_K(\underset{\sim}{x}_1 \ldots \underset{\sim}{x}_K)\psi_{M-K}(\underset{\sim}{x}_{K+1} \ldots \underset{\sim}{x}_M) + \text{h.c.}] \tag{4.15a}$$

$$H_2 = g \sum_{M=2}^{\infty} \int d^M \underset{\sim}{x} \, [\psi_M^\dagger \chi_M + \chi_M^\dagger \psi_M] \tag{4.15b}$$

$$H_3 = g \sum_{M=4}^{\infty} \int d^M \underset{\sim}{x} \sum_{K=2}^{M-2} [\chi_M^\dagger \chi_K \chi_{M-K} + \text{h.c.}] \tag{4.15c}$$

$$H_4 = g^2 \sum_{M=4}^{\infty} \int d^M x \sum_{K=2}^{M-2} \sum_{L=0}^{M-K} [\chi_M^\dagger \psi_K(\underset{\sim}{x}_L \ldots \underset{\sim}{x}_{L+K})\chi_{M-K} + \text{h.c.}] \tag{4.15d}$$

$$H_5 = \frac{g^2}{2} \sum_{M=4}^{\infty} \sum_{K=2}^{M-2} \sum_{J=1}^{K} \sum_{\substack{L=K+1 \\ L\geq J+2 \\ L\leq M+J-2}}^{M} \chi_K^\dagger \chi_{M-K}^\dagger \chi_{J+M-L}(\underset{\sim}{x}_1 \ldots \underset{\sim}{x}_J \underset{\sim}{x}_{L+1} \ldots \underset{\sim}{x}_M) \chi_{L-J}(\underset{\sim}{x}_{K+1} \ldots \underset{\sim}{x}_L \underset{\sim}{x}_{J+1} \ldots \underset{\sim}{x}_K). \tag{4.15e}$$

There are two noteworthy features about Eqs. (4.14) and (4.15). First, all of the lattice spacing dependence factors out: aP^- is independent of a, so that the ratio of P_1^- to P_0^- is independent of a. This is the principle that fixes the powers of a in Eq. (4.5). Secondly, although there are five distinct terms, all their coefficients are inter-related. This is because there are only two local alterations that can be made in a piece of string and their relative probability amplitudes are related by Eq. (4.6).

Next we would like to show, in the context of a simple calculation, the role of the critical dimension in the interacting theory: If D=26 the continuum limit of the zero loop S-matrix is finite with P^- given by Eqs. (4.11), (4.14) and (4.15). To illustrate this point, we shall calculate the lowest order transition amplitude for a ground state closed string to change to a ground state open string.

We must first normalize our free single string ground states so that they have the proper continuum limit:

$$\langle \underset{\sim}{P}', N | \underset{\sim}{P}, M \rangle = \delta(\underset{\sim}{P}' - \underset{\sim}{P}) \frac{\delta_{MN}}{aT_o} \tag{4.16}$$

where $|\underset{\sim}{P},M\rangle$ is a single free closed or open string ground state with M units of P^+. Thus in the continuum limit the RHS of (4.16) will be $\delta(\underset{\sim}{P}'-\underset{\sim}{P})\delta(P^{+'}-P^+)$. We must next work out the proportionality factors in

$$\langle 0 | \psi_M[\{\underset{\sim}{x}\},\tau] \, | \underset{\sim}{P}, M, \text{closed}\rangle = \frac{Z^{\text{closed}}}{\sqrt{aT_o}} \psi^{\text{closed}}_{\underset{\sim}{P},M} (\{\underset{\sim}{x}\},\tau) \tag{4.17a}$$

$$\langle 0 | \chi_M[\{\underset{\sim}{x}\},\tau] \, | \underset{\sim}{P}, M, \text{open}\rangle = \frac{Z^{\text{open}}}{\sqrt{aT_o}} \psi^{\text{open}}_{\underset{\sim}{P},M} (\{\underset{\sim}{x}\},\tau) \tag{4.17b}$$

where ψ^{closed} and ψ^{open} are single string ground state wave functions normalized to $\delta(\underset{\sim}{P}'-\underset{\sim}{P})$. Since we are working to lowest order (zero loops), we may work out the Z's from the commutation relations (4.8):

$$\begin{aligned} M\psi^{\text{closed}}_{\underset{\sim}{P},M} (\{\underset{\sim}{x}\}) &= \int d\underset{\sim}{y} \sum_i \prod_k \delta(\{\underset{\sim}{x}\}-\{\underset{\sim}{y}\})\psi^{\text{closed}}_{\underset{\sim}{P},M} (\{\underset{\sim}{y}\}) \\ &= \int dy \, \langle 0|\psi_M(\{\underset{\sim}{x}\})\psi^\dagger_M(\{\underset{\sim}{y}\})|0\rangle \psi^{\text{closed}}_{\underset{\sim}{P},M} (\{\underset{\sim}{y}\}) \\ &= \left(Z^{\text{closed}}\right)^2 \int d\underset{\sim}{y} \int d\underset{\sim}{Q} \, \psi^{\text{closed}}_{\underset{\sim}{Q},M}(\{\underset{\sim}{x}\})\psi^{*\text{closed}}_{\underset{\sim}{Q},M}(\{\underset{\sim}{y}\})\psi^{\text{closed}}_{\underset{\sim}{P},M}(\{\underset{\sim}{y}\}) \\ &= \left(Z^{\text{closed}}\right)^2 \psi^{\text{closed}}_{\underset{\sim}{P},M} \end{aligned}$$

in other words

$$Z^{\text{closed}} = \sqrt{M} \, . \tag{4.18a}$$

Analogously,

$$Z^{open} = 1. \tag{4.18b}$$

So finally, we have[5]

$$\begin{aligned}
&\langle \underset{\sim}{P},M \text{ open}|P_1^-|\underset{\sim}{Q},N \text{ closed}\rangle = \\
&= \frac{g}{a} \int d^M x \sum_K \langle \underset{\sim}{P},M \text{ open}|\chi_K^\dagger(\{\underset{\sim}{x}\})\psi_K(\{\underset{\sim}{y}\})|\underset{\sim}{Q},N \text{ closed}\rangle \\
&= \delta_{MN}\left(\frac{g}{a}\right)\left(\frac{\sqrt{N}}{aT_o}\right) \int d^M x\, \psi^{open*}_{\underset{\sim}{P},M}(\{\underset{\sim}{x}\})\psi^{closed}_{\underset{\sim}{Q},M}(\{\underset{\sim}{x}\}) \\
&\underset{M\to\infty}{\sim} \left(\frac{\delta_{MN}}{aT_o}\right) \delta(\underset{\sim}{P}) \frac{\sqrt{M}}{a} KM^{-d/16} \\
&\underset{M\to\infty}{\sim} \delta(P^+-Q^+)\delta(\underset{\sim}{P}) \frac{T_o}{P^+} K\, M^{3/2-d/16}
\end{aligned} \tag{4.19}$$

where d is the number of transverse dimensions. We see that the right hand side of Eq. (4.19) is finite and non-zero if and only if d=24 (D=26).

We shall close with a brief discussion of the "classical", by which mean zero loop, approximation to interacting string theory. We expect this limit to involve large field strengths. To discuss this limit systematically we scale the fields:

$$\hat{\psi} \equiv g^2\psi$$

$$\hat{\chi} \equiv g\chi$$

so that $[\hat{\psi},\hat{\psi}^\dagger] = O(g^4)$ and $[\hat{\chi},\hat{\chi}^\dagger] = O(g^2)$. Thus we expect the classical limit to be $g^2 \to 0$. In this limit, Heisenberg's equation for $\hat{\psi}$ and $\hat{\chi}$ become (very schematically)

$$a\mathcal{D}^{closed}\,\hat{\psi} = \hat{\psi}\hat{\psi} + \int \hat{\psi}^\dagger\hat{\psi} \tag{4.20a}$$

$$a\mathcal{J}^{open}\,\hat{\chi} = \hat{\chi}\hat{\chi} + \hat{\psi} + \hat{\psi}\hat{\chi} + \int\hat{\chi}^{\dagger}\hat{\chi} + \int\hat{\psi}^{\dagger}\hat{\chi} + \int\hat{\chi}^{\dagger}\hat{\chi}\hat{\chi}. \tag{4.20b}$$

Eqs. (4.20) describe the zero loop approximation to interacting string theory. $\hat{\psi}$ and $\hat{\chi}$ are c number multivariable fields. In principle, one might hope to study them to search for a non-trivial vacuum of the theory. Notice that the closed string field equation (4.20a) does not involve the open string in this limit. The open string equation, on the other hand, involves a closed string background.

Clearly the classical equations are formidible. Of course $\psi=\chi=0$ is a solution which leads to the usual perturbation theory; but non-perturbative phenomena are encoded in Eqs. (4.20). One might hope, for example, to learn something by studying these equations with the lattice regularization in place. Of course, the bosonic string is probably of academic interest only, and we should really apply such studies to the super string.

REFERENCES

1. Nambu, Y., Lectures at the Copenhagen Summer Symposium (1970); see also Goto, T., Prog. Theor. Phys. 46. 1560 (1971).

2. Goddard, P., Goldstone, J., Rebbi, C., and Thorn, C. B., Nucl. Phys. B56, 109 (1973).

3. Polyakov, A. M., Phys. Lett. B 103, 71 (1981).

4. Goto, T., Prog. Theor. Phys. 46 1560 (1971); Mansouri, J. and Nambu, Y., Phys. Lett. 38B, 375 (1972).

5. Giles, R. and Thorn, C. B., Phys. Rev. D16, 366 (1977).

6. Thorn, C. B., "The Theory of Interacting Relativistic Strings", University of Florida Theoretical Physics preprint UFTP-85-8.

7. Goddard, P. and Thorn, C. B., Phys. Lett. 40B, 235 (1972).

8. Brower, R. C., Phys. Rev. D6, 1655 (1972).

9. Thorn, C. B., Nucl. Phys. B248, 551 (1984).

10. Friedan, D., Qiu, Z., and Shenker, S., Phys. Rev. Lett. 52, 1575 (1984); Goddard, P., Kent, A., and Olive, D., Phys. Lett. 152, 88 (1985).

11. Thorn, C. B., Vertex Operators in Mathematics and Physics - Proceedings of a Conference Nov. 10-17, 1983, pp. 411-415. Publications of the Mathematical Sciences Research Institute #3, Springer-Verlag, 1984.

12. Mandelstam, S., Nucl. Phys. B64, 205 (1973); Nucl. Phys. B69, 77 (1974); Nucl Phys. B69, 413 (1974).

13. Kaku, M. and Kikkawa, K., Phys. Rev. D10, 1110 (1974).

THE INTERACTING-STRING PICTURE AND FUNCTIONAL INTEGRATION*

S. Mandelstam
Department of Physics, University of California, Berkeley, California

1. INTRODUCTION

I shall begin my lecture by comparing the interacting string picture, to be described here, with other approaches to string theory.

a) Interaction of strings with external fields: (Fig.1)

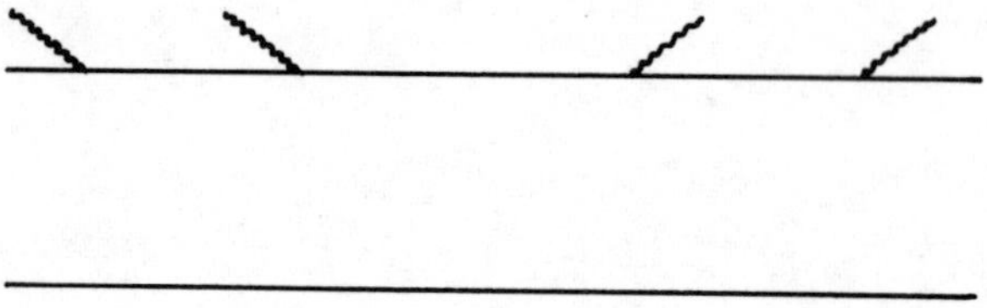

Fig. 1. Interaction of a string with external fields.

Most of the old dual-model calculations were performed by using the picture shown in Fig. 1, where a string emits or absorbs quanta of a separate system, described by an ordinary quantum field. The technique used was explicit manipulation of the algebra of operators which create or destroy excitations of the string. Even the pre-string approach to dual-resonance models could be reformulated in such a picture, as was pointed out independently by Nambu, Nielsen and Susskind [1].

Of course, one is not really interested in interactions of strings with external fields, one wishes to consider interactions of strings with themselves. In fact, starting from the picture of strings interacting with an external field, one was able to obtain a perturbative S-matrix theory of strings interacting with themselves. Higher-order

*This work supported by the National Science Foundation under Research Grant NO.PHY81-18547.

terms were obtained from lower order terms by explicit summation over intermediate states. The calculations were fairly lengthy; the technique, known as "sewing legs", was developed to a fine art. It was not obvious before completing the calculations that n-loop amplitudes with all the required properties would exist; it appeared rather miraculous that they did. In fact, the failure to obtain a suitable one-loop amplitude for $d \neq 26$ was the first indication that consistent models could not be constructed for smaller values of d. (Lovelace had previously shown that ghosts would exist in a covariant model with $d > 26$).

In spite of the complexity of the calculations, Alessandrini and Amati, and Kaku and Yu [2], independently calculated the n-loop amplitude for models current at the time, which possessed ghosts. The calculations for the ghost-free models were never completed, but the result was conjectured, and the much simpler interacting string calculations confirm the conjecture. The method has not been applied to superstrings beyond one loop.

We shall not attempt to give references to old dual-model calculations, but we should mention that most calculations in the ordinary Bose-string model, as well as calculations of tree amplitudes in the Neveu-Schwarz model without fermions, were first performed in this framework.

b) Interacting-string picture.

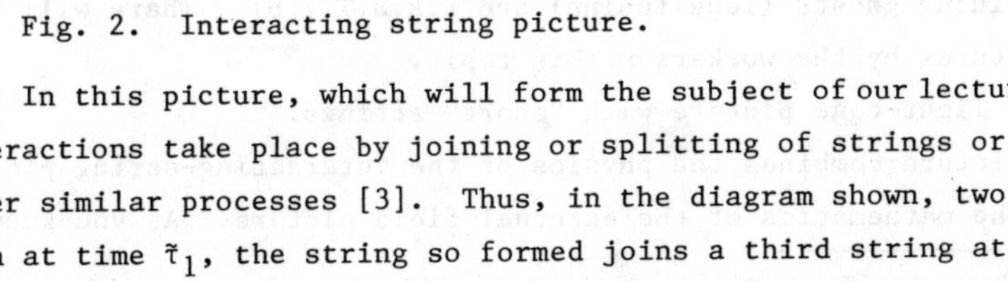

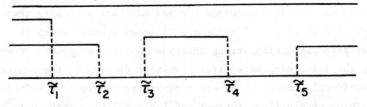

Fig. 2. Interacting string picture.

In this picture, which will form the subject of our lectures, interactions take place by joining or splitting of strings or by other similar processes [3]. Thus, in the diagram shown, two strings join at time $\tilde{\tau}_1$, the string so formed joins a third string at $\tilde{\tau}_2$, the string splits at time $\tilde{\tau}_3$, and so on. Fig. 2 represents a single-loop process, but tree diagrams or n-loop diagrams can be drawn in a similar way. Closed-string processes can be similarly described, as we shall see in more detail in the following section.

The technique used in the calculations is functional integration over all world sheets traced out by the string, and no summation over intermediate states is necessary. The mathematics required involves functions associated with potential theory of Riemann surfaces; elliptic functions (for one-loop amplitudes) and automorphic functions (for n-loop amplitudes). We shall give a short outline of automorphic functions in sec. 7*. Once one is familiar with them the calculations, even for n-loop processes, are straightforward.

The interacting-string picture was originally worked out in the light-cone frame, and we shall use it for all our calculations. The Lorentz invariance of the theory is not manifest, but it can be proved provided $d=d_{crit}$ [4]. A knowledge of the free string in light-cone coordinates, as treated in Thorn's first lecture, or in the original article by Goddard, Goldstone, Rebbi and Thorn [5], will be assumed. Even though we use the light-cone frame, the S-matrix for the Bose string will be obtained in a manifestly Lorentz-invariant form. We believe that we can obtain a superstring S-matrix for which Lorentz invariance, while not "manifest" in the technical sense, is fairly evident. The problem of obtaining scattering amplitudes which are simultaneously manifestly supersymmetric and manifestly Lorentz invariant (for d=4, n>2 or d=10, all n) is a well-known problem which is not specifically associated with string theory

During the last year, much effort has been devoted to the construction of a manifestly covariant interacting-string theory in a Hilbert space containing ghosts (longitudinal and B.R.S.T.)[6]. There will be several lectures by the workers on this topic.

c) The light-cone picture with "short" strings.

This picture combines the physics of the interacting-string picture with the mathematics of the external-field picture. As you know

---

* The functions involved are not actually elliptic or automorphic functions, but functions related to them such as θ-functions and prime forms.

from Thorn's lecture, the "length" of a string in the light-cone frame is the momentum in the + direction. The process shown in Fig. 1 may therefore be re-interpreted as a tree diagram corresponding to Fig. 2, with all strings but two having infinitesimal P^+. One can thereby apply the techniques of the approach a) to an interacting-string picture. The method can also be adapted to single-loop processes. If the number of external particles is sufficiently small, one can take a Lorentz frame where all particles but two have infinitesimal P^+, so that the restriction does not involve a loss of generality. Many of the calculations by Green and Schwarz, including all their single-loop calculations, have been performed in this picture.

The disadvantage of the method is that no-one knows how to apply it to n-loop processes ($n \geq 2$), even in principle. For such processes the more general interacting-string picture b) appears necessary.

d) Functional integration over world sheets.

An *ansatz* first made by Gervais and Sakita [7] expresses the string S-matrix as a functional integration over two-dimensional surfaces (world-sheets traced out by the string). The necessity for the critical dimension was not evident in the original work of Gervais and Sakita, but Polyakov [8] reformulated the *ansatz* in such a way that the critical dimension is singled out. Interactions as such are not introduced explicitly; the order of the perturbation series corresponds to the number of holes or handles on the surface.

The theories of Gervais and Sakita and of Polyakov are not *manifestly* quantum-mechanical theories of the usual type. If the S-matrix were known to be unitary, this would not be a defect of such theories. However, at present the only suggested method of proving the unitarity of the S-matrix is to establish the equivalence of the approach with one of the other approaches outlined above. In the critical dimension the functional integral of the interacting-string picture is probably equivalent to that of the G.S. and Polyakov pictures with a specific choice of parametrization. Such a precise equivalence remains to be proved in general, however. It is manifest at the tree level, and Polchinski [9] has calculated the single-loop amplitude using Polyakov's prescription. He found the same formula as that calculated by manifestly unitary methods. One would like to unify the Polyakov method with the interacting-string method, light-cone or covariant, in such a way that unitarity is manifest. Progress in this direction, with the covariant interacting-string picture, will be reported in the lectures of Friedan and Shenker.

In the following section we shall set up the interaction Hamiltonian of the Bose string; in other words, we shall define the vertex functions. The method of functional integration will be applied to perturbative calculations in sec. 3. We shall perform the functional integration and obtain the S-matrix as an (ordinary) integral of a function which depends on the Neumann functions and the determinant of the Laplacian of a Riemann surface. The formula is known as the analog model. Tree amplitudes will be treated in secs. 4 and 5; after a change of variables we shall evaluate the Neumann functions (sec.4) and the determinant of the Laplacian (sec. 5), thereby deriving an explicit formula for tree amplitudes. Sec. 5, as well as subsequent starred paragraphs, may be omitted by the reader who does not wish to follow all details. In sec. 6 and 7 we shall carry out similar calculations for single-loop amplitudes and, in somewhat less detail, for n-loop amplitudes.

We shall confine the discussion in the present article to Bose strings. For fermionic strings a general formula corresponding to that of sec. 3 is easily derived, but a technical complication has so far impeded the explicit calculation of scattering amplitudes in what we believe to be their simplest form. We are hopeful that methods for solving this problem are now at hand, but they have not yet been worked out. In a companion paper to the present article, which appears in the first section of the seminars of this proceedings, we shall give a general treatment of the present methods applied to the fermionic string, including plausibility arguments for finiteness at the n-loop level.

2. BOSE-STRING VERTICES

We are regarding the string theory as an ordinary quantum-mechanical Hamiltoninan system. It is first quantized; the degrees of freedom will be the displacements of the string, not the operators that create or destroy strings. Interactions will be specified by stating explicitly the matrix element of the Hamiltonian between a given state of m strings and a given state of n strings, where m and n are either 1 or 2. In other words, we write

$$H = H_{free} + H_{int}, \qquad (2.1)$$

where H_{free} is the G.G.R.T. Hamiltonian described by Thorn, and the matrix elements of H_{int} will be stated explicitly here.

H_{int} will only contain two-, three- or four-string interactions; even the last is absent in a pure closed-string system. In this respect string theory is much simpler than any other theory of gravity, where the Hamiltonian, if expanded in a series of n-particle terms, contains an infinite number of such terms.

We begin by examining the vertex for the splitting or joining of strings (Fig. 3).

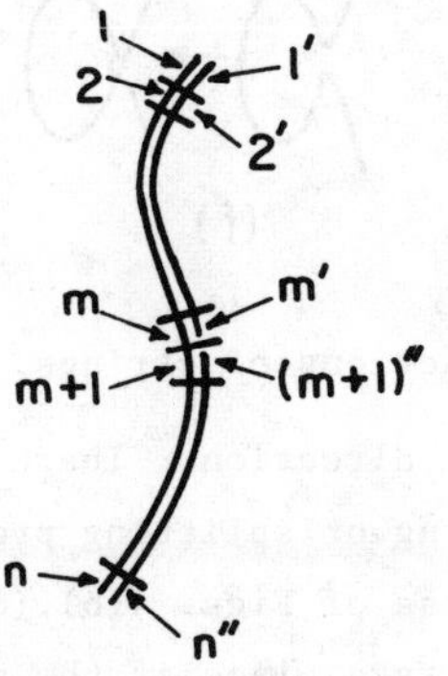

Fig. 3. Vertex for the splitting or joining of strings.

Our fundamental assumption is that the final strings occupy the same position as the initial string, i.e., *the vertex operator is only non-zero if the final and initial strings coincide along their entire lengths*. Thus, if we divide the strings into infinitesimal elements in the manner shown, the vertex operator will be:

$$g\delta(X_1-X_2')\ \delta(X_2-X_2')\ \cdots\ \delta(X_m-X_m')\ \ \delta(X_{m+1}-X_{m+1}'')\ \cdots\ \delta(X_n-X_n'') . \tag{2.2}$$

Unprimed, singly primed and doubly primed co-ordinates are associated with the incoming string, the first outgoing string and the second outgoing string respectively. The delta functions are (d-2)-delta functions in all transverse co-ordinates.

Other possible interaction terms, involving open or closed strings, are shown in Fig. 4. Fig. 4(a) is of course the same as Fig. 3. Each

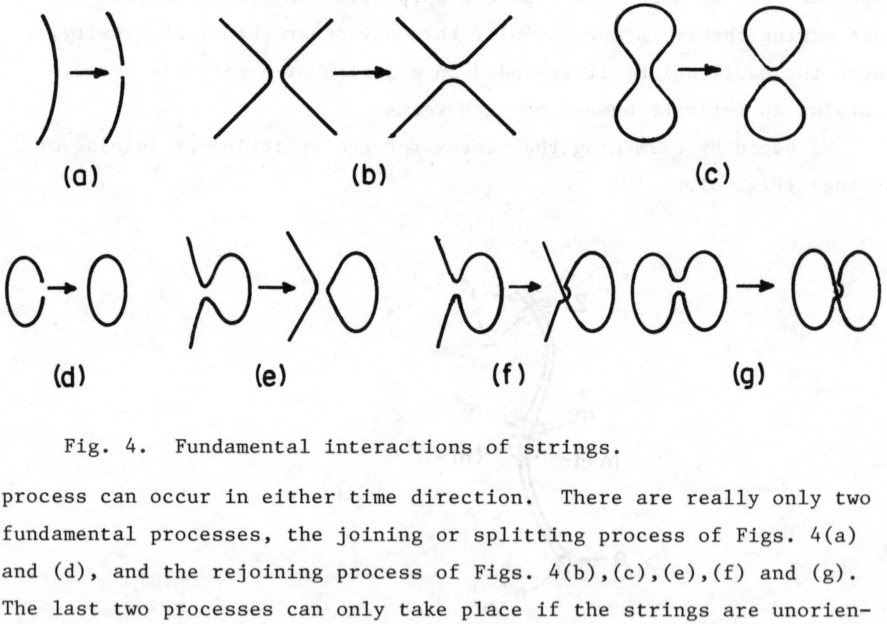

Fig. 4. Fundamental interactions of strings.

process can occur in either time direction. There are really only two fundamental processes, the joining or splitting process of Figs. 4(a) and (d), and the rejoining process of Figs. 4(b),(c),(e),(f) and (g). The last two processes can only take place if the strings are unoriented, i.e., if the string wave-function remains unchanged when the direction is reversed.

The vertex operators associated with all the interactions of Fig. 4 are obtained by generalizing (2.2) in the obvious way, and we need not write them down explicitly.

As we are working in the light-cone frame, it is necessary to prove that the interaction Hamiltonian described above leads to a Lorentz invariant theory. (In the covariant formulation, it is similarly necessary to prove ghost decoupling). We shall not reproduce the proof here; we direct the interested reader to ref. 4.

One necessary condition for Lorentz invariance is that the interaction be local, i.e., that the joining-splitting or rejoining interaction be independent of the configuration of the string away from the joining point. As a consequence, the possibility of the process shown in Fig. 4(a) implies that of the process shown in Fig. 4(d). A theory with only open strings is not consistent; a string must be able to close. Furthermore, we shall see in sec. 4 that a Lorentz-invariant system with open strings must contain both the joining-splitting and the rejoining interaction, and that the coupling constant for the latter must be equal to the square of that for the former. It is possible to obtain a pure closed-string system where the joining-splitting interaction is absent, and where the only vertices are those represented by Fig. 4(c) and, if the string is unoriented, by Fig. 4(g). (At the moment we are ignoring the extra restrictions imposed by absence of tachyons and anomalies.)

3. FUNCTIONAL INTEGRATION

Functional-Integration Formula for Strings.

The procedure of functional integration provides an efficient method for the perturbative calculation of scattering amplitudes in string theory. A general process is represented by Fig. 2. The horizontal τ-axis represents the light-cone time; since we are calculating S-matrix elements, *all* string diagrams will extend between $\tau=-\infty$ and $\tau=\infty$. The vertical σ-axis represents $\frac{1}{2\pi}$ times the P^+ momentum of that portion of the string below it. Conservation of P^+ enables us to draw the diagram with horizontal lines as shown. The d-2 transverse directions are at right angles to the paper. The transverse co-ordinates are of course discontinuous across the horizontal lines, since these lines bound different strings.

We begin by representing the propagation of free strings, between interactions, by functional integration. The formula may thus be represented symbolically in the form (with imaginary time):

$$S = \int d\tilde{\tau}_2 \cdots d\tilde{\tau}_{N-2} \sum_{int} \sum_{int'} \cdots \langle f | \int \mathcal{D} X^i(\sigma,\tau)\, \mathrm{Exp}\{-\int_{\tilde{\tau}_{N-2}}^{\infty} L d\tau\} | int \rangle$$

$$\times \langle int | v | int' \rangle \langle int' | \int \mathcal{D} X^i(\sigma,\tau)\, \mathrm{Exp}\{-\int_{\tilde{\tau}_{N-3}}^{\tilde{\tau}_{N-2}} L dt\} | int'' \rangle$$

$$\times \langle int'' | \cdots | i \rangle. \tag{3.1}$$

All functional integrations are products of factors, one for each string present between the interaction times. The amplitude $\langle int|v|int'\rangle$ is the matrix element of the vertex operator. We perform the integral $\int d\tilde{\tau}_2 \cdots d\tilde{\tau}_{N-2}$ over all interaction times but one, since the integration over the last interaction time simply gives us energy conservation. (The number of particles is denoted by N, so that the number of interactions is N-2.)

We now recall that the vertex operator v is a delta function in the X co-ordinates along the string. In other words, the X co-ordinates at the time $\tilde{\tau}_r+\varepsilon$ is the same as that at $\tilde{\tau}_r-\varepsilon$. Since we have to sum over all intermediate states, i.e., over all values of the X co-ordinates at the interaction time, we may simply forget about the interactions and *perform a single functional integral over the whole string diagram Fig. 2.*

We thus obtain the following formula for the S-matrix element between ground states of momentum P_r:

$$S = \int d\tilde{\tau}_2 \cdots d\tilde{\tau}_R\; B(\tilde{\tau}_1, \cdots, \tilde{\tau}_R), \tag{3.2a}$$

$$B = g^N \mathcal{N} \int \mathcal{D} X^i(\sigma,\tau)\, \mathrm{Exp}\{i \sum_r (\pi\alpha_r)^{-1} P_r^i \int d\sigma X^i(\sigma,\tau_1)$$

$$- \int d\sigma d\tau\, \mathcal{L}(\sigma,\tau) - \sum_r P_r^- \tau_1\}, \tag{3.2b}$$

$$\mathcal{L} = \frac{1}{4\pi} \left\{ \left(\frac{\partial X^i}{\partial \tau}\right)^2 + \left(\frac{\partial X^i}{\partial \sigma}\right)^2 \right\}. \tag{3.2c}$$

The first term in the exponent comes from the wave-function of the incoming and outgoing states. The indices i and r identify the component and the string respectively, and $\pi\alpha_r$ is the "length" of the string. All momenta P^i, P^- and $\alpha(=2P^+)$ are defined negatively for outgoing strings. The time τ_1 is the initial or final time, i.e., $\tau_1=-\infty$ for incoming strings and $+\infty$ for outgoing strings. More precisely, τ_1 is a constant, not necessarily the same for each string, and we are interested in the limit $\tau_1 \to +\infty$ or $-\infty$. The integration $\int d\sigma X(\sigma,\tau)$ is of course carried out over the length of the string, in the negative direction when α is negative. For simplicity we have suppressed the wave-functions of the incoming and outgoing strings except in their dependence on the barycentric co-ordinates $(\pi\alpha_r)^{-1}\int d\sigma X^i(\sigma,\tau_1)$; such wave-functions will affect the result only through a wave-function normalization factor (if the external particles are in their ground states). The last term in the exponent of (3.2b) arises because the S-matrix is defined (roughly) as the transition amplitude from which the factors Exp $(P^- t)$ of the external states have been removed.

We emphasize that the functional integral in (3.2b) depends on the shape of the string diagram through the boundary conditions. The normal derivative of X must be zero along the solid lines, and X can be discontinuous across these lines, since the term $(\partial X^i/\partial\sigma)^2$ in $\mathcal{L}$ connecting points on different sides of the solid lines is absent.

The Analog Model

The functional integral (3.2b) only involves the variables of integration $X^i(\sigma,\tau)$ linearly and quadractically in the exponent. It is therefore a Gaussian and may be performed explicitly. We recall the formula (for ordinary integration)

$$\int d\xi_1 \cdots d\xi_M \text{ Exp } \{A^T\xi - \tfrac{1}{2}\xi^T B\xi\} = |\tfrac{B}{\pi}|^{-\frac{1}{2}} \text{ Exp } \{\tfrac{1}{2}A^T B^{-1} A\}, \quad (3.3)$$

where A and ξ are column matrices and B is a square matrix.

We therefore require the reciprocal and the determinant of the Laplacian operator (with a negative sign), subject to the boundary conditions that the normal derivative be zero at the boundary (Neumann boundary conditions). In fact the reciprocal of the Laplacian with Neumann boundary does not exist owing to the presence of a zero mode, but we define the following function, known as the Neumann function:

$$-\left\{\left(\frac{\partial}{\partial\sigma}\right)^2+\left(\frac{\partial}{\partial\tau}\right)^2\right\} N(\sigma,\tau;\sigma',\tau')=-2\pi\delta(\sigma-\sigma')\delta(\tau-\tau'), \quad (3.4a)$$

$$\frac{\partial}{\partial n} N(\sigma,\tau;\sigma',\tau') = f(\sigma,\tau), \quad (3.4b)$$

where $\partial/\partial n$ is the normal derivative on the boundary, and f is an arbitrary function on the boundary which is independent of σ' and τ'. If there is no boundary, as is the case for closed-string diagrams, we may use the definition (which could also be used when there is a boundary)

$$-\left\{\left(\frac{\partial}{\partial\sigma}\right)^2+\left(\frac{\partial}{\partial\tau}\right)^2\right\} N(\sigma,\tau;\sigma',\tau')=-2\pi\delta(\sigma-\sigma')\delta(\tau-\tau') + f(\sigma,\tau), \quad (3.4c)$$

where f is a function independent of σ' and τ'. From the conservation-of-momentum equation $\sum_r P_r^i=0$ it may easily be checked that the extra terms on the right of (3.4) do not affect the validity of the formula for Gaussian integration, and that the result is independent of the function f. The determinant, of which we shall defer consideration, must also be modified.

We therefore obtain the following formula for the functional integral (3.2b):

$$B = g^{N-2}\,\mathfrak{N}|\Delta|^{-(d-2)/2}\ \mathrm{Exp}\ \{\tfrac{1}{2}\sum_{r,s}\ \pi^{-2}\ \alpha_r^{-1}\ \alpha_s^{-1}\ P_r^i P_s^i$$

$$\times \int d\sigma d\sigma' N(\sigma,\tau_1;\sigma',\tau_1) - \sum_r P_r^- \tau_1\} \quad . \quad (3.5)$$

The operator Δ is $-\frac{1}{2\pi^2}$ times the Laplacian, and $\mathfrak{N}$ is the Feynman normalization factor.

We may combine the second term in the Laplacian with the first by making use of the formula

$$\tau=(2\pi)^{-1}\sum_r \int_r d\sigma' N(\sigma,\tau;\sigma',\tau'), \quad (3.6)$$

where the σ' integration is over the rth string. Eq.(3.6) follows from the fact that both sides satisfy Laplace's equation and have normal derivations of -1 for $\tau=\tau_i$, 1 for $\tau=\tau_f$, and zero at the ends of the string. Hence

$$-\sum_r P_r^- \tau_1 = -(2\pi)^{-1} \sum_{r,s} P_r^- \int_s d\sigma' N(\sigma,\tau_1;\sigma',\tau_1), \qquad (3.7)$$

with σ on the rth string and σ' on the sth string. The right-hand side of (3.7) is independent of σ, so that we may integrate over σ and divide by $\pi\alpha_r$. Recalling that $\alpha_s = 2P_s^+$, we may write:

$$-\sum_r P_r \tau_1 = -\sum_{r,s} P_r^- P_s^+ \pi^{-2} \alpha_r^{-1} \alpha_s^{-1} \int_r d\sigma \int_s d\sigma' N(\sigma,\tau_1;\sigma',\tau_1) \qquad (3.8)$$

Eq. (3.5) thus becomes:

$$B = g^{N-2} \mathcal{N} |\Delta|^{-(d-2)/2} \operatorname{Exp} \{-\tfrac{1}{2} \sum_{r,s} \pi^{-2} \alpha_r^{-1} \alpha_s^{-1} P_r \cdot P_s \times \int_r d\sigma \int_s d\sigma' N(\sigma,\tau_1;\sigma',\tau_1)\} \quad , \qquad (3.9)$$

where $P_r \cdot P_s$ *is a d-vector product.*

Since the τ_1's approach $\pm\infty$, the whole of an incoming or outgoing string may be regarded as a single point in the calculation of the Neumann functions. This will become more evident when we examine conformal transformations in the following section. The function $N(\sigma,\tau_1;\sigma',\tau_1)$ will thus depend on the strings involved but, if $r \neq s$, they will not depend on the precise value of σ or σ'. We may therefore omit the integration over σ; the range of integration will simply cancel the factors $(\pi\alpha)^{-1}$. The terms with $r=s$ require a separate discussion; we shall see in the following section that they are simply normalization terms. We thus obtain our final general result:

$$B = g^{N-2} \mathcal{N} |\Delta|^{-(d-2)/2} \mathcal{M} \operatorname{Exp} \{-\sum_{r>s} P_r \cdot P_s \, N(\rho_r,\rho_s)\} \qquad (3.10)$$

where ρ_r represents any point on the rth external string, and

$$\mathcal{M} = \prod_r \operatorname{Exp} \{-\frac{\mu^2}{2} (\pi\alpha_r)^{-2} \int_r d\sigma \int_r d\sigma' N(\sigma,\tau_1;\sigma',\tau_1)\}. \qquad (3.11)$$

The symbol μ^2 is the square of the mass of the ground state, equal to $-(d-2)/24$. (We refer to Brink's lecture for a proof of this fact.) The formula (3.10) is the analog model of Fairlie and Nielsen.

We notice that all the momentum dependence of (3.10) is in the last factor.

It is evident that our treatment also applies to closed-string diagrams with the vertex Fig. 4(c). For the rejoining interaction, the integral $\int d\tilde{\tau}$ of (3.2a) will be replaced by a double integral over the interaction time and the position of the interaction point. The formula (3.10) will remain unchanged.

4. THE KOBA-NIELSEN FORMULA

The expression (3.10) is perfectly general; it applies to trees and n-loop diagrams. The calculation of the functions N and Δ will of course depend on the number of loops. In this section we shall confine ourselves to tree diagrams, and we begin with the function N.

The technique used will be that of conformal transformation, since the Neumann function is conformally invariant and is known for suitable regions. We combine the variables τ and σ into a single complex variable

$$\rho = \tau + i\sigma. \tag{4.1}$$

We now conformally transform the string diagram onto the upper half-plane (Fig. 5). The formula for doing so is a particularly simple

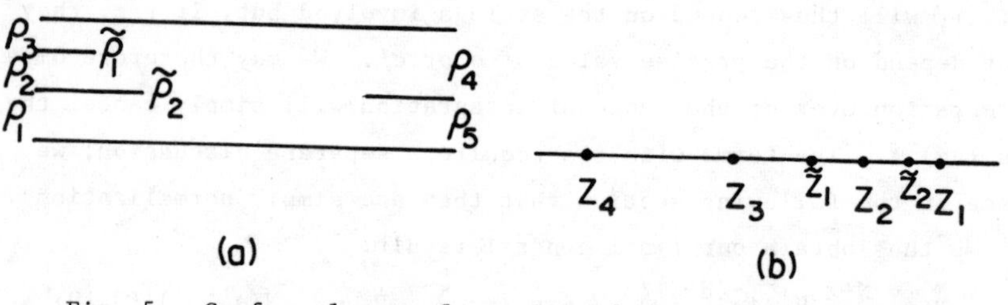

Fig. 5. Conformal transformation of the string diagram onto the upper half-plane.

case of the Schwarz-Christoffel transformation:

$$\rho = \sum_{r=1}^{N-1} \alpha_r \, \ell n(z - Z_r), \tag{4.2}$$

where Z_r is the point on the real axis onto which the rth string at $\tau = \pm\infty$ transforms. We have taken $Z_N = \infty$. It is easy to see that the transformation (4.2) does have the correct properties. Thus, as z decreases from ∞ to Z_1, ρ will decrease from ∞ to −∞. When z passes through the point Z_1, ρ will jump by iα. It will then increase until z reaches a certain value $\tilde{z}_2$, following which it will decrease to $-\infty$ as z approaches Z_2, and so on. Notice that we define the Z's as the points onto which the incoming and outgoing strings transform, the $\tilde{z}$'s, as the points onto which the joining points $\tilde{\rho}$ transform. For later use we observe that the $\tilde{z}$'s are given by the equation

$$\frac{\partial \rho}{\partial z}\Big|_{z=\tilde{z}} = 0, \tag{4.3a}$$

i.e.,

$$\sum \frac{\alpha_r}{\tilde{z}-Z_r} = 0. \tag{4.3b}$$

The Neumann function for the upper half-plane is

$$N(z,z') = \ln|z-z'| + \ln|z-z'^*|. \tag{4.4}$$

It is easily verified that this function does satisfy the defining equations.

The variables Z_r are not uniquely defined by the string diagram, since we may combine the conformal transformation (4.2) with a one-to-one conformal transformation of the upper half-plane onto itself. The most general such transformation is

$$z' = \frac{Az+B}{Cz+D} \tag{4.5a}$$

where A,B,C,D are real constants with AD-BC>0. We may multiply A,B,C and D by the same constant without changing (4.5a), and we usually fix the parameters by requiring that

$$AD-BC = 1. \tag{4.5b}$$

The transformation (4.4) is known as a projective transformation, a Möbius transformation or a linear fractional transformation. We can use it to fix three of the Z's at arbitrary values in a given cyclic order; if we take $Z_N = \infty$, we may fix two further Z's. One often takes $Z_1 = 0$, $Z_{N-1} = 1$, $Z_N = \infty$.

In principle we could invert Eq. (4.2) to find z in terms of ρ, use Eq. (4.4) to find $N(\rho,\rho')$, and substitute in Eq. (3.11). However, (4.2) cannot be inverted in terms of elementary functions, and it is much easier to change the variables of integration in (3.2a). For fixed α's, the shape of the string diagram depends on the $\tilde{\tau}$'s or on the Z's (with three of the latter fixed at arbitrary values), so that these sets of variables are functions of one another. We may accordingly change the variables $\tilde{\tau}_2 - \tilde{\tau}_1, \cdots, \tilde{\tau}_{N-2} - \tilde{\tau}_1$ to $Z_2, \cdots, Z_{N-2}$. The ρ_r's in (3.10) then become replaced by Z_r's, and we may substitute (4.4) in (3.10) and (3.2) to obtain the formula:

$$S = g^N \int_{Z_1}^{\infty} d\, Z_2 \cdots \int_{Z_{N-3}}^{Z_{N-1}} d\, Z_{N-2}\, \mathcal{N} |\Delta|^{-(d-2)/2} \mathcal{M}' \left|\frac{\partial \tilde{\tau}}{\partial Z}\right| \times \prod_{s<r<N} (Z_r - Z_s)^{-2P_r \cdot P_s} \qquad (4.6)$$

The factor $|\partial\tilde{\tau}/\partial Z|$ is of course the Jacobian of the transformation from the Z's to the $\tilde{\tau}$'s. We have set the variables Z_1 and Z_{N-1} at fixed values. The factor $\mathcal{M}'$ includes the factors $(Z_r - Z_s)^{-2P_r \cdot P_s}$ with r=N, $Z_N \to \infty$ the effect of these factors is to reverse the sign in the exponent of the factor in (3.11) with r=N.

It remains to evaluate the momentum-independent factor $\mathcal{N} |\Delta|^{-(d-2)/2} \mathcal{M} |\partial\tilde{\tau}/\partial Z|$. In our original work [3] we avoided a direct calculation of this factor. Following a procedure often used by Feynman, we used the functional integral to exhibit explicitly the dependence of the amplitude on certain parameters (in this case on the P's), and then chose the values of the parameters in such a way that the calculation was easily performed by other methods. We also made use of the known Lorentz invariance of the result. By a judicious choice of the Lorentz frame and of the values of the P's, we could obtain a configuration where the only relevant intermediate state between interactions was the ground state, so that the summation over states was trivial. On fixing the values of Z_1, Z_{N-1} and Z_N at 0, 1 and ∞ respectively, we found that the four momentum-independent factors in (4.6) simply gave us the relativistic factors $\prod (2P_r^+)^{-\frac{1}{2}}$. Since we made use of Lorentz invariance, the result was only true in the critical dimension.

The disadvantage of this method of circumventing the calculation of $|\Delta|$ is that it is not applicable beyond one loop. For n-loop diagrams (n≥2), there exists no sufficiently general configuration where the calculation is simply performed by other methods. We shall not therefore reproduce our calculations here, even though they are not very difficult. Instead, we shall outline a direct calculation of $|\Delta|$ (and of the other momentum-independent factors) in the following section.

If we fix Z_1, Z_{N-1} and Z_N at general values, we obtain the Koba-NIelsen formula for the covariant amplitude A (i.e., the amplitude without the factor $(2P^+)^{-\frac{1}{2}}$.

$$A = g^{N-2}\int_{Z_1}^{\infty} d\, Z_2 \ldots \int_{Z_{N-3}}^{Z_{n-1}} dZ_{N-2}\ (Z_N-Z_{N-1})(Z_{N-1}-Z_1)(Z_M-Z_1) \times \prod_{s<r\leq N} (Z_r-Z_s)^{-2P_r\cdot P_s} \tag{4.7}$$

Eq. (4.7) is justified by observing that A remains unchanged if all the Z's are subjected to the same projective transformation and that it reduces to (4.6), without the momentum-independent factors, if Z_1, Z_{N-1} and Z_N are fixed at the values 0, 1 and ∞ respectively. The integral is also independent of the choice of the Z's to be kept fixed, provided that the first three factors are replaced by the differences between the fixed Z's. The amplitude is therefore symmetric under cyclic interchange of the external particles, as it should be.

We notice that the integration in (4.7) diverges if $Z_r=Z_s$ and $2P_r\cdot P_s\geq 1$, i.e., $(P_r+P_s)^2\geq 1$. One then defines the amplitude by analytic continuation in the variable $(P_r+P_s)^2$, and one finds without difficulty that the amplitude has poles in $(P_r+P_s)^2$ at all integers ≥ -1. One finds similar poles in the variables $s_I = (\sum_{r\in I}P_r)^2$, where I is any set of consective r's. Such poles arise from the region where all the Z_r's, $r\in I$, approach one another or equivalently, by projective invariance, where the Z_r's with $r\in I$ are separated by a large distance from those with $r\notin I$. This region corresponds to the region in the string diagram, Fig. 5(a), where the interactions involving the particles r with $r\in I$ are separated by a large time interval from those involving the particles with $r\notin I$. The poles are therefore those associated with single-particle intermediate states in the Channel I.

In Fig. 5, the three Z's with positive α, corresponding to incoming strings, are adjacent to one another, as are the two Z's with negative α. Other configurations are possible. Thus, if the string diagram in Fig. 6(a) is mapped onto the upper half-plane, the Z's

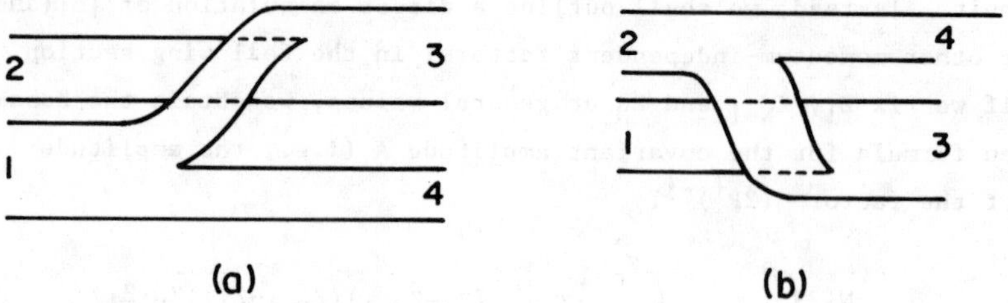

Fig. 6. String diagrams topologically similar to one with a rejoining interaction.

would be in the order Z_1, Z_3, Z_2, Z_4, with $\alpha_1,\alpha_2>0,\alpha_3,\alpha_4<0$. In this case eq. (4.3) would not always yield real values for the two joining points, i.e., not all real values of Z in the above cyclic order can be obtained by conformally transforming Fig. 6 onto the upper half-plane. Fig. 6(a) is topologically similar to a diagram where strings 1 and 2 go into 3 and 4 by a single rejoining interaction (Fig. 4(b)), as well as to Fig. 6(b)*. Any set of values of Z in the above cyclic order can be obtained by conformally transforming one and only one of the three diagrams onto the upper half-plane. The Koba-Nielsen integral (4.7), with the Z's in the above cyclic order, yields the sum of the scattering amplitudes corresponding to the three diagrams.

One of the three diagrams in question may be conformally equivalent to another with different values of the α's. The proof of Lorentz invariance of the S-matrix requires a conformal transformation between diagrams with different values of the α's, and it would therefore fail for any of the three diagrams taken separately. It is indeed fairly evident that the Koba-Nielsen integral for each of the three diagrams is not Lorentz invariant; the range of the Z integrations foreach diagram depends on the α's and therefore on the P^+'s. The Koba-Nielsen amplitude for the sum of the diagrams will be Lorentz invariant only if the amplitudes for individual diagrams all

*Interchange of incoming and outgoing particles is not allowed in the topology.

have the same overall constant g^2. Since the diagrams have two joining -splitting interactions or one rejoining interaction, *the coupling constant for the rejoining interaction is equal to the square of that for the joining-splitting interaction.* String models thus possess only one independent dimensionless coupling constant. Of course, this argument does not exclude the possibility that open strings, and the joining-splitting interaction, may be completely absent.

The analog of the Koba-Nielsen formula for closed strings, the Virasoro-Shapiro formula, can be derived in the same way. The boundariless closed-string diagram is conformally transformed onto the whole complex plane instead of the upper half-plane, with the $\tau=\pm\infty$ ends of the strings transforming onto *complex* points Z_r. Formulas (4.2) and (4.3) remain unchanged, with the "lengths" of the strings being $2\pi\alpha_r$. The original integral $\int d\tilde{\tau}$ in (3.2a) is now replaced by a double integral $\int d\tilde{\tau}\, d\tilde{\sigma}$, or $d^2\tilde{\rho}$, so that the integral $\int dZ_r$ after change of variables will be replaced by $\int d^2Z_r$ (defined as $\int dX_r dY_r$). Eq. (4.4) for the Neumann function in the z plane is replaced by the simpler equation

$$N(z,z') = \ln(z-z'). \tag{4.8}$$

Our final formula is then:

$$-4i\pi(ig^2/4\pi)^{N-2}\int d^2Z_2 \ldots \int d^2Z_{N-2} |Z_N - Z_{N-1}|^2 |Z_{N-1} - Z_1|^2 \; |Z_N - Z_1|^2 \times \prod_{r>s} |Z_r - Z_s|^{-P_r \cdot P_s}. \tag{4.9}$$

The diagrams analogous to Figs. 5 and 6 are no longer topologically distinct; the amplitudes corresponding to them can be added, and the ranges of integration in (4.7) cover the whole complex Z-plane.

5. THE DETERMINANT OF THE LAPLACIAN FOR TREE DIAGRAMS

Regularization.

We now outline the evaluation of the momentum-independent factors of (4.6) and, in particular, of the determinant of the Laplacian. We must first decide how to regularize the determinant, since the Laplacian is a singular operator. We shall in fact consider the determinant of the Laplacian and the Feynman normalization $\mathcal{N}$ together.

The boundary of the open-string diagram, Fig. 2, has infinite curvature at the interaction points. In the corresponding closed-string diagram, the surface itself has infinite curvature at the interaction points. We shall consider such points as limiting cases of configurations where the boundary or surface has large but finite curvature over a small region. As the curvature approaches infinity, $|\Delta|$ will split into the product of a finite factor which depends on the shape of the diagram, and a divergent factor which depends only on the curvature near the joining point. The latter factor will be absorbed into the definition of the coupling constant and will be unimportant for our purposes. It might be objected that such renormalization, if infinite, would destroy the finiteness properties that certain string models are claimed to possess in perturbation theory. However, the renormalization occurs at the tree level and is free of higher-order corrections, and is therefore harmless; it cannot give rise to anomalies and is unrelated to the β-function. In fact, Thorn has shown in his lattice model that there is no net infinite renormalization of the coupling constant at the tree level. The reader is referred to his lectures.

The logarithm of Δ, expressed as an integral operator, has the following behavior when its two arguments approach one another on a flat surface:

$$(\ell n\Delta)(\rho,\rho') = \frac{1}{\pi}|\rho-\rho'|^{-2} + \text{non-singular terms.} \tag{5.1}$$

In calculating $|\Delta| = \mathrm{Exp}\,\{\mathrm{Tr}\,\ell n\Delta\}$, we shall drop the contribution from the first term on the right of (5.1). Our regularization is probably equivalent to ζ-function regularization, used by mathematicians. By performing it we subtract an infinite term proportional to the area of the world sheet from the action or, in other words, we subtract an infinite term proportional to the length of the string from the energy. The sum of the lengths of the strings participating in a reaction, i.e., the sum of the P^+'s, is conserved, and the term subtracted has no physical significance. Such a regularization has previously been used in the calculation of the mass of the ground state by Brink and

Nielsen [10], presented by Brink to this conference.

On a curved surface, the right-hand side of (5.1) has an extra logarithmic singularity proportional to the curvature. This term is also subtracted out; since the surface is flat away from the interaction points, the subtraction is simply a coupling-constant renormalization.

The logarithm of the Feynman normalization factor $\mathcal{N}$ is proportional to the area of the world sheet, with possible extra contributions at the interaction points. One may therefore include it in the term to be subtracted out, and we need pay no further attention to the Feynman normalization.

At the boundary of an open string, there will be further singular contributions from the determinant of the Laplacian and from the Feynman normalization. The Brink-Nielsen calculation, which makes a cut-off in mode space, shows that the boundary terms combine with the surface terms to give a result proportional to the area. Finally, extra singular terms from the boundary near a joining point are renormalizations of the coupling constant for the joining interaction.

We should emphasize that the regularization in question is a regularization of the *Feynman functional integral*. It has nothing to do with a regularization of possible ultra-violet or infra-red infinities.

Another question that must be settled before the determinant of the Laplacian is defined is that of the zero mode. We shall replace the zero eigenvalue of the Laplacian by $1/\mathcal{A}$, where $\mathcal{A}$ is the area of the surface. The eigenvalue then behaves correctly under an overall change of scale. We shall also add an overall factor $1/2\pi$. It is not difficult to check that this is the correct prescription for the Feynman functional integral with *Neumann* boundary conditions for the zero mode of the string at the $\tau = \pm\infty$ ends of the diagram. The factor $(2\pi\mathcal{A})^{\frac{1}{2}}$ in $|\Delta|^{-\frac{1}{2}}$ arises from integrating over the co-ordinates of the zero-momentum wave-functions of the initial and final strings, with a transition amplitude given by the exponential of minus the classical action.

We have omitted the wave-functions of the higher modes in our functional-integration formulas. The amplitudes associated with such modes die off exponentially with time, and they will contribute at most a wave-function normalization factor, possibly α-dependent, whatever boundary conditions we use. The normalization factor may be determined by calculating the two-point S-matrix in zero-order perturbation theory. With open strings we shall find that there is no wave-function normalization provided that, in the calculation of $\ell n|\Delta|$, we take the average of Neumann and Dirichlet boundary conditions for the higher modes at $\tau = \pm\infty$. With closed strings the situation is simpler; there is no wave-function normalization whether we use Neumann or Dirichlet boundary conditions.

Calculation of $|\Delta|$

For tree diagrams the determinant of the Laplacian, like the Neumann function, can be calculated by making conformal transformations. Since all tree diagrams are conformally equivalent to the upper half-plane (or the whole plane for closed strings), they are conformally equivalent to one another. A knowledge of the behavior of $|\Delta|$ under a conformal transformation is therefore sufficient to perform the calculation.

The change of $|\Delta|$ under a conformal transformation has been calculated by McKean and Singer [11]. Their work has been applied to the Polyakov string model by Alvarez [12] and by Durhuus, Nielsen, Olesen and Petersen [13]. We quote the following formula, from ref. 12, for the change in the regularized determinant under a conformal transformation:

$$\delta(\ell n|\Delta|) = -\frac{1}{6\pi}\left\{\tfrac{1}{2}\int d^2z\sqrt{\hat{g}}\;\hat{g}^{ab}\frac{d\sigma}{dz^a}\frac{d\sigma}{dz^b} + \int d\hat{\ell}\;\hat{k}\sigma + \tfrac{1}{2}\int d^2z\sqrt{\hat{g}}\hat{R}\sigma\right\} \tag{5.2}$$

"Hatted" and unhatted symbols are quantities before and after the conformal transformation respectively. The variable σ is the logarithm of the scale change*:

$$|g| = |\hat{g}|e^{2\sigma} \tag{5.3}$$

* It may perhaps be worth while to point out that, in the present section, σ is *not* the imaginary part of ρ. The conflicting notation is unfortunately standard.

The first and last integrals are over the surface, the second over the boundary, $\hat{k}$ is the curvature of the (old) boundary and $\hat{R}$ is the scalar curvature of the (old) surface.

We shall use (5.2) to find the change in $|\Delta|$ under a conformal transformation *from* the upper half z-plane *to* the open-string diagram.

We use Eq. (4.2) to transform from the z-plane to the string diagram (i.e., we take $Z_N = \infty$); we then remove small semi-circular regions surrounding the points Z_r, $r \neq N$, as well as the region $|z| > R, R \to \infty$, surrounding Z_N. The first term in the braces of (5.2) is evaluated by partial integration; since $d^2\sigma/d\rho^2$ is zero except for a singular contribution at the interaction points, we may write:

$$\frac{1}{12\pi} \int d^2\rho (d^2\sigma/d\rho^2) = \frac{1}{12\pi} \int d\ell \sigma d_n \sigma, \tag{5.4}$$

where $\int d\ell$ is taken over small circles surrounding the interaction points and over the transforms of the boundaries surrounding the points $Z_r (1 \le r \le N)$, and d_n is the outward-directed normal derivative. The quantity $\ell n|\Delta|$ thus receives contributions from two sources.

i) The integrals (5.4) taken over the curves surrounding the joining points.

ii) The integrals (5.4) and the second term in braces of (5.2) taken along curves that exclude the external particles, as well as the effect of the excision of the regions surround the external particles on $\ell n|\Delta|$ in the z-plane. We first examine the contribution (i).

Near a joining point,

$$\rho - \tilde{\rho} = \tfrac{1}{2} c(z-\tilde{z})^2 \qquad c = \frac{\partial^2 \rho}{\partial z^2}. \tag{5.5}$$

Hence

$$e^\sigma = \left|\frac{\partial \rho}{\partial z}\right| = c|z-\tilde{z}|,$$

$$\sigma = \ell n\, c + \ell n\, |z-\tilde{z}| = \tfrac{1}{2} \{\ell n 2|\rho-\tilde{\rho}| + \ell n\, c\},$$

$$\sigma\, d_n\, \sigma = -\tfrac{1}{4}\, |\rho-\tilde{\rho}|^{-1}\, \ell n\, \{2|\rho-\tilde{\rho}| + \ell n\, c\},$$

$$-\frac{1}{12\pi} \int d\ell\, \sigma d_n \sigma = \frac{1}{24}\, (\ell n\, 2r + \ell n\, c), \qquad (5.6)$$

where r is the radius of the circle surrounding the joining point. If the boundary near the joining point is smoothed out in the manner discussed above, the first term on the right of (5.6) will give rise to a contribution that depends on the manner of smoothing but not on the shape of the string diagram, and it can be absorbed into the definition of the coupling constant. The second term gives a contribution of $(1/24)(d^2\rho/dz^2)$ to $\ell n\, |\Delta|$ so that,

$$|\Delta|_1^{-(d-2)/2} = \Pi' \left|\frac{\partial^2\rho}{\partial z^2}\right|^{-\frac{d-2}{48}}, \qquad (5.7)$$

where the subscript 1 indicates that we are only taking the contribution from the joining points, and the product is over all joining points. Throughout the paper, the product Π' will denote the product over all joining points.

The contribution from external particles is calculated in a similar way. The boundary of our region in the z-plane has a right-angle bend where the small semi-circle that excludes one of the Z's meets the real axis; the string diagram has right-angle bends at the ends of the string at $\tau = \pm\infty$. Normally such a corner would give a contribution to the second term of (5.2). The sign of this contribution is opposite depending on whether we use Neumann or Dirichlet boundary conditions at $\tau = \pm\infty$; it is of course taken for granted that we use Neumann boundary conditions at the ends of the string. As we have already stated, we shall average over Neumann and Dirichlet boundary conditions and thereby eliminate the term in question. Such a complication does not occur with closed strings.

We first notice that, in the z-plane

$$\ell n|\Delta|_z = \frac{1}{6}\, \{\sum_{r=1}^{N-1} \ell n\, \varepsilon_r - \ell n\, R\} + c_1 N + c_2, \qquad (5.8)$$

where ε_r is the radius of the semi-circle with center Z_r. The constants c_1 and c_2 may be dropped, since they may be absorbed into coupling constant and wave-function normalizations, the latter being fixed at unity by the calculation of the zero-order two-point function. (In fact, $2c_1 + c_2 = 0$.) Eq. (5.8) may easily be proved by applying projective transformations to change the values of the ε's and applying (5.2); one uses the fact that the result must be a sum of independent terms from different r's (if we use Neumann boundary conditions for the zero string mode at $\tau = \pm\infty$).

Near one of the Z_r's Eq. (2) can be approximated by the equation:

$$\rho = \alpha_r \,\ell n(z - Z_r), \tag{5.8}$$

so that

$$e^\sigma = \left|\frac{\partial\rho}{\partial z}\right| = |\alpha_r||z-Z_r|^{-1} = |\alpha_r|/\varepsilon_r,$$

$$\sigma = \ell n(|\alpha_r|/\varepsilon_r) = \ell n(|\alpha_r|-\mathrm{Re}\,\rho/\varepsilon_r,$$

$$\sigma\, d_n\sigma = \varepsilon_r^{-1}\,\ell n(|\alpha_r|/\varepsilon_r),$$

$$-\frac{1}{12\pi}\int d\ell\,\sigma d_n\sigma = -\frac{1}{12}\,\ell n(|\alpha_r|/\varepsilon_r, \tag{5.9a}$$

$$-\frac{1}{6\pi}\int d\hat{\ell}\,\hat{k}\sigma = \frac{1}{6}\,\ell n(|\alpha_r|/\varepsilon_r). \tag{5.9b}$$

Adding (5.9a), (5.9b) and the appropriate term of (5.8), we find that the contribution from the semi-circle with center Z_r is:

$$\ell n|\Delta|_{2r}^{-(d-2)/2} = -\frac{d-2}{24}\,\ell n\,|\alpha_r| + \varepsilon_r). \tag{5.10}$$

We can combine the term (5.10) with the contribution from the rth string to the factor $\mathcal{M}$, defined in (3.10). The quantity $\ell n\mathcal{M}$ is equal to $(d-2)/48$ times the average of $N(\rho,\rho')$, i.e., of $2\,\ell n\,|z-z'|$, over the rth string. Apart from a constant wave-function renormalization term this is just $(d-2)\,\ell n\,\varepsilon_r/24$, so that

$$\mathcal{M}_r\,\ell n\,|\Delta|_{2r}^{-(d-2)/2} = -\frac{d-2}{24}\,\ell n|\alpha_r|.\quad r \neq N \tag{5.11a}$$

The contribution from the semi-circle r=R may be calculated in a similar way; we find

$$\mathcal{M}'_N \, \ell n \, |\Delta|_{2N}^{-(d-2)/2} = -\frac{d-2}{24} \, \ell n \, |\alpha_N| . \tag{5.11b}$$

Combining (5.7), (5.11a) and (5.11b), and recalling that our regularized determinant includes the factor $\mathcal{N}$, we obtain finally:

$$\mathcal{N} |\Delta|^{-(s-2)/2} \mathcal{M}' = \{ \Pi' \left| \frac{\partial^2 \rho}{\partial z^2} \right|^{-\frac{1}{2}} \prod_{r \neq N} |\alpha_r|^{-1} |\alpha_N| \}^{(d-2)/24} \tag{5.12}$$

The right-hand side of (5.12) is unity when N=2, so that our wave-functions are correctly normalized.

Jacobian of the Conformal Transformation

We now turn to the last of the momentum-independent factors , i.e., the Jacobian $|\partial\tilde{\tau}/\partial Z|$. A direct calculation of this factor may be made if N=4, and probably also for general N. Such a direct calculation may well be prohibitively difficult with n-loop amplitudes, and we shall therefore use analyticity arguments to prove that the product

$$\left| \frac{\partial \tau}{\partial Z} \right| \quad \Pi' \left| \frac{\partial^2 \rho}{\partial z^2} \right|^{-\frac{1}{2}} \tag{5.13}$$

is indepndent of Z.

We regard (4.2) as defining ρ as a function of z when the quantities Z_r are complex; in other words, we consider a closed-string diagram. There are two possible sources of singularities of (5.13) as a function of the Z_r's; a pair of joining points or a pair of Z_r's may approach one another. In the first case we easily find that

$$\left| \frac{\partial \tilde{\tau}}{\partial Z} \right| \approx c_1 \delta^{\frac{1}{2}}, \quad \Pi' \frac{\partial^2 \rho}{\partial z^2} \approx c_2 \delta, \tag{5.14}$$

where δ is the distance between the two $\tilde{z}$'s. In the second case there will be a single $\tilde{z}$ near the pair of coinciding Z_r's, and

$$\left| \frac{\partial \tilde{\tau}}{\partial Z} \right| \approx c_3 \delta'^{-1}, \qquad \Pi' \frac{\partial^2 \rho}{\partial z^2} = c_4 \delta'^{-2}, \tag{5.15}$$

where δ' is the distance between the two Z's. In neither case does the expression (5.13) have singularities, so that it is independent of the Z's.

To calculate the value of (5.13) as a function of the α's, we take a configuration where $Z_1=0$, $Z_{r+1}/Z_r>>1$, $Z_N=\infty$. In the corresponding ρ-plane configuration the difference between all interaction times is large. There will be one $\tilde{z}$ of the order of magnitude of each Z_r, let us call it $\tilde{z}_r$. We may put $\tilde{z}_r-Z_s=\tilde{z}_r$ if s<r and $= \infty$ if s>r from (4.3b) we find that

$$\tilde{z}_r = \frac{\gamma_{r-1}}{\gamma_r} Z_r, \tag{5.16}$$

where

$$\gamma_r = \sum_{s\leq r} \alpha_r \tag{5.17}$$

We then find from (4.2) that

$$\tilde{\rho}_r = \gamma_r \, \ln Z_r + \text{const.} \tag{5.18}$$

and hence that

$$\left|\frac{\partial\tilde{\rho}}{\partial Z}\right| = \prod_{r=2}^{N-2} \gamma_r Z_r^{-1} . \tag{5.19}$$

(We recall that the variables are $\tilde{\rho}_r-\tilde{\rho}_{N-1}$ and Z_r, $2\leq r\leq N-2$.) It is also easily found by differentiating (4.3b) that

$$\frac{\partial^2\rho}{\partial\tilde{z}_r^2} = \gamma_r^3 \; \gamma_{r-1}^{-1} \; \alpha_r^{-1} \; Z_r^{-2}, \tag{5.20}$$

and hence

$$\left|\frac{\partial\tilde{\tau}}{\partial Z}\right| \; \prod{}' \; \left|\frac{\partial^2\rho}{\partial z^2}\right|^{-\frac{1}{2}} = \prod_{r=1}^{N-1} \; \alpha_r^{\frac{1}{2}}\alpha_N^{-3/2} Z_{N-1}, \tag{5.21}$$

since $\gamma_2 = \alpha_1$, $\gamma_{N-1} = \alpha_N$.

On multiplying (5.12) by (5.21) we obtain the result quoted in the last section, proivded d=26. We notice that we obtain the correct α-dependence only if d=26, and that the α's are not trivial normalization factors - a point that has been stressed by Thorn.

Closed Strings

The calculation of $|\Delta|$ for closed-string diagrams proceeds in the same way. All semi-circles of integration, both near the interaction points and near the external particles, become replaced by circles, and all contributions to $|\Delta|$ become replaced by their absolute squares. The Jacobian $|\partial\tilde{\tau}/\partial Z|$, too, becomes replaced by its absolute square, since the integrals are now $\int d^2\tilde{\rho}$ or $\int d^2Z$. Thus all cancellations occur as before, and the momentum-independent factors simply give a contribution $\prod_r \alpha_r^{-1}$.

The amplitude must now be mulitplied by a factor $\prod_r \ell_r$, where ℓ_r is the length of the rth string, since the interaction can take place anywhere on the incoming and outgoing strings. We must also include a normalization factor $\prod_r \ell_r^{-\frac{1}{2}}$ to take into account the arbitrariness in the origin of the σ co-ordinate. We are left with a factor $(4\pi)^{N/2}$ $\prod_r (2P_r^+)^{-\frac{1}{2}}$. As with open strings, we know that the wave-functions are correctly normalized, so that the factor $(4\pi)^{N/2}$ is not a wave-function normalization factor. We saw in the previous section that the rejoining coupling constant, as defined by the three-point function, must be ig^2; the factor i is due to the fact that the integration in the rejoining term of the open-string diagram is in the Imρ direction. We thus require a coupling constant renormalization of $i\ (4\pi)^{-3/2}$, so that we have an overall factor of $-4i\pi(i/4\pi)^{N-2}$

6. SINGLE-LOOP DIAGRAMS

A single-loop diagram cannot be conformally transformed onto the upper half-plane (or the whole plane,) as it is not simply connected.

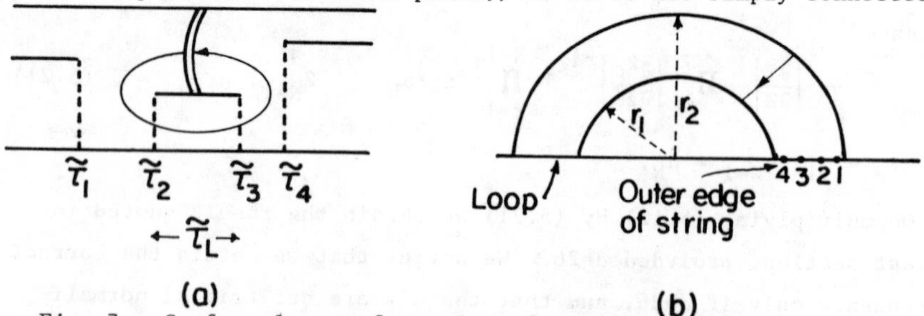

Fig. 7. Conformal transformation of one-loop uiagram

Instead, it is cut along the double line of Fig. 7(a) and is transformed onto a semi-annulus. The whole boundary of the string, excluding the middle line, is transformed onto the segment of the positive real axis, while the middle line (the "loop") is transformed onto the segment of the negative real axis. The left and right sides of the double line are transformed onto the outer and inner semi-circles respectively, and points on the two semi-circles with the same polar angle are identified.

The z-plane Neumann functions, besides satisfying (3.4), must be unchanged under multiplication by $r_1/r_2 \equiv w$. The logarithms of the Jacobi θ-functions [14], with some simple extra factors, satisfy our requirements. It is easy to derive the infinite series for the Neumann functions without prior knowledge of θ-functions; the method will be useful for generalization to the n-loop case. Thus*

$$N(z,z') = \ln |\Psi(z'/z,\ w)| + \ln|\Psi(z'^*/z,w)| \tag{6.1}$$

where

$$\ln\Psi(x,w) = \ln(1-x)-\tfrac{1}{2}\ln x + \sum_{n=1}^{\infty}\{\ln(1-w^n x)+\ln(1-w^n/x) -2\ln(1-w^n)\}+\ln^2 x/2\ln w. \tag{6.2}$$

The leading term is $\ln|z-z'|+\ln|z-z'^*|$ as before. The substitution $z\to wz$ replaces each term in the infinite series by its successor or predecessor, "end effects" at n=0 being taken care of by the second and last terms, so that the *function is periodic under the transformation* $z\to wz$. It is easily checked that (6.1) satisfies the equation (3.4). Its only singularity in the semi-annulus is at $z=z'$, but there will of course be further singularities at the equivalent points $w^{\pm n}z'$.

The last term in the braces of (6.2) is arbitrary; it has been inserted so that all corrections to the leading term vanish when $z=z'$. Any such corrections would affect the factor $\mathcal{M}$, Eq.(3.11). We should notice, however, that the first term in (6.20)is $\ln(z-z')-\ln z$ instead of $\ln(z-z')$ as it was for trees. Hence

$$\mathcal{M}_{loop} = \mathcal{M}_{tree} \prod_r Z_r^{-1} . \quad (d=26) \tag{6.3}$$

It is convenient to define the variables usually used in the theory of elliptic functions:

---

*We follow the notation of ref. 15.

$$\zeta = \ln z/2i\pi \qquad \xi = \ln x/2i\pi = \zeta' - \zeta \qquad \tau = \ln w/2i\pi. \quad (6.4)$$

The function Ψ is then given by the equation

$$\Psi(x,w) = -2i\pi \, \mathrm{Exp}\,(i\pi\xi^2/\tau)\theta_1(\xi|\tau)\theta_1{}'(0|\tau), \qquad (6.5)$$

where θ_1 is the first Jacobi θ-function [14].

The conformal transformation from the z-plane to the ρ-plane is given by a formula similar to (4.2), with functions analogous to (5.1) replacing the logarithm:

$$\rho = \sum_{r=1}^{N} \alpha_r \ln\{\Psi(Z_r/z,w)\} + \text{const.}$$

$$= \sum_{r=1}^{N} \alpha_r \ln\{\Psi'(Z_r/z,w)\} - \Sigma\alpha_r \ln \ln Z_r/\ln w + (\text{const.})', \quad (6.6)$$

where $\ln \Psi'$ is given by the right-hand side of (6.2) without the last term. The verification that the positive segment of the real z-axis transforms onto the outer boundary of the string diagram proceeds in the same way as for trees. On the negative segment of the real axis, only the second sum in (6.6) has a non-zero imaginary part. Thus

$$i\pi\alpha_L = (i\,\mathrm{Im}\rho)_{\mathrm{Loop}} = i\pi\,\Sigma\,\alpha_r \ln Z_r/\ln w. \qquad (6.7)$$

We observe that the negative segment of the real z-axis does map onto a horizontal line in the ρ-plane as required; furthermore, we obtain an expression for $\pi\alpha_L$, the length of the string below the loop.

In our method of defining the conformal transformation we fixed the origin and the point at infinity in the z-plane, so that the transformation from ρ to z cannot longer be combined with an arbitrary projective transformation of the z's, as it could for trees. It still has a one-parameter arbitrariness corresponding to a scale change of z.

An important difference between trees and loops is that all tree diagrams are conformally equivalent to one another, since they can all be conformally transformed onto the complex plane, whereas all loop diagrams are not. Two semi-annuli (Fig. 7(b)) with different values of w,i.e., of r_1/r_2, cannot be conformally transformed onto one another. A one-loop diagram is thus characterized by a single conformal invariant w. As with trees, we shall transform our ρ-plane variables of integration to z-plane variables, and w will be one such variable. In the ρ-plane the variables of integration will be the N-1 differences between the $\tilde{\tau}$'s, together with α_L. The Z-plane variables will be N-1 Z's (since the scale is arbitrary) and w.

*Determinant of the Laplacian

To find the value of $|\Delta|$ it is no longer sufficient to investigate conformal transformations, since all one-loop diagrams are not conformally equivalent. We must find how the determinant of the Laplacian operator for the semi-annulus of Fig. 7, with the boundary conditions that f(z)=f(wz), varies as a function of w; we then conformally transform the semi-annulus onto the string diagram and employ the procedure we used for trees.

In previous calculations [16, 17] the problem was treated as a quantum-mechanical string with periodic boundary conditions in the time. The sum over states in the calculation of the action could easily be performed. Here we shall evaluate the determinant explicitly. Our method will be to find the change of $|\Delta|$ under the non-conformal transformation:

$$z' = z + \varepsilon z\, \ell n|z|/w\,\ell n\, w. \tag{6.8}$$

The transformation has the property that it converts the equation $z_2 = wz_1$ into $z_2' = (w + \varepsilon)z_1'$, i.e., it increases w by ε.

On using the formula $\delta\, \ell n|\Delta| = \mathrm{Tr}(\delta\Delta)\Delta^{-1}$, we find that the change of $\ell n|\Delta|$ caused by the change (6.16) is:

$$\delta \ell n|\Delta| = -\frac{\varepsilon}{\pi w \ell n w}\,\mathrm{Tr}\,\Big\{\frac{1}{z^*}\frac{\partial}{\partial z}\Big(z\frac{\partial}{\partial z}\Big) + \frac{1}{z}\frac{\partial}{\partial z^*}\Big(z^*\frac{\partial}{\partial z^*}\Big)$$

$$+2(1+2\ell n|z|)\frac{\partial^2}{\partial z \partial z^*}\Big\}N(z,z'). \tag{6.9}$$

The right-hand side of (6.9), divided by ε, is just $d\ell n|\Delta|/dw$. We express N as the sum of four terms (without absolute-value signs) by (6.1), and use the series (6.2) for Ψ. It can be shown that the change in the term $\ell n|z-z'|+\ell n|z-z'^*|$ just gives the change in the regularization terms, and we omit it. On inserting the other terms, we find that

$$d\ell n|\Delta|/dw = -\frac{1}{\pi \ell n w |z|^2} \mathrm{Tr}\Big\{-\sum_{n=1}^{\infty}\Big[\frac{2w^{n-1}}{(1-w^n)^2} + \frac{w^{n-1}}{(1-z^*z^{-1}w^n)^2} + \frac{w^{n-1}}{(1-zz^{*-1}w^n)^2} + \frac{1}{w\ell n w}\Big]\Big\}. \quad (6.10)$$

(We have set $z' = z$ inside the trace.) To obtain the trace, we integrate between semi-circles of radii w z and z. Thus

$$d\ell n|\Delta|/dw = -2\sum_{n=1}^{\infty}\frac{w^{n-1}}{(1-w^n)^2} + \frac{1}{w\ell n w}. \quad (6.11)$$

As far as we can see, the easiest way of integrating the infinite series in (6.10) is to expand each term in a power series, integrate, and recombine the terms. We then find that

$$|\Delta|_z = 2\sum_n \ell n(1-w^n) + \ell n|\ell n w| + \text{const.} \quad (6.12a)$$

The constant can be found by taking the limit of small w. In this case only the last term in (6.2) is significant and, more important, the dependence on the polar angle may be neglected, so that we have a one-dimensional problem with periodic boundary conditions. The determinant of the operator $\frac{d^2}{dz^2}$ (z real) may easily be calculated, either directly or by going back to the Feynman formalism. We find

$$\text{const.} = -\ell n\pi \quad (6.12b)$$

The value of $|\Delta|$ in the ρ-plane can now be calculated by multiplying the exponential of (6.12a) by (5.7) and (5.11a), since the derivation of the last two factors is equally valid for trees and for loops. There is an additional contribution from (5.4) applied to the double line in Fig. 8(a); the factor is easily seen to be

$$w^{1/12}. \quad (6.13)$$

Notice that we do *not* obtain a contribution from the second term on the right of (5.2) applied to the semi-circles in Fig. 7 (b), since the semi-circles are not boundaries. Combining (5.7), (5.11a), (6.3). and (6.12), we obtain the result:

$$\mathcal{N}|\Delta|_\rho^{-(d-2)/2}\mathcal{M} = \prod_r Z_r^{-1}\, \{\Pi' \left|\frac{\partial^2\rho}{\partial z^2}\right|^{-\frac{1}{48}} \prod_r |\alpha_r|^{-\frac{1}{24}} w^{-\frac{1}{24}}$$

$$\times (\pi/|\ell n w|)^{\frac{1}{2}} \prod_n (1-w^n)^{-1}\}^{d-2}. \qquad (6.14)$$

We shall defer the examination of the Jacobian $|\partial\tilde{\tau}/\partial Z|$ until we we have treated closed loops, since we shall require the analytic properties of the amplitude when the variables become complex. We shall find an equation similar to (5.13),

$$\left|\frac{\partial(\tilde{\tau},\alpha_L)}{\partial(Z,w)}\right| = (w|\ell n w|)^{-1}\, Z_N \left\{\Pi\left|\frac{\partial^2\rho}{\partial z^2}\right|'\right\}^{\frac{1}{2}} \Pi\, \alpha_r^{\frac{1}{2}}, \qquad (6.15)$$

where Z_N is the variable over which we do not integrate.

Open-string One-loop Amplitude and its Singularities:

We can now insert (6.1), (6.2), (6.14) and (6.15) into our general formula (3.9) to obtain the result:

$$A = g^N\int_0^1 dw \int_{wZ_N}^{Z_N} dZ_1 \int_{Z_1}^{Z_N} dZ_2 \cdots \int_{Z_{N-2}}^{Z_N} dZ_{N-1}\, w^{-2}(\pi/|\ell n w|)^{13}$$

$$\times \prod_{n=1}^{\infty} (1-w^n)^{-24} \prod_{r=1}^{N-1} Z_r^{-1} \prod_{s<r} \{\Psi(Z_s/Z_r,w)\}^{-2P_r\cdot P_s}. \qquad (6.16)$$

The amplitude (6.16) has the expected one-particle poles and, as before, they arise from the region of integration where the Z_r's with $r \in I$ or with $r \notin I$ approach one another. When both sets of Z's approach one another we have a double pole due to a self-energy insertion. In addition, the amplitude has an imaginary part, i.e., considered as a function of the complex variable s_I it has a cut on the real s_I-axis. The cut arises from the region $w \to 0$ of the integration or, in the ρ-plane, the region where the time interval between the loop interactions ($\tilde{\tau}_L$ in Fig. 7(a)) becomes large. The imaginary part is thus the unitarity contribution of the two-particle intermediate state. In fact the integration in (6.16) diverges when $w \to 0$, and it is not difficult to check that the divergence is due to the tachyons and "photons" in the two-particle intermediate state.

Finally, the integrand of (6.16) is highly singular at the point w=1, and the amplitude is therefore divergent. We shall not discuss the divergence of the single-loop amplitude in these lectures, we direct the reader to Ref. 16 (and many others). Suffice it to say that the divergence can be analyzed by making a Jacobi transformation from w=1 to w=0; we shall treat such transformations at the end of this section. One then finds that the divergence can be absorbed into a coupling-constant renormalization and a "slope" renormalization, i.e., a renormalization of the overall mass scale.

The divergence may be interpreted in two ways. Since the region $w \to 1$ corresponds to the limit where the two loop interactions take place within a very short time interval, the divergence can be regarded as a typical ultra-violet divergence. Alternatively, we may observe that Fig. 7(a) is topologically the same as the diagram for the process $1+2 \to C+3+4$, where C is a closed string emitted by the process Fig. 4(e), and τ_f for the string C is *finite*. In other words, string C is emitted and disappears into the vacuum. The closed-string spectrum includes a massless scalar particle, the "dilaton", and the divergence occurs because the propagator $(P^2)^{-1}$ of the disappearing particle must be evaluated at its pole.

Closed-string Loop

Fig. 8(a) represents a closed-string loop. In this case the

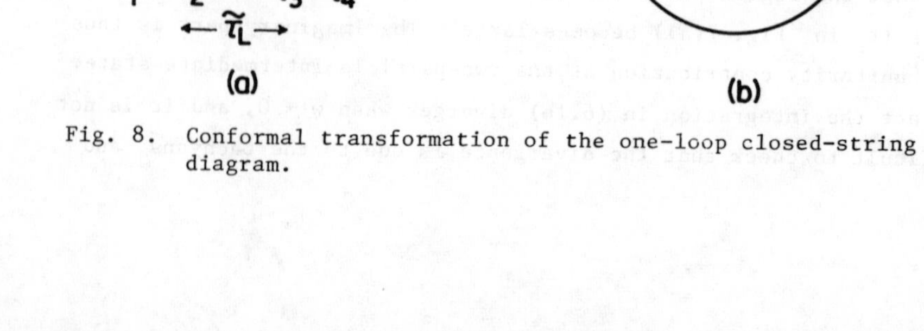

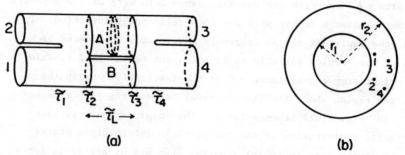

Fig. 8. Conformal transformation of the one-loop closed-string diagram.

string diagram is broken along the double dashed line and conformally transformed onto the whole annulus of Fig. 8 (b), with the left and right sides of the dashed line transformed onto the outer and inner circle respectively. The interactions can take place anywhere along the strings or, in other words, the strings above and below the loop may twist by an arbitrary angle before rejoining. Hence, in the z-plane, we identify points on the inner and outer circles whose polar angles differ by a fixed amount, i.e., we identify points z and wz, where w is a *complex* variable with $|w|=r_1/r_2$. The closed-string loop diagram, which is a Riemann surface with one handle, is thus characterized by one complex conformal invariant w.

The Neumann function is now

$$N(z,z') = \ell n|\chi(z'/z,w)| \tag{6.17}$$

where

$$\ell n\, \chi(x,w) = \ell n(1-x)-\tfrac{1}{2}\ell n x + \sum_{n=1}^{\infty} \{\ell n(1-w^n x) + \ell n(1-w^n/x) - 2\,\ell n(1-w^n)\} + \ell n^2|x|/2\ell n|w| \tag{6.18}$$

The function χ differs from Ψ by the modulus signs in the last term. As a result, χ is not analytic, and N satisfies Eq. (3.4c) rather than (3.4a). It is easily checked that the Neumann function is periodic under the transformation $x\to wx$, as required.

The conformal transformation from z to ρ is given by the formula

$$\rho = \sum_{r=1}^{N} \alpha_r\, \ell n\{\chi'(Z_r/z,w)\} - \sum \alpha_r \ell n z \ell n|Z_r|/\ell n|w|, \tag{6.19}$$

where $\ell n\,\chi'$ is given by the right-hand side of (6.18) without the last term. The value of α_L is given by a formula similar to (6.7):

$$2i\pi\alpha_L = (i\ \mathrm{Im}\rho)_{\mathrm{loop}} = -2i\pi\Sigma\alpha_r \ell n|Z_r|/\ell n|w|, \tag{6.20}$$

while the net twist (in units of α_L), i.e., the change of the $\mathrm{Im}\rho$ co-ordinate on going around the loop, is given by the formula

$$\beta_L = \Sigma\alpha_r\{\mathrm{Im}\ \ell n\ Z_r - \ell n\ |Z_r|\,\mathrm{Im}\ \ell n\ w/\ell n|w|\}. \tag{6.21}$$

If the parameters w and Z_r are gradually varied, we may reach a point where the time interval τ_L becomes infinitely short, but where the twisting angles are finite. Such a point is *not* a singular point. On varying the parameters further, the double dashed line in Fig. 8(a) becomes replaced by a line *surrounding* the loop.

Eqs. (6.19),(6.20) and (6.21) are not analytic in all the parameters. This is because the string diagram is characterized by N-1 complex interaction times and the the two real variables α_L and β_L, while its conformal transform is characterized by N complex variables- N-1 Z's and w.

*We calculate the determinant of the Laplacian in exactly the same way as for trees. Eq. (6.14) is replaced by the equation,

$$\mathcal{N}|\Delta|_\rho^{-(d-2)/2} \mathcal{M} = \prod_r |Z_r^{-2}| \{ \prod' |\frac{\partial^2 \rho}{\partial z^2}|^{-\frac{1}{24}} \prod_r |\alpha_r|^{-\frac{1}{12}} |w|^{-\frac{1}{12}}$$
$$\times (2\pi/|\ell n\ |w||^{-\frac{1}{2}} \prod_n |1-w^n|^{-2}\}^{d-2} . \qquad (6.22)$$

*Now let us calculate the Jacobian $|\partial\tilde{\tau}/\partial Z|$. The procedure is the same for open and closed strings; in the former case we allow our variables to be complex and restrict them to be real-valued at the end of the calculation, in the latter we take the absolute square of our result.

*As before we use arguments based on the analyticity, but the last term in (6.18) is not analytic in the Z_r's or in w. We therefore replace (6.19) by the equation

$$\rho = \sum_{r=1}^{N} \alpha_r\ \ell n\ \{\chi'\ (Z_r/z,\ w)\} - \alpha_L \ell n z, \qquad (6.23)$$

where α_L, instead of being given by (6.20), is treated as a fixed parameter. The change of the ρ co-ordinate on going round the loop is

$$\beta_L = \Sigma \alpha_r\ \ell n\ \ Z_r - \alpha_L\ \ell n\ w. \qquad (6.24)$$

We regard β_L as a fixed, purely imaginary parameter. We shall begin by examining the expression

$$\left|\frac{\partial\tilde{\rho}}{\partial Z}\right| \; \Pi' \; \left|\frac{\partial^2\rho}{\partial z^2}\right|^{-\frac{1}{2}} \tag{6.25}$$

where the derivatives in the Jacobian are to be taken with α_L and β_L fixed, and w varies with the Z's so that (6.24) remains true. Later we shall extend the Jacobian to include the dependent variables α_L and β_L and the independent variable w.

*The reasoning of sec. 5 can again be used to show that (6.25) has no singularities as a function of the Z's when w is kept fixed; the fact that the Z's are restricted by Eq. (6.24) does not affect the validity of our arguments. We wish to use the well-known elliptic-function theorem that a function analytic in an annulus and periodic at the boundaries is a constant. Eq. (6.24) interferes with the periodicity condition, so we assume that $\alpha_r = N_r\alpha$, where the N's are integers. The periodicity condition $Z_r \to w^{N_r n_r} Z_r$, with the n's integers such that $\Sigma\alpha_r n_r = 0$, is consistent with (6.24), and it is easily verified that (6.25) remains unchanged under this transformation. The expression is thus independent of the Z_r's when w is kept fixed. By continuity arguments the restriction $\alpha_r = N_r\alpha$ can now be removed.

*We next examine the w dependence. We can show without difficulty that (6.25) is invariant under a modular transformation (to be defined in the next sub-section), and any function of w which has modular invariance and which is holomorphic in a fundamental region of the modular group is known to be a constant. We shall see in the next sub-section that there exists a fundamental region within the circle $|w|=1$ which does not touch the boundary, and the only point within such a region where there may possibly be a singularity is the point w=0.

*An explicit calculation at w=0 is easily performed. Eq. (6.23) then reduces to the equation

$$\rho = \sum \alpha_r \, \ell n\, (z - Z_r) - \alpha_L \, \ell n\, z. \tag{6.26}$$

The approximation is valid as long as any ratio Z_r/Z_s is much greater than w, a restriction that can always be achieved consistently with (6.24). Eq. (6.26) is identical to (4.2), with r ranging from 0 to

$N + 1$, $\alpha_0 = -\alpha_L$, $\alpha_{N+1} = \alpha_L$. The calculation leading to (5.21) is thus valid, so that

$$\left|\frac{\partial\tilde{\rho}}{\partial Z}\right| \prod{}' \left|\frac{\partial^2\rho}{\partial z^2}\right|^{-\frac{1}{2}} = \left| \prod_{r=1}^{N} \alpha_r^{\frac{1}{2}} \alpha_L^{-1} Z_N \right|. \tag{6.27}$$

Hence the expression (6.25) is free of singularities at w=0, and Eq. (6.27) is valid for all values of w and the Z's.

*We recall that the variables α_L and β_L are kept fixed in the Jacobian on the right of (6.27). We actually require the Jacobian $\pi|\partial(\tilde{\rho},\alpha_L)/\partial(Z,w)|$ (open strings) or $\frac{1}{2}(2\pi)^2|\partial(\rho,\tilde{\rho}^*,\alpha_L,\beta_L)/\partial(Z,Z^*,w,w^*)|$ (closed strings). The extra factor $\frac{1}{2}$ is to correct for double counting due to interchange of the two points at which the string splits to form the loop [9]. Now it follows from (6.20) and (6.21) that:

$$2\pi \frac{\partial\alpha_L}{\partial|w|} = 2\pi \frac{\alpha_L}{|w|\ell n|w|}, \quad \frac{\partial\alpha_L}{\partial(\mathrm{Arg}\ w)} = 0, \quad 2\pi \frac{\partial\beta_L}{\partial(\mathrm{Arg}\ w)} = -2\pi\alpha_L. \tag{6.28}$$

For open strings we therefore obtain an extra factor $\pi\alpha_L/w|\ell n\ w|$, in agreement with the result previously quoted. For closed strings:

$$\tfrac{1}{2}(2\pi)^2 \left|\frac{\partial(\tilde{\rho},\rho^*,\alpha_L,\beta_L)}{\partial(Z,Z^*,w,w^*)}\right| \left|\frac{\partial^2\rho}{\partial z^2}\right|^{-1} = 2\pi^2 \left| \prod_{r=1}^{N} \alpha_r \quad Z_N^2 /w^{-2}|\ell n|w|| \right|. \tag{6.29}$$

Collecting together Eqs. (6.17),(6.18),(6.22) and (6.29) and inserting them into (3.10), we obtain the final result:

$$A = -\tfrac{1}{4}i\,4\pi(ig^2/4\pi)^N \int dw \int dZ_1 \ldots \int dZ_{N-1} |w|^{-4}(2\pi/|\ell n|w||)^{13}$$
$$\times \prod_{n=1}^{\infty} |1-w^n|^{-48} \prod_{r=1}^{N-1} |Z_r^{-2}| \prod_{s<r} |\chi(Z_s/Z_r,w)|^{-P_r\cdot P_s}. \tag{6.30}$$

The Z's must be integrated over the range $|wR|<|Z_r|<|R|$, where R is arbitrarily chosen.

Modular Invariance

We now have to investigate the range of the w-integration in (6.30). We begin by defining the modular group. It is convenient to transform from the variables z (or x) and w to ζ(or ξ) and τ, defined in (6.4). Functions which were singly periodic under the transformation $x \to wx$ are now doubly periodic under the transformations

$$\xi \to \xi + 1 \quad \xi \to \xi + \tau. \tag{6.31}$$

The annulus of Fig. 8(b), broken along the line $x = e^{y\,\ln w}$ $(0<y<1)$ becomes transformed onto the parallelogram with corners 0, 1, τ, $1 + \tau$. Breaking the annulus corresponds to breaking the string diagram along a line surrounding the loop.

Instead of the fundamental periods 1 and τ of (6.31), we may take a different pair, such as 1 and $\tau + 1$. Also, we may interchange the sides and rescale the parallelogram so that its new comers are 0, 1, $-\frac{1}{\tau}, -\frac{1}{\tau} + 1$. The two transformations are

$$\tau \to \tau + 1 \qquad \tau \to -\frac{1}{\tau}. \tag{6.32}$$

These transformations generate the *modular* group:

$$\tau' = \frac{A\tau+B}{C\tau+D} \quad A,B,C,D \text{ integral} \quad AD-BC=1. \tag{6.33a}$$

The reader may easily verify that the variables ξ transform as follows:

$$\xi' = \frac{\xi}{C\tau+D}. \tag{6.33b}$$

We may break the string diagram, Fig. 8(a), along the double dashed line and along a line surrounding the loop. By successively applying the transformations (6.19) and (6.4), we may then map it conformally onto the parallelogram. If the twisting angle is gradually changed by 2π, the period τ will become $\tau + 1$. If we perform the conformal transformation and then rotate and rescale the parallelogram as described above, the period τ will become $-1/\tau$. We may therefore replace the parameter τ associated with the string diagram by any value obtained from it after a modular transformation.

Since the Jacobi θ-functions are defined by their periodicity properties, they should transform simply when τ undergoes a modular transformation. In fact [14]

$$\theta_1(\xi'|\tau') = \varepsilon(C + D)^{\frac{1}{2}} \operatorname{Exp}\{i\pi C\xi^2/(C\tau+D)\}\theta_1(\xi|\tau), \tag{6.34}$$

where ε is a phase factor with $\varepsilon^8=1$. The partition function $\prod(1-w^n)$ is related to $\theta_1'(0)$ and it, too, has simple transformation properties:

$$w'^{\frac{1}{24}} \prod_n (1-w'^n) = \varepsilon^{\frac{1}{3}} (C\tau+D)^{\frac{1}{2}} w^{\frac{1}{24}} \prod_n (1-w^n). \quad (6.35)$$

The variables w and τ are of course related by (6.4).

Eq. (6.34) implies that the formula (6.19) for the conformal transformation from ρ to z is unaltered when all variable z, Z_r and w undergo a modular transformation.

A *fundamental region* of the modular group is defined as a region in the upper half τ-plane such that no two points in it are related by a non-trivial modular transformation, but any point in the upper half-plane may be obtained from a point in the region by a modular transformation. A particular fundamental region is defined by the inequalities:

$$-\tfrac{1}{2} < \mathrm{Re}\,\tau < \tfrac{1}{2}, \qquad |\tau| > 1, \quad (6.36a)$$

i.e,

$$|\ell n\, w| > 2\pi \quad (6.36b)$$

Since the right-hand side of Eq. (6.19) remains unchanged when the variables undergo a modular transformation, it follows that any configuration of the string diagram may be conformally transformed onto an annulus (or rectangle) with w (or τ) in a fundamental region. Hence *we must restrict the w-integrative of (6.30) to one fundamental region. In particular, we may choose the fundamental region (6.36).* This condition was first stated by Shapiro.

We may of course transform the string diagram onto any fundamental region, and the contribution from a given range of the integration variables cannot depend on which region we choose. The integrand of (6.30), together with the measure of integration, must therefore be invariant under a modular transformation.

In fact, we may divide (6.30) into several factors and assert that each is separately invariant under modular transformations. The Neumann functions are invariant under conformal transformations. We must include the factor $\mathcal{M}$ with the Neumann functions; it is not difficult to see that $\mathcal{M}/(\ell n|w|)^N$ is invariant. Thus

$$\prod_{s<r} |\chi(Z_s/Z_r,w)|^{-P_r \cdot P_s} (\ell n|w|)^N \tag{6.37a}$$

is invariant. The determinant of the Laplacian in the z-plane is invariant, provided we include the factor $|w|^{-1/12}$ from the conformal transformation. Hence the factor

$$|w|^{-2}(2\pi/|\ell n|w||)^{12} \prod_{n=1}^{\infty} |1-w^n|^{-48} \tag{6.37b}$$

is invariant. Finally, the factors for transforming the determinant of the Laplacian to the ρ-plane, as well as the product of the Jacobian $|\partial\tilde{\rho}/\partial\tilde{Z}|$ and the measure of integration are invariant. We must of course exclude the factors $(\ell n|w|)^N$, which come from $\mathcal{M}$. The factor

$$d^2w\, d^2Z, \ldots d^2Z_{N-1}\, |w|^{-2}(\ell n|w|)^{-N-1} \prod_{r=1}^{N-1} |Z_r^{-2}|$$

is therefore invariant. All these invariances may be explicitly verified using Eqs. (6.34) and (6.35); we emphasize, however, that it is not necessary to do so in order to establish the validity of our assertions.

As with open strings, the configuration where the time interval τ_L (Fig. 8(a)) is large corresponds to the limit $w \to 0$. The configuration where the loop shrinks to a point, i.e., where the ρ-interval, real and imaginary, between the interactions becomes small, can no longer correspond to the limit $w \to 1$, as this point is not in the fundamental region. It also corresponds to the limit $w \to 0$, but the transformation $z \to wz$ now represents the double dashed line in Fig. 8(b). Whether the limit $w \to 0$ corresponds to τ_1 being infinitely large or infinitely small depends on the values of the Z's; at the changeover point the length of one of the intermediate strings and τ_L will both be small, with the latter much larger than the former.

If a loop shrinks to a point and one of the intermediate strings becomes small, with the ratio of the dimensions remaining finite, w will not approach zero, but all the Z_r's will approach one another. Whenever a loop shrinks to a point all Z's will be in a region which, on a logarithmic scale, is small compared to the area of the annulus.

The region where all Z's approach one another does give rise to a divergence of the closed-string loop amplitude, and the divergence can be absorbed into a remormalization of the mass scale. The closed-string loop diagram is topologically similar to a tadpole diagram; a closed string emits a second closed string, which splits into a closed loop. We may again associate the divergence with the propagator of the disappearing dilaton.

7. THE n-LOOP AMPLITUDE

Conformal Transformation onto the z-Plane

Before treating the n-loop amplitude we shall indicate how the integrand of the one-loop amplitude may be rewritten in projectively invariant form. We may subject the z's of the previous section to an arbitrary fixed projective transformation. The relation $z \to wz$ itself becomes replaced by a projective transformation, while the circles with radii r_2, wr_2 become replaced by non-concentric circles. For convenience we shall suppose that they are outside one another. The string diagram is thus transformed onto the region exterior to two circles, the circumferences of which are related by a projective transformation. Points on the two circumferences which transform into one another by the projective transformation are identified.

We now consider an n-loop string diagram such as Fig. 9(a); for convenience we have drawn a two-loop diagram, but the discussion is completely general. Our treatment will be carried out within the

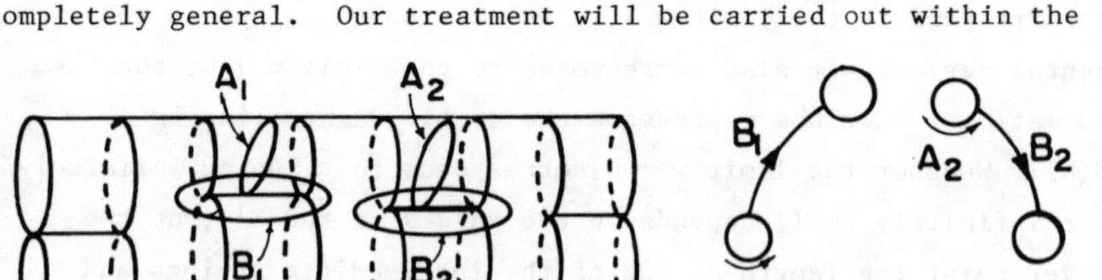

Fig. 9. Conformal transformation of an n-loop diagram onto the complex plane.

framework of closed strings. Any loop diagram, which is a Riemann surface with n handles, has a fundamental group generated by n A-cycles $A_1,\ldots A_n$ and n B-cycles $B_1,\ldots, B_n$. The cycle A_r intersects B_r once, otherwise no cycles intersect. Generators with these properties can be chosen in an infinite number of ways. We shall take the A-cycles to be the strings above the loops, and the B-cycles to be cycles surrounding the loops (Fig. 9(a)).

We now break the string diagram along the A-cycles, and map it conformally onto the region of the complex plane exterior to n pairs of circles; an A-cycle along which the diagram is broken maps onto a pair of circles in the z-plane. With each pair of circles we associate a complex projective transformation T_r $(1\leq r\leq n)$ that maps the circumferences onto one another. Points which transform into one another under the projective transformations are identified. The A-cycles thus *surround* the circles, the B-cycles take us *between* circles. The existence of a conformal transformation with the correct properties will be discussed shortly.

The n projective transformations $T_1,\ldots T_n$ and their reciprocals may be multiplied by one another in all combinations, with the restriction that two factors T_r and T_r^{-1} are not adjacent to one another, since such products would merely duplicate other products. We thus obtain an infinite group of projective transformations which represents the fundamental group of the string diagram. The projective transformations in the group will be denoted by $V_1, V_2,\ldots, V_i\ldots$, where the numbering is chosen arbitrarily. In the sequel, T_r will always denote one of the generators of the group, and the subscripts r or s $(1\leq r, s\leq n)$ will identify the generator. V_i will denote any projective transformation in the group, and the subscript i $(1\leq i<\infty)$ will identify the transformation.

Information about groups of projective transformations and automorphic functions may be found in the very readable book by Ford [18]. The encyclopaedic treatise of Fricke and Klein [19] may be used as a reference for matter not contained in Ford's book, and Chapter 12 of ref. 20 is a useful source of formulas and their proofs. For general properties of Riemann surfaces, ref. 21 may be consulted.

The generic projective transformation may written in the form

$$W: \quad \frac{x'-x_1}{x'-x_2} = w\,\frac{x-x_1}{x-x_2}\ . \qquad (7.1)$$

By interchanging x_1 and x_2 if necessary we can make $|w| \leq 1$, and we shall always do so. The points x_1 and x_2 are left unchanged by the transformation, and are known as *invariant points* or *limit points*. If $x \neq x_1$ or x_2,

$$\lim_{m\to\infty} W^m x = x_1, \quad \lim_{m\to\infty} W^{-m} x = x_2. \qquad (7.2)$$

Each projective transformation V_i will possess a pair of limit points. The set of limit points will have different properties depending on the original generators T_r. It may fill the whole plane, in which case the group is uninteresting. With certain groups there are regions of the z-plane which are free of limit points, but the set of limit points is not discrete. The classification of such groups according to the properties of their sets of limit points is complicated and is not of interest in the present context. We shall be interested in groups *whose limit points form a discrete set*. Such groups are called *Schottky* groups.

A fundamental region of a group of projective transformations is defined in the same way as a fundamental region of the modular group. (We may note in passing that the modular group is a group of projective transformations but it is not a Schottky group; its limit points cover the whole real axis.) It has been proved that any Schottky group with n generators has, as a fundamental region, the exterior of 2n topological circles (Fig. 8(b)). In order to be sufficiently general we have to drop the requirements that the circles in Fig. 8(b) be geometrical circles.

For future reference, we remark that each of the limit points z_{1r} and z_{2r} associated with the transformation T_r lies within one of the circles associated with this transformation. The transformation T_r will take any point outside the circle containing z_{2r} into that containing z_{1r}. Any transformation V_i whose leftmost factor is T_r will take a point in the fundamental region into the circle containing z_{1r}.

An important feature of Schottky groups is that all their elements may be written in the form (7.1), with $|w|$ strictly less than 1. As i becomes large (with any labelling), the parameter w associated with V_i approaches zero so that V_i takes any point, except its limit point z_2, to a point close to its limit point z_1.

Given generators $T_1,\ldots,T_n$ corresponding to a string diagram, we can construct alternative generators in two ways. We may subject them all to a given projective transformation, i.e., we may define $T_r' = A^{-1}T_rA$, where A is a projective transformation. We may also take a new set of A- and B- cycles which generate the fundamental group; the resulting transformation of the T's generalizes the modular transformation of the single-loop diagram.

A fundamental theorem conjectured by Poincaré and Klein and proved by Koebe [18] asserts that *any Riemann surface of genus n can be conformally transformed onto the complex plane with 2n holes in the manner described above. The transformation is unique up to an overall projective transformation and a modular transformation.*

Each generating projective transformation, as well as the arbitrary overall projective transformation, possesses three complex parameters. An n-loop string diagram, or, generally, a Riemann surface with n handles, is thus characterized by 3n-3 complex conformal invariants, a famous result of Riemann. One must of course integrate over the 3n-3 variables, as well as over the Z's, in the calculation of n-loop amplitudes.

The parametrization of Riemann surfaces up to conformal equivalence by Schottky groups is explicit, and the complex structure is manifest. There is no known explicit critierion for determining when a group generated by n projective transformations is a Schottky group. The lack of explicit knowledge of the Schottky region in the 3n-3 dimensional parameter space is not in itself a problem when calculating the n-loop amplitudes. One must integrate the 3n-3 parameters over one fundamental region of the appropriate modular group, and any such fundamental region lies within the the Schottky region except at known points.

In analogy with our treatment of the single-loop closed-string diagram, we have conformally transformed the n-loop diagram onto a multiply connected region of the z-plane. An alternative method would be to break the string diagram along the A-cycles and the B-cycles, and to transform it onto a simply connected region in the z-plane bounded by arcs of 2n circles. Instead of having a group generated by n complex projective transformations we would have a non-Schottky group generated by 2n real projective transformation. Groups of real projective transformations, known as Fuchsian groups, have been studied much more extensively by mathematicians than groups of complex transformations (Kleinian groups).

Automorphic and Related Functions

Before constructing the Neumann function we shall have to construct certain functions which were trivial in the one-loop case. The function $\ln z$ had the property that it changed by $2i\pi$ when we traversed the A-cycle of Fig. 8(b) and by $\ln w$, or $2i\pi\tau$, when we traversed the B-cycle. The term involving $\ln^2 z$ in the Neumann function was chosen because $\ln z$ had this property and the quantity τ, which was required in the definition of the modular group, was essentially the ratio of the two periods of $\ln z$. We require analytic functions in the n-loop z-plane with similar properties, i.e, they must change at most by constants when z traverses an A- or B-cycle.

A function with the required properties may be defined as follows:

$$v_r(z) = \sum_i{}^{(r)} \ln \frac{z-V_i z_{1r}}{z-V_i z_{2r}} \tag{7.3}$$

The points z_{1r} and z_{2r} are the two limit points of the transformation T_r, and the summation $\sum^{(r)}$ denotes a summation over all values of i *except* those values for which V_i, when expressed as a product of the T's, has a factor T_r or T_r^{-1} at its right-hand end. An extra such factor would not change the value of the term, so that we would count the same term an infinite number of times if we were to sum over *all* i.

We shall defer the question of the convergence of the series on the right of (7.3) for the moment, but we can at least see at this point that the higher terms do become small. For i large, the operator V_i takes any point except its invariant point z_{1i} onto a point near z_{2i}, so that the argument of the logarithm approaches unity.

To prove that the function v_r has the required properties it is convenient to subtract the constant $v_r(a)$:

$$v_r(z)-v_r(a) = \sum_i^{(r)} \ell n \frac{(z-V_i z_{1r})(a-V_i z_{2r})}{(z-V_i z_{2r})(a-V_i z_{1r})} . \qquad (7.4)$$

In the cross-ratio which forms the argument of the logarithm one can easily see that the operator V_i may be shifted onto the first term of each factor:

$$v_r(z)-v_r(a) = \sum_i^{(r)} \ell n \frac{(V_i^{-1}z-z_{1r})(V_i^{-1}a-z_{2r})}{(V_i^{-1}z-z_{2r})(V_i^{-1}a-z_{1r})} . \qquad (7.5)$$

We first examine the behavior of (7.5) when z traverses an A-cycle , i.e., when it goes once round a circle of Fig. 9(b). The points z_{1r} and z_{2r} lie within the circles associated with the transformation T_r. The point $V_i^{-1}z$ lies within one of the circles associated with its leftmost member, which by hypothesis is not T_r. Hence, if $V_i \neq I$, the points $V_i^{-1}z$ and z_{1r}, and $V_i^{-1}z$ and z_{2r}, lie within different circles and remain so as z traverses an A-cycle. Consequently the terms in (7.5) with $V_i \neq I$ do not change when z traverses the cycle. The term with $V_i= I$ changes by $2i\pi$ as z traverses an A-cycle surrounding the point z_{1r} in the counter-clockwise direction, by $-2i\pi$ if it traverses the A-cycle surrounding the point z_{2r}, and by zero if it traverses any other A-cycle. We use the symbols δ_{Ar} and δ_{Br} to denote the change in a function when z traverses the A or B cycle. Thus:

$$\delta_{As}\, v_r(z) = 2i\pi\delta_{rs} . \qquad (7.6)$$

Now let us examine the behavior of $v_r(z)$ when z traverses the sth B-cycle, with $s \neq r$. We must apply the transformation T_s to z, so that the right-hand side of (7.5) becomes

$$\sum_i^{(r)} \ell n \frac{(V_i^{-1}\, T_s z-z_{1r})(V_i^{-1}a-z_{2r})}{(V_i^{-1}\, T_s z-z_{2r})(V^{-1}a-z_{1r})} . \qquad (7.7)$$

We rewrite (7.7) in the form

$$\sum_i{}^{(r,s)} \sum_{n=-\infty}^{\infty} \ell n \frac{(V_i^{-1}\, T_s{}^{n+1}z - z_{1r})(V_i^{-1}\, T_s{}^n a - z_{2r})}{(V_i^{-1}\, T_s{}^{n+1}z - z_{2r})(V_i^{-1}\, T_s{}^n a - z_{1r})}, \quad (7.8)$$

where the summation $\Sigma^{(r,s)}$ excludes values of r and s such that V_i has $T_r^{\pm 1}$ as its rightmost member or $T_s^{\pm 1}$ as its leftmost member. By relabelling $T_s{}^{n+1}z$ as $T^n z$, we recover the original expression (7.5). However, we cannot drop the left-over terms with $n=\pm\infty$, since the terms in the individual series of (7.4) involving z or a separately do *not* become small as i or n becomes large. The change in $v_r(z)$ will thus be:

$$\sum_i{}^{(r,s)} \ell n \frac{(V_i^{-1}\, T_s{}^{\infty}z - z_{1r})(V_i^{-1}\, T_s{}^{-\infty}z - z_{2r})}{(V_i^{-1}\, T_s{}^{\infty}z - z_{2r})(V_i^{-1}\, T_s{}^{-\infty}z - z_{1r})}\,. \quad (7.9)$$

The point $T_s{}^{\infty}z$ and $T_s{}^{-\infty}z$ are just z_{1s} and z_{2s}, so that:

$$\delta_{Bs}\, v_r(z) = 2i\pi\tau_{rs}, \quad (7.10)$$

where

$$\tau_{rs} = \frac{1}{2i\pi} \sum_i{}^{(r,s)} \ell n \frac{(z_{1s} - V_i z_{1r})(z_{2s} - V_i z_{2r})}{(z_{1s} - V_i z_{2r})(z_{2s} - V_i z_{1r})}, (r\neq s). \quad (7.11a)$$

When r=s the term in (7.5) with $V_i^{-1} = T_s{}^n$, $n\neq 0$, does not occur, as it would violate the restriction that V_i not have $T_r^{\pm 1}$ as its rightmost member. The terms in (7.8) with $V_i = I$, $n\neq 0$ therefore do not occur. We must accordingly omit the term with $V_i = I$ from (7.11a) and, in its place, we must include the change of the term in (7.5) with $V_i = I$. This is just

$$\ell n \frac{(T_r z - z_{1r})(z - z_{2r})}{(T_r z - z_{2r})(z - z_{1r})}\,,$$

which may easily be shown to be equal to $\ell n\, w_r$, w_r being the w-parameter (Eq. (7.11) of T_r. Hence

$$\tau_{rr} = \frac{1}{2i\pi} \{\ell n\, w_r + \sum_{i\neq I}{}^{(r,s)} \ell n \frac{(z_{1r} - V_i z_{1r})(z_{2r} - V_i z_{2r})}{(z_{1r} - V_i z_{2r})(z_{2r} - V_i z_{1r})}\}\,. \quad (7.11b)$$

We may notice that the terms in the summation involving V_i and its inverse are equal.

The functions $v_r(z)$ thus have the required properties, and it may be shown that linear combinations of them are there only such functions. Instead of the single period τ associated with elliptic functions, we have a symmetric *period matrix* whose elements are defined by (7.11).

We can now introduce functions analogous to the χ's or χ''s of the one-loop problem. We define:

$$\ell n\phi'(z,z') = \ell n(z-z') + \sum_{i\neq I}{}' \ \ell n \frac{(z-V_i z')(z'-V_i z)}{(z-V_i z)(z'-V_i z')} . \qquad (7.12)$$

The notation Σ' indicates that we restrict the summation to one of each pair V_i, V_i^{-1}, the two terms being equal.

The change in ϕ' when its variables traverse an A-cycle is trivial:

$$\delta_{Ar,z}\ \phi'(z,z') = 0, \qquad (7.13)$$

where the symbol $\delta_{Ar,z}$ indicates that the variable z is traversing the A-cycle. In eq. (8.13) we assume that the path traced out by z does not enclose z'.

To find the change in ϕ' under the transformation $z \to Tz$, it is convenient to define the function:

$$\Pi(z,z_1';z'z_1) = \sum_i \ell n \frac{(z-V_i z')(z_1'-V_i z_1)}{(z-V_i z_1)(z_1'-V_i z')} . \qquad (7.14)$$

By following the reasoning used in determining the transformation properties of v_r, we find that:

$$\Pi(T_r z,z_1';z',z_1) - \Pi(z,z_1';z',z_1) = v_r(z') - v_r(z). \qquad (7.15)$$

Let us now take the limit of (7.14) when z approaches z_1 and z' approaches z_1'. The two terms $-\ell n(z-z_1)$ and $-\ell n(z'-z_1')$ must of course be subtracted off; the other terms reproduce the series (7.12) with each term counted twice. Hence,

$$\ell n\ \phi'(z,z') = \lim_{z\to z_1,\ z'\to z_1} \tfrac{1}{2}\{\Pi(z,z_1';z',z_1) + \ell n(z-z_1) + \ell n(z'-z_1')\}. \tag{7.16}$$

Now, from (7.15) and the analogous equation for $z_1 \to T_r\ z_1$, we may write:

$$\begin{aligned}\Pi(T_r z,z_1';z',T_r z_1) &= \Pi(T_r z,\ z_1'\ ;z',z_1) + v_r\ (z_1') - v_r\{T_r(z)\}\\ &= \Pi(z,z_1';z',z) + v_r(z_1') + v_r(z')-2v_r(z)\\ &\quad -2i\Pi\ \tau_{rr},\end{aligned} \tag{7.17}$$

from (7.15) and (7.10). Substituting (7.16) in (7.15), we find that

$$\delta_{Br,z}\ \ell n\phi'(z,z') = -\ v_r(z) + v_r(z') - \tfrac{1}{2}i\pi\tau_{rr} + \tfrac{1}{2}i\pi\ -\ell n\ (C_r z + D_r). \tag{7.18}$$

The constants C_r and D_r are of course the C- and D-constants (Eq.(4.5)) associated with the transformation T_r, and the last term on the right of (7.18) results from applying the transformation T_r to the extra logarithms on the right of (7.16).

*The term $\frac{1}{2}\pi$ on the right of (7.16) is not important for our purposes; in fact, we have not been precise in our definition of the $2i\pi$ terms in the logarithms, though they can easily be defined unambiguously, The $i\pi$ term (mod $2i\pi$) arises as follows: if we consider two cases of (7.15) where the rth B-cycle does and does not separate the points z' and z_1, the right-hand side will differ by a term $2i\pi$, since Π contains terms $\ell n\ (z-z')$ and $-\ell n\ (z-z_1)$. Now it is easy to see that, if the B-cycle traversed by the point z_1 in (7.17) separates $T_r z$ and z_1', the B-cycle traversed by z will not separate z' and z_1, and *vice versa*. All other $2i\pi$ terms will be the same for the two transformations. Therefore we obtain an odd number of $i\pi$ terms after dividing by 2. Another point worth noticing in this connection is that the expression $C_r z+D_r$ in (7.18) may be rewritten $w^{-\frac{1}{2}}(1-w)(z-z_{2r})/(z_{1r}-z_{2r})$. As the term $-\frac{1}{2}i\pi\tau_{rr}$ on the right of (7.18) contains a term $-\frac{1}{2}\ell n w$, (Eq. 7.11b)), there are no net terms involving $\frac{1}{2}\ell n w$, and therefore no ambiguity of $i\pi$ from this source.

The function ϕ' is known as the *prime form*. If it were not for the last termson the right of (7.18), it would be the analog of the Jacobi θ-function; as usual, the function $v_r(z)$ replaces $\ell n z$ and the period matrix replaces the period.

We can now obtain the analog of the function χ of the last section:

$$\ell n\phi(z,z') = \ell n\phi'(z,z') - (2\pi)^{-1} \sum_{r,s} \mathrm{Re}\{v_r(z)-v_r(z')\} \times \{(\mathrm{Im}\tau)^{-1}\}_{rs} \mathrm{Re}\{v_s(z)-v_s(z')\}, \quad (7.19)$$

where $(\mathrm{Im}\tau)^{-1}$ denotes the matrix inverse of the imaginary part of the period matrix.

The Neumann function is defined in the same way as previously:

$$N(z,z') = \ell n|\phi(z,z')|. \quad (7.20)$$

It satisfied the equation

$$\delta_{Br,z} N(z,z') = -\ell n|C_r z + D_r|. \quad (7.21)$$

Although the right-hand side is not zero, it depends only on z and not on z′, which is sufficient for our purpose. It is certainly possible to redefine

$$N'(z,z') = N(z,z') + f(z) + f(z') \quad (7.22)$$

and to choose the f's, which are not required to be analytic, so that $\delta_{Br,z} N' = 0$. (Actually the f's can be made analytic, N′ would then be the logarithm of what is known as the prime *function*.) N′ will still satisfy Eq. (3.4c). If the function N′ is then inserted into (3.8), the terms involving the f's will cancel due to momentum conservation. In fact, our Neumann function is slightly more convenient than the one-loop function in that, when z approaches z′ ,

$$N(z,z') = \ell n|z-z'| + O(z-z'). \quad (7.23)$$

*The series in (7.12) (more strictly its derivative) is a particular case of a *Poincaré series* [18]. (The term Poincaré θ-function is sometimes used but might lead to confusion, since the Poincaré series are not the analogues of the Jacobi θ-function.) It remains to discuss the convergence of (7.12), as well as of other similar series such as (7.3) and (7.11). The series are known to converge absolutely provided that the 3n-3 parameters defining the group of projective transformations satisfy certain inequalities [19]. By examining the functions within the region of convergence one can obtain the information necessary to determine the singularity structure and finiteness properties of the n-loop amplitude. It is also known that the functions defined by the series can be continued as analytic functions throughout the Schottky region. The question of convergence throughout the Schottky region remains open, and negative results have been obtained with regard to *absolute* convergence in certain cases.

There do exist other methods of defining the functions τ_{rs}, v_n and ϕ as ratios of convergent series, and then fitting parameters to satisfy certain conditions. We shall regard the functions as defined by analytic continuation if necessary.

Formula for the n-Loop Amplitude

We shall not give the derivation of the momentum-independent functions in the formula for the n-loop amplitude. We shall take as our variables of integration the Z's of the external particles and the parameters z_{1s}, z_{2s} and w_s associated with the transformation T_s. Three of the Z's, z_1's or z_2's will be fixed at arbitrary values.

The formulas for the momentum-independent factors are similar to to those in the one-loop amplitude, the main differences being the following:

i) The product $\prod_n (1-w^n)$ now becomes replaced by the product $\prod''$ $(1-w_i)$, where the product $\prod''$ denotes a product over all *conjugacy classes* in the group of projective transformations, an element and its inverse being together counted only once. (All members of the same conjugacy class have the same w.)

ii) The factor $2\pi/|\ell n|w||$ becomes replaced by $|\mathrm{Im}\ \tau|^{-1}$, the determinant of the imaginary part of the period matrix.

iii) There is an extra factor $\prod_r |z_{1r}-z_{2r}|^{-4}$; such a factor is required by projective invariance.

The formula for the n-loop amplitude is therefore:

$$A = 4^{1-n}(-i\pi)(ig^2/4\pi)^{N+2n-2} \int \prod_{r,s} d^2Z_r d^2z_{1s} d^2z_{2s} d^2w_s$$

$$\times (d^2z_a d^2z_b d^2z_c)^{-1} \left|(z_a-z_b)(z_b-z_c)(z_a-z_c)\right|^2$$

$$\times \prod_s |w_s(z_{1s}-z_{2s})|^{-4} |\mathrm{Im}\tau|^{-13} \prod_{i\neq I}{}'' |1-w_i|^{-48}$$

$$\times \prod_{r>r'} |\phi(Z_r, Z'_r)|^{-P_r \cdot P_{r'}}, \tag{7.24}$$

where z_a, z_b, z_c are three arbitrary Z's, z_1's or z_2's.

We have carried out our discussion in terms of closed strings, but open strings may be handled in the same way. The circles of Fig. 9 become replaced by semi-circles with the real axis as their diameter, and all variables Z_r, z_{1s}, z_{2s}, w_s are real. In this case the Poincaré series as we have written them are known to converge throughout the Schottky region [18]. Amplitudes with both open and closed strings may also be treated. Open-string loops are represented in Fig. 9b by semi-circles with the real axis as diameter and real parameters, closed-string loops by circles in the upper half-plane and complex parameters. The Z's for open-string external particles are real, those for closed-string external particles are complex with Im Z ≥ 0.

Modular Invariance

As with single-loop diagrams, we may select different generators of the fundamental group. The functions defined in this section will transform according to simple rules when we change the generators, and the loop amplitude will remain unchanged. There are four types of transformations which generate the modular group:

i) $A_r \leftrightarrow A_{r'}$, $B_r \leftrightarrow B_{r'}$ ii) $B_r \to B_r + A_r$

iii) $B_r \to B_r + A_{r'}$, $B_{r'} \to B_{r'} + A_r$ iv) $B_r \to A_r$ $A_r \to -B_r$,

where, in each case, r and r′ have specific values and all other cycles remain unchanged. In cases i), ii) and iii) the change of the period matrix is obvious; in iv), where an A-cycle is redefined, it is a straightforward exercise to define new v_n's as linear combinations of the old so as to satisfy (7.6), and then to calculate the period matrix. Thus:

$$\text{i)}\tau'_{rs} = \tau'_{sr} = \tau_{r's} \quad \tau'_{r's} = \tau'_{sr'} = \tau_{rs}, \qquad \text{ii)}\tau'_{rr'} = \tau_{rr}+1$$

$$\text{iii)}\tau'_{rr'} = \tau'_{r'r} = \tau_{rr'}+1, \qquad \text{iv)}\tau'_{rr} = -\tau^{-1}_{rr}$$

$$\tau'_{rs} = \tau'_{sr} = \tau_{rs}\tau^{-1}_{rr} \;(s\neq r),\; \tau'_{ss'} = \tau_{ss'} - \tau_{sr}\tau_{rs'}\tau^{-1}_{rr} \;(s,s'\neq r). \tag{7.25}$$

The subscripts s and s′ can take arbitrary values, and all periods not mentioned remain unchanged. The transformations (7.25) generate the Siegel modular group whose elements are again given by (6.32a), where τ is now the period matrix and A,B,C,D are matrices with integral entries such that AD-BC=DA-CB=1.

There do not exist explicit transformation formulas for the parameters of the T_r's corresponding to the above transformation formulas for the period matrix. One has to calculate the period matrix from the T_r's; two different sets of T_r's which lead to period matrices related to one another by a transformation of the Siegel modular group are modular equivalent. To calculate how the z's transform, one would have to calculate the v_r's for the equivalent sets of T_r's; the transformed z's are those for which the new v_r's are the correct linear combinations of the old. The transformation law for the z's is not of course required in the calculation of the loop amplitudes.

The equation corresponding to Eq. (6.34) holds for our functions ϕ, the new variables being defined in terms of the old according to the remarks just made. Owing to our definition of the ϕ's, the factors in (6.34) preceding the exponential will be absent. Though there is no precise equivalent of (6.35), the determinant factor equivalent to (6.37b) will still be invariant. We may repeat the reasoning of the single-loop case to show that our amplitude has modular invariance and that we must integrate over one fundamental region of the modular group.

Singularities or divergences in the amplitude can arise from configurations where two interactions come close to one another, when one approaches $\tau=\pm\infty$, or when one of the intermediate strings becomes infinitely short. In all cases the analysis of the behavior of the Schottky parameters is the same as in the one-loop case. If one of the interactions of a loop occurs either at $\tau=\pm\infty$ or near an external-particle interaction and the A- and B-cycles are as in Fig. 9(a), w approahces zero. If two loop interactions approach one another, the cycles must be drawn as in Fig. 10. The

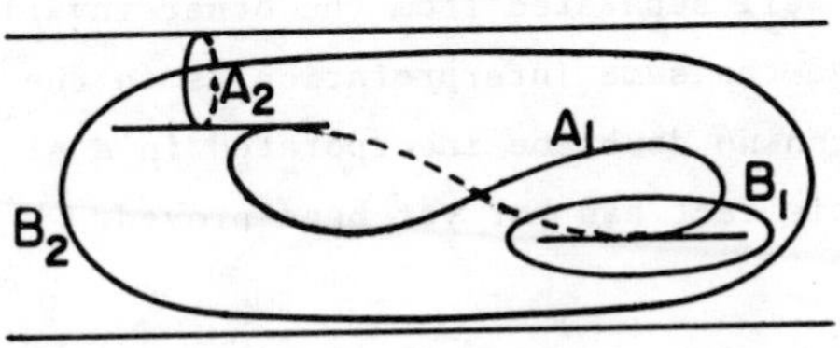

Fig. 10. Convenient cycles for closely-spaced loop interactions.

cycle 2 is not affected by the coincidence of the interactions, the w-parameter associated with the cycle 1 approaches zero. If a loop shrinks to a point with neither string becoming short, the A- and B-cycles of Fig. 9(a) are interchanged; w then approaches zero. If two loops together shrink into a small region, with the ratio of all dimensions remaining finite, one again uses the cycles of Fig. 10, with B_2 and $-A_2$ interchanged. In this case the cycle 1 is not affected by the shrinking, while the w-parameter for the cycle 2 approaches zero.

If an intermediate string becomes infinitely short, the two invariant points associated with its projective transformation will approach one another; w will approach zero if time difference between the interactions producing the string remains finite, but, if the time difference becomes of the order of the string length, w will remain finite. In the latter case we may take, as one of our fundamental generators T_r, the product of the projective transformation associated with the shrinking string and another projective transformation. The w-parameter associated with the new T_r will then approach zero.

In all cases, one of the w-parameters approaches zero. By making a modular transformation if necessary, one can take it to be the w-parameter associated with one of the generators. Hence there exists a fundamental region whose only singular points are those where the w's assoiated with the fundamental T_r's become small. The integrand of the scattering amplitudes will also have singularities when two or more Z's approach one another.

The only divergences not associated with unitarity arise from the configuration where the Z's, and possibly some invariant points, occur in a region which is well separated from the other invariant points. The singularities have the same interpretation as in the one-loop amplitude, and they can no doubt be incorporated in a slope-renormalization factor, but this fact has not yet been proved.

REFERENCES

1. Nambu, Y., Proc. Intern. Conf. on Symmetries and Quark Models, Wayne State University (1969).
   Nielsen, H. B., Proc. 15th Intern. Conf. on High-Energy Physics, Kiev, (1970).
   Susskind, L., Nuovo Cimento 69A, 457 (1970).

2. Alessandrini, V., Nuovo Cimento 2A, 321 (1971).
   Alessandrini, V. and Amati, D., Nuovo Cimento 4A, 793 (1971).
   Kaku, M. and Yu, L. P., Phys. Lett. 33B, 166 (1970); Phys. Rev. D3 2992, 3007, 3020 (1971).
   Lovelace, C., Phys. Lett. 32B, 703 (1970).

3. Mandelstam, S., Nucl. Phys. B64, 205 (1973).

4. Mandelstam, S., Nucl. Phys. B83, 413 (1974).

5. Goddard, P., Goldstone, J., Rebbi, C. and Thorn, C. B., Nucl. Phys. B56, 109 (1973).

6. Kato, M. and Ogawa, K., Nucl. Phys. B212, 443 (1983).
   Siegel, W., Phys. Lett. 151B 391, 396 (1985).
   Friedan, D., Martinec, E. and Shenker, S., Chicago preprint EFI85-32.
   Banks, T. and Peskin, M. E., SLAC preprint PUB-3740.
   Siegel, W. and Zwiebach, B., Berkeley preprint UCB-PTH-85/30.

7. Gervais, J-L. and Sakita, B., Phys. Rev. Lett. 30, 716 (1973).

8. Polyakov, A. M., Phys. Lett. 103B, 207, 211 (1981).

9. Polchinski, J. Texas preprint U.T.T.G. 13-85.

10. Brink, L. and Nielsen, H. B., Phys. Lett. 45B, 332 (1973).

11. McKean, H. and Singer, I. M., J. Diff. Geom. 1, 43 (1973).

12. Alvarez, O., Nucl. Phys. B216, 125 (1983).

13. Durhuus, B., Nielsen, H. B., Olesen, P. and Petersen, J. L., Nucl. Phys. B196, 498 (1982).
    Durhuus, B., Olesen, P. and Petersen, J. L., Nucl. Phys. B198,157; B201, 176 (1982).

14. Erdélyi, A. *et al.*, *Higher Transcendental Functions*. (Bateman Manuscript Project), vol. II Krieger, Malabar, 1981.

15. Schwarz, J., Phys. Reports 89, 224 (1982).

16. Mandelstam, S., Phys. Reports 13 260 (1974).

17. Arfaei, H. Nucl. Phys. 85, 535 (1975).

18. Ford, L. R., *Automorpnic Functions* (Chelsea, New York, 1951).

19. Fricke, R. and Klein, F. *Vorlesungen über die Theorie der Automorphen Funktionen* (2 Vols.) Teubner, Leipzig, 1926.

20. Baker, H. F., *Abel's Theorem and the Allied Theory*. Cambridge, 1897.

21. Ahlfors, L. and Sario, L., *Riemann Surfaces*. Princeton, 1960. Springer, G., Introduction to Reimann Surfaces. Addison-Wesley, Reading, Mass. 1957.

DIFFERENTIAL GEOMETRY IN STRING MODELS

Orlando Alvarez

Department of Physics
and
Lawrence Berkeley Laboratory
University of California
Berkeley, California 94720

1 Introduction

In this article we review the differential geometric approach to the quantization of strings. In a seminal paper, Polyakov [1] demonstrated the connection between the trace anomaly and the critical dimension. His insight into the role played by the Faddeev-Popov ghosts has been instrumental in much of the subsequent work on the quantization of strings. These lectures are an elaboration of Polyakov's work. We will discuss the differential geometry of two dimensional surfaces and its importance in the quantization of strings. The path integral quantization approach to strings [2] will be carefully analyzed to determine the correct effective measure for string theories. Our choice of measure for the path integral will be determined by differential geometric considerations. Once the measure is determined, the manifest diffeomorphism invariance of the theory will have to be broken by using the Faddeev-Popov *ansatz* . The gauge fixed theory will be studied in detail with emphasis on the role of conformal and gravitational anomalies [3]. In our analysis we will see that the path integral formulation of the gauge fixed theory requires summing over all the distinct complex structures on the manifold. This will lead us into an elementary discussion of Teichmüller space [4].

The author apologizes to colleagues he may not have referenced adequately. Any omissions are unintentional. The author is not completely familiar with the early literature.

2 Classical Theory

The path integral approach to quantized strings [2] postulates that the partition function is given by

$$Z = \sum_{\text{surfaces}} \exp(-\text{area}) . \tag{2.1}$$

The suggestion that the action is given by the area of the surface is due to Nambu [5], Goto [6], and Nielsen [7] and Susskind [32]. In the above equation one is supposed to sum over surfaces with all topologies. I will only discuss the case of closed orientable surfaces.

Let M be a fixed two dimensional manifold. Surfaces with the topology of M are described by maps

$$X : M \to \mathbf{R}^d . \tag{2.2}$$

We will always work in an Euclidean field theory. The area of the surface is given by

$$A = \int_M d^2z \sqrt{\det \gamma_{ab}} , \tag{2.3}$$

where

$$\gamma_{ab} = \frac{\partial X^\mu}{\partial z^a} \frac{\partial X^\mu}{\partial z^b} . \tag{2.4}$$

is the induced metric on the surface:

$$ds^2 = dX^\mu \, dX^\mu = \frac{\partial X^\mu}{\partial z^a} \frac{\partial X^\mu}{\partial z^b} dz^a \, dz^b . \tag{2.5}$$

The classical equations of motion are

$$\frac{\partial}{\partial z^a} \sqrt{\det \gamma} \, \gamma^{ab} \frac{\partial X^\mu}{\partial z^b} = 0 . \tag{2.6}$$

The nonlinearity of the action leads to difficulties in quantization. Polyakov [1] proposed that in the quantum theory one replace the area action by an equivalent classical action that depends on an additional intrinsic metric g_{ab}. The action used by Polyakov had been previously proposed by Brink, Di Vecchia and Howe [8], and by Deser and Zumino [9]. The new action is a functional of both the intrinsic metric g_{ab} and the coordinates X:

$$I[X,g] = \int_M d^2z \, \frac{1}{2}\sqrt{\det g}\, g^{ab} \frac{\partial X^\mu}{\partial z^a}\frac{\partial X^\mu}{\partial z^b} \,. \tag{2.7}$$

We will refer to the above as the *string action.*

The *string action* has the following obvious classical invariances:

1. Translations in physical spacetime, $X \to X + \text{constant}$. Physical rotations are also symmetries.
2. Invariance under diffeomorphisms of M. Heuristically one refers to these as z–reparametrizations.
3. Conformal (Weyl) invariance under a local rescaling of the metric given by $g_{ab}(z) \to g_{ab}(z)\exp[2\tau(z)]$.

There are two immediate consequences of item 3. Firstly, the energy momentum tensor T^{ab} is *automatically* traceless: $g_{ab}T^{ab} = 0$. This is true whether or not the equations of motion hold. Secondly, the equations of motion can determine the metric up to an arbitrary conformal factor.

The classical equations of motion are determined by the variational principle

$$\begin{aligned} \delta I = &\int_M d^2z\, \sqrt{g}\,\delta X^\mu\, \Delta X^\mu \\ &- \frac{1}{2}\int_M d^2z\,\sqrt{g}\,\delta g_{ab}\, T^{ab} \,, \end{aligned} \tag{2.8}$$

where the Laplacian Δ and the energy momentum tensor T^{ab} are defined by:

$$\begin{aligned} \Delta &= -\frac{1}{\sqrt{g}}\frac{\partial}{\partial z^a}\sqrt{g}g^{ab}\frac{\partial}{\partial z^b} \\ T_{ab} &= \partial_a X^\mu \partial_b X^\mu - \frac{1}{2} g_{ab} g^{cd}\partial_c X^\mu \partial_d X^\mu \,. \end{aligned} \tag{2.9}$$

The classical equations are

1. $\Delta X^\mu = 0$

2. $T_{ab} = 0$.

The T_{ab} equation cannot determine g_{ab} since the combination $g_{ab}g^{cd}$ is conformally invariant. The general solution to the energy-momentum equation is

$$g_{ab} = f(X, g, z)\gamma_{ab} \,, \tag{2.10}$$

where $f > 0$ is arbitrary. Inserting the above into $\Delta X = 0$ leads to the minimal area equations (2.6).

3 Quantum Theory

The principle of renormalizability tells us that the action should be constructed out of covariant expressions in X^μ and g_{ab} which are of dimension two or less and which respect the Euclidean invariance of physical spacetime. The most general such action consists of three terms:

$$\begin{aligned} I[X^\mu, g_{ab}] = A\,\frac{1}{2}\int_M d^2z\,\sqrt{g}\,g^{ab}\partial_a X^\mu \partial_b X^\mu \\ + B\int_M d^2z\,\sqrt{g} \\ + C\frac{1}{4\pi}\int_M d^2z\,\sqrt{g}\,R\,. \end{aligned} \tag{3.1}$$

The first term in the above is just the classical action we had previously discussed. The second term is a cosmological constant. The addition of such a term to the classical theory leads to inconsistent equations of motion. The third term is a topological invariant called the Euler characteristic χ:

$$\chi(M) = \frac{1}{4\pi}\int_M d^2z\,\sqrt{g}\,R\,. \tag{3.2}$$

It is formally a total derivative and therefore it does not contribute to the equations of motion. The Euler characteristic $\chi(M)$ is related to the topology of the manifold via the relation

$$\chi(M) = 2 - 2h - b\,, \tag{3.3}$$

where h is the number of handles on the surface and b is the number of boundaries.[1]

We have to evaluate the quantum partition function defined by

$$Z = \sum_{\text{topologies}} \int [dg][dX] \exp(-I[X,g])\,. \tag{3.4}$$

Since the action is invariant under the diffeomorphism group, it will be necessary to implement the Faddeev-Popov procedure. We will use a very geometrical approach in constructing the Faddeev-Popov determinants. The approach is based on the Riemannian geometry of the space of vector potentials. Let $\mathcal{C}$ be the configuration space of strings:

$$\mathcal{C} = \{X : M \to \mathbf{R}^d\}\,. \tag{3.5}$$

$\mathcal{C}$ has a natural metric![2] If X and X' are nearby maps then the infinitesimal vector field[3] between them δX has a norm given by

$$\| \delta X \|^2 = \int_M d^2z\, \sqrt{g}\, \delta X^\mu(z)\, \delta X^\mu(z)\,. \tag{3.6}$$

The measure $[dX]$ is the associated Riemann-Lebesque volume on the infinite dimensional space $\mathcal{C}$ with the above metric.

Example 3.1 *In an ordinary Riemannian space of finite dimension N with metric $ds^2 = G_{AB} dY^A dY^B$, the volume element is given by $d^N Y \sqrt{G}$.*

[1] In this article we are only considering surfaces without boundaries.

[2] We are looking for an expression which is covariant. There are many such expressions but locality and simplicity lead to a *natural* metric. We will write down the unique covariant expression which does not involve derivatives of the metric or derivatives of the X field, respects the translational invariance in X, and does not involve an arbitrary function.

[3] The quantities $\delta X^\mu(z, \bar{z})$ are the components of the tangent vector in $\mathcal{C}$.

We note that X^μ enters the action only through the conformally invariant kinetic energy term. The measure $[dX]$ is not conformally invariant since the metric (3.6) is not conformally invariant. In general the quantum partition function is not conformally invariant and there will be a trace anomaly:

$$\left\langle g_{ab} T^{ab} \right\rangle \neq 0 \,. \tag{3.7}$$

Note that since the action is quadratic in X^μ, the path integral is Gaussian. The functional integral over X is easy to perform with the result

$$\int d^d X_0 \; N \; (\det{}'\Delta)^{-d/2} \,, \tag{3.8}$$

where X_0 is the zero eigenvalue mode of the Laplacian, prime denotes the determinant with the zero eigenvalue removed and N is a normalization factor. The reader is referred to references [10], [16] for technical details which will not be crucial in our discussion.

The space of metrics $\mathcal{M}$ has a very interesting property. This space is defined as[4]

$$\mathcal{M} = \{ \;\; \mathbf{g} \;\; | \;\; \mathbf{g} \text{ is a metric on } M \} \,. \tag{3.9}$$

We demonstrate that $\mathcal{M}$ is a convex set. Note that a metric $\mathbf{g}$ is a positive definite quadratic form. Is it easy to verify that if $\mathbf{g}^0$ and $\mathbf{g}^1$ are metrics then the straight line

$$\mathbf{g}^t = (1-t)\mathbf{g}^0 + t\mathbf{g}^1 \tag{3.10}$$

for $0 \leq t \leq 1$ is also a positive definite quadratic form. The convexity of $\mathcal{M}$ implies that the topology of the space of metrics is trivial. Convexity states that given two metrics, the straight line connecting them is a metric. Therefore, there cannot be any holes. Furthermore, the space $\mathcal{M}$ is not compact. This follows easily from the global conformal transformation $g_{ab} \to \lambda g_{ab}$ as $\lambda \to \infty$.

The space of metrics, $\mathcal{M}$, has a natural metric.[5] An arbitrary tangent

[4] Boldface will be used to denote a tensor without referring to a coordinate frame.

[5] Remarks similar to the ones in footnote 2 apply to this situation. In this case there is no analog of translational invariance. The covariance and the absence of derivatives leads to an almost unique expression

vector $\delta\mathbf{g}$ at a point $\mathbf{g}$ in $\mathcal{M}$ has components $\delta g_{ab}(z,\bar{z})$. It is convenient to decompose the tangent vector into its traceless part and its trace part. The tensor

$$G_{ab}{}^{cd} = \frac{1}{2}\left(\delta_a^c\delta_b^d + \delta_a^d\delta_b^c - g_{ab}g^{cd}\right) \tag{3.11}$$

is the projector onto the space of symmetric traceless tensors. Since the symmetric traceless tensors are orthogonal to the pure trace tensors, one can do an orthogonal decomposition on the tangent vector $\delta\mathbf{g}$. This means that the most general positive definite quadratic form is given by

$$\| \delta\mathbf{g} \|^2 = \int_M d^2z\,\sqrt{g}\,\left(G^{abcd} + u g^{ab}g^{cd}\right)\delta g_{ab}\,\delta g_{cd}\;, \tag{3.12}$$

where $u > 0$ is an arbitrary constant.[6] This metric is not conformally invariant. The measure $[dg]$ is the associated Riemann-Lebesque measure.

The reason for the above decomposition is that the symmetric traceless part of $\delta\mathbf{g}$ transforms as helicity ± 2 and the trace part as helicity zero under the two dimensional rotation group. These three quantities may be taken as the three independent components of $\delta\mathbf{g}$:

$$\delta g_{ab} = \delta h_{ab} + 2g_{ab}\delta\tau\;, \tag{3.13}$$

where δh_{ab} is the symmetric traceless part and $\delta\tau$ is the trace part. Inserting the above into the expression for the metric leads to the result

$$\begin{aligned}\| \delta\mathbf{g} \|^2 &= \int_M d^2z\,\sqrt{g}\,G^{abcd}\delta h_{ab}\,\delta h_{cd} \\ &\quad + 16u\int_M d^2z\,\sqrt{g}\,(\delta\tau)^2\;,\end{aligned} \tag{3.14}$$

One immediately has the result that

$$[dg] = [dh][d\tau]\;. \tag{3.15}$$

One has to be a bit careful in the interpretation of the three preceeding equations. There is no coordinate system labelled by h and τ. The preceeding equations all involve the decomposition of a tangent vector in an

---

[6]In general u may be an arbitrary positive function. There is no reason for introducing such an *ad hoc* function.

This figure illustrates the bundle structure of the space of configurations $\mathcal{C} \times \mathcal{M}$. The above would be a principal bundle if the group Diff(M) *acted freely. Configurations of the type* (X, g) *are identified if they are related by a diffeomorphism. All the points on the heavy line are diffeomorphically equivalent. The quotient space of equivalence classes is obtained by identifying "physically equivalent" configurations. The equivalence class of the thick line is denoted by the bullet.*

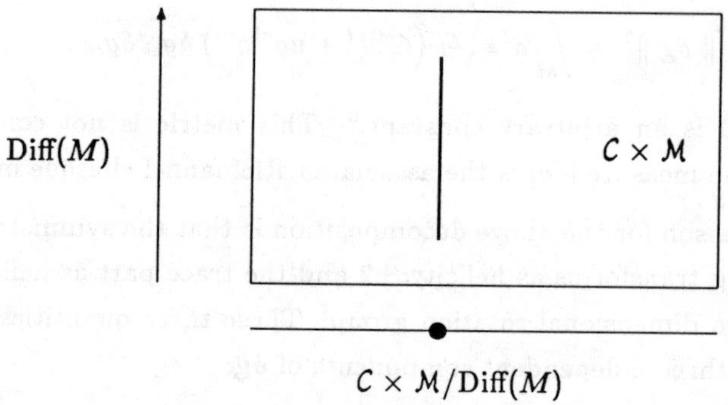

Figure 1: Bundle Structure of the Configuration Spaces

unspecified orthonormal frame. In general such a frame is not integrable and does not lead to a coordinate system. A more careful treatment may be found in reference [11].

The group of diffeomorphisms of M, Diff(M), acts on both g and X. Since the action and the measures are invariant under the action of Diff(M), "physically equivalent" configurations are counted manyfold. One has to gauge fix in one way or another. The configuration space over which one is integrating is the product space $\mathcal{C} \times \mathcal{M}$. The space of physical configurations is the quotient space $(\mathcal{C} \times \mathcal{M})/\mathrm{Diff}(M)$. This is the space of gauge equivalent configurations. Since the measure and the action are invariant under the diffeomorphism group, the theory can be reduced to a theory on the quotient space. The rest of this section is dedicated to formulating the reduction.

Not all diffeomorphisms are connected to the identity. If ϕ_0 and ϕ_1 are in the same connected component then $\phi_0^{-1} \circ \phi_1$ is in the connected component of the identity $\mathrm{Diff}_0(M)$. It follows that the full diffeomorphism group may be thought of as a discrete transformation in combination with an element of $\mathrm{Diff}_0(M)$. There may exist anomalies associated with these discrete diffeomorphisms [12]. We will concentrate on how $\mathrm{Diff}_0(M)$ affects the path integral but we will make some passing remarks on the discrete transformations.

Let δV^a be the infinitesimal generator of a diffeomorphism in the identity component. The strategy is to trade the symmetric traceless deformation δh_{ab} (two arbitrary functions) for the vector field δV^a (two arbitrary functions). Such a term would lead to a Jacobian in the measure from the change of variables:

$$[dg] = [dh][d\tau] = [dV][d\sigma] \text{ Jacobian} . \tag{3.16}$$

Since the action is invariant under the diffeomorphism group, one can perform the $[dV]$ integral obtaining the volume of the gauge group. In this way, one is effectively integrating over a single physical configuration with the invariance of the action manifesting itself as the volume of the gauge group.

If δV^a is the infinitesimal generator of a diffeomorphism in $\mathrm{Diff}_0(M)$, then the induced variation on the metric is given by the Lie derivative expression:

$$\delta g_{ab} = \nabla_a(\delta V_b) + \nabla_b(\delta V_a) . \tag{3.17}$$

We will trade δh_{ab} and $\delta\tau$ for δV^a and $\delta\sigma$ where the latter are defined by the equation:

$$\delta g_{ab} = \nabla_a(\delta V_b) + \nabla_b(\delta V_a) + 2g_{ab}\,\delta\sigma . \tag{3.18}$$

By projecting out the trace and the traceless piece one finds:

$$\begin{aligned} \delta h_{ab} &= 2G_{ab}{}^{cd}\nabla_c(\delta V_d) + 2g_{ab}\,\delta\sigma \\ &= P(\delta V)_{ab} ; \\ 2\delta\tau &= 2\delta\sigma + g^{cd}\nabla_c(\delta V_d) . \end{aligned} \tag{3.19}$$

The operator P maps vectors into symmetric traceless tensors. Each of these spaces is a rank two real vector bundle. The change of variable is implemented by the usual Jacobi prescription:

$$[d\tau][dh] = [d\sigma][dV]\left|\frac{\partial(\tau,h)}{\partial(\sigma,V)}\right| . \tag{3.20}$$

The Jacobian may be written as:

$$\left|\frac{\partial(\tau,h)}{\partial(\sigma,V)}\right| = \left|\det\begin{pmatrix} 1 & * \\ 0 & P\end{pmatrix}\right| . \tag{3.21}$$

The $*$ denotes an operator which is irrelevant since the matrix is upper triangular. The Jacobian is easily computed:

$$|\det P| = \left|\det(P^\dagger P)\right|^{1/2} . \tag{3.22}$$

The Jacobian is positive as required by the change of variables formula. The operator $P^\dagger$ is the adjoint of P. Note that $P^\dagger$ maps symmetric traceless tensors into vector fields via the formula:

$$(P^\dagger \delta h)^a = -\nabla_b(\delta h^{ba}) \tag{3.23}$$

The measure $[dV]$ is defined by the norm

$$\| \delta V \|^2 = \int_M d^2z \sqrt{g}\, g_{ab}\, \delta V^a\, \delta V^b \tag{3.24}$$

which is not conformally invariant.

At this stage we can finish our formal calculation of the partition function (3.4). Rewriting Z in terms of the new variables leads to the expression:

$$Z = N \sum_{\text{topologies}} \int [dV][d\sigma](dX_0)\, \left[\det(P^\dagger P)\right]^{1/2} (\det{}'\Delta)^{-d/2} . \tag{3.25}$$

If we formally define the volume of the connected diffeomorphism group by

$$\Omega_{\text{diff}\,0} = \int [dV] \tag{3.26}$$

then the partition function may be written as

$$Z = N \sum_{\text{topologies}} \int [d\sigma](dX_0)\, \Omega_{\text{diffo}} \Big[\det(P^\dagger P)\Big]^{1/2} (\det{}'\Delta)^{-d/2} \,. \qquad (3.27)$$

The volume of the diffeomorphism group depends on the conformal factor since the measure $[dV]$ is not conformally invariant.

We note that the above prescription is correct except for several technical details which are intimately related to the following questions:

1. What happens if $P(\delta V) = 0$?

2. Are all δh_{ab} induced by a δV^a?

A vector field ξ satisfying the equation $P\xi = 0$ is called a conformal Killing vector field (CKV). The defining equation may be written as

$$\nabla_a \xi_b + \nabla_b \xi_a - g_{ab} g^{cd} \nabla_c \xi_d = 0 \,. \qquad (3.28)$$

This equation is equivalent to the conformal Killing equation

$$\mathcal{L}_\xi\, g_{ab} = \lambda g_{ab} \,, \qquad (3.29)$$

where λ is given by $\nabla^a \xi_a$. Equation (3.19) tells us that a diffeomorphism generated by a conformal Killing vector is equivalent to a change in the conformal factor σ. Since each deformation of the metric is counted only once we see that such a diffeomorphism must be omitted. The zero modes of P must be omitted and therefore the correct Faddeev-Popov determinant is the square root of $\det{}' P^\dagger P$. A consequence of the exclusion of the zero modes of P is that the gauge fixed integral has a residual gauge invariance under diffeomorphisms generated by the conformal Killing vectors.

A second error in our derivation of (3.27) is that we implicitly assumed that all deformations δh_{ab} were expressible as infinitesimal gauge transformations (3.17). This is not true in general. There exist traceless deformations of the metric which are not equivalent to infinitesimal diffeomorphisms. This means that there are additional degrees of freedom besides the conformal factor σ which remain after extracting the gauge group. Fortunately, the new degrees of freedom are specified by a finite number of parameters. The new degrees of freedom satisfy the differential equation $\nabla_a \, \delta h^{ab} = 0$. These deformations are both traceless and transverse.

The above remarks are best understood by examining Figure 2. This figure illustrates the relationship of the space of vector fields to the space of traceless deformations of the metric. The situation depicted in the figure is identical to Witten's discussion of the index in supersymmetric theories [13]. Consider the direct sum of the space of vector fields and the space of traceless metrical deformations. An element of this direct sum may be represented as a matrix

$$\begin{pmatrix} \delta V \\ \delta h \end{pmatrix} . \tag{3.30}$$

Define a Hermitian "supersymmetry" operator Q by

$$Q = \begin{pmatrix} 0 & P^\dagger \\ P & 0 \end{pmatrix} , \tag{3.31}$$

and the operator $(-1)^F$ by

$$(-1)^F = \begin{pmatrix} 1 & 0 \\ 0 & -1 \end{pmatrix} , \tag{3.32}$$

where F is "fermion" number. One easily verifies that Q and $(-1)^F$ anti-commute. In analogy with supersymmetry one can define a Hamiltonian by $H = Q^2$. There is a pairing of a "bosonic state" with a "fermionic state" for each *positive* eigenvalue of H. There is no such correspondence for the states with zero "energy". The Witten index $\mathrm{Tr}(-1)^F$ is identical to the *analytic index* of the operator P defined by[7]

$$I(P) = \dim\ker P - \dim\ker P^\dagger \, . \tag{3.33}$$

The analytic index is invariant under reasonable perturbations of the operator P. Simply put, if a perturbation of P lifts a "bosonic" zero eigenvalue

[7] The *kernel* of the operator P is defined to be the set of solutions to the equation $PV = 0$. This space is finite dimensional for all the operators we will consider in this article.

The two boxes on the left represent the space of all vector fields on the manifold. The two boxes on the right are the traceless deformations of the metric. The operators P and $P^\dagger$ map between these spaces. On the spaces orthogonal to the kernels the mappings by the operators are one-to-one and onto. The $\ker P$ *is the space of global conformal Killing vector fields. The* $\ker P^\dagger$ *is the space of Teichmüller deformations.*

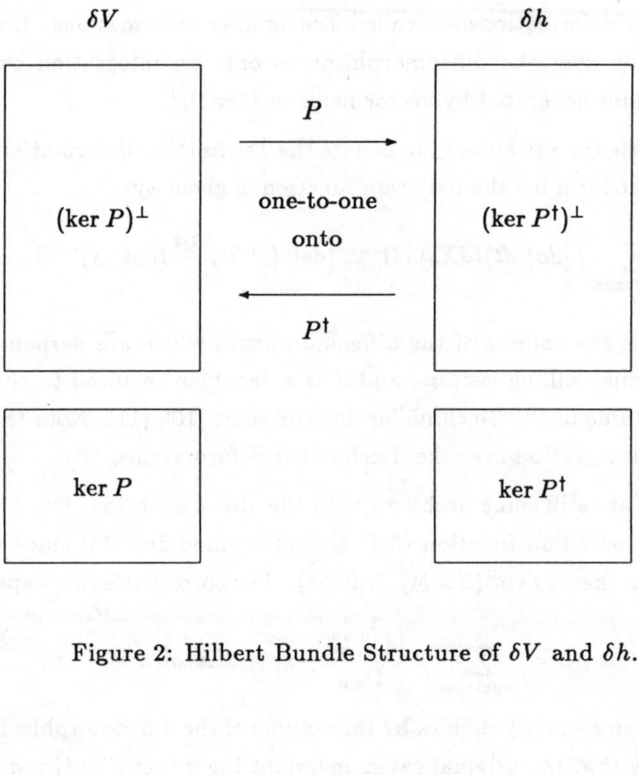

Figure 2: Hilbert Bundle Structure of δV and δh.

of H then it must also raise a "fermionic" zero eigenvalue of H. The theorem of Atiyah and Singer [14] shows that the analytic index is actually a topological invariant.

The discussion of the previous paragraph shows that we can trade a deformation of the metric in $(\ker P^\dagger)^\perp$ for an infinitesimal diffeomorphism in $\ker P$. Since we have to integrate over all metrics, the procedure we used to isolate the volume of the gauge group is actually incomplete. We are still required to integrate over metrical deformations which belong to $\ker P^\dagger$. These deformations are called Teichmüller deformations. Likewise, the integration over the diffeomorphisms is only an integration over the diffeomorphisms generated by vector fields in $(\ker P)^\perp$.

If we use the variables t_ω to denote the Teichmüller deformations then a more correct form for the partition function is given by

$$Z = N \sum_{\text{topologies}} \int [d\sigma](dt)(dX_0) J \Omega^\perp_{\text{diff}\,0} \left[\det{}'(P^\dagger P)\right]^{1/2} (\det{}'\Delta)^{-d/2} , \quad (3.34)$$

where $\Omega^\perp_{\text{diff}\,0}$ is the volume of the diffeomorphisms which are perpendicular to the conformal Killing vectors, and J is a Jacobian required to correctly count the volume of the Teichmüller deformations [10], [11]. Note that one performs an integration over the Teichmüller deformations.

There are still some problems with the above equation. Our starting point for the partition function (3.4) should be modified [11] since we are interested in a theory over $(\mathcal{C} \times \mathcal{M})/\text{Diff}(M)$. The correct starting expression is

$$Z = \sum_{\text{topologies}} \int \frac{[dg][dX]}{\Omega_{\text{diff}}} \exp(-I[X,g]) . \quad (3.35)$$

This expression correctly divides by the volume of the diffeomorphism group in such a way that the original gauge invariant theory on $\mathcal{C} \times \mathcal{M}$ reduces to a theory on the quotient $(\mathcal{C} \times \mathcal{M})/\text{Diff}(M)$. Taking this into account, we see that the partition function on the quotient is given by

$$Z = N \sum_{\text{topologies}} \int [d\sigma](dt)(dX_0) \frac{\Omega^\perp_{\text{diff}\,0}}{\Omega_{\text{diff}\,0}} \frac{\Omega_{\text{diff}\,0}}{\Omega_{\text{diff}}} J \left[\det{}'(P^\dagger P)\right]^{1/2} (\det{}'\Delta)^{-d/2}. \quad (3.36)$$

There are two remarks one can make about the above expression. The ratio $\Omega_{\text{diff}\,0}/\Omega^{\perp}_{\text{diff}\,0}$ is essentially Ω_{CKV}, the volume of the conformal Killing vectors.[8] This is just a reflection of the residual gauge symmetry. The ratio $\Omega_{\text{diff}}/\Omega_{\text{diff}\,0}$ counts the number of distinct path connected components of the diffeomorphism group.[9] The latter reflects the invariance of the naive partition function (3.4) under the full diffeomorphism group.

Even the above expression is not completely correct. There are some ultra-technical details [11] that arise due to our choice of coordinates. The careful analysis by Moore and Nelson shows that everything works out in the end.

Let us summarize the main points of our discussion. It is possible to implement the Faddeev-Popov procedure in string models. There are a finite number of Teichmüller deformations which must be included in the evaluation of the partition function. These deformations become very important when one studies multi-loop contributions to the partition function. In fact the number of real Teichmüller parameters is 0 for a sphere, 2 for a torus and $6(h-1)$ for a surface with more than one handle. The Teichmüller deformations will be discussed in more detail in a later section. The volume of conformal Killing vectors plays an important role in the case of a sphere leading to a vanishing contribution to the cosmological constant [11].

4 The Complex Tensor Calculus

Since every orientable two dimensional manifold is a Riemann surface, many of the differential geometric properties are best explored by using a differential geometric formalism in unity with complex analysis. The epsilon tensor[10] ϵ_{ab} defines a linear transformation on the tangent space $J^a{}_b = \epsilon^a{}_b$

[8] The factorization of volumes in an infinite dimensional space is not a very intuitive concept. For simplicity consider a cube with sides of length 1/2 in a ten dimensional space. If one thinks of the cube as the product of a seven cube and a three cube then the respective volume of each is 1/128 and 1/8. Augment the dimension and see what happens.

[9] The path connected components of the diffeomorphism group are denoted by $\pi_0(\text{Diff}(M))$.

[10] ϵ_{ab} is defined by $\epsilon_{12} = \sqrt{g}$.

with the property $J^2 = -1$. This object defines an integrable complex structure [15]. It is useful to combine this observation with a theorem of Gauss.

Theorem 4.1 *Locally one can choose a coordinate system on M such that the metric takes the simple form $ds^2 = \exp(2\sigma)\ (dx^2 + dy^2)$. Such a coordinate system is called conformal or isothermal.*

If one now defines $z = x+iy$ and $\bar{z} = x-iy$ then one sees that in a conformal coordinate system the operator J is simply multiplication by $\sqrt{-1}$. The partial derivative operators are defined by

$$\begin{aligned} \partial_z &= \frac{\partial}{\partial z} = \frac{1}{2}\left(\frac{\partial}{\partial x} - i\frac{\partial}{\partial y}\right) , \\ \partial_{\bar{z}} &= \frac{\partial}{\partial \bar{z}} = \frac{1}{2}\left(\frac{\partial}{\partial x} + i\frac{\partial}{\partial y}\right) . \end{aligned} \tag{4.1}$$

The Cauchy-Riemann equations are simply $\partial_{\bar{z}} f = 0$. Note that in this coordinate system the metric may be written as

$$ds^2 = \frac{1}{2}\exp(2\sigma)\ (dz \otimes d\bar{z} + d\bar{z} \otimes dz) . \tag{4.2}$$

In terms of the z and $\bar{z}$ coordinates the metric is off diagonal:

$$\begin{pmatrix} g_{zz} = 0 & g_{z\bar{z}} = (1/2)\exp(2\sigma) \\ g_{\bar{z}z} = (1/2)\exp(2\sigma) & g_{\bar{z}\bar{z}} = 0 \end{pmatrix} . \tag{4.3}$$

Note that the inverse metric has $z\bar{z}$ component given by $g^{z\bar{z}} = 2\exp(-2\sigma)$.

The following elementary observation is crucial to the subsequent ideas we will develop. Since we have a metric we can go to a local orthonormal frame. Since the irreducible representations of SO(2) = U(1) are one dimensional and characterized by an integer, *the helicity*, we can define the tensors of rank n to be those that under a rotation by θ transform as

$$T \to \exp(in\theta)\ T . \tag{4.4}$$

Example 4.1 *The helicity* ± 1 *vectors are given by* $V^1 \pm iV^2$. *The helicity* ± 2 *tensors correspond to a symmetric traceless tensor* h^{ab} *with components* $h^{11} \pm ih^{12}$.

The rank n tensors[11] are defined by

$$T \rightarrow \left(\frac{\partial z'}{\partial z}\right)^n T , \tag{4.5}$$

under the *analytic* change of coordinate $z \rightarrow z'$. It is only necessary to consider tensors which have only z indices because the metric can be used to trade a $\bar{z}$ index for a z index.

Example 4.2 $T_{\bar{z}}{}^{z} = g_{\bar{z}z} T^{zz}$.

We note that a tensor such as $T^z{}_z$ is a scalar. Consequentially, the only thing that matters is the difference between the number of upper indices and the number of lower indices. The space of helicity n tensors will be denoted by $\mathcal{T}^n$.

The vector bundles $\mathcal{T}^n$ are complex line bundles. On a Riemann surface there is always a *canonical* line bundle which may be taken to be the holomorphic tangent bundle T_cM. A holomorphic vector field is of the type $\xi = \xi^z \partial_z$ where the component is an arbitrary function of z and $\bar{z}$. The canonical bundle always exists. The spaces $\mathcal{T}^n$ may be identified with the n-fold tensor product of T_cM if $n > 0$. If $n < 0$ they may be identified with the n-fold product of the dual bundle T_c^*M. On any Riemann surface one can always define the positive chirality spin bundle S_+ to be the square root of the holomorphic tangent bundle T_cM. In other words, the transition functions are the square root of the transition functions for the tangent bundle:

$$\psi \rightarrow \sqrt{\left(\frac{\partial z'}{\partial z}\right)} \; \psi . \tag{4.6}$$

One is immediately led to ask the question whether it is possible to define the seventh root of the holomorphic tangent bundle. The answer is no. The

[11] The convention used here is the opposite of the convention use by Friedan [16].

above discussion seems to apply to an arbitrary power of the holomorphic tangent bundle. The above argument is a local argument and therefore incomplete. The global argument involves a little knowledge of characteristic classes [15]. The characteristic class associated with the holomorphic tangent bundle is the Chern class $c_1(T_cM)$ which when integrated over the manifold M gives an integer which is the Euler characteristic $\chi(M)$ of the manifold. If a line bundle E is the p-th power of the holomorphic tangent bundle then its Chern class is given by $c_1(E) = p \cdot c_1(T_cM)$. The integral of $c_1(E)$ over the manifold must be an integer! This is a fundamental result in the theory of characteristic classes. Since the Euler characteristic is even, we see that we can always construct the square root bundle, *i.e.*, the spinors. In general one cannot construct a seventh root bundle. On a manifold with eight handles one can construct the seventh root bundle. The only *canonical*, *i.e.*, generic, constructions of a fractional power of the holomorphic tangent bundle are the spinor bundle S_+ and the associated negative chirality bundle S_-.

One can define connections which commute with holomorphic changes of variable. Define $\nabla^z_n : \mathcal{T}^n \to \mathcal{T}^{n+1}$ by

$$\nabla^z_n T = g^{z\bar{z}} \partial_{\bar{z}} T \,, \tag{4.7}$$

and a second operator $\nabla^n_z : \mathcal{T}^n \to \mathcal{T}^{n-1}$ by

$$\begin{aligned} \nabla^n_z T &= \left(g^{z\bar{z}}\right)^n \partial_z \Big((g_{z\bar{z}})^n \, T \Big) \,. \\ &= \Big(\partial_z + 2n(\partial_z \sigma) \Big) T \,. \end{aligned} \tag{4.8}$$

These operators are the standard Christoffel connections expressed in a "helicity" basis. These operators belong to an infinite chain which connects the different $\mathcal{T}$ spaces.

$$\cdots \underset{\nabla^{-1}_z}{\overset{\nabla^z_{-2}}{\rightleftarrows}} \mathcal{T}^{-1} \underset{\nabla^{0}_z}{\overset{\nabla^z_{-1}}{\rightleftarrows}} \mathcal{T}^{0} \underset{\nabla^{1}_z}{\overset{\nabla^z_{0}}{\rightleftarrows}} \mathcal{T}^{1} \underset{\nabla^{2}_z}{\overset{\nabla^z_{1}}{\rightleftarrows}} \cdots \tag{4.9}$$

In the standard inner product one can show that the raising and lowering

∇ are adjoints of each other:

$$(\nabla_n^z)^\dagger = -\nabla_z^{n+1} \; . \tag{4.10}$$

The differential operator P is very simple when expressed in the complex basis. The infinitesimal diffeomorphism[12] generated by V^a corresponds to vector fields V^z and V_z with $V^z = (V^{\bar{z}})^*$. Similarly, the symmetric traceless tensor h_{ab} corresponds to h_{zz} and h^{zz} with $h^{zz} = (h^{\bar{z}\bar{z}})^*$. The equation PV corresponds to the pair of equations:

$$PV = \begin{cases} \nabla_1^z V^z \\ \nabla_z^{-1} V_z \end{cases} . \tag{4.11}$$

Likewise, the equation $P^\dagger h$ corresponds to the pair of equations:

$$P^\dagger h = \begin{cases} \nabla_z^2 h^{zz} \\ \nabla_{-2}^z h_{zz} \end{cases} . \tag{4.12}$$

The equation for a conformal Killing vector field $PV = 0$ is equivalent to $\nabla_1^z V^z = 0$. From the definition (4.7) of ∇_1^z it follows that the latter equation is $\partial_{\bar{z}} V^z = 0$. In other words, V^z is a global analytic vector field.[13] *Conformal Killing vector fields are just the global analytic vector fields.* If M has the topology of S^2 then the analytic vector fields correspond to the infinitesimal generators of SL(2, C), the automorphism group of the sphere. Likewise, one can show that $V^{\bar{z}}$ is an anti-analytic vector field.

The equation for the Teichmüller deformations $P^\dagger h = 0$ is equivalent to $\nabla_{-2}^z h_{zz} = 0$. This is equivalent to $\partial_{\bar{z}} h_{zz} = 0$. Teichmüller deformations correspond to global analytic tensor fields h_{zz}. The object $h_{zz}\, dz \otimes dz$ is called a quadratic differential.

The theory of elliptic operators states that the ∇ operators have finite dimensional kernels. The celebrated Riemann-Roch theorem relates

[12] We will drop the δ notation from V and h.

[13] The terminology is a bit confusing. We nave defined a holomorphic vector field to be one of the type $V^z \partial_z$ where V^z was arbitrary. Holomorphic refers only to the transformation law. A global analytic vector field is a holomorphic vector field with analytic components in each local coordinate chart.

the number of conformal Killing vectors and the number of Teichmüller deformations.

Theorem 4.2 Riemann-Roch: *The number of real conformal Killing vectors fields minus the number of real Teichmüller deformations is given by* $3\chi(M) = 6 - 6h$.

The counting of conformal Killing vectors is confusing. Sometimes people count the number of real parameters other times the number of complex parameters. We will elaborate on the counting when we discuss the Atiyah-Singer index theorem and the Teichmüller space.

There are several differential geometric formulas which will be useful. The Riemann curvature scalar is given by

$$R = -2\exp(-2\sigma)\,\partial^2\sigma\,. \tag{4.13}$$

The Ricci identity for the curvature tensor is

$$(\nabla_z^{n+1}\,\nabla_n^z - \nabla_{n-1}^z\,\nabla_z^n)T = \frac{n}{2}RT \tag{4.14}$$

for $T \in \mathcal{T}^n$.

5 Determinants and the Index Theorem

The evaluation of the partition function (3.36) requires the computation of several determinants. In this section we will demonstrate that the determinants may be evaluated with the use of the Atiyah-Singer index theorem [14]. We work at a naive level and avoid the complications introduced by Teichmüller deformations, conformal Killing vectors, and zero modes [10], [16]. Our main concern will be the conformal and the gravitational anomalies.

The watered down partition function is given by

$$Z = \int [d\sigma](\det\Delta)^{-d/2}(\det\nabla_1^z)(\det\nabla_z^{-1}) \tag{5.1}$$

where σ is the conformal factor of the metric and the Laplacian Δ is defined by

$$\begin{aligned}\Delta &= -\frac{1}{\sqrt{g}}\partial_a\sqrt{g}g^{ab}\partial_b \\ &= -\nabla^a\nabla_a \\ &= -(\nabla^z_{-1}\nabla^0_z + \nabla^1_z\nabla^z_0)\,.\end{aligned} \tag{5.2}$$

We remark that the integrand is manifestly real. The determinant of the self-adjoint Laplacian Δ is real. The Faddeev-Popov determinants constitute the Jacobian in the change of variables from traceless deformations of the metric to infinitesimal diffeomorphisms. Changes of variables require the absolute value of the determinant of the Jacobian matrix. In fact, $(\det \nabla^z_1)$ is the complex conjugate of $(\det \nabla^{-1}_z)$. Each of the determinants has a gravitational anomaly but the combined anomaly vanishes. Simply put, each determinant is complex with opposite phase. In addition, each determinant has a trace anomaly of *identical* value that contributes to the real part of the determinant. The miracle of (5.1) is that if $d = 26$ then the contribution to the conformal anomaly from the scalars cancels the contribution from the Faddeev-Popov ghosts.

In the Hamiltonian approach to closed string models one easily sees that the left movers are independent of the right movers. This is a key observation in the construction of the heterotic string. The same thing can be seen in the path integral approach. The fundamental observation is that by using the Ricci identity (4.14) one may rewrite the Laplacian (5.2) in the following forms:

$$\begin{aligned}\Delta &= -2\nabla^z_{-1}\nabla^0_z \\ &= -2\nabla^1_z\nabla^z_0\,.\end{aligned} \tag{5.3}$$

Therefore $\det\Delta$ factorizes into a "left-moving" piece and a "right-moving" piece. In fact, we will see that up to irrelevant infinite factors, $\det\nabla^z_{-1} = \det\nabla^z_0$, $\det\nabla^1_z = \det\nabla^0_z$, and $(\det\nabla^0_z)^* = \det\nabla^z_0$.

Consequently, the integrand of the partition function (5.1) may be written as a product

$$\left[(\det \nabla_z^0)^{-d/2}(\det \nabla_z^{-1})\right]\left[(\det \nabla_{-1}^z)^{-d/2}(\det \nabla_1^z)\right] \tag{5.4}$$

involving independent left movers and independent right movers.

We will see that all local anomalies cancel independently in the left and the right moving sectors if $d = 26$. This is an immediate consequence of the family's index theorem.

Firstly, we note that if S_+ and S_- are the positive and negative chirality spin bundles, then group theory immediately tells us that

$$\begin{aligned} \mathcal{T}^n &= S_+ \otimes S_+^{2n-1}, \\ \mathcal{T}^{n-1} &= S_- \otimes S_+^{2n-1}. \end{aligned} \tag{5.5}$$

The operator ∇_z^n is just the chiral Dirac operator $\mathcal{D}_V$ for chiral spinors coupled to the vector bundle $V = S_+^{2n-1}$. The operator ∇_z^n has an interpretation that transcends conformal coordinates! It is the chiral Dirac operator twisted with the vector bundle V, an operator which can be defined in an arbitrary coordinate system.

Let $W[g]$ be the vacuum functional. For the type of theories we are considering, the prototypical vacuum functional is of the form

$$W[g] = \log \det \nabla_z^n. \tag{5.6}$$

We will use a trick to evaluate the conformal anomaly. We observe that in two dimensions the conformal anomaly and the ordinary gauge anomaly differ by a factor of $\sqrt{-1}$.

In field theory, one is interested in the U(1) anomaly associated with the gauge transformation of the covariant derivative:

$$\partial_a + iA_a \to \partial_a + iA_a + i\partial_a\alpha\,. \tag{5.7}$$

The operator

$$\nabla_z^n = \partial_z + 2n(\partial_z\sigma) \tag{5.8}$$

looks like the gauge transformation in (5.7) except for a factor of i. This will allow us to use the index theorem to determine the dependence of $\log\det\nabla_z^n$ on σ. The strategy is to use the family's index theorem to calculate $\delta W/\delta\sigma$ and integrate to get the σ dependence of W.

To calculate the ordinary abelian gauge anomaly in two dimensions one uses the following prescription [17], [18] (We take the gauge field $\mathcal{A}$ to be anti-hermitian. The associated field strength will be denoted by $\mathcal{F} = d\mathcal{A}$):

1. Take an appropriate polynomial in $\mathcal{F}$ which is a four form (2 dimensions higher than the space-time dimensionality). For example, consider $\mathcal{F}^2 = \mathcal{F}\wedge\mathcal{F}$.
2. Write $\mathcal{F}^2 = d\omega$ where $\omega = \mathcal{A}\mathcal{F}$.
3. Make the substitution $\mathcal{A}\to\mathcal{F}+v$, $\mathcal{F}\to\mathcal{F}$ in ω, where v is the gauge transformation parameter. Under this substitution, the transformation law for ω is $\omega\to\omega+v\mathcal{F}$. The anomaly is given by $\delta W = \int v\mathcal{F}$.

We will use the family's index theorem to determine the appropriate polynomial including the correct normalization [19]. An appropriate choice [31] of v leads to either the gravitational, the Lorentz, or the conformal anomaly. We will only discuss the last case.[14]

Consider the chiral Dirac operator coupled to a vector bundle V on an even dimensional manifold X:

$$\mathcal{D}_V : S_+\otimes V \to S_-\otimes V. \tag{5.9}$$

More explicitly, $\mathcal{D}_V$ is given by

$$\mathcal{D}_V = \frac{1-\gamma_5}{2}\gamma^\alpha e_\alpha^a\left(\partial_a + \omega_a^{\alpha\beta}\frac{1}{8}[\gamma_\alpha,\gamma_\beta] + \mathcal{A}_a\right), \tag{5.10}$$

where $e_\alpha^a\partial_a$ defines an orthonormal frame, $\omega_a^{\alpha\beta}$ is the Cartan connection, and $\mathcal{A}_a$ is the anti-hermitian connection on V. The Cartan curvature $\mathcal{R}^{\alpha\beta}$ and

[14] In the non-abelian case the anomaly is given by $\int v\, d\mathcal{A}$. Since we are assuming that the theory has no net gravitational anomaly, the conformal anomaly will be covariant. This simplifies calculations a bit.

the field strength $\mathcal{F}$ are defined by

$$\mathcal{R} = d\omega + \omega^2 , \qquad \mathcal{F} = d\mathcal{A} + \mathcal{A}^2. \tag{5.11}$$

According to Atiyah and Singer, the index $\mathcal{I}(\mathcal{D}_V)$ may be obtained from the generating function

$$\mathcal{I}(\mathcal{D}_V) = \int_X \sqrt{\det\left(\frac{\mathcal{R}/4\pi}{\sinh \mathcal{R}/4\pi}\right)} \operatorname{tr}\exp\left(\frac{i\mathcal{F}}{2\pi}\right), \tag{5.12}$$

where one picks out the differential form corresponding to the dimension of X. We will be interested in both differential forms of degree two and of degree four. Expanding the equation above one finds

$$\mathcal{I}(\mathcal{D}_V) = \int_X \left[\frac{i}{2\pi}\operatorname{tr}\mathcal{F} + \left(\frac{\dim V}{192\pi^2}\operatorname{tr}\mathcal{R}^2 - \frac{1}{8\pi^2}\operatorname{tr}\mathcal{F}^2\right)\right]. \tag{5.13}$$

The ordinary index on the two dimensional world sheet M is given by

$$\mathcal{I}(\mathcal{D}_V) = \frac{i}{2\pi}\int_M \operatorname{tr}\mathcal{F}. \tag{5.14}$$

For the operator ∇_z^n, the vector bundle V is equal to S_+^{2n-1}, the tensors of helicity $(2n-1)/2$. The curvature two form on helicity one objects (tangent vectors) is $i\mathcal{R}_{12}$. Therefore, the curvature two form on V is given by

$$\mathcal{F} = \frac{2n-1}{2} i\mathcal{R}_{12}. \tag{5.15}$$

Inserting this into (5.14) we obtain the celebrated Riemann-Roch theorem

$$\begin{aligned} \mathcal{I}(\nabla_z^n) &= -\frac{2n-1}{2}\cdot\frac{1}{2\pi}\int_M \mathcal{R}_{12} \\ &= -\frac{2n-1}{2}\cdot\frac{1}{2\pi}\cdot\frac{1}{2}\int_M R\sqrt{g}d^2z \\ &= -\frac{2n-1}{2}\chi(M), \end{aligned} \tag{5.16}$$

where $\chi(M)$ is the Euler characteristic of M.

To calculate the gauge anomaly one needs the four form part of (5.13). The correctly normalized polynomial obtained from the family's index theorem is

$$2\pi i \cdot \left[\frac{\dim V}{192\pi^2} \operatorname{tr} \mathcal{R}^2 - \frac{1}{8\pi^2} \operatorname{tr} \mathcal{F}^2 \right], \tag{5.17}$$

the factor of $2\pi i$ arises since the index theorem gives an integer and one is interested in the phase of the determinant which must be single valued when one goes around any loop in the gauge group.

Applying the family's index theorem to ∇_z^n requires the substitution of

$$\begin{aligned} \dim V &= 1 \\ \operatorname{tr} \mathcal{R}^2 &= -2\mathcal{R}_{12}^2 \\ \operatorname{tr} \mathcal{F}^2 &= \left(\frac{2n-1}{2} i \right)^2 \mathcal{R}_{12}^2 \end{aligned} \tag{5.18}$$

into (5.17). Trivial algebra leads to the normalized polynomial

$$i \frac{1 + 6n(n-1)}{24\pi} \mathcal{R}_{12}^2 \,. \tag{5.19}$$

This formula is universal. It is the same for gravitational, Lorentz, or conformal anomalies. The only catch in using the above for conformal anomalies involves signs as will be explained later.

For the left moving sector, the anomaly associated with

$$(\det \nabla_z^0)^{-d/2} (\det \nabla_z^{-1}) \tag{5.20}$$

may be evaluated by using the universal expression

$$\left[-\frac{d}{2} \cdot 1 + 13 \right] i \frac{\mathcal{R}_{12}^2}{24\pi} = \frac{26-d}{2} i \frac{\mathcal{R}_{12}^2}{24\pi} \,. \tag{5.21}$$

If $d = 26$ then there will be no anomalies in the left moving sector or in the right moving sector.

It is also worthwhile to consider both sectors simultaneously as in Table 1. Note that the gravitational contributions have opposite signs for the left and right moving sectors whereas the conformal contributions have

	Left	Right
gravitational	$26-d$	$d-26$
conformal	$26-d$	$26-d$

Table 1: Anomaly Contributions

the same signs. In the case of gravitational anomalies there is a cancellation between the corresponding members of the left and the right moving sectors. Even if $d \neq 26$, the total contribution to gravitational anomalies vanishes and one is left with a conformal anomaly. Only in $d = 26$ does a complete anomaly cancellation occur within a left moving sector or a right moving sector.

We now turn to the superstring. In this case one has a bosonic Faddeev-Popov ghost for the gravitino. The combination of determinants that enters the partition function for the left moving sector is

$$(\det \nabla_z^0)^{-d/2}(\det \nabla_z^{1/2})^{d/2} \cdot (\det \nabla_z^{-1})(\det \nabla_z^{-1/2})^{-1}, \tag{5.22}$$

where $\nabla_z^{1/2}$ is the chiral Dirac operator on the Majorana fermions and $\nabla_z^{-1/2}$ is the operator that acts on the chiral gravitinos. The combined anomaly for the above is

$$\left[-\frac{d}{2} + \left(-\frac{1}{2}\right)\frac{d}{2} + 13 - \frac{11}{2}\right] i\frac{\mathcal{R}_{12}^2}{24\pi} = \frac{30-3d}{4}\frac{i\mathcal{R}_{12}^2}{24\pi}. \tag{5.23}$$

All anomalies cancel within the left moving sector if $d = 10$.

In the standard closed string model or in the supersymmetric version, one can move away from the critical dimension with a penalty [1],[20]. One has to introduce the Liouville field theory or the supersymmetric Liouville field theory in order to "compensate" for the conformal anomaly. The gravitational anomalies cancel between the left and the right moving sectors in all dimensions. One then has to face the problem of the dynamics of the Liouville field theory.

In the heterotic model things are more complicated. Consider the standard heterotic string [21] with a bosonic left moving sector and a supersymmetric right moving sector. The critical dimension of the left moving sector is $d_{\text{left}} = 26$, and of the right moving sector is $d_{\text{right}} = 10$. Away from the critical dimensions, one has both a conformal anomaly and a gravitational anomaly. The gravitational anomaly cannot be cancelled by the addition of a local counterterm. Therefore, the heterotic string has a gravitational anomaly away from the critical dimensions. Integrating the conformal anomaly leads to some modification of the Liouville field theory.

Consider a field theory where all the conformal anomalies and all the gravitational anomalies cancel. The functional integral (5.1) still involves an integration over the conformal factor σ. Since the integrand is independent of σ, the result of the σ integration is simply the volume Ω_{Weyl} of the Weyl group $\mathcal{W}$. In such a situation, the physical configuration space is not $\mathcal{C} \times \mathcal{M}/\text{Diff}(M)$ but $(\mathcal{C} \times \mathcal{M})/(\mathcal{W} \odot \text{Diff}(M))$, where $\odot$ denotes the semi-direct product. In conformally invariant field theories, gauge anomalies, gravitational anomalies and also the conformal anomaly must cancel. In such theories, the starting expression for the partition function involves a division by the volume of the group $\mathcal{W} \odot \text{Diff}(M)$ instead of the volume of the diffeomorphism group.

Using the prescription explained at the beginning of this section, we can use (5.19) to evaluate the trace anomaly. Remember that $\mathcal{R}^2_{12}$ becomes $v\mathcal{R}_{12}$, where v is the gauge parameter for a gauge anomaly. Since $\nabla^1_z = \partial_z + (2\partial_z\sigma)$ we see that $v = i2\delta\sigma$, where the extra i is required since we have absorbed some factors of i in obtaining (5.19):

$$\begin{aligned} \delta \log \det \nabla^n_z &= \pm \int (2i\delta\sigma) \frac{1 + 6n(n-1)}{24\pi} i\mathcal{R}_{12} \\ &= \mp \frac{1 + 6n(n-1)}{24\pi} \int (\delta\sigma) R\sqrt{g} d^2 z \,. \end{aligned} \tag{5.24}$$

We include the plus or minus sign due to ambiguities that result from using the index theorem. A simple one loop calculation for ∇^0_z shows that the correct sign is (-1) in the above. The final result is that

$$\delta \log \det \nabla^n_z = -\frac{1 + 6n(n-1)}{24\pi} \int d^2 z \sqrt{g} R(\delta\sigma). \tag{5.25}$$

This is the standard formula for the trace anomaly in two dimensions [20]. It may be integrated to obtain the σ functional dependence of $\det \nabla^n_z$.

Note that $1 + 6n(n-1) = 6(n-1/2)^2 - 1/2$, and therefore (5.25) is invariant under a reflection about $n = 1/2$. In other words, $n \to 1-n$ leaves (5.25) invariant. We immediately have the result that $\det \nabla^n_z = \det \nabla^{1-n}_z$. Since $-\nabla^z_{-n}$ is the adjoint of ∇^{1-n}_z, it follows that $(\det \nabla^n_z) = (\det \nabla^z_{-n})^*$.

The real part of the gravitational anomaly may be eliminated by a counter term. The discussion of the previous paragraphs shows that left and right movers contribute with opposite sign to the physically relevant imaginary part of the gravitational anomaly. The trace anomaly is real and therefore the left and right movers contribute equally.

It is straightforward to integrate the previous functional differential equation. The result is

$$\begin{aligned} \log \det \nabla^n_z &= -\frac{1+6n(n-1)}{24\pi} \int_M d^2z\, (\partial\sigma)^2 \\ &+ \Lambda^2 \int_M d^2z \sqrt{g} \\ &+ \text{independent of } \sigma\,. \end{aligned} \tag{5.26}$$

In the above Λ is an ultraviolet cutoff. The cosmological term is present if one uses a realistic regularization. The last term is the "integration constant" for the differential equation. Polchinski [22] discusses the possibility of setting the net cosmological constant zero to obtain a conformally invariant field theory.

The index theorem can also be used to discuss conformal and gravitational anomalies in background fields in which a factorization such as (5.22) occurs. In particular one can analyze certain theories at one loop. Typically what happens is that the determinant gets modified due to the background. The background field may be due to embedding the theory in a curved space or due to the presence of external gauge sources.[15] Without a background

[15] There are two curvatures in this scenario. The first is the intrinsic curvature of the Riemannian manifold M. The second is the curvature of the non-linear sigma model one is studying. The latter will be called the background curvature.

field the vector bundle V is simply S_+^{2n-1}. In the presence of a background field it becomes $V = S_+^{2n-1} \otimes E$ where E is some vector bundle that the background field acts on. The covariant derivative ∇_z^n gets replaced by a new covariant derivative

$$\nabla_z^n \to \nabla_z^n + A_z \, , \tag{5.27}$$

where A is the background field. The universal expression (5.17) remains the same but the field strength $\mathcal{F}$ has to be interpreted as having two contributions:

$$\mathcal{F} = \frac{2n-1}{2} i \mathcal{R}_{12} + F \, , \tag{5.28}$$

where $F = dA + A^2$. Inserting the above into (5.17) leads to the following normalized polynomial:

$$\begin{aligned} &\frac{3(2n-1)^2 - \dim E}{48\pi} \, i \, \mathcal{R}_{12}^2 \\ &+ \frac{2n-1}{4\pi} \mathcal{R}_{12} \operatorname{tr} F \\ &- \frac{1}{4\pi} i \operatorname{tr}\left(F^2\right) . \end{aligned} \tag{5.29}$$

This is a very useful expression that simultaneously gives information about the conformal anomalies, the gravitational anomalies and the gauge anomalies. One immediately learns that even in the absence of a background field there is a gauge anomaly in a U(1) current due to the intrinsic curvature of M. This follows from using our recipe and setting the background gauge fields to zero at the end of the calculation. The U(1) anomaly is proportional to $(2n-1)$R. This fact has been used in the construction of the fermionic vertex [23].

If tr F vanishes then the conformal anomaly lacks a contribution from the external fields. Such a contribution can also be made to vanish by an appropriate choice of the U(1) charges for all the different ∇_z^n that occur.

Finally we remark that the absence of gauge anomalies requires tr $F^2 = 0$ for the various determinants that occur. F can have a contribution from both a background curvature and a background gauge field. The condition found by Candelas *et al.* [24] and further discussed by Bagger *et al.* [25]

follows easily from this discussion. In heterotic like theories one needs

$$\operatorname{tr} G^2 - \operatorname{tr} \Omega^2 = 0 \,. \tag{5.30}$$

G is the background field strength for left moving sector and Ω is the embedding curvature two form for the right moving sector. In deriving the above we note that the anomalies for left movers and right movers have opposite sign and that the field strengths are anti-hermitian. Equation (5.29) has more uses [26] which will be discussed elsewhere.

6 Teichmüller Deformations

We have seen that Teichmüller deformations play an important role in string theory. An understanding of Teichmüller space is crucial for understanding multiloop string amplitudes. Our study will exploit both differential geometry and complex variables theory. Both viewpoints are complementary and illustrate different approaches to the same problem.

It is probably best to begin by making several general remarks about compact Riemann surfaces M without boundaries. There is a very powerful classification theorem which characterizes all Riemann surfaces [4].

Theorem 6.1 *A compact boundaryless Riemann surface with h handles is analytically equivalent to either: (1) sphere if $h = 0$, (2) torus = plane/lattice if $h = 1$, or (3) U/G where U is the upper half plane and G is a crystallographic subgroup of $SL(2,\mathbf{R})$ if $h > 1$.*

A corollary of this theorem is that there are essentially three distinct simply connected Riemann surfaces up to holomorphic equivalence: the sphere, the plane, and the upper-half plane. The corollary follows from the basic result of covering space theory which equates $\pi_1(M)$ with the group of deck transformations on the simply connected universal covering space.

Each of the universal covering spaces is a classic model for elliptic, Euclidean and hyperbolic geometries. Each of these manifolds has a canonical metric of constant curvature. The sphere has the standard positive curvature metric inherited from embedding it in $\mathbf{R}^3$. The plane has the Euclidean metric. The upper half plane has the constant negative curvature Poincaré metric

$$ds^2 = \frac{dz\, d\bar{z}}{(\mathrm{Im}\, z)^2} \,. \tag{6.1}$$

The existence of these three special metrics guarantees the existence of at least one constant curvature metric on any Riemann surface.

In the case of a torus the existence of a constant curvature metric is trivial. A lattice Γ is a collection of points

$$\Gamma = \{n_1\omega_1 + n_2\omega_2\} \,, \tag{6.2}$$

where n_1 and n_2 are integers, and ω_1 and ω_2 are complex numbers such that $\mathrm{Im}\,\omega_1/\omega_2 \neq 0$. A torus is the quotient space $\mathbf{C}/\Gamma$. Since the Euclidean metric is invariant under translations, it projects to a metric on the quotient $\mathbf{C}/\Gamma$.

An analogous argument works for surfaces with more handles. The crucial observation is that the Poincaré metric is invariant under the action of the automorphism group of the upper half plane $SL(2,\mathbf{R})$. If G is a crystallographic subgroup of $SL(2,\mathbf{R})$ then the Poincaré metric projects to a metric on the quotient U/G.

The existence of constant curvature metrics on Riemann surfaces interests us for a variety of reasons. The existence of a constant curvature metric on a Riemann surface is a global generalization of the local conformal flatness of Riemann surfaces. This global theorem agrees very nicely with the Gauss-Bonnet theorem:

$$\chi(M) = \frac{1}{4\pi}\int_M d^2z\,\sqrt{g}\,\mathrm{R} = 2 - 2h \,. \tag{6.3}$$

We note that for a constant curvature metric the sign of the curvature is determined by the Gauss-Bonnet theorem.

Our next concern is the uniqueness of the constant curvature metric which we analyze by linearization. Under an arbitrary variation δg_{ab} of the metric, the change in the scalar curvature is given by

$$\delta R = -\frac{1}{2} g^{ab}\, \delta g_{ab}\, R + \nabla^a \nabla^b \delta g_{ab} - \nabla^c \nabla_c \left(g^{ab}\, \delta g_{ab} \right) \ . \tag{6.4}$$

This equation is more transparent when written in local complex coordinates:

$$\delta R = -2(\delta\sigma) R + 2\Delta(\delta\sigma) + \nabla^z \nabla^z\, \delta g_{zz} + \nabla^{\bar z} \nabla^{\bar z}\, \delta g_{\bar z \bar z} \ . \tag{6.5}$$

Since the decomposition of the variation of the metric into trace and traceless pieces is an orthogonal decomposition we see that deformations in the Teichmüller directions leave the curvature invariant. This happens whether or not the reference metric has constant curvature.

The Teichmüller deformations count the number of "inequivalent" constant curvature metrics on a surface. This may be seen for the case $h > 1$ by looking at the linearization equation above and noticing that $\Delta - R$ is a positive operator for constant negative R.

Teichmüller realized that the symmetric traceless deformations of the metric correspond to infinitesimal changes in the complex structure of the manifold. In fact, all the inequivalent complex structures may be determined from the Teichmüller deformations. The proof of the latter statement is beyond the scope of these lectures but the former statement may be understood using the following simple argument. It was previously mentioned that a complex structure on a Riemannian manifold may be understood in terms of a linear transformation J on the tangent bundle satisfying $J^2 = -1$. It is clear that $J^a{}_b = g^{ac} \epsilon_{cb}$ is conformally invariant since ϵ_{ab} contains a factor of $\sqrt{g}$. Therefore the complex structure is preserved under conformal transformations. In fact, under a generic variation of the metric, the change in the conformal structure is determined entirely by the symmetric traceless deformations of the metric:

$$\delta J^a{}_b = -\epsilon_{cb}\, G^{acef}\, \delta g_{ef} \ . \tag{6.6}$$

Some of the complex structures generated by a generic variation of the metric are not truly new. From Figure 2 we see that most of the deformations of the complex structure may be associated with infinitesimal diffeomorphisms of the manifold. Only the Teichmüller deformations are not associated with diffeomorphisms. The Teichmüller deformations generate the new complex structures. In fact, the number of distinct complex structures is labelled by a finite number of parameters. This may be seen since the Teichmüller deformations are solutions to the elliptic equation $\nabla^{z}_{-2}\, h_{zz} = 0$. A theorem in elliptic operator theory guarantees that the kernel of the operator ∇^{z}_{-2} is finite dimensional.

It is remarkable that the number of independent complex structures is related to the number of inequivalent constant curvature metrics that one can put on the surface.

The Riemann-Roch theorem (5.16) provides more insight into the dimensionality of Teichmüller space. The number of solutions to the equation $\nabla^{z}_{-2}\, h_{zz} = 0$ minus the number of solutions to the equation $\nabla^{-1}_{z}\, V_z = 0$ is given by $-3\chi(M)/2 = 3(h-1)$ where h is the number of handles. The equation $\nabla^{-1}_{z}\, V_z = 0$ is equivalent to the statement that $V^{\bar{z}}$ is an anti-analytic vector field. The associated analytic vector field is defined by $V^z = (V^{\bar{z}})^*$. Therefore the Riemann-Roch theorem relates the number of Teichmüller deformations to the number of infinitesimal automorphisms of the Riemann surface.

Thus far everything has been very differential geometric in nature. Let us look at the complex analysis viewpoint. In a local complex coordinate system one can study the quadratic differentials $h_{zz}\, dz \otimes dz$. Since these objects are covariant under holomorphic changes of coordinates they define global quadratic differentials. One can then ask if there are any global analytic quadratic differentials. The original Riemann-Roch theorem is not formulated in terms of an index of a differential operator. It is a statement about globally analytic "n-th order" differentials. Teichmüller realized that there is a connection between the quadratic differentials and the number of complex structures. The differential geometric approach shows that there is a connection between quadratic differentials and deformations of the metric.

Riemann conjectured [27] that the number of distinct complex structures on a surface with $h > 1$ handles was described by $6(h-1)$ *real* parameters. The Riemann-Roch theorem says that these parameters may be taken as $3(h-1)$ complex parameters. One can give the following intuitive but not rigorous argument. A Riemann surface with $h \geq 1$ handles is topologically equivalent to a sphere with $2h$ pairwise identified circular holes. The three parameters that characterize each hole are the position and the radius. Therefore, the Riemann surface is described by $6h$ parameters. This is not completely accurate since an automorphism of the sphere would lead to the same Riemann surface. Also, the the Riemann surface is realized on the sphere up to an automorphism of the Riemann surface. The number of distinct complex structures is given by $6h - 6 + \alpha$ parameters, where α is the number of connected automorphisms of the surface. This argument can be made rigorous. Any Riemann surface with handles $h \geq 1$ can be canonically mapped into the extended complex plane [28] in such a way that there are $4h$ semi-infinite horizontal slits with appropriate edges identified. The ends of the slits of an identified pair have the same abscissa. The number of parameters that describe the canonical region is therefore $2 \cdot 3h$. The number of automorphisms of the extended complex plane is six. Therefore, the number of inequivalent Riemann surfaces is given by $6(h-1) + \alpha$ real parameters.

We can calculate the number of Teichmüller deformations by using the index theorem. If the manifold has the topology of a sphere then there are three independent global analytic vector fields. This is best seen by stereographically projecting the sphere onto the extended complex plane and noticing that the only normalizable analytic vector fields are 1, z and z^2, corresponding to the infinitesimal generators of $SL(2, C)$ transformations. The Riemann-Roch theorem then asserts that there are no Teichmüller deformations.

In the case of the torus the situation is slightly different. There is one complex Teichmüller parameter. This is an easy consequence of the classical theorem that states that the only global analytic functions on a torus are the constant functions. To prove this classical theorem we note that a torus may be represented in the plane by parallelograms which are identified by the discrete translations which define the lattice. Since the torus is compact, an analytic function must be bounded on the torus. The analytic function on the torus defines a global periodic analytic function on the complex plane. Therefore, the periodic analytic function is bounded in the entire complex plane and must be constant by Liouville's theorem. Global analytic vector fields V^z or Teichmüller deformations h_{zz} must be constant on a torus. There is one complex Teichmüller parameter and there is one complex parameter that describes the infinitesimal analytic diffeomorphism.

A different proof may be made by noticing that the connected analytic automorphisms of the plane (not the extended plane) are given by $z \to az + b$. The connected analytic diffeomorphisms which project down smoothly to the torus $\mathbf{C}/\Gamma$ are just $z \to z + b$.

The last proof generalizes to the case of more than one handle. The fundamental theorem states that the surface M may be identified with U/G where G was a crystallographic subgroup of $\mathrm{SL}(2,\mathbf{R})$. An analytic automorphism of M is an analytic diffeomorphism of U. The analytic automorphisms of M correspond to the automorphisms of U which project smoothly to U/G. It is immediately clear that the number of real parameters describing the infinitesimal diffeomorphisms is less than or equal to three since the dimension of $\mathrm{SL}(2,\mathbf{R})$ is three. It is non-trivial to show [4] that there are no infinitesimal $\mathrm{SL}(2,\mathbf{R})$ transformations which project to U/G. The Riemann-Roch theorem provides the desired result that the number of independent Teichmüller deformations is $3(h-1)$.

Teichmüller space is techically defined to be $t(M) = \mathcal{M}/(\mathcal{W} \odot \mathrm{Diff}_0(M))$. The symmetric traceless deformations δh_{ab} are the tangent vectors to the Teichmüller space. The *moduli* space of the manifold is defined to be $m(M) = \mathcal{M}/(\mathcal{W} \odot \mathrm{Diff}(M))$. Since $\mathcal{M}$ has no topology, the exact homotopy sequence shows that

$$\pi_1(t(m)) = \pi_0\,(\mathcal{W} \odot \mathrm{Diff}_0(M)) = 0\,, \tag{6.7}$$

because the Weyl group $\mathcal{W}$ is contractible. Therefore, Teichmüller space is contractible. Similarly, applying the exact homotopy sequence to the moduli space yields

$$\begin{aligned}\pi_1(m(M)) &= \pi_0(\mathcal{W} \odot \text{Diff}(M)) \\ &= \pi_0(\text{Diff}(M)) \ .\end{aligned} \tag{6.8}$$

In general the moduli space is topologically non-trivial. Since the space of physical configurations requires division by the full diffeomorphism group, one can have global anomalies [12].

In the Polyakov quantization of a conformally invariant string, the integration over metrics reduces to an integration over Teichmüller space. This is equivalent to summing over the complex structures on the Riemann surface. The importance of modular invariance was realized early in the history of string theory. Projective invariance played a key role in understanding the Koba-Nielsen formula. Alessandrini [29] realized that automorphic forms would play a central role in multiloop amplitudes. Polyakov's approach provides a way of determining the integration measure for the modular parameters. In some sense, the differential geometric approach is replaced by a complex analytic viewpoint. This raises interesting questions. Can one give an entirely complex analytic formulation of conformally invariant field theories? What is the complex analytic principle that determines the integration measure for the modular parameters?

The following may be of use in answering the questions. The differential geometric path integral formulation contains the principle of the composition of amplitudes [30].[16] There must be an analogue in the complex analytic formulation. The measure for the modular parameters must be chosen such that the amplitudes satisfy the composition rule.

Acknowledgments

The author acknowledges the Alfred P. Sloan Foundation for their support. This work was supported in part by the National Science Foundation under Contract PHY81-18547; and by the Director, Office of High Energy and Nuclear Physics of the U.S. Department of Energy under Contract DE-AC03-76SF00098.

[16] There is no satisfactory proof of this statement for string models.

References

1. A.M. Polyakov, *Phys. Lett.* **103B**, 207(1981).

2. J.L. Gervais and B. Sakita, *Phys. Rev.* **D4**, 2291(1971).

3. L. Alvarez-Gaumé and E. Witten, *Nucl. Phys.* **B234**, 269(1984).

4. W. Abikoff, *The Real Analytic Theory of Teichmüller Space*, Springer-Verlag, Berlin, 1980.

5. Y. Nambu, in *Symmetries and Quark Models*,, edited by R. Chand, Gordon and Breach, New York, 1970.

6. T. Goto, *Prog. Theor. Phys.* **46**, 126(1980).

7. H.B. Nielsen, "15th Inter. Conf. of High Energy Physics", Kiev 1970.

8. L. Brink, P. Di Vecchia and P. Howe, *Phys. Lett.* **65B**, 471(1976).

9. S. Deser and B. Zumino, *Phys. Lett.* **65B**, 369(1976).

10. O. Alvarez, *Nucl. Phys* **B216**, 125(1983).

11. G. Moore and P. Nelson, Harvard preprint, HUTP–85/A057.

12. E. Witten, "Global Anomalies in String Theories", to appear in *Proc. of the Argonne Symp. on Geometry, Anomalies and Topology*, Argonne, IL. March 28-30, 1985.

13. E. Witten, *Nucl. Phys.* **B202**, 253(1982)

14. M.F. Atiyah and I.M. Singer, *Ann. Math.* **87**, 485(1968).

15. S.S. Chern, *Complex Manifolds without Potential Theory*, Springer-Verlag, New York, 1979.

16. D.H. Friedan, "Introduction to Polyakov's String Model", in *Les Houches Summer School, 1982*, edited by J.B. Zuber and R. Stora, North Holland, Amsterdam, 1984.

17. B. Zumino, "Chiral Anomalies and Differential Geometry", in *Les Houches Summer School, 1983*, edited by B.S. DeWitt and R. Stora, North Holland, Amsterdam, 1984.

18. M.F. Atiyah and I.M. Singer, *Proc. Nat. Acad. Sci.* **81**, 2597(1984).

19. O. Alvarez, I.M. Singer and B. Zumino, *Comm. Math. Phys.* **96**, 409(1984).

20. A.M. Polyakov, *Phys. Lett.* **103B**, 211(1981).

21. D.J. Gross, J.A. Harvey, E. Martinec and R. Rohm, *Phys. Rev. Lett.* **54**, 502(1985), *Nucl. Phys.* **B256**, 253(1985), and Princeton preprint Print–85–0694.

22. J. Polchinski, Texas preprint, UTTG–13–85.

23. D. Friedan, E. Martinec and S. Shenker, Chicago preprint, EFI–85–32.

24. P.Candelas, G.T. Horowitz, A. Strominger and E. Witten, ITP preprint, NSF–ITP–84–170.

25. J. Bagger, D. Nemeschansky and S. Yankielowicz, SLAC preprint, SLAC–PUB–3588.

26. O. Alvarez, unpublished.

27. J. Dieudonné, *History of Algebraic Geometry*, Wadsworth, Monterey, 1985.

28. M. Shiffer and D.C. Spencer, *Functionals of Finite Riemann Surfaces*, Princeton Univ. Press, Princeton, 1954.

29. V. Alessandrini, *Nuovo Cimento* **2A**, 321(1971).

30. R.P. Feynman and A. Hibbs, *Quantum Mechanics and Path Integrals*, McGraw-Hill, New York, 1965.

31. W.A. Bardeen and B. Zumino, *Nucl. Phys.* **B244**, 421(1984).

32. L. Susskind, *Nuovo Cimento* **69A**, 457(1970).

Introduction to Two Dimensional Conformal and Superconformal Field Theory

Stephen H. Shenker
Enrico Fermi and James Franck Institutes
and Department of Physics
University of Chicago, Chicago, Illinois 60637

Some of the basic properties of conformal and superconformal field theories in two dimensions are discussed in connection with the string and superstring theories built from them. In the first lecture the stress-energy tensor, the Virasoro algebra, highest weight states, primary fields, operator product coefficients, bootstrap ideas, and unitary and degenerate representations of the Virasoro algebra are discussed. In the second lecture the basic structure of superconformal two dimensional field theory is sketched and then the Ramond Neveu-Schwarz formulation of the superstring is described. Some of the issues involved in constructing the fermion vertex in this formalism are discussed.

1. Two Dimensional Conformal Field Theory

In these two lectures I will describe some aspects of conformal and superconformal two dimensional field theories and the string and superstring theories built from them. The lectures are meant to be read in conjunction with the notes of Daniel Friedan[1]. A large number of other contributions to this volume also discuss related issues.

In this first lecture I will discuss some of the rich structure present in two-dimensional conformally invariant field theories. My approach follows that of Belavin, Polyakov, and Zamolodchikov[BPZ][2].

The central object in the study of such theories is the stress energy tensor $T_{\alpha\beta}$. Euclidean invariance (we work in Euclidean signature throughout) implies conservation, and scale invariance implies tracelessness.

$$\partial_\alpha T_{\alpha\beta} = 0 \qquad T_{\alpha\alpha} = 0 \tag{1.1}$$

These equations imply the Cauchy-Riemann equations for the combinations

$$T(z) = T_{11} - T_{22} + iT_{12} \qquad \bar{T}(\bar{z}) = T_{11} - T_{22} - iT_{12} \tag{1.2}$$

Correlation functions of $T(\bar{T})$ are meromorphic (anti-meromorphic) functions.

Analyticity, bose symmetry, the spin-2 nature of $T_{\alpha\beta}$ and the requirement that $T_{\alpha\beta}$ generate translations and scale transformations force the operator product of T with itself to have the form

$$T(z)T(w) \sim \frac{c}{2}\frac{1}{(z-w)^4} + \frac{2T(w)}{(z-w)^2} + \frac{\partial T}{\partial w}\frac{1}{(z-w)} + \cdots \tag{1.3}$$

The number c is a parameter of the given theory related to the trace anomaly[3]. $\bar{T}$ obeys an analogous equation with the same value of c.

The most useful operator interpretation for a scale invariant theory is radial quantization[4]. Dilation plays the role of time translation, radial ordering that of time ordering. In such a quantization the "time zero surface" can be chosen as the unit circle and operators defined by integrating fields on this surface. In particular we can define operators that generate conformal transformations. In two dimensions there are many such transformations since any analytic mapping is conformal. A basis for the infinitesimal conformal transformations is

$$z \to z + \epsilon_n z^{n+1} \qquad \epsilon_n \text{ complex.} \tag{1.4}$$

The operator generating such a transformation is $\epsilon_n L_n + \bar{\epsilon}_n \bar{L}_n$ where

$$L_n = \frac{1}{2\pi i}\oint dz\, z^{n+1} T(z)\,, \qquad \bar{L}_n = \frac{1}{2\pi i}\oint d\bar{z}\, \bar{z}^{n+1} \bar{T}(\bar{z}). \tag{1.5}$$

This quantization gives the following hermiticity properties:

$$L_n^+ = L_{-n}\,,\ \bar{L}_n^+ = \bar{L}_{-n}\,. \tag{1.6}$$

Note that $n = 0$, ϵ real in (1.4) is a dilation so $L_0 + \bar{L}_0$ is the radial "hamiltonian". The commutation relations of these operators can be determined[3] from the operator product expansion (1.3)

$$[L_m, L_n] = (m-n)L_{m+n} + \frac{c}{12}(m^3 - m)\delta_{m,-n}. \tag{1.7}$$

The $\bar{L}$'s obey the same algebra as the L's, and the L's and $\bar{L}$'s commute. The algebra (1.7) is called the Virasoro algebra[5]. The first term on the RHS is just what we would expect from the composition of transformations (1.4). The second "central term" with coefficient c is the consequence of the projective representation of the transformations in the Hilbert space of the quantum field theory.

We see that scale invariance and a local conserved stress-energy tensor imply the existence of an infinite dimensional algebra that acts on the state space of the theory. This infinite dimensional algebra is the source of much of the magic in two dimensional conformally invariant field theories. In particular the state space will decompose into irreducible representations of this algebra which will be very large. In fact there are certain theories whose entire Fock space consists of a finite number of such representations. The properties of all states in an irrep are related to each other purely by the Virasoro algebra.

What do these representations look like? Choose the hermitian operators L_0 and $\bar{L}_0$ diagonal;

$$L_0|h> = h|h>\,,\ \ \bar{L}_0|\bar{h}> = \bar{h}|\bar{h}> \qquad h, \bar{h} \text{ real.} \tag{1.8}$$

(We will usually suppress the $\bar{L}$ algebra since its properties are analogous to the L algebra. The state space is built up of tensor products of irreps of the two commuting algebras.) Equation (1.7) shows that for $n > 0$ L_{-n} raises L_0 by n units, L_{+n} lowers L_0 by n units. We can then construct "highest weight" representations in the standard fashion: choose an eigenstate of L_0, $|h>$ that is annihilated by all the L_{+n}. Then the set of states formed by applying products of the $\{L_{-n}\}$ is a representation space for the algebra. These states are conveniently organized into levels by L_0 eigenvalue. $L_{-1}|h>$ is at level 1; $L^2_{-1}|h>$, $L_{-2}|h>$ are at level 2; and so on. This set of states is called a Verma module; h is the weight of the irrep, $|h>$ is a "highest weight" state. In dual theory, it is called a physical state (if $h=1$).

Such states exist in a quantum field theory. The hamiltonian, $L_0+\bar{L}_0$ is lowered by L_{+n}. The lowest "energy" state, the vacuum $|0>$, must be annihilated by all the L_{+n} - it is a highest weight state. Scale invariance implies that L_0 annihilates $|0>$ as well; $L_0|0>=0$. This implies translation invariance, $L_{-1}|0>= 0$, since

$$|| L_{-1}|0>||^2 = <0|L_{+1}L_{-1}|0> = <0|2L_0|0> = 0\,. \tag{1.9}$$

We should stop for a minute and contrast the situation in two dimensions with that in higher dimensions. There the conformal algebra is finite dimensional, the analogous object in two dimensions is the closed subalgebra $L_0, L_{\pm1}$, $\bar{L}_0, \bar{L}_{\pm1}$ called $SL_2(C)$ that we have shown annihilates the vacuum. These generate the infinitesimal fractional linear transformations . The irreps of $SL_2(C)$ are much smaller than those of Virasoro. There are in general an infinite number of SL_2 irreps in each Virasoro irrep; the Virasoro algebra ties together their behavior.

There is more in a quantum field theory than states; there are local field operators as well. In a $2-d$ conformally invariant quantum field theorythere are special fields called primary fields[2] characterized by their operator products with $T(z)$

$$T(z)\phi(w,\bar{w}) \sim \frac{h\,\phi(w)}{(z-w)^2} + \frac{1}{(z-w)}\frac{\partial\phi}{\partial w} \tag{1.10}$$

The absence of higher order poles distinguishes primary from other scaling fields. The coefficients are fixed by the requirement that T generate dilations and translations.

Equation (1.10) gives the following commutation relations

$$[L_n, \phi(w)] = h(n+1)w^n\phi + w^{n+1}\partial_w\phi\ . \tag{1.11}$$

Observing that $\bar{L}_n$ will obey a similar formula with h replaced by $\bar{h}$ and specializing to $n = 0$ serves to identify $h + \bar{h}$ as the scaling dimension of ϕ and $h - \bar{h}$ as its euclidean spin ($L_0 - \bar{L}_0$ generates rotations). The commutation relations (1.11) are also the conditions that ϕ must obey to be a vertex operator in dual theory (along with a constraint on h).

From (1.11) and the highest weight properties of the vacuum we find, for $n > 0$,

$$L_n\, \phi(0)|0> = 0\,; \qquad L_0\, \phi(0)|0> = h\phi(0)|0> \tag{1.12}$$

So $\phi(0)|0> = |h>$ is also a highest weight state with weight h. Primary fields are those that create highest weight states from the vacuum.

What kind of fields create descendants? It is easy to see from (1.5) that

$$L_{k_n} \ldots L_{k_1}\phi(0)|0> \tag{1.13}$$

is created by a field $\phi^{\{k\}}(z)$ at $z = 0$ where

$$\phi^{\{k\}}(z) = L_{k_n}(z) \ldots L_{k_1}(z)\phi(z) \tag{1.14}$$

and

$$L_k(z) = \frac{1}{2\pi i}\oint dw\ (w - z)^{k+1}T(w)\ . \tag{1.15}$$

What are these fields? It is clear that

$$L_{-1}(z)\phi(z) = \frac{\partial\phi}{\partial z}\ . \tag{1.16}$$

Other descendant fields are composites of the stress-energy tensor with ϕ. The properties of all these fields and not just the derivatives (1.16) are organized by the Virasoro algebra.

We need to know more about fields than their action on the vacuum. We need to know their matrix elements between arbitrary states. This information will be contained in three point correlation functions. Let $\{O_\alpha\}$ be arbitrary fields, primary or descendant.

$$\langle O_\gamma(z_2) O_\beta(z_1) O_\alpha(0)\rangle \tag{1.17}$$

contains the information we seek. But we can evaluate (1.17) using the operator product expansion

$$O_\beta(z) O_\gamma(0) \sim \sum_\delta C_{\beta\gamma\delta} O_\delta z^{-h_\beta-h_\gamma+h_\delta} \tag{1.18}$$

where the $C_{\beta\gamma\delta}$ are the operator product coefficients and the $\bar{z}$ dependence is supressed. So the three point function (1.17) is

$$\sum_\delta C_{\beta\gamma\delta} z_1{}^{-h_\beta-h_\gamma+h_\delta} \langle O_\gamma(z_2) O_\delta(0)\rangle\,. \tag{1.19}$$

The matrix element information is related to the operator product coefficients and completely determined two point functions. It is useful to notice that the two point function of descendants of different primaries vanishes (in fact $SL(2)$ primary is enough to ensure this.)

The key simplification is that the operator product coefficients of descendants are determined by those of the primaries. Imagine a three point function of two primaries ϕ_a, ϕ_b and one descendant of primary ϕ_c

$$< 0|\phi_a(z_2)\phi_b(z_1) L_{k_n} \ldots L_{k_1} \phi_c(0)|0 > \tag{1.20}$$

By using the commutation relations (1.11) and the highest weight properties of the vacuum we can relate (1.20) to derivatives of the three point function of the primaries. So operator product coefficients of descendants can be determined from those of the primaries mechanically.

The complete information necessary to specify a conformally invariant two dimensional quantum field theory is the Virasoro highest weight representation content $\{h_a, \bar{h}_a\}$ and the operator product coefficients C_{abc} between the primary fields creating them. In string theory these operator product coefficients are just the three point couplings between particles created by the "vertex operators" or primary fields.

There are constraints on this information. For instance, consider the four point function

$$\langle \phi_a(z_1)\phi_b(z_2)\phi_c(z_3)\phi_d(z_4)\rangle \, . \tag{1.21}$$

Take $z_1 \to z_2$ and $z_3 \to z_4$ and use the operator product expansion to evaluate. We find (schematically)

$$\langle \phi_a\phi_b\phi_c\phi_d\rangle = \sum_{\alpha\beta} C_{ab\alpha}C_{cd\beta}\langle O_\alpha O_\beta\rangle \tag{1.22}$$

Where α, β range over all fields, primaries and descendants. Now evaluate another way, taking $z_1 \to z_3$ and $z_2 \to z_4$. We find

$$\langle \phi_a\phi_b\phi_c\phi_d\rangle = \sum_{\alpha\beta} C_{ac\alpha}C_{bd\beta}\langle O_\alpha O_\beta\rangle \tag{1.23}$$

and similarly for $z_1 \to z_4$ and $z_2 \to z_3$. The equality of these expressions is a necessary consistency requirement. It is just the constraint of duality in the corresponding string four point amplitude. The sums over descendants can be done, in principle mechanically, turning this requirement into algebraic equations on primary field operator product coefficients. The in general difficult technical problem of evaluating the descendant contribution is just the calculation of the conformal blocks of BPZ[2]. These equations provide strong constraints on conformal quantum field theories. Polyakov[6] originally proposed using them in a "conformal bootstrap" to solve for conformally invariant systems in d dimensions. The dramatic simplification in two dimensions[2] is that there are many fewer fields to consider – just the primaries. The difficulties are first the technical one of evaluating the conformal blocks and second the possibility of a large number of coupled primaries.

BPZ noticed that there are certain special c, h values where things simplify enormously. Gervais and Neveu[7] also found these values. One way to see this structure is to study the constraint imposed by requiring that the state space be a Hilbert space i.e., having a positive inner product[8]. The basic constraint is that each vector in each Verma module must have non-negative norm squared. Equation (1.9) already tells us that $L_{-1}|h>$ is not an acceptable state unless $h \geq 0$. At higher levels in a Verma module we need to check the matrix of inner products of the basis vectors, the contravariant form introduced in reference 9, and verify that there is no negative eigenvalue. An important tool is the Kac[10] formula for the determinant of this matrix of inner products.

As an example consider the second level. We have a basis of two states, $L_{-1}^2|h>$ and $L_{-2}|h>$ and hence a two by two matrix of inner products. There is no negative eigenvalue for $c \geq 1, h \geq 0$ (a result true at all levels), but for $c < 1$ there is a region of h where one occurs. The boundary curve of this region is given by

$$h = \frac{1}{16}(5 - c \pm \sqrt{(c-1)(c-25)}) \tag{1.24}$$

and on it there is a zero eigenvalue, corresponding to an eigenvector

$$|null>= (L_{-1}^2 + \alpha L_{-2})|h> \qquad \alpha = \frac{3}{4h+2} \tag{1.25}$$

that satisfies

$$< v|null >= 0 \tag{1.26}$$

for all vectors $|v>$ *. It is not hard to see that the first time a null vector appears it corresponds to a highest weight state within the Verma module.

BPZ pointed out a remarkable property of such null vectors. Let

$$|h>= \phi(0)|0> \tag{1.27}$$

be a highest weight state and have a null vector at second level. Then the n point function

$$< 0|\phi(z_1)\ldots\phi(z_n)(L_{-1}^2 + \alpha L_{-2})\phi(0)|0> \tag{1.28}$$

---

*For vectors within the Verma module this is immediate. For vectors outside it follows if the state space is composed entirely of highest weight irreps. Unitarity is sufficient for this. I thank C. Thorn for raising this point and D. Friedan for the unitarity observation.

vanishes because the vector is null. But (1.28) can also be evaluated by commuting the Virasoro operators to the right using (1.11). This gives a linear differential equation that the n point function must satisfy! This property of degenerate representations (i.e., those containing null vectors) of infinite dimensional algebras is very important. For example, Knizhnik and Zamolodchikov[11] have used this idea in the Kac-Moody context to calculate correlation functions of sigma models with Wess-Zumino term at their fixed point[12].

The full unitarity analysis[8] shows that all unitary representations for $c < 1$ are degenerate, giving a physical motivation for studying the degenerate case. In fact the unitary representations occur only at a discrete set of points given by:

$$c = 1 - \frac{6}{m(m+1)} \qquad m = 2, 3, 4, \ldots \tag{1.29}$$

$$h = \frac{(p(m+1) - qm)^2 - 1}{4m(m+1)} \qquad p = 1 \ldots m-1 \qquad q = 1 \ldots p \tag{1.30}$$

This discrete series contains a number of interesting two dimensional statistical mechanical critical points. The unitarity result in combination with the differential equation technique and effective methods to solve the equations[13] reduces the study of conformally invariant two dimensional quantum field theories with $c < 1$ to a finite algorithm. Goddard, Kent and Olive[14] have shown that all elements of the list (1.29,1.30) are in fact unitary by constructing manifestly unitary realizations for all of them.

The first nontrivial member of the series is at $c = 1/2$. The allowed h values are $0, 1/16, 1/2$. A Majorana fermion has $c = 1/2$ and provides a realization of the $0, 1/2$ representations. The full representation content of that theory is $(\psi = \psi_1 + i\psi_2)$

$$\begin{array}{cc} \textit{field} & (h, \bar{h}) \\ \psi & (\tfrac{1}{2}, 0) \\ \bar{\psi} & (0, \tfrac{1}{2}) \\ \psi\bar{\psi} & (\tfrac{1}{2}, \tfrac{1}{2}) \\ I & (0, 0) \end{array} \tag{1.31}$$

The nonvanishing operator product coefficients are: $(\epsilon = \psi\bar{\psi})$

$$\begin{aligned} \psi\psi &\sim I \\ \bar{\psi}\bar{\psi} &\sim I \\ \psi\bar{\psi} &\sim \epsilon \\ \epsilon\epsilon &\sim I \end{aligned} \tag{1.32}$$

The magnitudes of the coefficients are all one. This then is the complete specification of this very simple theory. Next lecture we will find a place for the thus far absent $h = 1/16$ representation.

2. Superconformal Field Theory and Superstrings

In this lecture I will describe some aspect of superconformal field theories in two dimensions and the fermionic string built from them. Again, this lecture should be read in conjunction with Daniel Friedan's notes[1]. Much of the material discussed here can be found in references 15–19.

We work in two dimensional superspace[20] and use complex coordinates $z, \theta, \bar{z}, \bar{\theta}$. We will often leave the $(\bar{z}, \bar{\theta})$ dependence implicit. The superconformal properties of a general theory are determined by properties of the super stress-energy tensor

$$T(z,\theta) = T_F(z) + \theta T_B(z) \tag{2.1}$$

where T_B is the ordinary bosonic stress tensor and T_F is its fermionic superpartner. T_F is a conformal primary field with $h = 3/2$, $\bar{h} = 0$. The moments of $T(z,\theta)$ form the supersymmetric extensions of the Virasoro algebra with operators G_n, L_n

$$\begin{aligned} T_F(z) &= \frac{1}{2}\sum_n z^{-n-3/2} G_n \\ T_B(z) &= \sum_n z^{-n-2} L_n \,. \end{aligned} \tag{2.2}$$

The commutation relations follow from the operator product

$$T(z_1,\theta_1)T(z_2,\theta_2) \sim \frac{\hat{c}/4}{z_{12}^3} + \frac{3}{2}\frac{\theta_{12}}{z_{12}}T(z_2,\theta_2) + \frac{1}{2}\frac{D_2T}{z_{12}} + \frac{\theta_{12}}{z_{12}}\partial_2 T \tag{2.3}$$

where $\hat{c}$ is the trace anomaly, normalized so that $\hat{c} = \frac{2}{3}c$, $z_{12} \equiv z_1 - z_2 - \theta_1\theta_2$ and $\theta_{12} = \theta_1 - \theta_2$. They are

$$\begin{aligned} [L_m, L_n] &= (m-n)L_{m+n} + \frac{\hat{c}}{8}(m^3 - m)\delta_{m,-n} \\ [L_m, G_n] &= (\frac{m}{2} - n)G_{m+n} \\ \{G_m, G_n\} &= 2L_{m+n} + \frac{\hat{c}}{2}(m^2 - \frac{1}{4})\delta_{m,-n}. \end{aligned} \tag{2.4}$$

We now discuss boundary conditions. First we introduce a new set of coordinates via the conformal mapping $w = \tau + i\sigma = \log z$. The z-plane maps onto a cylinder with "time" τ and periodic "space" σ. Dilation in z is translation in τ so radial quantization on the plane is just conventional quantization on the cylinder. There are two natural boundary conditions for the fermionic fields on the cylinder, period or anti-periodic. We refer to the periodic case as Ramond boundary conditions, anti-periodic as Neveu-Schwarz. Translated back to the z-plane (remembering $\phi(dz)^h(d\bar{z})^h$ is conformally invariant) antiperiodic fermi (half-integer h) fields are single valued; periodic fermi fields are double valued, $\psi \to -\psi$ along a cut beginning at the origin. From (2.2) we see that the mode numbers n for fermionic fields are integers in Ramond and half integers in Neveu-Schwarz. The algebra (2.4) with n integral is called the Ramond algebra[21], with n half integral the Neveu-Schwarz algebra[22]. The full superconformal field theory contains both Ramond and Neveu-Schwarz sectors.

Notions of highest weight states and descendants apply in this situation is well. The vacuum is a highest weight state for the reasons discussed in the last lecture and is in the NS sector, as we shall see. Other highest weights in the NS sector are created by superconformal primary fields (often called conformal superfields) acting on the vacuum. A highest weight state in the R sector must be created by an operator that changes the boundary conditions of the fermionic fields as it acts on the NS vacuum. Such an operator can be visualized as the endpoint of a cut in the fermionic fields and is called a spin field[15]. Superfields take NS to NS and R to R; spin fields take NS to R and R to NS.

We now begin a discussion of the Ramond Neveu-Schwarz fermionic string[21,22]. It is constructed out of D free superfields[20] $\mathbf{X}^\mu$ coupled to $2-d$ supergravity[23,24]. We gauge fix to superconformal gauge. In the critical dimension $D=10$ this leaves the free superconformal field theory of the $\mathbf{X}^\mu$ and the Faddeev-Popov ghosts[24,25,26]. The component expansion of $\mathbf{X}^\mu$ is given by

$$\mathbf{X}^\mu(z,\theta,\bar{z},\bar{\theta}) = X^\mu + \theta\psi^\mu + \bar{\theta}\bar{\psi}^\mu + \theta\bar{\theta}F^\mu \tag{2.5}$$

where X^μ is a free massless scalar, ψ^μ and $\bar{\psi}^\mu$ are left and right handed Majorana-Weyl fermions and F is auxiliary. Free field equations of motion let us work with chiral superfields

$$\mathbf{X}^\mu(z,\theta) = X^\mu(z) + \theta\psi^\mu(z) \tag{2.6}$$

with two point function

$$\langle \mathbf{X}^\mu(z_1,\theta_1)\,\mathbf{X}^\nu(z_2,\theta_2)\rangle = -g^{\mu\nu}\log z_{12}. \tag{2.7}$$

In this example

$$T(z,\theta) = -\frac{1}{2}D\mathbf{X}D^2\mathbf{X} \tag{2.8}$$

where D is the super-covariant derivative

$$D = \frac{\partial}{\partial\theta} + \theta\frac{\partial}{\partial z}\,. \tag{2.9}$$

Vertex operators V that create physical states in the Neveu-Schwarz sector are superconformal primary fields. The amplitude is formed by integrating vertex operator correlation functions with the invariant measure $dzd\theta^*$. In order that

$$\int dz d\theta\, V(z,\theta) \tag{2.10}$$

*In the closed Type II string there will be an integral over $\bar{z},\bar{\theta}$ as well. In the heterotic string[27] there is no $\bar{\theta}$ and only an integration on $\bar{z}$.

be superconformal invariant V must have conformal weight $h = 1/2$. A first example of a vertex operator is

$$e^{ip\mathbf{X}} . \tag{2.11}$$

This exponential has conformal weight $h = p^2/2$ so $p^2 = 1$. This means $m^2 = -1$; this vertex operator creates a tachyon. A second example is

$$\kappa^\mu D\mathbf{X}^\mu e^{ip\mathbf{X}} . \tag{2.12}$$

This is a superconformal primary if $p \cdot \kappa = 0$ and has $h = 1/2$ if $p^2 = 0$. This vertex operator creates a massless particle. Lorentz properties show that it corresponds to a massless vector. Note that the θ integral of the tachyon vertex is odd under $(-1)^F$ where F is sheet fermion number, while the integral of the massless vector vertex is $(-1)^F$ even. We can eliminate the tachyon in this theory by projecting on $(-1)^F$ even operators. This projection is called the GSO projection[28].

We now turn to the Ramond sector. Here it is more convenient to look at states at first, rather than the operators that create them. Fermionic fields have integer moding so the expansion of $\psi^\mu(z)$ is

$$\psi^\mu(z) = \sum_{n \epsilon \mathbf{Z}} \psi_n^\mu z^{-n-\frac{1}{2}} \tag{2.13}$$

The operators ψ_n^μ obey commutation relations

$$\begin{aligned} \{\psi_m^\mu, \psi_n^\nu\} &= -\delta_{m+n} g^{\mu\nu} \\ [L_0, \psi_m^\mu] &= -m\psi_m^\mu \end{aligned} \tag{2.14}$$

First focus on the zero modes ψ_0^μ that commute with the "hamiltonian" L_0. The first part of equation (2.14) restricted to the zero modes forms a Clifford algebra. Because the zero modes commute with L_0 its eigenspaces must form representations of the Clifford algebra. In the lowest energy sector, in particular, ψ_0^μ are just realized as gamma matrices, and the lowest energy states form a space-time spinor. It is in this way that space-time spinor indices enter the $R-NS$ string. $(-1)^F$ anticommutes with ψ_0^μ so $(-1)^F$ acts like γ^{D+1} on ground states. The masses of these states are determined by their L_0 eigenvalue. On the z-plane in the R sector translation invariance is clearly absent - there is a cut emanating from the origin. As we saw last lecture this means that L_0 cannot be zero on the vacuum. L_0 can be computed from the ψ propagator on the cut plane using, e.g.,

$$\begin{aligned} < 0|L_0|0 >= < 0|\frac{1}{2}[L_{+1}, L_{-1}|0 >\sim \\ \oint dz \oint dw\ z^2 < T(z)T(w) > \end{aligned} \tag{2.15}$$

(See reference 1). The result is

$$< 0|L_0|0 >= \frac{D}{16}. \tag{2.16}$$

This h value of 1/16 (per Ramond fermion) gives a realization of the 1/16 unitary representation at $c = 1/2$ we saw last time.

We need to consider the ghost system[24,25,26] as well. The superconformal gauge fixing gives rise to ghost superfields

$$\begin{aligned} B(z,\theta) = \beta(z) + \theta b(z) \\ C(z,\theta) = c(z) + \theta\gamma(z) \end{aligned} \tag{2.17}$$

where b, c are the anticommuting ghosts of the ordinary string with h values of 2 and -1. β and γ are their superpartners with h values of 3/2 and $-1/2$. The trace anomaly values of these systems are as follows:

$$\begin{aligned} c_X &= D \\ c_\psi &= D/2 \\ c_{b,c} &= -26 \\ c_{\beta,\gamma} &= +11 \end{aligned} \tag{2.18}$$

When $D = 10$, $c_{TOT} = 0$ and the theory is consistent. The ground state energies in the Ramond sector are as follows: (β, γ are sheet spinors so their boundary conditions shift with sector)

$$\begin{array}{cc} X^\mu & P^2/2 \\ \psi^\mu & D/16 \\ b, c & -1 \\ \beta, \gamma & 6/16 \end{array} \tag{2.19}$$

So when $D = 10$ the physical state condition $(L_0)_{TOT} = 0$ implies $P^2 = 0$, i.e., massless space-time spinors. This result has a simple interpretation in terms of sheet supersymmetry. The algebra (2.4) contains the relation

$$G_0^2 = L_0 - \frac{\hat{c}}{16} \tag{2.20}$$

This is the supersymmetry relation $Q^2 = H$ where $Q = G_0$ and $H = L_0 - \hat{c}/16$. A supersymmetric vacuum $G_0|0>= 0$ has L_0 eigenvalue $\hat{c}/16$. So, assuming sheet supersymmetry is not broken (true because we are just dealing with free field theories)

$$(L_0)_{TOT} = \frac{\hat{c}_{TOT}}{16} \tag{2.21}$$

From (2.18) $\hat{c}_{TOT} = 0$ if $D = 10$ so we have massless particles there. $(-1)^F$ acts like γ^{11} on ground states so the net space-time chirality is just Witten's $(-1)^F$ index[29] for Ramond sheet supersymmetry. The nonvanishing of this index provides a criterion in more general superconformal theories corresponding to string compactifications for the presence of massless fermions.

We saw in the Neveu-Schwarz sector that the GSO projection was useful to eliminate the tachyon. In the Ramond sector on the massless states it is just a chirality projection $\gamma^{11} = +1$. The resulting massless states are a massless vector and a massless Majorana-Weyl spinor– a $N = 1$ $d = 10$ super-multiplet[28]. This projected theory is the superstring.

We note that the GSO projection is necessary to preserve modular invariance on the torus if we want both Ramond and Neveu-Schwarz sectors in the theory. Summing over different σ boundary conditions combined with $\sigma \leftrightarrow \tau$ interchange (a modular transformation) implies summing over τ boundary conditions. This sum just implements the GSO projection.

To compute scattering amplitudes of fermions we need to know the vertex operators that create the space-time spinor states. The matter part of this operator was already constructed in the earlier epoch of string theory[30]. This vertex operator must act on the vacuum which is in the Neveu-Schwarz sector† and create a state in the Ramond sector. This tells us that the fermion vertex operator must change the boundary conditions of the ψ^μ. It opens a cut changing e.g., cylinder antiperiodic to periodic boundary conditions. This is parallel to the construction of the Ising disorder variable out of fermions[31]. The $h = 1/16$ value for one fermion is related to the Ising (dis)order exponent. This is the simplest example of a "spin" field[15].

We now review an intuitively appealing way to compute correlations of such operators. Consider[32] the case $D = 2$. A cut across which $\psi^1 \to -\psi^1$ and $\psi^2 \to -\psi^2$ can be represented as an $SO(2)$ gauge field concentrated along the cut with field strength only at the ends of the cut. Changing the position of the cut is just a gauge transformation‡. The field strength is adjusted to give a π phase for parallel transport around the endpoint. The spin field is like a point magnetic vortex on the world sheet. The GSO projection eliminates operators that see the "string" of this vortex. The projected theory is local.

We now can bosonize[33] the two Majorana equal one Dirac fermion and rewrite the gauge field coupling to the fermion current j_α

$$\begin{aligned}(\psi^1, \psi^2) &\to \phi \\ j_\alpha &\to \epsilon_{\alpha\beta}\partial_\beta\phi \\ \int d^2\xi A_\alpha j_\alpha &\to \int d^2\xi \phi \epsilon_{\alpha\beta}\partial_\beta A_\alpha \\ &= \int d^2\xi \phi B\end{aligned} \tag{2.22}$$

where B is the field strength, $B = \frac{1}{2}\delta(\xi - z) - \frac{1}{2}\delta(\xi - w)$. We see that we only need to compute the ratio of the partition function of a free scalar field with point sources to that of one without, i.e., a correlation function of exponentials of free scalars. These will just be power laws. In fact the answer for the two point

†Here this vacuum is a null state that does not correspond to a physical particle.

‡The gauge field can actually be coupled to the anomalous left or right handed fermion currents alone. Moving the cut locally involves gauge transformations in regions of zero field strength where the anomaly vanishes.

function is

$$\frac{1}{|z-w|^{1/4}} \tag{2.23}$$

Because we have the full $z, \bar{z}$ dependence here this gives $h = 1/8 = 2/16$, as advertised.

For $D = 10$, roughly speaking, we can do the same thing[17]. Group the 10 fermions into 5 groups of 2 and then bosonize. We can then make a vertex operator representation[34] of the spin fields as[§]

$$S_\alpha = e^{i\vec{\alpha}\cdot\vec{\phi}} \tag{2.24}$$

where $\vec{\phi} = (\phi_1 \dots \phi_5)$, $\vec{\alpha} = (\pm\frac{1}{2} \dots \pm\frac{1}{2})$. The 32 possibilities fill out a $D = 10$ Majorana spinor. The conformal weight h of S_α is $|\vec{\alpha}|^2/2 = 10/16$. Correlation functions can also be computed using the $SO(9,1)$ Kac-Moody algebra generated by currents $\psi^\mu\psi^\nu$ and the null vector differential equation in the Virasoro-Kac-Moody semi-direct product[17].

This operator has $h = 10/16$, not $h = 1$ necessary for the fermion vertex operator to have the correct conformal properties: thatthe integral $\int dz V_F$ should be conformally invariant. The resolution of this problem involves the ghosts, as suggested by Goddard and Olive. This is reasonable since the spinor ghosts change boundary condition at the cut as well as the ψ^μ. There should be a spin field for the ghosts included in the vertex as well. The Ramond ghost ground state of energy 6/16 should be created by a conformal field with that weight. If we denote that spin field by Σ we can construct a candidate $h = 1$ massless fermion vertex as

$$V_F \sim u^\alpha \Sigma S_\alpha e^{ipX}, \qquad p\cdot\gamma u = 0,\ p^2 = 0\,. \tag{2.25}$$

This turns out to be about half right. The full working out of these ideas is described in references 1,18,19.

The structure of superconformal field theories described above is quite general. It tells us how to construct spin fields and hence fermion vertex operators in

[§] We must really include a cocycle here.

general superconformal theories, for instance those describing compactified superstrings. In fact the first indication of the general structure came from a different direction: a study of the unitary representations of the Ramond and Neveu-Schwarz algebras. We[15] found that there were discrete series of unitary representations for each algebra with the same values of c but different values of h. This strongly suggested the existence of theories containing both types of representations in the way described above. This has been verified explicitly[15,35] for the first two elements of the discrete series. The first one corresponds to the Ising model with vacancies–a system that has already been realized in the laboratory!

2.1 *

Acknowledgements I would like to thank J. Cohn, E. Martinec, and Z. Qiu for enjoyable collaboration and valuable discussions. I would also like to thank P. Di Vecchia, L. Dixon, J. Harvey, C. Thorn, C. Vafa and A. Zamolodchikov for enlightening remarks. I am especially grateful to D. Friedan for countless conversations about all aspects of this subject. This work was supported in part by DOE grant DE-FG02-84ER-45144, NSF grant PHY-8451285 and the Alfred P. Sloan foundation.

REFERENCES

1. D. Friedan, lecture notes in this volume.

2. A.A. Belavin, A. M. Polyakov, and A. B. Zamolodchikov, *Nuc. Phys.* B241 (1984) 333.

3. D. Friedan, in 1982 Les Houches summer school *Recent Advances in Field Theory and Statistical Mechanics*, J. B. Zuber and R. Stora, eds., North-Holland(1984).

4. cf. S. Fubini, A. J. Hanson, and R. Jackiw, *Phys. Rev.* D7 (1973) 1732.

5. M. Virasoro, *Phys. Rev.* D1 (1970) 2933; I. M. Gelfand and D. B. Fuchs, *Functs. Anal. Prilozhen* 2 (1968) 92; in the context of string theory, the Schwinger term in the conformal algebra was discovered by J. Weis, unpublished (cf. S. Fubini and G. Veneziano, *Ann. Phys.* 63 (1971) 12).

6. A. M. Polyakov,, *Zh. Eksp. Teor. Fiz.* 57 (1969) 271; [*Sov. Phys. JETP* 30 (1970) 151].

7. J-L. Gervais and A. Neveu, *Nuc. Phys.* B199 (1982) 50; B209 (1982) 125; B224 (1983) 320; B238 (1984) 125; B238 (1984) 301;

8. D. Friedan, Z. Qiu, and S.H. Shenker, *Phys. Rev. Lett.* 52 (1984) 1575; in *Vertex Operators in Mathematics and Physics*, J. Lepowsky et.al. (eds), Springer-Verlag (1984).

9. R. Brower, *Phys. Rev.* D6 (1972) 1655; P. Goddard and C. Thorn, *Phys. Lett.* 40B (1972) 235.

10. V. G. Kac in Proc. Int. Cong. Math., Helsinki, 1978 and in *Group Theoretical Methods in Physics*, edited by W. Beiglbock, A. Bohm, Lecture Notes in Physics Vol. 94 (Springer-Verlag, New York, 1979), p. 441; B. L. Feigin and D. B. Fuchs, *Funkts Anal. Prilozhen.* 16 (1982) 47 [*Funct. Anal. Appl.* 16 (1982) 114].

11. V. Knizhnik and A. B. Zamolodchikov, *Nuc. Phys.* B247 (1984) 83.

12. E. Witten, *Comm. Math. Phys.* 92 (1984) 451.

13. B. L. Feigin and D. B. Fuchs, Moscow preprint (1983); V. L. Dotsenko and V. Fateev, *Nuc. Phys.* B240 (1984) 312.

14. P. Goddard, A. Kent and D. Olive, DAMTP preprint (1985).

15. D. Friedan, Z. Qiu and S.H. Shenker, *Phys. Lett.* 151B (1985) 37.

16. M. Bershadsky, V. Knizhnik, and M. Teitelman, *Phys. Lett.* 151B (1985) 31.

17. J. Cohn, D. Friedan, Z. Qiu and S.H. Shenker, in preparation.

18. D. Friedan, E. Martinec, and S.H. Shenker, *Phys. Lett.* 160B (1985) 55.

19. D. Friedan, E. Martinec, and S.H. Shenker EFI preprint 85-89, November 1985.

20. L. Brink and J. O. Winnberg, *Nuc. Phys.* B103 (1976) 445.

21. P. Ramond, *Phys. Rev.* D3 (1971) 2415.

22. A. Neveu and J. Schwarz, *Nuc. Phys.* B31 (1971) 86.

23. L. Brink, P. DiVecchia, and P. Howe, *Phys. Lett.* 65B (1976) 471; S. Deser and B. Zumino, *Phys. Lett.* 65B (1976) 369.

24. A.M. Polyakov, *Phys. Lett.* 103B (1981) 207, 211.

25. E. Martinec, *Phys. Rev.* D28 (1983) 2604.

26. M. Kato and K. Ogawa, *Nuc. Phys.* B212 (1983) 443; S. Hwang, *Phys. Rev.* D28 (1983) 2614;K. Fujikawa, *Phys. Rev.* D25 (1982) 2584.

27. D. Gross, J. Harvey, E. Martinec, and R. Rohm, *Phys. Rev. Lett.* 54 (1985) 502,, *Nuc. Phys.* B256 (1985) 253, and Princeton preprint (July 1985).

28. F. Gliozzi, D. Olive, and J. Scherk, *Nuc. Phys.* B122 (1977) 253.

29. E. Witten, *Nuc. Phys.* B202 (1982) 253.

30. C. Thorn, *Phys. Rev.* D4 (1971) 1112; J. Schwarz, *Phys. Lett.* 37B (1971) 315; E. Corrigan and D. Olive, *N. Cim.* 11A (1972) 749; E. Corrigan and P. Goddard, *N. Cim.* 18A (1973) 339; L. Brink, D. Olive, C. Rebbi, and J. Scherk, *Phys. Lett.* 45B (1973) 379; D. Olive and J. Scherk, *Nucl. Phys.* B64 (1973) 334; J. Schwarz and C. C. Wu, *Phys. Lett.* 47B (1973) 453; E. Corrigan, P. Goddard, D. Olive, and R. Smith, *Nucl. Phys.* B67 (1973) 477.

31. L. P. Kadanoff and H. Ceva, *Phys. Rev.* B3 (1971) 3918.

32. A. Luther and I. Peschel, *Phys. Rev.* B12 (1975) 3908; J. B. Zuber and C. Itzykson, *Phys. Rev.* D15 (1977) 2875.

33. cf. S. Coleman, *Phys. Rev.* D11 (1975) 2088.

34. M. Halpern, *Phys. Rev.* D12 (1975) 184; T. Banks, D. Horn and H. Neuberger, *Nucl. Phys.* B108 (1976); I. Frenkel and V. Kac, *Invent. Math.* 62 (1980) 23; G. Segal, *Comm. Math. Phys.* 80 (1981) 301; P. Goddard and D. Olive, in *Vertex Operators in Mathematics and Physics*, J. Lepowsky et.al. (eds), Springer-Verlag (1984).

35. D. Friedan and S.H. Shenker, in preparation.

Notes on String Theory and Two Dimensional Conformal Field Theory

Daniel Friedan

Enrico Fermi Institute and Department of Physics
University of Chicago, Chicago, Illinois 60637

The basic problem in covariant first quantized string theory is to construct the world surface of the string as a local two dimensional conformally invariant quantum field theory. The problem divides in two parts. A conformal field theory is completely defined by the operator product expansions of its quantum fields, which can be determined at arbitrarily small distance. So the first task is to describe the local structure of the world surface. Once the conformal field theory is defined by its local properties, its global behavior can be checked to determine the consistency of the string loop expansion.

These notes are about the superconformal invariance of the world surface of supersymmetric string. The main topics are the construction of the vertex operators for emission of spacetime fermions and the demonstration of spacetime supersymmetry in the covariant first quantization. Only the local structure of the world surface is described; explicit global information is given only for the two sphere, in order to calculate tree amplitudes. The tree amplitudes illustrate how global facts such as spacetime supersymmetry and BRST invariance are obtained from local information coded in operator products of chiral fields and, in particular, conformal currents. The translation from local to global information is based on the analyticity of chiral quantum fields in two dimensions. The chiral fields on the string world surface include the super stress-energy tensor, the Fadeev-Popov ghost fields and their anomalous currents, the BRST superconformal current, and the conformal current for spacetime supersymmetry.

These notes are meant to be read in conjunction with the lectures of Stephen Shenker[1], and describe work done with him, Joanne Cohn, Emil Martinec and Zongan Qiu[2-6]. Only a few references are given, and then only to relatively recent work. The references are definitely not meant to convey the history of the subject. A more complete introduction to the literature can be found in reference 5. Some of the ideas of conformal field theory and covariant bosonic string theory are discussed in reference 7 from the point of view which is taken here.

Many of the arguments and calculations in these notes are presented rather telegraphically. The industrious reader might treat the gaps as exercises or problems.

Section 1 is a sketch of the general strategy of covariant first quantization; section 2 develops the most basic properties of super Riemann surfaces; section 3 sketches superconformal field theory; section 4 describes the superconformal world surface of fermionic string and the superconformal ghosts; section 5 is a general discussion of two dimensional free tensor quantum fields satisfying first order equations of motion; section 6 applies the general results of section 5 to the superconformal ghosts and constructs the BRST current; and section 7 constructs the fermion vertex and the spacetime supersymmetry current.

1. INTRODUCTION

A theory of gravity, such as string theory, should at least provide a manifestly Lorentz covariant scheme for calculating scattering amplitudes in flat spacetime. Covariant first quantization of strings could also be useful as a step towards understanding the underlying structure of string.

A manifestly relativistic first quantization of string can be carried out using the language of two dimensional conformal quantum field theory to describe sums over world surfaces of first quantized strings. The analog in particle theory is the relativistic calculation of scattering amplitudes in first quantization by representing Feynman diagrams as sums over particle world lines (joined at interaction vertices).

1.1 Covariant quantization of bosonic strings

The basic ideas of covariant first quantization of strings are realized in the bosonic theory[8]. A world surface is given by its location in spacetime, $x^\mu(z,\bar{z})$, and by an intrinsic metric $g_{ab}(z,\bar{z})$ on the parameter space of the complex variable z, with line element $ds^2 = g_{zz}dz^2 + g_{z\bar{z}}dzd\bar{z} + g_{\bar{z}z}d\bar{z}dz + g_{\bar{z}\bar{z}}d\bar{z}^2$. The intrinsic metric makes it possible to write a sum over world surfaces $\int dx\, dg\, e^{-S(g,x)}$ which is both local in parameter space and invariant under reparametrizations, and whose action

$$S(g,x) = \int d^2z\sqrt{g}\,\left(\mu^2 + \lambda R^{(2)} + g^{ab}\partial_a x^\mu \partial_b x_\mu + \cdots\right) \tag{1.1.1}$$

can be expanded in powers of the two dimensional derivatives.

The reparametrizations of the world surface act as a gauge group in the functional integral over surfaces. A natural gauge fixing condition is $g_{ab} = \rho(z,\bar{z})g^{(m)}_{ab}$. where $g^{(m)}_{ab}$ is some background metric. In this gauge the integral over metrics becomes an integral over the conformal factors $\rho(z,\bar{z})$ and over the conformal classes of metrics, represented by a collection of background metrics $g^{(m)}_{ab}$ which are indexed by a finite number of moduli $m = (m^1, m^2, \ldots)$. The conformal classes of two dimensional surfaces are the Riemann surfaces.

A Fadeev-Popov determinant is introduced into the functional integral because of the gauge fixing. The determinant is calculated by a Grassmann integral over conjugate ghost fields $b(z), c(z)$ which are chiral fermion fields on the world surface, of spins 2 and -1 respectively, corresponding to variations of the gauge condition and to infinitesimal reparametrizations of the world surface. The gauge fixed functional integral has the form

$$\sum_{topologies} e^{-4\pi\lambda(Euler\#)} \int_{moduli} dm \int_{fields} dx\, db\, dc \\ \exp\left\{-\int d^2z\left(\partial x\bar{\partial}x + b\bar{\partial}c + \bar{b}\partial\bar{c}\right)\right\} \tag{1.1.2}$$

when the action is written in conformal coordinates $(z,\bar{z})$ with $g^{(m)}_{zz} = 0$, $g^{(m)}_{z\bar{z}} = \frac{1}{2}$, and interactions of dimension > 2 are dropped from the two dimensional action because they are irrelevant (nonrenormalizable) in the continuum limit of parameter space. The coefficient $e^{-4\pi\lambda}$ is the string coupling constant. In the sum over surfaces, the Euler number indexes the string loop expansion.

Note that the conformal factor ρ is left out of 1.1.2. The classical action in 1.1.2 is independent of ρ, but this conformal invariance does not persist in the two dimensional quantum field theory of x^μ, b, c if there is a net conformal anomaly, which always happens except in the critical dimension $d = 26$. In the critical dimension ρ drops from the surface dynamics, leaving 1.1.2. In noncritical spacetime dimensions the ρ field must be dynamical, but as yet no acceptable quantum dynamics for ρ has been formulated for $2 \leq d \leq 25$.

In the critical dimension $d = 26$, the vanishing of the conformal anomaly means that the x^μ, b, c quantum field theory depends only on the conformal class of the surface, and its partition function transforms as a density on moduli space, so that the integral 1.1.2 over moduli makes sense (locally in moduli space).

1.2 Scattering amplitudes

To calculate a Greens function of N strings, let the sum over topologies in 1.1.2 range over surfaces with N boundary components and fixed wave functionals on the boundary values, representing N external strings. The boundaries can be pictured as holes in a compact Riemann surface without boundary. The radii of the holes are N of the (real) moduli of the original surface. The integrals over radii near zero produce poles in the external spacetime momenta, and the N point scattering amplitudes are the residues at these poles. The amplitudes can thus be calculated as functional integrals over surfaces with N infinitesmal holes, and particular boundary conditions at the holes. The locations of the holes are the remaining moduli for the boundaries. The infinitesmal holes can be represented as local quantum fields on the world surface, called ***vertex operators***. The scattering amplitudes have the form

$$\begin{aligned} G(p_1, \ldots, p_B) = \sum_{topologies} \int_{moduli} dm \int d^2 z_1 \cdots d^2 z_N \\ Z(m) \left\langle V_1(p_1, z_1) \cdots V_N(p_N, z_N) \right\rangle_m \end{aligned} \quad (1.2.1)$$

where $Z(m)$ is the partition function of the x^μ, b, c system (including the string coupling) on the compact Riemann surface without boundary whose moduli are m, and $\langle\cdots\rangle_m$ is the correlation function on the surface. The contribution from the simplest topology, the two sphere, gives the tree amplitudes.

The reparametrization invariance of the original functional integral means that the integrals over the z_i should be conformally invariant, which implies that the vertex operators should have quantum dimension 1 in z and also 1 in $\bar{z}$. The simplest examples are the exponentials $e^{ik\cdot x}$, $\frac{1}{2}k^2 = 1$, which are the vertex operators for the tachyonic states of the bosonic string. The duality properties of the string amplitudes are manifest in 1.2.1; the factorization of amplitudes is expressed by the operator product expansion of vertex operators. For example, the leading singularity in the operator product of tachyon vertex operators,

$$e^{ik\cdot x}(z_1)\, e^{ik'\cdot x}(z_2) \sim (z_1 - z_2)^{k\cdot k'}\, e^{i(k+k')\cdot x}(z_2)\,, \tag{1.2.2}$$

yields a tachyon pole in the intermediate momentum $k + k'$ at $\frac{1}{2}(k + k')^2 = 1$, coming from the integration over z_1 near z_2.

1.3 Unitarity

Scattering amplitudes calculated by this prescription are manifestly Lorentz covariant, but not obviously unitary. The demonstration of unitarity has two parts. First, the tree amplitudes must be shown free of ghosts; and, second, the sum over surfaces of nontrivial topology must be shown to produce loop corrections consistent with the tree amplitudes.

In the covariant first quantization, the states of the two dimensional field theory on the cylinder, subject to the residual constraints of reparametrization invariance, are the physical states of a single string. This identification of the states is formally apparent the Schrödinger picture of the two dimensional field theory on the cylinder, where the states are wave functionals on circles in spacetime. The Hilbert space of the two dimensional field theory has an indefinite metric, because of the Lorentz signature of spacetime and the Fermi statistics of the ghosts.

The original covariant approach worked only with the matter fields x^μ. The physical states are defined by the gauge conditions $L^z_{+n}\,|phys\rangle = 0$ and the mass-shell condition $(L^z_0-1)\,|phys\rangle = 0$, in terms of the Virasoro operators L^z_n generating the residual gauge algebra of conformal transformations. The gauge and mass-shell conditions are equivalent to the conformal invariance of the integral over locations of vertex operators in 1.1.2. The metric on the physical states can be shown nonnegative, the unphysical states can be shown to decouple from the physical states in tree amplitudes The problem with the classical approach is that the matter sector by itself is conformally anomalous, so its partition function must be corrected to become a density on moduli space, and the unphysical states do not manifestly decouple in the loop corrections.

1.4 The BRST current

An alternative approach uses the BRST quantization of the world surface[9]. The fermionic BRST charge is defined in the combined matter-ghost system, satisfying $Q^\dagger_{BRST} = Q_{BRST}, Q^2_{BRST} = 0$. The physical states are the invariant states, $Q_{BRST}\,|phys\rangle = 0$, modulo the null invariant states $Q_{BRST}\,|state\rangle$. The metric on physical states can be shown to be positive, if only by showing that the two definitions of the physical states are equivalent.

The decoupling of physical states can be shown by writing the BRST variations of fields as contour integrals of a conformal (chiral) current:

$$(\delta_{BRST}\Phi)(w) = \frac{1}{2\pi i}\oint_{C_w} dz\; j_{BRST}(z)\,\Phi(w) \qquad (1.4.1)$$

where C_w is a simple contour surrounding w. The physical vertex operators $V(z)$ are the BRST invariant fields, where invariance means that $(\delta_{BRST}V)(z)$ is a total derivative, so that the integral over z vanishes. Equation 1.4.1 is equivalent to an operator product formula. This is a local property of the conformal field theory which obtains inside all correlation functions, on all surfaces.

The decoupling of null states can be shown by considering the correlation function of $N-1$ physical vertex operators and one $BRST$-null vertex operator.

Write the null vertex in the form 1.4.1. Deform the contour to surround the each of the $N-1$ physical vertices, giving a sum of total derivatives, each of which vanishes after integration over the locations of the vertices. This argument depends on j_{BRST} being conserved on any surface, which means that it must be a conformal current, a conformal field of weight $(1,0)$, i.e., of weight 1 in z and 0 in $\bar{z}$. The expectation values of vertex operators on arbitrary surfaces must be $BRST$ invariant, which means checking that contour integrals of j_{BRST} vanish up to total derivatives, as long as the contour surrounds no vertex operators.

One way to show that the loop corrections are consistent with the tree amplitudes is to use a representation of the Riemann surfaces in which all the curvature of the intrinsic metric $g_{ab}^{(m)}$ is concentrated at isolated points[10]. The functional integrals are explicit sums over states, and the moduli play the role of Schwinger parameters. Since the conformal symmetry is anomaly-free, any representation of the surfaces is equivalent to any other, so a more suitable representation can be used for calculation (for example, letting the $g_{ab}^{(m)}$ be the constant curvature metrics). It would be good to have a less mechanical method for showing unitarity in the BRST formalism, perhaps employing relations between functional integrals over Riemann surfaces of different topologies.

1.5 Supersymmetric strings

The covariant first quantization of supersymmetric strings[11] follows the pattern of the bosonic theory. The main difference is that the gauge symmetry of the fermionic world surface is two-dimensional superconformal invariance. The integral over Riemann surfaces turns into an integral over super Riemann surfaces (see section 2).

These notes concentrate on the local properties of the superconformal field theory of the superstring world surface and on the calculation of tree amplitudes. The main problem is to construct the superconformal BRST current and the algebra of BRST invariant vertex operators for the bosonic and fermionic modes of the string[4,5]. Given the superconformal covariance of the BRST current and the vertex operators, the remaining issues in the calculation of loop corrections are construction of BRST invariant expectation values for arbitrary surfaces, and proof of the finiteness of the integrals over moduli[12].

This covariant approach is not manifestly supersymmetric. The supersymmetry of the amplitudes is demonstrated by constructing conformal currents $j_\alpha(z)$ whose charges generate the spacetime supersymmetry algebra[4,5]. Contour integrals of $j_\alpha(z)$ give the supersymmetry variations of vertex operators:

$$(\delta_\alpha V)(w) = \frac{1}{2\pi i} \oint_{C_w} dz\, j_\alpha(z)\, V(w)\,. \tag{1.5.1}$$

Integrating j_α around a contractible contour gives zero. On the other hand, a trivial contour can be deformed into a sum of contours, one surrounding each vertex operator. Thus the supersymmetry variation of any correlation function vanishes, even before the integral over the positions of vertex operators and over moduli. Therefore the amplitudes are supersymmetric. In calculating loop corrections, the crucial issue is whether contour integrals of j_α vanish for contractible contours on non-simply connected surfaces.

2. Super Riemann Surfaces

This section is about tensor analysis on superconformal manifolds of one complex dimension, the super Riemann surfaces. In the covariant first quantization, the world surfaces of fermionic strings are super Riemann surfaces. The point of view taken in this section has evolved from references 13 and 2-5.

2.1 Super coordinates

A one dimensional complex supermanifold is locally described by an ordinary complex coordinate z and an anticommuting coordinate θ, $\theta^2 = 0$, making a complex super coordinate $\mathbf{z} = (z, \theta)$. The superderivative is the square root of the ordinary derivative:

$$D = \frac{\partial}{\partial\theta} + \theta\frac{\partial}{\partial z} \qquad D^2 = \frac{\partial}{\partial z}\,. \tag{2.1.1}$$

A super analytic function is a solution of $\overline{D} f = 0$, and consists of two ordinary analytic functions: $f(\mathbf{z}) = f_0(z) + \theta f_1(z)$, with f_0 commuting with θ and f_1 anticommuting with θ.

2.2 Superconformal transformations

A super analytic map $\mathbf{z} \to \tilde{\mathbf{z}}(\mathbf{z}) = (\tilde{z}(z,\theta), \tilde{\theta}(z,\theta))$ transforms the superderivative according to

$$D = (D\tilde{\theta})\widetilde{D} + (D\tilde{z} - \tilde{\theta} D\tilde{\theta})\widetilde{D}^2 . \tag{2.2.1}$$

A super analytic map is called a superconformal transformation when the superderivative transforms homogeneously:

$$D = (D\tilde{\theta})\,\widetilde{D}\,, \tag{2.2.2}$$

i.e., if

$$D\tilde{z} - \tilde{\theta} D\tilde{\theta} = 0\,. \tag{2.2.3}$$

2.3 Super Riemann surfaces

It follows from 2.2.2 that a composition of superconformal transformations, $\mathbf{z} \to \tilde{\mathbf{z}} \to \tilde{\tilde{\mathbf{z}}}$, is also a superconformal transformation. A super Riemann surface can thus be defined as a collection of superconformal coordinate patches, that is, as a set of super coordinate neighborhoods patched together by superconformal transformations. A super coordinate neighborhood is just an ordinary neighborhood in z.

Consider special super Riemann surfaces for which the patching transformations are all of the form $\mathbf{z} \to (\tilde{z}, \tilde{\theta}) = (\tilde{z}(z), \mu(z)\,\theta)$, so that the ordinary patching transformations $z \to \tilde{z}$ make an ordinary Riemann surface, and θ behaves like an ordinary tensor. The superconformal condition 2.2.2 becomes $\partial\tilde{z} = \mu(z)^2$. Thus the $\mu(z)$ are transition functions for a line bundle over the ordinary Riemann surface whose square is the canonical line bundle. The canonical line bundle is the bundle of $(1,0)$ forms; its transition functions are $\partial\tilde{z}$. A square root of the canonical bundle is called a bundle of half-forms. On Riemann surfaces half-forms are spinors. Thus ordinary Riemann surfaces with spin structures are special cases of super Riemann surfaces.

2.4 Superconformal tensor fields

In a composition of superconformal transformations, $\mathbf{z} \to \tilde{\mathbf{z}} \to \tilde{\tilde{\mathbf{z}}}$, the super jacobians obey

$$D\tilde{\tilde{\theta}} = (D\tilde{\theta})\,(\widetilde{D}\tilde{\tilde{\theta}})\,. \tag{2.4.1}$$

This composition law allows a super differential $\mathbf{dz}$ to be defined by the transformation law

$$\mathbf{d\tilde{z}} = (D\tilde{\theta})\,\mathbf{dz} \qquad \frac{\mathbf{d\tilde{z}}}{\mathbf{dz}} = D\tilde{\theta}\,. \tag{2.4.2}$$

Then superconformal tensor fields $\phi(\mathbf{z})$, can be defined by the condition that $\phi(\mathbf{z})\mathbf{dz}^{2h}$ be superconformally covariant, where h is called the *weight* or *dimension* of ϕ. This means that

$$\phi(\mathbf{z})\mathbf{dz} = \tilde{\phi}(\tilde{\mathbf{z}})\mathbf{d\tilde{z}}, \qquad \phi(\mathbf{z}) = \tilde{\phi}(\tilde{\mathbf{z}})(D\tilde{\theta})^{2h}\,. \tag{2.4.3}$$

The superconformal tensor fields are the analogues of ordinary conformal tensor fields $\phi(z)$, of weight or dimension h, for which $\phi(z)dz^h$ is conformally covariant, i.e., for which $\phi(z) = \tilde{\phi}(\tilde{z})(d\tilde{z}/dz)^h$.

The component fields of $\phi(\mathbf{z}) = \phi_0(z) + \theta\phi_1(z)$ consist of an ordinary conformal field of weight h, ϕ_0, and an ordinary conformal field of weight $h + 1/2$, ϕ_1. When ϕ is a quantum field, its Fermi/Bose statistics are the statistics of ϕ_0, opposite to the statistics of ϕ_1.

Globally defined superconformal tensor fields have weights which are either integer or half-integer. But in what follows it will be useful to manipulate quantum tensor fields whose weights are not integer or half-integer, and so can only be defined locally. Globally defined fields can be constructed as products of locally defined fields.

2.5 Superconformal vector fields

Infinitesimal superconformal transformations $\mathbf{z} \to \tilde{\mathbf{z}} = \mathbf{z} + \delta\mathbf{z}(\mathbf{z})$, transform superconformal tensor fields by the infinitesimal version of 2.4.3,

$$\phi = \tilde{\phi} + \delta_v\phi \qquad \delta_v\phi = (v\partial + \tfrac{1}{2}DvD + h\partial v)\,\phi \qquad v(\mathbf{z}) = \delta z + \theta\,\delta\theta\,, \tag{2.5.1}$$

written in terms of the superconformal vector field $v(\mathbf{z})$, which is itself a superconformal tensor field of weight -1. The commutation relations of the Lie algebra of infinitesimal superconformal transformations is the same as the commutation relations of the Lie derivatives δ_v:

$$\delta_{[v,w]} = [\delta_v,\, \delta_w]\,, \qquad [v, w] = v\partial w - w\partial v + \tfrac{1}{2} DvDw\,. \tag{2.5.2}$$

2.6 Super contour integrals

Integration over the anticommuting coordinate θ is given by

$$\int d\theta\, \theta = 1\,, \qquad \int d\theta\, 1 = 0\,. \tag{2.6.1}$$

The super contour integral is the ordinary contour integral over z combined with the integral over θ , that is,

$$\oint_C \mathbf{dz}\, \omega(\mathbf{z}) = \oint_C dz \int d\theta\, \omega(\mathbf{z}) = \oint_C dz\, \omega_1(z)\,. \tag{2.6.2}$$

The use of the super differential $\mathbf{dz}$ is justified by considering the behavior of the super contour integral under superconformal transformations:

$$\oint_{\tilde{C}} \mathbf{d\tilde{z}}\, \tilde{\omega}(\tilde{\mathbf{z}}) = \oint_C \mathbf{dz}\, \tilde{\omega}(\tilde{\mathbf{z}}(\mathbf{z}))\,. \tag{2.6.3}$$

A dimension $\frac{1}{2}$ superconformal tensor field is called a ***superconformal current.*** By 2.6.3, the super contour integrals of superconformal currents are invariant under superconformal transformations. Also, if $f(\mathbf{z})$ is a regular super analytic function in a domain bounded by C, then $\oint_C \mathbf{dz}\, Df = 0$.

2.7 Indefinite integrals and Cauchy formulas

Define the indefinite integral

$$f(\mathbf{z}_1, \mathbf{z}_2) = \int_{\mathbf{z}_2}^{\mathbf{z}_1} \mathbf{dz}\, \omega(\mathbf{z}) \tag{2.7.1}$$

by

$$f(\mathbf{z}_2, \mathbf{z}_2) = 0\,, \qquad D_1 f(\mathbf{z}_1, \mathbf{z}_2) = \omega(\mathbf{z}_1)\,. \tag{2.7.2}$$

The natural coordinates for super translation invariant functions on the plane are

$$\theta_{12} = \theta_1 - \theta_2 = \int_{\mathbf{z}_2}^{\mathbf{z}_1} \mathbf{dz} \qquad z_{12} = z_1 - z_2 - \theta_1\theta_2 = \int_{\mathbf{z}_2}^{\mathbf{z}_1} \mathbf{dz} \int_{\mathbf{z}_2}^{\mathbf{z}} \mathbf{dz}'$$
$$D_1 z_{12} = \theta_{12} = D_2 z_{12} \qquad D_1\theta_{12} = 1 = -D_2\theta_{12}\,. \tag{2.7.3}$$

A super analytic function $f(\mathbf{z})$ can be expanded in a power series around $\mathbf{z}_2$:

$$\begin{aligned} f(\mathbf{z}_1) &= \sum_{n=0}^{\infty} \frac{1}{n!} (z_{12})^n\, \partial_2^n\, (1 + \theta_{12} D_2)\, f(\mathbf{z}_2) \\ &= f(\mathbf{z}_2) + \theta_{12} D_2 f + z_{12}\partial_2 f + \cdots\,. \end{aligned} \tag{2.7.4}$$

The super Cauchy formulas are

$$\frac{1}{2\pi i}\oint_{C_2} \mathbf{dz}_1\, z_{12}^{-n-1} = 0 \qquad \frac{1}{2\pi i}\oint_{C_2} \mathbf{dz}_1\, \theta_{12}\, z_{12}^{-n-1} = \delta_{n,0} \tag{2.7.5}$$

where C_2 is a simple contour winding once around z_2. Combining 2.7.4 and 2.7.5,

$$\begin{aligned} \frac{1}{2\pi i}\oint_{C_2} \mathbf{dz}_1\, f(\mathbf{z}_1)\, \theta_{12}\, z_{12}^{-n-1} &= \frac{1}{n!}\, \partial_2^n\, f(\mathbf{z}_2) \\ \frac{1}{2\pi i}\oint_{C_2} \mathbf{dz}_1\, f(\mathbf{z}_1)\, z_{12}^{-n-1} &= \frac{1}{n!}\, \partial_2^n\, D_2\, f(\mathbf{z}_2)\,. \end{aligned} \tag{2.7.6}$$

2.8 Periods and moduli

On a topologically nontrivial super Riemann surface, the indefinite integral 2.7.1 is defined only up to the super periods of ω, $\int_{\mathbf{z}}^{\mathbf{z}} \mathbf{dz}'\, \omega(\mathbf{z}')$. The theory of super Jacobian varieties and super theta functions should parallel and organize the theory of ordinary theta functions.

The super moduli of super Riemann surfaces are the variations of the patching transformations which define the super Riemann surface, modulo superconformal transformations of the coordinate neighborhoods[14]. Infinitesimally the moduli are the superconformal vector fields on the overlaps of coordinate neighborhoods, modulo differences of superconformal vector fields on the neighborhoods themselves. This describes the first cohomology group of the super Riemann surface with coefficients in the superconformal vector fields. This cohomology group is realized as the $(-\frac{1}{2},\frac{1}{2})$ forms modulo the image of $\bar{D}$ acting on the $(-\frac{1}{2},0)$ forms. Super integration on the super Riemann surface identifies the dual space of the infinitesimal moduli space with space of superconformal tensors of weight $\frac{3}{2}$ (see equation 3.2.5 below). For genus $g > 1$, the Riemann-Roch theorem (see section 5.8 below) and the vanishing theorem for superconformal fields of negative weight, give $2(g-1)$ as the number of weight $\frac{3}{2}$ conformal fields and $3(g-1)$ as the number of weight 2 conformal tensor fields. Therefore the super moduli space has $3(g-1)$ ordinary complex dimensions and $2(g-1)$ anticommuting complex dimensions.

The bosonic dimension of the super moduli space is exactly the dimension of the moduli space of ordinary Riemann surfaces with spin structure. Therefore the super moduli space consists of the ordinary moduli space of Riemann surfaces with spin structures, plus $2(g-1)$ fermionic coordinates lying in a vector bundle over the ordinary moduli space.

For the generic compact Riemann surface of genus g there are 2^{2g} spin structures, corresponding to all the possible sign changes of half-forms transported around the $2g$ non-trivial cycles on the surface. When the corresponding Riemann surface has nontrivial automorphisms, there are fewer spin structures. The space of spin structures is thus a 2^{2g}-sheeted covering of ordinary moduli space, branched at the singular points. This covering can be realized by the 2^{2g} first order theta functions of integer characteristic which are the partition functions of free Majorana-Weyl fermions in the various spin structures.

3. SUPERCONFORMAL FIELD THEORY

3.1 Conformal fields and operator product expansions

In a conformal field theory[8,7,15,16] the *primary* or *conformal* fields are conformal tensors of weight $(h, \bar{h})$, or, equivalently, scaling dimension $h+\bar{h}$ and spin $h-\bar{h}$. Chiral fields have $\bar{h} = 0$ (or $h = 0$), so they are analytic fields (or antianalytic fields). The operator product expansions of conformal fields,

$$\phi_i(z_1)\, \phi_j(z_2) \sim \sum_k (z_1 - z_2)^{h_k - h_i - h_j}\, C_{ijk}\, \phi_k(z_2)\,, \tag{3.1.1}$$

stand for identities which obtain in every correlation function of ϕ_i and ϕ_j:

$$\langle \cdots \phi_i(z_1)\, \phi_j(z_2) \cdots \rangle = \sum_{k,n} (z_1 - z_2)^{h_k - h_i - h_j + n} C_{ijk} \langle \cdots [\phi_k]_n(z_2) \cdots \rangle\,, \tag{3.1.2}$$

where the notation $[\phi_k]_n$, $[\phi_k]_0 = \phi_k$, stands for a sum over the descendent fields of ϕ_k on level n with coefficients which depend only on the weights $h_{i,j,k}$ (see Shenker's lectures[1] for an explanation of descendent fields). The identity 3.1.2 holds for z_1 near z_2 and, by analytic continuation, for all z_1 and z_2. For nonchiral fields the sum in 3.1.2 should include factors $(\bar{z}_1 - \bar{z}_2)^{h_k - h_i - h_j + m}$.

There is an expectation value $\langle \cdots \rangle_m$ for each Riemann surface, but the operator product identities are independent of the surface. The operator product expansions are local properties of the conformal field theory. They can be regarded as defining the quantum field theory, since 3.1.2 can be used to reconstruct the correlation functions from the operator product expansions.

In a superconformal field theory the primary superfields, the superconformal fields, are superconformal tensor fields on super Riemann surfaces. A two dimensional quantum field theory can be invariant under superconformal transformations in $\mathbf{z}$ alone or in both $\mathbf{z}$ and $\bar{\mathbf{z}}$. It suffices to consider $\tilde{\mathbf{z}}$ alone, since the discussion of local superconformal invariance in $\bar{\mathbf{z}}$ is parallel and independent.

The partition function $Z(m)$ depends on the super moduli parametrizing the super Riemann surfaces. The superconformal fields obey operator product expansions analogous to 3.1.1, with z_{12} and θ_{12} taking the place of $z_1 - z_2$. In power counting, θ_{12} counts as $z_{12}^{1/2}$.

3.2 The super stress-energy tensor

The fundamental quantum field in a superconformal field theory is the super stress-energy tensor

$$T(\mathbf{z}) = T_F(z) + \theta T_B(z). \tag{3.2.1}$$

$T(\mathbf{z})$ is a chiral superfield of dimension 3/2. T_B is the ordinary stress-energy tensor (dimension (2,0)); T_F is its super partner (dimension (3/2,0)). $T(\mathbf{z})$ generates the superconformal transformations 2.5.1 by

$$\delta_v \phi(\mathbf{z}_2) = \frac{1}{2\pi i} \oint_{C_2} d\mathbf{z}_1 \; v(\mathbf{z}_1) \, T(\mathbf{z}_1) \, \phi(\mathbf{z}_2) \,, \tag{3.2.2}$$

where C_2 is a simple contour winding once around z_2. As usual, this identity holds within correlation functions.

By the super Cauchy formulas 2.7.6, the transformation law 3.2.2 is equivalent to the operator product expansion

$$T(\mathbf{z}_1) \, \phi(\mathbf{z}_2) \sim \frac{\theta_{12}}{z_{12}^2} \, h \, \phi(\mathbf{z}_2) + \frac{1/2}{z_{12}} D_2\phi + \frac{\theta_{12}}{z_{12}} \partial_2 \phi + \cdots \tag{3.2.3}$$

where the ommitted terms are nonsingular. Only the singular part of the operator product expansion contributes to the contour integral.

The super stress-energy tensor is itself an anomalous superconformal field of weight 3/2:

$$\begin{aligned} T(\mathbf{z}_1) \, T(\mathbf{z}_2) \quad &\sim \quad \frac{\hat{c}}{4} \frac{1}{z_{12}^3} + \frac{3}{2} \frac{\theta_{12}}{z_{12}^2} T(\mathbf{z}_2) + \frac{1}{2} \frac{1}{z_{12}} D_2 T + \frac{\theta_{12}}{z_{12}} \partial_2 T \\ \delta_v T \quad &= \quad (v\partial + \tfrac{1}{2}(Dv)D + \tfrac{3}{2}\partial v) \, T + \tfrac{1}{8} \hat{c} \, \partial^2 D v \,. \end{aligned} \tag{3.2.4}$$

There are at least two approaches to deriving 3.2.4. In the first approach, the subleading singularities in $T(\mathbf{z}_1) \, T(\mathbf{z}_2)$ are determined by 3.2.2, which fixes the dimension of $T(\mathbf{z})$ to be 3/2, which fixes the two point function, which determines the leading singularity, up to a constant $\hat{c}$. The second approach is to determine the form of 3.2.4, up to the arbitrary number $\hat{c}$, by symmetries of the quantum field theory on the plane: Euclidean invariance, supersymmetry and scale invariance. The second argument applies especially to superymmetric critical phenomena[16,3]. The coefficient $\hat{c}$ of the anomaly, the central term in the operator product, is the fundamental characteristic number of a two dimensional superconformal field theory.

The super stress-energy tensor represents the infinitesimal variations of the super moduli:

$$\frac{\partial}{\partial m^i} \log Z(m) = \int \mathbf{d}^2\mathbf{z}\, \bar{f}_i(z,\bar{z})\, \langle T(\mathbf{z})\rangle_m + \text{c.c.} \tag{3.2.5}$$

where $\bar{f}_i$ is a $(-1,\frac{1}{2})$ form representing the infinitesmal variation of m. One way to calculate $\langle T(\mathbf{z})\rangle_m$ is to take the expectation value of 3.2.4:

$$\langle T(\mathbf{z}_1)\, T(\mathbf{z}_2)\rangle_m \sim \tfrac{1}{4}\hat{c}\, z_{12}^{-3} + \tfrac{3}{2}\theta_{12} z_{12}^{-2}\, \langle T(\mathbf{z}_2)\rangle_m\,. \tag{3.2.6}$$

3.3 The global superconformal group $\widehat{\mathbf{SL}}_2$

The effect of a finite superconformal transformation $\mathbf{z} \to \tilde{\mathbf{z}}$ on the super stress-energy tensor is computed by requiring the operator product expansion 3.2.4 to hold in both $\mathbf{z}$ and $\tilde{\mathbf{z}}$:

$$T(\mathbf{z}) = \tilde{T}(\tilde{\mathbf{z}})(D\tilde{\theta})^3 + \tfrac{1}{4}\hat{c}S(\mathbf{z},\tilde{\mathbf{z}}) \tag{3.3.1}$$

where $S(\mathbf{z},\tilde{\mathbf{z}})$ is the super Schwarzian derivative

$$S(\mathbf{z},\tilde{\mathbf{z}}) = \frac{D^4\tilde{\theta}}{D\tilde{\theta}} - 2\frac{D^3\tilde{\theta}\, D^2\tilde{\theta}}{D\tilde{\theta}\ \ D\tilde{\theta}} = \partial(\log D\tilde{\theta})\, D(\log \partial(-1/D\tilde{\theta}))\,. \tag{3.3.2}$$

On the sphere, the globally defined superconformal vector fields are of the form

$$v(\mathbf{z}) = (v_{-1} + v_0 z + v_1 z^2) + \theta(\hat{v}_{-1/2} + \hat{v}_{1/2} z)\,, \tag{3.3.3}$$

forming the super Lie algebra Osp(2, 1). The correlation functions of superconformal fields are Osp(2, 1) invariant, because at large distance the correlation functions of the super stress-energy tensor are of the form of its two-point function,

$$\langle T(\mathbf{z})\cdots\rangle \underset{z\to\infty}{\sim} O(z^{-3} + \theta z^{-4})\,, \tag{3.3.4}$$

which implies that contour integrals of $v(\mathbf{z})T(\mathbf{z})$ vanish at infinity for the vector fields 3.3.3.

The vector fields 3.3.3 are exactly the solutions of $\partial^2 Dv = 0$, which is the infinitesmal form of the super Schwarzian derivative 3.3.2. By equation 3.3.1, the super Schwarzian derivative obeys the composition law

$$S(\mathbf{z}, \tilde{\tilde{\mathbf{z}}}) = S(\mathbf{z}, \tilde{\mathbf{z}}) + (D\tilde{\theta})^3 S(\tilde{\mathbf{z}}, \tilde{\tilde{\mathbf{z}}}) \tag{3.3.5}$$

implying that $S(\mathbf{z}, \tilde{\mathbf{z}}) = 0$ for all the global superconformal transformations which can be made from successive infinitesimal transformations, i.e., the connected component of the identity in the global superconformal super group. The solutions of $S(\mathbf{z}_1, \tilde{\mathbf{z}}_1) = 0$ are

$$\tilde{\theta}_1 = \theta_0 + \theta_{12}/z_{12} \qquad \tilde{z}_1 = z_0 + (\alpha + \theta_1\theta_0)/z_{12} \tag{3.3.6}$$

where the parameters of the transformation are $\mathbf{z}_0 = (z_0, \theta_0)$, $\mathbf{z}_2 = (z_2, \theta_2)$ and α. This group is $\widehat{SL}_2$, a supersymmetric extension of the ordinary global conformal group SL_2 of fractional linear maps $x \to (az+b)/(cz+d)$.

The Lie algebra of ordinary conformal vector fields on the cylinder, or the punctured plane, is the the complexification of the Lie algebra of Diff(S^1), which is the group of diffeomorphisms of the circle. Diff(S^1) can be identified with the conformal transformations of the punctured plane which satisfy the reality condition $\tilde{z}(1/z^*) = 1/\tilde{z}(z)^*$.

The superconformal vector fields on cylinder or the punctured plane form two super Lie algebras, the Ramond and Neveu-Schwarz algebras, corresponding to the two spin structures (boundary conditions) on the cylinder. The superconformal algebras are the complexifications of the super Lie algebras of the two groups of super diffeomorphisms of the circle, $\widehat{\text{Diff}}_\pm(S^1)$. The distinction between the two super groups only appears when there is a strong enough topology on the group to distinguish the two boundary conditions on the circle which define trivial and nontrivial $O(1)$ spinors. The super Schwarzian derivative is the globally invariant generator of the second cohomology group of the super group $\widehat{\text{Diff}}_\pm(S^1)$, just as the ordinary Schwarzian derivative is the SL_2 invariant generator for the two cohomology of Diff(S^1)[17].

3.4 Operator interpretation

The simplest operator interpretation of a conformal field theory is given by the radial quantization. It is constructed from the correlation functions on the sphere or, equivalently, on the plane or the infinite cylinder. If $z = e^w = e^{\tau + i\sigma}$ is the standard complex coordinate for the plane, so that w is the standard coordinate on the cylinder, then correlation functions on the sphere or the plane or the cylinder are interpreted as vacuum expectation values of τ-ordered products

$$\langle \phi(z_1) \cdots \rangle = \langle 0 | \, \tau \{ \phi(z_1) \cdots \} \, | 0 \rangle \tag{3.4.1}$$

where the operator valued fields are τ-ordered by putting fields of large $|z|$ to the left and fields of small $|z|$ to the right.

On the cylinder there are two spin structures, given by periodic or antiperiodic boundary conditions in the σ direction. Thus the Hilbert space of the radial quantization divides into two sectors; the Neveu-Schwarz (NS) sector, in which the spinor fields are single valued on the plane (but double valued on the cylinder because of the factor $dz^{1/2}$) and the Ramond (R) sector in which the spinor fields are single valued on the cylinder (but double valued on the plane). The superconformal fields are block diagonal in the NS $\oplus$ R decomposition of the Hilbert space. Vacuum expectation values are single valued in the plane, so the vacuum state $|0\rangle$ is in the NS sector.

A highest weight state $|S\rangle$ in the R sector is an ordinary conformal highest weight state, so it corresponds to some ordinary conformal field $S(z)$. This conformal field is called a ***spin*** field. It is block off-diagonal in the NS$\oplus$R decomposition. A spin field $S(z)$, acting on the vacuum in the NS sector, creates the highest weight state $|S\rangle = S(0)\,|0\rangle$ in the R sector.

Integrating over all super Riemann surfaces includes summing over all spin structures. On the torus there are four spin structures, two boundary conditions in each of two directions. The partition function is a Hilbert space trace. Picture one direction as euclidean time and the other as space. The sum over spatial boundary conditions is the sum over NS and R sectors in the trace. Summing over boundary conditions for the spinor fields introduces a projection operator $\frac{1}{2}+\frac{1}{2}\Gamma$ in the trace, where the chirality operator Γ commutes with integer spin fields and anticommutes with half-integer spin fields. On surfaces of genus >1 the sum over spin structures provides in each loop a sum over R and NS sectors and a chiral projection.

3.5 Superconformal generators

An infinitesmal superconformal transformation of the punctured plane corresponds to a superconformal vector field $v(\mathbf{z})$, analytic away from the origin. The transformation is generated by the operator

$$T_{[v]} = \frac{1}{2\pi i}\oint_{C_0} d\mathbf{z}\, v(\mathbf{z})T(\mathbf{z}) \tag{3.5.1}$$

where C_0 winds once around the origin, making a "space-like hypersurface" in the radial quantization. Commutation relations of operators can be represented in terms of τ-ordered products,

$$\begin{aligned}
[T_{[v]},\, \phi(\mathbf{z}_2)] &= \frac{1}{2\pi i}\oint_{C_{0,2}-C_0} d\mathbf{z}_1\, v(\mathbf{z}_1)\, T(\mathbf{z}_1)\, \phi(\mathbf{z}_2) \\
&= \frac{1}{2\pi i}\oint_{C_2} d\mathbf{z}_1\, v(\mathbf{z}_1)\, T(\mathbf{z}_1)\, \phi(\mathbf{z}_2) \\
&= \delta_v \phi(\mathbf{z}_2)
\end{aligned} \tag{3.5.2}$$

where $C_{0,2}$ winds around z_2 and the origin, e.g. $|z|>|z_2|$, and C_0 winds around the origin but not z_2, e.g. $|z_2|>|z|>0$. The deformation of contours is justified by the analyticity of the τ-ordered products for $z_2 \neq 0, z_1$. The contours and deformations can always be chosen so as to miss any other quantum fields which might be present in the correlation function.

The contour integral argument shows that the commutation relations of the generators $L_{[v]}$ are encoded in the singular part of the operator products of $T(\mathbf{z})$.

This argument is quite general. The singular parts of the operator product expansions of analytic (chiral) fields, with themselves and with other quantum fields, are equivalent to their commutation relations. The equivalence is realized by the contour argument. Even for nonchiral fields, the singular operator product expansions are equivalent to commutation relations, but in the absence of analyticity the contour integral must be replaced by a principal part interpretation of the singularity. The advantages of operator products is that they are independent of the Hilbert space interpretation, they obtain on arbitrary two dimensional surfaces, and they are easily calculated from functional integral representations of correlation functions.

3.6 Operator products of component fields

Expanding 3.2.3, 3.2.4 in component fields gives

$$\begin{aligned}
T_B(z_1)\,T_B(z_2) &\sim \frac{3\hat{c}/4}{(z_1-z_2)^4} + \frac{2}{(z_1-z_2)^2}\,T_B(z_2) + \frac{1}{z_1-z_2}\,\partial_2 T_B \\
T_B(z_1)\,T_F(z_2) &\sim \frac{3/2}{(z_1-z_2)^2}\,T_F(z_2) + \frac{1}{z_1-z_2}\,\partial_2 T_F \\
T_F(z_1)\,T_F(z_2) &\sim \frac{\hat{c}/4}{(z_1-z_2)^3} + \frac{1/2}{z_1-z_2}\,T_B(z_2) \\
T_B(z_1)\,\phi_0(z_2) &\sim \frac{h}{(z_1-z_2)^2}\,\phi_0(z_2) + \frac{1}{z_1-z_2}\,\partial_2\phi_0 \\
T_B(z_1)\,\phi_1(z_2) &\sim \frac{h+1/2}{(z_1-z_2)^2}\,\phi_1(z_2) + \frac{1}{z_1-z_2}\,\partial_2\phi_1 \\
T_F(z_1)\,\phi_0(z_2) &\sim \frac{1/2}{z_1-z_2}\,\phi_1(z_2) \\
T_F(z_1)\,\phi_1(z_2) &\sim \frac{h}{(z_1-z_2)^2}\,\phi_0(z_2) + \frac{1/2}{z_1-z_2}\,\partial_2\phi_0\,.
\end{aligned} \tag{3.6.1}$$

3.7 Mode expansions

The fields expand in Laurent (Fourier) series:

$$\begin{aligned}
T_F(z) &= \sum_n z^{-n-3/2}\,\tfrac{1}{2}G_n \qquad & \phi_0(z) &= \sum_n z^{-n-h}\,\phi_{0,n} \\
T_B(z) &= \sum_n z^{-n-2}\,L_n \qquad & \phi_1(z) &= \sum_n z^{-n-h-1/2}\,\phi_{1,n}\,.
\end{aligned} \tag{3.7.1}$$

The powers of z are such that, when $z \to \log z$ takes the plane to the cylinder, the covariance of, for example, $\phi_0(z)\,(dz)^h$, implies that the $\phi_{0,n}$ are the Fourier coefficients of ϕ_0 on the cylinder. A component field of integer weight h is always indexed by integers n. A field of half-integer weight h is indexed by integers n in the R sector and by half-integers n in the NS sector. Euclidean time reversal on the cylinder corresponds to $z \to \tilde{z} = 1/\bar{z}$. The adjoint of a field is given by $(\phi(z)dz^h)^\dagger = \phi^\dagger(\tilde{z})d\tilde{z}^h$. The reality of the super stress-energy tensor implies

$$L_m^\dagger = L_{-m} \qquad G_m^\dagger = G_{-m}\,. \tag{3.7.2}$$

3.8 Commutation relations of normal modes

Commutation relations are derived from the operator product expansions by representing modes as contour integrals, then deforming contours. An anticommuting parameter ϵ is introduced in order to express anticommutation relations as commutation relations, whatever the statistics of the field ϕ. The commutation relations are

$$\begin{aligned}
[L_m,\ \phi_0(z)] &= z^{m+1}\partial\phi_0 + h(m+1)z^m\phi_0(z) \\
[L_m,\ \phi_1(z)] &= z^{m+1}\partial\phi_0 + (h+\tfrac{1}{2})(m+1)z^m\phi_0(z) \\
[\epsilon G_m,\ \phi_0(z)] &= \epsilon\, z^{m+1/2}\phi_1(z) \\
[\epsilon G_m,\ \phi_1(z)] &= \epsilon[z^{m+1/2}\partial\phi_0 + 2(m+\tfrac{1}{2})h\, z^{m-1/2}\phi_0(z)]
\end{aligned} \tag{3.8.1}$$

$$\begin{aligned}
[L_m,\ T_F(z)] &= z^{m+1}\partial T_F + \tfrac{3}{2}(m+1)z^m T_F(z) \\
[L_m,\ T_B(z)] &= z^{m+1}\partial T_B + 2(m+1)z^m T_B(z) + \tfrac{1}{8}\hat{c}(m^3-m)z^{m-2} \\
[G_m,\ T_F(z)]_+ &= z^{m+1/2}T_B(z) + \tfrac{1}{4}\hat{c}(m^2-\tfrac{1}{4})z^{m-3/2} \\
[G_m,\ T_B(z)] &= z^{m+1/2}\partial T_F + 3(m+\tfrac{1}{2})z^{m-1/2}T_F(z)
\end{aligned} \tag{3.8.2}$$

$$\begin{aligned}
[L_m,\ \phi_{0,n}] &= [(h-1)m-n]\phi_{0,m+n} \\
[\epsilon G_m,\ \phi_{0,n}] &= \epsilon\phi_{1,m+n} \\
[\epsilon G_m,\ \phi_{1,n}] &= \epsilon[(2h-1)m-n]\phi_{0,m+n}
\end{aligned} \tag{3.8.3}$$

$$\begin{aligned}
[L_m, L_n] &= (m-n)L_{m+n} + \tfrac{1}{8}\hat{c}(m^3 - m)\delta_{m+n,0} \\
[L_m, G_n] &= (\tfrac{1}{2}m - n)G_{m+n} \qquad (3.8.4)\\
[G_m, G_n]_+ &= 2L_{m+n} + \tfrac{1}{2}\hat{c}(m^2 - \tfrac{1}{4})\delta_{m+n,0}
\end{aligned}$$

The conformal generators L_n form the Virasoro algebra; the superconformal generators G_n, L_n form the Ramond algebra (integer n) and the Neveu-Schwarz algebra (half-integer n). $\widehat{SL}_2$ is generated by $G_{-1/2}$, $G_{1/2}$, L_{-1}, L_0, L_1. In particular, L_0 ($+\bar{L}_0$) is the generator of dilations, which is the hamiltonian in radial quantization. The mode expansions are arranged so that

$$[L_0, \phi_n] = -n\,\phi_n \qquad (3.8.5)$$

as long as the form 3.7.1 is used for the mode expansion of $\phi(z)$.

3.9 Highest weight states and conformal fields

A ground state for the superconformal algebra is a state $|h\rangle$ which is annihilated by all the lowering operators L_{+n}, G_{+n} and has eigenvalue h for L_0. In mathematical terminology the ground states are called highest weight states, because mathematicians usually call $-h$ the weight of the state.

In the Ramond sector G_0 commutes with L_0 and therefore acts on the ground states. G_0 anticommutes with the chirality operator Γ. If $h \neq \hat{c}/16$ then $G_0^2 \neq 0$ and the ground states come in pairs of opposite chirality. If $h = \hat{c}/16$ then $G_0\,|h\rangle = 0$ (this is a bit subtle in a nonunitary theory). Then $|h\rangle$ is a supersymmetric ground state for the Ramond system on the cylinder with supersymmetry generator G_0, and is not necessarily paired with a state of opposite chirality. The Witten index of the Ramond sector is the net chirality of the $h = \hat{c}/16$ states. In a unitary system $G_0^2 \geq 0$ so all $h = \hat{c}/16$ states are heighest weight states.

In an ordinary conformal field theory there is a one to one correspondence between conformal fields $\phi(z)$ of conformal weight h and highest weight states $|h\rangle$ (for the L_n) with eigenvalue $L_0 = h$. The correspondence is $|h\rangle = \phi(0)\,|0\rangle$. Given the conformal field ϕ, the state $\phi(0)\,|0\rangle$ exists and is nonzero, because correlation functions of $\phi(z)$ are finite at $z = 0$, and because no quantum field can annihilate the vacuum (in the unitary two dimensional field theory associated with euclidean spacetime). The highest weight condition on $|h\rangle$ then follows from the commutation relations of $\phi(z)$ with the conformal generators. Conversely, given the highest weight state $|h\rangle$, completeness of the field algebra implies that there is a quantum field $\phi(z)$ which has a matrix element between the vacuum and $|h\rangle$. By subtracting other quantum fields it can be assured that ϕ creates no states of energy less than h. Then $|h\rangle = \phi(0)\,|0\rangle$, because $z = 0$ corresponds to $\tau = -\infty$. The highest weight condition on $|h\rangle$ implies the operator product 3.6.1, first only for $z_2 = 0$ and acting on the vacuum, then as an operator statement because no fields annihilate the vacuum, and then for all z_2 because of two dimensional translation invariance.

In superconformal field theories, the NS heighest weight states $|h\rangle$ correspond to superconformal fields $\phi(\mathbf{z}) = \phi_0(z) + \theta\phi_1(z)$, where $\phi_0(0)\,|0\rangle = |h\rangle$ and $\phi_1(0)\,|0\rangle = G_{-1/2}\,|h\rangle$. The highest weight conditions for the superconformal algebra correspond to the operator product 3.2.3 between T and ϕ. The highest weight states $|h\rangle$ in the R sector correspond to the spin fields. These are pairs $S_\pm(z)$ of conformal fields such that

$$\begin{aligned} &S_+(0)\,|0\rangle = |h\rangle \qquad S_-(0)\,|0\rangle = G_0\,|h\rangle \\ &T_F(z_1)\,S_+(z_2) \sim \tfrac{1}{2}(z_1 - z_2)^{-3/2}\,S_-(z_2) \\ &T_F(z_1)\,S_-(z_2) \sim \tfrac{1}{2}(h - \tfrac{1}{16}\hat{c})(z_1 - z_2)^{-3/2}\,S_+(z_2)\,. \end{aligned} \tag{3.9.1}$$

The Ramond supersymmetry is unbroken only if $h = \hat{c}/16$. Then it is possible to have $G_0\,|h\rangle = 0$, $S_-(z) = 0$ and a nonzero Witten index.

Note that T_F and $S_\pm$ are not mutually local. The field theory containing spin fields $S_\pm$ becomes local only in combination with the chiral projection $\Gamma = 1$. Both the sum over sectors and the chiral projection are accomplished by the sum over spin structures. The projection eliminates T_F and the other half-integer fields, and eliminates one spin field in every chiral pair.

4. THE FERMIONIC STRING

The world surface of fermionic string is described by a two dimensional superconformal field theory, consising of a matter superfield, X^μ, $\mu = 1, \ldots, d$ which gives the location of the world surface in spacetime; and ghost superfields B, C which arise from fixing the superconformal gauge on the world surface[11].

The radial quantization of this superconformal field theory gives the one string Hilbert space. Both matter and ghost systems are indefinite metric field theories (although the matter sector becomes unitary when the spacetime metric has euclidean signature); the positive metric of the string Hilbert space only appears after the BRST condition is imposed on the states of the two dimensional field theory.

4.1 Matter fields

The action and equation of motion of the matter superfield (in flat spacetime) are

$$S_{matter} = \frac{1}{2\pi} \int d^2z \, d\theta d\bar{\theta} \, \overline{D} X^\mu D X_\mu \, , \qquad \overline{D} D \, X^\mu = 0 \, , \tag{4.1.1}$$

$$X^\mu(\mathbf{z}, \bar{\mathbf{z}}) = X^\mu(\mathbf{z}) + X^\mu(\bar{\mathbf{z}}) \, , \qquad X^\mu(\mathbf{z}) = x^\mu(z) + \theta \psi^\mu(z) \, , \tag{4.1.2}$$

after eliminating the auxiliary field $F^\mu = \partial_\theta \partial_{\bar{\theta}} X^\mu$ by its equation of motion. The action of the component fields is

$$S_{matter} = \frac{1}{2\pi} \int d^2z \, \left(\bar{\partial} x^\mu \partial x_\mu - \psi \bar{\partial} \psi - \bar{\psi} \partial \bar{\psi} \right) \, . \tag{4.1.3}$$

This is the model appropriate to the type II superstrings. For heterotic strings there is no $\bar{\theta}$ and an additional $E_8 \times E_8$ or $SO(32)$ chiral current algebra in $\bar{z}$. For type I superstrings the world surfaces include the nonorientable surfaces (which are not globally complex), and have boundaries; and there are gauge degrees of freedom on the boundaries. In any case, the present discussion is only concerned with the superconformal aspect of the world surface, so only the (z, θ) sector is discussed.

The matter chirality operator Γ is defined by $[\Gamma, x^\mu] = 0$, $[\Gamma, \psi^\mu]_+ = 0$. In the type II theory there are two separate chirality operators, Γ and $\overline{\Gamma}$, and two projections.

4.2 Superconformal ghosts

The field C is the ghost for infinitesimal transformations of the super world surface; it has weight -1 and Fermi statistics. The field B is the ghost for infinitesimal variations of the superconformal gauge condition and is conjugate to C; it has weight 3/2 and Bose statistics. The action and equations of motion are

$$S_{ghost} = \frac{1}{\pi}\int d^2z\, d\theta d\bar{\theta}\, B\overline{D}C\,, \qquad \overline{D}B = 0 = \overline{D}C\,. \tag{4.2.1}$$

The dimensions and statistics of the component fields are

$$\begin{array}{llll} B(\mathbf{z})=\beta(z)+\theta b(z), & \beta:\ h=3/2\ \text{(Bose)}, & b:\ h=2 & \text{(Fermi)} \\ C(\mathbf{z})=c(z)+\theta\gamma(z), & c:\ h=-1\ \text{(Fermi)}, & \gamma:\ h=-1/2 & \text{(Bose)} \end{array} \tag{4.2.2}$$

The action for the component fields is

$$S_{ghost} = \frac{1}{\pi}\int d^2z\, \left(b\bar{\partial}c + \beta\bar{\partial}\gamma\right). \tag{4.2.3}$$

The chirality operator Γ commutes with b and c and anticommutes with β and γ.

4.3 Two-point functions

All correlation functions of free fields are determined by the two-point functions. For the time being the following expressions should be regarded as the singular parts of the two point functions on any surface. Later it will become apparent that these are also the exact two point functions on the sphere or plane. They are derived from the action using the identities $\partial\bar{\partial}\ln|z|^2 = \pi\delta^2(z)$, $\bar{\partial}\, z^{-1} = \pi\delta^2(z)$.

$$\begin{array}{llll} X^\mu(\mathbf{z}_1)\,X^\nu(\mathbf{z}_2) \sim -g^{\mu\nu}\ln z_{12} & B(\mathbf{z}_1)\,C(\mathbf{z}_2) \sim & \theta_{12}z_{12}^{-1} & \sim C(\mathbf{z}_1)\,B(\mathbf{z}_2) \\ x^\mu(z_1)\,x^\nu(z_2) \sim -g^{\mu\nu}\ln(z_1-z_2) & c(z_1)\,b(z_2) & \sim (z_1-z_2)^{-1} \sim & b(z_1)\,c(z_2) \\ \psi^\mu(z_1)\,\psi^\nu(z_2) \sim -g^{\mu\nu}(z_1-z_2)^{-1} & \gamma(z_1)\,\beta(z_2) & \sim (z_1-z_2)^{-1} \sim & -\beta(z_1)\,\gamma(z_2) \end{array} \tag{4.3.1}$$

4.4 Stress-energy tensors

The stress-energy tensor (and central charge) has contributions from the matter fields and from the ghosts: $T = T^X + T^{gh}$, $\hat{c} = \hat{c}^X + \hat{c}^{gh}$. One way to find the stress-energy tensor is to use the known form of the operator product expansions with the free fields. Write T as the most general superfield of dimension 3/2 bilinear in the fields and neutral in all conserved charges, and then fix the unknown numerical coefficients by the operator products. For example, write $T^{gh} = a_1 C\partial B + a_2 DCDB + a_3 \partial CB$ and then calculate operator products with B, C by making partial contractions.

Products of fields at coincident points usually need renormalization. In free field theory all divergences come from self-contractions, so bilinears in the free fields have finite connected correlation functions, thus finite singular operator product expansions. For calculations of finite parts of operator products a simple systematic regularization of bilinears is simply to subtract the singular part of the operator product of the free fields. On the sphere this means simply omitting all self-contractions. Here is a sample calculation of one contribution to the $T\,C$ operator product:

$$\begin{aligned} C\partial_1 B\ C(\mathbf{z}_2) &\sim -\langle \partial_1 B\, C(\mathbf{z}_2)\rangle\ C(\mathbf{z}_1) \\ &\sim \theta_{12} z_{12}^{-2}\ C(\mathbf{z}_1) \\ &\sim \theta_{12} z_{12}^{-2}\ C(\mathbf{z}_2) + \theta_{12} z_{12}^{-1}\ \partial_2 C\,. \end{aligned} \tag{4.4.1}$$

For the rest of the calculations, keep in mind that C, DB and DX obey Fermi statistics, and take advantage of the identity $\theta_{12}^2 = 0$. The results are

$$\begin{aligned} T^X &= -\frac{1}{2} DX^\mu \partial X_\mu \\ T^{gh} &= -C\,\partial B + \frac{1}{2} DC\,DB - \frac{3}{2}\partial C\,B\,. \end{aligned} \tag{4.4.2}$$

$$\begin{aligned} T_F^X &= -\tfrac{1}{2}\psi_\mu \partial x^\mu & T_B^X &= -\tfrac{1}{2}\partial x^\mu \partial x_\mu - \tfrac{1}{2}\partial\psi^\mu \psi_\mu \\ T_F^{gh} &= -c\partial\beta - \tfrac{3}{2}\partial c\beta + \tfrac{1}{2}\gamma\, b & T_B^{gh} &= c\partial b + 2\partial c b - \tfrac{1}{2}\gamma\partial\beta - \tfrac{3}{2}\partial\gamma\beta \end{aligned} \tag{4.4.3}$$

Once the coefficients in T are fixed, the $T\,T$ operator products are calculated by partial contractions; the central terms are the double contractions. For example,

$$T^X(\mathbf{z}_1)\,T^X(\mathbf{z}_2)\sim\tfrac{1}{4}\left(\langle D_1X^\mu\,D_2X^\nu\rangle\,\langle\partial_1X_\mu\,\partial_2X_\nu\rangle+\langle D_1X^\mu\,\partial_2X_\nu\rangle\,\langle\partial_1X_\mu\,D_2X^\nu\rangle\right)$$
$$T^{gh}(\mathbf{z}_1)\,T^{gh}(\mathbf{z}_2)\sim\langle\partial_1B\,C(\mathbf{z}_2)\rangle\,\langle C(\mathbf{z}_1)\,\partial_2B\rangle+\cdots\,. \tag{4.4.4}$$

The results are

$$\hat{c}^X=d \qquad \hat{c}^{gh}=-10 \qquad \hat{c}=d-10\,. \tag{4.4.5}$$

The critical dimension $d=10$ is determined by the condition that the combined matter ghost system be free of conformal anomaly. The combined two dimensional quantum field theory then depends only on the super conformal class of the world surface.

4.5 Mode expansions

The mode expansions of the free fields are

$$\begin{aligned}
\partial x^\mu &= \textstyle\sum_n z^{-n-1}a_n^\mu & x^\mu(z) &= \tfrac{1}{2}q^\mu+a_0^\mu\ln z+\textstyle\sum_{n\neq0}\dfrac{z^{-n}}{-n}\,a_n^\mu\\
\psi^\mu(z) &= \textstyle\sum_n z^{-n-1/2}\,\psi_n^\mu & &\\
b(z) &= \textstyle\sum_n z^{-n-2}\,b_n & \beta(z) &= \textstyle\sum_n z^{-n-3/2}\,\beta_n\\
c(z) &= \textstyle\sum_n z^{-n+1}\,c_n & \gamma(z) &= \textstyle\sum_n z^{-n+1/2}\,\gamma_n
\end{aligned} \tag{4.5.1}$$

The superconformal generators are

$$\begin{aligned}
G_n^X &= \textstyle\sum_k -\psi_{n-k}^\mu a_k^\mu\\
L_n^X &= \textstyle\sum_k -\tfrac{1}{2}a_{n-k}^\mu a_k^\mu+\tfrac{1}{2}(\tfrac{1}{2}n-k)\psi_{n-k}^\mu\psi_k^\mu\\
G_n^{gh} &= \textstyle\sum_k (3n-k)c_{n-k}\beta_k+\gamma_{n-k}b_k\\
L_n^{gh} &= \textstyle\sum_k (k-2n)c_{n-k}b_k+(\tfrac{3}{2}n-k)\gamma_{n-k}\beta_k\,.
\end{aligned} \tag{4.5.2}$$

There is implicit normal ordering of the quadratic expressions for L_0 in 4.5.2 (see the discussion of renormalization in section 4.4 and of ground state energies in section 4.7). The indices n of the modes L_n, a_n, b_n and c_n are integers. In the NS sector the indices n of the modes G_n, ψ_n^μ, γ_n, and β_n are half-integers ($n\in\mathbf{Z}+\tfrac{1}{2}$).

In the R sector they are integers ($n \in \mathbf{Z}$). The behavior of the fields under $z \to 1/\bar{z}$ and hermitian conjugation gives

$$\begin{aligned} (a^\mu_n)^\dagger &= -a^\mu_n & b^\dagger_n &= b_n & c^\dagger_n &= c_n \\ (\psi^\mu_n)^\dagger &= \psi^\mu_n & \beta^\dagger_n &= -\beta_n & \gamma^\dagger_n &= \gamma_n\,. \end{aligned} \tag{4.5.3}$$

4.6 (Anti-)commutation relations

The (anti-)commutation relations of the modes are calculated from contour integrals of the two point functions, and depend only on the singular parts. Only the nonzero relations are written here:

$$\begin{aligned} [a^\mu_m,\, a^\nu_n] &= -g^{\mu\nu}\, m\, \delta_{m+n} & [a^\mu_0,\, q^\nu] &= -g^{\mu\nu} \qquad p^\mu = i a^\mu_0 \\ [\psi^\mu_m,\, \psi^\nu_n]_+ &= -g^{\mu\nu}\delta_{m+n} & [c_m,\, b_n]_+ &= \delta_{m+n} = [\gamma_m,\, \beta_n] \end{aligned} \tag{4.6.1}$$

$$\begin{aligned} [G_m,\, a^\mu_n] &= -n\psi^\mu_{m+n} & [L_m,\, a^\mu_n] &= -n a^\mu_{m+n} \\ [G_m,\, \psi^\mu_n]_+ &= a^\mu_{m+n} & [L_m,\, \psi^\mu_n] &= (-\tfrac{1}{2}m - n)\psi^\mu_{m+n} \\ [G_m,\, c_n]_+ &= \gamma_{m+n} & [L_m,\, c_n] &= (-2m - n)c_{m+n} \\ [G_m,\, b_n]_+ &= (2m - n)\beta_{m+n} & [L_m,\, b_n] &= (m - n)b_{m+n} \\ [G_m,\, \gamma_n] &= (-3m - n)c_{m+n} & [L_m,\, \gamma_n] &= (-\tfrac{3}{2}m - n)\gamma_{m+n} \\ [G_m,\, \beta_n] &= b_{m+n} & [L_m,\, \beta_n] &= (\tfrac{1}{2}m - n)\beta_{m+n} \end{aligned} \tag{4.6.2}$$

4.7 Matter ground states and zero modes

The zero mode algebra of the matter system is generated by the total spacetime momentum operator $p^\mu = i\, a^\mu_0$ and, in the R sector, the fermionic zero modes ψ^μ_0. From 4.6.1, the ψ^μ_0 zero modes satisfy the anticommutation relations of the spacetime γ-matrices:

$$[\psi^\mu_0,\, \psi^\nu_0]_+ = -g^{\mu\nu}\,. \tag{4.7.1}$$

The matter ground states are the states annihilated by all the lowering operators a^μ_{+n}, ψ^μ_{+n}. In the NS sector there is one ground state $|k\rangle$ for each momentum eigenvalue $p^\mu = k^\mu$. In the R sector the ground states can be written $|k,\alpha\rangle = |k\rangle \otimes |\alpha\rangle$ where $|k\rangle$ is the ground state of x^μ and $|\alpha\rangle$ is a ground state of ψ^μ. By

4.7.1, the states $|\alpha\rangle$ form a Dirac spinor, with indices $\alpha = 1 \cdots 2^{d/2}$. Summarizing,

$$\begin{aligned} a_0^\mu \,|k\rangle &= ik^\mu \,|k\rangle & \psi_0^\mu \,|\alpha\rangle &= \gamma^{\mu\alpha}{}_\beta \,|\beta\rangle & G_0^X \,|k,\alpha\rangle &= -i\not{k}\,|k,\alpha\rangle \\ L_0^x \,|k\rangle &= \tfrac{1}{2}k^2 \,|k\rangle & L_0^\psi \,|\alpha\rangle &= \tfrac{1}{16}d\,|\alpha\rangle & L_0^X \,|k,\alpha\rangle &= (\tfrac{1}{2}k^2 + \tfrac{1}{16}d)\,|k,\alpha\rangle \ . \end{aligned} \tag{4.7.2}$$

The L_0^ψ eigenvalue follows from the Ramond supersymmetry algebra $L_0 = G_0^2 + d/16$ and the supersymmetry of the Ramond ground states $|0,\alpha\rangle$.

The $\widehat{SL}_2$ invariant matter vacuum $|0\rangle$ is the zero momentum ground state of the NS sector. The full Hilbert space of the two dimensional field theory is generated from the ground states by the raising operators a_{-n}^μ, ψ_{-n}^μ. Thus the NS sector contains only states which transform as Lorentz vectors, i.e., are spacetime bosons; and the R sector contains only states which transform as Lorentz spinors, and so are spacetime fermions.

The chirality operator Γ is normalized to be $+1$ on the vacuum, thus $\Gamma = 1$ on the NS ground states (since $[\Gamma, q] = 0$), and on the excited states $\Gamma = (-1)^F$, the fermion parity. Γ acts on the R ground states $|\alpha\rangle$ as $\pm\gamma_{d+1}$, because it anticommutes with the ψ_0^μ. The choice of sign is conventional and immaterial except in the type II theories where $\Gamma = \gamma_{d+1} = \pm\overline{\Gamma}$ are the two inequivalent possibities.

If the projection $\Gamma = 1$ were made only on the matter sector, it would produce a theory with spacetime chirality (except for the $\Gamma = -\overline{\Gamma}$ type II theory). But the Γ projection acts on the combined matter ghost system, so there will only be spacetime chirality after projection if the ghost states do *not* come in chiral pairs, i.e., if the Witten index of the ghost Ramond sector is nonzero.

The ground states $|k\rangle$ of the NS sector are created from the vacuum by the superfields $e^{ik\cdot X}$, because they are superconformal fields which create momentum

k^μ and have weight $\frac{1}{2}k^2$:

$$\begin{aligned} DX^\mu(\mathbf{z}_1)\, e^{ik\cdot X}(\mathbf{z}_2) &\sim \langle DX^\mu(\mathbf{z}_1)\, X^\nu(\mathbf{z}_2)\rangle\, (ik_\nu)\, e^{ik\cdot X} \sim \theta_{12} z_{12}^{-1}\, (-ik^\mu)\, e^{ik\cdot X} \\ T^X(\mathbf{z}_1)\, e^{ik\cdot X}(\mathbf{z}_2) &\sim \tfrac{1}{2}k_\sigma k_\tau\, \langle D_1 X^\mu\, X^\sigma(\mathbf{z}_2)\rangle\, \langle \partial_1 X_\mu\, X^\tau(\mathbf{z}_2)\rangle\, e^{ik\cdot X}(\mathbf{z}_2) + \cdots \\ &\sim \left(\tfrac{1}{2}k^2\, \theta_{12} z_{12}^{-2} + \tfrac{1}{2} z_{12}^{-1} D_2 + \theta_{12} z_{12}^{-1}\, \partial_2\right) e^{ik\cdot X} . \end{aligned} \tag{4.7.3}$$

The spin fields $S_\alpha(z)$ are the conformal fields which create the ground states $|\alpha\rangle$ of ψ in the R sector:

$$|\alpha\rangle = S_\alpha(0)\ |0\rangle \qquad |k,\alpha\rangle = S_\alpha e^{ik\cdot X}(0)\ |0\rangle\ . \tag{4.7.4}$$

The Ramond supersymmetry of $|\alpha\rangle$, equation 4.7.2, implies that $S_\alpha(z)$ has weight $d/16 = 5/8$.

4.8 SO(10) current algebra

The spin field $S_\alpha(z)$ can be constructed from the $SO(10)$ chiral current algebra of the ψ system[6]:

$$\begin{aligned} j^{\mu\nu}(z) &= \psi^\mu \psi^\nu(z) \\ j^{\mu\nu}(z) j^{\sigma\tau}(w) &\sim (z-w)^{-2}(g^{\mu\tau} g^{\nu\sigma} - \mu{\leftrightarrow}\nu) + (z-w)^{-1} \\ &\qquad \times g^{\mu\sigma} j^{\nu\tau}(w)\, (1-\mu{\leftrightarrow}\nu)(1-\sigma{\leftrightarrow}\tau)\, . \end{aligned} \tag{4.8.1}$$

The current algebra determines the entire theory because the stress-energy tensor T_B^ψ is generated by the currents[18]:

$$j^{\mu\nu}(z)\, j_{\mu\nu}(w) \sim \frac{d-d^2}{(z-w)^2} + 2(d-1)(\partial\psi^\mu)\psi^\mu(w) \tag{4.8.2}$$

$$T_B^\psi(z) = \frac{-1/4}{d-1}\, j^{\mu\nu} j_{\mu\nu}(z)\, . \tag{4.8.3}$$

The Sugawara stress-energy tensor 4.8.3 is renormalized by subtracting the leading singularity in the operator product 4.8.2.

Adopt the spinor conventions

$$\begin{aligned} &[\gamma_\mu, \gamma_\nu]_+ = -g_{\mu\nu} && \epsilon_{\alpha\beta} = -\epsilon_{\beta\alpha} && \epsilon_{\alpha\beta}\,\epsilon^{\beta\gamma} = \delta_\alpha^\gamma \\ &u^\alpha = \epsilon^{\alpha\beta} u_\beta && u_\alpha = \epsilon_{\alpha\beta} u^\beta && \bar{A}_\beta^\alpha = \epsilon^{\alpha\gamma} A_\gamma^\delta \epsilon_{\delta\beta} \\ &\gamma_\mu^{\alpha\beta} = \gamma_\mu^{\beta\alpha} && \gamma_{[\mu}\gamma_{\nu]}{}^{\alpha\beta} = \gamma_{[\mu}\gamma_{\nu]}{}^{\beta\alpha} && \bar{\gamma}_\mu = -\gamma_\mu\,. \end{aligned} \tag{4.8.4}$$

The fermion field ψ^μ is completely determined from the current algebra by the operator product

$$j^{\mu\nu}(z)\,\psi^\sigma(w) \sim \frac{1}{z-w}(g^{\mu\sigma}\psi^\nu - g^{\nu\sigma}\psi^\mu)(w) \tag{4.8.5}$$

and S_α is determined by

$$j^{\mu\nu}(z)\,S_\alpha(w) \sim \frac{1}{z-w}\,\tfrac{1}{2}\,\gamma^{[\mu}\gamma^{\nu]}{}_\alpha^\beta\,S_\beta(w)\,, \tag{4.8.6}$$

in the sense that all of their operator products are determined by the currents.

The operator products (for $d = 10$) are (writing S_α for $S_\alpha(0)$, ψ^μ for $\psi^\mu(0)$):

$$\begin{aligned} \psi^\mu(z)\,\psi^\nu &\sim -g^{\mu\nu}\,(z-w)^{-1} \\ \psi^\mu(z)\,S_\alpha &\sim (z-w)^{-\frac{1}{2}}\gamma^\mu{}_\alpha^\beta\,S_\beta \\ \psi^\mu(z)\,S^\alpha &\sim (z-w)^{-1/2}(-\gamma^\mu{}_\beta^\alpha)\,S^\beta \\ \psi^\mu\psi^\nu(z)\,S_\alpha &\sim (z-w)^{-1}\tfrac{1}{2}\gamma^{[\mu}\gamma^{\nu]}{}_\alpha^\beta\,S_\beta \\ \psi^\mu\psi^\nu(z)\,S^\alpha &\sim (z-w)^{-1}(-\tfrac{1}{2})\gamma^{[\mu}\gamma^{\nu]}{}_\beta^\alpha\,S^\beta \\ S_\alpha(z)\,S_\beta &\sim z^{-5/4}\epsilon_{\alpha\beta} + z^{-3/4}\gamma^\mu_{\alpha\beta}\,\psi_\mu + z^{-1/4}\tfrac{1}{2}\gamma^\mu\gamma^\nu{}_{\alpha\beta}\,j_{\mu\nu} \\ S^\alpha(z)\,S^\beta &\sim z^{-5/4}(-\epsilon^{\alpha\beta}) + z^{-3/4}\gamma_\mu^{\alpha\beta}\psi^\mu + z^{-1/4}\tfrac{1}{2}\gamma_\mu\gamma_\nu{}^{\alpha\beta}\,j^{\mu\nu} \\ S^\alpha(z)\,S_\beta &\sim z^{-5/4}\delta_\beta^\alpha + z^{-3/4}\gamma_\mu{}_\alpha^\beta\psi^\mu + z^{-1/4}\tfrac{1}{2}\gamma_\mu\gamma_\nu{}_\beta^\alpha\,j^{\mu\nu} \\ S_\alpha(z)\,S^\beta &\sim z^{-5/4}(-\delta_\beta^\alpha) + z^{-3/4}\gamma_\mu{}_\alpha^\beta\psi^\mu + z^{-1/4}\tfrac{1}{2}\gamma_\mu\gamma_\nu{}_\alpha^\beta j^{\mu\nu}\,. \end{aligned} \tag{4.8.7}$$

The coefficients are given by the following arguments. The $\psi^\mu\,\psi^\nu$ operator product and the leading term in the $S_\alpha\,S_\beta$ operator product are fixed by $SO(10)$ invariance up to normalizations. The $\psi\ S$ operator product is obtained by requiring consistency of $j^{\mu\nu} S_\alpha$ with $\psi^\mu \psi^\nu S_\alpha$. The ψ contribution to $S\,S$ is found by evaluating $\langle\psi(z)\, S(w)\, S\rangle \propto w^{-1/2} z^{-3/4}(z-w)^{-1/2}$ in the two limits $z \to w$ and $w \to 0$. Similarly, the $j^{\mu\nu}$ contribution is determined by evaluating $\langle j\, S\, S\rangle$.

Note the fractional powers of z in the operator products 4.8.7, appropriate to the fractional dimension of S_α. Note also that $\Gamma\,|\alpha\rangle = \gamma_{11}\,|\alpha\rangle$ implies $\Gamma S_\alpha \Gamma^{-1} = (\gamma_{11} S)_\alpha$, which is inconsistent with the $S\,S \sim \gamma\psi$ operator product. Thus the spin fields of the matter sector do not by themselves form a local quantum field theory, and the chirality operator Γ acting in the matter sector alone is not an automorphism of the local algebra of spin fields. These difficulties are resolved by combining S_α with the spin fields of the superconformal ghosts.

The advantage of the current algebra approach is its manifest Lorentz invariance. The spin fields S_α can also be realized explicitly as ordinary vertex operators, that is, as exponentials of free chiral scalar fields[1-6]. The vertex operator construction is not manifestly Lorentz invariant, but it allows explicit calculation of correlation functions, on any surface. On the other hand, the current algebra is useful for obtaining the first few coefficients in operator product expansions. It can be used to find correlation functions[18,6], but not easily, except in the simplest situations.

It will be useful to know some subleading terms in the operator products when leading terms vanish. Assume that the indices in the following operator products are contracted with spinors u^α, v^β and a vector k_μ satisfying $\bar{u}\not{k}v = 0$, and use the shorthand $\gamma^\mu_{\alpha\beta} = 0$ for this situation. Then

$$\psi^\mu(z)\, S_\alpha \sim z^{1/2}\, \psi^\mu_{-1} S_\alpha \tag{4.8.8}$$

where $\psi^\mu_{-1} S_\alpha(z)$ is the conformal field corresponding to the state $\psi^\mu_{-1}\,|\alpha\rangle$. In general,

$$\psi^\mu \psi^\nu(z)\, S^\beta \sim z^{-1} \tfrac{1}{2} \gamma^{[\nu\beta\gamma} \gamma^{\mu]}_{\gamma\delta} S^\delta + \gamma^{[\nu\beta\gamma} \psi^{\mu]}_{-1} S_\gamma \tag{4.8.9}$$

Thus, when $\gamma^\mu_{\alpha\beta} = 0$, the following is a finite product,

$$\gamma_{\nu\alpha\beta} \psi^\mu \psi^\nu S^\beta = (1 - \tfrac{1}{2} d) \psi^\mu_{-1} S_\alpha\,, \tag{4.8.10}$$

and

$$\psi^\mu(z)\, S_\alpha \sim (z-w)^{\frac{1}{2}}(1-\tfrac{1}{2}d)^{-1}\gamma_{\nu\alpha\beta}\psi^\mu\psi^\nu S^\beta\,. \tag{4.8.11}$$

4.9 N=2 supersymmetry of the ghosts

As an aside, it might be interesting that the superconformal ghost system has an additional supersymmetry. Combined with the manifest $N=1$ superconformal invariance, this gives $O(2)$ extended superconformal symmetry. The fundamental $O(2)$ superconformal multiplet in two dimensions consists of the stress-energy tensor, two dimension 3/2 conformal fields, and the dimension 1 current of the $O(2)$ symmetry. Under an $N=1$ superconformal subalgebra these fields split into the $N=1$ super stress-energy tensor $T(\mathbf{z})$ and a dimension 1 superconformal field $J(\mathbf{z})$. The condition on $J(\mathbf{z})$ which gives the closure of the $N=2$ algebra is

$$J(\mathbf{z}_1)\,J(\mathbf{z}_2) \sim \tfrac{1}{2}\hat{c}\,z_{12}^{-1} + 2\theta_{12}\,z_{12}^{-1}\,T(\mathbf{z}_2)\,. \tag{4.9.1}$$

In the superconformal ghost system,

$$J = 2(DB)C + 3B(DC) \tag{4.9.2}$$

is a dimension 1 superconformal field which satisfies 4.9.1. Thus the ghost system has an $N=2$ superconformal symmetry.

5. First Order Free Fields

The component fields b, c and β, γ of the superconformal ghosts are special cases of free fields satisfying first order equations of motion. This section discusses the general case[8,11,19,2,4,5].

5.1 Fields, action, modes, two-point functions

Let $\mathbf{b}(z)$ and $\mathbf{c}(z)$ be conjugate conformal fields:

$$S = \frac{1}{\pi}\int d^2z\; \mathbf{b}\bar{\partial}\mathbf{c} \qquad \bar{\partial}\mathbf{b} = \bar{\partial}\mathbf{c} = 0$$
$$\text{weight}(\mathbf{b}) = \lambda \qquad \text{weight}(\mathbf{c}) = 1-\lambda\,. \tag{5.1.1}$$

The conformal ghosts b, c have $\lambda = 2$; their superpartners β, γ have $\lambda = 3/2$. The basic facts are

$$\epsilon = \begin{cases} +1 & \text{Fermi statistics} \\ -1 & \text{Bose statistics} \end{cases} \qquad \delta = \begin{cases} 0 & \text{NS sector} \\ \frac{1}{2} & \text{R sector} \end{cases}$$

$$\mathbf{c}(z)\,\mathbf{b}(w) \sim \frac{1}{z-w} \qquad \mathbf{b}(z)\,\mathbf{c}(w) \sim \frac{\epsilon}{z-w}$$

$$\mathbf{b}(z) = \sum_{n\in\delta-\lambda+\mathbf{Z}} z^{-n-\lambda}\,\mathbf{b}_n \qquad \mathbf{c}(z) = \sum_{n\in\delta+\lambda+\mathbf{Z}} z^{-n-(1-\lambda)}\,\mathbf{c}_n \tag{5.1.2}$$

$$\mathbf{c}_m\mathbf{b}_n + \epsilon\,\mathbf{c}_m\mathbf{b}_n = \delta_{m+n} \qquad \mathbf{b}_n^+ = \epsilon\mathbf{b}_{-n} \qquad \mathbf{c}_n^\dagger = \mathbf{c}_{-n}\,.$$

$$Q = \epsilon(1-2\lambda) \qquad \lambda = \tfrac{1}{2}(1-\epsilon Q)\,.$$

In the NS sector the fields are single valued on the plane; in the R sector they are double valued. Strictly speaking, the R sector should be present only for $\lambda \in \frac{1}{2}+\mathbf{Z}$. When $\lambda \in \mathbf{Z}$ the case $\delta = \frac{1}{2}$ is a *twisted* sector. In 5.1.2 the fields are operators on an indefinite metric Hilbert space.

5.2 The stress-energy tensor

The stress-energy tensor is determined by the weights of $\mathbf{b}$ and $\mathbf{c}$:

$$T_B^{\mathbf{bc}} = -\lambda\mathbf{b}\partial\mathbf{c} + (1-\lambda)\,\partial\mathbf{b}\,\mathbf{c} = \tfrac{1}{2}(\partial\mathbf{b}\mathbf{c} - \mathbf{b}\partial\mathbf{c}) + \tfrac{1}{2}\epsilon Q\partial(\mathbf{b}\mathbf{c})\,. \tag{5.2.1}$$

As usual, $T_B^{\mathbf{bc}}$ is renormalized by subtracting the singular part of the $\mathbf{b}$, $\mathbf{c}$ operator products. Double contractions give the conformal anomaly

$$\begin{gathered} T_B^{\mathbf{bc}}(z)\,T_B^{\mathbf{bc}}(w) \sim \tfrac{1}{2}c^{\mathbf{bc}}(z-w)^{-4} \\ c^{\mathbf{bc}} = -\epsilon(12\lambda^2 - 12\lambda + 2) = \epsilon(1-3Q^2)\,. \end{gathered} \tag{5.2.2}$$

The mode expansions are:

$$\begin{gathered} L_m^{\mathbf{bc}} = \textstyle\sum_k \left[k - (1-\lambda)m\right]\mathbf{b}_{m-k}\,\mathbf{c}_k = \sum_k \epsilon(k-\lambda m)\,\mathbf{c}_{m-k}\,\mathbf{b}_k \\ [L_m^{\mathbf{bc}}, \mathbf{b}_n] = (-(1-\lambda)m - n)\mathbf{b}_{m+n} \qquad [L_m^{\mathbf{bc}}, \mathbf{c}_n] = (-\lambda m - n)\mathbf{c}_{m+n} \end{gathered} \tag{5.2.3}$$

For the superconformal ghosts,

$$\begin{array}{lllll} b,\ c: & \epsilon = 1, & \lambda = 2, & Q = -3, & c^{bc} = -26 \\ \beta,\ \gamma: & \epsilon = -1, & \lambda = \frac{3}{2}, & Q = 2, & c^{\beta\gamma} = 11 \end{array} \tag{5.2.4}$$

5.3 The U(1)-current

The action 5.1.1 has a chiral $U(1)$ symmetry whose chiral current is

$$\mathbf{j}(z) = -\mathbf{bc} = \epsilon\,\mathbf{cb} = \sum_n z^{-n-1}\mathbf{j}_n \qquad \mathbf{j}_n = \sum_k \epsilon\,\mathbf{c}_{n-k}\mathbf{b}_k$$

$$\begin{aligned} \mathbf{j}(z)\,\mathbf{b}(w) &\sim (-1)\,(z-w)^{-1}\,\mathbf{b}(w) & \mathbf{j}(z)\,\mathbf{c}(w) &\sim (+1)\,(z-w)^{-1}\,\mathbf{c}(w) \\ [\mathbf{j}_m, \mathbf{b}_n] &= -\mathbf{b}_{m+n} & [\mathbf{j}_m, \mathbf{c}_n] &= +\mathbf{c}_{m+n}\,. \end{aligned}$$

$$\mathbf{j}_0 = \text{charge operator} \qquad \text{charge}(\mathbf{b}) = -1 \qquad \text{charge}(\mathbf{c}) = +1$$

$$\mathbf{j}(z)\;\mathbf{j}(w) \sim \epsilon\,(z-w)^{-2} \qquad [\mathbf{j}_m, \mathbf{j}_n] = \epsilon\, m\,\delta_{m+n}\,. \tag{5.3.1}$$

The algebra of the chiral current and the stress-energy tensor is anomalous:

$$\begin{aligned} T^{\mathbf{bc}}(z)\,\mathbf{j}(w) &\sim Q\,(z-w)^{-3} + (z-w)^{-2}\,\mathbf{j}(z) \\ [L^{\mathbf{bc}}_m, \mathbf{j}_n] &= -n\mathbf{j}_{m+n} + \tfrac{1}{2}Q\,m(m+1)\delta_{m+n} \end{aligned} \tag{5.3.2}$$

so $\mathbf{j}(z)$ is scale and translation covariant ($m = 0, -1$) but *not* conformally covariant. The anomaly coefficient Q can be interpreted as a background charge on the sphere (see also equations 5.4.2 and5.6.3 below):

$$\mathbf{j}_0^\dagger = -[L^{\mathbf{bc}}_{-1}, \mathbf{j}_1]^\dagger = -[L^{\mathbf{bc}}_1, \mathbf{j}_{-1}] = -\mathbf{j}_0 - Q\,. \tag{5.3.3}$$

There is no normal ordering ambiguity in $\mathbf{j}_m$ for $m \neq 0$, therefore

$$\mathbf{j}_m^\dagger = -\mathbf{j}_{-m} - Q\,\delta_{m,0}\,. \tag{5.3.4}$$

5.4 The Fermi/Bose sea

A Fermi/Bose sea is a state $|q\rangle$ which splits the normal modes:

$$\begin{aligned} \mathbf{b}_n\,|q\rangle &= 0 \qquad n > \epsilon q - \lambda \\ \mathbf{c}_n\,|q\rangle &= 0 \qquad n \geq -\epsilon q + \lambda \end{aligned} \tag{5.4.1}$$

where $q \in \mathbf{Z}$ for the NS sector and $q \in \frac{1}{2} + \mathbf{Z}$ for the R sector. By 5.1.2, the only nonzero inner products are

$$\langle q - Q\,|q\rangle = 1\,. \tag{5.4.2}$$

The two point function in the sea $|q\rangle$ is

$$\langle \mathbf{c}(z)\,\mathbf{b}(w)\rangle_q = \sum_{m,n} z^{-m-(1-\lambda)} w^{-n-\lambda} \langle q-Q|\,\mathbf{c}_m \mathbf{b}_n\,|q\rangle = \left(\frac{z}{w}\right)^{\epsilon q} \frac{1}{z-w} \tag{5.4.3}$$

from which,

$$\begin{aligned}
\langle \mathbf{j}(z)\,\mathbf{j}(w)\rangle_q &= \epsilon(z-w)^{-2}\\
\left\langle T^{\mathbf{bc}}(z)\,\mathbf{j}(w)\right\rangle_q &= Q(z-w)^{-3} + (z-w)^{-2}\frac{q}{z}\\
\left\langle T^{\mathbf{bc}}(z)\,T^{\mathbf{bc}}(w)\right\rangle_q &= (z-w)^{-4}\tfrac{1}{2}c^{\mathbf{bc}} + (z-w)^{-2}\epsilon q(Q+q)\frac{1}{zw}
\end{aligned} \tag{5.4.4}$$

$$\begin{aligned}
L^{\mathbf{bc}}_{+n}\,|q\rangle &= 0 \qquad \langle \mathbf{j}(z)\rangle_q = qz^{-1} \qquad \left\langle T^{\mathbf{bc}}(z)\right\rangle_q = \tfrac{1}{2}\epsilon q(Q+q)z^{-2}\\
\mathbf{j}_{+n}\,|q\rangle &= 0 \qquad \mathbf{j}_0\,|q\rangle = q\,|q\rangle \qquad L^{\mathbf{bc}}_0\,|q\rangle = \tfrac{1}{2}\epsilon q(Q+q)\,|q\rangle\,.
\end{aligned} \tag{5.4.5}$$

Thus the Bose/Fermi sea $|q\rangle$ has charge q, and it is apparent that 5.4.2 expresses the presence of a background charge Q. An SL_2 invariant state has $L_0 = 0$, so the only candidates are $|0\rangle$ and $|-Q\rangle$. Only a neutral state can be translation invariant, so $|0\rangle$ is the unique SL_2 invariant state.

Each Fermi sea can be obtained from any other by applying a monomial in the fields **b**, **c**. But this is not true for the Bose seas. The Bose seas $|q\rangle$ generate inequivalent representations of the **b**, **c** algebra.

5.5 The U(1) stress-energy tensor

Define the $U(1)$ stress-energy tensor by

$$T^{\mathbf{j}}(z) = \tfrac{1}{2}\epsilon\left(\mathbf{j}(z)^2 - Q\partial\mathbf{j}\right), \tag{5.5.1}$$

subtracting the singularity in the **j**, **j** operator product 5.3.1. The linear term in 5.5.1 is designed so that $T^{\mathbf{j}}_B$ and $T^{\mathbf{bc}}_B$ will have the same commutation relations with **j**:

$$\begin{aligned}
T^{\mathbf{j}}(z)\,\mathbf{j}(w) &\sim (z-w)^{-3}Q + (z-w)^{-2}\mathbf{j}(z)\\
T^{\mathbf{j}}(z)\,T^{\mathbf{j}}(w) &\sim (z-w)^{-4}\tfrac{1}{2}c^{\mathbf{j}} + (z-w)^{-2}2T^{\mathbf{j}}(w) + (z-w)^{-1}\partial_w T^{\mathbf{j}}
\end{aligned} \tag{5.5.2}$$

$$c^{\mathbf{j}} = 1 - 3\epsilon Q^2 = \begin{cases} c^{\mathbf{bc}} & \text{Fermi statistics} \\ c^{\mathbf{bc}} + 2 & \text{Bose statistics.} \end{cases} \tag{5.5.3}$$

In the Fermi case the $U(1)$ current algebra gives the complete dynamics ($T_B^{\mathbf{j}} = T_B^{\mathbf{bc}}$), but in the Bose case it does not. For Bose systems, define[20]

$$T^{[-2]}(z) = T^{\mathbf{bc}}(z) - T^{\mathbf{j}}(z) \,. \tag{5.5.4}$$

By 5.3.2 and 5.5.2, $T^{[-2]}(z)$ commutes with $\mathbf{j}(w)$ and therefore with $T^{\mathbf{j}}(w)$, and $T^{[-2]}(z)$ generates a conformal algebra with central charge $c^{[-2]} = -2$.

5.6 Bosonization

Use the $U(1)$ current to define a chiral scalar field $\phi(z)$:

$$\mathbf{j}(z) = \epsilon\partial\phi(z) \qquad \phi(z) = \epsilon\int^{z} dw\,\mathbf{j}(w)$$

$$\mathbf{j}(z)\,\phi(w) \sim (z-w)^{-1} \qquad \phi(z)\,\phi(w) \sim \epsilon\,\ln(z-w)$$

$$\mathbf{j}(z)\,e^{q\phi(w)} \sim q(z-w)^{-1}\,e^{q\phi}(w) \qquad [\mathbf{j}_0,\, e^{q\phi(w)}] = q\,e^{q\phi(w)}$$

$$T^{\mathbf{j}}(z)\,e^{q\phi(w)} \sim \left[\tfrac{1}{2}\epsilon q(Q+q)\,(z-w)^{-2} + (z-w)^{-1}\,\partial_w\right] e^{q\phi(w)} \tag{5.6.1}$$

$$\text{charge}(e^{q\phi}) = q \qquad \text{weight}(e^{q\phi}) = \tfrac{1}{2}\epsilon q(Q+q)\,.$$

$$e^{q\phi(z)}\quad e^{q'\phi(w)} \sim (z-w)^{\epsilon q q'}\,e^{q\phi(z)+q'\phi(w)}$$

$$e^{q\phi(0)}\,|0\rangle = |q\rangle \tag{5.6.2}$$

The soliton operator $e^{q\phi}$ shifts the Fermi/Bose sea level by q units of charge. Equation 5.4.2 gives

$$\langle 0|\,e^{-Q\phi(z)}\,|0\rangle = 1 \tag{5.6.3}$$

which again shows the need for charge $-Q$ to absorb the background charge Q on the sphere.

The $U(1)$ current can be fermionized in terms of fundamental solitons $e^{\pm\phi}$. In the Fermi case the fundamental solitons are exactly the original $\mathbf{b}$, $\mathbf{c}$ fields

$$\mathbf{b}(z) = e^{-\phi(z)} \qquad \mathbf{c}(z) = e^{\phi(z)} \qquad \text{(Fermi statistics)}\,. \tag{5.6.4}$$

The $e^{\pm\phi} e^{q\phi}$ operator products, given by equation 5.6.1, can be compared with 5.4.1 to confirm 5.6.2 for the Fermi systems. In the Bose case the $U(1)$ solitons cannot give the original fields, because of the missing central charge -2 and because the soliton fields $e^{\pm\phi(z)}$ are fermionic, while **b**, **c** are bosonic.

5.7 The c = -2 system

Define

$$\eta = \partial c\, e^{-\phi} \qquad \partial\xi = \partial b\, e^{+\phi}. \tag{5.7.1}$$

$\eta(z)$ and $\xi(z)$ are conjugate free fermion fields of conformal weights 1 and 0 respectively:

$$\begin{gathered} \eta(z)\,\xi(w) \sim (z-w)^{-1} \sim \xi(z)\,\eta(w) \\ \eta(z) = \textstyle\sum_n z^{-n-1}\eta_n \qquad \xi(z) = \sum_n z^{-n-1}\xi_n \qquad [\eta_m,\, \xi_n]_+ = \delta_{m+n}\,. \end{gathered} \tag{5.7.2}$$

η and ξ commute with ϕ and have

$$\lambda^{\eta\xi} = 1 \qquad Q^{\eta\xi} = -1 \qquad c^{\eta\xi} = -2\,. \tag{5.7.3}$$

Thus every first order Bose system consists of its own $U(1)$ current algebra along with the $\lambda = 1$ first order Fermi system η, ξ. The Bose fields can be written

$$\mathbf{b} = e^{-\phi}\partial\xi \qquad \mathbf{c} = e^{\phi}\eta \qquad \text{(Bose statistics)}\,. \tag{5.7.4}$$

and again the operator products of the exponentials confirm 5.6.2.

The η, ξ system contains its own chiral $U(1)$ current, which gives a second chiral scalar $\chi(z)$:

$$\partial\chi = \eta\xi \qquad \chi(z)\,\chi(w) \sim \ln(z-w) \qquad \eta = e^{-\chi} \qquad \xi = e^{\chi}\,. \tag{5.7.5}$$

The zero-mode algebra $[\eta_0,\, \xi_0] = 1$ forces the ground state of the η, ξ system to be twofold degenerate. The two ground states are the SL_2 invariant $|0\rangle$ and its hermitian conjugate $|-Q\rangle_{\eta\xi}$. Since $\xi = e^{\chi}$,

$$\langle 0|\, \xi(z)\, |0\rangle_{\eta\xi} = \langle 0|\, \xi_0\, |0\rangle_{\eta\xi} = 1\,. \tag{5.7.6}$$

But in the construction 5.7.4 for the original fields **b**, **c** only $\rho = \partial\xi$ appears. Therefore the $c = -2$ system is really η, ρ and not η, ξ. In the η, ρ system, it is consistent to fix $\eta_0 = 0$. Then the η, ρ system has a unique ground state:

$$\langle 0| \cdots |0\rangle_{\eta\rho} = \langle 0|\, \xi_0 \cdots |0\rangle_{\eta\xi} \, . \tag{5.7.7}$$

5.8 The chiral scalar and Riemann-Roch

The anomalous operator product of the stress-energy tensor with the chiral $U(1)$ current, equation 5.3.2, is equivalent to the anomalous conservation law for the chiral current:

$$\bar\partial \mathbf{j}(z) = \epsilon\, \bar\partial \partial \phi(z) = \tfrac{1}{8} Q R^{(2)} \sqrt{g} \tag{5.8.1}$$

where $R^{(2)}\sqrt{g}$ is the two dimensional scalar curvature density. The operator product 5.3.2 is derived from the conservation law 5.8.1 by differentiating with respect to the two-metric to get a Ward identity, and using the Ward identity to determine the singular part of the operator product expansion[7].

The anomalous conservation law is the equation of motion for ϕ, derived from the action

$$S(\phi) = \frac{1}{2\pi} \int d^2 z \, \left(-\epsilon \bar\partial \phi \partial \phi - \tfrac{1}{4} Q \sqrt{g} R^{(2)} \phi \right) . \tag{5.8.2}$$

Note that the action is well-behaved for Fermi systems ($\epsilon = 1$) if $\phi \to i\phi$. The exponentials then take the familiar form $e^{iq\phi}$.

To find the background charge on an arbitrary Riemann surface, note that the action 5.8.2 gives expectation values $\left\langle e^{q\phi} \right\rangle = 0$ unless $q + q^{back} = 0$ with the background charge given by

$$q^{back} = Q \frac{1}{8\pi} \int d^z \, \sqrt{g}\, R^{(2)} = Q(1-g) \tag{5.8.3}$$

where g is the genus of the Riemann surface, and $2(1-g)$ is the Euler number given by the Gauss-Bonnet formula

$$\frac{1}{4\pi} \int d^z \, \sqrt{g}\, R^{(2)} = 2(1-g) \, . \tag{5.8.4}$$

The sphere has $g = 0$, which gives the background charge $q^{back} = Q$, as already seen in the operator representation. The background charge is related to the number of solutions of the equations of motion 5.1.1:

$$\# \text{ of } \mathbf{b} \text{ solutions} - \# \text{ of } \mathbf{c} \text{ solutions} = \epsilon q^{back} = (1-2\lambda)(g-1) \tag{5.8.5}$$

which is the Riemann-Roch formula.

6. The Superconformal Ghosts

6.1 Bosonization

Specializing the constructions of the previous section to the superconformal ghosts,

$$\begin{aligned} &\beta = e^{-\phi}\partial\xi = e^{-\phi+\chi}\partial\chi && \gamma = e^{\phi}\eta = e^{\phi-\chi} \\ &\xi = e^{\chi} && \eta = e^{-\chi} \\ &b = e^{-\sigma} && c = e^{\sigma}\,. \end{aligned} \tag{6.1.1}$$

The properties of the chiral scalars are

$$\begin{aligned} &\phi(z)\,\phi(w) = -\ln(z-w) && Q^{\phi} = 2 && c^{\phi} = 13 && \mathrm{wt}(e^{q\phi}) = -\tfrac{1}{2}q(q+2) \\ &\chi(z)\,\chi(w) = +\ln(z-w) && Q^{\chi} = -1 && c^{\phi} = {-2} && \mathrm{wt}(e^{q\chi}) = +\tfrac{1}{2}q(q-1) \\ &\sigma(z)\,\sigma(w) = +\ln(z-w) && Q^{\sigma} = -3 && c^{\sigma} = -26 && \mathrm{wt}(e^{q\sigma}) = +\tfrac{1}{2}q(q-3) \end{aligned} \tag{6.1.2}$$

The total ghost charge is ϕ-charge plus σ-charge; the ghost charge operator is $j_0^{\phi} + j_0^{\sigma}$. The inequivalent representations of the β, γ algebra are indexed by the $\phi + \chi$-charge, since β and γ both commute with $j_0^{\phi} + j_0^{\chi}$. One way of picturing this extra quantum number is to fermionize the current $\beta\gamma$, giving *two* charged solitons, $e^{\pm\phi}$ and $e^{\pm\chi}$. The $e^{\pm\chi}$ solitons are free, but the $e^{\pm\phi}$ solitons cannot be free, since their dimensions do not add up to 1.

6.2 Spin fields

The integer weight ghost fields b, c, η, $\partial\xi$ are not affected by the spin fields, so the ghost contribution to the spin fields comes only from ϕ. The exponentials $e^{q\phi}$ are components of superfields (NS operators) for $q \in \mathbf{Z}$, and spin fields (R operators) for $q \in \frac{1}{2} + \mathbf{Z}$. Note that these spin fields, like the S_α of the matter sector, have fractional weights and do not by themselves form a local algebra of fields. Only the combined ϕ, ψ^μ system will have a local algebra of spin fields.

The integer weight ghost fields all commute with the chirality operator Γ. The identity has even chirality, as does the corresponding state, the SL_2 invariant vacuum $|0\rangle_\phi$. The solitons $e^{\pm\phi}$ are odd, so

$$\Gamma\,|q\rangle_\phi = (-1)^q\,|q\rangle_\phi \qquad q \in \mathbf{Z}\,. \tag{6.2.1}$$

For the Ramond states, up to a conventional choice of overall sign,

$$\Gamma\,|q\rangle_\phi = (-1)^{q+1/2}\,|q\rangle_\phi \qquad q \in \tfrac{1}{2} + \mathbf{Z}\,. \tag{6.2.2}$$

In terms of the spin fields,

$$\Gamma e^{q\phi}\Gamma^{-1} = e^{q\phi}\begin{cases}(-1)^q & q \in \mathbf{Z}\\ (-1)^{q+1/2} & q \in \frac{1}{2}+\mathbf{Z}\end{cases} \tag{6.2.3}$$

which is obviously inconsistent with the operator products of the exponentials, in the same way that $\Gamma S_\alpha \Gamma^{-1} = (\gamma_{11} S)_\alpha$ is inconsistent with the operator products of S_α. Only in the combined ϕ, ψ^μ system can the chirality operator be extended to the spin fields, with the $\Gamma = 1$ projection giving a local field theory.

The spin field $e^{\phi/2}$, of weight $-5/8 = \hat{c}^{BC}/16$, corresponds to the unique state $|1/2\rangle_\phi \otimes |0\rangle_{\sigma\chi}$ of unbroken Ramond supersymmetry in the ghost system. The Witten index is thus $+1$ in the Ramond sector of the ghosts. This is responsible for spacetime chirality in the covariant formulation of the fermionic string theory. If the ghost states were all paired in chirality, then the $\Gamma = 1$ projection would produce only states of paired spacetime chirality.

6.3 The BRST current

The BRST supercurrent is

$$J_{BRST} = DC(C\,DB - \tfrac{3}{4}\,DC\,B) \tag{6.3.1}$$

where normal ordering is done with respect to the $\widehat{SL}_2$ invariant state, in which $\langle B(\mathbf{z}_1)\,C(\mathbf{z}_2)\rangle = \theta_{12}/z_{12}$. In Feynman diagrams for correlation functions on the sphere which involve J_{BRST}, no self-contractions are included. The BRST charge is

$$\epsilon\, Q_{BRST} = \frac{1}{2\pi i}\oint dz\, d\theta\ \epsilon\, J_{BRST}(\mathbf{z}) \qquad Q^{\dagger}_{BRST} = Q_{BRST}\,. \tag{6.3.2}$$

The BRST current is completely specified by three conditions:

1. the BRST transformation laws of the superconformal matter fields of weight h:
$$[\epsilon Q_{BRST},\, \Phi_{matter}] = [\,\epsilon C + \tfrac{1}{2}D(\epsilon C)D + h\partial(\epsilon C)]\,\Phi_{matter}\,, \tag{6.3.3}$$

2. the transformation laws of the ghost fields:
$$\begin{aligned}[\epsilon Q_{BRST},\, C] &= \epsilon\left(C\partial C - \tfrac{1}{4}\,DC\,DC\right)\\ [\epsilon Q_{BRST},\, B] &= -\epsilon T\end{aligned} \tag{6.3.4}$$

3. the requirement that J_{BRST} be an anomaly free supercurrent, i.e. a superconformal field of weight 1/2:

$$T(\mathbf{z}_1)\,J_{BRST}(\mathbf{z}_2) \sim \tfrac{1}{2}\,\theta_{12}z_{12}^{-2}\,J_{BRST}(\mathbf{z}_2) + \tfrac{1}{2}z_{12}^{-1}D_2J_{BRST} + \theta_{12}z_{12}^{-1}\,\partial_2 J_{BRST}\,. \tag{6.3.5}$$

The last condition ensures that J_{BRST} is analytic (conserved) on any world-surface, and that its contour integrals are conformal invariants. From properties 1–3 it follows that

$$Q^2_{BRST} = 0\,, \tag{6.3.6}$$

because, by the BRST transformation laws, Q^2_{BRST} commutes with all the matter fields and C, and $[Q^2_{BRST},\, B] = -[T,\, Q_{BRST}]_+ = 0$ because J_{BRST} is a conformal supercurrent. Therefore Q^2_{BRST} commutes with all the fields and must be a multiple of the identity. But it has total ghost charge +2 while the identity is neutral. Therefore $Q^2_{BRST} = 0$.

The procedure for finding (or verifying) equation 6.3.1 for J_{BRST} is to write the most general superfield of total ghost charge +1, and then to fix its coefficients by evaluating the operator products needed to verify equations 6.3.3-6.3.5, using the two-point function θ_{12}/z_{12}. Only if $d = 10$ is it possible to satisfy all of the defining properties simultaneously. Rewriting Q_{BRST} in the form

$$\epsilon Q_{BRST} = \frac{1}{2\pi i}\oint dz\, d\theta\, \epsilon \left(CT^{[X]} + \tfrac{1}{2}CT^{[BC]}\right) \tag{6.3.7}$$

makes it easy to derive 6.3.3. From 6.3.3 it is obvious that any $h = 1/2$ superconformal field in the matter sector is a BRST invariant vertex operator. These vertex operators are enough to give the complete S-matrix of the spacetime bosons (the NS sector). It is also useful to rewrite Q_{BRST} again:

$$\begin{aligned}
Q_{BRST} &= Q^{(0)}_{BRST} + Q^{(1)}_{BRST} + Q^{(2)}_{BRST} \\
Q^{(0)}_{BRST} &= \frac{1}{2\pi i}\oint dz\, (cT_B^{[X\beta\gamma]} - c\partial c b) \\
Q^{(1)}_{BRST} &= \frac{1}{2\pi i}\oint dz\, \tfrac{1}{2}\gamma\psi_\mu \partial x^m u \quad = \frac{1}{2\pi i}\oint dz\, \tfrac{1}{2}e^{\phi-\chi}\psi_\mu\partial x^\mu \\
Q^{(2)}_{BRST} &= \frac{1}{2\pi i}\oint dz\, \tfrac{1}{4}\gamma^2 b \quad = \frac{1}{2\pi i}\oint dz\, \tfrac{1}{4}e^{2\phi-2\chi-\sigma}
\end{aligned} \tag{6.3.8}$$

6.4 *BRST* invariant expectation values

The vacuum $|0\rangle$ is the SL_2 invariant state (for all of the fields). All charge operators annihilate the vacuum. The vacuum expectation values

$$\langle \cdots \rangle_Q = \langle 0| \cdots |0\rangle \tag{6.4.1}$$

are the correlation functions of the conformally invariant quantum field theory on the sphere (or plane or cylinder). These expectation values vanish unless they contain ghost operators which exactly soak up the ghost background charges. In particular, $\langle 1 \rangle_Q = 0$. The disadvantage of these expectation values is the background b, c charge. It requires that three of the vertex operators contain a factor of c. In tree amplitudes this is not a problem of principle, since the BRST quantization treats all operators of the matter – ghost system equally. But it is an inconvenience, because the asymmetry of the distribution of the b, c charges among the vertices obscures the duality of the scattering amplitudes, and because it would be attractive to have the full S-matrix entirely in terms of the x, ψ, ϕ system. In loops the problem is serious, since there are no vertex operators with the negative b, c charge needed to neutralize the background.

Let

$$\langle -Q| = \langle 0|\, e^{3\sigma-2\phi}(\infty) \qquad \langle -Q \mid 0\rangle = 1 \tag{6.4.2}$$

be the state conjugate to the vacuum, in which the ghost background charges have been neutralized. Note that $e^{3\sigma-2\phi}$ is a conformal field of weight 0. The vacuum and its hermitian conjugate are both BRST invariant

$$Q_{BRST}\,|0\rangle = 0 \qquad \langle -Q|\, Q_{BRST} = 0\,. \tag{6.4.3}$$

The vacuum is invariant because the conformal generators commute with Q_{BRST} and the SL_2 invariant state is unique. To show that its conjugate is invariant, first show directly from

$$\begin{aligned} b_n|0\rangle &= 0 \;\; n \geq -1 \qquad & c_n|0\rangle &= 0 \;\; n \geq 2 \\ \beta_n|0\rangle &= 0 \;\; n \geq -1/2 \qquad & \gamma_n|0\rangle &= 0 \;\; n \geq 3/2 \end{aligned} \tag{6.4.4}$$

that

$$Q^0_{BRST}\; e^{3\sigma(0)}\,|0\rangle = 0\,. \tag{6.4.5}$$

Then use version 6.3.8 for Q_{BRST}, and the standard operator products of exponentials.

Both the vacuum expectation value $\langle\cdots\rangle_Q$ and the expectation value

$$\langle\cdots\rangle_0 = \langle -Q|\cdots|0\rangle = \langle e^{3\sigma-2\phi}(\infty)\cdots\rangle_Q \tag{6.4.6}$$

are BRST invariant. The advantage of $\langle\cdots\rangle_0$ is charge neutrality; $\langle 1\rangle_0 = 1$. On higher genus Riemann surfaces, there is a manifestly BRST invariant expectation value $\langle\cdots\rangle_{Q(1-g)}$ with background charge $Q(1-g)$. For calculating loop amplitudes, the problem is to screen the background charge to get a BRST invariant, neutral expectation value on an arbitrary Riemann surface[12].

The correlation functions on the sphere of exponentials of the chiral scalars ϕ, χ, σ are calculated using two-point functions in the simplest possible form, $\pm \ln(z-w)$, ommitting self-contractions exactly as if there were no background charge. The only effect of the background charge is to determine which expectation values of exponentials are nonzero, namely those which neutralize the background charges.

Note that these *BRST*-invariant expectation values are for the small algebra of η, $\rho = \partial\xi$, ***not*** for the large algebra of η, ξ which includes the ξ_0 zero mode. In the large algebra, the neutralizer of the background charge is $e^{3\sigma-2\phi+\chi}$, but

$$[Q_{BRST}, e^{3\sigma-2\phi+\chi}] \neq 0\,. \tag{6.4.7}$$

This is a key point in the construction of the fermion vertex.

7. The Fermion Vertex and Spacetime Supersymmetry

The object is to construct BRS invariant vertex operators for spacetime fermions, and to construct a two dimensional chiral current for spacetime supersymmetry. The vertex operators must be spin fields in the matter ghost system so they should combine the spin fields S_α of the ψ^μ system and the spin fields $e^{\pm\phi/2}$ of the β, γ system. The fermion vertex operator should be a fermion field on the world surface in order that the fermion amplitudes have the antisymmetry properties appropriate to spacetime Fermi statistics.

It is enough to construct the vertex for massless fermions, since the scattering amplitudes of all the other states appear as residues of the massless fermion amplitudes at poles in the intermediate momenta. In the language of two dimensional field theory, the vertex operators for the massless fermions generate through their operator products the algebra of vertex operators for all physical states.

7.1 $V_{-1/2}$

It would be simplest if the fermion vertex did not couple to the ordinary conformal ghosts b, c. The fermion vertex operator must then anticommute (up to a total derivative) with each of $Q^{(0,1,2)}_{BRST}$ because each has a different b, c charge. The vertex operator should be an ordinary conformal field of weight 1 in order to anticommute with $Q^{(0)}_{BRST}$.

The simplest candidate for the matter part of the massless fermion vertex is $u^\alpha S_\alpha e^{ik\cdot x}$, which has dimension $5/8 + k^2/2 = 5/8$. The ghost sector must supply the missing 3/8 weight:

$$V_{-1/2} = u^\alpha e^{-\phi/2} S_\alpha e^{ik\cdot x} . \tag{7.1.1}$$

$V_{-1/2}$ is an ordinary conformal field of weight 1 if $k^2 = 0$. For invariance under $Q^{(1)}_{BRST}$ u will have to satisfy the massless Dirac equation $\not{k} u = 0$.

The $\Gamma = 1$ projection requires that u^α be left handed. Henceforth the convention will be that S_α is left handed, i.e., $\Gamma S_\alpha \Gamma^{-1} = S_\alpha$, and S^α is right handed. After the chiral projection, $V_{-1/2}(z)$ becomes a local fermionic field. It is fermionic because

$$V_{-1/2}(z)\, V_{-1/2}(w) \sim (z-w)^{-1} u^\alpha u^\beta \gamma^\mu_{\alpha\beta} e^{-\phi} \psi_\mu e^{2ik\cdot x}(w) , \tag{7.1.2}$$

which is odd under $z \leftrightarrow w$.

Because $V_{-1/2}$ is a conformal field of weight 1,

$$[Q^{(0)}_{BRST}, V_{-1/2}]_+ = \partial(c V_{-1/2}) . \tag{7.1.3}$$

The spacetime Dirac equation $\not{k} u = 0$ implies $[Q^{(1)}_{BRST}, V_{-1/2}(z)]_+ = 0$ because

$$e^{\phi-\chi} \psi_\mu \partial_z x^\mu\, V_{-1/2}(w) \sim (z-w)^{-1} (-i\not{k} u)_\alpha e^{\frac{1}{2}\phi - \chi} S^\alpha e^{ik\cdot x} \sim (z-w)^0 . \tag{7.1.4}$$

The last piece of the BRST invariance, $[Q^{(2)}_{BRST}, V_{-1/2}(z)]_+ = 0$, follows from the nonsingularity of the operator product expansion

$$e^{2\phi-2\chi} b(z)\, V_{-1/2}(w) \sim (z-w)^{+1} e^{\frac{3}{2}\phi - 2\chi} b u^\alpha S_\alpha e^{ik\cdot x}(w) . \tag{7.1.5}$$

Combining the three pieces gives BRST invariance of $V_{-1/2}$:

$$[Q_{BRST}, V_{-1/2}]_+ = \partial(c V_{-1/2}) . \tag{7.1.6}$$

7.2 $V_{1/2}$

$V_{-1/2}$ cannot be the entire fermion vertex operator because it has nonzero ϕ charge. The correlation functions $\langle V_{-1/2}(z)\cdots\rangle_0$ on the sphere all vanish by charge conservation. The correlation functions $\langle V_{-1/2}(z)\cdots\rangle_Q$ vanish except for the four point function. This difficulty can be avoided if there is a second fermion vertex, $V_{1/2}(z)$, having the opposite ϕ charge. $e^{\phi/2}$ has dimension $-5/8$ and odd chirality, so the spin field with even chirality, $e^{\phi/2}S^\alpha$ has dimension 0 and is righthanded in spacetime. To get even chirality, weight 1 and lefthandedness in spacetime, write a vertex of the form $e^{\frac{1}{2}\phi}u^\alpha\gamma^\mu_{\alpha\beta}S^\beta\partial x_\mu e^{ik\cdot x}$. The question now becomes BRST invariance.

A BRST invariant vertex operator $V_{1/2}$ can be constructed from $V_{-1/2}$ using the extension of the matter-ghost system which contains the field $\xi(z)$. Recall that the ghost system contains $\partial\xi$, but not ξ itself. Thus $\xi V_{-1/2}$ is not in the matter-ghost system, but $[Q_{BRST}, \xi V_{-1/2}]$ *is*, because the commutation with Q_{BRST} can absorb the zero mode of ξ. Since $Q^2_{BRST} = 0$, the commutator is automatically BRST invariant. Normally, commutation with Q_{BRST} gives vertex operators for BRST-exact states, which are null and decouple from physical states. But here the algebra of fields has been expanded so that all BRST invariant states are BRST-exact in the large algebra. Thus any BRST-closed state in the small algebra can be represented as a commutator with Q_{BRST} in the large algebra.

To be precise, define $V_{1/2}$ by

$$\begin{aligned} V_{1/2} &= 2[Q_{BRST}, \xi V_{-1/2}] - \partial(2c\xi V_{-1/2}) \\ &= 2[Q^{(1)}_{BRST}, V_{-1/2}] + \tfrac{1}{2}be^{3\phi/2-\chi}u^\alpha S_\alpha e^{ik\cdot x} \,. \end{aligned} \tag{7.2.1}$$

The total derivative is subtracted because it contains ξ, whereas $V_{1/2}$ should be in the small algebra. This modification does not affect BRST invariance because the BRST commutator is still a total derivative:

$$[Q_{BRST}, V_{1/2}]_+ = -\partial[Q_{BRST}, 2c\xi V_{-1/2}]_+ \,. \tag{7.2.2}$$

The term in 7.2.1 containing $b(z)$ will never contribute to correlation functions because neither $V_{-1/2}$ nor $V_{1/2}$ contains $c(z)$. So $V_{1/2}$ might as well be defined as

$2[Q^{(1)}_{BRST}, V_{-1/2}]$, which can be calculated using the operator product 4.8.11:

$$V_{1/2} = e^{\phi/2} u^\alpha \gamma^\mu_{\alpha\beta} (\partial x_\mu + \tfrac{i}{4} k\cdot\psi\psi_\mu) S^\beta e^{ik\cdot x} . \tag{7.2.3}$$

7.3 Scattering amplitudes

The two fermion vertex operators, $V_{-1/2}$ and $V_{1/2}$, give tree level fermion scattering amplitudes by formulas of the form

$$A(1,\ldots,N) = \int dz_1 \cdots dz_N \langle V_{q_1} \cdots V_{q_N} \rangle_0 \, , \tag{7.3.1}$$

but it must be shown that these formulas do not depend on the choice of ϕ charges $q_i = \pm\frac{1}{2}$, $\sum q_i = 0$. This will be done in section 7.4, by showing that the expectation value in equation 7.3.1 is invariant under rearrangements of the q_i.

First note that, given this rearrangement lemma, the fermion amplitudes factorize, as they should, on the Neveu-Schwarz amplitudes for spacetime bosons:

$$\begin{aligned} V_{-1/2}(z_1)\, V_{1/2}(z_2) &\sim -2[Q^{(1)}_{BRST}, V_{-1/2}(z_1)\, \xi V_{1/2}(z_2)] \\ &\sim 2[Q^{(1)}_{BRST}, (z_1 - z_2)^{-1+k_1\cdot k_2} u_1^\alpha u_2^\beta \gamma^\mu_{\alpha\beta} \xi e^{-\phi} \psi_\mu e^{i(k_1+k_2)\cdot x}(z_2)] \\ &\sim (z_1 - z_2)^{-1+k_1\cdot k_2}\, u_1^\alpha u_2^\beta \gamma^\mu_{\alpha\beta} \\ &\qquad \times (\partial x_\mu + i(k_1+k_2)\cdot\psi\psi_\mu)\, e^{i(k_1+k_2)\cdot x}(z_2) \, , \end{aligned} \tag{7.3.2}$$

so the integral over z_1 near z_2 gives a pole at $(k_1+k_2)^2 = 0$ whose residue is a NS massless vector boson vertex operator of the form $V_0(z_2) = \int d\theta \cdot DXe^{ik\cdot X}$.

On the other hand, the operator products $V_{1/2}\, V_{1/2}$ and $V_{-1/2}\, V_{-1/2}$ factorize the amplitudes on vertex operators $V_{\pm 1}$ with ϕ charge ± 1. For example, $V_{-1/2}\, V_{-1/2} \sim V_{-1} = e^{-\phi}\psi e^{ik\cdot x}$. But any factorization on $V_{\pm 1}$ is exactly equivalent to a factorization on V_0, since each pair of fermion vertices which are brought close together can be taken to be of opposite charge by the rearrangement lemma. Therefore $V_{\pm 1}$ are alternative forms of the massless boson vertex. Equation 7.3.2 shows explicitly that V_0 is derived from V_{-1} exactly as $V_{1/2}$ is derived from $V_{-1/2}$.

A more complicated calculation shows that V_{+1} is derived in the same way from V_0. In the classical formulation of the fermionic string, the physical states corresponding to the vertex operators V_{-1} and V_0 were discussed as two equivalent "pictures" for the string states, although, because the ghosts were missing, there was no weight 1 vertex operator V_{-1}.

It is clear now that there are infinitely many pictures, corresponding to the infinitely many Bose seas which give the inequivalent representations of the superconformal ghost fields β, γ. The infinite number of equivalent vertex operators for each physical state are derived by the picture changing operation $V_{q+1} = [Q_{BRST}, \xi V_q]$, and by its inverse. The rearrangement lemma given below for $V_{\pm 1/2}$ can be generalized to show the equivalence of all the pictures for both fermions and bosons.

7.4 The rearrangement lemma

The object is to show that

$$\left\langle \cdots V_{1/2}(z) \cdots V_{-1/2}(w) \cdots \right\rangle_0 = \left\langle \cdots V_{-1/2}(z) \cdots V_{1/2}(w) \cdots \right\rangle_0 . \tag{7.4.1}$$

The idea is to use the equivalence of expectation values in the small and large algebras and the contour integral form of the BRST transformation:

$$\begin{aligned} \langle \cdots A(z) \cdots [\epsilon Q_{BRST}, \xi A(w)] \cdots \rangle_{\partial\xi} &= \frac{1}{2\pi i} \oint_{C_w} dz' \, \langle \xi(\infty) \cdots A(z) \cdots j_{BRST}(z') \, \xi(w) A(w) \rangle \\ &= \frac{1}{2\pi i} \oint_{C_z} dz' \, \langle \cdots \xi(z) A(z) \cdots j_{BRST}(z') \, \xi(w) A(w) \rangle \\ &= \langle \cdots [\epsilon Q_{BRST}, \xi A(w)] \cdots A(z) \cdots \rangle_{\partial\xi} . \end{aligned} \tag{7.4.2}$$

The contour can be deformed by BRST invariance of the expectation value and of the operators represented by ellipses. The ξ field can be moved because its dimension is zero and only its zero mode participates in the expectation value.

7.5 Spacetime supersymmetry

The spacetime supersymmetry current $q_\alpha(z)$ is simply the fermion vertex at zero momentum. It takes the forms

$$\cdots \quad q_{\alpha,-1/2}(z) = e^{-\phi/2} S_\alpha \qquad q_{\alpha,1/2} = e^{\phi/2} \gamma^\mu_{\alpha\beta} S^\beta \partial x_\mu \qquad \cdots . \tag{7.5.1}$$

in the various pictures. $q_\alpha(z)$ is a BRST invariant dimension 1 conformal field, so its contour integral

$$Q_{\alpha,q} = \frac{1}{2\pi i} \oint dz\, q_{\alpha,q}(z) \tag{7.5.2}$$

is invariantly defined. The operator product 7.3.2 gives

$$[Q_{\alpha,-1/2}, Q_{\beta,1/2}]_+ = \gamma^\mu_{\alpha\beta} p_\mu \tag{7.5.3}$$

where p_μ is the spacetime momentum operator, and $[Q_{\alpha,1/2}, V_{-1/2}] = V_0$, i.e.,

$$[Q_{\alpha,1/2}, e^{-\phi/2} S_\beta e^{ik\cdot x}]_+ = \gamma^\mu_{\alpha\beta} \psi_\mu e^{ik\cdot x} , \tag{7.5.4}$$

showing that Q_α is the spacetime supersymmetry generator. Since Q_α commutes with the screening operator $e^{3\sigma-2\phi}$ for the background charge, the contour argument shows that the expectation values $\langle \cdots \rangle_0$ on the sphere are invariant under supersymmetry. Thus the tree amplitudes are supersymmetric. All that is needed to show supersymmetry of the loop expansion, since $q_\alpha(z)$ is a conformal current, is to show that the screening charges on arbitrary Riemann surfaces preserve spacetime supersymmetry.

REFERENCES

1. S.H. Shenker, this volume.

2. D. Friedan, Z. Qiu, and S. Shenker, *Proc. of the Santa Fe Meeting of the APS Div. of Particles and Fields, October 31 – November 3, 1984*, T. Goldman and M. Nieto (eds.), World Scientific (1985).

3. D. Friedan, Z. Qiu, and S. Shenker, *Phys. Lett.* 151B (1985) 37.

4. D. Friedan, E. Martinec, and S. Shenker, *Phys. Lett.* 160B (1985) 55.

5. D. Friedan, E. Martinec, and S. Shenker, *Conformal Invariance, Supersymmetry and String Theory*, Fermi Institute preprint EFI 85-89 and Princeton preprint, submitted to Nucl. Phys. B.

6. J. Cohn, D. Friedan, Z. Qiu, and S. Shenker, EFI preprint 85-90 (presented in reference 2).

7. D. Friedan, in 1982 Les Houches summer school *Recent Advances in Field Theory and Statistical Mechanics*, J-B. Zuber and R. Stora (eds.), North-Holland (1984).

8. A.M. Polyakov, *Phys. Lett.* 103B (1981) 207.

9. K. Fujikawa, *Phys. Rev.* D25 (1982) 2584; M. Kato and K. Ogawa, *Nuc. Phys.* B212 (1983) 443; and S. Hwang, *Phys. Rev.* D28 (1983) 2614.

10. See Mandelstam's lectures in this volume.

11. A.M. Polyakov, *Phys. Lett.* 103B (1981) 211; elaborated in E. Martinec, *Phys. Rev.* D28 (1983) 2604.

12. E. Martinec, in preparation.

13. D. Friedan and P. Windey, *Nucl. Phys.* B235[FS11] (1984) 395.

14. Cf. R.C. Gunning, *Lectures on Riemann Surfaces,* Princeton Univ. Press (1966).

15. A.A. Belavin, A. M. Polyakov, and A. B. Zamolodchikov, *Nuc. Phys.* B241 (1984) 333.

16. D. Friedan, Z. Qiu, and S. Shenker in *Vertex Operators in Mathematics and Physics*, J. Lepowsky et.al. (eds.), Springer-Verlag (1984); *Phys. Rev. Lett.* 52 (1984) 1575.

17. G. Segal, *Comm. Math. Phys.* 80 (1981) 301;

18. V. Knizhnik and A. B. Zamolodchikov, *Nuc. Phys.* B247 (1984) 83.

19. A. Chodos and C. Thorn, *Nuc. Phys.* B72 (1974) 509; B. L. Feigin and D. B. Fuchs, Moscow preprint (1983) and *Funct. Anal. Appl* 16 (1982) 114; V. L. Dotsenko and V. Fateev, *Nuc. Phys.* B240 (1984) 312; C. Thorn, *Nucl. Phys.* B248 (1985) 551.

20. Cf. P. Goddard and D. Olive, *Nucl. Phys.* B257[FS14] (1985) 83.

AN INTRODUCTION TO KAC-MOODY ALGEBRAS AND THEIR PHYSICAL APPLICATIONS

P. Goddard
Department of Applied Mathematics and Theoretical Physics,
University of Cambridge, Silver Street, Cambridge CB3 9EW
U.K.

and

D. Olive

Blackett Laboratory, Imperial College, London SW7 2BZ

ABSTRACT

Kac-Moody algebras, the physical applications and results on their representation theory are surveyed. The Sugawara construction of the Virasoro algebra associated with a Kac-Moody algebra is described and it is used to produce the full discrete series of representations of the Virasoro algebra. The quark model construction of representations of Kac-Moody algebras is also described. Conditions necessary for the equivalence of two-dimensional σ-models to free fermion theories are derived.

1. Introduction and Background

1.1 Introduction

Since physics is the study of symmetry in nature and group theory the mathematical analysis of symmetry it follows that group theory naturally provides appropriate mathematical tools for theoretical physics. Indeed the use of finite groups are infinite groups which are finite dimensional (e.g. the rotation group) is well established. An important feature of nature is locality (or causality) and it seems that the theory of affine Kac-Moody algebras (and their associated Virasoro algebras) provides an extremely powerful yet natural framework for unifying the concepts of symmetry and locality. These algebras are infinite dimensional and so would exponentiate into infinite dimensional groups but at present it is easier to study them as algebras, thereby regarding them as the second stage generalisation of angular momentum theory beyond ordinary finite dimensional Lie algebra theory. They nevertheless constitute the class of infinite dimensional algebra in which the dimension diverges to infinity in the most controlled way.

These lectures are intended to complement others that we have given on the subject (GODDARD; OLIVE). These latter notes will therefore constitute background and parallel reading for these lectures which deal with two related topics much studied in the physics literature in the last year or so, namely the construction of Virasoro generators from Kac-Moody generators by Sugawara's construction (section 2) and the use of the "quark model" to construct Kac-Moody, and hence Virasoro generators (section 3). The treatment will elaborate and synthesise the results of a series of papers written by us, together with A. Kent and W. Nahm.

We shall consider the affine, untwisted Kac-Moody algebra $\hat{g}$ which has commutation relations:

$$[T^i_m, T^j_n] = if^{ij\ell} T^\ell_{m+n} + k\delta^{ij} m\, \delta_{m+n,0}, \tag{1.1}$$

where the central (c-number) term k commutes with all the T^i_m. The suffices m and n take integer values and $f^{ij\ell}$ are the totally antisymmetric structure constants of the compact Lie algebra g, an orthonormal basis for whose generators is given by T^i_0 (i=1,...dim g). More discussion of (1.1), the structure of the root system of $\hat{g}$ and its representation theory can be found in our Srni notes (GODDARD; OLIVE) as well as in the mathematical literature (see e.g. KAC).

Associated with $\hat{g}$ is a VIRASORO algebra with generators L_m satisfying:

$$[L_m, T^j_n] = -n\, T^j_{m+n}, \tag{1.2}$$

$$[L_m, L_n] = (m-n)\, L_{m+n} + \frac{c}{12} m(m^2-1)\delta_{m+n,0}, \tag{1.3}$$

where c is another central (c-number) term.

Equations (1.1)-(1.3) constitute the system to be studied in these lectures. They comprise the semi-direct product of the Kac-Moody algebra $\hat{g}$ with the Virasoro algebra. One of the key questions of physical interest will concern the possible allowed values of the c-numbers k and c and their interrelation, and this will depend on the physical applications we have in mind as now reviewed.

1.2 Survey of Physical Applications

There are at least three disparate physical applications of the algebraic structure above, two quite old, namely the current algebra theory of the 1960's and the string theory of particle interactions of the early 1970's. The third is recent and more surprising, the theory of the behaviour of spin systems on two dimensional lattices at their critical temperature at which phase transitions occur. The feature common to all those is the connection with conformally invariant quantum field theories in two dimensions. The fact that we talk about two rather than four dimensions is the price paid for an algebraic structure which is tractable according to present knowledge. Nevertheless it is not such a limitation from the physical point of view since the string theories are currently thought to be the most realistic theory of unified particle interactions and since experimentalists can readily make and study two dimensional substances with lattice structure.

To see the connection with current algebras let us first define the Kac-Moody "field":-

$$T^j(z) = \sum_{m \in \mathbb{Z}} z^{-m} T^j_m , \tag{1.4}$$

where z is a complex variable usually considered in the vicinity of the unit circle. Now define the "current"

$$J^j(\xi) = \frac{\hbar}{2\pi R} T^j(e^{i\xi/R}) . \tag{1.5}$$

In terms of this, the Kac-Moody algebra (1.1) reads:

$$[J^i(\xi), J^j(\eta)] = i\hbar f^{ij\ell} J^\ell(\xi)\delta(\xi-\eta) + \frac{i\hbar^2 k}{2\pi}\delta^{ij}\frac{\partial}{\partial\xi}\delta(\xi-\eta) \tag{1.6}$$

This is recognizable as a "current algebra" relation (see ADLER and DASHEN) with the c-number term appearing as the derivative of a delta-function and hence as a Schwinger term. The current J has period $2\pi R$ in its argument ξ yet R does not appear explicitly in (1.6) and so can be as large as we choose. The appearance of Planck's constant, $\hbar$ squared in the SCHWINGER term emphasises that that term is a second order quantum effect and in fact comes from a loop Feynman diagram. The integer suffix in (1.1) is seen to be $R/\hbar$ times the momentum conjugate to x, quantised because of the periods in ξ.

Equations (1.6) can arise in at least two apparently different two dimensional quantum field theories. In the theory of free massless "quarks" (or fermions) the commutation relations of currents $(j^{t,i}(t,x),\ j^{x,i}(t,x))$, $(i=1,\ldots \dim g)$ bilinear in the quarks are calculated in the usual way. The Schwinger term occurs in the commutator of $j^{t,i}$ with $j^{x,j}$ only. When $j^{t,i}$ and $j^{x,i}$ are added (or subtracted) to obtain the light cone components $j^{+,i}(j^{-,i})$ it is found that $j^{+,i}$, commutes with $j^{-,j}$ and that each individually satisfies (1.6). Since the quarks are free and massless their equations of motion imply that $j^{+,i}$ depends on t and x in the combination of $t-x$ only. Likewise $j^{-,i}$ depends on $t+x$ only. Hence the current algebra constitutes two commuting copies of (1.6) satisfied by $j^{+,i}(t-x)$ and $j^{-,i}(t+x)$.

The energy momentum tensor $\theta^{\mu\nu}$ is the standard one bilinear in free massless fermions and hence traceless as well as symmetric. Thus there are only two independent components which can be taken to be θ^{++} and θ^{--} depending on $t-x$ and $t+x$ by virtue of the conservation law. The Dirac, Schwinger commutation relations take the form of two commuting Virasoro algebras with θ^{++} related to L_n (and θ^{--} to $\bar{L}_n$) by an analogue of eqns.(1.4) and (1.5). Altogether we have two commuting copies of the semidirect product of the Virasoro algebra with $\hat{g}$.

When we try to construct the current algebra (1.6) with bosonic fields instead of fermionic ones we find that the most promising possibility occurs when we consider dim g fields constrained to lie on the manifold of the Lie group G obtained by exponentiating g. Unfortunately the commutator of $j^{x,i}$ with $j^{x,j}$ vanishes instead of yielding a term proportional to $f^{ij\ell} j^{t,\ell}$ as in the fermionic case and as a consequence it is impossible to obtain the $\hat{g}$ Kac-Moody algebra unless we heed the recent observation of WITTEN and add to the usual G invariant kinetic energy a new term called the Wess-Zumino term. When this is added as prescribed by Witten yet again we obtain two commuting copies of the semidirect product of a Virasoro algebra with $\hat{g}$.

When z is related to space and time as explained above, i.e. identified with $\exp(i(t \pm x)/R)$, it follows that the Virasoro generators $L_o, \bar{L}_o$ are proportional to $(H \pm P)$, the translation generators on the light cone, i.e.

$$L_o \sim (H+P)/2 \ , \qquad (1.7)$$

and hence, in a quantum field theory, intrinsically positive according to the usual desiderata.

In string theory (see the review edited by JACOB) z is related to the parameters τ and σ on the string world sheet in a similar way so that $2\pi R$ is now simply the length of the string measured in σ. A similar sort of algebraic structure to eqns.(1.1)-(1.3) arises with the Kac-Moody generators now corresponding to the vertex operators for particle emission. Again L_o is intrinsically positive. It is well known from the constructions in string theory that the value of the c-number, c occurring in the Virasoro algebra (1.3) is given by:

$$c = \text{number of bosons} + \frac{1}{2}\,\text{number of fermions} \ , \qquad (1.8)$$

and thus quantised. In quantum field theory these constructions correspond to the energy momentum tensor for free, massless, real bosons and fermions. Thus the spectrum (1.8) is a necessary condition for a conformal quantum field theory to be equivalent to a free theory.

Lattice systems automatically possess a fundamental scale, the lattice spacing, but this effectively diverges at the critical temperature leaving a theory which is scale invariant, and local (because of the nearest neighbour interactions) and hence conformally invariant as all conformal currents are conserved by virtue of the traceless energy momentum tensor.

In two dimensions the algebra of conformal transformations is infinite dimensional (the sum of two commuting Virasoro algebras with generators L_n and $\bar{L}_n$) as can be seen by considering the infinitesimal conformal transformation

$$\delta_n z = \varepsilon_n z^{n+1} \qquad (\varepsilon_n \text{ small})$$

generated by L_n. If we identify the Cartesian coordinates x,y via

$$z = x + i y \tag{1.9}$$

instead of via $\exp(i(t+x)/R)$ as before we see, putting $n=0$ that, if ε_0 is real, δ_0 is a scale transformation and if it is imaginary δ_0 is a rotation. We conclude that $L_0 + \bar{L}_0$ is the scale or dilatation generator D and $L_0 - \bar{L}_0$ the rotation generator S. Hence instead of (1.7), we have

$$L_0 \sim (D+S)/2 . \tag{1.10}$$

This identification has sometimes been made in quantum field theory. In either case (1.9) or (1.10) L_0 has the fundamental property that its spectrum has to be positive.

The realisation that the Virasoro algebra, originally found in string theory is also relevant to conformally invariant quantum field theory has a long history. See GERVAIS and SAKITA; FERRARA, GATTO and GRILLO; FUBINI, HANSON and JACKIW; FRIEDAN; BELAVIN, POLYAKOV and ZAMOLODCHIKOV.

Finally let us mention that there are other applications of Kac-Moody algebras in which k vanishes, yielding what is called a loop algebra (for example in soliton theory). These are fundamentally different from the applications above with $L_0 \geq 0$, as we shall now see, and will not be further discussed in these lectures.

1.3 The Permitted values of k,c and h

We have explained that in the physical applications considered L_0 is fundamentally positive:

$$L_0 \geq 0 . \tag{1.11}$$

In addition we usually require that we have a state space with positive definite scalar product and hermiticity conditions:

$$T_n^{i\dagger} = T_{-n}^i \; ; \; L_n^\dagger = L_{-n} . \tag{1.12}$$

Such representations are called unitary and only these will be considered henceforth.

According to the commutation relations (1.2) and (1.3) the action of T^i_n or L_n $(n \geq 1)$ on an eigenstate of L_o, lowers its eigenvalue by n. In view of (1.11) this procedure must stop eventually and hence there must be states $|\psi\rangle$ satisfying

$$L_n|\psi\rangle = 0 , \tag{1.13}$$

$$T^i_n|\psi\rangle = 0 \quad n \geq 1 \tag{1.14}$$

These states must form a representation of $g + \{L_o\}$ and this representation is irreducible precisely when that of the complete algebra $\hat{g} + \{L_o\}$ is. Thus, given irreducibility, all the states $|\psi\rangle$ have the same L_o eigenvalue, h say, which indeed has to be positive as

$$0 \leq \| L_{-1}|\psi\rangle \| = \langle\psi|L_1 L_{-1}|\psi\rangle$$

$$= \langle\psi|[L_1, L_{-1}]|\psi\rangle = 2\langle\psi|L_o|\psi\rangle = 2h\||\psi\rangle\|$$

The physical requirement that L_o be positive (1.11) has implied that we have what is perversely called a "highest weight" representation of $\hat{g}$ and the unitary such possibilities have been classified by mathematicians. See KAC and our Srní lectures (GODDARD, OLIVE) for a sketch.

In such positive L_o, unitary representations it is possible to establish a remarkable theorem concerning the permitted values of the c numbers k and c appearing in (1.1) and (1.2):

<u>Theorem on permitted c-numbers</u>

(i) $2k/\psi^2$ is a non-negative integer (the level, x); (1.15)

(ii) either $c \geqslant 1$ or $c = 1 - 6/(m+2)(m+3)$, $m=0,1,2,3,\ldots$, (1.16)

Further if m=0 (so c=0) only the trivial representation with T^i_m and L_n equal to zero exists. Likewise if x=0, T^i_m vanishes.

Part (i) of this theorem is well known and proofs can be found in KAC, and also GODDARD, OLIVE. The result means that k is quantised in units determined by the size of the structure constants of g, conveniently taken to be $\psi^2/2$ when ψ is the highest root of g. We could simplify matters by choosing $\psi^2 = 2$ but as explained later this condition cannot be maintained simultaneously for an algebra and a subalgebra in general so we prefer to allow ourselves flexibility.

The Virasoro generators L_n (unlike the T^i_m) are intrinsically normalised by the fact that their structure constants are integers and thus c is likewise intrinsically normalised. Part (ii) of the theorem concerning the allowed values of c, is a recent, surprising discovery due to FRIEDAN, QIU and SHENKER. It states that the spectrum of c resembles that of a scattering problem with a threshold at c=1, with a discrete sequence of bound states accumulating there.

The result (part(ii)) was established in an indirect way, by tracing zeros of the Kac determinant (of matrices of scalar products of states $L_{-1}|\psi\rangle$; $L_{-2}|\psi\rangle$, $(L_{-1})^2|\psi\rangle$; etc when $|\psi\rangle$ satisfies (1.13)). This argument was incomplete in that it failed to establish the existence of unitary representations of (1.3) for all the values of c in (1.14). This will be shown by different methods in section 2.6.

The possible eigenvalues of L_o exceed by a positive integer (≥ 0) of the eigenvalue h for a state satisfying (1.13). FRIEDAN, QIU and SHENKER also showed that there were only a finite number of possible h values for the discrete sequence of c values less than unity, and that there were rational numbers:

$$h = h_{p,q} \equiv \frac{[(m+3)p-(m+2)q]^2-1}{4(m+2)(m+3)} \; ; \; 1 \leq p \leq m-1; \; 1 \leq q \leq p \qquad (1.17)$$

For example

if $c = 0$, $h = 0$;

$c = \frac{1}{2}$, $h = 0, 1/16, 1/2$;

$c = 7/10$, $h = 0, 3/80, 1/10, 7/16, 3/5, 3/2$.

Physically these results are of great importance since the eigenvalues of $L_o + \bar{L}_o$ yield by (1.10) the dimensions of possible fields are hence critical exponents. These are measurable and found to be rational. This is now explained by (1.17) which is potentially a consequence of the representation theory of the Virasoro algebra. FRIEDAN, QIU and SHENKER compared (1.17) with the critical exponents of known lattice models and concluded that $c = \frac{1}{2}$ for the Ising model, $c = 7/10$ for the tricritical Ising model and so on.

The only non-trivial value of c common to the sequence (1.16) and equation (1.8) is $c = \frac{1}{2}$. This tells us that the Virasoro algebra for the Ising model can be realised by the energy momentum tensor for a single, real, free, massless fermion. Thus a new light is shed on a result of ONSAGER, originally found forty years ago. One question to be dealt with here concerns the corresponding construbtion for higher terms in the sequence (1.4) of discrete c values.

The purpose of this introductory section has been to set the background and motivate the physical importance of the representation theory of the Virasoro algebra. The reader is urged to consult the references cited above for more information.

2. Sugawara's Construction of the Virasoro Algebra

2.1 Sugawara's Construction for g Simple

We have written down in (1.1) to (1.3), the semidirect product of the Kac-Moody and Virasoro algebras as if the generators were totally independent but in fact there is a construction of the Virasoro algebra in terms of bilinears in the Kac-Moody generators which it will be the purpose of these notes to explain and develop. The idea arose naturally in the theory of current algebras (GELL-MANN; GELL-MANN and NE'EMAN; ADLER and DASHEN) where it was argued that the full dynamics of the theory should be formulated in terms of currents. This means that the energy momentum tensor should be expressed in terms of currents and it was realised in 1968 by SUGAWARA that this was possible if this tensor was bilinear in currents and if the SCHWINGER term (often ignored hitherto) was taken into account. This idea was originally applied in four space time dimensions but it was soon realised that it worked most neatly in two dimensions. Since the currents correspond to the T^i_m, and the energy momentum tensor to L_m, Sugawara's construction in the present language reads

$$\mathcal{L}(z) \equiv \sum_{n \in \mathbb{Z}} z^{-n} \mathcal{L}_n = \frac{1}{2k + Q_\psi} {}^{\times}_{\times} \sum_{i=1}^{\dim g} T^i(z)\, T^i(z) {}^{\times}_{\times} \tag{2.1}$$

or, equivalently:-

$$\mathcal{L}_n = \frac{1}{2k + Q_\psi} \sum_{m \in \mathbb{Z}} {}^{\times}_{\times} \sum_{i=1}^{\dim g} T^i_{m+n}\, T^i_{-m} {}^{\times}_{\times} \tag{2.2}$$

The first statement emphasises that two quantum field operators $T^i(z)$ (see (1.5)) are multiplied together at the same point. In order to avoid a singularity and obtain a quantity with finite matrix elements in a highest weight representation it is necessary to introduce a normal ordering denoted by the double crosses whereby the T^i_m with positive suffices are moved to the right of those with negative suffices (since, by (1.14), $T^i_n, n \geq 1$ are "like" destruction operators). The singularity is exhibited by the Wick contraction following from this definition and (1.1).

$$T^i(z)\, T^i(\zeta) = {}^{\times}_{\times} T^i(z)\, T^i(\zeta) {}^{\times}_{\times} + \frac{kz\zeta}{(z-\zeta)^2}, \quad |z| > |\zeta| . \tag{2.3}$$

Notice that Sugawara's construction is like a Kac-Moody generalisation of the quadratic Casimir operator for g. Notice also that the prefactor is not $1/(2k)$ (as Sugawara thought), but is subtly "renormalised" to be $(2k + Q_\psi)^{-1}$ where Q_ψ is the quadratic Casimir in the adjoint representation of g (so that ψ denotes its highest weight). The necessity for this prefactor is seen by calculating the L_m, T^i_n commutator, paying attention to the normal ordering. Given the bilinear form of $\mathcal{L}$ in terms of T^i_m a simple determination of the coefficient was given by KNIZHNIK and ZAMOLODCHIKOV. Let the highest weight states (1.12) belong to a representation of g with matrix representation t^i. Then

$$T^i_o|\Psi\rangle = |\Psi\rangle\, t^i \tag{2.4}$$

If $1/\beta$ is the prefactor, we have by (1.12)

$$\mathcal{L}_{-1}|\Psi\rangle = \sum_{i=1}^{\dim g} \beta^{-1} T^i_{-1}|\Psi\rangle t^i\ .$$

Acting on this equation with T^j_1 and using its commutation relations (1.1) and (1.2) with T^i_{-1} and $\mathcal{L}_{-1}$ yields $\beta = 2k + Q_\psi$.

Finally it has to be checked that $\mathcal{L}$ indeed satisfies (1.3) and that the c number is actually

$$c_g = \frac{2k \dim g}{2k + Q_\psi} = \frac{x \dim g}{x + \tilde{h}(g)}\ , \tag{2.5}$$

where x is the level $2k/\psi^2$ mentioned in section (1.3) and $\tilde{h}(g) = Q_\psi/\psi^2$ is called the dual Coxeter number. We shall prove it is an integer and list its values later.

The check is done by using (1.2) in the explicit form (2.2) for $\mathcal{L}_n$. The c number arises in restoring the normal ordering after the commutation, using (1.1).

In fact the states of an irreducible representation of $\hat{g}$ with the lowest possible L_o value, namely those annihilated by T^i_n $(n > 0)$ (and hence by L_n $n > 0$) form an irreducible representation of g where the quadratic Casimir operator is given by $Q_\psi = \sum_{i=1}^{\dim g} (t^i)^2$ [see (2.4)]. Then the eigenvalue of L_o on these states is

$$L_o \to Q_\psi/(2k + Q_\psi) \geq 0 \tag{2.6}$$

The results described here form the culmination of a long series of papers in the physics literature: CALLAN, DASHEN and SHARP; SUGAWARA; SOMMERFIELD; COLEMAN, GROSS and JACKIW; BARDAKCI and HALPERN; DELL'ANTONIO, FRISHMAN and ZWANZIGER; and DASHEN and FRISHMAN, ending with the correct prefactor for level 1 representations of SU(N). More recently the correct general formula (2.1) has been given by KNIZHNIK and ZAMOLODCHIKOV; GODDARD and OLIVE (1985); and TODOROV. Applications in string theory were discussed by NEMESCHANSKY and YANKIELOWICZ and JAIN, SHANKAR and WADIA. The Sugawara formula has also appeared in the mathematical literature (SEGAL; FRENKEL; KAC; GOODMAN and WALLACH).

Finally notice that Sugawara's construction is automatically unitary in the sense (1.12) if the Kac-Moody generators are. Further $\mathcal{L}_o$ is then positive.

2.2 Sugawara's construction for g not simple

If the Lie algebra g is semisimple, i.e. $g = g_1 + g_2 + \ldots$ then the relevant construction is simply

$$\mathcal{L}^{g} = \mathcal{L}^{g_1} + \mathcal{L}^{g_2} + \ldots \tag{2.7}$$

where $\mathcal{L}^{g_i}$ denotes (2.1) for g_i. The c number for $\mathcal{L}^{g}$ is also obtained additively.

$$c_g = c_{g_1} + c_{g_2} + \ldots \tag{2.8}$$

The result is valid even if g is not semisimple. Thus if g is abelian its structure constants vanish, $Q_\psi = 0$, and we have simply

$$c_g = \dim g = \operatorname{rank} g \ . \tag{2.9}$$

A physical example of this case is VIRASORO's original construction in string theory. The quantities P^i_m of string theory generate the Kac-Moody algebra:

$$[P^i_m, P^j_n] = mk\, \delta_{m+n,0}\, \delta^{ij} \ , \tag{2.10}$$

(usually with $k=1$ chosen) and Sugawara's construction (2.1) reduces to Virasoro's construction

$$L(z) = \frac{1}{2k} \mathop{\times}\limits_{\times} \sum_{i=1}^{\operatorname{rank} g} P^i(z)^2 \mathop{\times}\limits_{\times} \ . \tag{2.11}$$

Here the group g is just real space $\mathbb{R}^d$ with rank and dimension d. That the c number is d was originally noted by WEIS.

This result also applies if space is compactified to a torus, $\mathbb{R}^d$ divided by a lattice, since it is still an abelian group.

2.3 Properties of c_g for g simple

We shall show that c_g is a rational number lying between $\dim g$ and $\operatorname{rank} g$, attaining its lower bound, $\operatorname{rank} g$, if and only if g is simply laced (i.e. has roots all of the same length) and a level 1 representation of $\hat{g}$ is considered. To do this we shall find an expression for Q_ψ in terms of the root system of g. Q_ψ was the quadratic Casimir in the adjoint representation of g and hence is defined in terms of structure constants of g :

$$Q_\psi \, \delta^{ij} = \sum_{m,n=1}^{\dim g} f^{imn} f^{jmn} . \tag{2.12}$$

It will be useful for later work to consider a more general real representation of g than the adjoint. Let it have antihermitian real generators M^i satisfying

$$[M^i, M^j] = f^{ijk} M^k , \quad M^{i*} = M^i . \tag{2.13}$$

Since g is simple (and compact)

$$\mathrm{tr}(M^i M^j) = -\kappa_M \, \delta^{ij} = -x_M \psi^2 \, \delta^{ij} \tag{2.14}$$

for some real, positive κ_M. Putting $i = j$ and summing from 1 to dim g yields

$$Q_M \dim M = \kappa_M \dim g \tag{2.15}$$

where Q_M is the quadratic Casimir, $-\sum_{i=1}^{\dim g} M^i M^i$. Of course in the adjoint representation Q_M equals Q_ψ. By summing instead over the rank g generators of the Cartan subalgebra we find

$$\kappa_M = \sum \mu^2 / (\mathrm{rank}\, g) \tag{2.16}$$

where the sum is over the weights of the representation M (with multiplicities included). The weights of the adjoint representation are the roots and when g is simple they have at most two distinct lengths. Let there be n_L long roots and n_S short roots respectively. Then

$$\dim g = n_L + n_S + \mathrm{rank}\, g \tag{2.17}$$

and

$$\left[\frac{\text{long root}}{\text{short root}}\right]^2 \equiv (L/S)^2 = 1, 2 \text{ or } 3$$

By (2.16)

$$\tilde{h} \equiv Q_\psi/\psi^2 = (n_L + (L/S)^{-2} n_S)/\mathrm{rank}\, g \tag{2.18}$$

so that the dual coxeter number $\tilde{h}$ is certainly rational and hence so is c_g (2.5).

Obviously c_g is less than $\dim g$ and we now show what is less obvious, that it is greater than or equal to $\operatorname{rank} g$. By (2.5), (2.17) and (2.18)

$$c_g - \operatorname{rank} g = \frac{n_L(x-1) + n_S(x-(L/S)^{-2})}{x+\tilde{h}} \geq 0 \tag{2.19}$$

as we consider level $x \geq 1$ and $(L/S)^{-2} \leq 1$ by definition.

What is more important is that c_g equals $\operatorname{rank} g$ if and only if $(L/S)^2 = 1$, i.e. g is simply laced, and $x=1$, i.e. the $\hat{g}$ representation has level 1. This conclusion remains true if g is semisimple in the sense that each component g_i is to be simply laced. These conditions are precisely those for the validity of the "vertex operator" representation of g and we shall show subsequently that this is no coincidence; there must be such a construction when c_g and $\operatorname{rank} g$ coincide.

Finally, introducing more Lie algebra theory, we shall show that the dual Coxeter number $\tilde{h}$ is an integer. If M generates an irreducible representation of g it has a unique highest weight λ, and then

$$Q_M = \lambda(\lambda + 2\rho)$$

where ρ is half the sum of positive roots of g. Hence

$$\tilde{h} = Q_\psi/\psi^2 = 1 + 2\rho.\psi/\psi^2$$

and ψ is the highest weight of the adjoint representation. Now ψ/ψ^2 is a co-root of g and can be expanded as an integer linear combination of the simple co-roots of g:

$$\psi/\psi^2 = \sum_{i=1}^{\operatorname{rank} g} m_i \alpha_i/(\alpha_i)^2$$

Then as $2\rho.\alpha_i/(\alpha_i)^2 = 1$,

$$\tilde{h} = 1 + \sum_{i=1}^{\operatorname{rank} g} m_i$$

and is clearly an integer. Its value can easily be calculated for simple groups: A_n : n+1, C_n : n+1, E_6 : 12, E_7 : 18, E_8 : 30, F_4 : 9, G_2 : 4 and SO(N) : N-2, $(N \geq 5)$.

2.4 The situation where g has a subalgebra $h \subset g$

In the first instance let us suppose that g and h are both simple. We can choose an orthonormal basis for g which includes as a subset an orthonormal basis for h, let us say $i = 1,2,\ldots \dim h$. We automatically obtain a Kac-Moody algebra $\hat{h}$ inheriting the same central term k as $\hat{g}$. But the $\hat{h}$ level may differ from that of $\hat{g}$ because the highest roots of g and h, ψ and ϕ, say, may have unequal lengths.

We may suppose $\hat{g}$ has level $2k/\psi^2$ equal to 1. Then the level of $\hat{h}$, $2k/\phi^2$, must equal 1, 2 or 3 etc. Hence ψ^2/ϕ^2 equals an integer greater than or equal to unity. In general the $\hat{h}$ level must be greater than or equal to the $\hat{g}$ level.

Sugawara's construction can be applied to both $\hat{g}$ and $\hat{h}$ to obtain Virasoro generators $\mathcal{L}^g$ and $\mathcal{L}^h$ respectively. Of course they have different prefactors and different c-numbers in general. We have, by (1.2)

$$[\mathcal{L}^g_m, T^j_n] = -n\, T^j_{m+n} \qquad j = 1 \ldots \dim g$$

$$[\mathcal{L}^h_m, T^j_n] = -n\, T^j_{m+n} \qquad j = 1 \ldots \dim h$$

Hence, subtracting

$$[\mathcal{L}^g_m - \mathcal{L}^h_m, T^j_n] = 0 \qquad j = 1 \ldots \dim h \tag{2.20}$$

and so, by (2.2)

$$[\mathcal{L}^g_m - \mathcal{L}^h_m, \mathcal{L}^h_n] = 0 \tag{2.21}$$

Thus $\mathcal{L}^g_m$ has been broken into two mutually commuting pieces:

$$\mathcal{L}^g_m = \mathcal{L}^h_m + K_m, \tag{2.22}$$

where K_m commutes with the $\hat{h}$ Kac-Moody algebra (2.20) and can be thought of as relating to the coset G/H. Further, since by (2.21)

$$[\mathcal{L}^g_m, \mathcal{L}^g_n] = [\mathcal{L}^h_m, \mathcal{L}^h_n] + [K_m, K_n]$$

we deduce that K_m, like $\mathcal{L}^g_m$ and $\mathcal{L}^h_m$, satisfies a Virasoro algebra and that the c-number is

$$c_K = c_g - c_h = \frac{2k \dim g}{2k + Q_\psi} - \frac{2k \dim h}{2k + Q_\phi} \tag{2.23}$$

Since the eigenvalues of $\mathcal{L}^g_o$ are bounded below, so are those of K_o. Therefore the highest weight representation of $\hat{g}$ must decompose into highest weight representations of the K_m Virasoro algebra so that in particular we must have c_K positive i.e.

$$c_K \geq 0 \tag{2.24}$$

with c_K vanishing if and only if K_m vanishes, according to the theorem of section (1.3). Of course if c_K is less than unity it must take one of the values in the discrete series (1.16), and in section (2.6) we shall make choices of h g so as to obtain the complete series (1.16) with explicitly unitary representations. Thus by subtracting two Virasoro algebras whose c numbers exceed unity (section 2.3), we can obtain a Virasoro algebra with $0 < c < 1$.

These results were due to GODDARD and OLIVE (1985) and GODDARD, KENT and OLIVE. Particular examples of related algebraic structures occur in the earlier work of BARDAKCI and HALPERN and of MANDELSTAM.

Generalisation to the cases when g and h are not simple is easily made using the results of section (2.2).

2.5 A quantum equivalence theorem and the vertex operator construction

A second use of the preceding argument concerns the deduction that the single numerical condition $c_g = c_h$ implies that

$$\mathcal{L}^g = \mathcal{L}^h \tag{2.25}$$

since if c_K vanishes then so does K in a highest weight representation by the theorem of section 1.3. We call (2.5) a quantum equivalence theorem since it establishes the equality of two apparently different operators which, in a conformally invariant quantum field theory correspond to components of the energy momentum tensors of two apparently different theories. Two theories with the same energy momentum tensor have the same Hamiltonian and so are indeed quantum equivalent.

This is a remarkably powerful result which we shall return to in section 3 when we study fermions. As an immediate application consider the Cartan subalgebra t of g which exponentiates to a maximal torus T subgroup of G. T is an abelian group, isomorphic to $\mathbb{R}^{\text{rank}\, g}/\Lambda_R(g)$ where $\Lambda_R(g)$ is the root lattice of g. We saw in section 2.3 that $c_K = c_g - c_t = c_g - \text{rank}\, g$ vanishes if and only if g is simply laced, and a level 1 representation of g is considered. Thus, when these two conditions are satisfies, the quantum equivalence theorem (2.25) states that, choosing $\psi^2 = 2$,

$$\frac{\text{rank}\, g}{2\dim g} \sum_{i=1}^{\dim g} {}^{\times}_{\times} T^i(z)\, T^i(z) {}^{\times}_{\times} = \tfrac{1}{2} \sum_{i=1}^{\text{rank}\, g} {}^{\times}_{\times} T^i(z)\, T^i(z) {}^{\times}_{\times} \,. \tag{2.26}$$

Now the $T^i(z)$, $i = 1,2,\ldots \text{rank}\, g$, appearing on the right hand side of (2.26) and corresponding to the Cartan subalgebra of g satisfy (2.10) (with k=1), according to (1.1). Thus the equivalence theorem (2.24) states that when g is simply laced and $x=1$, the Sugawara construction (2.2) equals Virasoro's construction (2.11) for a string moving on the maximal torus T of G. This suggests that it must be possible to construct the Kac-Moody generators E^{α}_m corresponding to the step operators E^{α} of g from the T^i_m ($i = 1, \ldots$ rank g) and indeed this is precisely what the vertex operator construction achieves. One obtains the FUBINI-VENEZIANO vector $Q^i(z)$ by integrating

$$iz \frac{dQ^i}{dz} = T^i(z) \qquad i = 1, \ldots, \text{rank}\, g ,$$

Then the E^{α}_m are the Laurent coefficients in the expansion of the vertex operator

$$\sum_{m \in \mathbb{Z}} z^{-m} \varepsilon(\alpha, T^i_0)\, E^{\alpha}_m = z : e^{i\alpha . Q(z)} : , \tag{2.27}$$

where the normal ordering is with respect to the bosonic oscillators which are the Laurent coefficients of $T^i(z)$, $i = 1, \ldots$, rank g, and $\varepsilon(\alpha, T^i_o)$ is a Klein transformation needed to correct some signs in commutation relations. See GODDARD and OLIVE (1984) for more information about this construction due to FRENKEL and KAC and SEGAL. The explicit verification of (2.26) is due to FRENKEL.

Two commuting copies of the above result (2.26) can be interpreted in terms of two dimensional conformally invariant quantum field theory. In section 1.2 we saw that an interesting example was the "Wess-Zumino" theory, with a field confined to the manifold of a Lie group G, and described by an action consisting of the usual kinetic term plus a Wess-Zumino term, each normalised to that the energy momentum tensor was indeed bilinear i.e. in the Sugawara form, in conserved currents satisfying the $\hat{g}$ Kac-Moody algebra. WITTEN showed that the level of $\hat{g}$ was equal to the only free parameter left in the action, the overall coefficient which had to be an integer to ensure the single valuedness of $\exp\{i \text{ Action}/\hbar\}$.

Equation (2.26) shows that if g is simply laced and the level is unity then the Wess-Zumino model is quantum equivalent to a free scalar field theory on T, the maximal torus of G. At first sight this is highly surprising because the two equivalent theories possess different numbers of independent fields. Yet the total number of degrees of freedom is infinite in each case. The scalar field on the torus T "feels" the non-abelian structure of G through the periodicity structure of the torus which is specified by the root lattice of g. [$T = \mathbb{R}^d / \Lambda_R(g)$].

The NAMBU action for the string can be regarded as a chiral model on the group $\mathbb{R}^d$ defined by flat space. The ghost free nature of the theory depends on the conformal symmetry of this action but then holds only in 26 dimensions (10 for the fermionic string). Actually it is the c-number of the Virasoro algebra which is critical and in flat space this equals the dimension (2,10). If we try to follow the Kaluza-Klein philosophy and let the string move on the manifold of $G \times \mathbb{R}^4$ we must add the Wess-Zumino term to the action in order to retain conformal symmetry and hence the no ghost theorem. Since we must choose g simply laced and level 1 because of the string vertex operators, we have, by the above results, that the critical c number equals the rank of g not its dimension. Insofar as supergravity is a special limit of the superstring theory this shows that the recent attempts to use Kaluza-Klein theory in supergravity were wrong, because the seven extra dimensions were apparently used in the wrong way.

We believe that the above comment on the relation between string theories and chiral models were due to WITTEN, as reported by FREUND. Later NEMESCHANSKY and YANKIELOWICZ made similar remarks.

2.6 The discrete sequence of Virasoro c-numbers less than unity

We saw (eq . (2.19)) that Sugawara's construction always yielded a Virasoro c-number c_g exceeding the rank of g and hence never less than unity. Nevertheless c less than unity could result from a judicious choice of $h \subset g$ (and level 1) in the construction $K = \mathcal{L}^g - \mathcal{L}^h$ of section 2.4. We now see that the complete sequence $c = 1 - 6/(m+2)(m+3)$ results from the choice of $G/H = Sp(m+1)/Sp(m) \times Sp(1)$ - with level 1. That $Sp(m) \times Sp(1)$ is a subgroup of $Sp(m+1)$ and inherits the same level that can be seen from the extended Dynkin diagram for the symplectic algebras.

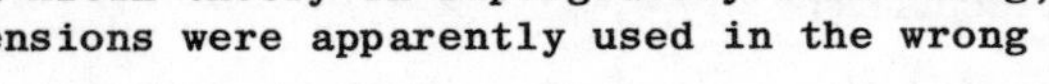

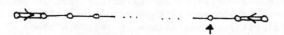

Deletion of the arrowed point leaves the Dynkin diagram for the desired subgroup. Since the deleted point corresponds to a short root it is evident that the highest roots of each factor of h have the same lengths as that of g . Hence as explained in section 2.4 the levels are all the same and hence all of level 1 if we choose that in the first place. Then by (2.5)

$$c_{Sp(m)} = \frac{\dim Sp(m)}{1 + \tilde{h}(Sp(m))} \tag{2.28}$$

$Sp(m)$ has long roots $\pm 2e_i$, $i = 1,2,\ldots,m$, short roots $\pm e_i \pm e_j$ $(i \neq j)$, and rank m. Hence $n_L = 2m$, $n_S = 2m(m-1)$ and $\dim Sp(m) = (2m+1)m$. Thus by (2.18) the dual Coxeter number $\tilde{h}$ equals

$$\tilde{h}_{(Sp(m))} = (n_L + n_S\,(L/S)^{-2})/m = (2m + m(m-1))/m = m+1 ,$$

as quoted at the end of section 2.3. Hence inserting in (2.28)

$$c_{Sp(m)} = \frac{m(2m+1)}{m+2} = 2m - 3 + 6/(m+2) ,$$

and so, as desired (GODDARD, KENT and OLIVE)

$$c_K = c_{Sp(m+1)} - c_{Sp(m)} - c_{Sp(1)} = 1-6/(m+2)(m+3)$$

We shall see in the next chapter that level 1 Sp(m) Kac-Moody algebra representations can be constructed in a unitary way by considering fermion fields in the defining representation of Sp(m). Partial results already exist on finding within the fermionic Fock space highest weight states of K_n corresponding to the critical exponents (1.17) (ALTSCHULER).

The relation between the above construction and the lattice models with the corresponding value of c is not yet clear but it is interesting that the defining representation of Sp(m) used is quaternionic since quaternions are essentially Pauli spin matrices.

It is possible to reformulate the above construction in terms of SU(2) groups, using results of the next section and this may also aid the physical interpretation. It has led to a proof that all the possible h values (1.17) can occur in unitary representations (GODDARD, KENT and OLIVE).

3. The Quark Model

3.1 Representations of Kac-Moody algebras bilinear in fermions

So far our discussion of unitary highest weight representations of (affine untwisted) Kac-Moody algebras has mainly been in general terms and without reference to any specific construction. In this section we shall construct explicit representations in the Fock space of N real fermi fields. This is basically the quark model current algebra representation (in two dimensions), in which currents are represented by bilinears in fermi fields,

$$T^i(z) = \sum_{n\in\mathbb{Z}} T^i_n(z) = \frac{i}{2} H^\alpha(z) M^i_{\alpha\beta} H^\beta(z) \tag{3.1}$$

where the fermi fields $H^\alpha(z)$, $1 \leq \alpha \leq N$, are either periodic or anti-periodic on the unit circle, and the real antisymmetric matrices M^i represent the algebra g, i.e. satisfy eqs.(2.13).

Since the fermi fields $H(z)$ are either periodic or antiperiodic, they have an expansion

$$H^\alpha(z) = \sum b^\alpha_r z^{-r} \tag{3.2}$$

where the sum over r is

<u>either</u> over $r \in \mathbb{Z}$ Ramond (R) periodic case , (3.3a)

<u>or</u> over $r \in \mathbb{Z}+\frac{1}{2}$ Neveu-Schwarz (NS) antiperiodic case , (3.3b)

corresponding to the RAMOND and NEVEU-SCHWARZ fields of string theory. The fermion annihilation and creation operators satisfy the hermiticity condition

$$b_r^{\alpha\dagger} = b_{-r}^{\alpha} \tag{3.4}$$

and the vacuum condition

$$b_r^{\alpha}\, \psi_o = 0\ , \qquad r > 0\ , \tag{3.5}$$

where the vacuum state ψ_o can be taken to be unique in the NS case but, in the R case, it has to be $2^{N/2}$ fold degenerate, N even, so that it provides a representation of the γ-matrix algebra

$$\{b_o^{\alpha}, b_o^{\beta}\} = \delta^{\alpha\beta} \tag{3.6}$$

[The degeneracy is $2^{(N-1)/2}$ if N is odd.]

The Kac-Moody generators defined by eq.(3.1) satisfy the algebra of eq.(1.1) with the level

$$x = 2k/\psi^2 = \kappa_M/\psi^2 \tag{3.7}$$

where κ_M is defined as in eq. (2.14),

$$tr(M^i M^j) = -\,\kappa_M\, \delta^{ij}\ , \tag{3.8a}$$

is related to the value of the quadratic Casimir operator in the representation M^i, Q_M, by eq. (2.15),

$$N\, Q_M = \kappa_M \dim g\ . \tag{3.8b}$$

The level x is called, in the representation theory of the finite-dimensional algebra g , the Dynkin index of the representation given by M^i.

The representation of $\hat{g}$ that we obtain in this way is in general decomposable into a number of irreducible components, each an irreducible highest weight unitary representation of $\hat{g}$. Such representations are characterised by the level x and the irreducible representation of g into which the vacuum or highest weight states, i.e. those satisfying (1.14), fall. If λ is the highest weight of this vacuum representation of g , then unitarity implies that

$$x\frac{\psi^2}{\alpha^2} = \frac{2k}{\alpha^2} \geq \left|\frac{2\lambda.\alpha}{\alpha^2}\right| \tag{3.9}$$

for all roots α. In fact the unitary representations of $\hat{g}$ are labelled by levels and highest weights of g , (x,λ), satisfying (3.9).

To establish (3.9) we rewrite (1.1) using a Cartan-Weyl basis for g, consisting of a Cartan subalgebra T^i, $1 \le i \le \text{rank}\, g$, and step operators E^α, where α ranges over the roots of g. Then $\hat{g}$ takes the form

$$[T^i_m, T^j_n] = \kappa_m \delta^{ij} \delta_{m,-n}, \quad 1 \le i,j \le \text{rank}\, g, \tag{3.10}$$

$$[T^i_m, E^\alpha_n] = \alpha^i E^\alpha_{m+n}, \tag{3.11}$$

$$[E^\alpha_m, E^\beta_n] = \varepsilon(\alpha,\beta)\, E^{\alpha+\beta}_{m+n}, \quad \alpha+\beta \text{ a root}, \tag{3.12a}$$

$$= \frac{2}{\alpha^2}(\alpha.H_{m+n} + mk\delta_{m,-n}), \quad \alpha = -\beta, \tag{3.12b}$$

$$= 0, \quad \text{otherwise}. \tag{3.12c}$$

Here each $\varepsilon(\alpha,\beta) = \pm 1$. From these commutation relations we have in particular

$$[E^\alpha_1, E^{-\alpha}_{-1}] = \frac{2}{\alpha^2}(\alpha.T_o + k) \tag{3.13}$$

so that $E^\alpha_1, E^{-\alpha}_{-1}, \frac{1}{\alpha^2}(\alpha.T_o + k)$ form an su(2) algebra, isomorphic to I_+, I_-, I_3. Thus any eigenvalue of $\frac{2}{\alpha^2}(\alpha.T_o + k)$ must be integral. Hence

$$\frac{2}{\alpha^2}(\alpha.\lambda + k) \ge 0 \tag{3.14}$$

for all roots α, from which (3.9) follows.

The condition (3.9) can be quite restrictive. For g = so(n), in level 1 representations of $\hat{g}$, the vacuum can only transform like the scalar, vector or spinor representations of so(n). For g = su(n), the vacuum representations of g occurring in level 1 representations of $\hat{g}$ are the scalar, n and $\bar{n}$.

Let us consider some specific examples of the fermionic representation (3.1) of $\hat{g}$. If we take M^i to be the n-dimensional representation of g = so(n) we obtain a level 1 representation of $\hat{g}$. The Neveu-Schwarz and Ramond cases yield inequivalent representations. In the Neveu-Schwarz case it decomposes into two irreducible components with the vacuum transforming under the scalar and vector representations of so(n), respectively. In the Ramond case, we again get two irreducible components if n is even but only one if n is odd; the vacuum representations correspond to the possible spinor representations of so(n).

The possible level 1 representations of the other classical groups can be obtained from the inclusions $su(n) \subset so(2n)$ and $sp(n) \subset su(2n) \subset so(4n)$. By taking tensor products of these and the level 2 representations of $so(n)$ given by the inclusion $so(n) \subset su(n) \subset so(2n)$, we can obtain all the representations of the classical groups satisfying (3.9).

3.2 Free fermions in two dimensions

We shall now consider in more detail how the representation discussed in the last section is obtained from the current algebra associated with real free massless fermion fields in one space and one time dimension. To describe such a field, we use a real representation of the two-dimensional γ-matrices:

$$\gamma_o = \begin{pmatrix} 0 & 1 \\ 1 & 0 \end{pmatrix} , \quad \gamma^1 = \begin{pmatrix} 0 & -1 \\ 1 & 0 \end{pmatrix} , \quad \gamma^5 \equiv \gamma^o\gamma^1 = \begin{pmatrix} 1 & 0 \\ 0 & -1 \end{pmatrix} , \tag{3.15}$$

and a field ψ with two real spinor components which, following WITTEN, we write as

$$\psi = \begin{pmatrix} \psi_- \\ \psi_+ \end{pmatrix} , \tag{3.16}$$

for reasons that will become apparent. The system is described by the action

$$S = \frac{1}{2} \int d^2x \, \bar{\psi} \, i\gamma^\mu \partial_\mu \psi \tag{3.17}$$

where $\bar{\psi} = \psi^T \gamma^o$. This leads to the Dirac equation of motion

$$\gamma^\mu \partial_\mu \psi = 0 \tag{3.18}$$

where, as usual, $\partial_\mu = \partial/\partial x^\mu$, $x^o = t$, $x^1 = x$. Eq.(3.18), when written out in components, is equivalent to

$$(\partial_o + \partial_1)\psi_- = (\partial_o - \partial_1)\psi_+ = 0 . \tag{3.19}$$

These Weyl equations say that ψ_+ and ψ_- are functions of $t+x$ and $t-x$ respectively:

$$\psi_+ \equiv \psi_+(t+x) , \quad \psi_- \equiv \psi_-(t-x) . \tag{3.20}$$

Each of ψ_+, ψ_- is a Weyl (i.e. an eigenvector of γ_5) and Majorana (i.e. real) spinor. (Such spinors only exist in space-time dimensions $8n+2$, where n is an integer; see GLIOZZI, OLIVE and SCHERK.)

For the moment let us consider only one of the two Weyl components of ψ, $\psi_+(t+x)$ say. Independently, for each of the Weyl components, we have canonical anticommutation relations

$$\{\psi(x),\psi(y)\} = \hbar\delta(x-y) \tag{3.21}$$

for $\psi \equiv \psi_+$, say, with $\{\psi_+(x),\psi_-(y)\} = 0$. Actually we need to consider an elaboration of this theory in which we have an internal symmetry index α taking values from 1 to N; that is N non-interacting copies of the real fermion theory we have been discussing. Then the anticommutation relations (3.21) are replaced by

$$\{\psi_\alpha(x),\psi_\beta(y)\} = \hbar\delta(x-y)\delta_{\alpha\beta} . \tag{3.22}$$

From these fields we can construct a current algebra associated with any algebra g which has an N-dimensional real antisymmetric representation. Then g is a subalgebra of the so(n) symmetry algebra of the theory. Associated with this symmetry are conserved currents

$$J^i_\mu = \frac{i}{2\sqrt{2}}\bar{\psi} M^i \gamma_\mu \psi . \tag{3.23}$$

It is convenient to define light-cone coordinates for vectors $v = (v^0, v^1)$ by

$$v^\pm = (v^0 \pm v^1)/\sqrt{2} , \quad v_\pm = (v_0 \pm v_1)/\sqrt{2} , \tag{3.24}$$

so that $v^\pm = v_\mp$ and, for two vectors v and w,

$$v^\mu w_\mu = v^+ w^- + v^- w^+ = v_+ w_- + v_- w_+ . \tag{3.25}$$

Then

$$J^i_\pm = \frac{i}{2} \psi^T_\pm M^i \psi_\pm \tag{3.26}$$

so that J^i_+ is only a function of $x^+ = (t+x)/\sqrt{2}$ and J^a_- is only a function of $x^- = (t-x)/\sqrt{2}$. Again we can consider independently $J^i_+(x^+)$ and $J^i_-(x^-)$ and they will commute with one another because ψ_+ and ψ_- anticommute. Let us denote either by $J^i(\xi)$. Then the canonical anticommutations imply the current algebra (1.6),

$$[J^i(\xi),J^j(\eta)] = if^{ij\ell}J^\ell(\xi)\delta(\xi-\eta) + \frac{i\kappa_M}{4\pi}\delta^{ij}\delta'(\xi-\eta)\hbar^2 . \tag{3.27}$$

Imposing periodicity as in eq.(1.5),

$$J^i_\mu(x,t) = J^i_\mu(x + 2\pi R,t) , \tag{3.28}$$

the dimensionless fields of the last section are obtained by

$$J^i(\xi) = \frac{\hbar}{2\pi R} T^i(e^{i\xi/R}) \tag{3.29}$$

and

$$\psi^\alpha(\xi) = \sqrt{\frac{\hbar}{2\pi R}} H^\alpha(e^{i\xi/R}) . \tag{3.30}$$

3.3 Two Virasoro algebras

In the context of section 3.2, there are two Virasoro algebras that can be defined naturally. We can employ the Sugawara construction of eq.(2.1), to obtain the Virasoro algebra $\mathcal{L}_n^g$ from the Kac-Moody algebra (3.1). Alternatively we use the energy-momentum tensor of the free fermion theory to produce a Virasoro algebra, or rather two commuting ones, as in section 1.1. If we concentrate on one of these, θ_{++}, we have, in terms of eq.(3.30), that it is proportional to

$$L(z) \equiv \sum L_{-n} z^n = \frac{i}{2} z \,{}^\circ_\circ \frac{dH}{dz} H {}^\circ_\circ + \varepsilon N \tag{3.31}$$

where

$$\varepsilon = 0 \text{ (NS case)} \quad \text{or} \quad \frac{1}{16} \text{ (R case)} ,$$

and the open dots denote normal ordering with respect to the fermi oscillators,

$${}^\circ_\circ b_r b_s {}^\circ_\circ = b_r b_s \qquad \text{if} \quad r < 0 , \tag{3.33a}$$

$$= \frac{1}{2}[b_r, b_s] \qquad \text{if} \quad r = 0 , \tag{3.33b}$$

$$= -b_s b_r \qquad \text{if} \quad r > 0 . \tag{3.33c}$$

The expressions (3.31) are familiar from string theory. They satisfy the Virasoro algebra (1.3) with $c = \frac{1}{2}N$.

We see from this that a single real fermi field (N=1) yields the first non-zero value of the FQS series, $c = \frac{1}{2}$. All of the corresponding values of h occur; the states $|0\rangle$, $b_{-\frac{1}{2}}|0\rangle$ in the NS case have $h = 0$ and $h = \frac{1}{2}$, respectively, whilst the vacuum in the R case has $h = \frac{1}{16}$. These values are relevant to the critical values of the Ising model.

The Virasoro algebra (3.31) satisfies

$$[L_m, b_r^\alpha] = - r\, b_{m+r}^\alpha \tag{3.34}$$

and hence

$$[L_m, T_n^i] = - n\, T_{m+n}^i \tag{3.35}$$

as in eq.(1.2). But we also have that

$$[\mathcal{L}^g_m, T^i_n] = -n\,T^i_{m+n} \,, \tag{3.36}$$

so that if, as in section 2.4, we introduce K_m by (GODDARD and OLIVE, 1985),

$$L_m = \mathcal{L}^g_m + K_m \,, \tag{3.37}$$

$$[K_m, T^i_n] = 0 \,. \tag{3.38}$$

From this it follows that

$$[K_m, \mathcal{L}^g_n] = 0 \,. \tag{3.39}$$

Thus eq.(3.37) expresses L_m as the sum of commuting Virasoro algebras.

The c-number in the Virasoro algebra $\mathcal{L}^g_n$ is by (2.5),

$$c_g = \frac{\kappa_M \dim g}{\kappa_M + Q_\psi} = \frac{NQ_M}{\kappa_M + Q_\psi} \tag{3.40}$$

and so the c-number for K_n is

$$c_K = \frac{N}{2}\left(1 - \frac{2Q_M}{\kappa_M + Q_\psi}\right) \tag{3.41}$$

which hence must be non-negative.

There are a number of situations in which we get interesting values for c_K. For the 7 dimensional irreducible representation of $g = so(3)$ or G_2 we get $c_K = \frac{7}{10}$. For the 6 dimensional representation of $sp(3) \oplus u(1)$ we get $c_K = \frac{4}{5}$.

There are also many cases in which we get $c_K = 0$, e.g. the n of $so(n)$ the $n \oplus \bar{n}$ of $u(n)$, the $2n$ of $sp(n) \oplus sp(1)$, the adjoint representation of any g. Now, since we have highest weight representations of L_n, $\mathcal{L}_n$, and since K_n, $\mathcal{L}_n$ commute, by eq.(3.37) we must have a highest weight representation of K_n. So, if $c_K = 0$, it follows that $K_n = 0$ because this is the only unitary highest weight representation with this c-number; in this case

$$L(z) = \mathcal{L}^g(z) \,. \tag{3.42}$$

So we see that

$$c_K = 0 \tag{3.43}$$

is the condition that the energy momentum tensor for the free fermion theory can be written in the Sugawara form associated with the algebra g . In the next section we shall see that it also has another interpretation, as the condition for equivalence of a principal G σ-model and a free fermion theory.

Finally in this section let us note that the fact that $c_K = 0$ for the N dimensional representation of so(N) means that

$$L_n = \mathcal{L}_n^{so(N)} \tag{3.44}$$

and hence

$$K_n = \mathcal{L}_n^{so(N)} - \mathcal{L}_n^{g} . \tag{3.45}$$

This shows that the construction of K_n in this section is not merely analogous to that of section 2.4 but is a particular case of it corresponding to $g = so(N)$.

3.4 Nonabelian bosonization

Our aim in this section is to develop an interpretation of eq. (3.42) as the condition for the equivalence of a principal G σ-model, with a suitable Wess-Zumino term, to a free fermion theory. The action for a principal G σ-model with a Wess-Zumino term takes the form

$$-\frac{1}{4\lambda^2}\int \mathrm{tr}(g^{-1}\partial_\mu g\, g^{-1}\partial^\mu g)\, d^2x + \nu\hbar\Gamma \tag{3.46}$$

where

$$\Gamma = \frac{1}{24\pi}\int_B \varepsilon^{\lambda\mu\rho}\mathrm{tr}(g^{-1}\partial_\lambda g\, g^{-1}\partial_\mu g\, g^{-1}\partial_\rho g)\, d^3y , \tag{3.47}$$

B is a three-dimensional manifold whose boundary is space-time and ν has to be an integer in order that the functional integral, defining the quantum theory, be single-valued (for further details see WITTEN, or the review of GODDARD).

The theory is conformally invariant if $\lambda^2 = 4\pi/\nu\hbar$. Then the equation of motion implies that the light-cone components

$$J_+ = g^{-1}\partial_+ g , \quad J_- = (\partial_- g)g^{-1} \tag{3.48}$$

are functions of x^+ only and x^- only, respectively;

$$J_\pm \equiv J_\pm(x^\pm) , \quad x^\pm = \frac{1}{\sqrt{2}}(t\pm x) . \tag{3.49}$$

Further the canonical commutation relations imply that these $J_\pm$ satisfy commuting "Kac-Moody algebras" with level ν . To make this more precise, we introduce an infrared cut-off by putting the theory onto a circle of radius R , though we may hope eventually to let $R \to \infty$. Then, if we set

$$-i\,[\frac{\nu R}{\sqrt{2}}]\,g^{-1}\partial_{+}g = t^{j}T^{j}(z)\ , \tag{3.50}$$

where $\{t^j\}$ denotes an orthonormal basis for the Lie algebra, the canonical commutation relations for the σ-model give

$$[T^{i}_{m},T^{j}_{n}] = if^{ij\ell}T^{\ell}_{m+n} + \frac{\nu}{2}\psi^{2}m\,\delta^{ij}\delta_{m,-n}\ . \tag{3.51}$$

(A similar statement can be made for the components of J_-.) We can represent J_+ (and so, of course, similarly J_-) by bilinears in fermion fields using eq.(3.1), but it is necessary that the Kac-Moody algebras have the same level, i.e.

$$\nu = \kappa_M/\psi^2 \tag{3.52}$$

But this condition is not sufficient for the σ-model to be equivalent to the free fermion theory; it is also necessary for the fermion energy momentum tensor of the theories to be equated under the isomorphism of the two Kac-Moody algebras.

In both models, the energy momentum tensor, $\theta_{\mu\nu}$, is traceless, i.e. $\theta_{+-} = 0$, so that it has independent components θ_{++},θ_{--}. For the σ-model

$$\theta_{++} = \frac{1}{2}J^{a}_{+}J^{a}_{+} \sim \mathcal{L}^{g}(z)\ , \tag{3.53}$$

whilst for the free fermion theory

$$\theta_{++} = \frac{i}{2}z\,{}^{\circ}_{\circ}\frac{dH}{dz}H{}^{\circ}_{\circ} \sim L(z)\ . \tag{3.54}$$

Thus what we also need for equivalence is eq.(3.42), at least up to quantum effects. Actually what one proves in general is that

$$\frac{1}{2}{}^{\times}_{\times}T(z)T(z){}^{\times}_{\times} = \frac{1}{2}{}^{\circ}_{\circ}T(z)T(z){}^{\circ}_{\circ} + Q_M\,L(z) \tag{3.55}$$

which relates the operation of normal ordering with respect to the components T^i_m of $T^i(z)$ to that of normal ordering with respect to the fermionic oscillators b^{α}_{r}. So in order to achieve (3.42) we need

$${}^{\circ}_{\circ}T(z)T(z){}^{\circ}_{\circ} = 0 \tag{3.56}$$

This happens exactly if $c_K = 0$. For if $c_K = 0$, $K_n = 0$ and, by (3.41),

$$\kappa_M + Q_\psi = 2Q_M\ , \tag{3.57}$$

so that

$$\mathcal{L}^{g}(z) = \frac{1}{2Q_M}{}^{\times}_{\times}T(z)T(z){}^{\times}_{\times} = L(z) \tag{3.58}$$

implying (3.56). On the other hand if eq.(3.56) holds, (3.57) follows because the normalisation of a Virasoro algebra is uniquely fixed. This in terms implies $c_K = 0$.

3.5 Symmetric spaces and fermion-boson equivalence

Hence we have seen that for the principal G σ-model Wess-Zumino term to be equivalent, at the conformal point $\lambda^2 = 4\pi\nu/\not{k}$, to the theory of free fermions transforming under the real representation M of G , it is necessary not only that the level κ_M/ψ^2 of the corresponding fermionic representation of $\hat{g}$ be equal to the coefficient, ν , of the Wess-Zumino term, but also that $\circ T(z)T(z)\circ$ vanishes, and further that this condition is equivalent to the vanishing of the Virasoro algebra K_n which in turn is equivalent to the vanishing of c_K , defined by eq.(3.41). We have pointed that this condition obtains for the adjoint representation of any group, the n-dimensional representation of so(n) , the $n \oplus \bar{n}$ of u(n) , etc. Other examples may be found quite easily by trial and error but now we derive a general and geometrical criterion (GODDARD, NAHM and OLIVE).

Firstly, it follows immediately from

$$\circ T^i(z)T^i(z)\circ = -\frac{1}{4} M^i_{\alpha\beta}M^i_{\gamma\delta}\ \circ H^\alpha(z)H^\beta(z)H^\gamma(z)H^\delta(z)\circ \tag{3.59}$$

that the condition (3.56) is equivalent to

$$M^i_{\alpha\beta}\ M^i_{\gamma\delta} + M^i_{\alpha\delta}\ M^i_{\beta\gamma} + M^i_{\alpha\gamma}\ M^i_{\delta\beta} = 0\ . \tag{3.60}$$

This condition can be rephrased as the condition that there should exist a group G' , with Lie algebra g' , such that $G' \supset G$ and G'/G is a symmetric space whose tangent space g'/g transforms under G under the representation M . This condition means that, if $g = \{t^i\}$, $g' = \{t^i\}\ \ \{p^a\} \equiv \{t^I\}$, where $tr(t^i p^a) = 0$ (so using capital Latin letters for g' , small Latin letters for g and Greek letters for the part of g' orthogonal to g),

$$[t^i, t^j] = if^{ij\ell}t^\ell\ , \tag{3.61a}$$

$$[t^i, p^\alpha] = if^{i\alpha\beta}p^\beta\ , \tag{3.61b}$$

$$[p^\alpha, p^\beta] = if^{\alpha\beta j}t^j\ , \tag{3.61c}$$

and

$$M^i_{\alpha\beta} = f^{i\alpha\beta} \tag{3.62}$$

Choosing the extra generators p^α orthogonal to g implies

$$f^{ij\alpha} = 0\ , \tag{3.63}$$

whereas the condition that G'/G is a symmetric case amounts to

$$f^{\alpha\beta\gamma} = 0 \ , \tag{3.64}$$

as indicated by eq.(3.61c).

We can now see that, if we have such a symmetric space, eq.(3.60) follows from the Jacobi identity

$$f^{IJK} f^{LMK} + f^{ILK} f^{MJK} + f^{IMK} f^{JLK} = 0 \ . \tag{3.65}$$

Taking $I = \alpha$, $J = \beta$, $L = \gamma$, $M = \delta$, the sum in practice only involves K ranging over small Latin letters. The argument can easily be reversed to show that g can be extended to g' if eq.(3.60) holds. Thus we have arrived at the result that a principal G σ-model with Wess-Zumino terms with integral coefficient ν can be equivalent to a theory of free fermions only if ν is the Dynkin index of a representation M of G given by a symmetric space G'/G . A virtue of this result is that symmetric spaces have been classified and tabulated; see e.g. the book of HELGASON, from which a table is quoted in GODDARD, NAHM and OLIVE.

One extension of these ideas is to consider adding to the system consisting of fermions transforming according to the representation g'/g of g , fermions in the adjoint representation of g . This system is manifestly supersymmetric if we bosonize the g'/g fermions. On the other hand it can be described <u>either</u> entirely in terms of bosons by bosonising the adjoint representation g fermions (which satisfy our equivalence criterion as they correspond to the symmetric space $g \times g/g$) <u>or</u> entirely in terms of fermions, which altogether form the adjoint representation of g' . It may well be that such supersymmetric systems, which have entirely bosonic descriptions, are of particular interest. They recall the early attempts by KRAEMMER and NIELSEN and by GODDARD to build the spinning string by using a twistable boson string. In this work the relation between the bosonic and fermionic degrees of freedom was made using the current algebra representation of Kac-Moody algebras but WITTEN's understanding of the Wess-Zumino term was lacking.

4. Review of further developments

In this last section we shall mention briefly some topics which we have not been able to treat in the previous sections.

In section 2.5 we stated that, following the work of FRENKEL and KAC, and of SEGAL, vertex operators can be used to construct all the level 1 representations of the Kac-Moody algebras $\hat{g}$ associated with simply laced algebras g (i.e. algebras of A, D or E type). FRENKEL also showed how to use these operators to construct fermion fields (if g is of D type), which are Ramond or Neveu-Schwarz fields. These ideas are developed further in GODDARD and OLIVE (1984). They complete the fermion-boson equivalence by inverting the map which expresses bosons as bilinears in fermions.

In general, these fermion fields have to transform under the vector representation of D_n except that, in the case $n=4$, where because of the triality properties of $D_4 = so(8)$, spinorial fermions can be constructed (GODDARD, OLIVE and SCHWIMMER). This property of so(8) underlies the calculations that enabled GREEN and SCHWARZ (1981) to explicitly establish the space-time supersymmetry, evident from the work of GLIOZZI, OLIVE and SCHERK, of the 10-dimensional string model. This same fact can be exploited to construct the level 1 representation of $\hat{E}_8$ out of fermions in an unconventional way. The fermions used are not all independent of one another as in the quark model representation. In a sense, they are interacting.

Yet more generalisations of the vertex operator construction are possible. One interesting possibility, developed by GODDARD and OLIVE (1984) is to regard the FRENKEL-KAC construction as a "transverse" or "light cone gauge" construction in the sense of string theory and consider instead a "covariant" version with contour integrals of the vertex operator associated with length squared 2 points of any integral lattice, with Euclidean, singular or Lorentzian metric, thereby obtaining simply laced finite dimensional Lie algebras, affine untwisted Kac-Moody algebras or Lorentzian algebras respectively. Nestings of different sorts of algebra inside each other correspond to the nestings of the different types of lattice. Particularly interesting lattices to consider are the self-dual even ones because of their relation to electromagnetic duality conjectures (GODDARD, NUYTS and OLIVE) and their associated modularity properties. The no-ghost theorem of string theory (BROWER; GODDARD and THORN) implies that only three such Lorentzian lattices are of interest, those in 10, 18 or 26 dimensions. The first two contain the weight lattices of E_8, $E_8 \times E_8$ and $spin(32)/\mathbb{Z}_2$ in an interesting way as explained by GODDARD and OLIVE (1984). After the dramatic discovery by GREEN and SCHWARZ (1984) that the latter two Lie algebras constituted the unique anomaly free gauge algebras for supersymmetric gauge theories in ten dimensions, these were further developed by GROSS, HARVEY, MARTINEC and ROHM.

We end by remarking that, although one of the remarkable and exciting features of the study of Kac-Moody and Virasoro algebra is how much of mathematics and physics becomes interrelated and unified, there is no doubt that much remains to be explored and discovered.

REFERENCES

ADLER S. AND DASHEN R.: Current Algebras and Applications to Particle Physics (Benjamin, New York 1968).

ALTSCHULER D.: Critical Exponents from Infinite-Dimensional Symplectic Algebras, University of Geneva preprint UGVA-DPT 1985/06-466.

BARDACKI K. AND HALPERN M.: Phys. Rev. D3 (1971), 2493.

BELAVIN A.A., POLYAKOV A.M. AND ZAMOLODCHIKOV A.B.: Nucl. Phys. B241 (19 333.

BROWER R.: Phys. Rev. D6 (1972) 1655.

CALLAN C.G., DAHSEN R.F. AND SHARP D.H.: Phys. Rev. 165 (1968) 1883.

COLEMAN S., GROSS D. AND JACKIW R.: Phys. Rev. 180 (1969) 1359.

DASHEN R. AND FRISHMAN Y.: Phys. Rev. D11 (1975) 278.

DELL'ANTONIO G.F., FRISHMAN Y. AND ZWANZIGER D.: Phys. Rev. D6 (1972) 988

FERRARO S., GRILLO A.F. AND GATTO R.: Nuovo Cimento 12A (1972) 959.

FRENKEL I.B.: Proc. Natl. Acad. Sci. U.S.A. 77 (1980) 6306; J. Funct. Anal. 44 (1981) 259.

FRENKEL I.B. AND KAC V.G.: Inv. Math. 62 (1980) 23.

FRIEDAN D.M., QIU Z. AND SHENKER S.: Phys. Rev. Lett. 52 (1984) 1575; Verte Operators in Mathematics and Physics; MSRI Publication No.3 (Springer 1984) p.491.

FUBINI S., HANSON A.J. AND JACKIW R.: Phys. Rev. D7 (1973) 1732.

FUBINI S. AND VENEZIANO G.: Nuovo Cimento 67A (1970) 29.

GELL-MANN M.: Phys. Rev. 125 (1962) 1067; Phys. Letters 8 (1964) 214.

GELL-MANN M. AND NE'EMAN Y.: The Eightfold Way (Benjamin, New York 1964).

GERVAIS J-L. AND SAKITA B.: Nucl. Phys. B34 (1971) 477.

GLIOZZI F., OLIVE D. AND SCHERK J.: Nuclear Phys. B122 (1977) 253.

GODDARD P.: Nuclear Phys. B116 (1976) 157.

GODDARD P.: Kac-Moody and Virasoro Algebras: Representations and Applications, DAMTP preprint 85/7.

GODDARD P., KENT A. AND OLIVE D.: Phys. Lett. 152B (1985) 88; Unitary representations of the Virasoro and Super Virasoro Algebras, DAMTP preprint 85/21.

GODDARD P., NUYTS J. AND OLIVE D.: Nucl. Phys. B125 (1977) 1.

GODDARD P., NAHM W. AND OLIVE D.: Symmetric Spaces, Sugawara's energy momentum tensor in two dimensions and free fermions; Imperial/TP/84-85/25.

GODDARD P. AND OLIVE D.: Vertex Operators in Mathematics and Physics, MSRI Publication No.3 (Springer 1984) p.51.

GODDARD P. AND OLIVE D.: Nuclear Phys. B257 [FS14] (1985) 226.

GODDARD P., OLIVE D. AND SCHWIMMER A.: Phys. Lett. 157B (1985) 393.

GODDARD P. AND THORN C.: Phys. Lett. 40B (1972) 235.

GOODMAN R. AND WALLACH N.R.: J. Reine. Angew. Math. 347 (1984) 69.

GREEN M. AND SCHWARZ J.: Nuclear Phys. B181 (1981) 502.

GREEN M. AND SCHWARZ J.: Phys. Lett. 149B (1984) 117.

GROSS D., HARVEY J.A., MARTINEC E. AND ROHM R.: Phys. Rev. Lett. 54 (1985) 502; Nucl. Phys. B256 (1985) 253.

HELGASON S.: Differential Geometry, Lie Groups, and Symmetric Spaces, (Academic Press 1978).

JACOB M. (Editor): Dual Theory (North Holland 1974).

JAIN S., SHANKAR R. AND WADIA S.: Conformal Invariance and String Theory in Compact Space:Bosons, Tata Institute Preprint TIFR/PH/85-3.

KAC V.G.: Infinite-dimensional Lie Algebras - An Introduction (Birkhäuser 1983).

KNIZHNIK V.G. AND ZAMOLODCHIKOV A.B.: Nuclear Phys. B247 (1984) 83.

KRAEMMER A.B. AND NIELSEN H.B.: Nuclear Phys. B98 (1975) 29.

MANDELSTAM S.: Phys. Rev. D7 (1973) 3763 and 3777; Phys. Rev. D11 (1975) 3026.

NAMBU Y.: Lectures for the Copenhagen Symposium 1970 (unpublished).

NEMESCHANSKY D. AND YANKIELOWICZ S.: Phys. Rev. Lett. 54 (1985) 620.

NEUEU A. AND SCHWARZ J.H.: Nucl. Phys. B31 (1971) 86; Phys. Rev. D4 (1971) 1109.

OLIVE D.I.: Kac-Moody Algebras: An Introduction for Physicists, Imperial preprint TP/84-85/14.

ONSAGER L.: Phys. Rev. 65 (1944) 117.

RAMOND P.: Phys. Rev. D3 (1971) 2415.

SCHWINGER J.: Phys. Rev. Lett. 3 (1959) 296.

SEGAL G.: Commun. Math. Phys. 81 (1981) 301.

SOMMERFIELD C.: Phys. Rev. 176 (1968) 2019.

SUGAWARA H.: Phys. Rev. 170 (1968) 1659.

TODOROV I.T.: Phys. Lett. 153B (1985) 77.

VIRASORO M.: Phys. Rev. D1(1969) 2933.

WEIS J.: unpublished (1970).

WITTEN E.: Commun. Math. Phys. 92 (1984) 455.

SUPERSTRINGS AND THE LIGHT-CONE GAUGE

Lars Brink

Institute of Theoretical Physics
Chalmers University of Technology
S-412 96 Göteborg
SWEDEN

1. INTRODUCTION

A basic feature of string theories is the geometric origin of the action of the free bosonic string, the Nambu-Hara-Goto action[1]. It was constructed as the area of the world surface traced out by the propagating string; an obviously reparametrization invariant action. To generalize this concept to spinning strings turned out to be not quite straightforward. Such strings demand (classically) Grassmann degrees of freedom and there does not seem to be a generalization of the area concept above to a world surface endowed with anticommuting degrees of freedom. The solution to this problem was to consider the theories as general relativity theories in two space-time dimensions[2] (the world-sheet), which, however, somewhat shadowed the original geometric idea.

It is certainly important to have a geometric understanding of strings, since the interaction, the splitting and joining of strings[3], are clearly geometric concepts. However, since progress in this direction has turned out to be slow it is natural to seek other ways of describing strings and their interactions. This is what I will do in these lectures. In fact, I will follow a rather extreme line; a line that will lead quickly to results, but will obscure the geometry. I will not present it as a fundamental principle, but just as a practical way which we can use while we are searching for a more fundamental starting point.

2. BOSONIC STRINGS

We start by considering a bosonic string described by its coordinate $x^\mu(\sigma,\tau)$. The world-sheet is spanned by a time-like variable τ and a space-like one σ which runs between 0 and π, and μ runs over the d-dimensional Minkowski space. Although we will start by considering a free string, we must construct a free such theory which could interact in a natural way. (In fact, one lesson we have learnt from string theory is that in a consistent theory, interactions are essentially built in if we just allow the free strings to split and join[3]).

The interacting theories that we aim at should satisfy

(i) Poincaré invariance
(ii) causality
(iii) unitarity.

In the standard treatment the Poincaré invariance is automatic; the unitarity is ensured by the gauge invariance of the theory, the reparametrization invariance, and causality is implemented by choosing the correct description of the interactions. Here we will start in another end. In order for a string to be able to interact causally, we demand that each point along the string carry the same time, i.e. the time coordinate should not depend on σ. This is certainly the first requirement to impose in order to have consistent interactions for extended objects. In a gauge invariant theory, it is enough that the points carry the same time up to a gauge transformation. We do have a choice here, however. According to the analysis of Dirac[4] we can either choose x^o, or one of the light-cone directions $x^\pm \equiv \frac{1}{\sqrt{2}}(x^o \pm x^{d-1})$ as the evolution parameter. (In the sequel we write the components of a vector A^μ as A^+, A^-, A^i and $\eta_{\mu\nu}A^\mu B^\nu = -A^+B^- - A^-B^+ + A^iB^i$). Since we are going to describe a massless string, i.e. a string with no explicit mass term in the free action, it is natural to work in the light-cone frame and choose x^+ as the "time". Our first assumption is then*

$$x'^{+} = 0 \quad , \tag{2.1}$$

i.e.

$$x^{+}(\sigma,\tau) = x^{+} + p^{+}\tau \quad , \tag{2.2}$$

where x^+ and p^+ are integration constants. We will interpret p^+ as the total momentum in the + -direction. In order to do so we have put a dimensionful constant to one. We will soon correct this fact.

To implement the third requirement above, we assume that all the propagating degrees of freedom lie in the transverse part of $x^{\mu}(\sigma,\tau)$, i.e. in $x^{i}(\sigma,\tau)$ where $i=1,\ldots,d-2$. Since we are going to describe freely propagating strings, it is natural to choose an action

$$S = -\frac{T}{2}\int_{\tau_i}^{\tau_f} d\tau \int_{0}^{\pi} d\sigma \; \eta^{\alpha\beta}\partial_{\alpha}x^{i}\partial_{\beta}x^{i} \quad , \tag{2.3}$$

with α and β taking the values τ and σ and the metric $\eta^{\alpha\beta}$ is chosen space-like. T is a proportionality factor (which ensures that x^i is a length) and will turn out to be the string tension.

The possible flaw with this action is that it is not explicitly Poincaré invariant. The invariance can only be proven by an explicit construction of the Poincaré generators. Before performing the analysis let us check the dynamical content of the action. When varying it to obtain the equations of motion some care has to be exercised be-

*We use the standard notation that $\dot{x} \equiv \frac{\partial x}{\partial \tau}$ and $x' \equiv \frac{\partial x}{\partial \sigma}$.

cause of the finite σ-interval. Disregarding possible surface terms, one obtains the equations of motion

$$\ddot{x}^i - x''^i = 0 \quad . \tag{2.4}$$

The canonical conjugate momentum density is

$$p^i(\sigma,\tau) = T\dot{x}^i(\sigma,\tau) \quad , \tag{2.5}$$

and the total momentum is

$$p^i = T \int_0^\pi d\sigma \, \dot{x}^i(\sigma,\tau) \quad . \tag{2.6}$$

By comparing with (2.2) we find that by choosing $T = \frac{1}{\pi}$, also p^+ defined by (2.2) satisfy (2.6). For the rest of the lectures we will make this choice.

We can now quantize the theory by the canonical commutators with $\hbar=1$).

$$\left[x^i(\sigma,\tau),\, p^j(\sigma',\tau)\right] = i\,\delta^{ij}\delta(\sigma-\sigma') \quad . \tag{2.7}$$

Returning to the surface terms we find two sets of boundary conditions which make the surface terms vanish:

(i) $x'^i(\sigma=0,\pi) = 0$, open strings:

$$x^i(\sigma,\tau) = x^i + p^i\tau + i \sum_{n\neq 0} \frac{1}{n}\alpha_n^{\ i} \cos n\sigma \, e^{in\tau} \tag{2.8}$$

(ii) x^i periodic , closed strings:

$$x^i(\sigma,\tau) = x^i + p^i\tau + \frac{i}{2} \sum_{n\neq 0} \frac{1}{n}\left(\alpha_n^{\ i} e^{-2in(\tau-\sigma)} + \tilde{\alpha}_n^{\ i} e^{-2in(\tau+\sigma)}\right) \quad . \tag{2.9}$$

The canonical commutators are satisfied if

$$[\alpha_n^{\ i},\alpha_m^{\ j}] = n\,\delta_{n+m,0}\,\delta^{ij} \tag{2.10}$$

$$[\tilde{\alpha}_n^{\ i},\tilde{\alpha}_m^{\ j}] = n\,\delta_{n+m,0}\,\delta^{ij} \quad . \tag{2.11}$$

The Hilbert space of states can be constructed by introducing a vacuum $|0\rangle$ and demanding

$$\alpha_n^{\ i}\,|0\rangle = 0 \qquad n>0$$

$$\tilde{\alpha}_n^{\ i}\,|0\rangle = 0 \qquad n>0 \quad . \tag{2.12}$$

The space of states is then constructed by letting α^i_{-n} and $\tilde{\alpha}^i_{-n}$, $n>0$, act as creation operators building up an explicitly positive-norm Hilbert space.

It remains to check the Poincaré invariance. The generators must be constructed out of $x^i(\sigma)$, $p^i(\sigma)$, x^+, p^+ and possibly a zero mode x^- conjugate to p^+. Since there is no such linear Lorentz transformation, they must be <u>non-linear</u>. We will first construct the representation of the generators classically and use Poisson brackets instead of the commutators (2.7). We will also use a notation with lower case letters for these generators and use capital letters eventually for second-quantized representations. In the light-cone frame the generator p^- is the Hamiltonian. Now

$$p^- \sim i\,\frac{\partial}{\partial x^+} = i\,\frac{\partial\tau}{\partial x^+}\frac{\partial}{\partial\tau} = \frac{1}{p^+}(i\,\frac{\partial}{\partial\tau}) \sim \frac{1}{p^+}\,H \quad , \tag{2.13}$$

where H is the Hamiltonian connected to the Lagrangian (2.3). The Hamiltonian density is

$$\mathcal{H} = \dot{x}^i p^i - \mathcal{L} = \frac{1}{2\pi}[\pi^2 p^{i^2} + x'^2] \quad . \tag{2.14}$$

Hence the translation part of the algebra is

$$p^{-} = \frac{1}{2\pi p^{+}} \int_{o}^{\pi} d\sigma[\pi^{2} p^{i^{2}} + x'^{i^{2}}] \tag{2.15}$$

$$p^{+} = p^{+} \tag{2.16}$$

$$p^{i} = \int_{o}^{\pi} p^{i}(\sigma) d\sigma \quad . \tag{2.17}$$

The Lorentz part we write as

$$j^{ij} = \int_{o}^{\pi} d\sigma(x^{i} p^{j} - x^{j} p^{i}) \tag{2.18}$$

$$j^{+i} = \int_{o}^{\pi} d\sigma(x^{+} p^{i} - x^{i} p^{+}) \tag{2.19}$$

$$j^{+-} = x^{+} p^{-} - x^{-} p^{+} \tag{2.20}$$

$$j^{-i} = \int d\sigma[x^{-}(\sigma) p^{i}(\sigma) - x^{i}(\sigma) p^{-}(\sigma)] \quad . \tag{2.21}$$

Here we have introduced $x^{-}(\sigma)$ which should be determined in terms of x^{i} and p^{i} apart from its zero mode x^{-}. We have introduced the generators one by one. The generator j^{ij} follows from the explicit transverse invariance in (2.3). The generators j^{+i} and j^{+-} are natural and are easily seen to be correct. Finally we determine j^{-i} first classically (to avoid ordering problems). A tedious calculation shows that the algebra is indeed satisfied if we choose

$$x'^{-}(\sigma) = \frac{\pi}{p^{+}} p^{i} x'^{i} \quad . \tag{2.22}$$

If we so check the quantum algebra we symmetrize the generators in order to keep the hermiticity. This affects the checking of the commutator $[j^{i-}, j^{j-}] = 0$, since j^{i-} is cubic. A detailed computation shows that the commutator is satisfied only if d=26[5] (the critical dimension).

Inserting the solution (2.8) into p^- (2.15) we obtain

$$p^- = \frac{p^{i^2}}{2p^+} + \frac{1}{p^+} \sum_{n=1}^{\infty} \alpha_{-n}^i \alpha_n^i \quad , \tag{2.23}$$

i.e.

$$p^2 = -2 \sum_{n=1}^{\infty} \alpha_{-n}^i \alpha_n^i = -m^2 \quad . \tag{2.24}$$

Reintroduce the dimensionful constant as $\alpha' = \frac{1}{2\pi T}$, the Regge slope. Then the correct expression is

$$\alpha' m^2 = \sum_{n=1}^{\infty} \alpha_{-n}^i \alpha_n^i \quad . \tag{2.25}$$

The mass-squared is hence built up by an infinite set "harmonic oscillator energies"[6]. For the quantum case each oscillator will contribute an energy due to zero-point fluctuations (which is due to the symmetrization above). The lowest mass level will then be

$$\alpha' m_o^2 = \frac{d-2}{2} \sum_{n=1}^{\infty} n \quad . \tag{2.26}$$

This is clearly a divergent sum which must be regularized. We do so by comparing the sum to the Riemann ζ-function[7]

$$\zeta(s) = \sum_{n=1}^{\infty} n^{-s} \quad . \quad \text{Re } s > 1 \quad . \tag{2.27}$$

This is a function that can be analytically continued to s = -1 and

$$\zeta(-1) = -\frac{1}{12} \quad . \tag{2.28}$$

In this way the infinite series has been regularized into

$$\alpha' m_o^{\ 2} = -\frac{d-2}{24} \quad . \tag{2.29}$$

Hence this term should be included in p^-.

There is also a more standard way[6] of obtaining this result by adding counterterms to the action (2.3), which amounts to renormalizing the speed of light, (which, of course, is another parameter of the theory). It is quite attractive that it is the quantum theory that demands a finite speed of light.

The really important consequence of Eq. (2.29) is that the lowest state is a tachyon. This really means that an interacting theory based on this bosonic string will not make sense. Also since the tachyon is the scalar state with no excitations of the higher modes, I find it hard to believe that there exists a consistent truncation in which the tachyon is left out.

Let us now redo the last analysis for closed strings. Inserting the solution (2.9) we find

$$\frac{\alpha'}{2} m^2 = \sum_{n=1}^{\infty} (\alpha_{-n}^{\ i}\alpha_n^{\ i} + \tilde{\alpha}_{-n}^{\ i}\tilde{\alpha}_n^{\ i}) \equiv N + \tilde{N} \quad . \tag{2.30}$$

Also in this sector we find a tachyon. For closed strings we get a further constraint. Consider again Eq. (2.22). Integrating it between 0 and σ we find $x^-(\sigma)$. Although $x^-(\sigma)$ is depending on the x^i's we must demand that it represent a component of $x^\mu(\sigma)$ and hence must be periodic. Then

$$\int_o^\pi d\sigma\, x'^- = 0 = \frac{1}{p^+}\int_o^\pi d\sigma\, \dot{x}^i x'^i = \frac{\pi}{p^+}(N-\tilde{N}) \quad , \tag{2.31}$$

i.e. classically $N=\tilde{N}$ and quantum mechanically we impose this condition on the physical states.

The formulation derived here is the "light-cone gauge" formulation of the Nambu-Hara-Goto string action[1]. In the usual description[5] one starts with a reparametrization and Poincaré invariant action. By specifying a gauge, one arrives at the formulation above. Such an approach which certainly is more general starts with a geometric action. This is quite different from the approach here, where we start with a non-covariant action which, however, explicitly describes free positive-norm states.

The Poincaré algebra spanned by the generators (2.15)-(2.21) with the constraint (2.22) contains all the information about the strings. This is typical for the light-cone gauge. By finding the non-linear representation we know the complete dynamics of the system, since the Hamiltonian is one of the generators. The regrettable thing is that so far we have not found a deductive way to find the generators, but have had to allow some guesswork and then check.

By describing the dynamics of free bosonic strings we have found that such a theory is much more constrained than a corresponding theory for point-particles. This is certainly a most wanted property, since one lesson we have learnt from modern gauge field theories for point particles is that seemingly whole classes of theories are theoretically consistent and only experiments can tell which theories Nature is using. In string theories we can entertain the hope that only one model is consistent and this then should be the theory of Nature!

3. SPINNING STRINGS

The representation of the Poincaré algebra found in the last section was found to lead to an inconsistent theory. To obtain a consistent one, we need to change the expression for p^- (2.15) so as to avoid tachyons. The most natural thing is to introduce a set of anticommuting harmonic oscillators such that their zero-point fluctuations compensate the ones from the commuting oscillators.

The first problem to solve is to determine in which representation of the transverse symmetry group SO(d-2), the new oscillators should be chosen. The x-coordinates belong to the vector representation. We can always try this representation also for the new set, which we shall do first, but we should keep in mind, that for certain values of d-2, there are other representations with the same dimension as the vector one.

Consider hence a set of anticommuting harmonic oscillators d_n^i satisfying

$$\{d_m^{\ i}, d_n^{\ j}\} = \delta_{n+m,o}\,\delta^{ij} \quad . \tag{3.1}$$

If the relevant mass formula for open strings is

$$\alpha' p^2 = -\Big(\sum_{n=1}^{\infty} \alpha_{-n}^{\ i}\alpha_n^{\ i} + \sum_{n=1}^{\infty} n\, d_{-n}^{\ i} d_n^{\ i}\Big) \quad , \tag{3.2}$$

we can deduce p^- from this expression. To write it in a coordinate basis we introduce two normal-mode expansions

$$\lambda^{1i} = \sum_{n=-\infty}^{\infty} d_n^{\ i}\, e^{-in(\tau-\sigma)} \tag{3.3}$$

$$\lambda^{2i} = \sum_{n=-\infty}^{\infty} d_n^{\ i}\, e^{-in(\tau+\sigma)} \quad , \tag{3.4}$$

such that

$$\{\lambda^{Ai}(\sigma,\tau),\ \lambda^{Bj}(\sigma',\tau)\} = \pi\delta^{AB}\delta^{ij}\delta(\sigma-\sigma') \quad . \tag{3.5}$$

The generator p^- can then be written (classically) as

$$p^- = \frac{1}{2\pi p^+} \int_o^\pi d\sigma(\pi^2 p^{i\,2} + x'^{i\,2} - i\lambda^{1i}\lambda'^{1i} + i\lambda^{2i}\lambda'^{2i}) \quad . \qquad (3.6)$$

Before trying to construct the remaining generators let us consider the dynamics following from (3.6). Since $p^- = \frac{1}{p^+} H$, the corresponding action is

$$S = -\frac{1}{2\pi} \int d\tau \int_o^\pi d\sigma[\eta^{\alpha\beta}\partial_\alpha x^i \partial_\beta x^i + i\bar{\lambda}^i \rho^\alpha \partial_\alpha \lambda^i] \quad , \qquad (3.7)$$

where we combine the two λ^i's into two-dimensional 2-component spinors and use the Majorana representation for the 2×2 Dirac matrices, here called ρ^α. To get a consistent theory the surface terms obtained upon variation of (3.7) must be zero. In the case of open strings the boundary conditions for the λ's are

$$\lambda^{1i}(o,\tau) = \lambda^{2i}(o,\tau) \qquad (3.8)$$

$$\lambda^{1i}(\pi,\tau) = \begin{cases} \lambda^{2i}(\pi,\tau) & (3.9a) \\ -\lambda^{2i}(\pi,\tau) & (3.9b) \end{cases} \quad .$$

In fact we have two choices. The first choice together with equations of motion gives the solutions (3.3) and (3.4). In this sector we know that there are no tachyons. The other choice results in expansions

$$\lambda^{1i} = \sum_r b_r^i e^{-ir(\tau-\sigma)} \qquad (3.10)$$

$$\lambda^{2i} = \sum_r b_r^i e^{-ir(\tau+\sigma)} \quad , \qquad (3.11)$$

where the index r takes all half-integer values. The b's satisfy the anticommutators

$$\{b_r^{\,i}, b_s^{\,j}\} = \delta_{r+s,o}\, \delta^{ij} \quad . \tag{3.12}$$

The (classical) mass-shell condition now reads

$$\alpha' m^2 = \sum_{n=1}^{\infty} \alpha_{-n}^{\,i} \alpha_n^{\,i} + \sum_{r=1/2}^{\infty} r b_{-r}^{\,i} b_r^{\,i} \quad . \tag{3.13}$$

Computing the contributions from the zero-point fluctuations we find

$$\begin{aligned} \alpha' m_o^{\,2} &= \frac{d-2}{2}\Big[\sum_{n=1}^{\infty} n - \sum_{n=1}^{\infty} (n-1/2)\Big] \\ &= \frac{d-2}{2} \sum_{n=1}^{\infty} n(1-\frac{1}{2}+1) \\ &\rightarrow -\frac{d-2}{16} \quad , \end{aligned} \tag{3.14}$$

when the sum is renormalized. Again we find tachyons! Hence this sector of the model is unphysical and basing an interacting theory upon it would lead to inconsistencies.

The states we have discovered spanned by the d- and b-oscillators together with the α's are in fact the spectrum of the Ramond[8)]-Neveu-Schwarz[9)] model. The states constructed out of d-oscillators all have to transform as fermions and constitute the states of the Ramond sector, while the ones constructed out of b-modes which are bosonic constitute the Neveu-Schwarz sector. Both sectors are needed in order to have a model with both fermions and bosons.

There is, in fact, a way to truncate the spectrum to avoid tachyons, which can be proven to be consistent with interactions[10)]. Consider the projector

$$P = \frac{1}{2}(1-(-1)^{\Sigma b_{-r} b_r}) \quad . \tag{3.15}$$

By demanding it be zero on physical states

$$P|\text{phys}\rangle = 0 \tag{3.16}$$

we obtain a tachyon free spectrum. A consistent interaction can be set up, if also the spinors are chosen to be of Majorana-Weyl type (the critical dimension is 10). This leads to the superstrings, which we will describe in the next section.

The Poincaré generators can be constructed and in the quantum case the algebra only works in d=10. We will not give them here, but will discuss them in the next section.

In the case of closed strings there are also two sectors depending on what boundary conditions are chosen. One sector is obtained if the λ's are periodic in σ. This leads to 2 sets of integer moded oscillators $d_n^{\,i}$ and $\tilde{d}_n^{\,i}$. The other sector is obtained by choosing the λ's antiperiodic, which leads to 2 sets of half-integer moded oscillators $b_r^{\,i}$ and $\tilde{b}_r^{\,i}$. This sector has a tachyon.

We have discussed the spinning string completely in the light-cone gauge. Also for this string theory, there exists a covariant reparametrization invariant action[2)], which when gauge fixed leads to the representation above.

4. SUPERSTRINGS

In the last section we added anticommuting degrees of freedom to cancel zero-point fluctuation. They transformed as the vector representation of SO(d-2). For d=3,4,6 and 10 we could also choose the lowest spinor representation, since it has the same dimension as the vector one. Since SO(d-2) is a compact group, the scalar product of two spinors is just the contracted sum as for vectors (which is the only product used in sect. 3). We can, then, take over all work in sect. 3. We only make the substitution

$$\lambda^{Ai} \to S^{Aa} \quad , \tag{4.1}$$

where A still is a 2-component spinor index and a is a d-2-component spinor index. This leads to the superstring theory, which was discussed in another formulation in the last section.

The action for the superstring theory is then[11)]

$$S = -\frac{1}{2\pi} \int d\tau \int_o^\pi d\sigma [\eta^{\alpha\beta} \partial_\alpha x^i \partial_\beta x^i + i \, \overline{S}^a \rho^\alpha \partial_\alpha S^a] \quad . \tag{4.2}$$

We know that this action leads to a sector (for open strings) which starts with massless particles as the lowest lying states. In this sector the solution to the equations of motion is

$$S_a^{\ 1} = \sum_{n=-\infty}^{\infty} S_n^{\ a} \, e^{-in(\tau-\sigma)} \tag{4.3}$$

$$S_a^{\ 2} = \sum_{n=-\infty}^{\infty} S_n^{\ a} \, e^{-in(\tau+\sigma)} \tag{4.4}$$

with the anticommutation rule

$$\{S_a^{\ A}(\sigma,\tau), S_b^{\ B}(\sigma',\tau)\} = \pi \, \delta_{ab} \delta^{AB} \, \delta(\sigma-\sigma') \tag{4.5}$$

$$\{S_n^{\ a}, S_m^{\ b}\} = \delta_{n+m,o} \delta^{ab} \quad . \tag{4.6}$$

The operators $S_{-n}^{\ a}$ with n positive are creation operators. They will take a bosonic state that is acts on to a fermionic one. Hence this sector will contain both bosons and fermions, in fact equally many of each kind at each mass level, building up supermultiplets at each level (as we will soon prove).

The other sector which follows by using the other set of boundary conditions corresponding to (3.8) and (3.9b), we know has tachyons. Furthermore the fermionic oscillators will be half-integer moded and there will not be an equal number of bosons and fermions at each mass level, thus ruining the possibility to have a supersymmetry. Since the first sector contains all we want, we simply decree, that we only use the boundary conditions (3.8) and (3.9a).

Similarly for closed strings we decree that we only use periodic boundary conditions. This leads to the following solutions to the equations of motion

$$S_a^{\ 1} = \sum_{n=-\infty}^{\infty} S_n^{\ 1a}\, e^{-2in(\tau-\sigma)} \tag{4.7}$$

$$S_a^{\ 2} = \sum_{n=-\infty}^{\infty} S_n^{\ 2a}\, e^{-2in(\tau+\sigma)} \quad . \tag{4.8}$$

The arduous task now is to check if there is a representation of the Poincaré algebra spanned on this string theory. In fact there is[12], and the marvellous fact is that it can also be extended to a super-Poincaré algebra! Since light-cone supersymmetry[13] might not be too familiar, let me first review it. The supersymmetry charge Q in ten dimensions decomposes into two SO(8) light-cone spinors $Q_+^{\ a}$ and $Q_-^{\ \dot{a}}$, where the indices $a, \dot{a} = 1,\ldots,8$ denote the two inequivalent 8-component spinors of SO(8). The algebra is

$$\{Q_+^{\ a}, Q_+^{\ b}\} = 2p^+ \delta^{ab} \tag{4.9a}$$

$$\{Q_-^{\ \dot{a}}, Q_-^{\ \dot{b}}\} = 2p^- \delta^{\dot{a}\dot{b}} \tag{4.9b}$$

$$\{Q_+{}^a, Q_-{}^{\dot{b}}\} = \sqrt{2}(\gamma_i)^{a\dot{b}} p^i \tag{4.9c}$$

For further notations, see Appendix.

Let me so write down the representation of the super-Poincaré algebra. Again I stress that there is some guesswork behind the construction of it. For the case of closed strings the algebra turns out to be an N=2 super-Poincaré algebra. The most general algebra is

$$p^+ = p^+ \tag{4.10a}$$

$$p^i = \int_o^\pi d\sigma\, p^i(\sigma,\tau) \tag{4.10b}$$

$$p^- = \frac{1}{2\pi p^+} \int_o^\pi d\sigma\, [\pi^2 p^{i^2} + x'^{i^2} - i(S^1 S'^1 - S^2 S'^2)] \tag{4.10c}$$

$$q_1{}^{+a} = \sqrt{\frac{2p^+}{\pi}} \int d\sigma\, S_1{}^a \tag{4.11a}$$

$$q_2{}^{+a} = \sqrt{\frac{2p^+}{\pi}} \int d\sigma\, S_2{}^a \tag{4.11b}$$

$$q_1{}^{-\dot{a}} = \frac{1}{\pi\sqrt{p^+}} \int d\sigma (\gamma^i S_1)^{\dot{a}} (\pi p^i - x'^i) \tag{4.11c}$$

$$q_2{}^{-\dot{a}} = \frac{1}{\pi\sqrt{p^+}} \int d\sigma (\gamma^i S_2)^{\dot{a}} (\pi p^i + x'^i) \tag{4.11d}$$

$$j^{ij} = \int_o^\pi d\sigma [x^i p^j - x^j p^i + \frac{1}{4\pi}(S^1 \gamma^{ij} S^1 + S^2 \gamma^{ij} S^2)] \tag{4.12a}$$

$$j^{+i} = \int_0^\pi d\sigma(x^+ p^i - x^i p^+) \tag{4.12b}$$

$$j^{+-} = x^+ p^- - x^- p^+ \tag{4.12c}$$

$$j^{-i} = \frac{1}{2}\int_0^\pi d\sigma[\{x^-(\sigma),p^i\} - \{x^i,p^-(\sigma)\}$$

$$- \frac{i}{4\pi\sqrt{\pi p^+}} (S^1\gamma^{ij}S^1(\pi p^j - x'^j) + S^2\gamma^{ij}S^2(\pi p^j + x'^j)) + 4\,\frac{p^i}{p^+}] \quad , \tag{4.12e}$$

where

$$x'(\sigma) = \frac{\pi}{p^+} p^i x'^i + \frac{i}{2p^+}(S^1 S'^1 + S^2 S'^2) \quad . \tag{4.13}$$

In fact this algebra is enough to cover all known string models.

(i) Type IIb superstrings: This is the full algebra (4.10)-(4.12) with periodic boundary conditions for the coordinates. Note that this is a chiral model, since the creation operators S^{1a}_{-n} and S^{2a}_{-n} create spinors of only one chirality.

(ii) Type IIa superstrings: The anticommuting coordinate $S_2{}^a$ can instead be chosen to transform as the other spinor representation, $S_2{}^{\dot{a}}$. Nothing is affected in the algebra since S_1 and S_2 are never contracted with each other. The action corresponding to (4.2) cannot be written in a two-dimensional covariant form, but who cares? This model is not chiral since the spinor states can be combined to Majorana states.

(iii) Type I superstrings: For open strings we must use the boundary conditions corresponding to (2.8) and (3.8) and (3.9a). Then $q_1^+=q_2^+$ and $q_1^-=q_2^-$ and the supersymmetry is reduced to an N=1 one. One can also perform this truncation for closed strings.

(iv) The bosonic strings: Put $S^1=S^2=0$. No supersymmetry, of course.

(v) The heterotic string[14]: Consider first the bosonic closed string. The generators can be written as a sum of two pieces, one built from the right-moving part of x and p in terms of α-oscillators and one in terms of the left-moving part in terms of the $\tilde{\alpha}$-oscillators. Both parts separately satisfy the algebra, and one can in principle set one part to zero. Consider so the full algebra (4.10)-(4.12). It is straightforward to see that in the terms where S^A couple to x and p, S^1 couple to the right-moving part and S^2 to the left-moving. (They are right-moving and left-moving resp.) A consistent truncation can now be made by putting, say the left-moving parts to zero. This reduces the algebra to an N=1 supersymmetry. The heterotic string is now constructed by putting together one right-moving superstring constructed as above and a 26-dimensional bosonic left-moving string. For all the details, see the lecture by David Gross.[15]

(vi) The spinning string: By performing triality transformations back to vectors λ^i such as $S^{1a}S^{1a} \to \lambda^{1i}\lambda^{1i}$ and $S^1\gamma^{ij}S^1 \to \lambda^{1i}\lambda^{1j}$ and similarly for S^2 one can easily read off the representation for the spinning string. No supersymmetry survives, of course.

Considering the superstring theories I, IIa and IIb we know we have a quantum theory of free strings where the lowest lying states are massless. Next we like to know what is the spin content of these massless states. For the bosonic string one can choose a scalar vacuum state |0> as the ground state. Putting all fermionic oscillators in (4.12) to zero we find this state to be indeed a scalar one. However, in the superstring case there is an extra piece in the zero mode part of J^{ij} (4.12)

$$s_o^{ij} = \frac{1}{4}(S_o^1\gamma^{ij}S_o^1 + S_o^2\gamma^{ij}S_o^2) \quad . \qquad (4.14)$$

Its effect on a vacuum state will be non-zero in general.

We should also realize that the massless level must be a supermultiplet. Such a one can be constructed from the anticommutator (4.9a) combined with the knowledge of (4.14). The generator q_+^a is real. In the 4-dimensional case it is customary to go from an SO(2) description to a U(1) one forming complex generators (with no Lorentz index) which then build up a Clifford algebra and we can define creation and annihilation operators from which we construct the supermultiplet. For d=10 we have so far used an SO(8) covariant notation. To decompose the generators into creation and annihilation operators we must break the covariance into SU(4) × U(1). This is the formalism we will use eventually for the field theories for open strings. Here, however, I will describe an alternative formalism, which uses the full SO(8) covariance. This method can equally well be used in d=4. Consider the zero mode part of J^{ij} (4.12a) which is the relevant part on vacuum states and start with open strings

$$J_o^{ij} = \ell^{ij} + \frac{1}{2} S_o\gamma^{ij}S_o \equiv \ell^{ij} + s_o^{ij} \quad . \qquad (4.15)$$

The last term in (4.15) is the spin contribution. If we try to start with a scalar vacuum $|0\rangle$ we need a constraint

$$s_o^{ij}\ |0\rangle = 0 \quad . \qquad (4.16)$$

By multiplying (4.16) by another S_o and using (4.16) and Fierz rearrangements we will find that (4.16) can only work for d=4.

For d=10 we have to try the next simplest thing. We take the vacuum to be a vector $|i\rangle$ with $\langle i|j\rangle = \delta^{ij}$. Then we must insist on a constraint

$$s_o^{ij} |k\rangle = \delta_{ik} |j\rangle - \delta_{jk} |i\rangle \quad . \tag{4.17}$$

This time we find when we bang a S_o on (4.17) that for the general state

$$\psi^{ia} \equiv q_+^{\ a} |i\rangle \quad , \tag{4.18}$$

with the decomposition

$$\psi^{ia} = \tilde{\psi}^{ia} + \frac{1}{8} \gamma^i \gamma^j \psi^j \quad , \tag{4.19}$$

where $\gamma^i \tilde{\psi}^{ia} = 0$

that

$$\tilde{\psi}^{ia} = 0 \ , \tag{4.20}$$

but that there is no constraint on

$$|\dot{a}\rangle \equiv \frac{1}{8} (\gamma^i q_+)^{\dot{a}} |i\rangle \quad . \tag{4.21}$$

with $\langle\dot{b}|\dot{a}\rangle = p^+ \delta_{\dot{a}\dot{b}}$. Checking the Lorentz properties of this state we find that

$$s_o^{ij} |\dot{a}\rangle = -\frac{1}{2} (\gamma^{ij})^{\dot{a}\dot{b}} |\dot{b}\rangle \quad . \tag{4.22}$$

Hence it transforms properly as a spinor. From the Fierz property (A.7)

$$q_+^{\ a} q_+^{\ b} = p^+ \delta^{ab} + \frac{1}{16} (\gamma^{ij})^{ab} \ q_+ \gamma^{ij} q_+ \quad , \tag{4.23}$$

we find that no other massless states can be constructed and the massless sector contains then one vector state and one spinor state which we recognize as the Yang-Mills multiplet in 10 dimensions. We can, of course, let both states transform according to some representation (such as the adjoint one) of some internal group. For type I theories, there is a standard way for including internal symmetry quantum numbers in scattering amplitudes introduced by Chan and Paton[16)] a long time ago. What one does is to associate a matrix $(\lambda_i)_{ab}$ with the i'th external string state and multiply the N-point scattering amplitude with a group theory factor $\mathrm{tr}(\lambda_1 \ldots \lambda_N)$. When checking that such factors factorize properly so as to not destroy the factorization properties of the scattering amplitudes one finds that the possible internal symmetry groups are SO(n), U(n) or Sp(2n)[17)]. Note that the exceptional groups are not possible.

It may be instructive to also check the first excited level. Here one can form 128 boson states $\alpha_{-1}{}^{i}\,|j\rangle$ and $S_{-1}{}^{a}\,|\dot{b}\rangle$ and 128 fermionic states $\alpha_{-1}{}^{i}\,|\dot{a}\rangle$ and $S_{-1}{}^{a}\,|i\rangle$. These form various reducible SO(8) multiplets which can be reassembled into SO(9) representation (since they are massive) using the Lorentz generators. Doing this one finds for the bosons the SO(9) representations 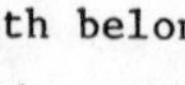 and 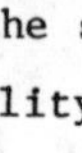 of dimensions 44 and 84 resp. The fermions form a single 128 dimensional Rarita-Schwinger SO(9) multiplet (like the $\tilde{\psi}^{ia}$ in (4.19)).

In the closed string case the massless spectrum is generated by two supersymmetry generators $q_+{}^1$ and $q_+{}^2$. If both belong to the same SO(8) representation (this means that they have the same chirality in 10 dimensions (type IIb), we can form complex generators

$$q_+{}^a = \frac{1}{\sqrt{2}}\,(q_+{}^1 + iq_+{}^2)^a \quad , \tag{4.24}$$

and hence construct creation and annihilation oeprators. For this case we can introduce a scalar vacuum, (which has to be complex, which can be seen from the supersymmetry transformations). Checking J_o^{ij} with the s_o^{ij} term as in (4.14) we see that the vacuum is indeed a scalar state. Choosing q_+^a as a creation operator we can form the following massless supermultiplet

$$| 0\rangle \sim \phi$$

$$q_+^a | 0\rangle \sim \phi^a$$

$$q_+^a q_+^b | 0\rangle \sim \phi^{ab}$$

$$\vdots$$

$$q_+^{a_1} \cdots q_+^{a_8} | 0\rangle \sim \phi^{a_1 \cdots a_8} \quad .$$

It is easily seen that this is a reducible multiplet. To obtain an irreducible representation we impose the conditions

$$(\phi^{a_1 a_2 \cdots a_{2N}})^* = \frac{1}{(8-2N)!} \varepsilon^{a_1 a_2 \cdots a_8} \phi^{a_{2N+1} \cdots a_8} \qquad (4.25)$$

$$(\phi^{a_1 a_2 \cdots a_{2N+1}})^* = \frac{1}{(7-2N)!} \varepsilon^{a_1 a_2 \cdots a_8} \phi^{a_{2N+2} \cdots a_8} \quad . \qquad (4.26)$$

Because of the triality properties of the representations of SO(8) we rewrite the states with vector indices. We now find the bosonic spectrum:

2 scalars ϕ

2 antisymmetric tensors $A^{ij} \sim (\gamma^{ij})^{ab} \phi^{ab}$

1 self-dual antisymmetric tensor $A^{ijk\ell}$

1 graviton g^{ij} ,

where the last two sets of states follow from $\phi^{a_1 \ldots a_4}$. The fermionic spectrum is:

2 spinors ϕ^{a}

2 Rarita-Schwinger states $\tilde{\phi}^{ia}$.

If q_+^1 and q_+^2 have opposite Weyl properties (type IIa) we cannot form creation operators in an SO(8) covariant way. We then have to follow the procedure of the N=1 case. We can start with a tensor state

$$|ij\rangle \sim |i\rangle \otimes |j\rangle \sim \phi^{ij}$$

and generate the states by stuttering the N=1 procedure

$$|a\rangle \otimes |i\rangle \sim \phi^{ai}$$

$$|i\rangle \otimes |\dot{a}\rangle \sim \chi^{\dot{a}i}$$

$$|a\rangle \otimes |\dot{a}\rangle \sim \phi^{a\dot{a}} .$$

The bosonic spectrum is then

1 graviton $g^{ij} = \phi^{(ij)} - \frac{1}{8}\phi^{kk}\delta^{ij}$

1 antisymmetric tensor $A^{ij} = \phi^{[ij]}$

1 scalar $\phi = \phi^{ii}$

1 vector $A^{i} = (\gamma^{i})^{a\dot{a}}\phi^{a\dot{a}}$

1 antisymmetric tensor $A^{ijk} = (\gamma^{ijk})^{a\dot{a}}\phi^{a\dot{a}}$.

The fermionic spectrum is

2 spinors $\dot{\phi}^a$, $\dot{\chi}^a$

2 Rarita-Schwinger states $\tilde{\phi}^{ai}$, $\tilde{\dot{\chi}}^{ai}$.

This technique could, of course, also have been used in the other case above.

We can also impose constraints on the N=2 spectrum to become an N=1 spectrum. These closed strings (type I) are important since they will be able to couple to the open strings. To find this spectrum we linearly combine $Q^1+Q^2 = Q$ and use the N=1 technique. The spectrum is then clearly

$|i\rangle \otimes |j\rangle$

$|i\rangle \otimes |a\rangle$

i.e.

1 graviton	g^{ij}
1 antisymmetric tensor	A^{ij}
1 scalar	ϕ
1 spinor	ψ^a
1 Rarita-Schwinger state	$\tilde{\psi}^{ai}$.

So far we have only discussed free string states. We could anticipate that all closed strings states also belong to some representation of some internal group. However, it will turn out that interaction is only possible if this representation is the trivial one.

Since the strings are extended objects in one dimension there is a possibility that they can carry an intrinsic orientation, like an "arrow" pointing in one direction along its length. When strings are oriented there are two distinct classical states for a single spatial configuration, corresponding to the two possible orientations. An open string is oriented, loosely speaking, if the end points are different, while a closed string is oriented, if one can distinguish a mode running one way around the string from a mode running the other way.

Hence the basic question is whether a string described by $x^{\mu}(\sigma)$, $S^{Aa}(\sigma)$ is the same as one described by $x^{\mu}(\pi-\sigma)$, $S^{Aa}(\pi-\sigma)$. Consider the closed string solution (2.9), (4.7) and (4.8). The replacement $\sigma \to \pi-\sigma$ corresponds to the interchanges $\alpha_n \leftrightarrow \tilde{\alpha}_n$, $S_n^1 \leftrightarrow S_n^2$. In the type II case the two strings connected by the interchanges are different, while for type I the constraints are just such as to make the two strings the same state. Hence we conclude that type II strings are oriented while type I closed strings are non-oriented. Heterotic strings are evidently oriented. The analysis here is classical but can be taken over to the quantum case, by considering matrix elements of the operators $x^{\mu}(\sigma)$ and $S^{Aa}(\sigma)$. The same conclusions are reached here.

For open strings we argued above that one can allow for an internal (global) symmetry (the remnant of the gauge symmetry in a covariant formalism). The interaction allows for letting the states transform as $\phi^a{}_b$ where the index a and b run over the fundamental representation and its complex conjugate resp. Hence $\phi^a{}_b$ transforms either as the adjoint representation or the singlet one. The intuitive way to interpret this is to say that each end carry a "quark" and an "antiquark". If, hence, the "quarks" are different from the "antiquarks" the string is oriented, otherwise not. Now these strings can join their ends to form closed strings, if they are singlets. Then the SU(N) strings will form oriented closed strings, while the SO(n) and Sp(2n) strings will form non-oriented closed strings. But these strings must be of type I, since the open strings are, and hence should be non-oriented. Thus we conclude by this non-rigorous argument that only the gauge groups SO(n) or Sp(2n) are possible[18].

The notation of orientability will be important for perturbation expansions. For oriented strings the interaction must be such that the orientations match up. This will make the perturbation expansion of type I strings different from type II ones.

We have discussed the superstrings so far in the light-cone gauge. This is enough to build an interacting theory as we will see in the next sections. However, it would be advantageous to have a covariant formalism. Such an action has been found by Green and Schwarz[19) but it is not yet clear if that action can be quantized covariantly[20).

5. FIELD THEORY FOR FREE SUPERSTRINGS

To describe the interactions among point-like particles, it is most often advantageous to go over to a second-quantized formalism, a quantum field theory. Consider a scalar particle described by its phase space, x^μ and p^μ. A field ϕ will associate a value to each operator in a commuting set of the phase space, say x^μ. The Lorentz generators act on a field $\phi(x^\mu)$ such that x^μ multiplies and $p^\mu \to -i \dfrac{\partial}{\partial x^\mu}$. The variable x^μ can now be interpreted as a c-number. (It can be taken as the eigenvalue of the operator x^μ). A second-quantized formalism can so be constructed by also introducing a momentum $\pi(x)$ canonically conjugate the $\phi(x)$ and then the generators of the Lorentz algebra can be written as expression

$$G \sim \int d^3x \; \pi(x) \; g \; \phi(x) \tag{5.1}$$

It is straightforward to write an action for $\phi(x)$ which describes any number of free, propagating particles. Interactions are introduced by adding polynomials in $\phi(x)$ which are Lorentz invariant.

We will now follow the same lines to construct a field theory for superstrings. We start by considering the super-Poincaré algebra for type IIb strings (4.10)-(4.12). We need to write it in terms of a maximal set of commuting coordinates. For the bosonic coordinates this is an easy task; we can simply choose to use $x^i(\sigma)$, x^-, x^+ and then represent $p^i(\sigma)$ as $-i\frac{\delta}{\delta x^i(\sigma)} \equiv \delta^i$, $p^+ = i\frac{\partial}{\partial x^-} \equiv i\partial_-$.

The original representation of the anticommuting coordinates is, however, not suitable since the S's do not anticommute with themselves. Instead we choose

$$\theta^a = \frac{1}{\sqrt{2p^+\pi}}(S_1 + iS_2) \qquad (5.2)$$

$$d^a = \sqrt{\frac{p^+}{2\pi}}(S_1 - iS_2) \qquad (5.3)$$

They satisfy

$$\{\theta^a(\sigma), \theta^b(\sigma')\} = \{d^a(\sigma), d^b(\sigma')\} = 0 \qquad (5.4)$$

$$\{\theta^a(\sigma), d^b(\sigma')\} = \delta^{ab}\,\delta(\sigma - \sigma') \quad . \qquad (5.5)$$

We can now use $\theta^a(\sigma)$ as the coordinate and represent $d^a(\sigma)$ as $\frac{\delta}{\delta\theta^a(\sigma)}$. It is not completely trivial to rewrite the algebra in terms of θ^a and d^a since they involve p^+ which has non-trivial commutations with generators j^{+-} and j^{i-}. The correct new representation is[21)22)]

$$p^- = h = \frac{i}{2\partial_-}\int d\sigma\left(\pi\delta^i\delta^i - \frac{1}{\pi}x'^i x'^i + dd'\frac{1}{\partial_-} - \theta\theta'\partial_-\right) \equiv \int d\sigma\, h(\sigma)$$

$$p^+ = i\partial_- \ , \qquad p^i = -i\int d\sigma\,\delta^i$$

$$j^{ij} = -i\int d\sigma\left(x^i\delta^j - x^j\delta^i + \frac{1}{2}\theta\gamma^{ij}d\right)$$

$$j^{+i} = -i\int d\sigma\left(x^+\delta^i + x^i\partial_-\right)$$

$$j^{+-} = x^{+}h - ix^{-}\partial_{-} - \frac{i}{4}\int d\sigma[\theta^{a}, d^{a}] + 2i$$

$$j^{-i} = -\frac{i}{2}\int d\sigma(\{x^{-}(\sigma),\delta^{i}\} - i\{x^{i}, h(\sigma)\}$$

$$+ \frac{1}{2\sqrt{\pi}}\left([(\gamma^{i}\theta)^{\dot{a}}, (\gamma^{j}d)^{\dot{a}}]\frac{\pi\delta^{j}}{\partial_{-}} - (d\gamma^{ij}d\frac{1}{\partial_{-}} - \theta\gamma^{ij}\theta\partial_{-})\frac{x'^{j}}{\partial_{-}}\right) - 4\frac{\delta^{i}}{\partial_{-}})$$

$$q_{1}^{+a} = i\sqrt{2}\int d\sigma\, \theta^{a}\partial_{-}\ , \quad q_{1}^{-\dot{a}} = -\frac{i}{\sqrt{\pi}}\int d\sigma[(\gamma^{i}\theta)^{\dot{a}}\,\pi\delta^{i} - (\gamma^{i}d)^{\dot{a}}\frac{x'^{i}}{\partial_{-}}]$$

$$q_{2}^{+a} = \sqrt{2}\int d\sigma\, d^{a}\ , \quad q_{2}^{-\dot{a}} = -\frac{1}{\sqrt{\pi}}\int d\sigma[(\gamma^{i}d)^{a}\frac{\pi\delta^{i}}{\partial_{-}} + (\gamma^{i}\theta)^{\dot{a}}\, x'^{i}]. \tag{5.6}$$

The $x^{-}(\sigma)$ occurring in j^{-i} is determined by

$$x'^{-}(\sigma) = -\frac{\pi}{\partial_{-}}(x'^{i}\delta^{i} + \theta' d)\ , \tag{5.7}$$

The appropriate field is

$$\Psi[x^{+},x^{-},x^{i}(\sigma),\theta^{a}(\sigma)]\quad ,$$

where the coordinates are c-numbers. This turns out to be a reducible representation. Define the Fourier transformed field

$$\hat{\Psi}[x,\lambda] = \int D^{8}\theta(\sigma)\, e^{\int d\sigma\, \lambda^{a}\theta^{a}}\, \Psi[x,\theta]\quad . \tag{5.8}$$

Impose so the constraint

$$\hat{\Psi}[x,\lambda] = \Psi[x,\theta]^{*}\quad , \tag{5.9}$$

where we identify $\theta^{*} = \frac{1}{p^{+}}\lambda$. This constraint hence relates Ψ^{*} to Ψ, which means that in a variation of fields in an action, Ψ^{*} should not be varied independently from Ψ. Furthermore we impose the constraint that any point on the string can be chosen as the origin, i.e.

$$\Psi[x(\sigma+\sigma_o), \theta(\sigma+\sigma_o)] = \Psi[x(\sigma), \theta(\sigma)] \quad . \tag{5.10}$$

Checking it in a mode basis shows that $N=\tilde{N}$ on a field.

The next step is to second-quantize the theory by imposing a commutator between the field and its conjugate momentum. In the light-cone gauge one can in fact essentially use the field itself or more precisely $\partial_-\Psi$ as the momentum. (Note that ∂_- is a space derivative.) Call a configuration $x^-, x^i, \theta^a \equiv \Sigma$. Then we choose

$$[\partial_-\Psi[\Sigma_1], \Psi[\Sigma_2]]_{x_1^+ = x_2^+} = -\frac{i}{2}\Delta^{17}[\Sigma_1,\Sigma_2] \quad , \tag{5.11}$$

where

$$\Delta^{17}[\Sigma_1,\Sigma_2] \equiv \delta(x_1^- - x_2^-) \int d\sigma_o \, \Delta^8[x_1^{\,i}(\sigma) - x_2^{\,i}(\sigma+\sigma_o)]$$
$$\times \Delta^8[\theta_1^{\,a}(\sigma) - \theta_2^{\,a}(\sigma+\sigma_o)] \quad . \tag{5.12}$$

The ∂_- in the commutator is essential to obtain the correct symmetry. It is only in the light-cone gauge that we can use a space derivative of the field as a momentum. In a covariant description, there is no such covariant expression.

Given the canonical commutation rule (5.11), one can represent the super-Poincaré generators as

$$G = i \int D\Sigma \, \partial_-\Psi[\Sigma] \, g \, \Psi[\Sigma] \quad , \tag{5.13}$$

with g as in (5.6). From the Hamiltonian p^- we can write down the equation of motion

$$\partial_+\Psi = -i\, h\, \Psi \tag{5.14}$$

and construct the action

$$S = \int \partial_-\Psi(\partial_+\Psi + i\, h\, \Psi)D\Sigma\, dx^+$$

$$= \int \partial_+\Psi\partial_-\Psi\, D\Sigma\, dx^+ + i \int H\, dx^+ \quad . \tag{5.15}$$

If we perform a similar analysis for the underlying point-particle field theory one can show that there is an infinity of possible representations of the super-Poincaré algebra[22)], apart from the one obtained by truncating the string to a point. These ones can be interpreted as corresponding to terms with $\Box^2$, $\Box^3$ etc in the action and are possible counterterms to the kinetic term. In an interacting quantum theory these terms will be generated unless there is a symmetry forbidding them. Such a symmetry does not seem to exist and this is a strong indication that the quantum theory does not make sense. In the superstring theory one can prove[22)] that <u>the representation (5.13) is unique</u> leaving no possible counterterm to the kinetic term.

The expressions (5.13) and (5.15) are quite formal and must be given a precise meaning. This can be done by relating them to infinite mode expansions. The coordinates and functional derivatives can be expanded as

$$x^i(\sigma) = x^i + \frac{i}{2} \sum_{n\neq 0} \frac{1}{n}(\alpha_n^i e^{2in\sigma} + \tilde{\alpha}_n^i e^{-2in\sigma})$$

$$\equiv x^i + \sum_{n=1}^{\infty} \frac{1}{\sqrt{n}} (x_n^i \cos 2n\sigma + \tilde{x}_n^i \sin 2n\,\sigma) \tag{5.16}$$

$$\delta^i(\sigma) = \frac{1}{\pi}\{\frac{\partial}{\partial x^i} + i \sum_{n\neq 0} (\alpha_n{}^i e^{2in\sigma} + \tilde{\alpha}_n{}^i e^{-2in\sigma})\}$$

$$\equiv \frac{1}{\pi}\{\frac{\partial}{\partial x^i} + \sum_{n=1}^{\infty} 2\sqrt{n}\,(\frac{\partial}{\partial x_n{}^i}\cos 2n\sigma + \frac{\partial}{\partial \tilde{x}_n{}^i}\sin 2n\sigma)\} \quad , \tag{5.17}$$

where

$$x_n{}^i = \frac{i}{2\sqrt{n}}(\alpha_n - \alpha_{-n} + \tilde{\alpha}_n - \tilde{\alpha}_{-n}) \tag{5.18}$$

$$\tilde{x}_n{}^i = -\frac{1}{2\sqrt{n}}(\alpha_n + \alpha_{-n} - \tilde{\alpha}_n - \tilde{\alpha}_{-n}) \tag{5.19}$$

and

$$\theta^a(\sigma) = \frac{1}{\sqrt{2p^+\pi}}\sum_{n=-\infty}^{\infty}(S_n{}^1 + iS_{-n}{}^2)e^{2in\sigma}$$

$$\equiv \frac{1}{\sqrt{\pi}}\sum_{n=-\infty}^{\infty}(\theta_n{}^1 + i\theta_{-n}{}^2)e^{2in\sigma}$$

$$\equiv \theta_o{}^a + \sum_{n=1}^{\infty}(\theta_n{}^a\cos 2n\sigma + \tilde{\theta}_n{}^a\sin 2n\sigma) \tag{5.20}$$

$$d^a(\sigma) = \sqrt{\frac{p^+}{2\pi}}\sum_{n=-\infty}^{\infty}(S_n{}^1 - iS_{-n}{}^2)e^{2in\sigma}$$

$$\equiv \frac{1}{\pi}\{\frac{\partial}{\partial \theta_o{}^a} + \sum_{n=1}^{\infty} 2(\frac{\partial}{\partial \theta_n{}^a}\cos 2n\sigma + \frac{\partial}{\partial \tilde{\theta}_n{}^a}\sin 2n\sigma)\} \quad , \tag{5.21}$$

where

$$\theta_n{}^a = \frac{1}{\sqrt{2p^+\pi}}(S_n{}^1 + S_{-n}{}^1 + iS_n{}^2 + iS_{-n}{}^2) \tag{5.22}$$

$$\tilde{\theta}_n^{\,a} = \frac{i}{\sqrt{2p^+\pi}}(S_n^{\,1} - S_{-n}^{\,1} - iS_n^{\,2} + iS_{-n}^{\,2}) \quad . \tag{5.23}$$

The coordinates $x_n^{\,i}$, $\tilde{x}_n^{\,i}$, $\theta_n^{\,a}$ and $\tilde{\theta}_n^{\,a}$ are harmonic oscillator coordinates. A field $\Psi[x(\sigma),\theta(\sigma)]$ can then be expanded in component fields using a complete set of harmonic oscillator wave functions $\psi_n(x)$ and corresponding wave functions $\chi_n(\theta)$ for the anticommuting variables

$$\Psi[x(\sigma),\theta(\sigma)] = \sum_{n_k,n'_\ell,m_s,m'_t} \phi_{(n_k,n'_\ell,m_s,m'_t)}(x,\theta_o)$$

$$\prod_k \psi_{n_k}(x_k) \prod_\ell \psi_{n'_\ell}(\tilde{x}_\ell) \prod_s \chi_{m_s}(\theta_s) \prod_t \chi_{m'_t}(\tilde{\theta}_t) \quad . \tag{5.24}$$

The functional measure in (5.15) is then defined such that when the integrations over the harmonic oscillator coordinates are performed the action is

$$S = \int d^{10}x\, d^8\theta [-\frac{1}{2} \sum_{\{n\}} \phi_{\{n\}}(x,\theta_o)(\Box - N_\phi)\phi_{\{n\}}(x,\theta_o)$$

$$+ \frac{i}{2} \sum_{\{m\}} \psi_{\{m\}}(x,\theta_o) \frac{\Box - M_\psi}{\partial_-} \psi_{\{m\}}(x,\theta_o)] \quad . \tag{5.25}$$

In this expression $\{n\}$ incorporates number labels for every bosonic and fermionic oscillator made with the understanding that the bosonic fields are included in the first sum and fermionic ones in the second sum. The action is hence a sum of ordinary kinetic actions for all component fields of the string.

When we turn to open strings and type IIa ones, the construction above cannot be used. As in the construction of the massless supermultiplets in sect. 4, we can find an SO(8) covariant formalism with vector or tensor superfields. These fields will, however, be highly reducible and have to satisfy quite complex constraints, which will make the formalism inaccessible. In a superfield formalism one should try to use scalar superfields to avoid a high redundancy in field components. The way to achieve it here is to break the explicit SO(8) invariance down to SU(4)×U(1)[23]. Consider two SO(8) spinors A^a and B^a. Then in SU(4) notation $A^a \to A_A, \bar{A}^A$, A=1,...,4, and similarly for B^a.

$$A^a B^a = \frac{1}{2}(A_A \bar{B}^A + \bar{A}^A B_A) \tag{5.26}$$

The chiral theories containing spinors S^{1a} and S^{2a} decompose as

$$S^{1a} \to S_A \; , \; \bar{S}^A \tag{5.27}$$

$$S^{2a} \to \tilde{S}_A \; , \; \bar{\tilde{S}}^A \tag{5.28}$$

and the type IIa case as

$$S^{1a} \to S_A \; , \; \bar{S}^A \tag{5.29}$$

$$S^{2\dot{a}} \to \tilde{S}^A \; , \; \bar{\tilde{S}}_A \quad . \tag{5.30}$$

Since the anticommutator is

$$\{S_A(\sigma), \bar{S}^B(\sigma')\} = \pi \, \delta_A{}^B \, \delta(\sigma-\sigma') \tag{5.31}$$

and similarly for the other S's, we use

$$\theta_A = \frac{1}{\sqrt{\pi p^+}} S_A \tag{5.32}$$

$$\frac{\delta}{\delta\theta_A} = \sqrt{\frac{p^+}{\pi}} \, \bar{S}^A \tag{5.33}$$

$$\tilde{\theta}_A = \frac{1}{\sqrt{\pi p^+}} \tilde{S}_A \tag{5.34}$$

$$\frac{\delta}{\delta\tilde{\theta}_A} = \sqrt{\frac{p^+}{\pi}}\, \bar{\tilde{S}}^A \tag{5.35}$$

as coordinates and derivatives and rewrite the algebra (4.10)-(4.12). Sometimes it is convenient to rewrite also the SO(8) vector x^i in SU(4)×U(1) notation by

$$x^i \rightarrow (x^I, x^R, x^L) \quad I = 1,\ldots,6$$

$$x^R = \frac{1}{\sqrt{2}}(x^7 + ix^8)$$

$$x^L = \frac{1}{\sqrt{2}}(x^7 - ix^8) \quad .$$

For the open strings a scalar field

$$\Phi^{ab}[x(\sigma),\theta(\sigma),\tilde{\theta}(\sigma)]$$

can be used, and for closed strings

$$\Psi[x(\sigma),\theta(\sigma),\tilde{\theta}(\sigma)] \quad .$$

The appropriate boundary conditions for the coordinates distinguish the various cases. Also for this case one finds the fields to be reducible and a "reality constraint" as in (5.9) must be imposed. Type I strings also have to satisfy a non-orientation constraint

$$\Phi^{ab}[x(\sigma),\theta(\sigma),\tilde{\theta}(\sigma)] = -\Phi^{ba}[x(\pi-\sigma),\tilde{\theta}(\pi-\sigma),\theta(\pi-\sigma)] \tag{5.36}$$

$$\Psi[x(\sigma),\theta(\sigma),\tilde{\theta}(\sigma)] = \Psi[x(-\sigma),\tilde{\theta}(-\sigma),\theta(-\sigma)] \quad . \tag{5.37}$$

Again one can introduce a canonical commutation relation and rewrite the super-Poincaré algebra in a second-quantized form and find the Hamiltonian and the Lagrangian. As in the previous case one can show that the representations are unique[22]. There are no possible counterterms (apart from the action itself) to the kinetic terms.

At this stage let us digest the notion of a quantum string. In (5.24) we have given each quantum state, that the string can appear in, a local point-like structure. One way of understanding this fact is to consider each point along the string as a separate point-like object, which can be excited to give a specific state. (This will not be a mass eigenstate and hence not a true state, but it is easier to comprehend than the harmonic modes, which are the true states.) However, when a string propagates all the states (or the points along the string) propagate coherently and we must consider an extended object. In the next section we will discuss interactions and consider three-string vertices. If we project each string to specific states, we will obtain three-point functions.

A further remark is that in a superfield $\Psi[x(\sigma),\theta(\sigma)]$, the Grassmann coordinates are more than just book-keeping devices. We cannot Taylor expand to get a finite number of fields $\Psi[x(\sigma)]$ as could be done for point-particle superfields.

6. INTERACTING FIELD THEORY FOR TYPE IIB SUPERSTRINGS

The light-cone gauge formulation is quite advantageous for the construction of interactions. The generator p^- is the Hamiltonian and we can simply ask if we can add interaction terms to it and still keep an algebra that closes. It turns out that all generators that transform the system out of the quantization plane, x^+=const., will have to contain interaction terms. These generators are called Hamiltonians or dynamical generators (in contrast to the linearly realized generators, which are called kinematical ones) in the notation of Dirac[4].

To obtain the precise form of the interaction terms we will work in two steps. Firstly, we will represent the generators in a functional form. A general form for the dynamical generators will be assumed and then the closure of the algebra has to be imposed. In this process we will be able to pin down the expressions to a unique form. Then to give the precise meaning of this form, we must go over to a mode basis to check that each coupling is properly defined. After that we can go back to a functional form again.

We write a general three-string contribution to a generator (in type IIb) as

$$G_3 = i \int D\Sigma_1 D\Sigma_2 \, \Pi[\Sigma_1,\Sigma_2] \partial_- \Psi[\Sigma_1+\Sigma_2] \, g(\sigma_1,\sigma_2) \, \Psi[\Sigma_1] \, \Psi[\Sigma_2], \quad (6.1)$$

where

$$\Pi[\Sigma_1,\Sigma_2] = \delta(x_1^- - x_2^-) \int d\sigma d\sigma' \, \delta^8(x_1^i(\sigma)-x_2^i(\sigma'))$$

$$\delta^8(\theta_1^a(\sigma)-\theta_2^a(\sigma')) \quad , \quad (6.2)$$

i.e. the configurations Σ_1 and Σ_2 have one point in common. This factor will be suppressed in the sequel. The configuration $\Sigma_1+\Sigma_2$ is the union of the configurations Σ_1 and Σ_2. To make the interaction local, the operator $g(\sigma_1,\sigma_2)$ acts at points σ_1 and σ_2 infinitesimally close to the point common to Σ_1 and Σ_2, the interaction point. When we transcribe to the mode basis we will find that convergence factors have to be inserted to damp singular behaviour near this point, but they are of no importance for the closure of the super-Poincaré algebra.

When trying to close the algebra, we will deal with commutators between two- and three string operators. If we let

$$A = i \int D\Sigma \, \partial_- \Psi[\Sigma] \int d\sigma \, a(\sigma) \, \Psi[\Sigma] \quad (6.3)$$

$$B = i \int D\Sigma_1 D\Sigma_2 \, \partial_- \Psi[\Sigma_1+\Sigma_2] b_1(\sigma_1) \, \Psi[\Sigma_1] b_2(\sigma_2) \, \Psi[\Sigma_2] \quad , \quad (6.4)$$

and use the fact that functional operators can be integrated partially between $\Psi[\Sigma_1]$ or $\Psi[\Sigma_2]$ and $\Psi[\Sigma_1+\Sigma_2]$ (with appropriate changes of signs), we get the result

$$[A,B] = i \int D\Sigma_1 D\Sigma_2\, \partial_-\Psi[\Sigma_1+\Sigma_2] \int d\sigma \{[b_1(\sigma_1),a(\sigma)]\, \Psi[\Sigma_1] b_2(\sigma_2)\, \Psi[\Sigma_2] + b_1(\sigma_1)\, \Psi[\Sigma_1]\, [b_2(\sigma_2),a(\sigma)]\, \Psi[\Sigma_2]\} \tag{6.5}$$

under the condition that $a(\sigma)$ does not contain ∂_-. The ∂_--structure must be handled carefully; to take a simple example:

$$[i \int D\Sigma\, \partial_-\Psi[\Sigma] \int d\sigma \frac{a(\sigma)}{\partial_-} \Psi[\Sigma]\ ,\ i \int D\Sigma_1 D\Sigma_2\, \partial_-\Psi[\Sigma_1+\Sigma_2]\, \Psi[\Sigma_1]\Psi[\Sigma_2]]$$

$$= i \int D\Sigma_1 D\Sigma_2\, \partial_-\Psi[\Sigma_1+\Sigma_2] \int d\sigma \frac{1}{\partial_-}\{\frac{a(\sigma)}{\partial_-} \Psi[\Sigma_1]\, \partial_-\Psi[\Sigma_2] + \partial_-\Psi[\Sigma_1]\, \frac{a(\sigma)}{\partial_-} \Psi[\Sigma_2]\} \quad . \tag{6.6}$$

One would expect the three-string vertex to contain terms with at most two transverse functional derivatives, since the three-point couplings involving massless states contain the three-graviton vertex, but here we will allow a general expression in the Hamiltonian.

We will follow the procedure used in constructing ordinary light-cone superfield theory. We start with the three-string part of the dynamical supersymmetry generators $Q_A^{-\dot{a}}$ and the Hamiltonian p^- and consider their mutual commutators as well as the ones with the kinematical generators. We introduce the notation

$$Q^{-\dot{a}}_{A_3} = i \int D\Sigma_1 D\Sigma_2\, \partial_-\Psi[\Sigma_1+\Sigma_2]\, q_A^{\dot{a}}(\sigma_1,\sigma_2)\, \Psi[\Sigma_1]\, \Psi[\Sigma_2] \tag{6.7}$$

$$H_3 = i \int D\Sigma_1 D\Sigma_2\, \partial_-\Psi[\Sigma_1+\Sigma_2]\, h(\sigma_1,\sigma_2)\, \Psi[\Sigma_1]\, \Psi[\Sigma_2] \quad . \tag{6.8}$$

The procedure now is to go through the various (anti-)commutators in order of simplicity. The commutators $\{Q_A^{+a}, Q_B^{-\dot{a}}\}$, $[Q_A^{+a}, H]$,

$[J^{+i}, Q_A^{-\dot{a}}]$ and $[J^{+i}, H]$ force the functional derivatives to enter only via $\underline{d}^a = \frac{d_1^{\ a}}{\partial_{-1}} - \frac{d_2^{\ a}}{\partial_{-2}}$ and $\underline{\delta}^i = \frac{\delta_1^{\ i}}{\partial_{-1}} - \frac{\delta_2^{\ i}}{\partial_{-2}}$. From $\{Q_1^{-\dot{a}}, Q_2^{-\dot{b}}\}$ it follows that x^i appears only through $\underline{x}^i = \frac{x_1^{\ i}}{\partial_{-1}} - \frac{x_2^{\ i}}{\partial_{-2}}$. From this anticommutator it also follows that there is no explicit θ-dependence in the functional operators. Furthermore, the commutator $[J^{+-}, Q_A^{-\dot{a}}]$ yields $\#\,\partial_- + \frac{1}{2}\,\#\,d = 3/2$.

With this knowledge one can attempt a more restricted form for the operator $q_A^{\dot{a}}(\sigma_1,\sigma_2)$. In the commutators $\{Q_A^{-\dot{a}}, Q_B^{-\dot{b}}\}$ the fact that they be proportional to $\delta^{\dot{a}\dot{b}}$ is a very strong restriction. By starting with the part in $q_A^{\dot{a}}$ with no spinor derivatives we can then work upwards in the number of spinor derivatives. After a long and painstaking calculation, where abundant use of SO(8) properties is utilized, one finds, in fact, a unique result for $q_A^{\dot{a}}$ and h.

$$h = \kappa \partial_-^{\ 3} \sum_{n=0,2,4,6,8} c^{ij}_{a_1 \ldots a_n} \underline{d}^{a_1} \ldots \underline{d}^{a_n} \underline{\bar{p}}^i p^j \frac{\partial_{-1}^{\frac{n}{2}} \partial_{-2}^{\frac{n}{2}}}{\partial_-^{\frac{n}{2}}} , \qquad (6.9)$$

where

$$\underline{p}^i = -i\,\underline{\delta}^i + \underline{x}'^i \quad , \quad \underline{\bar{p}}^i = i\,\underline{\delta}^i + \underline{x}'^i \qquad (6.10)$$

$$C^{ij} = \delta^{ij}$$

$$C^{ij}_{a_1a_2} = -\frac{i}{2}\gamma^{ij}_{ab}$$

$$C^{ij}_{a_1a_2a_3a_4} = -\frac{1}{4!}\, t^{ij}_{a_1a_2a_3a_4}$$

$$C^{ij}_{a_1\cdots a_6} = \frac{i}{2\cdot 6!}\,\varepsilon_{a_1\cdots a_8}\,\gamma^{ij}_{a_7a_8}$$

$$C^{ij}_{a_1\cdots a_8} = \frac{1}{8!}\,\varepsilon_{a_1\cdots a_8}\,\delta^{ij} \quad . \tag{6.11}$$

The matrices used are defined by

$$u^{ia}_{abc} = -\gamma^{ij}_{[ab}\gamma^{j}_{c]\dot{a}} \quad , \quad t^{ij}_{abcd} = \gamma^{ik}_{[ab}\gamma^{jk}_{cd]} \quad . \tag{6.12}$$

What remains to be proven is to find the exact ∂_--structure. This can be obtained from the commutators with J^{i-}. This is a very long and intricate calculation, which should be done to check the whole algebra. To find just the ∂_--structure one can try a simpler method such as checking a four-point amplitude. This has been done in the open-string case[23].

To be complete, let us now investigate the Hamiltonian in the mode basis. Consider the Hamiltonian (6.8) with h given by (6.9). Let us introduce the following integration

$$\int D\Sigma_3\ \Delta^{17}[\Sigma_3-\Sigma_1-\Sigma_2] = 1 \tag{6.13}$$

to make the expression symmetric in the strings 1, 2 and 3. It is here advantageous to rescale the σ-variables to be proportional to p^+. In these new variables the length of string 3 is the lengths of string 1 + string 2, and the condition (6.13) just says that strings 1+2 lie on top of string 3 at the interaction or that string 3 just splits into strings 1 and 2 such that their lengths in σ match up. This is clearly a very geometric interaction! Unfortunately, in this formulation there is the operator $h(\sigma_1,\sigma_2)$, acting at the point of splitting, which somewhat shadows the pure geometric interpretation of splitting. To perform the explicit integrations over all higher modes is an arduous task. Let us instead consider the expression

$$E = \kappa \int D\Sigma_1 D\Sigma_2 D\Sigma_3 \; \Delta^{17}[\Sigma_3-\Sigma_1-\Sigma_2] \; \Psi[\Sigma_1] \; \Psi[\Sigma_2] \; \Psi[\Sigma_3] \quad . \tag{6.14}$$

If one expands all coordinates in their normal modes and integrate over all non-zero modes the result is formally

$$E = \int \prod_{r=1}^{3} d^9x_r d^8\theta_r \; \delta^{17}(z_1-z_2)\, \delta^{17}(z_2-z_3)$$
$$\sum_{\{n^{(1)},n^{(2)},n^{(3)}\}} C(\{n^{(1)},n^{(2)},n^{(3)}\}) \prod_{r=1}^{3} \phi_{\{n^{(r)}\}}(z_r) \quad , \tag{6.15}$$

where $\{n^{(r)}\}$ is short for the infinite set of mode numbers $\{n_k^{(r)}, n'^{(r)}_\ell, m_s^{(r)}, m'^{(r)}_t\}$ in Eq. (5.24) and z_r is short for (x_r,θ_r). This is a horrendous expression and in order to avoid this morass of indices we define the number-basis vector by

$$|V\rangle = \sum_{\{n^{(1)},n^{(2)},n^{(3)}\}} C(\{n^{(1)},n^{(2)},n^{(3)}\}) \; |n^{(1)},n^{(2)},n^{(3)}\rangle \quad . \tag{6.16}$$

A coupling of three specific fields is then given by

$$C_{N_1,N_2,N_3} = \langle N_1,N_2,N_3 \,|V\rangle \quad , \tag{6.17}$$

where $|N_1,N_2,N_3\rangle$ is a Fock space vector with excitation number given by the three sets of integers N_1, N_2 and N_3. From $|V\rangle$ one can in principle obtain the starting expression E.

To obtain $|V\rangle$ by direct computation is quite tedious. A faster way is to notice that if we insert $x_3{}^i(\sigma) - x_1{}^i(\sigma) - x_2{}^i(\sigma)$ in the integrand of expression (6.14) we get zero because of the Δ-functional. Following through the steps from (6.14) to (6.16) we deduce that $|V\rangle$ has to satisfy

$$(x_3{}^i(\sigma) - x_1{}^i(\sigma) - x_2{}^i(\sigma))\ |V\rangle = 0 \tag{6.18}$$

$$(\theta_3{}^a(\sigma) - \theta_1{}^a(\sigma) - \theta_2{}^a(\sigma))\ |V\rangle = 0 \quad . \tag{6.19}$$

These conditions determine $|V\rangle$ up to certain overall factors that do not involve the oscillators. Such factors will be determined eventually by the continued analysis. By setting up a general expression for $|V\rangle$ in terms of creation operators, the conditions (6.18) and (6.19) leads to

$$|V\rangle = \exp\,(E_\alpha + E_\theta)\ |0\rangle\ \delta^{17}(z_1{}^o - z_2{}^o)\delta^{17}(z_2{}^o - z_3{}^o) \tag{6.20}$$

with $z^o = (x,\theta_o)$, the zero modes and

$$E_\alpha = \frac{1}{2}\{\sum_{r,s=1}^{3}\ \sum_{m=1}^{\infty}(\alpha_{-m}^{(r)}\ \overline{N}_{mn}^{rs}\ \alpha_{-n}^{(s)} + \tilde{\alpha}_{-m}^{(r)}\ \overline{N}_{mn}^{rs}\ \tilde{\alpha}_{-n}^{(s)})$$

$$+ \mathbb{P}\sum_{r=1}^{3}\ \sum_{m=1}^{\infty}\ \overline{N}_m^r(\alpha_{-m}^{(r)} + \tilde{\alpha}_{-m}^{(r)}) - \frac{\tau_o}{\alpha}\ \mathbb{P}^2\} \tag{6.21}$$

$$E_\theta = \frac{1}{2} \sum_{r,s=1}^{3} \sum_{m,n=1}^{\infty} \frac{1}{\alpha_r} (\theta^{1(r)}_{-m} (C\bar{N}^{rs})_{mn} \theta^{1(s)}_{-n} + \theta^{2(r)}_{-m} (C\bar{N}^{rs})_{mn} \theta^{2(s)}_{-n})$$

$$+ \frac{i}{2} \alpha (\sum_{r=1}^{3} \sum_{m=1}^{\infty} \theta^{1(r)}_{-m} \frac{C}{\alpha_r} \bar{N}^r_m)(\sum_{s=1}^{3} \sum_{n=1}^{\infty} \theta^{2(s)}_{-n} \frac{C}{\alpha_s} \bar{N}^s_n)$$

$$- \Lambda \sum_{r=1}^{3} \sum_{m=1}^{\infty} \frac{1}{\alpha_r} (\bar{N}^r C)_m (e^{-i\pi/4} \theta^{1(r)}_{-m} + e^{i\pi/4} \theta^{2(r)}_{-m}) \quad , \qquad (6.22)$$

with $\theta^{1,2}_{-m}$ defined in (5.20) and where

$$\alpha_r = 2p^+_r$$

$$\alpha = \alpha_1 \alpha_2 \alpha_3$$

$$\tau_o = \sum_{r=1}^{3} \alpha_r \ln \alpha_r$$

$$\mathbb{P}^i = \alpha_1 p_2{}^i - \alpha_2 p_1{}^i$$

$$\Lambda^a = \alpha_1 \frac{\partial}{\partial \theta_{o\,2}{}^a} - \alpha_2 \frac{\partial}{\partial \theta_{o\,1}{}^a}$$

$$C_{mn} = m\, \delta_{m,n}$$

$$\bar{N}^{rs}_{mn} = - \frac{mn\alpha}{n\alpha_r + m\alpha_s} \bar{N}^r_m \bar{N}^s_n$$

$$\bar{N}^r_m = \frac{1}{\alpha_r} \frac{(-1)^{m+1}}{m!} \frac{\Gamma(m(1+\frac{\alpha_{r+1}}{\alpha_r}))}{\Gamma(1-m\frac{\alpha_{r+1}}{\alpha_r})} \qquad (6.23)$$

We can now consider the Hamiltonian (6.8) with $h(\sigma_1,\sigma_2)$ as in (6.9). By use of the δ-functional (6.13) h can be symmetrized in an expression $h(\sigma_1,\sigma_2,\sigma_3)$. Following the steps (6.14) to (6.16) we find that it corresponds to a mode-basis vertex vector

$$|H\rangle = h(\sigma_1,\sigma_2,\sigma_3)\ |V\rangle \quad , \tag{6.24}$$

with h expressed in terms of oscillators. If the operators in h is commuted through the exponential in $|V\rangle$, one finds that

$$p_1^i(\sigma)\ |V\rangle \underset{\sigma\to\sigma_1}{\to} \frac{1}{\pi}(\sigma_1-\sigma)^{-1/2}\ Z^i\ |V\rangle \tag{6.25}$$

$$d_1^a(\sigma)\ |V\rangle \underset{\sigma\to\sigma_1}{\to} \frac{1}{\pi}(\sigma_1-\sigma)^{-1/2}\ Y^a\ |V\rangle \quad , \tag{6.26}$$

where $Z^i\ |V\rangle$ and $Y^a\ |V\rangle$ are vectors of finite norm.

Finally we can now transform back to the functional expression to write the correctly normalized three-string hamiltonian as (6.8) and (6.9) where we let

$$\underline{p}^i(\sigma) \to \sqrt{\sigma_1-\sigma}\ \underline{p}^i(\sigma)$$

$$\underline{\bar{p}}^i(\sigma) \to \sqrt{\sigma_1-\sigma}\ \underline{\bar{p}}^i(\sigma)$$

$$\underline{d}^a(\sigma) \to \sqrt{\sigma_1-\sigma}\ \underline{d}^a(\sigma) \tag{6.27}$$

and consider the full expression in the limit $\sigma\to\sigma_1$.

To obtain a further check on the Hamiltonian, we may consider the couplings of three massless particles. They are most easily obtained from the vertex vector $|H\rangle$ (6.24) by taking its matrix elements with three ground states. Alternatively one lets $\sigma\to 0$ properly to recover point-like particles in (6.8) and (6.9). One can show that these expressions correspond to the cubic couplings of N=2 supergravity.

7. OTHER STRING INTERACTION

In order to construct interactions the burden is to find non-linear representations of the super-Poincaré algebra using second-quantized functional fields. For open strings it is also rather straightforward to construct the three-string interaction. The natural interaction to try is when two end-points on two strings join to make up one string. We start with the field representation of sect. 5 and attempt a Hamiltonian of the form

$$H_3 = i \int D\Sigma_1 D\Sigma_2 \, \Pi[\Sigma_1,\Sigma_2] h(\sigma_1,\sigma_2)$$
$$\mathrm{Tr}[\partial_- \Phi[\Sigma_1+\Sigma_2]\Phi[\Sigma_1]\Phi[\Sigma_2]] \quad , \tag{7.1}$$

where the trace is taken over the SO(N) or Sp(2N) indices of the fields Φ. This time

$$\Pi[\Sigma_1,\Sigma_2] = \delta^8(x_1{}^i(\pi\alpha_1) - x_2{}^i(0))$$
$$\times \delta^4(\theta_{A_1}(\pi\alpha_1) - \theta_{A_2}(0))$$
$$\delta^4(\tilde{\theta}_{A_1}(\pi\alpha_1) - \tilde{\theta}_{A_2}(0)) \quad , \tag{7.2}$$

where we used σ's such that the length of a string is $\pi\alpha \equiv 2p^+\pi$.

Again one can grind through the algebra to find a unique answer for the Hamiltonian with h given by

$$h = g\,\partial_- \sum_{n=0,2,4} C^i{}_{A_1,\ldots,A_n} \frac{\delta}{\delta\theta_{A_1}} \cdots \frac{\delta}{\delta\theta_{A_n}} \underline{P}^i \frac{\partial_{-1}{}^{n/2}\partial_{-2}{}^{n/2}}{\partial_-{}^{n/2}} \tag{7.3}$$

with notations as in (6.9). (For further information on notations, see Appendix).

$$c^i = \frac{1}{\sqrt{2}} \qquad \text{for } i=L \text{ otherwise zero} \quad ,$$

$$c^i{}_{AB} = \rho^I{}_{AB} \qquad \text{for } i=I \text{ otherwise zero} \quad ,$$

$$c^i{}_{ABCD} = \frac{\sqrt{2}}{3}\,\varepsilon_{ABCD} \qquad \text{for } i=R \text{ otherwise zero} \quad , \tag{7.4}$$

As in the case of the closed string a detailed investigation in the mode-basis shows that certain convergence factors must be inserted, namely for each operator $\frac{\delta}{\delta\theta_A}$ or $\underline{p}^i$ a factor $\lim_{\sigma\to\pi\alpha_1} (\pi\alpha_1-\sigma)^{1/2}$ must multiply the operator.

The construction of the open-string vertex can be used also to construct closed-string vertices. These ones are obtained by a "stuttering process" by direct products of open-string vertices. Since we have seen that left-going modes and right-going modes are separated in all generators, we can build up the closed string vertex by a product of two open string vertices, one containing left-going modes and one right-going modes. This can also be translated into the functional form as a product of left-going operators and right-going operators.

In this way we know the three-string interactions. Is there higher-string interactions as there are higher-point interactions for point-particles? The answer is most probably no! One can check that the four-string interaction term one expects from the commutator $\{\dot{Q}^{-a}_{A_3}, \dot{Q}^{-b}_{B_3}\}$ is indeed zero[23] (apart from a possible non-zero term in the forward direction) and that an explicit computation of a four-particle amplitude using only three-string vertices yields a Lorentz-invariant expression. Hence the algebra closes with only the three-string terms. Whether it is still possible to include higher-string interactions has not yet been excluded but it is highly unlikely.

The two types of interactions introduced are both local. To have a consistent theory we must demand that the interactions can occur as soon as two end-point touch or two intermediate points meet.

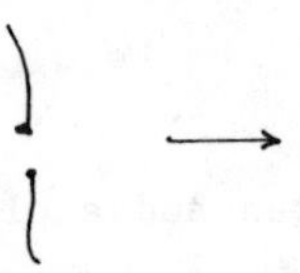

strength g

Fig 7.1 Two end-points meet to join.

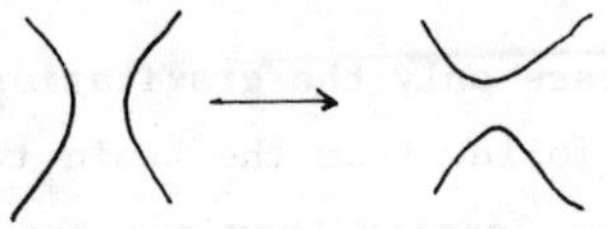

strength κ

Fig 7.2 Two internal points meet to exchange segments.

In the case of an open string theory with a natural three-string "Yang-Mills" coupling as in (7.3), one string can curl up to a closed string leading to a coupling

$$g\ \Psi\Phi^{aa}\ .$$

Similarly for the gravity coupling of (6.9) two open strings can touch, exchange segments and scatter into two new strings giving a coupling

$$\kappa\ \Phi^4\ .$$

Note that this coupling is different from an ordinary 4-string coupling where 1→3 and 3→1 can occur.

With the gravity coupling an open string can also decay into a closed string and an open string giving a coupling

$\kappa\ \Psi\Phi^2$.

Finally in type I theories both an open and a closed string can double on itself to open up to a new open or closed string leading to couplings

$\kappa\phi^2$, $\kappa\ \Psi^2$.

Note that in the type II case only the gravitational interaction is possible. All couplings must follow from the basic two three-string interactions derived above, since locally they are the same as either of the two. Some of the couplings have indeed been checked to agree with this requirement[23].

The checking of the super-Poincaré invariance at the interaction level has only been made for the classical theory. Checking the full quantum theory can only further constrain the theory. At the loop level there will be a condition $\kappa \sim g^2$ emerging. Furthermore, anomalies can emerge in the quantum theory for the chiral theories. Remarkably for type I strings with gauge group SO(32)[24] and for the type IIb strings[25] there are no anomalies at the one loop level although they contain chiral fermions!

All the functional actions for the respective strings are unique[22]. This can also be proven for the heterotic string[26]. Hence there are no possible counterterms other than possibly the action itself. If one can prove that all quantum corrections respect the super-Poincaré invariance, then the theories are at least renormalizable. Mandelstam has shown that for superstrings the S-matrix is covariant[27], so it should at least be renormalizable. To finally settle the question of finiteness, more insight into higher loops is needed.

At this stage there are five possible models which have passed all consistency checks so far. I think there is a realistic chance that further insights into the quantum theories can further reduce this number and perhaps only one model is perfectly consistent. Such a scheme should then give the model in a space $M_4 \times K$, where M_4 is the four-dimensional Minkowski space and K an internal space. If this is the only solution, then Nature better use it!

APPENDIX

Some Notations and Conventions

The algebra of SO(8) has three inequivalent real eight-dimensional representations, one vector and two spinors. We use 8-valued indices i,j, ... corresponding to the vector, a,b, ... corresponding to one spinor and $\dot{a}$, $\dot{b}$, ..., corresponding to the other spinor. Dirac matrices $\gamma^i_{a\dot{a}}$ may be regarded as Clebsch-Gordan coefficients for combining the three eights into a singlet. A second set of matrices $\tilde{\gamma}^i_{\dot{a}a}$ is also introduced. We choose

$$\tilde{\gamma} = \gamma^T \tag{A.1}$$

$$\{\gamma^i, \tilde{\gamma}^j\} = 2\,\delta^{ij} \quad . \tag{A.2}$$

The 16 × 16 matrices

$$\begin{bmatrix} 0 & \gamma^i_{a\dot{a}} \\ \tilde{\gamma}^i_{\dot{b}b} & 0 \end{bmatrix}$$

form a Clifford algebra. We also define

$$\gamma^{ij}_{ab} = \frac{1}{2}[\gamma^i_{a\dot{a}}\,\tilde{\gamma}^j_{\dot{a}b} - \gamma^j_{a\dot{a}}\,\tilde{\gamma}^i_{\dot{a}b}] \quad . \tag{A.3}$$

These matrices are seen to be antisymmetric in a and b using (A.1).

We can also define

$$\gamma^{ij}_{\dot{a}\dot{b}} = \frac{1}{2}[\tilde{\gamma}^{i}_{\dot{a}a}\, \gamma^{j}_{a\dot{b}} - \tilde{\gamma}^{j}_{\dot{a}a}\, \gamma^{i}_{a\dot{b}}] \tag{A.4}$$

which in a similar fashion is antisymmetric in $\dot{a}$ and $\dot{b}$.

To span the whole 8 × 8 dimensional matrix spaces we also define

$$\gamma^{ijk\ell}_{ab} \equiv (\gamma^{[i}\, \tilde{\gamma}^{j}\, \gamma^{k}\, \tilde{\gamma}^{\ell]})_{ab} \tag{A.5}$$

$$\gamma^{ijk\ell}_{\dot{a}\dot{b}} \equiv (\tilde{\gamma}^{[i}\, \gamma^{j}\, \tilde{\gamma}^{k}\, \gamma^{\ell]})_{\dot{a}\dot{b}} \quad . \tag{A.6}$$

These matrices are symmetric.

The general Fierz formula is

$$M_{ab} = \frac{1}{8}\,\delta_{ab}\ \mathrm{tr}\, M - \frac{1}{16}\,\gamma^{ij}_{ab}\ \mathrm{tr}(\gamma^{ij}M)$$

$$+ \frac{1}{384}\,\gamma^{ijk\ell}_{ab}\ \mathrm{tr}(\gamma^{ijk\ell}M) \quad . \tag{A.7}$$

In the case of SU(4), the six-vector can be obtained as the antisymmetric tensor product of two 4's or two $\overline{4}$'s. The corresponding Clebsch-Gordan coefficients (or Dirac matrices) are denoted $\rho^{I}{}_{AB}$ and ρ^{IAB}. They are normalized as usual so that

$$\rho^{IAB}\,\rho^{J}{}_{BC} + \rho^{JAB}\,\rho^{I}{}_{BC} = 2\,\delta^{A}{}_{C}\,\delta^{IJ} \quad . \tag{A.8}$$

We also define

$$\rho^{IJ}{}_{A}{}^{B} = \frac{1}{2}(\rho^{I}{}_{AC}\,\rho^{JCB} - \rho^{J}{}_{AC}\,\rho^{ICB}) \quad . \tag{A.9}$$

REFERENCES

1) Nambu, Y., Lectures at Copenhagen Symposium, unpublished (1970)
Hara, O., Progr. Theor. Phys. 46, 1549 (1971)
Goto, T., Progr. Theor. Phys. 46, 1560 (1971).

2) Brink, L., Di Vecchia, P. and Howe, P.S., Phys. Lett. 65B, 471 (1976)

Deser, S. and Zumino, B., Phys. Lett. 65B, 369 (1976).

3) Mandelstam, S., Nucl. Phys. B64, 205 (1973).

4) Dirac, P.A.M., Rev. Mod. Phys. 26, 392 (1949).

5) Goddard, P., Goldstone, J., Rebbi, C. and Thorn, C.B., Nucl. Phys. B56, 109 (1973).

6) Brink, L. and Nielsen, H.B., Phys. Lett. 45B, 332 (1973).

7) This method was suggested by F. Gliozzi, unpublished.

8) Ramond, P.M., Phys. Rev. D3, 2415 (1971).

9) Neveu, A. and Schwarz, J.H., Nucl. Phys. B31, 86 (1971); Phys. Rev. D4, 1109 (1971).

10) Gliozzi, F., Scherk, J. and Olive, D.I., Phys. Lett. 65B, 282 (1976); Nucl. Phys. B122, 253 (1977).

11) Green, M.B. and Schwarz, J.H., Phys. Lett. 109B, 444 (1982).

12) Green, M.B. and Schwarz, J.H., Nucl. Phys. B181, 502 (1981).

13) Brink, L., Lindgren, O. and Nilsson, B.E.W., Nucl. Phys. B212, 401 (1983).

14) Gross, D.J., Harvey, J.A., Martinec, E. and Rohm, R., Phys. Rev. Lett. 54, 502 (1985); Nucl. Phys. B256, 253 (1985).

15) Gross, D.J., lectures in this volume.

16) Paton, J. and Hong-Mo, C., Nucl. Phys. B10, 519 (1969).

17) Marcus, N. and Sagnotti, A., Phys. Lett. 119B, 97 (1982).

18) Schwarz, J.H., Phys. Rep. 69, 223 (1982).

19) Green, M.B. and Schwarz, J.H., Phys. Lett. 136B, 367 (1984).

20) Bengtsson, I. and Cederwall, M., Institute of Theoretical Physics, Göteborg 84-21 (1984).

21) Green, M.B., Schwarz, J.H. and Brink, L., Nucl. Phys. B219, 437 (1983).

22) Bengtsson, A.K.H., Brink, L., Cederwall, M. and Ögren, M., Nucl. Phys. B254, 625 (1985).

23) Green, M.B. and Schwarz, J.H., Nucl. Phys. B243, 475 (1984).

24) Green, M.B. and Schwarz, J.H., Phys. Lett. 149B, 117 (1984).

25) Alvarez-Gaumé, L. and Witten, E., Nucl. Phys. B234, 269 (1984).

26) Brink, L., Cederwall, M. and Green, M.B., to be published.

27) See S. Mandelstam, this volume.

LECTURES ON SUPERSTRINGS

Michael B. Green,

Dept. of Physics, Queen Mary College, University of London,

Mile End Road, London E1 4NS, U.K.

Superstring theories have not yet been formulated in terms of a single compelling principle such as that of general relativity. However, enough is now known about the structure of these theories to justify the optimism that certain of them might be consistent quantum theories that unify gravity and the other forces. The fact that the quantum consistency of superstring theories restricts the possible ten-dimensional unifying symmetry groups to be $E_8 \times E_8$ or SO(32) (or (Spin 32)/Z_2 which has the same algebra as SO(32)) is a novel development in particle physics. The case of $E_8 \times E_8$ is particularly interesting since, in the process of compactification from ten to four dimensions, it can break to a realistic chiral symmetry group describing all the observed interactions and the spectrum of the known particles.

The subject is still in a somewhat primitive state by comparison with the sophistication of our understanding of conventional "point" field theory. Results have been arrived at by a variety of techniques which I shall survey in these lectures.

In the first lecture I will start with the description of the dynamics of a free classical superstring moving in a flat super-space-time background. In order to avoid problems associated with covariant quantization in this formulation I shall describe the first-quantized superstring theory in the light-cone gauge. This provides the basis for calculating the spectra of the various types of superstring theories and is adequate for most perturbation theory calculations around a background space with no Riemann curvature.

The second lecture will survey the formulation of the second-quantized interacting field theory of superstrings in the light-cone gauge. This involves fields which create and destroy complete strings and which are therefore functionals of the string configurations. The use of the light-cone gauge is presumably an undesirable feature since it obscures much of the geometric structure of the theory. However, for the moment this is the only more or less complete understanding of interacting super string field theory we have.

The third lecture will summarize the present status of the one-loop calculations. I will also present, in some detail, the calculation of the Yang-Mills anomaly in the open-string theory and demonstrate its cancellation for the group SO(32).

I COVARIANT SUPERSTRING DYNAMICS

(a) The Covariant Action

The most geometrically appealing formulation of superstring theories begins with a generalization of the Nambu-Goto action[1] of bosonic string theory (namely, the area of the world-sheet swept out as the string moves through space-time) to super-space-time[2]. The other formulation, based on the observation that the spectrum of the "spinning string" theory[3] can be truncated to be supersymmetric in ten-dimensional space-time[4], does not incorporate space-time supersymmetry manifestly.

As a string moves through space-time it sweeps out a world-sheet that is parametrized by a time-like parameter τ and a spacelike parameter σ. The coordinates of the superstring map the world-sheet into superspace . These are the space-time coordinates $X^\mu(\sigma,\tau)$ (where $\mu = 0,1,..,D-1$ is a space-time index in D dimensions) and, in general, two Grassmann (anticommuting) coordinates $\theta^{Aa}(\sigma,\tau)$ which are space-time spinors (A=1,2) and the spinor index $a = 1,2,...2^{D/2}$). In ten dimensions these spinors will be taken to be both Majorana (i.e. real in the Majorana representation of the Dirac gamma matrices) and to satisfy the chirality (Weyl) constraints

$$(1 + \eta^A \gamma_{11})^{ab}\, \theta^{Ab} = 0 \tag{1.1}$$

where $\eta^A = \pm 1$ and $\gamma_{11} = \gamma^0 \gamma^1 \ldots\ \gamma^9$. The Dirac gamma matrices γ^μ satisfy $\{\gamma^\mu,\gamma^\nu\} = -2\eta^{\mu\nu}$ where the ten-dimensional Minkowski metric $\eta^{\mu\nu} = \mathrm{diag}(-1,1,..,1)$. When $\eta^1 = -\eta^2$ the theory has no net chirality whereas when $\eta^1 = \eta^2$ the theory is chiral. The fact that both the Majorana and the chirality conditions can be imposed simultaneously is a special property of ten dimensions (more generally of 2 mod 8 dimensions). More generally, the classical theory will also make sense in dimensions D = 3 (with Majorana spinors), 4 (with Majorana or Weyl spinors) and 6 (with Weyl spinors) but since the case D = 10 is of particular interest in the quantum theory I will use notation appropriate to that dimension. The super-Poincaré transformations on which these theories are based are (suppressing spinor indices)

$$\delta\theta^A = \tfrac{1}{4}\, \omega_{\mu\nu}\, \gamma^{\mu\nu}\, \theta^A + \epsilon^A \tag{1.2}$$

$$\delta X^\mu = \omega^\mu{}_\nu\, X^\nu + a^\mu + i\bar{\epsilon}^A \gamma^\mu \theta^A \tag{1.3}$$

where $\omega_{\mu\nu}$ and a_μ are infinitessimal parameters of the Poincaré group while $\epsilon^A \equiv \epsilon^{Aa}$ are the two 32-component Grassmann spinor parameters which also satisfy Majorana and Weyl conditions. [In six dimensions θ^A is chiral but not Majorana and the term $i\bar{\epsilon}^A\gamma^\mu\theta^A$ is replaced by $i(\bar{\epsilon}^A\gamma^\mu\theta^A - \bar{\theta}^A\gamma^\mu\epsilon^A)$.] For the case of the heterotic string[5] there is only one (Majorana-Weyl) spinor coordinate.

The natural covariant action for a relativistic point particle is the length of the world-line traversed as it moves in space-time. In the bosonic string theory this action is generalized[1] to the area of the world-sheet. The fact that this is invariant under arbitrary reparametrizations of the world-sheet ($\sigma \to \tilde{\sigma}(\sigma,\tau)$ and $\tau \to \tilde{\tau}(\sigma,\tau)$) is crucial for the consistency of the theory. The action for the superstring consists of several terms

$$S = S_1 + S_2 \quad (+\ S_3 \text{ in the case of the heterotic string}). \tag{1.4}$$

The first term is the obvious guess for a supersymmetric generalization of the bosonic action. It is convenient to write this in the form that invokes a two-dimensional metric tensor $g^{\alpha\beta}(\sigma,\tau)$ so that the action looks like two-dimensional general relativity on the world-sheet (with the coordinates X^μ and θ^{Aa} being scalars under two-dimensional reparametrizations of the world-sheet)

$$S_1 = -\frac{T}{2}\int \eta_{\mu\nu}\sqrt{-g}g^{\alpha\beta}\,\Pi^\mu_{\ \alpha}\Pi^\nu_{\ \beta}\,d^2\xi \tag{1.5}$$

(where $d^2\xi = d\sigma\, d\tau$ and the indices $\alpha,\beta = \sigma,\tau$) and

$$\Pi^\mu_\alpha = \partial_\alpha X^\mu - i\bar{\theta}^A\gamma^\mu\partial_\alpha\theta^A \tag{1.6}$$

when θ^A is a Majorana spinor (otherwise the term $i\bar{\theta}^A\gamma^\mu\partial_\alpha\theta^A$ is replaced by $\frac{1}{2}\, i(\bar{\theta}^A\gamma^\mu\partial_\alpha\theta^A - \partial_\alpha\bar{\theta}^A\gamma^\mu\theta^A)$). The string has been taken to be moving in flat Minkowski space-time ($\eta^{\mu\nu}$ is the D-dimensional Minkowski metric). Recent considerations of strings moving in curved space-time backgrounds involve the replacement of $\eta^{\mu\nu}$ by the background metric $G^{\mu\nu}(X)$ (as well as the addition of other terms) in which case the resulting non-linear σ model is only a consistent string theory for very special spaces. S_1 is manifestly invariant under the global transformations of eqs.(1.2) and (1.3) (since Π^μ_α is manifestly supersymmetric) as well as under arbitrary reparametrizations of σ and τ. The metric, $g^{\alpha\beta}(\sigma,\tau)$, is an auxiliary field which (at least in the classical theory) can be eliminated by replacing it in the action by the solution of its equations of motion ($g_{\alpha\beta} = f(\sigma,\tau)\, \Pi^\mu_\alpha \Pi_{\mu\beta}$ where $f(\sigma,\tau)$ is an arbitrary function). The resulting expression is the "area" of the world-sheet in superspace. The form of the action S_1 in eq. (1.5) is a generalization to superstrings of the action of ref. 6 and used by Polyakov[7] in discussing the quantization of the bosonic and spinning string theories.

By itself S_1 does not define a conformally-invariant quantum theory, presumably because of the terms cubic and quartic in the coordinates. However, it is possible to add another term, S_2, to the action to remedy this where

$$S_2 = -iT \int \eta_{\mu\nu}\, \epsilon^{\alpha\beta} \left\{ \partial_\alpha X^\mu \, (\bar\theta^1 \gamma^\nu \partial_\beta \theta^1 - \bar\theta^2 \gamma^\nu \partial_\beta \theta^2) - i\bar\theta^1 \gamma^\mu \partial_\alpha \theta^1 \, \bar\theta^2 \gamma^\nu \partial_\beta \theta^2 \right\} d^2\xi \quad . \tag{1.7}$$

This term is also manifestly invariant under reparametrizations of σ and τ due to the presence of the two-dimensional Levi-Cevita tensor density, $\epsilon^{\alpha\beta}$. The fact that S_2 is also invariant under the global supersymmetry transformations is not so manifest. To verify this consider, for simplicity, the case of the heterotic superstring obtained by setting $\theta^2 = 0$ in eq.(1.7). The variation under supersymmetry transformations is then given by substituting the ϵ variation from eqs. (1.2) and (1.3)

$$\delta_\epsilon S_2 = \int (\bar\epsilon \gamma^\mu \partial_\alpha \theta \, \bar\theta \gamma_\mu \partial_\beta \theta + \text{total derivatives})\, d\sigma\, d\tau \, . \tag{1.8}$$

It is straightforward to verify that this vanishes by using the identity (proved by using Fierz transformations)

$$\bar{\epsilon}\gamma^{\mu}\lambda_{[1}\bar{\lambda}_{2}\gamma^{\mu}\lambda_{3]} = 0 \tag{1.9}$$

where $\lambda_1 = \theta$, $\lambda_2 = \partial_\tau\theta$, $\lambda_3 = \partial_\sigma\theta$ and [ ] denotes antisymmetrization. This is the same identity that is used in proving the supersymmetry of supersymmetric Yang-Mills theories in dimensions D = 3, 4, 6 and 10 (with the appropriate kind of spinors in each of these dimensions). There is therefore a restriction on the possible space-time dimensionality in the classical theory. We shall see that the critical dimension in the quantum theory is D = 10 which means that the excitations of the string are purely transverse in ten dimensions. [If it is possible to define consistent quantum theories in 3, 4 or 6 dimensions it is presumably necessary to account for new longitudinal modes in the manner suggested by Polyakov[7].]

The relative coefficient of S_1 and S_2 in eq.(1.5) and (1.7) is uniquely determined by requiring that the total action have extra local symmetries which lead to it describing a conformally invariant two-dimensional theory. In particular there is a local fermionic invariance (analogous to that for the superparticle discussed in ref. 8). It is useful to introduce projection operators

$$P_{\pm} = \tfrac{1}{2}\left[g^{\alpha\beta} \pm \frac{\epsilon^{\alpha\beta}}{\sqrt{-g}}\right] \tag{1.10}$$

which project onto the self-dual and anti-self-dual pieces of two-dimensional vectors. If any two-vector V_α satisfies $V^\alpha \equiv V^\alpha_\pm = P^{\alpha\beta}_\pm V_\beta$ then the two components are equal (for + sign) or equal in magnitude but opposite in sign (for − sign). The action $S_1 + S_2$ is invariant under fermionic transformations with Grassmann parameters $\kappa^{1\alpha}$ and $\kappa^{2\alpha}$,

$$\delta_\kappa\theta^A = 2i\gamma.\Pi_\alpha\kappa^{A\alpha} \quad , \qquad \delta_\kappa X^\mu = i\bar{\theta}^A\gamma^\mu\delta_\kappa\theta^A \ ,$$

$$\delta_\kappa(\sqrt{-g}g^{\alpha\beta}) = -16\sqrt{-g}(\bar{\kappa}^{1\alpha}\partial^\beta_-\theta^1 + \bar{\kappa}^{2\alpha}\partial^\beta_+\theta^2) \tag{1.11}$$

The parameters $\kappa^{1\alpha}$ and $\kappa^{2\alpha}$ have suppressed space-time spinor indices and are self-dual ($\kappa^{1\alpha} \equiv \kappa_{+}^{1\alpha}$) and anti-self-dual ($\kappa^{2\alpha} \equiv \kappa_{-}^{2\alpha}$) respectively. The proof of the κ invariance of S involves the same Fierz identity used in the proof of its ϵ invariance. These supersymmetries are reminiscent of two-dimensional supersymmetry although the parameters are not two-dimensional spinors and there is no two-dimensional gravitino field.

The action S possesses further local invariance under bosonic transformations which can be discovered by trying to close the algebra of the κ transformations. These bosonic transformations are

$$\delta_\lambda \theta^A = \sqrt{-g}\partial_\alpha \theta^A \lambda^{A\alpha} , \qquad \delta_\lambda X^\mu = i\bar{\theta}^A \gamma^\mu \delta_\lambda \theta^A ,$$

$$\delta_\lambda (\sqrt{-g} g^{\alpha\beta}) \quad . \tag{1.12}$$

The parameters satisfy $\lambda^{1\alpha} = \lambda_{+}^{1\alpha}$, $\lambda^{2\alpha} = \lambda_{-}^{2\alpha}$ respectively. For the heterotic string where there is just one superspace spinor $\theta(\sigma,\tau)$ the action contains, in addition to S_1 and S_2, the third term S_3 that incorporates the internal quantum numbers of (Spin 32)/Z_2 (which has the same Lie algebra as SO(32)) or $E_8 \times E_8$. This term may be written in terms of a two-dimensional chiral fermion field Ψ^I ($I = 1,2,..,32$) as

$$S_3 = \frac{1}{2}\, i\, T \int \bar{\Psi}^I (1-\rho_3) e^\alpha_A \rho^A \partial_\alpha \Psi^I \, d^2\xi \tag{1.13}$$

where ρ^A are the two-dimensional Dirac matrices and $\rho_3 = \rho_1\rho_2$. The zweibein e^α_A defines the metric by $g^{\alpha\beta} = e^\alpha_A e^{\beta A}$. The consistency of the heterotic superstring when one-loop corrections are incorporated requires that SO(32) or $E_8 \times E_8$ be symmetries of S_3. This is acheived by arranging that Ψ^I transform as a 32-component vector of SO(32) or as the (16,16) representation of the SO(16) × SO(16) subgroup of $E_8 \times E_8$. Since this very appealing idea is discussed in great detail elsewhere in this workshop I will not describe it in any detail in these lectures.

The geometrical interpretation of the term S_2 has been clarified[9] by writing it in terms of the supersymmetric one-form (for example, for the case of the heterotic string)

$$\Omega^M \equiv (\Omega^\mu, \Omega^a) \tag{1.14}$$

where

$$\Omega^\mu = dX^\mu - i\bar{\theta}\gamma^\mu d\theta \equiv \Omega^\mu_\alpha \, d\xi^\alpha \tag{1.15}$$

$$\Omega^a = d\theta^a \equiv \Omega^a_\alpha \, d\xi^\alpha \tag{1.16}$$

and $d\xi^\alpha \equiv (d\tau, d\sigma)$. In this notation the action S_1 is written as

$$S_1 = -\frac{1}{2} T \int \sqrt{-g}\, g^{\alpha\beta}\, \eta_{\mu\nu} \Omega^\mu_\alpha \Omega^\nu_\beta \tag{1.17}$$

S_2 can be written as a kind of Wess-Zumino term by formally extending the dimension of the world-sheet to a three-dimensional space (in the style of Witten[10]) and introducing the manifestly super-Poincaré invariant three-form

$$\Omega^3 = -i(C\gamma^\mu)_{ab}\, \Omega_\mu\, \Omega^a\, \Omega^b \tag{1.18}$$

so that

$$S_2 = -\frac{1}{2} T \int \Omega^3 \tag{1.19}$$

The earlier expression for S_2 (eq.(1.7)) is recovered by using the fact that $\Omega^3 = d\Omega^2$ where $\Omega^2 = -idX^\mu \wedge \bar{\theta}\gamma_\mu d\theta$ in the case of the heterotic superstring and a generalization involving two θ's in the case of type II superstring theories.

The situation is reminiscent of a non-linear σ-model defined on a group manifold in which the addition of a Wess-Zumino term leads to a free theory for a certain value of the relative couplings of the two terms[11]. In that case the action $S = S_1/g + S_2$ has a beta function which has a zero at a special value of g at which the theory is conformally invariant. The Wess-Zumino term can also be interpreted in terms of a torsion which parallelizes the curvature at the point at which the beta function vanishes[12]. A similar interpretation is possible in the case of the superstring theories (where the σ-model defines a mapping of the two-dimensional world-sheet into N = 1 or N = 2 superspace).

The covariant action S has recently been generalized (for N = 1 theories) to describe a curved gravitational background[13]. This involves using a curved metric instead of $\eta^{\mu\nu}$ in S_1 and coupling the using a three-form that arises in ten-dimensional supergravity[14] to generalize Ω^3 in S_2. A similar construction has also been carried out for N = 2 theories[15].

(b) Equations of Motion

In type I or type II theories the equation of motion for $g^{\alpha\beta}$ arises only from the S_1 term in the action and gives the two-dimensional Einstein equation

$$\Pi^{\mu}_{\alpha}\,\Pi_{\mu\beta} - \frac{1}{2}\, g_{\alpha\beta}\, g^{\gamma\delta}\, \Pi^{\mu}_{\gamma}\, \Pi_{\mu\delta} = 0 \ . \tag{1.20}$$

[For the heterotic string there is an additional term arising from the variation of the zweibein in S_3.] This equation expresses the vanishing of the two-dimensional energy-momentum tensor which is traceless and symmetric and has two independent components. In a conformal gauge, defined by

$$g^{\alpha\beta} = e^{\phi}\, \eta^{\alpha\beta} \tag{1.21}$$

where $\eta^{\alpha\beta} = \begin{bmatrix} -1 & 0 \\ 0 & 1 \end{bmatrix}$, eq. (1.20) describes two independent constraints on the coordinates. In terms of the quantities $\Pi^{\mu}_{\pm}$

$$\Pi^{\mu}_{\pm} \equiv (\, \Pi^{\mu}_{\tau} \pm \Pi^{\mu}_{\sigma}\,)/\sqrt{2} \tag{1.22}$$

(the components of $P_{\mp\alpha\beta}\Pi^{\mu\beta}$) these "Virasoro" constraints are

$$\Pi_{+}\cdot\,\Pi_{+} = 0 = \Pi_{-}.\Pi_{-} \qquad . \tag{1.23}$$

The equations for $X^{\mu}(\sigma,\tau)$, determined from the action, can be written in a conformal gauge as

$$\partial_{\alpha}(\partial^{\alpha}X^{\mu} - 2i\bar{\theta}^{1}\gamma^{\mu}\partial^{\alpha}_{+}\theta^{1} - 2i\bar{\theta}^{2}\gamma^{\mu}\partial^{\alpha}_{-}\theta^{2}) = 0 \tag{1.24}$$

(where $\partial^{\alpha}_{\pm} \equiv P^{\alpha\beta}_{\mp}\,\partial_{\beta}$). The θ^{Aa} equations are

$$\gamma.\Pi_{-}\partial_{+}\theta^{1} = 0 = \gamma.\Pi_{+}\partial_{-}\theta^{2} \tag{1.25}$$

where $\partial_{\pm} \equiv (\partial_{\tau}\pm\partial_{\sigma})/\sqrt{2}$. In deriving these equations from the action various boundary conditions have been imposed which eliminate surface terms. For closed strings these conditions impose periodicity of the coordinates in σ. For open strings the conditions require

$$\theta^1 = \theta^2 \; , \qquad \Pi^{\mu}_{\sigma} = 0 \text{ at the endpoints } \sigma = 0, \pi. \tag{1.26}$$

From the transformations in eq.(1.2) it is clear that the open-string supersymmetry is truncated to N = 1 since eq.(1.26) requires $\epsilon^1 = \epsilon^2$.

The covariant quantization of this action is hampered by the fact that there are additional phase-space constraints involving the momenta, p^A_{θ}, conjugate to θ^A. Defining

$$p^A_{\theta} = \frac{\delta S}{\delta \theta^A} \tag{1.27}$$

it is easy to see that p_{θ} is related to functions of X^{μ}, P^{μ} and θ^A. These constraints are mixtures of first and second class constraints which must be separated before quantization. Despite some progress[16] it seems probable that covariant quantization will involve introducing extra variables into the action to relax the constraints[17]. I will take the pragmatic route of passing to the light-cone gauge in which the super-Poincaré invariance is not manifest but the quantum theory is easy to formulate.

(c) The Light-Cone Gauge

The parameters, $\kappa^{A\alpha}$ have the same number of independent components as θ^A (after allowing for the fact that they are (anti) self-dual). However, it follows from the transformation laws in eq.(1.11) that only half the components of the θ's can be gauged away because the operators $\gamma.\Pi_{\pm}$ are nilpotent (since $(\gamma.\Pi_{\pm})^2 = (\Pi_{\pm})^2 = 0$ by use of eq.(1.23)). By a suitable choice of κ^A the fermionic gauge invariance can be used to choose

$$\frac{1}{2}\,(\gamma^- \gamma^+)^{ab}\, \theta^{Ab} = 0 \tag{1.28}$$

where $\gamma^{\pm}$ are the light-cone Dirac matrices which satisfy $(\gamma^+)^2 = 0 = (\gamma^-)^2$ so that $\frac{1}{2}(\gamma^+\gamma^-)$ and $\frac{1}{2}(\gamma^-\gamma^+)$ are projection operators. The $\pm$ components of any 10-vector, V^μ, are defined by $V^\pm = (V^0 \pm V^9)/\sqrt{2}$. Each spinor, θ, satisfying eq.(1.28) has eight independent components, half as many as a general Majorana-Weyl spinor. With this choice of κ gauge the equation of motion for $X^+(\sigma,\tau)$ in the conformal gauge, eq. (1.24) becomes $\partial^2 X^+(\sigma,\tau) = 0$. Just as in the original development of the light-cone gauge treatment of the bosonic theory in ref. 18 this allows the choice of a special parametrization, known as the light-cone gauge, in which the "time" coordinate, $X^+(\sigma,\tau)$, takes a common value for all values of σ

$$X^+(\sigma,\tau) = x^+ + \frac{1}{\pi T} p^+\tau \quad . \tag{1.29}$$

[The tension T will often be set equal to $1/\pi$ from now on.] The equations of motion are particularly simple in this gauge. From eqs.(1.24) and (1.25) we have

$$\partial^2 X^I = 0 \qquad \text{with } I = 1,2,..,8 \quad , \tag{1.30}$$

$$\partial_+\theta^1 \equiv \frac{1}{\sqrt{2}}(\partial_\tau + \partial_\sigma)\,\theta^1 = 0 = \partial_-\theta^2 \equiv \frac{1}{\sqrt{2}}(\partial_\tau - \partial_\sigma)\,\theta^2 \quad . \tag{1.31}$$

The last two equations are the components of the two-dimensional Dirac equation

$$\rho.\partial\theta = 0 \tag{1.31$'$}$$

where the two-component spinor θ is defined by

$$\theta = \begin{bmatrix} \theta^{1a} \\ \theta^{2a} \end{bmatrix} \tag{1.32}$$

In passing to the light-cone gauge the ten-dimensional scalars, θ^1 and θ^2, which were independent world-sheet scalars in the covariant action have become the two components of a world-sheet spinor.

This is a consequence of the way in which super-Poincaré transformations act in the light-cone gauge. Since the light-cone gauge conditions (eqs.(1.28) and (1.29)) are not covariant they are altered by certain super-Poincaré transformations. In order to ensure that the gauge conditions are unaltered in the transformed frame the transformations must be supplemented by compensating (super) reparametrizations. The net effect is that under these compensated transformations the θ's transform as two-dimensional spinors. The Virasoro constraint equations $\Pi_+^2 = 0 = \Pi_-^2$ can be explicitly solved to express X^- in terms of the coordinates X^I and θ^A by substituting eqs.(1.28) and (1.29) into these constraint equations. The result is

$$\dot{X}^- = \frac{1}{p^+}(\dot{\underline{X}}^2 + \acute{\underline{X}}^{\,2}) + \frac{i}{\sqrt{2}}\bar{\theta}^2\gamma^-\partial_+\theta^2 + \frac{i}{\sqrt{2}}\bar{\theta}^1\gamma^-\partial_-\theta^1 \qquad (1.33)$$

$$\acute{X}^- = \frac{1}{p^+}(\dot{\underline{X}}.\acute{\underline{X}}) + \frac{i}{\sqrt{2}}\bar{\theta}^2\gamma^-\partial_+\theta^2 - \frac{i}{\sqrt{2}}\bar{\theta}^1\gamma^-\partial_-\theta^1 \qquad (1.34)$$

where $\underline{X} \equiv X^I$ and $\dot{\underline{X}} \equiv \partial_\tau\underline{X}$ and $\acute{\underline{X}} \equiv \partial_\sigma\underline{X}$.

It is convenient to introduce an SO(8) spinor notation in which any sixteen-component Majorana-Weyl spinor ψ^a $(a = 1,2,..,16)$ is written as the sum of two inequivalent SO(8) spinors

$$\psi = \tfrac{1}{2}\gamma^-\gamma^+\psi + \tfrac{1}{2}\gamma^+\gamma^-\psi \qquad (1.35a)$$

$$\equiv \psi^a + \psi^{\dot{a}} \qquad (1.35b)$$

where the superscripts a and ȧ in the last line take the values 1,2,..,8. The physical modes of the type I or type II theories are now represented by the SO(8) vector

$$X^I(\sigma,\tau) \qquad (1.36a)$$

and two SO(8) spinors, written in the two-dimensional spinor notation as

$$\theta^a(\sigma,\tau) \equiv \begin{bmatrix} \theta^{1a}(\sigma,\tau) \\ \theta^{2a}(\sigma,\tau) \end{bmatrix} \tag{1.36b}$$

(while the heterotic string has only a single SO(8) spinor that can be thought of as eight chiral two-dimensional spinors). In the non-chiral type IIa theory one of the two SO(8) spinors is a dotted spinor while the other is an undotted spinor.

The light-cone gauge equations can be deduced from a light-cone gauge action[19]

$$S = -\frac{1}{2}T\int_0^\pi d\sigma \int d\tau \left\{ \eta^{\alpha\beta}\partial_\alpha X^I \partial_\beta X^I + \frac{1}{2}ip^+\bar{\theta}^a\rho.\partial\theta^a \right\} \quad . \tag{1.37}$$

A bar on top of a spinor now denotes *two-dimensional* conjugation (i.e. $\bar{\theta} \equiv \theta\rho^0$). This action is invariant under the supersymmetry transformations

$$\delta\theta^a = \eta^a + \gamma^I_{a\dot{b}}\rho.\partial X^I \epsilon^{\dot{b}} \tag{1.38}$$

$$\delta X^I = \frac{i}{p^+}\bar{\epsilon}^{\dot{a}}\tilde{\gamma}^I_{\dot{a}b}\theta^b \tag{1.39}$$

where the Grassmann parameters η^a and $\epsilon^{\dot{a}}$ are both SO(8) spinors and world-sheet spinors

$$\eta^a \equiv \begin{bmatrix} \eta_1^a \\ \eta_2^a \end{bmatrix} \qquad \epsilon^{\dot{a}} \equiv \begin{bmatrix} \epsilon_1^{\dot{a}} \\ \epsilon_2^{\dot{a}} \end{bmatrix} \tag{1.40}$$

Together they build up the 32 components of the two supercharges.

The matrices $\gamma^I_{a\dot{b}}$ and $\tilde{\gamma}^I_{\dot{b}a}$, defined by

$$\gamma^I_{a\dot{b}}\tilde{\gamma}^J_{\dot{b}c} + \gamma^J_{a\dot{b}}\tilde{\gamma}^I_{\dot{b}c} = 2\delta^{IJ}\delta_{ac} \tag{1.41a}$$

$$\tilde{\gamma}^I_{\dot{a}b}\gamma^J_{b\dot{c}} + \tilde{\gamma}^J_{\dot{a}b}\gamma^I_{b\dot{c}} = 2\delta^{IJ}\delta_{\dot{a}\dot{c}} \tag{1.41b}$$

are related to the 16 × 16 Dirac matrices of the chiral ten-dimensional theory by

$$\gamma^I = \begin{bmatrix} 0 & \gamma^I_{\cdot ab} \\ \tilde{\gamma}^I_{ba} & 0 \end{bmatrix} . \tag{1.42}$$

The solutions of the equations of motion (eqs.(1.30) and (1.31)) can be written as Fourier expansions so that, for example, the open string coordinates satisfying the boundary conditions of eq. (1.26) (i.e. $\theta^1 = \theta^2$ and $\partial_\sigma X^I = 0$ at the endpoints) are given by

$$\theta^{1a}(\sigma,\tau) = \sum_{-\infty}^{\infty} \theta^a_n \, e^{-in(\tau-\sigma)} \qquad \theta^{2a}(\sigma,\tau) = \sum_{-\infty}^{\infty} \theta^a_n \, e^{-in(\tau+\sigma)} \tag{1.43}$$

$$X^I(\sigma,\tau) = x^I + p^I\tau + i\sum_{n=1}^{\infty} \frac{1}{n}(\alpha^I_n e^{-in\tau} - \alpha^I_{-n} e^{in\tau}) \cos n\sigma \tag{1.44}$$

(where $\alpha^I_{-n} \equiv \alpha^{I*}_n$ and $\theta^a_{-n} \equiv \theta^{a*}_n$). The momentum conjugate to $X^I(\sigma,\tau)$ is given by

$$P^I(\sigma,\tau) = \frac{\delta S}{\delta \dot{X}^I} = \frac{1}{\pi} \sum_{-\infty}^{\infty} \alpha^I_n \, e^{-in\tau} \cos n\sigma \quad . \tag{1.45}$$

Poisson brackets can be obtained from the action in the usual way and I will immediately transcribe them to (anti)-commutator brackets (setting $\hbar = 1$) which gives

$$[\, P^I(\sigma,\tau) \; , \; X^J(\sigma',\tau)] = -i \, \delta^{IJ} \, \delta(\sigma - \sigma') \tag{1.46}$$

$$\{\, \theta^a(\sigma,\tau) \; , \; \theta^b(\sigma',\tau) \,\} = \frac{2\pi}{p^+} \delta^{ab} \, \delta(\sigma-\sigma') \quad . \tag{1.47}$$

Substituting the mode expansions into these expressions gives the relations

$$[\, \alpha^I_m \; , \; \alpha^J_n \,] = m \, \delta^{IJ} \, \delta_{m+n,0} \tag{1.48}$$

$$\{\, \theta^a_m \; , \; \theta^b_n \,\} = \frac{2}{p^+} \delta^{ab} \, \delta_{m+n,0} \quad . \tag{1.49}$$

The generators of all the super-Poincaré transformations can be represented in terms of the bosonic oscillator modes $\underline{\alpha}_n$ and their fermionic partners, θ_n[20]. I will not describe this here in detail as I shall want to use a slightly different formalism in the treatment of superstring field theory in the next lecture. Suffice it to note that the hamiltonian operator, h (which is the operator conjugate to X^+ so that $h \equiv \int_0^\pi d\sigma\, P^-(\sigma,\tau)$) is given by $h = \frac{1}{\pi}\int_0^\pi d\sigma\, \dot{X}^-$ which can be obtained from the expression for X^- (eq. (1.34)) and is given in terms of the modes by

$$h \equiv p^- = \frac{1}{p^+}\left(N + \frac{\underline{p}^2}{2}\right) \quad . \tag{1.50}$$

where

$$N = \sum_{n=1}^{\infty}\left(\underline{\alpha}_{-n}^{\dagger} \cdot \underline{\alpha}_n + \frac{1}{2}\, p^+ n \theta_n^{\dagger a}\, \theta_n^a\right) \tag{1.51}$$

Notice that the normal ordering constants cancel between Bose and Fermi modes at each value of n. This is not the case in the heterotic string theory. The mass of any open-string state is given by

$$(\text{Mass})^2 = 2p^+p^- - \underline{p}^2 = 2N \quad . \tag{1.52}$$

The massless ground states in the open-string sector form the Yang-Mills supermultiplet. These are

an SO(8) vector $|i\rangle$ of gauge bosons

an SO(8) spinor $|a\rangle$ of fermions

Similar arguments for the closed-string theories lead to two independent sets of modes α_n^I, θ_n^a and $\tilde{\alpha}_n^I$, $\tilde{\theta}_n^a$ corresponding to waves running around the string in either direction. The states of type II closed string theories have masses determined by

$$(\text{Mass})^2 = 4\ (N+\tilde{N}) \tag{1.53}$$

subject to the constraint

$$N = \tilde{N} \tag{1.54}$$

which follows by requiring that $X^-(\sigma,\tau)$ is periodic (by integrating eq. (1.34) from $\sigma = 0$ to $\sigma = \pi$). The ground states of the type IIb theory form the massless D = 10 chiral N = 2 supergravity multiplet consisting of

$|i\rangle \otimes |\tilde{j}\rangle$, $|a\rangle \otimes |\tilde{b}\rangle$ 128 boson states

$|i\rangle \otimes |\tilde{a}\rangle$, $|a\rangle \otimes |\tilde{i}\rangle$ 128 fermion states

where $|\ \rangle$ and $|\tilde{\ }\rangle$ indicate the open-string states in each oscillator space. [In the IIa theory the two spaces have fermion ground states of opposite type, $|a\rangle$ and $|\tilde{\dot{a}}\rangle$.]

The type I closed-string states are obtained as a truncation of the type IIb theory by symmetrizing the states between the two types of spaces. This halves the number of ground states which now form the massless supermultiplet of N=1 supergravity in ten dimensions.

More details of the SO(8) formalism are given by Brink in his contribution to this workshop.

(d) SU(4) × U(1) Formalism[21].

The fact that the Θ's are self-conjugate (and therefore do not anti-commute with themselves) means that they are simultaneously "position" and "momentum" variables. This will not be satisfactory for formulating a field theory of superstrings in which the fields are functions of the coordinates. For the type IIb theory it is possible to take the complex combinations $\theta^{1a} + i\theta^{2a}$ and $\theta^{1a} - i\theta^{2a}$, etc. to be position and momentum variables[22] but this does not adapt to N = 1 theories (type I or heterotic). A more satisfactory resolution is to form complex combinations of the components of a single spinor which amounts to breaking the manifest SO(8) symmetry down to SU(4) × U(1) by the identifications (for the type IIb theory)

$$\begin{aligned} \theta^{1a} &\to \theta_A \,,\ \lambda^A \\ \theta^{2a} &\to \tilde{\theta}_A \,,\ \tilde{\lambda}^A \\ 8 &\to \bar{4}_{1/2} \,,\ 4_{-1/2} \end{aligned} \tag{1.55}$$

where the last line indicates the SU(4) × U(1) content and an upper index indicates an SU(4) spinor whereas a lower index indicates an SU(4) anti-spinor (and $\theta^{\bar{A}} \equiv \theta_A$). [For the type IIa theory the SU(4) indices on the spinors is changed to θ_A, $\tilde{\theta}^A$, λ^A and $\tilde{\lambda}_A$.] Similarly

$$\begin{aligned} X^I &\to X^i \,,\ X^L \,,\ X^R \\ 8 &\to 6_0 \,,\ 1_1 \,,\ 1_{-1} \end{aligned} \tag{1.56}$$

where $i = 1,2,..,6$ labels the 6 of SU(4) and $X^L \equiv (X^7+iX^8)/\sqrt{2}$, $X^R \equiv (X^7-iX^8)/\sqrt{2}$ are SU(4) singlets. The fact that an SU(4) subgroup is picked out amounts to treating the six transverse dimensions differently from the other two. Since we hope that six dimensions will compactify in the end such a formalism is certainly adequate for our purposes.

In writing the mode expansions of the coordinates in the interacting theory it will prove useful to use a normalization of the parameter σ such that it spans the region

$$0 \leqslant \sigma \leqslant 2\pi|p^+| \equiv \pi|\alpha| \tag{1.57}$$

where $\alpha = 2p^+$ as introduced by Mandelstam[23]. The anticommutation relations in eq. (1.47) are now given by

$$\{ \lambda^A(\sigma,\tau) , \theta_B(\sigma',\tau) \} = \delta^A{}_B \delta(\sigma-\sigma') = \{ \tilde{\lambda}^A(\sigma,\tau) , \tilde{\theta}_B(\sigma,\tau) \} \quad (1.58)$$

For diversity, I will consider the mode expansions for the type IIb theory in this subsection. These are

$$X^I = x^I + p^I\tau + \frac{1}{2} i \sum_{n\neq 0} \frac{1}{n}(\alpha^I_n e^{2in(\tau+\sigma)/|\alpha|} + \tilde{\alpha}^I_n e^{2in(\tau-\sigma)/|\alpha|}) \quad (1.59a)$$

$$\theta_A = \frac{1}{\alpha} \sum_{-\infty}^{\infty} Q_{An} e^{-2in(\tau-\sigma)/|\alpha|} \quad (1.59b)$$

$$\tilde{\theta}_A = \frac{1}{\alpha} \sum_{-\infty}^{\infty} \tilde{Q}_{An} e^{-2in(\tau+\sigma)/|\alpha|} \quad (1.59c)$$

$$\lambda^A = \frac{1}{\pi|\alpha|} \sum_{-\infty}^{\infty} Q^A_n e^{-2in(\tau-\sigma)/|\alpha|} \quad (1.59d)$$

$$\tilde{\lambda}^A = \frac{1}{\pi|\alpha|} \sum_{-\infty}^{\infty} \tilde{Q}^A_n e^{-2in(\tau+\sigma)/|\alpha|} \quad (1.59e)$$

(and $P^I(\sigma,\tau) = \frac{\dot{X}^I}{\pi}(\sigma,\tau)$ as before). The α_n and $\tilde{\alpha}_n$ modes satisfy the commutation relations of eq. (1.48) whereas the Q_n's and $\tilde{Q}_n$'s satisfy

$$\{ Q^A_m , Q_{Bn} \} = \alpha\, \delta_{m+n,0}\, \delta^A{}_B = \{ \tilde{Q}^A_m , \tilde{Q}_{Bn} \} \quad (1.60)$$

with the other anticommutators vanishing.

Each sixteen-component supercharge generator breaks into two SO(8) spinors, q^a and $q^{\dot{a}}$, where the undotted piece generates a linear transformation of the coordinates whereas the dotted piece is more complicated since it also performs a compensating κ transformation to restore the gauge condition. Each undotted SO(8) supercharge splits into two SU(4) spinors which are integrals of charge densities which may be determined by the Noether method to be given by

$$q_{1A}(\sigma) = \epsilon(\alpha)\, \theta_A(\sigma) \qquad q_{2A}(\sigma) = \epsilon(\alpha)\, \tilde{\theta}_A(\sigma)$$

$$q^A_1(\sigma) = \lambda^A(\sigma) \qquad q^A_2(\sigma) = \tilde{\lambda}(\sigma) \quad (1.61)$$

(where $\epsilon(\alpha) = \mathrm{sign}(\alpha)$) and hence

$$\{ q_1^A(\sigma) , q_{1B}(\sigma') \} = \delta_B^A \, \delta(\sigma-\sigma') = \{ q_2^A(\sigma) , q_{2B}(\sigma') \} . \qquad (1.62)$$

When integrated these relations give the piece of the N = 2 supercharge algebra associated with the linearly realized supersymmetries. The dotted SO(8) spinors split into SU(4) spinors which will be denoted by q^-_{1A}, q_1^{-A}, q^-_{2A} and q_2^{-A}. These are represented by integrals of quadratic functions of the coordinates. For example,

$$q_1^{-A} = \int_0^{\pi|\alpha|} \{ \sqrt{2}\rho_i^{AB} (P^i - \frac{\acute{X}^i}{\pi}) \theta_B + 2\pi\epsilon(\alpha)(P^L - \frac{\acute{X}^L}{\pi})\tilde{\lambda}^A \} \, d\sigma \qquad (1.63)$$

where the matrices ρ_i^{AB} are Clebsch-Gordon coefficients for SU(4) normalized so that

$$\rho_i^{AB} \rho_{jBC} + \rho_j^{AB} \rho_{iBC} = \delta_{ij} \, \delta_C^A . \qquad (1.64)$$

The formulae for the other q^-'s can be found in ref. 21. The rest of the anticommutation relations of the supercharge algebra can be obtained from them. These include

$$\{ q_1^{-A} , q^-_{1B} \} = 2\delta_B^A \, h_{cl} = \{ q_2^{-A} , q^-_{2B} \} \qquad (1.65)$$

$$\{ q_1^A , q_1^{-B} \} = \sqrt{2} \, \rho_i^{AB} \, p^i = \{ q_2^A , q_2^{-B} \} \qquad (1.66)$$

$$\{ q_{1A} , q^-_{1B} \} = \sqrt{2} \, \rho_{iAB} \, p^i = \{ q_{2A} , q^-_{2B} \} \qquad (1.67)$$

The subsidiary condition $N = \tilde{N}$ must be used in verifying the closure of the algebra. The hamiltonian, h_{cl}, which appears in eq.(1.65) is given by

$$h_{cl} = \frac{1}{\pi}\int_0^{\pi|\alpha|} d\sigma \, \{\epsilon(\alpha)(\pi^2 \underline{P}^2 + \underline{\acute{X}}^2) - 2\pi i(\theta_A \acute{\lambda}^A + \tilde{\theta}_A \acute{\tilde{\lambda}}^A)\} \tag{1.68}$$

$$= \frac{4}{\alpha}(N + \tilde{N}) + \frac{\underline{P}^2}{\alpha} \tag{1.69}$$

where

$$N = \sum_{n \neq 0} \left(\frac{1}{2}\alpha^I_{-n} \cdot \alpha^I_n + \frac{n}{\alpha} Q_{-nA} Q^A_n\right) \tag{1.70}$$

$$\tilde{N} = \sum_{n \neq 0} \left(\frac{1}{2}\tilde{\alpha}^I_{-n} \cdot \tilde{\alpha}^I_n + \frac{n}{\alpha} \tilde{Q}_{-nA} \tilde{Q}^A_n\right) \tag{1.71}$$

The Lorentz generators $J^{\mu\nu}$ may be represented in analogous fashion and the closure of the Lorentz algebra verified, including the notorious term $[J^{I-}, J^{J-}]$. This term only vanishes for the ten-dimensional theory obtained from the covariant action. If the superstring theories in D = 3, 4, or 6 dimensions are assumed to have purely transverse excitations in the light-cone gauge this commutator is found not to vanish[24] (at least with the usual definitions of the Lorentz generators). This reinforces the observation that these theories are not in their critical dimensions.

In this SU(4) formalism the massless ground states are described by wave functions that depend on θ_0 and $\tilde{\theta}_0$ which are the wave functions of the corresponding super-Yang-Mills or supergravity point field theories. The excited states are constructed as before by operating on the ground states with the creation operators.

The states of the type II theories are tensor products of states in the untilde and tilde Fock spaces obtained by applying α_{-n}, $\tilde{\alpha}_{-n}$, Q_{-n} and $\tilde{Q}_{-n}$ to the massless ground state N = 2 supergravity multiplet. The subsidiary condition $N = \tilde{N}$ must also be enforced.

The type I closed-string states are obtained from the type IIb states by symmetrizing them in the oscillators α and $\tilde{\alpha}$ as well as Q and $\tilde{Q}$. This truncates the supersymmetry to N = 1.

The open-string algebra has one ten-dimensional supercharge (since $\Theta_0 = \tilde{\Theta}_0$ due to the boundary conditions) given by averaging the N = 2 supercharges so that any of the various SU(4) components of the supercharges is given by

$$q = \frac{1}{2}(q_1 + q_2) \tag{1.72}$$

The open-string states are created by just one set of α_n and Q_n modes operating on the ground-state Yang-Mills supermultiplet.

Rules for calculating scattering amplitudes can be formulated in the first-quantized formalism by inventing vertices for emitting on-shell ground states which are consistent with the constraints of supersymmetry. This was how many of the calculations were initially performed both for tree diagrams[20] and for one-loop diagrams[42].

II LIGHT-CONE-GAUGE FIELD THEORY OF SUPERSTRINGS

The first-quantized superstring theory is adequate for calculations in perturbation theory around flat space-time in the critical dimension. The amplitudes for the bosonic and spinning string theories have been constructed in the light-cone gauge[23] by a generalization of Feynman's path-integral formalism, as sums over all possible connected world-sheets that join the incoming and outgoing strings. The interactions occur at the ponts where the world-sheet splits. The "Born" (or "tree") diagrams are given by summing over those surfaces which have no handles attached and no holes cut out (in the case of an open string theory where the boundary of a hole corresponds to the world-line of a string endpoint). Adding handles or holes corresponds to the higher order corrections (associated with loop diagrams, for example). The functional treatment of the covariant formulation of string theory was suggested[25] some time ago but not studied in depth until recently[7]. The functional formulation of the perturbation theory diagrams for superstring theories is the subject of intensive study both in the light-cone gauge[26] and in a covariant formulation.

Just as in any point particle theory this first-quantized formulation is probably not appropriate for understanding non-perturbative aspects, such as the compactification of the theory. In order to develop our understanding of these theories it will be necessary to study them in the language of string field theory in which string fields create and destroy complete strings. There is, presumably, something like a geometrical formulation of the theory that generalizes the familiar description of general relativity in terms of the geometry of space-time. Although there has been progress towards a supercovariant, gauge-invariant interacting string field theory[27], this is only incompletely understood for the moment.

The interacting quantum field theory of strings has, however, been formulated in the light-cone gauge, both for the older string theories[28] and superstring theories[21,22,29]. This is not completely satisfying since it is likely that any deep geometric structure in the theory will be obscured by the choice of a special gauge. Nevertheless it may be adequate for certain purposes, such as studying compactified solutions in which four dimensions are flat (including the + and - directions). This field theory formalism also provides a systematic way of generating the light-cone-gauge superstring perturbation theory diagrams including certain local operators that have to be inserted at the interaction points where the world-sheet splits. [Such operators do not occur in the bosonic theory.]

In this lecture I will outline the way in which the superstring fields are defined and the construction of the free field representation of the super-Poincaré algebra. I will then show how the interaction terms are almost uniquely specified by demanding that the algebra be represented non-linearly on the fields in the interacting theory. The whole discussion is a generalization of conventional point field theory. However, one particularly striking general feature of the light-cone gauge field theory of closed strings is the fact that the interactions are simply cubic in the string fields (in contrast with point field theories containing the Einstein-Hilbert ≈action which involve infinite orders of interactions between the fluctuations of the metric). This is intuitively plausible from the fact that the string interactions correspond to critical points on the world-sheet, which generically involve the splitting of one string into two or the joining of two strings into one. I will show how this is required by the consistency of the supersymmetry algebra in the interacting field theory. Most of this lecture is based on material contained in ref. (21).

(a) Light-Cone-Gauge Superstring Fields

A string field is a scalar *functional* of the light-cone string superspace coordinates. In the light-cone gauge it is often useful to use a Fourier transform with respect to X^- so that the momentum p^+ is a variable ($p+ = i\partial/\partial X^-$). A string field is then a function of $\Theta(\sigma)$, $\tilde{\Theta}(\sigma)$, $\underset{\sim}{X}(\sigma)$, X^+ and p^+ (or α) which means that it is a function of the configuration of the whole string and is not an explicit function of σ. By convention a field operator with $p^+ > 0$ will be a creation operator whereas one with $p^+ < 0$ will be an annihilation operator. A closed-string field will be denoted by

$$\Psi[\underset{\sim}{X}(\sigma),\Theta_A(\sigma),\tilde{\Theta}_A(\sigma),\alpha,X^+]$$

for a type IIb (i.e. chiral N=2) field (the non-chiral type IIa field is $\Psi[\underset{\sim}{X}(\sigma),\Theta_A(\sigma),\tilde{\Theta}^A(\sigma),\alpha,X^+]$ i.e. it is a function of $\tilde{\Theta}$ in a 4 instead of one in a $\bar{4}$). An open-string field is denoted by

$$\Phi[X(\sigma),\theta_A(\sigma),\tilde{\theta}_A(\sigma),\alpha,X^+]$$

These fields describe an infinite set of ordinary point fields, one for every state of excitation of the string. The open-string field is also a matrix in an internal symmetry group. By direct analogy with ordinary scalar field theory in light-cone coordinates[30] the action S is related to the field-theory hamiltonian H by

$$S = \int D^{16}Z\ dX^+\ dX^-\ (\partial_+\Psi\partial_-\Psi + \mathrm{tr}\partial_+\Phi\partial_-\Phi) - \int dX^+\ H \tag{2.1}$$

Note that $\partial_\pm = \partial/\partial X^\pm$ so that the action is linear in time derivatives (if H is independent of them). In eq. (2.1) $D^{16}Z$ denotes the functional integration over the eight transverse coordinates $X(\sigma)$ and eight Grassmann coordinates $\theta_A(\sigma)$, $\tilde{\theta}_A(\sigma)$ i.e.

$$D^{16}Z \equiv D^8X^I(\sigma)\ D^4\theta_A(\sigma)\ D^4\tilde{\theta}_A(\sigma) \tag{2.2}$$

which will be interpreted as the infinite product of the differentials of the modes of the coordinates.

For closed strings it is important to impose the condition that the superfield Ψ does not depend on the origin of the σ parameter

$$\Psi[X^I(\sigma+\sigma_0),\theta_A(\sigma+\sigma_0),\tilde{\theta}_A(\sigma+\sigma_0)] = \Psi[X^I(\sigma),\theta_A(\sigma),\tilde{\theta}_A(\sigma)] \tag{2.3}$$

where σ_0 is an arbitrary constant. For infinitessimal σ_0 this equation becomes

$$\int_0^{\pi\alpha} d\sigma \left\{ X'^I\frac{\delta}{\delta X^I} + \theta'_A\frac{\delta}{\delta\theta_A} + \tilde{\theta}'_A\frac{\delta}{\delta\tilde{\theta}_A} \right\} \Psi = 0 \tag{2.4}$$

This is exactly the same condition as $N-\tilde{N} = 0$ obtained earlier (eq. (1.54)) and in this context is to be imposed as a (first class) constraint on the string fields in the action. The massless type IIb states, in particular, are described by the ground state component superfield $\psi(x,\theta_{oA},\theta_{oA},\alpha,x^+)$ which has 2^8 terms when expanded in powers of Θ, corresponding to the states of type IIb supergravity.

Type I closed string fields satisfy the additional constraint that expresses the fact that they are unoriented

$$\Psi[X^I(\sigma),\Theta_A(\sigma),\tilde{\Theta}_A(\sigma)] = \Psi[X^I(-\sigma),\tilde{\Theta}_A(-\sigma),\Theta_A(-\sigma)] \qquad (2.5)$$

The symmetry under $\sigma \leftrightarrow -\sigma$ and $\Theta \leftrightarrow \tilde{\Theta}$ expressed by eq. (2.5) implies (from the mode expansions in eqs. (1.59)) that the physical states are symmetric under the interchange of (α_n,Q_n) and $(\tilde{\alpha}_n,\tilde{Q}_n)$ as required for type I closed strings.

Open strings carry quantum numbers at their endpoints associated with the defining representation of one of the classical groups SO(n), U(n) or Sp(2n) which appear to be the only ones that can be incorporated in this manner[31] in the classical theory. In the quantum theory anomalies rule out all groups except SO(32). In this ("Chan-Paton") scheme[32] the indices a and b of Φ_{ab} are associated with either endpoint of the string and the particular group G represented in this way depends on the conditions imposed on the Φ. These conditions are designed so that the massless ground states lie in the adjoint representation of G. For example, the group SO(n) is obtained by requiring Φ_{ab} to be real with a,b = 1,2,...,n and imposing the condition

$$\Phi_{ab}[X^I(\sigma),\Theta_A(\sigma),\tilde{\Theta}_A(\sigma)] = -\Phi_{ba}[X^I(\pi|\alpha|-\sigma),\tilde{\Theta}_A(\pi|\alpha|-\sigma),\Theta_A(\pi|\alpha|-\sigma)] \qquad (2.6)$$

This relates a superfield to one with reversed orientation which (as follows from the mode expansions of eqs. (1.59)) identifies states which differ by the substitutions $\alpha^I_n \to (-1)^n\alpha^I_n$ and $\Theta_{An} \to (-1)^n\tilde{\Theta}_{An}$. This means that the even mass levels (with masses satisfying $(\text{mass})^2/2\pi T = 0,2,4...$) have component fields satisfying $\Phi_{ab} = -\Phi_{ba}$ and therefore lie in the adjoint representation. The odd levels (with $(\text{mass})^2/2\pi T = 1,3,5,..$) have components satisfying $\Phi_{ab} = \Phi_{ba}$.

(b) Free Field Theory of Superstrings

For every generator of the super-Poincaré algebra g, which was expressed in terms of the coordinates and momenta in the first lecture, we now associate a field theory generator G made out of the string superfields. In general this generator will be a complicated function of the fields which I shall assume can be expanded as a series

$$G = G_2 + G_3 + \dots \qquad (2.7)$$

where G_2 is quadratic in fields, G_3 is cubic and so on. Actually, this counting works for purely closed-string theories or for the purely open-string part of the type I theories. However, in the terms involving mixtures of open and closed strings a closed string field counts as two powers of an open string field. The G's are constructed to satisfy the same algebra as the g's.

In the free theory all the G's are quadratic and have the form

$$G_2 = \int_0^\infty \alpha d\alpha \int D^{16}Z \; (\Psi_{-\alpha}\; g\; \Psi_\alpha + \mathrm{tr}\Phi_{-\alpha}\; g\; \Phi_\alpha) \qquad (2.8)$$

where the dependence of the fields on α (i.e. on p^+) is explicitly displayed (and reality of the fields implies $\Psi_{-\alpha} = \Psi_\alpha{}^*$ and $\Phi_{-\alpha} = \Phi_\alpha{}^\dagger$). The form of eq. (2.8) generalizes that of conventional scalar field theory expressed in light-cone coordinates. [Note that the normalization of the string fields implicit in this equation differs by a power of $\sqrt{\alpha}$ from that adopted by some authors.] The G's satisfy an algebra that is isomorphic to the algebra of the g's as will be seen by use of Poisson brackets.

The momenta conjugate to the fields, Π_Ψ and Π_Φ, are defined by

$$\Pi_\Psi \equiv \frac{\delta S}{\delta \partial_+ \Psi} = \partial_- \Psi \qquad \Pi_\Phi \equiv \frac{\delta S}{\delta \partial_+ \Phi} = \partial_- \Phi \qquad (2.9)$$

These are phase-space constraints since the momenta are proportional to "spatial" derivatives of the fields. These constraints lead to a simple modification of the canonical Poisson brackets as is well-known in scalar light-cone field theory[30]. Even though all the arguments of this lecture are at the level of the classical string field theory (i.e. I will not treat the fields as quantum operators so that the ordering of the fields will never matter) I shall describe these constrained Poisson brackets as if they were equal time quantum commutation relations by including appropriate powers of i. These relations, analogous to those of scalar point field theory can be deduced simply from the requirement that the G's defined by eq. (2.8) form a closed algebra. For closed strings this gives

$$[\Psi[1]\ ,\ \Psi[2]] = \frac{\delta(\alpha_1+\alpha_2)}{\alpha_2}\int_0^{\pi|\alpha_2|}\frac{d\sigma_0}{\pi|\alpha_2|}\,\Delta^{16}[Z_1(\sigma)-Z_2(\sigma+\sigma_0)] \qquad (2.10)$$

where the integral over σ_0 takes into account the constraint of eq. (2.3) and $\Delta^{16}(Z)$ represents the product of an infinite number of Dirac delta functions, one for each mode of X^I, Θ_A and $\tilde{\Theta}_A$. For type I closed strings there is also a projection on the right-hand side which imposes the constraint of eq. (2.5). For open strings the commutation relations which are consistent with the group theory constraints are (when the internal group is SO(32)),

$$[\Phi_{ab}[1]\ ,\ \Phi_{cd}[2]] = \frac{\delta(\alpha_1+\alpha_2)}{2\alpha_2}\Big[\delta_{ab}\delta_{cd}\Delta^{16}(Z_1(\sigma)-Z_2(\sigma)) - \delta_{ac}\delta_{bd}\Delta^{16}(Z_1(\sigma)-Z_2(\pi|\alpha_2|-\sigma))\Big] \qquad (2.11)$$

with analogous expressions for the other possible groups.

From these commutation relations it is easy to check that the G_2's (defined in eq. (2.8) generate the appropriate changes in the fields,

$$[G_2,\Psi] = g\Psi \qquad \text{and} \qquad [G_2,\Phi] = g\Phi \qquad (2.12)$$

For example, the time dependence or the fields is generated by the hamiltonian H_2, with h being given (eq. (1.69)) by the sum of harmonic oscillator hamiltonians (and their Grassmann equivalents). The time (i.e. X^+) dependence of the free field theory is therefore determined by the equation of motion

$$i \frac{\partial \Psi}{\partial X^{+}} = h \Psi \tag{2.13}$$

with a similar equation for the open string field. A string field can be expanded as a sum over a complete set of eigenfunctions of h in the form (for open strings)

$$\Phi[\chi(\sigma),\Theta_A(\sigma),\tilde{\Theta}_A(\sigma)] =$$

$$\sum_{n_k^I m_{As}} \Phi_{n_k^I m_{As}}(x,\Theta_0) \prod_k \prod_I H_{n_k^I}(x_k^I) \prod_s \prod_A \Xi_{m_{As}}(\Theta_{As}) \tag{2.14}$$

where H and Ξ are the eigenfunctions of the number operators occurring inside h (and the numbers n_k^I and m_{As} are the occupation numbers of the levels labelled by (k,I) for the Bose modes, x_k^I, and (s,A) for the Fermi modes, Θ_{As}). The coefficients $\Phi_{\{n\}}$ are ordinary point superfields which are either fermionic or bosonic depending on whether the level {n} has an even or odd number of Θ excitations. By substituting this expansion into the free field action we can rewrite it as an infinite sum of ordinary point field free field actions (suppressing group theory indices)

$$S = \int_0^\infty d\alpha \int dx\, d\Theta_0 \sum_{\{n\}} \left\{ \frac{1}{2}\Phi_{-\alpha\{n\}}(\Box - N_{\{n\}})\Phi_{\alpha\{n\}} + \frac{1}{\alpha}\Psi_{-\alpha\{n\}}(\Box - N_{\{n\}})\Psi_{\alpha\{n\}} \right\} \tag{2.15}$$

where the fermionic superfields have been denoted $\Psi_{\{n\}}$ to distinguish them from the bosonic ones.

(c) Interacting Superstring Fields.

The choice of the light-cone gauge spoils the manifest super-Poincaré invariance of a theory. As a result, in the interacting light-cone-gauge field theory certain of the super-Poincaré generators act non-linearly on the fields. These are the generators that incorporate compensating gauge transformations which restore the light-cone gauge conditions, namely, J^{+-}, J^{I-}, $Q^{-}{}_{A}$, $\tilde{Q}^{-}{}_{A}$, Q^{-A}, $\tilde{Q}^{-A}$ and P^{-} (which is the hamiltonian, H). The form of these generators can be determined uniquely by demanding that they satisfy the super-Poincare algebra (and assuming that they contain no more than two spatial derivatives). In fact, requiring the supersymmetry sub-algebra to be satisfied already determines the form of the generators almost uniquely. The appropriate equations are

$$\{ Q^{-A} , Q^{-}_{B} \} = 2 \delta^{A}{}_{B} H \tag{2.16}$$

$$\{ Q^{-A} , Q^{-B} \} = 0 = \{ Q^{-}{}_{A} , Q^{-}{}_{B} \} \tag{2.17}$$

$$\vdots$$

By substituting the series expansions (eq. (2.7)) for the Q's and H into these equations we can satisfy them order by order in the string fields (allowing for the comment after eq. (2.7) concerning terms involving mixtures of open and closed string fields).

(d) The Cubic Interaction of Open Superstrings.

In order to illustrate the method I will begin with the term in the interaction hamiltonian which describes the splitting of one open string into two or the joining of two open strings into one as illustrated in fig. 2.1a.

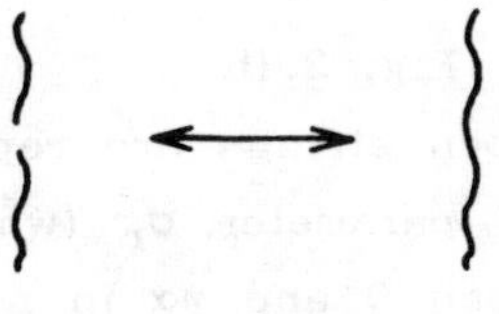

Fig. 2.1a

We want to solve the equations like (2.16) and (2.17) by substituting the series expansions (eq. (2.7)) for the generators. The lowest order involves only the quadratic pieces of the generators and the equations are satisfied by the free field theory expressions. At the next order eqs. (2.16) and (2.17) give (for terms involving purely open-string or purely closed-string fields)

$$\{ Q_3^{-A}, Q_{2B}^{-} \} + \{ Q_2^{-A}, Q_{3B}^{-} \} = 2\, \delta_B^A H_3 \qquad (2.18)$$

$$\vdots$$

Needless to say the rest of the superalgebra, which involves those generators that generate linear transformations and do not get interaction corrections, provides strong constraints on the form of H_3, and the $Q_{\bar{3}}$'s. Together with eqs. (2.18) these constraints almost completely determine these generators. The small remaining ambiguity involves overall powers of α (i.e. p^+) in the expressions for the interactions. In principle, these could be fixed by using the Lorentz generators J^{I-} and J^{+-} (but in practice it is easier to fix them by treating a special case with external ground-state particles).

The interaction can best be represented on a plot of the σ–τ space first introduced by Mandelstam[23] (with the interaction taken at $\tau = 0$).

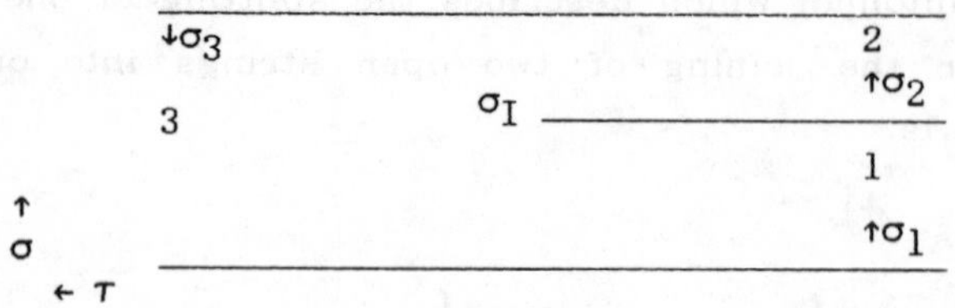

Fig. 2.1b

The boundaries of the open strings are represented by solid lines. Each string has its own parameter σ_r (where $r = 1,2,3$ labels the string) which runs between 0 and $\pi\alpha$ in a sense that depends on whether the string is incoming or outgoing as indicated in the figure. The α's are chosen to be positive for incoming strings (1 and 2) and negative for outgoing ones (3). The coordinates on each string, $X^{(r)}(\sigma_r)$, $\theta^{(r)}(\sigma_r)$ and $\tilde{\theta}^{(r)}(\sigma_r)$ (setting $\tau=0$), have mode expansions like eqs. (1.59) on each string. It is useful to use one parameter, $\sigma = -\pi\alpha_3 - \sigma_3 = \sigma_1 = \pi\alpha_1 + \sigma_2$, and to define $X_r(\sigma) \equiv X^{(r)}(\sigma_r)$ (with the understanding that X_r vanishes outside of the region in σ in which string r is defined) with similar definitions for the Grassmann coordinates.

One physical ingredient that is built into the form of the interaction hamiltonian is that the coordinates should be continuous at the interaction time. This means that at $\tau = 0$ we must enforce the conditions

$$\sum_{r=1}^{3} \epsilon_r Z_r(\sigma) = 0 \tag{2.19}$$

where $\epsilon_r = \text{sign}(\alpha_r)$ and $Z_r(\sigma) \equiv (X_r(\sigma), \theta_r(\sigma), \tilde{\theta}(\sigma))$. This is incorporated into the expression for any of the generators by a delta functional. The general form of a generator is therefore

$$G_3 = g\int (\prod_{r=1}^{3} d\alpha_r D^{16}Z_r)\delta(\Sigma\alpha_r)\Delta^{16}[\Sigma\epsilon_r Z_r]\hat{G}(\sigma_I)\,\text{tr}(\Phi[1]\Phi[2]\Phi[3]) \tag{2.20}$$

where $\hat{G}$ is an operator that acts on the fields at the interaction point σ_I. The notation Φ[r] denotes the field as a functional of the coordinates of string r. The expression for G_3 can also be written in terms of the Fourier transform of the field $\tilde{\Phi}$ in which case the coordinate *continuity* Δ funtional is replaced by one expressing the *conservation* of momenta, $\underset{\sim}{P}_r(\sigma)$, $\lambda_r^A(\sigma)$, $\tilde{\lambda}_r^A(\sigma)$ at the interaction time.

This form of the interaction terms is not only motivated by the physical picture of the strings interacting at their endpoints but also guarantees that the linearly realized symmetries are correctly incorporated. These are the symmetries generated by J^{IJ}, J^{I+}, p^I, Q^A and Q_A which do not get interaction corrections.

In the case of the bosonic string theory considered in ref. 28 the form of the interaction was guessed to be like eq. (2.20) without the operator $\hat{G}$ (and obviously with no fermionic variables) but with certain extra factors in the integration measure. This expression was shown to agree with that obtained by Mandelstam in ref.23 (who had obtained the vertex by considering the Feynman path integral for the infinite time process shown in the fig. 2.1 and then factored off the external legs). The major difference in the case of the superstring theories (as well as in the spinning string theory) is the occurrence of the operator $\hat{G}$ at the interaction point.

The expression for G_3 can be written in terms of the component fields by substituting the expansion of eq. (2.14) into eq. (2.20) and integrating over the non-zero modes of the coordinates. This gives an expression of the form

$$G_3 = \int \left(\prod_{r=1}^{3} d\alpha_r d^8 x_r d^4 \theta_{0r} \right) \sum_{\{n_1, n_2, n_3\}} C_{n_1 n_2 n_3} \Phi_{n_1} \Phi_{n_2} \Phi_{n_3} \tag{2.21}$$

where the coefficients $C_{n_1 n_2 n_3}$ are (non-local) functions which determine the interactions between states of arbitrary occupation numbers, denoted $\{n_r\}$ for string r. These coefficients can be written in terms of states of definite occupation numbers for each string as

$$C_{n_1 n_2 n_3} = \langle n_1 | \otimes \langle n_2 | \otimes \langle n_3 | G \rangle \tag{2.22}$$

where $|G\rangle$ is a "vertex" ket vector defined in the tensor product of the Fock spaces of the three strings (and $\langle n_r|$ is the bra vector for string r in a state with occupation numbers $\{n_r\}$). This relation can be inverted to give

$$| G \rangle = \sum_{n_1 n_2 n_3} C_{n_1 n_2 n_3} |n_1\rangle \otimes |n_2\rangle \otimes |n_3\rangle \tag{2.23}$$

For all the generators this can be written in the form

$$|G\rangle = \hat{G}\,|V\rangle \tag{2.24}$$

where $\hat{G}$ is the oscillator-basis representation of the operator $\hat{G}(\sigma_I)$ that occurred in eq. (2.20) and the vertex $|V\rangle$ is the oscillator-basis representation of the Δ functional in eq. (2.20). It is given by the exponential of a quadratic form in the creation operators for each string, α^r_{-n} and θ^r_{-n} acting on the product of the three ground states. The explicit expression for $|V\rangle$ is given in ref. 21 (and generalizes that of the bosonic string theory which only involved the α modes).

The representation of the theory in the oscillator basis gives well-defined expressions with no singular operators. This makes it convenient for deriving the explicit expressions for the $\hat{G}$'s by solving the equations like eq. (2.18). In the position (or momentum) basis there are delicacies in defining various operators (such as the momentum, $\underline{p}(\sigma)$) in the vicinity of the interaction point since they are singular and must be regulated. This singular behaviour is essential for there to be a non-trivial interaction. However, (as stressed in ref. (36)) in the oscillator basis it is not at all obvious that the theory is local on the world-sheet. This locality is an essential ingredient in understanding duality which therefore appears to be an accidental miracle in the oscillator basis. The position-space representation of the interactions can be reconstructed in the form given in eq. (2.20) where the operator $\hat{G}$ is given as a polynomial in the functional derivative

$$Z^J = -\,i\,\sqrt{\pi}\,\lim_{\epsilon\to 0}\sqrt{\epsilon}\,\frac{\delta}{\delta X^J(\sigma_I-\epsilon)} \tag{2.25}$$

and a Grassmann coordinate

$$Y_A = \lim_{\epsilon\to 0}\sqrt{\epsilon}\,[\theta_A(\sigma_I-\epsilon) + \tilde{\theta}_A(\sigma_I-\epsilon)]/\sqrt{2} \tag{2.26}$$

where the factors of $\sqrt{\epsilon}$ cancel the singular behaviour of the operators near the interaction point. It does not matter on which string these operators are defined - they all give the same limiting answer. In terms of these operators the generators, $\hat{G}$, appearing in the supercharges are given by

$$\hat{Q}^{-}_{3A} = Y_A \tag{2.27}$$

$$\hat{Q}^{-A}_{3} = \frac{2}{3} \epsilon^{ABCD} Y_B Y_C Y_D \tag{2.28}$$

$$H_3 = \frac{1}{\sqrt{2}} Z^L + \frac{1}{2} (\rho^i)^{AB} Z^i Y_A Y_B + \frac{1}{6\sqrt{2}} Z^R \epsilon^{ABCD} Y_A Y_B Y_C Y_D \tag{2.29}$$

The fact that the expression for the interaction term in the hamiltonian is linear in Z (i.e. in the momentum $P(\sigma_I)$) is reminiscent of the usual Yang-Mills cubic interaction. In fact, it is easy to isolate the term involving just the zero modes (the usual light-cone gauge Yang-Mills superfields) in the interaction from eq. (2.20). This term has the form of the familiar light-cone gauge cubic super-Yang-Mills interaction but smeared out by a gaussian factor (due to the fact that the centres of masses of the three strings do not coincide). In the limit of infinite string tension the interaction reduces to that of usual point super-Yang-Mills field theory.

(e) Open - Closed Superstring Interaction.

The fact that the interaction of fig. 1 is local on the string means that the same joining or splitting process can take a single open string into a closed string and vice versa.

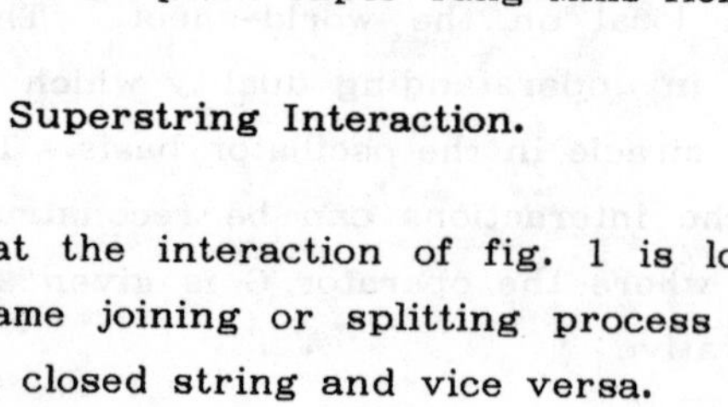

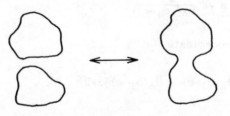

Fig. 2.2

The expression for this interaction, which involves $g\Psi \mathrm{tr}\Phi$, is again deduced by determining its contribution to the supersymmetry generators[21]. [It was also discussed in detail in the bosonic theory using a lattice regulator in the σ-τ plane[34].] As expected on the grounds of locality the expressions for the generators involve the same local operator $\hat{G}(\sigma_I)$ acting at the joining or splitting point as in the cubic open-string interaction.

(f) Closed Supertring Interactions.

Two oriented (type II) closed strings can interact by touching at a point

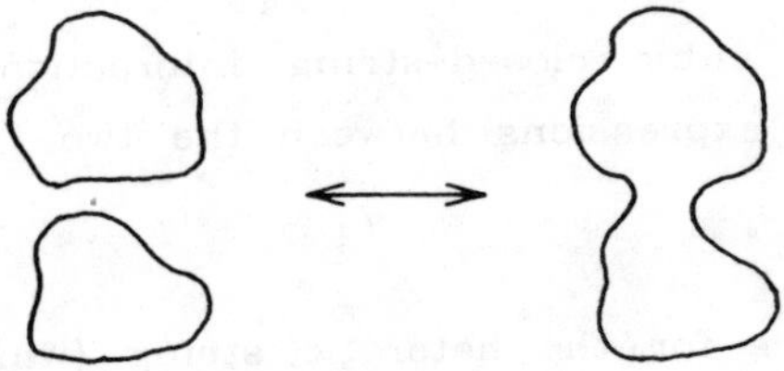

Fig. 2.3

This is a local interaction which is cubic in the closed-string fields (i.e. contains Φ^3) and is of gravitational strength κ (since it reduces to the cubic gravitational coupling in the low energy limit). It has a form that is very simply related to the cubic open-string interaction described above. The operator acting on the fields in the interaction terms for any of the generators is the product of two open-string factors. In the oscillator basis these generators are given by tensor products of two factors where one factor is made of the untilde modes and the other of the tilde ones.

$$| \, Q_1^{-A} \rangle = | \, Q^{-A} \rangle \otimes | \, \tilde{H} \rangle \tag{2.30}$$

$$| \, Q_2^{-A} \rangle = | \, H \rangle \otimes | \, \tilde{Q}^{-A} \rangle \tag{2.31}$$

$$| Q^-_{1A}\rangle = |Q^-_A\rangle \otimes | \tilde{H}\rangle \tag{2.32}$$

$$| Q^-_{2A}\rangle = | H\rangle \otimes | \tilde{Q}^-_A\rangle \tag{2.33}$$

$$| H\rangle_{closed} = | H\rangle \otimes | \tilde{H}\rangle \tag{2.34}$$

where the right-hand sides of these equations involve the open-string expressions of eqs. (2.27) - (2.29). The closure of the closed-string algebra follows simply from the fact that the open-string algebra closes. The expressions for the generators in terms of the fields can be deduced from these expressions in a similar manner to the open string case.

The type I cubic closed-string interaction is obtained by symmetrizing these expressions between the two types of oscillator space.

The generators for the heterotic string (which has only one supercharge) are obtained by substituting the (twenty-six dimensional) bosonic string vertex $|\tilde{V}\rangle$ instead of the right-hand factors in eqs.(2.30)-(2.34) and compactifying the zero modes in sixteen of the dimensions on a suitable hypertorus[35]. This involves a subtle issue concerning the fact that these internal bosonic coordinates are constrained to be right-moving (or left-moving) on the world-sheet and therefore do not commute.

(g) Other Interactions

The interaction illustrated in fig. 2.3 is the only contribution to the oriented closed-superstring theories (types II or heterotic). For the (unoriented) type I theories there are a number of other interaction terms in which two internal points touch. The existence of all these terms is obviously necessary to ensure the closure of the superalgebra.

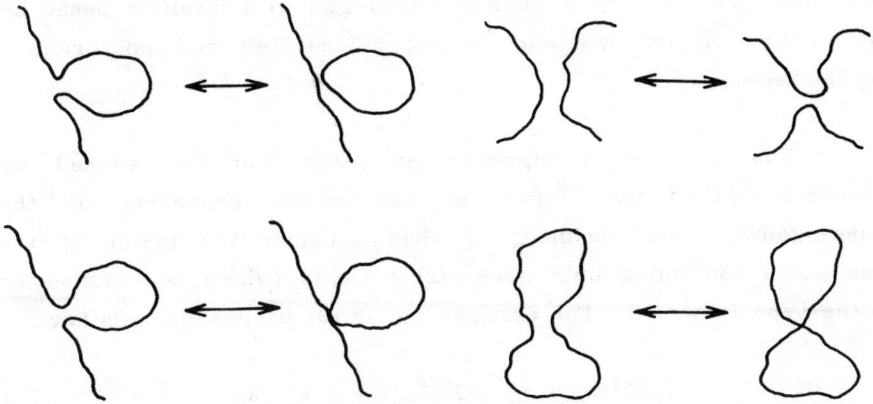

Fig. 2.4

All these terms are of strength κ and they all involve the *same* local touching interaction. [The last two were inadvertently omitted from ref. 21 but included in ref. 36.] Since the interaction is local along the string the form of the expressions for each of these terms is determined by the cubic closed-string coupling which involves the same touching interaction. This should be checked explicitly by verifying the closure of the super-Poincaré algebra including these terms.

(h) The Absence of Higher Order Interactions

The seven interaction terms depicted in the figures are the only ones needed to satisfy the closure of the algebra. None of these interactions is of higher order than cubic in closed-string fields and there is just one term, the $(\Phi)^4$ term, which is quartic in open-string fields. Notice that this only contributes to the scattering of two incoming strings into two outgoing ones (and not to the $1\to3$ or $3\to1$ processes) so that it does not contain the whole Yang-Mills contact term. This quartic term also only contributes to one of the possible group theory trace factors in a four-particle amplitude. The fact that these are the only interactions conforms with intuition based on the picture of splitting and joining of oriented and non-oriented world-sheets.

The absence of higher-order terms can be deduced by considering the next terms in the series expansion of the superalgebra. For simplicity, I shall consider the pieces of the generators containing only open-string fields (which is a subsector of the type I theory). For example, eq. (2.16) implies the relation

$$\{ Q_3^{-A}, Q_{3B}^{-} \} + \{ Q_2^{-A}, Q_{4B}^{-} \} + \{ Q_4^{-A}, Q_{2B}^{-} \} = 2\ \delta_B^A\ H_4 \tag{2.35}$$

Notice that in the *quantum* field theory of strings there could, in principle, also be corrections to the right-hand side arising from effects of normal ordering the string fields. The arguments presented here all treat the string fields *classically*. In order to show that the quartic terms are absent we would have to show that

$$\{ Q_3^{-A}, Q_{3B}^{-} \} \tag{2.36}$$

vanishes. Since we know that there is a quartic open-string interaction that contributes to a particular group theory factor the easiest way to investigate this anticommutator is to consider four-particle matrix elements between states of definite quantum numbers which exclude that factor. Fig. 2.5 depicts the diagrams that contribute with a ("Chan-Paton") factor $tr(\lambda_1\lambda_2\lambda_3\lambda_4)$ (where λ_r is a matrix in the fundamental representation of the internal symmetry group representing the quantum numbers of the particle r) in which case the possible intermediate states are a one-string state and a three-string state. The internal lines in these diagrams represent the complete sums over intermediate states (and not propagators). The labelling of the vertices indicates the occurrence of the local fermionic operators ($\hat{Q}^{-A}$ or $\hat{Q}_{\bar{B}}$) acting at the interaction points ($\sigma = \sigma_A$ and $\sigma = \sigma_B$).

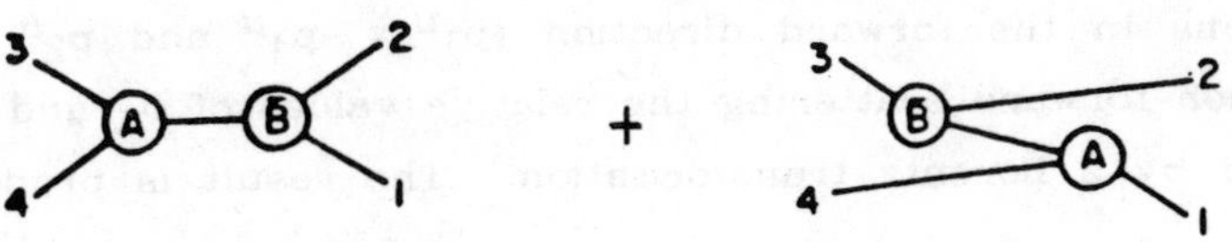

$+ \; A \leftrightarrow \bar{B}$

Fig. 2.5

In the σ-τ plane the two interactions occur at the same value of τ, so *both* the diagrams shown explicitly in fig. 2.5 are represented in parameter space by

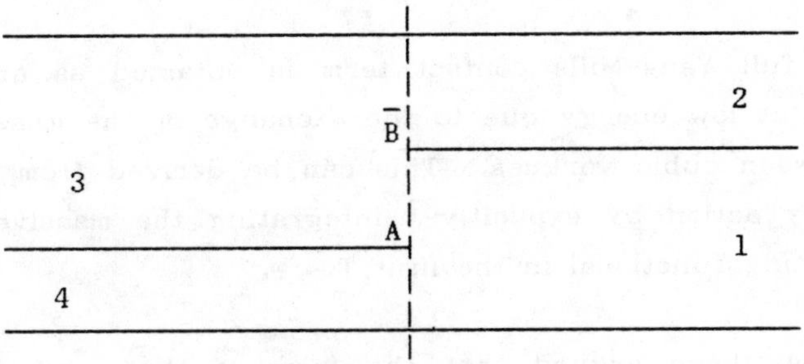

Fig. 2.6

The two cases in fig. 2.5 differ only in the order of the fermionic factors in the vertices. However,

$$\{ \hat{Q}^{-A}(\sigma_A) \; , \; \hat{Q}^{-}_{B}(\sigma_B) \} = 0 \qquad (2.37)$$

Therefore the matrix element of the anticommutator vanishes.

This reasoning breaks down if the points A and B coincide. This

only happens in the forward direction ($p_1^\mu = -p_4^\mu$ and $p_2^\mu = -p_3^\mu$) since for non-forward scattering the relative values of σ_A and σ_B can be changed by a Lorentz transformation. The result is proportional to

$$\delta^{10}(p_1 + p_4)\, \delta^4(\alpha_1\theta_1 + \alpha_4\theta_4)\, \delta^{10}(p_2 + p_3)\, \delta^4(\alpha_2\theta_2 + \alpha_3\theta_3) \qquad (2.38)$$

which suggests that the anticommutator (2.36) is proportional to the identity operator. It seems to imply an extra term in H_4 of a rather peculiar nature that only contributes to the disconnected piece of the S matrix and hence does not affect tree diagrams. Similar considerations should also apply yo the closed-string theories.

The full Yang-Mills contact term is obtained as an effective interaction at low energy due to the exchange of the massive string states between cubic vertices. This can be derived from the string field theory action by explicitly[37] integrating the massive fields in the generating functional in the limit $T \to \infty$.

It has been argued that the form of the cubic interaction between either open or closed strings derived by the methods used in this section is unique (up to powers of p+ which can only be determined by using the full Lorentz algebra). This suggests that there are no possible counterterms other than the action itself which are consistent with super-Poincare invariance. This in turn is suggestive of finiteness or renormalizability[38]. However, the analysis has only been carried out at the level of the classical superstring theory and it ignores quantum effects that we know can give rise to anomalies in the super-Poincaré algebra.

III LOOP AMPLITUDES AND ANOMALIES

The string perturbation expansion is the generalization to string theories of the Feynman diagram expansion of point field theory. The diagrams can be thought of as representing functional integrals over all possible world-sheets joining initial and final strings including surfaces of non-trivial topology which have handles attached or holes cut out. In this lecture I will first review in outline the divergence structure of the one-loop calculations. I will then show that superstring theories are gauge invariant at the level of tree diagrams as a prelude to describing the calculation of the one-loop Yang-Mills anomaly in the type I theory. Requiring the absence of this anomaly[39] restricts the gauge group to SO(32). The one-loop diagrams are also finite for this choice of group[36].

The question of checking gauge invariance in string theories and the rules for setting up the calculation using the covariant operator formalism are well known and straightforward. The difficulties in the evaluation of the anomaly involve subtleties that have apparently caused some confusion in other covariant formulations of the problem (such as the functional formalism).

(a) Loops of Open Strings (Type I theories)

The one-loop open-string amplitude is given by a sum of contributions from path integrals over orientable and non-orientable surfaces. The contributions from planar oriented surfaces to the amplitude can be represented by any of the diagrams in fig. 3.1. The first two diagrams look like a box diagram and a self-energy diagram respectively. The equivalence of these ways of representing the amplitude illustrates the duality property of string theories. The box diagram is divergent due to the infinite sum over massive states circulating around the loop and due to the high loop momentum. It is often very interesting to represent the amplitude in the third way shown in fig. 3.1(c) which is obtained by another distortion of the world-sheet. The diagram has the form of an open-string tree diagram with a closed string emitted into the vacuum at zero momentum k^μ.

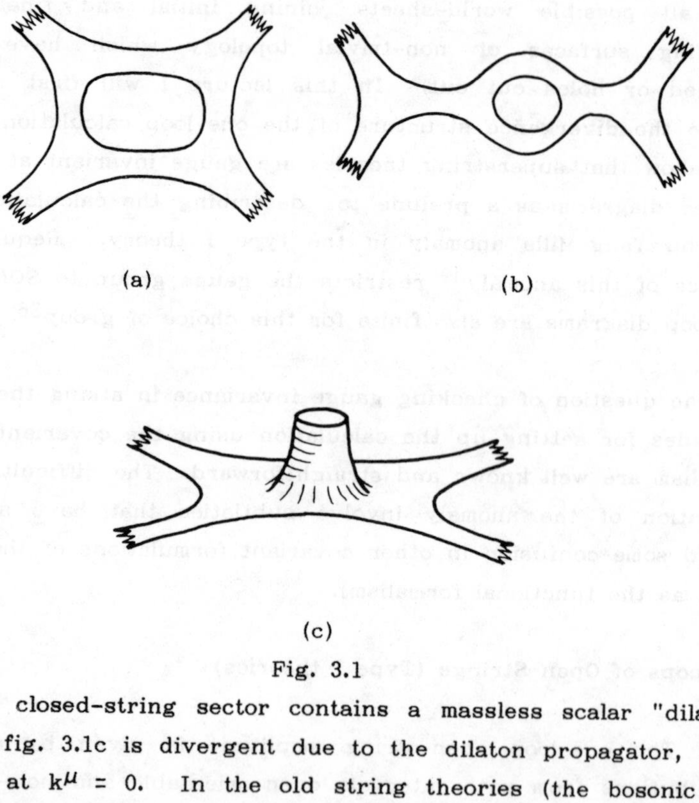

(a) (b)

(c)

Fig. 3.1

Since the closed-string sector contains a massless scalar "dilaton" state the fig. 3.1c is divergent due to the dilaton propagator, $1/k^2$, evaluated at $k^\mu = 0$. In the old string theories (the bosonic and spinning string theories) there is also a tachyon state in the emitted closed-string tadpole (which makes the divergence appear even worse in the usual representation of the loop due to the use of an invalid integral representation). The form of the divergence, first discussed in detail in ref. 40 for the bosonic theory, is that of the derivative of a tree diagram with respect to the string tension[41].

The amplitude of fig. 3.1 can also be represented by an annulus in parameter-space by an appropriate choice of the two-dimensional metric as illustrated in fig. 3.2. In the expression for the amplitude the inner radius, r, of the annulus is to be integrated from 0 to 1 (with the outer radius fixed equal to 1). The residual Möbius invariance of the integrand can then be used to fix one of the external particles at z = 1 on the outer boundary with the positions, z_r, of the N-1 other ones being integrated around the outer boundary.

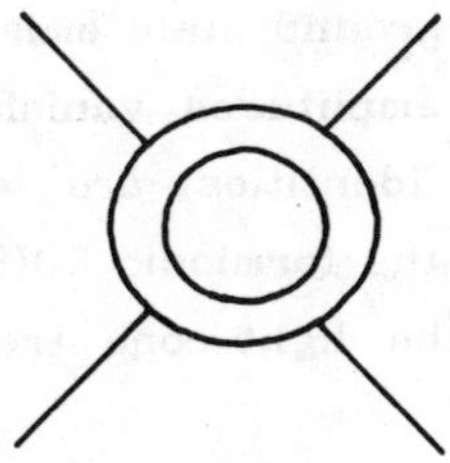

Fig. 3.2

The other divergent diagrams are represented by non-orientable surfaces (Möbius strips).

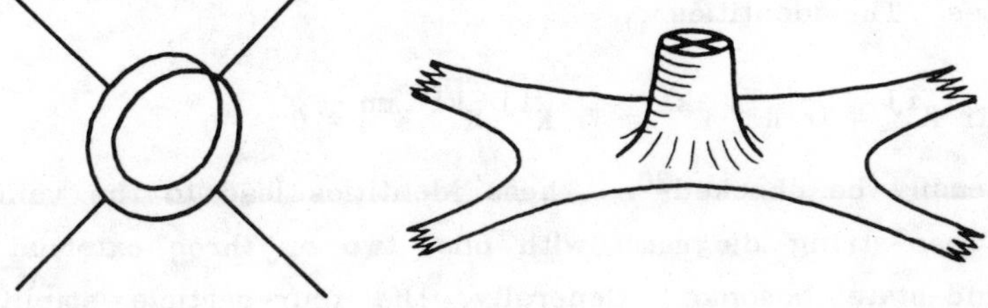

(a) (b)

Fig. 3.3

Fig. 3.3(b) illustrates a divergence due to the emission of a closed string into the vacuum at zero momentum, this time via a non-orientable cross cap (a boundary with diametrically opposite points identified).

Open strings carry quantum numbers of an internal symmetry group at their endpoints in the manner first proposed by Chan and Paton[32]. The only groups that can be included in this way at the level of the open-string tree diagrams are the classical groups SO(n), Sp(2n) and U(n)[31] (with U(n) excluded when considering the coupling of open to closed strings).

The open-string one-loop amplitudes in superstring theories (with four external ground state bosons) were considered in ref. 36,42 using the light-cone-gauge ground state emission vertices obtained in ref. 20. Certain of these amplitudes vanish due to supersymmetry trace identities. These identities are expressed in terms of supertraces of functions of the fermionic SO(8) spinor zero modes, θ_0^a, (a = 1,2,..8) which enter the light-cone treatment and satisfy (eq. 1.49)

$$\{ \theta_0^a, \theta_0^b \} = \frac{2}{p^+}\delta^{ab} \tag{3.1}$$

The vertices for emitting ground-state on-shell bosons in the light-cone gauge involve operators $K^{ij} = \frac{1}{4}\,\theta_0\gamma^{ij}\theta_0$ which act like spin operators on the ground state super-Yang-Mills vector and spinor indices. The identities

$$\mathrm{Tr}\, K^{ij} = \mathrm{Tr}\, K^{ij}\, K^{kl} = \mathrm{Tr}\, K^{ij}\, K^{kl}\, K^{mn} = 0 \tag{3.2}$$

can easily be checked[20]. These identities lead to the vanishing of the open-string diagrams with one, two or three external on-shell ground-state bosons. Generally, the four-particle amplitude has divergences from both the planar orientable loop (fig. 3.1) and the non-orientable loop (fig. 3.3) but these cancel when the symmetry group is SO(32)[39]. This result has recently been generalized to certain one-loop N-particle amplitudes for $N > 4$[43]. [There is an ambiguity in the way the separate divergences are regularized in ref. 36 that raises doubts about the generalization of the results to higher loop diagrams.]

The last class of one-loop diagrams built from open strings has orientable world-sheets with external particles attached to both boundaries as, for example, in fig. 3.4.

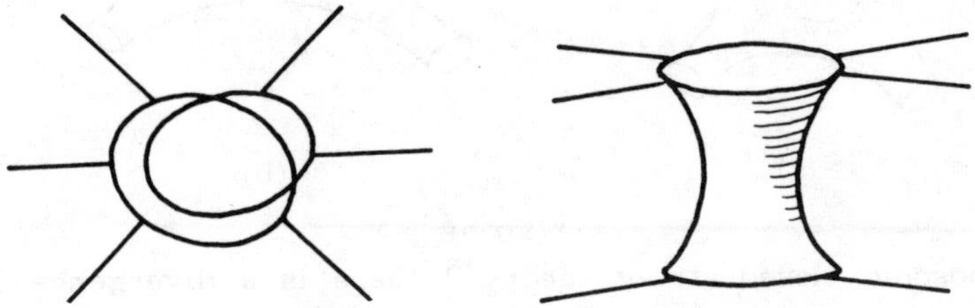

(a) (b)

Fig. 3.4

This contribution is not divergent. However, it generally has a branch cut in the Mandelstam invariant of the channel with vacuum quantum numbers that leads to a violation of unitarity. In 1971 it was realized by Lovelace[44] that this cut became a pole in D = 26 dimensions (for the bosonic theory) if he made the (correct) assumption that in that dimension an extra set of states decoupled by virtue of the Virasoro gauge conditions. [This corresponds to the condition for transversality of the physical Hilbert space.] The pole is interpreted as a closed string bound state formed by the joining of the ends of a single open string. [This was originally proposed as the origin of the Pomeranchuk Regge pole in hadronic physics.]

(b) Closed-String Loop Amplitude

In the case of the type IIa, type IIb and heterotic superstring theories the perturbation series only involves orientable closed surfaces with handles but no boundaries. [The type I closed-string amplitudes are represented by non-orientable as well as orientable surfaces with holes cut out as well as handles attached. None of these have yet been calculated.] An N-particle one-loop amplitude can be represented by

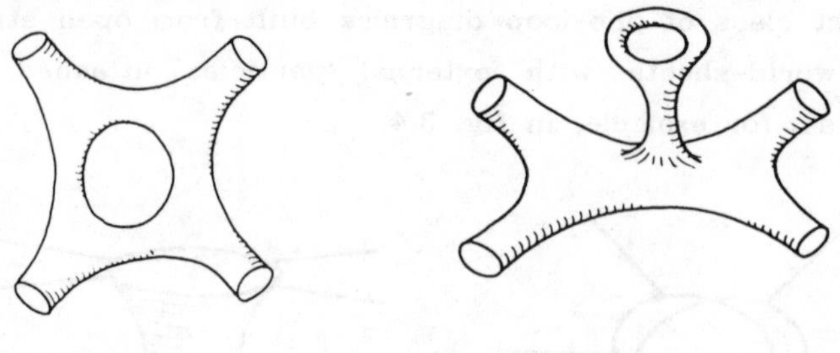

(a) (b)

Fig. 3.5

In the bosonic closed-string theory[45] there is a divergence in this one-loop diagram. This can once again be associated with the emission into the vacuum of a tadpole which disappears into the vacuum via a closed-string loop as in fig. 3.5(b). By an appropriate choice of world-sheet metric the σ-τ parameter space can be mapped into a torus. The expression for the amplitude involves complex integrations over the ratio, τ, of the moduli as well as the positions, ν_r, of N-1 of the external particles which are integrated over the whole surface of the torus (with ν_N held fixed). In D = 26 dimensions the integrand is a modular function (which means that it is invariant under the transformations $\tau \rightarrow \tau + 1$ and $\tau \rightarrow -1/\tau$) and the τ integration must be restricted to the fundamental region of the modular group (in other dimensions there is a violation of unitarity).

The divergence of the expression for the amplitude arises from the endpoint of the integrations at which all the external particles come together on the surface of the torus (i.e. all the ν_r are equal) irrespective of the value of τ. In this limit the tadpole neck in fig. 3.3(b) has vanishingly small radius. The divergence is again associated with the emission of a dilaton at zero momentum. The residue of the divergence is proportional to the coupling of an on-shell dilaton to a torus. This residue is also divergent in the bosonic string theory due to the tachyon circulating around the loop (which is *not* an ultra-violet divergence).

In the case of type II superstring theories the supersymmetry identities in eq. (3.2) cause the one, two and three-particle amplitudes to vanish[42]. This suggests that the N-particle amplitude should be finite in these theories since the residue of the dilaton tadpole now vanishes. The four-particle amplitude was found to be finite by explicit calculation[19]. This has also now shown to be true for the heterotic superstring theories[5].

(c) Multi-Loop Amplitudes

There are no firm results yet about the divergences of multi-loop diagrams but the subject is under intensive study. The oriented closed-string theories are easiest to consider because there is only one diagram at any order in perturbation theory (at L loops its world-sheet is topologically a sphere with L handles). Assuming that the only possible divergences arise from the emission of dilatons into the vacuum there is a new potential divergence at each order. For example, fig. 3.6 shows two representations of the two-loop diagram.

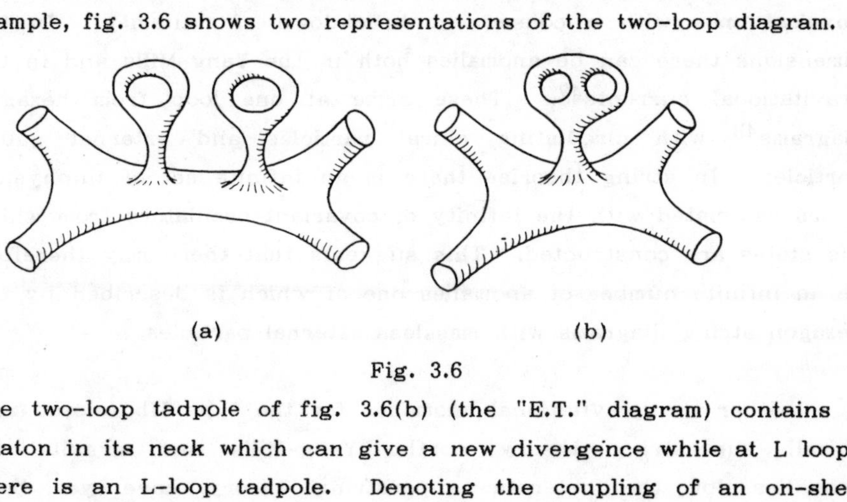

(a) (b)

Fig. 3.6

The two-loop tadpole of fig. 3.6(b) (the "E.T." diagram) contains a dilaton in its neck which can give a new divergence while at L loops there is an L-loop tadpole. Denoting the coupling of an on-shell dilaton to an L-loop torus by the condition for finiteness is therefore

 = 0 (3.3)

This equation is also the condition that supersymmetry be unbroken. In perturbation theory around ten-dimensional Minkowski space supersymmetry cannot be broken in perturbation theory unless there are Lorentz anomalies (since the gravitini do not have the right partners with which to form a massive states). This means that

$$\text{absence of anomalies} \leftrightarrow \text{finiteness}$$

at any order in perturbation theory[46]. The question of whether the theories are actually free of infinities (or of anomalies) requires explicit calculations of L-loop diagrams which is at present the subject of extensive study[26].

(d) One-Loop Chiral Gauge Anomalies (in type I theories).

The occurrence of gauge anomalies in any theory is a disaster since they result in the coupling of unphysical gauge modes to physical ones with a consequent breakdown of unitarity. In ten dimensions there can be anomalies both in the Yang-Mills and in the gravitational currents[47]. These arise at one loop from hexagon diagrams[48] with circulating chiral particles and external gauge particles. In string theories there is an infinite set of unphysical modes associated with the infinity of covariant oscillators from which the states are constructed. This suggests that there may therefore be an infinite number of anomalies one of which is described by the hexagon string diagrams with massless external particles.

Neither the gravitational anomalies for the type IIb (closed and chiral) superstring theory northe Yang-Mills and gravitational anomalies for the heterotic superstring theory have yet been evaluated explicitly. However, the finiteness of the one-loop amplitudes makes it very plausible that the anomalies vanish by the arguments mentioned above. Further evidence for the absence of anomalies in these theories is that they reduce to anomaly-free point field theories at low energy.

I will now describe the calculation of the Yang-Mills anomaly in type I theories in some detail. Even though the type I theories may not, in the end, be phenomenologically interesting or even perhaps, consistent, the calculation illustrates techniques that may be of interest. To begin with review the proof of the gauge invariance of the tree diagrams in the covariant formulation of the spinning string theory.

Gauge Invariance of Tree Diagrams.

I will begin with a review of the proof of the gauge invariance of the N-particle tree diagrams in the covariant approach to the theory. As an example consider a fermionic string emitting ≈ground-state vector particles with momenta k_r and polarizations ζ_r satisfying the mass-shell condition

$$k_r^2 = 0 \tag{3.4}$$

and the transversality condition

$$\zeta_r . k_r = 0 \tag{3.5}$$

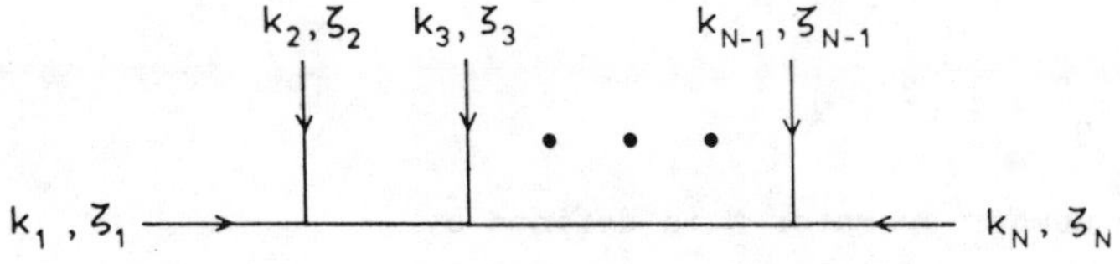

Fig. 3.7

The states are defined in terms of covariant bosonic and fermioni and fermionic oscillator modes, α_n^μ and d_n^μ where

$$[\, \alpha_m^\mu \, , \, \alpha_n^\nu \,] = m\, \delta_{m+n,0}\, \eta^{\mu\nu} \tag{3.6}$$

$$\{\, d_m^\mu \, , \, d_n^\nu \,\} = \delta_{m+n,0}\, \eta^{\mu\nu} \, . \tag{3.7}$$

The amplitude is given by the expression (ignoring the group theory factor)

$$T_N = \langle k_1 | V(k_2,\varsigma_2,1)\ \Delta \ldots V(k_{N-1},\varsigma_{N-1},1)\ \frac{1}{2}(1+\Gamma_{11})|k_N\rangle \tag{3.8}$$

where the fermionic string propagator, Δ, and the vertex for emitting an on-shell vector particle of momentuom k_r and polarization ς_r, $V(k_r,\varsigma_r,z)$, are constructed from the momentum operator, $P^\mu(z)$, and the Ramond field[3], $\Gamma^\mu(z)$. The variable $\ln z/2\pi i$ is the proper time, τ, (i.e. $z = 1$ is $\tau = 0$) and the particles are attached to the end of the string $\sigma = 0$. These operators are defined by

$$P^\mu(z) = \sum_{-\infty}^{\infty} \alpha_n z^n \equiv \frac{dX^\mu(z)}{dz} \tag{3.9}$$

$$\Gamma^\mu(z) = \gamma^\mu + i\sqrt{2}\ \gamma_{11} \sum_1^{\infty}(d_n^\mu z^{-n} + d_n^\mu z^n) \tag{3.10}$$

The propagator is defined by

$$\Delta = (F_0)^{-1} = \frac{F_0}{L_0} \tag{3.11}$$

where

$$F_0 = \frac{1}{i\sqrt{2}} \int \frac{dz}{2\pi iz}\ \Gamma(z).P(z) \tag{3.12}$$

and

$$L_0 = F_0^{\ 2} = \frac{1}{2} P^2 + N \tag{3.13}$$

where the number operator N is defined by

$$N = \sum_{n=1}^{\infty} (\alpha_n^\dagger.\alpha_n + n\, d_n^\dagger.d_n) \tag{3.14}$$

The vertex for emitting string r at $\tau = \ln z/2\pi i$ is defined by

$$V(k_r,\varsigma_r,1) = g\varsigma_r.\Gamma(1)\ V_0(k_r,1) \tag{3.15}$$

where $V_0(k_r,1)$ is the usual factor of $\exp\{ik_r.X(1)\}$. The factor of

$\frac{1}{2}(1+\Gamma_{11})$ in eq. (3.8) is the generalized chirality projection operator introduced[4] to obtain a space-time supersymmetric theory. Γ_{11} is the string generalization of γ_{11} and defined by $\Gamma_{11} = \gamma_{11} (-1)^{\Sigma d^{11}_{-n} \cdot d_n}$ so that Γ_{11} anti-commutes with the matrices $\Gamma^{\mu}(z)$. It also anti-commutes with the propagator and with the vertex so that the factors of $\frac{1}{2}(1+\Gamma_{11})$ associated with each internal line can all be moved to the right-hand end of the tree as in eq. (3.8).

Gauge invariance can be checked by coupling the logitudinal mode of any given external gauge particle by substituting its momentum, k_r, for its polarization, ζ_r, in the vertex. The rth vertex becomes

$$V(k_r, k_r, 1) \equiv k_r.\Gamma(1)\, e^{ik_r.X(1)}$$

$$= i\sqrt{2}\left[F_0 , e^{ik_r.X(1)} \right] \tag{3.16}$$

The two terms in the commutator have a factor of F_0 which cancels the propagator one side or the other of the vertex, leaving the product of two factors of V_0 from adjacent vertices. In string theories the product of two vertices in a tree with no propagator between them is singular and this causes the tree to vanish (essentially because of a factor $\exp\{-k_r.k_{r+1}\ \Sigma 1/n\}$ which vanishes). As a result of this "cancelled propagator argument" the tree amplitude is proved to be gauge invariant. In usual point field theory the terms with cancelled propagators do not vanish and need to be compensated by the presence of contact terms in the action.

The cancelled propagator argument must be used with extreme caution in loop amplitudes since a zero may be compensated by an infinity from the loop. This is what happens in the loop anomaly calculation where the result is finite.

The One-Loop Anomaly.

Further details of this section may be found in ref. 39. The first anomalous diagrams in open-string theories are hexagon diagrams with external massless gauge particles (open-string ground states) and circulating chiral fermions.

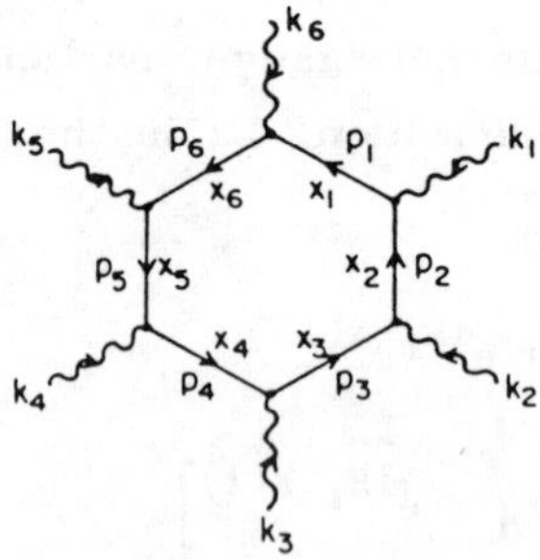

Fig. 3.8

The planar diagram (which is like fig. 3.1 but with six external states) has a group theory Chan-Paton factor

$$G^{planar} = n\, tr(\lambda_1 \ldots \lambda_6) \tag{3.17}$$

The λ's are matrices in the fundamental representation of SO(n), Sp(2n) or U(n) and the factor of n comes from the trace around the inner boundary which gives tr(1). The full amplitude must also be Bose symmetrized.

Only the components in the amplitude with an odd number of Γ_{11} factors will contribute to the anomaly. These give

$$T = \int d^{10}p \; Tr\{\Delta V(k_1,\zeta_1,1)\Delta \ldots \Delta V(k_6,\zeta_6,1)\Gamma_{11}\} \tag{3.18}$$

where the propagator Δ is defined in eq. (3.11) and the vertex $V(k_r,\zeta_r,1)$ is defined in eq. (3.16). An overall constant and the group theory factor have been omitted from eq. (3.18). In writing this expression factors of $(1+\Gamma_{11})$ have been moved around the loop so that only one power of Γ_{11} remains and is arbitrarily placed next to the vertex for particle 6. Just as in ordinary point field theory anomaly calculations this is only a formal manoeuvre since T is divergent. However, the calculation is easier in this asymmetric configuration with only one power of Γ_{11} and the answer can easily be related to the result of the symmetric calculation.

The next step is to regulate T. In ref. 39 two different regulators were considered one of which was analogous to a Pauli–Villars regulator while the other was analogous to a Fujikawa regulator. They gave different expressions for the anomaly although they reduced to the same low-energy limit. This may merely reflect the fact that the theory is sick if there is any non-zero anomaly. I will describe the "Pauli-Villars" (P.V.) method here. Introducing a P.V. propagator

$$\Delta_m = (F_0 - im)^{-1} = \frac{F_0 + im}{L_0 + m^2} \tag{3.19}$$

in T instead of Δ defines T_m where $T-T_m$ is the regulated amplitude. The regulator mass, m, will be taken to infinity at the end of the calculation. Gauge invariance can be checked for the regulated amplitude by substituting k_r for ζ_r successively on each line. Due to the asymmetric position of Γ_{11} the only non-trivial case is that of line 6 (since the cancelled propagator argument works on the other legs in the finite, regulated expression). The vertex for line 6 can be rewritten using

$$k_6.\Gamma(1)V_0(k_6,1)\Gamma_{11} = i\sqrt{2}\left\{F_0, V_0(k_6,1)\,\Gamma_{11}\right\} \tag{3.20}$$

in T and

$$k_6.\Gamma(1)V_0(k_6,1)\Gamma_{11} = i\sqrt{2}\left\{F_0-im,\ V_0(k_6,1)\Gamma_{11}\right\}$$
$$-2\sqrt{2}\ m\ V_0(k_6,1)\Gamma_{11} \tag{3.21}$$

in T_m. In each case the term in { } involves an inverse propagator that leads to a vanishing contribution (recalling that $T-T_m$ is finite). The only non-zero contribution to the anomaly therefore comes from the last term in eq. (3.21) which leads to the anomaly A of the form

$$A = m\int d^{10}p\ \mathrm{Tr}\left\{\frac{F_0+im}{L_0+m^2}V(k_1,\varsigma_1,1)\ldots V(k_5,\varsigma_5,1)\frac{F_0+im}{L_0+m^2}V_0(k_6,1)\Gamma_{11}\right\} \tag{3.22}$$

The Dirac trace can be performed easily due to the γ_{11} which eats up ten γ^{μ} factors out of the eleven which are present in the integrand of eq. (3.22) (one in each of the six F_0's and one in each of the five V's). The result is

$$A = im^2\ \epsilon(\varsigma,k)\int d^{10}p\mathrm{Tr}\left\{\frac{1}{L_0+m^2}V_0(k_1,1)\ldots\frac{1}{L_0+m^2}(-1)^{\Sigma d_{-n}\cdot d_n}\right\} \tag{3.23}$$

where

$$\epsilon(\varsigma,k) \equiv \epsilon_{\mu_1\ldots\mu_5\nu_1\ldots\nu_5}\ \varsigma_1^{\mu_1}\ldots\ \varsigma_5^{\mu_5}\ k_1^{\nu_1}\ldots\ k_5^{\nu_5} \tag{3.24}$$

is proportional to the usual anomaly in point field theory in ten dimensions. The integral is analogous to standard bosonic open-string loop integrals. It is usual to represent each propagator by

$$(L_0+m^2)^{-1} = \int_0^1 dx\ x^{L_0+m^2-1} \tag{3.25}$$

where the x's are labelled as in fig. 3.8. The trace in eq. (3.23) is reexpressed by using

$$x^{L_0}\ V_0(k,1) = V_0(k,x)\ x^{L_0} \tag{3.26}$$

Using the variables

$$\rho_i = x_1 \ldots x_i \quad , \qquad \rho_6 = \omega \tag{3.27}$$

gives

$$A = m^2 \, \epsilon(\varsigma,k) \int \Pi \frac{dx_i^i}{x_i} \omega^{m^2} \int d^{10}p \, \mathrm{Tr} \left\{ \omega^{L_0} V_0(k_1,\rho_1) \ldots \right.$$
$$\left. \ldots V_0(k_6,\rho_6) \, (-1)^{\Sigma d_{-n} \cdot d_n} \right\} \tag{3.28}$$

In the limit of large m this expression vanishes exponentially fast unless ω is close to 1. The region near $\omega = 1$ can be examined by a change of variables (a Jacobi imaginary transformation) to

$$z_i = \exp \left\{ 2\pi i \frac{\ln \rho_i}{\ln \omega} \right\} \tag{3.29}$$

where $z_6 = 1$ and the z_i's are ordered around the unit circle in the complex plane. The oscillator arithmetic in eq. (3.28) can be elegantly described by using an identity for $\int d^{10}p \, \mathrm{Tr}\{\ldots\}$ (proved in ref. 49) which recasts this expression into a form involving vertices which are functions of z_r giving

$$A = m^2 \, \epsilon(\varsigma,k) \int \Pi \frac{dz_i}{z_i} \int_0^1 d\omega \, \omega^{m^2-1}$$

$$\times \; \langle p = 0 | \, \mathrm{Tr} \left\{ \exp\left(\frac{4\pi^2 N}{\ln \omega} \right) V_0(z_1,k_1) \ldots V_0(z_6,k_6) \right\} | p = 0 \rangle \tag{3.30}$$

where the trace is now over non-zero modes only. Recalling that the number operator N has integer eigenvalues we see that as m gets large only the term in the trace with N = 0 survives. Furthermore, since

$$\int_0^1 d\omega \, \omega^{m^2-1} = m^{-2} \tag{3.31}$$

the result is finite and is given by

$$A = \epsilon(\zeta,k) \int \prod_{i=1}^{5} \frac{dz_i}{z_i} \langle 0|V_0(z_1) \ldots V_0(z_6)|0\rangle$$

$$= \epsilon(\zeta,k) \int \prod_{i=1}^{5} \frac{dz_i}{z_i} \prod_{i<j} (z_j - z_i)^{k_i . k_j} \tag{3.32}$$

At low momenta (i.e. when $k_i.k_j \ll 1$ in the units with $T = 1/\pi$) the integral in eq. (3.32) is a constant and the result reduces to the usual hexagon diagram anomaly of point field theory. The integral contains all the "stringiness" which includes non-trivial dynamical structure, the significance of which is not clear.

In addition to the planar diagrams we must add the diagrams with twists on internal propagators. Those with an odd number of twists are Möbius strips and only have one boundary. They are associated with group theory factors like

$$G^{\text{Möbius}} = \pm \operatorname{Tr}(\lambda_1 \ldots \lambda_6) \tag{3.33}$$

with the + sign for USp(2n) and the - sign for SO(n) (for U(n) there is no contribution from these non-orientable diagrams - this case is, in any case, known to be inconsistent at the level of loop amplitudes and need not be considered here[31]. These signs, associated with a twist, can be understood from the symmetry properties of the group generators expressed in terms of the fundamental representation. In contrast to eq.(3.17) there is no factor of n in this expression because there is only one boundary on a Möbius strip. However, there are 32 diagrams with an odd number of twists on internal propagators and apart from the group theory factor, they contribute with the same weight to the anomaly as in the planar case (after Bose symmetrization). The calculation is very similar to the earlier one and will not be repeated here. Once again, the low energy limit of the contribution to the anomaly from these terms coincides with the calculation of a hexagon diagram in the massless field theory.

After adding the planar and Möbius strip contributions, the anomaly has a factor of

$(n + 32)$ for $USp(2n)$

(3.34)

$(n - 32)$ for $SO(n)$

The last class of diagrams is those with an even number of twists such that the world sheet is an annulus with particles attached to both the inner and the outer boundaries such as that shown in fig. 3.4(a). Elementary group theory shows that for the groups SO(n) or USp(n) there must be an even number of particles attached to each boundary of the world-sheet. The contribution shown in fig. 3.8 has a group theory factor of $tr(\lambda_1\lambda_2\lambda_3\lambda_4)tr(\lambda_5\lambda_6)$. The anomaly calculation is similar to the earlier one with the important distinction that the exponential factor inside eq. (3.30) now becomes

$$\exp\left\{\frac{4\pi^2(8N + k_5.k_6)}{8\ln\omega}\right\} \tag{3.35}$$

which vanishes for any value of $N \geqslant 0$ (with $k_5.k_6 \geqslant 0$ by a suitable continuation, if necessary). This result at first seems rather surprising since the low-energy limit of zero is zero! This does not coincide with the (non-zero) contribution to the anomaly arising from the corresponding hexagon diagrams in the massless effective field theory. The resolution to this paradox lies in the fact that in the low energy limit fig. 3.4(a) gives another anomalous contribution in addition to the usual hexagon loop. This arises from the closed-string bound states in the 5-6 channel which can be seen by distorting the diagram into the cylinder of fig. 3.4(b). These states include the massless supergravity multiplet which survives the low energy limit in which massive states decouple. At low energy it is clear from fig. 3.4(b) that there is a contribution from a *tree* diagram with the exchange of the supergravity multiplet. This tree contains an anomalous piece arising from the exchange of the antisymmetric tensor field, $B_{\mu\nu}$.

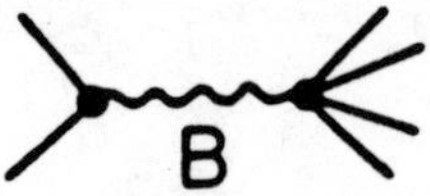

Fig. 3.9

The coupling between $B_{\mu\nu}$ and the pair of gauge particles at one end of the tree is the usual coupling present in the much-studied system of D = 10 super-Yang-Mills coupled to supergravity[50] (which I shall refer to as the "minimal" theory). However, there is, in addition, an *anomalous* coupling between the $B_{\mu\nu}$ field and four gauge particles that is not included in the "minimal" theory.

The total anomaly is therefore given entirely by the planar and non-orientable diagrams. From eq. (3.34) we see that the anomaly cancels for the group SO(32). Furthermore, the cancellation can be understood in the language of the low-energy point field theory as a cancellation between the usual one-loop quantum effect and a classical effect due to the presence of an anomalous term in the action.

(f) Anomaly Cancellations in the Low-Energy Theory

As yet there has not been an explicit calculation of gravitational anomalies in any string theory or of the Yang-Mills anomaly in the heterotic string theory. However the argument for the cancellation of the Yang-Mills anomaly based on the low-energy field theory generalizes to gravitational and mixed anomalies. The requirement that all these anomalies should cancel by the addition of local terms to the action is very restrictive. Details of how these restrictions arise are given in ref. 39 and I shall only quote the results here. The gravitational anomalies can only be cancelled if the Yang-Mills group has dimension 496 (i.e. there are 496 species of chiral fermions in the Yang-Mills sector). The Yang-Mills anomaly and the mixed Yang-Mills gravitational anomalies can only be cancelled if the gauge group has the property that for an arbitrary matrix, F, in the adjoint representation

$$\mathrm{Tr}\ F^6 = \frac{1}{48}\ \mathrm{Tr}\ F^2\left\{(\mathrm{Tr}\ F^4 - \frac{1}{300}\ (\mathrm{Tr}\ F^2)^2\right\} \tag{3.36}$$

These conditions only hold for the algebras of SO(32) and $E_8 \times E_8$ (apart from the apparently trivial cases of $U(1)^{496}$ and $E_8 \times U(1)^{248}$). This suggested that there ought to be a superstring theory with the gauge algebra of $E_8 \times E_8$ as well as SO(32). The heterotic superstring is just such a theory. It actually has (Spin 32)/Z_2 instead of SO(32) but these have the same Lie algebra.

The absence of the gravitational and Yang-Mills anomalies does not imply the vanishing of anomalies in the divergence of the supercurrent[51]. For completeness it would be satisfying to calculate these anomalies explicitly in superstring theories or merely in the low energy point field theory limits.

The absence of infinitessimal gauge anomalies is a pre-requisite for the consistency of the theory. It is also necessary for anomalies associated with "large" ten-dimensional gauge transformations, which are not continuously connected to the identity, to be absent. This has been shown[52] to be the case for the two interesting gauge groups, at least when space-time is taken to be a ten-sphere, S^{10}.

The restriction to these two groups in the heterotic theory can be seen by considering an arbitrary loop diagram. In a closed-string theory consistency requires that the expression for a loop diagram is the integral of a modular function. This restricts the possible groups of the heterotic string to the ones associated with rank 16, self-dual, even lattices[5]. Requiring modular invariance of the loop amplitudes for closed strings is equivalent to requiring invariance of the theory under "large" reparametrizations of the world-sheet i.e. reparametrizations not continuously connected to the identity[53].

REFERENCES

(1) Y. Nambu, Lectures at the Copenhagen Symposium, 1970; T. Goto, Progr. Theor. Phys. **46** (1971) 1560.

(2) M.B. Green and J.H. Schwarz, Phys. Lett. **136B** (1984) 367; Nucl. Phys. **B243** (1984) 285.

(3) P.M.Ramond, Phys. Rev. **D3** (1971) 2415; A. Neveu and J.H. Schwarz, Nucl. Phys. **B31** (1971) 86; Phys. Rev. **D4** (1971) 1109.

(4) F. Gliozzi, J. Scherk and D.I. Olive, Nucl. Phys. **B122** (1977) 253

(5) D.J. Gross, J.A. Harvey, E. Martinec, and R. Rohm, Phys. Rev. Lett. **54** (1985) 502; Nucl. Phys. **B256** (1985) 253; "Heterotic String Theory II. The Interacting Heterotic String", Princeton preprint (June 1985).

(6) S. Deser and B. Zumino, Phys. Lett. **65B** (1976) 369. L. Brink, P. Di Vecchia and P.S. Howe, Phys. Lett. **65B** (1976) 471

(7) A.M. Polyakov, Phys. Lett. **103B** (1981) 207; Phys. Lett. **103B** (1981) 211.

(8) W. Siegel, Phys. Lett. **128B** (1983) 397.

(9) M. Henneaux and L. Mezincescu, Phys. Lett. **152B** (1985) 340; T.L. Curtright, L. Mezincescu, C.K. Zachos, Argonne preprint ANL-HEP-PR-85-28 (1985).

(10) E. Witten, Nucl. Phys. **B223** (1983) 422.

(11) E. Witten, Comm. Math. Phys. **92** (1984) 455.

(12) T. Curtright and C. Zakhos, Phys. Rev. Lett. **53** (1984) 1799.

(13) E. Witten, "Twistor-Like Transform in Ten Dimensions", Princeton preprint (May, 1985).

(14) B. Nielsen, Nucl. Phys. **B188** (1981) 176.

(15) M.T. Grisaru, P. Howe, L. Mezincescu, B. Nilsson, P.K. Townsend, D.A.M.T.P. Cambridge preprint (1985).

(16) I. Bengtsson and M. Cederwall, Göteborg preprint (1985). T. Hori and K. Kamimura, U. of Tokyo preprint (1984).

(17) W. Siegel, Berkeley preprint UCB-PTH-85/23 (May, 1985).

(18) P. Goddard, J. Goldstone, C. Rebbi and C.B. Thorn, Nucl. Phys.**B56** (1973) 109.

(19) M.B. Green and J.H. Schwarz, Phys. Lett. **109B** (1982) 444.

(20) M.B. Green and J.H. Schwarz, Nucl. Phys. **B198** (1982) 252.

(21) M.B. Green and J.H. Schwarz, Phys. Lett. **140B** (1984) 33; Nucl. Phys. **B243** (1984) 475.

(22) M.B. Green, J.H. Schwarz and L. Brink, Nucl. Phys. **B219** (1983) 437.
(23) S. Mandelstam, Nucl. Phys. **B64** (1973) 205; **B69** (1974) 77.
(24) L. Brink and M.B. Green (unpublished).
(25) C.S. Hsue, B. Sakita and M.A. Virasoro, Phys. Rev. **D2** (1970) 2857
J.- L. Gervais and B. Sakita, Phys. Rev. **D4** (1971) 2291
(26) S. Mandelstam, Proceedings of the Niels Bohr Centennial Conference, Copenhagen, Denmark (May, 1985); Proceedings of the Workshop on Unified String Theories, ITP Santa Barbara, 1985.
A. Restuccia and J. G. Taylor, King's College, London preprints (June, 1985).
(27) W. Siegel, Phys. Lett. **149B** (1984) 157, 162;
Banks and M. Peskin, Proceedings of the Argonne-Chicago Symposium on Anomalies, Geometry and Topology (World Scientific, 1985); SLAC preprint SLAC-PUB-3740 (July,1985);
D. Friedan, E.F.I. preprint EFI 85-27 (1985);
A. Neveu and P.C. West, CERN preprint CERN-TH 4000/85;
K. Itoh, T. Kugo, H. Kunitomo and H. Ooguri, Kyoto preprint(1985)
W. Siegel and B. Zweibach, Berkeley preprint UCB-PTH-85/30(1985);
M. Kaku and K. Lykken, CUNY preprint (1985);
M. Kaku, Osaka University preprints OU-HET 79 and 80 (1985);
S. Raby, R. Slansky and G. West, Los Alamos preprint (1985).
(28) M. Kaku and K. Kikkawa, Phys. Rev. **D10** (1974) 1110, 1823;
E. Cremmer and J.-L. Gervais, Nucl. Phys. **B76** (1974) 209;
J.F.L. Hopkinson R.W. Tucker and P.A. Collins, Phys. Rev. **D12** (1975) 1653.
(29) M.B. Green and J.H. Schwarz, Nucl. Phys. **B218** (1983) 43.
(30) E. Tomboulis, Phys. Rev. **D8** (1973) 2736.
(31) J.H. Schwarz, Proceedings of Johns Hopkins Workshop on Current Problems in Particle Theory (Florence, 1982) 233;
N. Marcus and A. Sagnotti, Phys. Lett. **119B** (1982) 97.
(32) H.M. Chan and J. Paton, Nucl. Phys. **B10** (1969) 519.
(33) C.B. Thorn, University of Florida UFTP-85-8.
(34) R. Giles and C.B. Thorn, Phys. Rev. **D16** (1976) 366.
(35) L. Brink, M. Cederwall and M.B. Green, Göteborg preprint, (1985).
(36) M.B. Green and J.H. Schwarz, Phys. Lett. **151B** (1985) 21.
(37) M.B. Green, Proceedings of the Argonne-Chicago Symposium on Algebra, Geometry and Topology (1985).
(38) A.K.H. Bengtsson, L. Brink and M. Cederwall, Nucl. Phys. **B254** (1985) 625.
(39) M.B. Green and J.H. Schwarz, Phys. Lett., **149B** (1984) 117;
M.B. Green and J.H. Schwarz, Nucl. Phys. **B225** (1985) 93.
(40) A. Neveu and J. Scherk, Phys. Rev. **D1** (1970) 2355.
(41) J.A. Shapiro, Phys. Rev. **D11** (1975) 2937;
M. Ademollo et al., Nucl. Phys. **B124** (1975) 461.
(42) M.B. Green and J.H. Schwarz, Nucl. Phys.**B198** (1982) 441;
(43) P.H. Frampton P. Moxhay and Y.J. Ng, Harvard preprint HUTP-85/A059 (1985);
L. Clavelli, U. of Alabama preprint (1985).
(44) C. Lovelace, Phys. Lett. **34** (1971) 500.

(45) J. A. Shapiro, Phys. Rev. D5 (1972) 1945.
(46) M.B. Green, "Developments in Superstring Theory",Caltech preprint CALT-68-1219 (1984), to be published in the volume in honour of E.S. Fradkin's 60th birthday.
(47) L. Alvarez-Gaumé and E. Witten, Nucl. Phys. B234 (1983) 269.
(48) P.H.Frampton and T.W.Kephart,Phys. Rev. Lett.,50(1983)1343;1347; P.K. Townsend and G. Sierra, Nucl. Phys. B222 (1983) 493.
(49) L. Clavelli and J.A.Shapiro, Nucl. Phys. B57 (1973) 490.
(50) A.H. Chamseddine, Phys. Rev. D24 (1981) 3065
E. Bergshoeff, M. de Roo, B. de Wit and P. van Nieuwenhuisen, Nucl. Phys. B195 (1982) 97;
G.F. Chapline and N.S. Manton, Phys. Lett. 120B (1983) 105.
(51) R. Kallosh, Phys. Lett. 159B (1985) 111.
(52) E. Witten, "Global Gravitational Anomalies", Princeton preprint (1985).
(53) E. Witten, Proceedings of the Argonne-Chicago Symposium on Anomalies, Geometry and Topology (1985).

THE HETEROTIC STRING

David J. Gross*

Joseph Henry Laboratories
Princeton University
Princeton, New Jersey 08544

I. INTRODUCTION

Traditional string theories, either bosonic or supersymmetric, came in two varieties, closed string theories and open string theories. Closed strings are neutral objects which describe at low energies gravity or supergravity. Open strings have geometrically invariant ends to which charges can be attached, thereby obtaining, in addition to gravity, Yang-Mills gauge interactions. Recently a new kind of string theory was discovered[1]--the heterotic string, which is a chiral hybrid of the closed superstring and the closed bosonic string, and which produces by an internal dynamical mechanism gauge interactions of a totally specified kind. Although this theory was found in an attempt to produce a superstring theory which would yield a low energy $E_8 \times E_8$ supersymmetric, anomaly free, gauge theory, as suggested by the anomaly cancellation mechanism of Green and Schwarz,[2] it fits naturally into the general framework of consistent string theories.

The consistency requirements on string theories are remarkably restrictive. To preserve the two dimensional reparametrization invariance of the string dynamics, which is responsible for the emergence of local gauge symmetries, we must take the dimension of space-time to be 10 (for supersymmetric strings) or 26 (for bosonic strings). Open

*Research supported in part by NSF Grant PHY80-19754.

string theories can be used to generate low energy Yang Mills symmetries by attaching labels (or charges) to the ends of the string. In the case of superstrings this can produce SO(N) or Sp(2N) gauge groups, but existence of one loop anomalies, which might render the theory consisent, require the unique gauge group SO(32).[2] Thus, to our knowledge, the only open superstring that yields a consistent perturbative expansion about flat space (i.e. is free of tachyons and ghosts, is unitary, Lorentz invariant and finite) is that of Green and Schwarz. The low energy limit of this theory describes N=1 supergravity with an SO(32) local gauge symmetry in ten dimensions.

Closed superstrings appear to be more unique. Since there is no invariant location where one could attach charges, they are "neutral" objects, which produce, at low energy, pure N=2 ten dimensional supergravity. Actually there is a discrete choice that must be made in setting up a closed superstring theory, resulting in the existence of two distinct theories. The ambiguity arises due to the different possibilities of choosing the chirality of the fermionic coordinates of a closed superstring. The fermionic degrees of freedom consist of a ten-dimensional Majorana-Weyl spinor, $S^a(\tau,\sigma)$, which has precisely eight physical degrees of freedom that match the eight transverse bosonic coordinates of the string. This spinor, regarded as a two-dimensional field on the string world sheet (σ,τ), can be decomposed into its right moving, $S^a(\tau-\sigma)$, and left moving, $S^a(\tau+\sigma)$, pieces. One then has the option of choosing the ten-dimensional chiralities of these spinors to be the same or to be opposite, resulting in a chiral or non-chiral theory respectively. The chiral theory, remarkably, yields an anomaly free N=2 chiral supergravity in ten dimensions.[3]

One possibility had been overlooked in the construction of closed string theories. All the physical degrees of freedom of closed oriented strings are two-dimensional massless fields and can be separated into right and left movers, which do not mix even when the strings interact. Therefore there is no apparent reason why one must insist on symmetry between these modes. In fact, in the chiral model described above, the symmetry is broken by choosing opposite chiralities for the right and left moving spinors. In the heterotic theory the symmetry is even more dramatically broken--the right- and left-moving modes are completely different. This idea can only work for closed strings. Open string boundary conditions mix the modes, a right mover when reflected off the boundary becomes a left mover. It also cannot work for nonorientable strings in which we identify the right- and left-movers, as in the case of "Type I" closed strings which couple to open strings. However, for Type II closed strings there is no apriori reason why the right and left movers must be identical--as long as each sector of the theory is separately consisent.

With this idea in mind we constructed the heterotic string--a combination of the right-movers of the ten-dimensional superstring and the left-movers of the 26-dimensional bosonic string.[1] Together we have ten full fledged coordinates as well as a right-moving Majorana-Weyl spinor coordinate appropriate for realizing N=1 supersymmetry and 16 left-moving bosonic coordinates which provide the arena for the gauge degrees of freedom.

Most of the consistency checks for the heterotic string (e.g. Lorentz invariance) follow automatically from the separate consistency of the right- and left-moving sectors. Supersymmetry (N=1) is assured by the supersymmetry of the right-movers alone, and the absence of tachyons in this sector eliminates the would-be tachyonic state of the left-moving bosonic sector from the physical Hilbert space. Most remarkably, the demand that the heterotic string be free of global (two-dimensional) diffeomorphism anomalies requires an (almost) unique compactification of the sixteen internal bosonic left-moving coordinates. These must be compactified on a special sixteen dimensional torus of determined radius, whose defining lattice is self dual. With such a compactification the heterotic string then contains, in addition to massless gravitons, 496 massless gauge bosons in the adjoint representation of either $E_8 \times E_8$ or Spin $(32)/Z_2$.[1,4]

At first sight the heterotic string appears awkward and strange; upon closer examination it has many attractive features. Unlike other closed string theories it is inherently chiral, containing only one

Majorana-Weyl-spinor coordinate. It produces gauge interactions, not by the ad hoc procedure of attaching charges to the ends of open strings, but via a generalized Kaluza-Klein mechanism. Being a closed string theory it has a much simpler structure than the alternative open superstrings. Among the rest, it contains a unique interaction, in contrast to the many (five) interactions of open strings, the world sheets that describe its time evolution are orientable closed surfaces so that at each order of perturbation theory there is but a single string diagram and thus no cancellations of separately divergent diagrams are required for heterotic loops to be finite. Most important, the early phenomenological hints that superstring theory might describe the real world are largely based on simple compactifications of the $E_8 \times E_8$ heterotic string.[5] Many of these preliminary clues are based on general properties of the $E_8 \times E_8$ heterotic string and are likely to persist in a more realistic solution to the theory.

II. THE FREE HETEROTIC STRING

Free string theories are constructed by the first quantization of an action which is given by the invariant area of the world sheet swept out by the string or by its supersymmetric generalization. The fermionic coordinates of superstrings can be described either as ten-dimensional spinors or by ten-dimensional fermionic vector fields. Similarly the sixteen, left-moving coordinates of the heterotic string can be described by 32 real fermionic coordinates, by the local coordinates on the group manifolds of $E_8 \times E_8$ or spin $(32)/Z_2$, or by sixteen bosonic coordinates. Here we shall take the right-moving fermionic coordinates to be described by a ten-dimensional spinor and the left-moving internal coordinates to be sixteen bosonic fields. This will have the advantage of making the ten-dimensional supersymmetry manifest and of yielding a rather physical picture of the left-moving internal space. The price we pay is that this formulism is only tractable in light cone gauge, so that we must relinquish manifest Lorentz invariance.

The manifestly supersymmetric action for the heterotic string is given by the Green-Schwarz action[6] for the right-movers plus the Nambu action for the left-movers. The superstring action is that of non-linear sigma model on superspace (the super-translation group manifold) with a Wess-Zumino term.[7] Consider an element of the supertranslation group, $h = \exp i(X\cdot P + \theta\cdot Q)$, where X_μ (μ=0,1,...,a) and θ^a (a ten-dimensional Majorana-Weyl spinor) are the superspace coordinates, P and Q are the generators of translations and supersymmetry translations. Then

$$\Pi_\alpha = h^{-1}(x)\,\partial_\alpha h(x) = (\partial_\alpha X^\mu - i\bar{\theta}\gamma^\mu\partial_\alpha\theta)\,P_\mu + \partial_\alpha\,\theta\cdot Q, \tag{1}$$

and defining $tr(P_\mu P_\nu) = \eta_{\mu\nu}$, $tr(Q_a Q_b) = 0$, $tr(Q_a Q_b P^\mu) = (\gamma^\mu C^{-1})_{ab}$, we have

$$S_R = -T/2\{\int d^2\xi\; e q^{\alpha\beta}\; tr\;\Pi_\alpha\Pi_\beta + \int_M d^3\xi\;\varepsilon^{\alpha\beta\gamma}\; tr\;\Pi_\alpha\Pi_\beta\Pi_\gamma$$

$$+ \int d^2\xi e\;\lambda^{++}\;(e^\alpha_+\;\Pi_\alpha)^2\}. \tag{2}$$

Here $\xi^\pm = \tau\pm\sigma$ are two dimensional light cone coordinates, $e^\alpha{}_a$ is the two-bein $(e^\alpha{}_a e^{\beta a} = g^{\alpha\beta})$. The second term is the Wess-Zumino term, where M is a three-dimensional manifold whose boundary coincides with the world sheet. The last term enforces the constraint that the coordinates are only right moving.

The left movers are given by a similar action

$$S_L = -T/2\int d^2\xi\; e[\tfrac{1}{2}\,g^{\alpha\beta}\partial_\alpha X^A\partial_\beta X^A + \lambda^{--}\;(e^\alpha_-\;\partial_\alpha X^A)^2], \tag{3}$$

where A=0,...25.

When expanded, the full action is recognizable as the Green-Schwarz action for right movers alone plus the Nambu action for 26 left movers. It is therefore invariant under ten-dimensional Poincare transformations, N=1 supersymmetry, which acts on the right movers alone, and 16-dimensional Poincare transformations of X^I (I=A-9=1,2,...16). In addition it is invariant under local two-dimensional reparametrizations and the local fermionic transformation of Green and Schwatrz.[6] These local symmetries enable us to choose a gauge--"light cone gauge"--where $g^{\alpha\beta} \sim \eta^{\alpha\beta}$, $X^+(\sigma,\tau) = \pi/T\, P^+\tau + x^+$, and $\gamma^+\theta = 0$. (For details see the lectures of M. Green, this volume). The resulting dynamics is given by the light cone action

$$S_{HeT} = -\frac{1}{2\pi}\int d^2\xi\, [(\partial_\alpha X^i)^2 + \frac{i}{2}\bar{S}(\partial_\tau + \partial_\sigma)S + (\partial_\alpha X^I)^2 + \lambda[(\partial_\tau - \partial_\sigma)X^I]^2], \tag{4}$$

where we have chosen units in which the string tension $T = \frac{1}{2\pi\alpha'} = \frac{1}{\pi}$. The physical degrees of freedom are now manifest--eight transverse coordinates $X^i (i=1\ldots8)$, a Majorana-Weyl-light cone right moving spinor S^a $((1+\gamma^{11})S^a = \gamma^+ S^a = 0)$, and sixteen left-moving coordinates $X^I (I=1,\ldots,16)$. The equations of motions are then

$$\partial^2 X^i = 0; \qquad (\partial_\tau + \partial_\sigma)S^a = 0; \qquad (\partial_\tau - \partial_\sigma)X^I = 0; \tag{5}$$

to which we must append the constraints that follow from the gauge fixing.

Canonical quantization of the transverse and fermonic coordinates is straightforward. Since we are dealing with closed strings, the fields are periodic functions of $0 \leq \sigma \leq \pi$ and can be expanded as

$$X^i(\tau-\sigma) = \frac{1}{2}x^i + \frac{1}{2}p^i(\tau-\sigma) + \frac{i}{2}\sum_{n\neq o}\frac{\alpha^i_n}{n}e^{-2in(\tau-\sigma)}$$

$$X^i(\tau+\sigma) = \frac{1}{2}x^i + \frac{1}{2}p^i(\tau+\sigma) + \frac{i}{2}\sum_{n\neq o}\frac{\tilde{\alpha}^i_n}{n}e^{-2in(\tau+\sigma)} \tag{6}$$

$$S^a(\tau-\sigma) = \sum_{n=-\infty}^{+\infty} S^a_n e^{-2in(\tau-\sigma)}, \tag{7}$$

where

$$[x^i, p^j] = i\delta^{ij},\ [\alpha^i_n, \alpha^j_m] = [\tilde{\alpha}^i_n, \tilde{\alpha}^j_m] = n\delta_{n+m,o}\,\delta^{ij}$$

$$[\alpha^n_i, \tilde{\alpha}^m_j] = 0$$

$$\{S^a_m, \bar{S}^b_n\} = (\gamma^+ h)^{ab}\,\delta_{n+m,o}. \tag{8}$$

The quantization of the left moving coordinates, X^I, is more subtle. We must impose the second class constraint (which is actually the equation of motion) that

$$\phi(\sigma,\tau) = (\partial_\tau - \partial_\sigma) X^I = 0. \qquad (9)$$

But this is not consistent with the cannonical commutator of X^I and its conjugate momentum $P^I = \frac{1}{\pi} \partial_\tau X^I$

$$[X^I(\sigma,\tau),\ P^J(\sigma',\tau)] = i\delta^{IJ}\ \delta(\sigma-\sigma'). \qquad (10)$$

We therefore modify the commutation relations, ala Dirac, defining

$$[A,B]_{DIRAC} = [A,B] - \sum_{\phi_i,\phi_j} [A,\phi_i]\ C_{ij}\ [\phi_j B],$$

where the sum runs over all constraints ϕ_i whose commutator is $[\phi_i,\phi_j] = (C^{-1})_{ij}$. The Dirac bracket is then consistent with the constraint, $[A,\phi_i] = 0$, which can be imposed as an operator identity. In our case we must modify Eq. (10) to read

$$[X^I(\sigma,\tau),\ P^J(\sigma',\tau)] = \frac{i}{2}\,\delta^{IJ}\ \delta(\sigma-\sigma').$$

$$[X^I(\sigma,\tau),\ X^J(\sigma',\tau)] = -\ \frac{i}{4}\,\delta^{IJ}\ \mathrm{Sgn}(\sigma-\sigma'). \qquad (11)$$

Therefore $X^I = X^I(\tau+\sigma)$ has the expansion

$$X^I(\tau+\sigma) = X^I + P^I(\tau+\sigma) + \frac{i}{2} \sum_{n\neq 0} \frac{\tilde{\alpha}^I_n}{n}\ e^{-2in(\tau+\sigma)}, \qquad (12)$$

where upon quantization

$$[\tilde{\alpha}^I_n,\ \tilde{\alpha}^J_m] = n\delta_{n+m,o}\ \delta^{IJ}$$

$$[X^I,\ P^I] = \frac{i}{2}\,\delta^{IJ}. \qquad (13)$$

Note the factor of 1/2 in the commutator of X^I and P^I, which implies that $2P^I$ is the generator of translations in the internal space. The allowed values of P^I must be restricted since X^I must be a periodic function of σ. They will be determined by the structure of the internal sixteen dimensional space.

In light-cone gauge $X^+(\tau,\sigma) = x^+ + p^+\tau$ and X^- is determined by solving the constraints that result from choosing a conformally flat metric on the world sheet. If we expand $X^-(\tau,\sigma)$ as

$$X^-(\tau,\sigma) = x^- + p^-\tau + \frac{i}{2}\sum_{n\neq o}\frac{1}{n}\left(\alpha_n^- e^{-2in(\tau-\sigma)} + \tilde{\alpha}_n^- e^{-2in(\tau+\sigma)}\right) \quad (14)$$

then α_n^- is given, as in the fermionic string, by ($n\neq o$)

$$\alpha_n^- = \frac{1}{p^+}\sum_n \alpha_m^i\, \alpha_{n-m}^i + \frac{1}{2p^+}\sum_m \left(m - \frac{n}{2}\right)\bar{S}_{n-m}\,\gamma^-\, S_m \quad (15)$$

and $\tilde{\alpha}_n^-$ is constructed, as in the bosonic string, as ($n\neq o$)

$$\tilde{\alpha}_n^- = \frac{1}{p^+}\sum_m \left(\tilde{\alpha}_m^i\,\tilde{\alpha}_{n-m}^i + \tilde{\alpha}_m^I\,\tilde{\alpha}_{n-m}^I\right). \quad (16)$$

(Note, in these formulas $\alpha_o^i = \tilde{\alpha}_o^i = \frac{1}{2}p^i$, $\tilde{\alpha}_o^I = P^I$.)
These same constraints determine p^-, the generator of τ translations conjugate to X^+, and thereby the mass operator $m^2 = 2p^+p^- - (p^i)^2$ of the string

$$\frac{1}{4}(\text{mass})^2 = N + (\tilde{N}-1) + \frac{1}{2}\sum_{I=1}^{16}(P^I)^2, \quad (17)$$

where $N(\tilde{N})$ are the normal ordered number operators for the right (left) movers

$$N = \sum_{n=1}^{\infty}\left(\alpha_{-n}^i\,\alpha_n^i + \frac{1}{2}n\,\bar{S}_{-n}\,\gamma^-\,S_n\right)$$

$$\tilde{N} = \sum_{n=1}^{\infty}\left(\tilde{\alpha}_{-n}^i\,\tilde{\alpha}_n^i + \tilde{\alpha}_{-n}^I\,\tilde{\alpha}_n^I\right). \quad (18)$$

The subtraction of -1 in (2.17) is due to the normal ordering of $\tilde{N}$, which is unnecessary in the case of the right movers due to fermion-boson cancellations. This subtraction can also be seen to be necessary to ensure ten dimensional Lorentz invariance. Finally the factor of $(P^I)^2/2$ comes from the internal momentum and winding number (see below) of the left movers.

In addition there is a further constraint that requires

$$N = \tilde{N} - 1 + \frac{1}{2}\sum_{I=1}^{16} (P^I)^2. \qquad (19)$$

This constraint has a simple physical explanation. Since there is no distinguished point on a closed string, we are free to shift the origin of the σ coordinate by an arbitrary amount Δ. This is achieved by the unitary operator

$$U(\Delta) \equiv e^{2i\Delta(N-\tilde{N}+1-1/2\sum_I (P^I)^2)}, \qquad (20)$$

which (recall that $[X^I, P^J] = \frac{i}{2}\delta^{IJ}$) satisfies $U(\Delta)\ F(\tau,\sigma)\ U^+(\Delta) = F(\tau,\sigma+\Delta)$, where F can be X^i, X^I or S^a. The operator $U(\Delta)$ must therefore equal the identity operator on the space of physical states, which must therefore satisfy eq. (19). Again the subtraction constant, -1, is due to the normal ordering of $\tilde{N}$ and is necessary for Lorentz invariance.

Now let us consider the internal left-moving coordinates X^I whose properties are responsible for the new features of the heterotic string. These coordinates can be thought of as parametrizing an internal space T. It is unlikely that T could be curved without producing an inconsistent theory, since the resulting two-dimensional field theory of X^I would be an interacting nonlinear σ model with conformal anomalies. Therefore we consider only flat internal manifolds, taking T to be a sixteen-dimensional torus.

Since closed strings contain gravity in their low energy limit one expects that a compactified closed string theory will contain massless vectors associated with the isometries of the compact space. In the case of a flat 16-dimensional torus this would yield the gauge bosons of $[U(1)]^{16}$. A remarkable feature of closed string theories is that for special choices of the compact space there will exist additional massless vector mesons. These are in fact massless solitons of the closed string theory. They combine with the Kaluza-Klein gauge bosons to fill out the adjoint representation of a simple Lie group whose rank equals the dimension of T. In the case of the heterotic string the structure of T is so severely limited that only two choices are consistent. These produce the gauge mesons of $Spin(32)/Z_2$ or $E_8 \times E_8$!

To see this take T to be the most general torus, which may be thought of as R^{16} modulo a lattice Γ generated by 16 basis vectors e_i^I (i=1...16). We identify the center of mass coordinate X^I with its translation by πe_i^I [$\sqrt{e_i{}^I \cdot e_i{}^I}$ is the diameter of the torus in the ith direction]

$$X^I \equiv X^I + \pi \sum_{i=1}^{16} n_i e_i{}^I \equiv X^I + \pi L^I \tag{21}$$

with n_i integers. On such a torus the allowed center of mass momenta P^I lie on the dual lattice, Γ^*, generated by $e_i{}^{*I}$ (i=1...16), defined by

$$\sum_{I=1}^{16} e_i{}^I \cdot e_j{}^{*I} = \delta_{ij} . \tag{22}$$

In other words $\exp(2\pi i \, P^I \cdot L^I)$, which represents a translation around T by the element L^I (recall that $2P^I$ generates translations), must equal one, so that

$$P^I = \sum_{i=1}^{16} m_i e_i^{*I} \quad (m_i = \text{integer}). \tag{23}$$

For the heterotic string, which is a periodic function of σ on T,

$$X^I = X^I + P^I\tau + L^I\sigma + \ldots = X^I(\tau+\sigma) = X^I + P^I(\tau+\sigma) + \text{oscillators} \tag{24}$$

represents a string configuration which winds around the torus n_i times in the ith direction, with momenta P^I which equals the winding number

L^I. This is in fact a soliton--a classical solution of the equations of motion which is characterized by topological quantum numbers (n_i) that classify the maps of the one sphere $(0<\sigma<\pi)$ onto T $(\pi_1(T) = Z^{16})$. Such solitons must be included in the spectrum of the interacting closed string, since a string with $L^I = 0$ can split into one with $+L^I$ and $-L^I$.

Next we recall the constraint of Eq. (19), which requires that the allowed values of $(P^I=L^I)^2$ are even integers. Therefore, since all (L^I+L^J) are allowed windings, $L^I \cdot L^J$ must be an integer (since $(L^I+L^J)^2$, $(L^I)^2$ and $(L^J)^2$ must all be even numbers). The winding numbers L^I must therefore lie on an integer, even lattice, so that the "metric" $g_{ij} = \sum_{I=1}^{16} e_i^{\ I} e_j^{\ I}$ is integer valued, and g_{ii} = even. The momenta P^I lie on the dual lattice Γ^*, which for integer Γ, contains Γ. [Since every vector in Γ, $V^I = \sum_{i=1}^{16} n_i e_i^{\ I}$, can be written as a vector in Γ^*, $V_i = \Sigma m_i e_i^{*I}$, with $m_i = \sum_{j=1}^{16} n_j g_{ji}$ = integer.] In general Γ^* contains more points than Γ, including some that are not of even length squared. For example if Γ = hypercubic lattice with spacing $\sqrt{2}$, Γ^* = hypercubic lattice with spacing $1/\sqrt{2}$. From a geometrical point of view we would, however, expect that all vectors in Γ^* are allowed momenta, in which case Γ^* must also be integer and even. But then $\Gamma \supset \Gamma^*$ and since $\Gamma^* \supset \Gamma$ the lattice must be self dual, $\Gamma = \Gamma^*$.

In that case

$$L^I = P^I + \sum_{i=1}^{16} n_i e_i^{\ I} \quad (n_i = \text{integer}), \tag{25}$$

and the diameters of the torus are all equal to $\sqrt{2} = \sqrt{\frac{2}{\pi T}}$.

Actually, for the free string we need not take Γ^* to be integer and even. But if we do so, the non geometric nature of the momenta would produce trouble at the interacting level where we would find two-dimensional (global) diffeomorphism anomalies that render string loops nonunitary or sick. More on this below.

Even self dual lattices are extremely rare. They only exist in 8n dimensions. In sixteen dimensions there are two such lattices, $\Gamma_8 \times \Gamma_8$ and Γ_{16}. The first is the direct product of two Γ_8's, where Γ_8 is the root lattice of the exceptional Lie algebra E_8; the second is the weight lattice of Spin(32)/Z_2, generated by the weights of the adjoint of SO(32) plus one of the spinor representations. Let us consider $\Gamma_8 \times \Gamma_8$, generated by e_i^I, the roots of $E_8 \times E_8$. In this case the torus T is the "maximal torus" of $E_8 \times E_8$, generated by $\exp iH_i$, where H_i constitute the Cartan subalgebra of $E_8 \times E_8$.

The appearance of these special tori and the consequent emergence of the gauge group G(G=$E_8 \times E_8$ or Spin(32)/Z) might seem mysterious. To those familiar with the theory of affine Lie (Kac-Moody) algebras the mathematical framework is familiar(see the lectures of Olive, this volume); nonetheless let us take a physicist's approach. Consider the massless states of the theory. These must satisfy $N = \tilde{N} - 1 + 1/2\ \Sigma(P_I)^2 = 0$. The bosonic states with $\tilde{N}=1$, $P^I=0$, consist of gravitons $(\tilde{\alpha}_{-1}^i|0\rangle)$ and 16 corresponding gauge bosons $(\tilde{\alpha}_{-1}^I|0\rangle)$. These massless vectors are expected--they arise from the $U(1)^{16}$ isometry of the torus. The corresponding conserved charges are simply the components of the internal (left-handed) momenta P^I. However there are additional massless vectors, solitons of the string theory, with winding numbers $L^I = P^I$ such that $(P^I)^2 = 2$. For the allowed self-dual lattices there are precisely 480 such vectors. The solitons have nonvanishing $U(1)^{16}$ charges, which can be identified with the non-zero weights of the adjoint representation of G, and they are massless vector mesons. It is well known that the only consisent theory of massless vector bosons with nonvanishing charges is a local gauge theory. But why couldn't the gauge group be $U(1)^{496}$? The reason is that the interactions will allow a soliton of charge, which equals winding number, P^I to break up into two solitons of charge $P^I + K^I$ and $-K^I$ respectively. Thus each gauge boson couples to every other one, and the corresponding gauge group must be G. To show this explicitly we should construct the generators of G in the Fock space of the string oscillators. This we shall do below when we discuss interactions, since these generators are simply the vertex operators for a string to emit a massless gauge boson of vanishing momenta.

Let us now consider the full spectrum of the heterotic string. The physical states are simply direct products of the Fock spaces $|\rangle_R \times |\rangle_L$ of the right moving fermionic string and the left moving bosonic string, subject to the constraint, which ensures that the masses are non negative,

$$(\text{mass})^2 = 8N \geq 0$$

$$N = \tilde{N} + \frac{1}{2}(P^I)^2 - 1. \qquad (26)$$

The right-handed ground state is annihilated by α^i_n and S^i_n $(n>0)$ and N. It consists of 8 bosonic states $|i\rangle_R$ and 8 fermionic states $|a\rangle_R$, which form a massless vector and spinor supermultiplet. The left-moving ground state consists of $\tilde{\alpha}^i_{-1}|0\rangle_L$ $(\tilde{N}=1,\ P^I=0)$ and $\tilde{\alpha}^I_{-1}|0\rangle_L$ $(\tilde{N}=1,\ P^I=0)$ and $|P^I,\ (P^I)^2=2\rangle_L$. The most general state

$$\Pi_i\, \alpha^{i}_{-n_i}\ \Pi_j\, S^{a_j}_{-m_j}\ \Pi_k\, \tilde{\alpha}^{k}_{-n_k}\ |\text{Ground State}\rangle,$$

can be decomposed into irreducible representations of D=10, N=1 supersymmetry and of the group G.

The demonstration that these states are indeed Lorentz invariant and supersymmetric is straightforward. It is a consequence of the fact that the generators of Lorentz and supersymmetry transformations act separately on the right and left movers, and that the left movers are Lorentz invariant in 26 dimensions and the right movers Lorentz invariant and N=1 supersymmetric in 10 dimensions.

The ground state is a direct product of $|i\rangle_R + |a\rangle_R$ with $\tilde{\alpha}^i_{-1}|0\rangle_L$ (which yields the N=1 D=10 supergravity multiplet), and with $\tilde{\alpha}^I_{-1}|0\rangle_L$ and $|(P^I)^2=2\rangle_L$ (which yields the N=1 Yang Mills supermultiplet in the adjoint of $E_8 \times E_8$). The higher mass states are easily assembled into SO(9)×G multiplets. For example the first massive level, with $(\text{mass})^2 = 8$, has N=1 and $(\tilde{N},\ (P^I)^2) = (2,0),\ (1,2)$ or $(0,4)$. We separately assemble the right and left moves into SO(9) multiplets and take direct products. The right movers contain $\alpha^i_{-1}|j\rangle_R$ and $S^a_{-}\ (b)_R$ which fill out the SO(9) representations <u>44</u> = and <u>84</u> = , as well as the fermion states $\alpha^{i}_{-1}|a\rangle_R$ and $S^{a}_{-1}|i\rangle_R$ which form the 128 of SO(a). In the left-moving sector we have the $(E_8\times E_8,\ SO(9))$ representations: ((1,1), $\underline{44}$) which contain $\tilde{\alpha}^{i}_{-1}\ \tilde{\alpha}^{j}_{-1}|0\rangle_L$, ((248,1) + (1,248), $\underline{9}$) which contains $\tilde{\alpha}^{i}_{-1}|(P^I)^2=2\rangle_L$, and ((3875,1) + (1,2875) + (248,248) + 2(1,1), $\underline{1}$) which contains $|(P^I)^2=4\rangle$. Altogether, at this level, we have 18,883,584 physical degrees of freedom!

At higher mass levels the number of states will increase rapidly, not only will we have the usual proliferation of higher spin states but also we will get even larger representations of $E_8\times E_8$. The number of states of mass M in the heterotic string increases as $d(M) \underset{M\to\infty}{\sim} \exp\beta(\beta_H M)$, with β_H given by $(2+\sqrt{2})\ \pi\ \sqrt{\alpha'}$, the mean of corresponding factors for the fermionic superstring and the 26 dimensional bosonic string.

III. INTERACTIONS

Ordinary field theories of particles allow for some arbitrariness in introducing interactions. Even in the case of gauge theories the form of the interactions are not totally fixed by the requirement of gauge symmetry; one must in addition impose other demands, such as renormalizability, to rule out higher derivative couplings. String theories do not seem to have any such ambiguities. The interactions that describe the splitting and joining of strings at points appear to be uniquely determined. Eventually this unique dynamics will be understood to be a consequence of the symmetry principles tht underly the second quantized field theory of strings. However it is also clear how we must introduce interctions among strings from the first quantized functional integral approach. In order to calculate the amplitude for string scattering we should sum over world surfaces that describe the splitting and rejoining of strings, weighted with the same exponential of the classical action that is used in describing the free propagation. The tree or Born approximation is then given by summing over world sheets where external string cylinders are attached to a sphere, whereas N loop amplitudes will be obtained by attaching the cylinders to a sphere with N handles. From the point of view of the first quantized theory the interactions of strings are introduced in a purely topological fashion, by enlarging the class of manifolds in which the two-dimensional fields, $X^i(\sigma,\tau)$ and $S^a(\sigma,\tau)$, propagate. This is one of the reasons why string theories are found to be renormalizable, since there is simply no way by which this interaction can be modified.

Just as free superstrings can be described in a variety of equivalent formulisms, so can their interactions. Here I shall only describe the interactions of heterotic strings in light cone gauge, with the interaction described by an opertor, acting on the direct product of single string Fock spaces, which yields the amplitude for a single string to split into two.[8] The advantage of the light cone gauge formulation is that there is a direct translation of this Fock space operator to a cubic interaction of a second quantized string field theory. The perturbative expansion of the S matrix is then given directly by $\tau=X^+$ ordered perturbation theory. [For a review of this approach see the lectures by Mandelstam, this volume.] Particularly simple are the vertices that describe the emission of a zero size pointlike string. (Recall, that when treating interacting strings we must choose their σ length to be proportional to P^+, so that the P^+ density is the same for all strings, and is locally conserved. Thus zero length corresponds to vanishing P^+.) With this vertex one can derive tree and one-loop amplitudes for the scattering of massless particles, the full three string vertex is only required beyond the one loop level.

The amplitude for string 2 and 3 to join and become string 1 is proportional to $\Pi_\sigma \delta\ [X^2(\sigma,\tau) - (X^2 + X^3)(\sigma,\tau)]$, which expresses the fact that string 1 coincides with string 2 plus 3 at the time of interaction. For a pointlike string 3, this operator is simply proportional to $I_{12} \int_0^\pi d\sigma\ \delta[X^1(\tau,\sigma) - X^3]$, where I_{12} constrains X^1 to equal X^2. Written in the Fock space of string 1 and 2, which we can identify, the vertex operator for the emission of a pointlike string with momentum k, at time τ is of the form

$$V(\tau,k,\lambda) = \int_0^\pi d\sigma\ W[\tau,k,\lambda]\ e^{ik\cdot X(\sigma,\tau)}, \qquad (27)$$

where W depends on the quantum numbers of the emitted particle, and we integrate over all points, σ, on string 1=2 where particle 3 is emitted. The S matrix amplitude can then be evaluated by sandwiching the operator

$$S = \int_{-\infty}^{+\infty} d\tau_1 \ldots d\tau_N \; T\big(V(\tau_1,k_1,\lambda_1)\ldots V(\tau_N,k_N,\lambda_N)\big) \tag{28}$$

between asymptotic states of string 1 and 2 (T denotes τ ordering), and summing over inequivalent τ orderings.

When we consider closed orientable strings the vertex operator V factorizes (inside the σ integral) into a product of V_R and V_L, which act on right and left moving modes separately. This is a consequence of the fact that the string coordinates are massless two dimensional free fields which propagate on orientable surfaces without interacting with each other. This operator is especially simple if we consider the special kinematics where the emitted massless particle has vanishing k^+ and k^-. This requires, since $k^2 = k^i \cdot k^i = 0$, that we continue to complex values of k^i, but greatly simplifies calculations due to the absence of the complicated dependent variables X^- and P^-.

The form factor W will be proportional to the wave function of the emitted massless particle, which can either belong to the N=1 supergravity multiplet or to th N=1 super Yang Mills multiplet. In light cone gauge these are described by the wave functions:

<u>Supergravity Multiplet</u>:

$\rho^{\mu\nu}(k)$: graviton (traceless, symmetric $\rho^{\mu\nu}$) + antisymmetric tensor (antisymmetric $\rho^{\mu\nu}$) + dilaton (the trace of $\rho^{\mu\nu}$), where

$$k_\mu \rho^{\mu\nu} = k_\nu \rho^{\mu\nu} = \rho^{+\nu} = \rho^{\mu +} = 0 \tag{29}$$

$U^{\mu a}(k)$: gravitino + massless spinor, where

$$U^{+a} = \not{k} U^\mu = k_\mu U^{\mu a} = \gamma^+ U^\mu = \tfrac{1}{2}(1-\gamma^{11})\, U = 0. \tag{30}$$

Yang Mills Multiplet:

$\rho_\mu^I(k)$ and $\rho_\mu^{K_I}(k)$: gauge bosons, neutral and charged, $(K^I)^2=2$,

$$k^\mu \rho_\mu = \rho^+ = 0. \tag{31}$$

$U^I(k)$, $U^{K_I}(k)$: gauginos, neutral and charged

$$\not{k} U = \gamma^+ U = \frac{1}{2}(1-\gamma^{11})\, U = 0 \tag{32}$$

Again, life is greatly simplified if we take the polarization tensors to be transverse in the frame where $k^+ = k^- = 0$.

The precise form of the operator W can be determined by a variety of considerations. In the conformal field theory approach the vertex operator must have conformal weight two and yield a representation of N=1 supersymmetry × G. Imposing these requirements fixes the form of W. Alternatively, the vertex for massless particles can be derived from the variation of the string action with respect to background massless fields. Thus the graviton vertex, say for the bosonic closed string theory, can be deduced by varying the action, $-1/4\pi \int d^2\xi\, g_{\mu\nu}(x)\, \partial_\alpha X^\mu\, \partial_\alpha X^\nu$, which describes the string embedded in a space-time with metric $g_{\mu\nu}$, with respect to $g_{\mu\nu}(y)$. This would yield the vertex for graviton emission at y,

$$V_{\mu\nu}(y) \sim \delta(y-x(\sigma,\tau))\, \partial_\alpha X^\mu(\sigma,\tau)\, \partial_\alpha X^\nu(\sigma,\tau),$$

from which we deduce that $W^{GRAV}(\tau,k) \sim \rho^{\mu\nu}(k) \frac{\partial X^\mu}{\partial(\tau+\sigma)} \frac{\partial X^\nu}{\partial(\tau-\sigma)}$. Finally, in the light cone approach, the full interaction Hamiltonian that describes the splitting of closed strings, can be determined to be the overlap δ-functional of the strings, plus an operator insertion at the point of interaction whose form is determining by the requirement of supersymmetry.[9] This construction can be carried out for closed bosonic and superstrings, and could easily be generalized to the heterotic string. For present purposes we can simply construct the heterotic vertices as products of the known right-moving superstring and left-moving bosonic string amplitudes.

Thus the vertices for emission of particles in the supergravity multiplet are given by ($k^+=0$),

$$V_B^{GRAV}(k) = \frac{4g}{\pi}\,\rho_{\mu\nu}(k) \int_0^\pi d\sigma\, B^\mu \tilde{P}^\nu\, e^{ik_\mu X^\mu(\tau,\sigma)}$$

$$V_F^{GRAV}(k) = \frac{4g}{\pi} \int_0^\pi d\sigma\, F^a \tilde{P}^\nu U^{a\nu}(k)\, e^{ik_\mu X^\mu(\tau,\sigma)} \qquad (33)$$

In the special kinematic frame, where $k^- = \rho^{-\mu} = \rho^{\mu-} = U^{-a} = 0$, we only require the transverse components of B^μ, P^μ, which are

$$B^i = P^i + \frac{1}{2} k^j R^{ij}$$

$$P^i = \frac{dX^i}{d(\tau-\sigma)} = \frac{1}{2} p^i + \sum_{n\neq 0} \alpha_n^i e^{-2in(\tau-\sigma)},$$

$$R^{ij} = \frac{1}{8} \bar{S}\gamma^{ij-} S$$

and

$$F^a = \frac{i}{2}(p^+)^{-1/2}\,[\bar{S}\gamma\cdot P - \frac{1}{6} : R^{ij} k^i \bar{S}\gamma^j :]^a \qquad (34)$$

comes from the right-moving gravitational vertex of the superstring, and

$$\tilde{P}^i = \frac{1}{2} p^i + \sum_{n\neq 0} \tilde{\alpha}_n^i e^{-2in(\tau+\sigma)}.$$

The constant, g, that appears in the normalization of the vertex has been chosen so tht it will equal the Yang Mills coupling constant.

The vertices for the emission of neutral gauge mesons are essentially those of the gravitational multiplet, with the index ν being replaced by the internal index I. Indeed these sixteen bosons can be regarded as Kaluza-Klein gauge bosons arising from the $U(1)^{16}$ isometry of the internal torus. We thus have (for $k^+=0$)

$$V_B^I = \frac{4g}{\pi} \rho_\mu^I(k) \int_0^\pi d\sigma\, B^\mu\, \tilde{P}^I\, e^{ik_\mu X^\mu(\tau,\sigma)}$$

and

$$V_F^I = \frac{4g}{\pi} \int_0^\pi d\sigma\, F^a\, \tilde{P}^I\, U^{aI}(k)\, e^{ik_\mu X^\mu(\tau,\sigma)} , \qquad (35)$$

The vertex for the emission of the "charged" gauge bosons (and their supersymmetric partners) is of different form. The left-moving state here corresponds to the <u>tachyon</u> ground state of the bosonic string with momentum (=winding number) K^I in the internal dimension. Therefore the vertex for emission of the vector supermultiplet labeled by K^I is:

$$V_B^{K^I}(k) = \frac{4g}{\pi} \rho_\mu(k) \int_0^\pi d\sigma\, B^\mu\, e^{ik_\mu X^\mu(\tau,\sigma)} \; : e^{2iK^I \cdot X^I(\tau+\sigma)} : \; C(k)$$

$$V_F^{K^I}(k) = \frac{4g}{\pi} \int_0^\pi d\sigma\, F^a U^a(k)\, e^{ik_\mu X^\mu(\tau,\sigma)} \; : e^{2iK^I \cdot X^I(\tau+\sigma)} : \; C(k) \qquad (36)$$

The factor $:e^{2iK^I \cdot X^I(\tau+\sigma)}:$ arises from the left-moving part of the "tachyon" vertex of the bosonic string. The factor of two in the exponential is the fact that $2K^I$ is the generator of translations in the internal left moving space . This term must be normal ordered (for clarity the hats to refer to operators in the string Fock space), since $(K^I)^2$ does not vanish.

$$:e^{2iK^I \cdot X^I(\tau+\sigma)}: = e^{-K^I \sum_{n=-\infty}^{-1} \frac{\tilde{\alpha}_n^I}{n} e^{-2in(\tau+\sigma)}} \; e^{2iK^I(x^I + p^I(\tau+\sigma))}$$

$$\cdot\; e^{-K^I \sum_{n=1}^{\infty} \frac{\tilde{\alpha}_n}{n} e^{-2in(\tau+\sigma)}} \qquad (37)$$

The factor of $e^{2iK^I \cdot X^I}$ in the vertex shifts the internal momentum of a given string state by the root vector K^I. The factor $C(K^I)$ is an operator "co-cycle" or "twist". It appears because this vertex creates a string soliton which corresponds to a nontrivial map of the circle onto the sixteen dimensional torus T. The vertex opertor need only be a projective representation of the group of compositions of these maps, and can differ from the translation operator by a cocycle C(K), which acting on states of internal momenta P^I yields

$$C(K)\ |P\rangle = \varepsilon(K,P)\ |P\rangle \tag{38}$$

as long as

$$C(K)\ C(L) = \varepsilon(K,L)\ C(K+L) \tag{39}$$

where $\varepsilon(K,L)$ is a pure phase. For the composition to be associative $\varepsilon(K,L)$ must satisfy

$$\varepsilon(K,L)\ \varepsilon(K+L,M) = \varepsilon(L,M)\ \varepsilon(K,L+M) \tag{40}$$

The groups of translations on the lattices Γ_{16} and $\Gamma_8\times\Gamma_8$ have such nontrivial projective representations, and we must use them to construct vertex operators that will yield representations of the algebra of G = SO(32) or $E_8\times E_8$.[10] An explicit representation of $\varepsilon(K,L)$ can be chosen so that

$$\varepsilon(K,L)\ \varepsilon(L,K) = (-)^{K\cdot L}$$

$$\varepsilon(K,0) = -\varepsilon(K,-K) = 1. \tag{41}$$

Since our vertices are not manifestly Lorentz invariant, supersymmetric or G invariant we must check that these symmetries are preserved by the interactions. The demonstration of Lorenz invariance and supersymmetry is identical to that given for the superstring and for the bosonic string, since the generators of these symmetries act separately on each factor of the heterotic vertices.

For example, consider supersymmetry. The heterotic string contains one supersymmetry, generated by

$$Q^a = i\sqrt{p^+}\,(\gamma^+ S_0)^a + 2i\,\frac{1}{\sqrt{p^+}}\sum_n (\gamma_i S_{-n})^a\,\alpha_n^i\,. \tag{42}$$

This acts only on the right-moving modes on the string, which are those of the fermionic superstring. One must then show that this generator transforms V_B into V_F, and vice versa. This is a consequence of

$$[\bar{\varepsilon}\cdot Q,\ \rho_\mu(k)\,B^\mu\,e^{ik_\nu\cdot X_R^\nu(\tau-\sigma)}] = F^a(k)\,\frac{1}{2}\,\gamma_{\mu\nu}\,k^\mu\rho^\nu(k)\varepsilon^a\,e^{ik_\nu X_R^\nu(\tau-\sigma)}$$

$$+\ \tau \text{ derivative term} \tag{43}$$

and

$$[\bar{\varepsilon}\cdot Q,\ F^a(k)U^a(k)\,e^{ik_\nu\cdot X_R^\nu(\tau-\sigma)}] = B^\mu(k)\cdot 2\bar{\varepsilon}\,(\gamma_\mu - \frac{k^\mu\gamma^-}{k^-})U(k)\,e^{ik_\nu\cdot X_R^\nu(\tau-\sigma)}$$

$$+\ \tau \text{ derivative term}), \tag{44}$$

which are not difficult to verify (at least for $k^+=0$).

This ensures that the on shell scattering amplitudes will be supersymmetric, since the τ derivatives can be integrated by parts and will not contribute to on shell matrix elements.

Finally let us examine the issue of invariance under the symmetry group $G = E_8\times E_8$ or Spin $(32)/Z_2$. We first must construct the operators which represent the generators of G in the string Fock space. We can deduce the form of these operators by considering the vertex for the emission of gauge bosons of zero momentum, which couple to the G charges. The generators in the Chevally basis consist of P^I, and

$$E(K^I) = \int \frac{dz}{2\pi i Z} :\ e^{2iK^I\cdot X^I(z)}\ :\ C(K^I), \tag{45}$$

(with $(K^I)^2 = 2$, $z = e^{2i(\tau+\sigma)}$). To prove G invariance of the interaction we must show that the vertices, V^I and V^{K^I}, yield an adjoint representation of G, in other words satisfy the commutation relations

$$[P^I,\ V^J_{B,F}] = 0$$

$$[P^I,\ V^{K^J}_{B,F}] = K^I\ V^{K^J}_{B,F}$$

$$[E(L^I),\ V^J_{B,F}] = L^J\ V^{L^I}_{B,F}$$

$$[E(L^I),\ V^{K^J}_{B,F}] = \begin{cases} \varepsilon(L,K)\ V^{L^I+K^I}_{B,F} & (L+K)^2=2 \\ L^I \cdot V^I_{B,F} & L+K = 0 \\ 0 & \text{otherwise} \end{cases} \tag{46}$$

The first two commutators are trivial, since P^I commutes with P^J and $[P^I,\ :e^{2iK^I \cdot X^I(\tau+\sigma)}:] = K^I :e^{2iK^I \cdot X^I(\tau+\sigma)}:$. The third commutator follows from the fact that

$$[:e^{2iL^I \cdot X^I(\tau+\sigma')}:,\ P^I(\tau+\sigma)] = \delta(\sigma-\sigma')\ L^I\ :e^{2iL^I X^I(\tau+\sigma)}: . \tag{47}$$

Finally, the last commutator is derived using the easily verified (for $|w|<|z|$)

$$:e^{2iK\cdot X(z)}:\ :e^{2iL\cdot X(w)}: = :e^{2i(K\cdot X(z)+L\cdot X(w))}:\ (z-w)^{K\cdot L}(wz)^{-K\cdot L/2} \tag{48}$$

and the properties of $C(K)$ and $\varepsilon(K,L)$. A by product of this proof is that the generators, P^I and $E(K^I)$, do indeed satisfy the commutation relations of the Lie algebra of G since, when $k^\mu=0$, the vertices are proportional to these generators. Consequently the physical heterotic string states form unitary representations of the algebra of G generated by P^I and $E(K)$, the interaction vertices as well as the S-Matrix amplitudes are G invariant.

IV. TREE AND LOOP AMPLITUDES

The scattering amplitudes of massless particles can be evaluated in the tree and one-loop order of perturbtion theory by using the vertex operators constructed above. In the case of N+2 tree amplitudes (the Born approximation) we must simply evaluate the S-operator, Eq. (28), between asymptotic, on shell, string states and sum over the (N-2)! time ordering of the interaction vertices. The matrix elements are easily evaluated using the simple algebra of harmonic oscillator creation and annihilation operators.

It is convenient to remove the explicit τ and σ dependence from the vertex operators and to absorb it in propagators. This is achieved by noting that the τ dependence of a vertex is generated by the light cone Hamiltonian (which equals p^+p^-)

$$H = \frac{p^2}{2} + 2N + 2(\tilde{N}-1) + \sum_{I=1}^{16} (P^I)^2 \tag{49}$$

and the σ dependence by U, defined in Eq. (20). Thus if $\tilde{V}(\tau,\sigma)$ is the vertex for the emission of a pointlike string at time τ and position σ,

$$V(k,\tau) = e^{iH\tau} \int_0^\pi \frac{d\sigma}{\pi} U(\sigma)\, \tilde{V}(\tau=\sigma=o)\, U^\dagger(\sigma)\, e^{-iH\tau}. \tag{50}$$

Performing the integration over the times, τ_i, and emission points σ_i, then yields a propagator (t = Euclidean time = $i\tau$)

$$\Delta = \int_0^\infty dt \int_0^\pi \frac{d\sigma}{\pi} e^{-Ht}\, U(\sigma) = \int_{|z|<1} d^2z |z|^{(P^2/4)-2}\, z^N\, \bar{z}^{\tilde{N}-1+\Sigma_I (P^I)^2/2}, \tag{51}$$

sandwiched between each vertex, so that the N+2 point amplitude is proportional to $\langle N+2|\, \tilde{V}_{N+1}(k_{N+1})\, \Delta \tilde{V}_N(K_N)\, \Delta \ldots \Delta\, \tilde{V}(k_2)|\, 1\rangle$. Note that the propagator is equal to $1/H\ \delta(N-\tilde{N}+1-\frac{1}{2}(P^I)^2)$ and thus projects onto states satisfying the constraint (26). The amplitude is then given as an integral over (N-3) complex variables Z_i. The sum over all τ ordering of the vertices has the effect of simply enlarging the Z_i integration regions. The oscillator algebra is easily performed using the formulism of coherent states, since the vertices are linear in the α_n's and create, when acting as the vacuum, coherent states, and the Z^N factors have a simple action on such states. The most tedious part of the calculation is dealing with the spin dependent factors in the amplitude.

To illustrate the form of the resulting amplitudes let us consider the scattering of four charged gauge bosons of momenta k_i, polarizations $\rho_i(k_i)$, and charges $K_i(K_i^2=2)$, with $\Sigma k_i = \Sigma K_i = 0$. This amplitude has the remarkably simple form

$$A(1,\ldots,4) = g^2 K(\rho_1,k_1;\ldots;\rho_4 k_4)\cdot\varepsilon$$

$$\cdot\frac{\Gamma(-1-\frac{u}{8}+\frac{U}{2})\,\Gamma(-1-\frac{s}{8}+\frac{S}{2})\,\Gamma(-1-\frac{t}{8}+\frac{T}{2})}{\Gamma(1+\frac{u}{8})\,\Gamma(1+\frac{s}{8})\,\Gamma(1+\frac{t}{8})} \tag{52}$$

where K is the kinematical factor familiar from the superstring

$$K = -\frac{1}{4}(st\,\rho_1\cdot\rho_3\rho_2\cdot\rho_4 + su\,\rho_2\cdot\rho_3\rho_1\cdot\rho_4 + tu\,\rho_1\cdot\rho_2\rho_3\cdot\rho_4)$$

$$+\frac{1}{2}s(\rho_1\cdot k_4\rho_3\cdot k_2\rho_2\cdot\rho_4 + \rho_2\cdot k_3\rho_4\cdot k_1\rho_1\cdot\rho_3 + \rho_1\cdot k_3\rho_4\cdot k_2\rho_2\cdot\rho_3$$

$$+\rho_2\cdot k_4\rho_3\cdot k_1\rho_1\cdot\rho_4) + \frac{1}{2}t(\rho_2\cdot k_1\rho_4\cdot k_2\rho_3\cdot\rho_1 + \rho_3\cdot k_4\rho_1\cdot k_2\rho_2\cdot\rho_4$$

$$+\rho_2\cdot k_4\rho_1\cdot k_3\rho_3\cdot\rho_4 + \rho_3\cdot k_1\rho_4\cdot k_2\rho_2\cdot\rho_1) + \frac{1}{2}u(\rho_1\cdot k_2\rho_4\cdot k_3\rho_3\cdot\rho_2$$

$$+\rho_3\cdot k_4\rho_2\cdot k_1\rho_1\cdot\rho_4 + \rho_1\cdot k_4\rho_2\cdot k_3\rho_3\cdot\rho_4 + \rho_3\cdot k_2\rho_4\cdot k_1\rho_1\cdot\rho_2)$$

and ε is a phase factor

$$\varepsilon(K_1,K_2,K_3,K_4) = \varepsilon(K_3,-K_4)\,\varepsilon(K_1,K_2)\,(-)^{K_1K_3}, \tag{53}$$

and s,t,u are the usual invariants: $s = (k_1+k_2)^2$, $t = (k_2+k_3)^2$, $u = (k_1+k_3)^2$; $s+t+u=o$; and S,T,U are corresponding lattice momenta invariants $S = (K_1+K_2)^2$, $T = (K_2+K_3)^2$, $U = (K_1+K_3)^2$; $S+T+U=8$.

It is easily established that ε is totally symmetric under interchange of any two labels, and thus the amplitude is totally symmetric. It contains poles which correspond to graviton as well as gauge boson exchanges. For example, if S=0 $(K_1=-K_2)$ or S=2 $(K_1K_2=-1)$ amplitude has a pole for S=0, corresponding to graviton or charge gauge boson exchange respectively. If, however, S=4 $(K_1 \cdot K_2=0)$, or S=6 $(K_1 \cdot K_2=1)$ or S=8$(K_1=K_2)$ the lowest mass poles are at S=8, 16 or 24 respectively, corresponding to the exchange of massive particles with $(\text{mass})^2 = 8[\frac{1}{2} S-1]$.

Since this single amplitude contains both gauge boson and graviton exchange we can use it to determine the relation between Newton's constant, $G_N = K^2/8\pi$ and the gauge coupling in 10 dimensions, g_{10}, since these appear as the residues of the graviton and gauge boson poles. In comparing these poles with standard field theory diagrams it is useful to use the fact that the kinemantical factor K simplifies at s=0,

$$K(k_i,\rho_i)_{s=(k_1+k_2)^2=0} = \frac{t}{8}\,[\rho^1_\alpha\rho^2_\beta\, V_{\alpha\beta\mu}(k_1,k_2,-k_1-k_2)] \cdot \cdot [\rho^3_\gamma\rho^4_\delta\, V_{\gamma\delta\mu}(k_3,k_4,-k_3-k_4)], \tag{54}$$

where $V_{\alpha\beta\gamma}(k_1,k_2,k_3)$ is the standard Yang-Mills vertex

$$V_{\alpha\beta\gamma}(k_1,k_2,k_3) = g_{\alpha\beta}(k_1-k_2)_\gamma + g_{\beta\gamma}(k_2-k_3)_\alpha + g_{\gamma\alpha}(K_3-k_1)_\beta. \tag{55}$$

One then derives that (in units where $\alpha' = 1/2$, $T = 1/\pi$)

$$g_{10} = g = 2K\ . \tag{56}$$

This relation between the gravitational and gauge couplings will survive compactification, since both g_{10} and K are related to their four dimensional values by the same factor of $\sqrt{V}$, where V is the volume of the compactified six-dimensional manifold. Thus we can determine the value of α' in terms of Newton's constant and the Yang-Mills coupling; i.e. $\alpha' = \frac{1}{2\pi T} = \frac{2K^2}{g^2}$. Since g is of order one, $\sqrt{\alpha'}$ is necessarily of order the Planck length!

Heterotic loop amplitudes, at least those involving external massless particles, are not more difficult to construct than trees. One simply "sews" together the external legs of a tree amplitude and sums over all states. Thus, schematically, the one loop amplitude is given by

$$A_{1\text{-loop}}(1,2,\ldots,N) \sim \mathrm{tr}[\Delta V(N) \ldots \Delta V(2)\, \Delta V(1)], \tag{57}$$

where the trace sums over all string states, labeled by bosonic and fermionic oscillator occupation numbers, space-time and internal momenta. An explicit evaluation of the one loop four gauge boson scattering amplitude is given in Ref. 8. Here I shall simply discuss two important issues that arise at the one loop level.

From the earliest days of the "dual-resonance" model one worried about the existence of ultraviolet divergences. Since the tree amplitudes contained particles of arbitrarily high spin one might have expected to find severe ultraviolet divergences in the loop amplitudes. However this was not the case. The one loop amplitude of the open bosonic string contained only a very simple divergence, which could be absorbed by a change in the parameters of the theory. This was interpreted as a sign that the dual resonance model was renormalizable, in the traditional sense of quantum field theory.[11] Subsequent analysis has shown however that these infinities do not arise from traditional ultraviolet divergences, but rather from infrared divergences associated with the existence of a massless, spin zero particle (the dilaton) which can have a non vanishing vacuum expectation value. Consequently the occurrence of infinities in string loops implies the instability of the perturbative vacuum state and not the non-renormalizability of the theory.

The reason this instability was often confused with coupling constant and slope renormalizability is that a shift of the vacuum expectation value of the dilaton field can be reabsorbed by a rescaling of the string tension α'. This fact, by the way, explains why string theories appear to depend on a dimensionless coupling constant, $g^2(\alpha')^{-3}$, that can be arbitrarily adjusted. The reason is that the existing construction of string theories do not determine the vacuum expectation value of the dilaton.[12] In the absence of a dynamical choice of vacuum state the theory is then characterized by a free parameter which distinguishes different (equally acceptable) vacua. Eventually we expect that the choice of vacuum will be unique, in which case the theory will contain no adjustable parameters.

The ultraviolet finiteness of string theories is in accord with the remarks made previously, where we noted that string interactions are introduced in a purely topological fashion and in the first quantized formulation just modify the topology of the manifolds on which the strings propagate. It is then clear that, as long as reparametrization invariance is maintained, no new counterterms can be tolerated --there are simply no other consistent forms of string interactions.

What is the source then of the infinities that sometimes appear in string loops? Consider the planar loop diagram for open strings. The world sheet for this process has the topology of a disc , to which the external particles are attached, with a hole cut out. It can be viewed as describing the rescattering of open strings (the box diagram), or equivalently as the emission of a closed string, which goes into the vacuum, from the tree diagram. This diagram does contain, for almost all open string theories, a divergence which arises when the diameter of the hole shrinks to zero. This region of parameter space describes, among the rest, the process in which the closed string which is emitted develops an infinitely long and infinitely narrow neck, thus corresponding to an on mass shell particle going into the vacuum. Since the closed string sector always contains (when constructed about flat space) a massless spin zero particle (the 'dilaton', ϕ) one would expect a divergence to arise from this process if the dilaton tadpole is non vanishing, since one would then be sitting on the dilaton pole. In fact one can show by explicitly factoring the loop amplitude into a tree $\times$ closed string propagator $\times$ closed string to vacuum amplitude, that the divergence arises from this source.[13] The divergence is therefore an indication of vacuum instability, $\langle vac|\phi|vac\rangle \neq 0$, and not of the existence of ultraviolet divergences in the theory. Nonetheless the instability of the vacuum in a string theory is a serious problem, since we only know how to construct these theories by perturbing about a stable vacuum state. The old open bosonic string, in fact, contained additional sources of vacuum instability, namely the existence of tachyons in the free string and the generation of a cosmological constant in one loop order. Open superstring theories are much healthier. These have no tachyons and supersymmetry prevents a cosmological constant from developing. Green and Schwarz have argued that the anomaly free O(32) superstring theory is also free of dilaton tadpole induced divergences.[14] This is due to a cancellation of the amplitudes for the closed string dilaton to go into the vacuum by attaching to a disc and a corresponding non-orientable diagram where it attaches to a cross cap. This cancellation is rather difficult to establish rigorously and it is not clear whether it would survive beyond one loop order. Closed string theories, especially supersymmetric ones, are even healthier. The old bosonic closed string theory was free of vacuum instability divergences at the tree level. This is because the dilaton tadpole diagram corresponds to attaching an infinitely thin cylinder to a sphere or equivalently evaluating

$$\langle vac|\phi|vac\rangle_{loop} \sim \int_{sphere} d^2\xi\, \partial_\alpha X^\mu(\xi)\, \partial_\alpha X^\mu(\xi)\, e^{iS}. \qquad (58)$$

Since the vertex operator for the dilaton, $(\partial_\alpha X^\mu)^2$, has conformal weight two, conformal invariance alone suffices to ensure (as long as it is maintained) that $\langle\phi\rangle = 0$. However at one loop order the dilaton can acquire a nonvanishing vacuum expectation value. Conformal invariance does not require that the matrix element of $(\partial X)^2$ evaluated on a torus vanish. Indeed, explicit evaluation of closed bosonic one-loop amplitudes shows that the only divergences that occur are due either to external particle mass renormalization or to emission of a dilaton from the closed string tree which then goes into a torus. These theories are also unacceptable since they contain tachyons, and a cosmological constant will develop at the one-loop level.

Supersymmetric closed string theories, the type II theories and the heterotic string, are the healthiest yet. Supersymmetry requires the vanishing of the one-loop dilaton tadpole as well as the one-loop mass and vertex renormalization (for massless external particles). Thus the one loop amplitudes, for massless external states, are completely finite. Furthermore we have every reason to expect that similar arguments of conformal invariance and/or supersymmetry will prevent dilaton tadpoles from developing to any order in perturbation theory. Thus we expect that the heterotic string (as well as the type II superstring) will prove to be finite to all orders in perturbation theory.

The second issue is that of global diffeomorphisms. The world sheet for closed string N-loop amplitudes have the topology of a sphere with N handles. In constructing these amplitudes we must sum over all such manifolds, up to reparametrizations or diffeomorphisms which are gauge symmetries of the covariant action of the string. In choosing the orthonormal and light-cone gauge conditions we have totally fixed the gauge, insofar as local diffeomorphisms are concerned. However global diffeomorphisms which cannot be reached continuously from the identity are not eliminated by the gauge choice and must be factored out by hand. The group of components of (orientation preserving) diffeomorphisms is called the modular group. In the operator approach that we are following we must show that the amplitude can be written as a sum over equal contributions from manifolds which are globally diffeomorphic, and then mod out by the modular group.

Consider for example the torus, which is the relevant manifold for the closed string one loop amplitude. A torus can be described as a parallelogram in the $z = 1/\pi\ (\sigma+i\tau)$ plane, where we identify sides according to $z\to z+1$ and $z\to z+T$. A general torus is thus defined by the complex parameter T, corresponding to a cylinder of circumference π (in the δ direction) and of length π Im T, which is joined at the ends after a rotation of one end by angle 2π Re T. However different values of T may correspond to the same torus. Thus if we shift T by one we obtain a torus which has been out along a circle of constant τ, rotated by 2π and rejoined. This is a global diffeomorphism of the torus. The full modular group is generated by the above $T\to T+1$, $T\to T/T-1$ (which corrsponds to cutting along a τ-circle and rotating one end by 2π

before rejoining) and $T\to -1/T$ (which corresponds to interchanging σ and τ). These generate SL(2,Z), i.e.

$$T \to \frac{aT+b}{cT+d} \quad \begin{matrix} a,b,c,d = \text{integers} \\ ad-bc = 1 \end{matrix} \tag{59}$$

Heterotic string amplitudes will involve an integration over a parameter T which labels the shape of the torus. Demanding modular invariance of the amplitudes, particularly under the transformation $T\to -1/T$ (which effectively interchanges σ and τ) is what determines the compactification of the internal coordintes and produces the gauge groups $E_8\times E_8$ or $Spin(32)/Z_2$. To illustrate how this works consider the vacuum one loop amplitude, i.e. with no external legs. This amplitude is simply given by tr $e^{-i\pi TH}$, describing a string propagating (in the τ direction) around a torus parametrized by T. H is the generator of τ translation, Eq. (49). Let us consider only the internal sixteen left moving bosonic center of mass degrees of freedom, so the relevant part of H is $\sum_{I=1}^{10} (P^I)^2$. The internal trace is easily evaluated to be proportional to the "partition function" of the lattice

$$f(T) = (T)^4 \sum_{L\varepsilon\Lambda} e^{-i\pi T(L\cdot L)}, \tag{60}$$

where the sum runs over all intenal momenta, L, which lie on the lattice Λ, defined in II, which characterizes the torus on which X^I are compactified. Thus $L^I = \Sigma n_i e_i^{\ I}$ with integer n_i and $L\cdot L = \sum_{i,j} n_i g_{ij} n_j$.

Modular invariance requires that we get the same answer if we think of the amplitude as describing strings propagating around the torus in the σ-direction, or that $f(T) = f(-1/T)$. To investigate how f(T) transforms under this modular transformation, consider

$$F(T,X) \equiv \sum_{L\varepsilon\Lambda} e^{-i\pi T(L-X)^2} \tag{61}$$

where X is a sixteen dimensional vector. F is a periodic function of X on $T^{16} = R^{16}/\Lambda$, in other words $F(T,X) = F(T,X\pm e_i)$. Therefore it has a Fourier transform

$$F(T,X) = \sum_{M\epsilon\Lambda^*} e^{-2i\pi M\cdot X}\,\tilde{F}(T,M) \qquad (62)$$

where M lies on the dual lattice, i.e. $M^I = \sum_{i=1}^{16} m_i e_i^{*I}$. We then evaluate $\tilde{F}(T,X)$ as ($\sqrt{|g|}$ = volume of the torus)

$$\tilde{F}(T,M) = \int_{T^{16}} \frac{d^{16}X}{\sqrt{|g|}}\, e^{2i\pi M\cdot X}\, F(T,X) \qquad (63)$$

When the expansion (62) is inserted into (63), the sum over $L\epsilon\Lambda$ extends the X integral to the whole of R^{16}, which can then be performed to yield

$$\tilde{F}(T,M) = \frac{1}{\sqrt{|g|}}\left(\frac{1}{T}\right)^8 e^{i\pi M^2/T}. \qquad (64)$$

When this is inserted into the Fourier expansion of $F(T,0) = f(T)/T^4$, we derive

$$f(T) = \frac{1}{\sqrt{|g|}}\left(-\frac{1}{T}\right)^4 \sum_{M\epsilon\Lambda^*} e^{-i\pi(-1/T)M\cdot M} = \frac{1}{\sqrt{|g|}}\, f^*\!\left(-\frac{1}{T}\right). \qquad (65)$$

Here f^* is the same lattice sum that appears in the definition of f, with Λ replaced by Λ^*. For modular invariance we then require that the lattice be self dual, $\Lambda=\Lambda^*$ ($|g|=1$ then follows automatically).

This completes the line of argument that consistency, or lack of anomalies, requires the special compactification of the internal left moving coordinates of the heterotic string on the maximal torus of $E_8\times E_8$ or $Spin(32)/Z_2$. Witten has given arugments that no new global diffeomorphism anomalies will arise beyond one-loop order.[15]

IV THE LOW ENERGY FIELD THEORY

So far we have been discussing the heterotic string embedded in ten dimensional flat space. However, if the theory is to have any relevance to the real world, we must find solutions to the theory in which six of the spatial dimensions are compactified. String theories automatically contain gravitons and gravitational interactions, thus the issue of which manifold the string lives in is one of string dynamics. The choice of an embedding space with some background metric, corresponds to the choice of a candidate vacuum state for the string. The perturbative construction of a string theory, embedded in a given manifold, should be regarded as the construction of quantum perturbation theory about a given classical solution. The starting point for constructing alternate solutions to quantum string theory is, therefore, the string classical equations of motion. If these are satisfied, and no tachyons exist in the spectrum of small string fluctuations, then one can expect a consistent perturbative expansion about this state.

A truly satisfactory form of string field theory is not yet available. One does know how to formulate the field theory of strings in light cone gauge, and one might attempt to construct nontrivial solutions of the classical field equations. The equations are quite formidable since they involve string functional fields. To date this avenue has not been explored with any degree of success. The existing solutions have been consructed by other means. It is likely that eventually the full power of string field theory will be required.

The first method, which is closest to the usual effective Lagrangian approach in point field theories, starts with the string theory tree approximation and systematically integrates out the massive modes to obtain an effective field theory describing the interactions of the massless modes. The effective Lagrangian is constructed order by order in $\alpha' P^2$, where P is a typical momentum of the massless particles. It should accurately describe, therefore, the low energy physics of the massless particles. The low energy effective field theories that arise from string theories have proved useful in many regards. The anomaly cancelation mechanism of Green and Schwarz, although discovered by stringy arguments, can be understood by counterterms which may be added to the low energy field theory. We can also use these field theories to look for nontrivial vacuum solutions.

A second approach starts with the first quantized string moving in background fields corresponding to coherent superpositions of string modes. The action for the string is that of a generalized nonlinear sigma model, where the target space is, might have nonvanishing curvature, torsion, etc... For example, the appropriate action for a bosonic closed string would be

$$S = -\frac{1}{2\alpha'}\int d^2\xi\sqrt{g}\ [G_{AB}(x)\ g^{\alpha\beta}\ \partial_\alpha X^A\ \partial_\beta X^B + B_{AB}(x)\ \varepsilon^{\alpha\beta}\ \partial_\alpha X^A\ \partial_\beta X^B + \alpha'\ R^{(2)}\ \phi(x)] \tag{66}$$

where G_{AB} (B_{AB}) is the metric field (torsion field) of the manifold in which the string propagates, which might be a nontrivial function of $X^A(\sigma,\tau)$. The last term represents the dilaton field $\phi(x)$, which couples to the two dimension curvature tensor.[17] One then constructs a first quantized (free) string theory using this action. This is more difficult, in the case of a X dependent metric, than it is for constant G_{AB}, since it requires solving a nontrivial, interacting, two dimensional field theory. In principle a consistent string theory can be constructed as long as the σ-model is <u>conformally invariant</u>. Conformal invariance is crucial; it is the symmetry which assures the decoupling of unphysical degrees of freedom and is responsible for the emergence of space-time gauge symmetries. Furthermore, as we argued above in the case of the dilaton tadpole, it is conformal invariance that ensures the vanishing of massless particle tadpoles, and thereby the stability of the string vacuum.[5]

Conformal invariance, in this approach, is the principle that yields the classical string equations of motion. The above action, for example, is naively conformally (scale) invariant, however anomalies will in general destroy this symmetry. These conformal anomalies can be studied using the ordinary machinery of the renormalization group. The nonlinear σ model is renormalizable with $G_{AB}(x)$ and $B_{AB}(x)$ representing continuous sets of dimensionless couplings. To assure conformal invariance one requires that the β-functions (which will be local functions of G_{AB}, B_{AB} and the dilaton field Φ) vanish. These β functions can, in principle, be calculated to any order in perturbation theory, thereby deriving a perturbative expansion of the classical string equations of motion.[16,17] [For a review of this approach see the lectures by Martinec, this volume.] Again, this perturbation theory is an expansion in powers of $\alpha'P^2$ or α'/R^2 where R is a typical size of the (compact) manifold described by G_{AB}.

The main advantage of this approach is tht it might enable one to derive results that are true to all orders of perturbation theory. For example, the best argument that the "Calabi-Yau" manifolds of SU(3) holonomy are exact solutions of the heterotic string equations of motion are based on the vanishing of the β function of the N=2 supersymmetric σ-model on Kahler, Ricci flat manifolds[18] [see the lecture by P. Ginsparg, this volume].

The perturbative expansion of the effective Lagrangian can be constructed more simply by direct methods, using the string S-matrix calculated about flat space. Here the idea is to consider the on-shell scattering amplitudes of the massless particles (the graviton, the antisymmetric torsion field, the gauge bosons, the dilaton, and their fermionic partners) in tree approximation. One then constructs an effective field theory Lagrangian, $\mathcal{L}_o$, that reproduces this S-matrix. This too can be done in a perturbative fashion. One first writes an effective Lagrangian that describes the free massless particles and their three point couplings. One then considers the four point scattering amplitude. The unitarity of the theory guarantees that the massless poles will be generated by the tree graphs of $_o$. What remains has no singularities for vanishing external momenta, and can thus be expanded in a power series in $\alpha'P^2$. One then adds new 4-point, local vertices to $\mathcal{L}_o$, order by order in α', to obtain a new effective Lagrangian. This procedure can then be repeated for the five, six, etc. particle amplitudes, thereby yielding, in principle, the effective Lagrangian to all orders. At each stage one can exploit the fact that the string theory has many local and global symmetries, and thus generate higher order terms that must emerge in higher orders as consequence of these symmetries.

The effective Lagrangian so constructed will not be unique. It is well known that a redefinition of fields (a point transformation) will not affect on shell S-matrix amplitudes. Thus if we have a Lagrangian, $L(\phi_i)$, constructed to yield the S-matrix for the particles described by the fields ϕ_i, the Lagrangian $L[\phi(\phi')] \equiv L'(\phi')$ will yield the same S-matrix. This is true, order by order in perturbation theory, as long as the transformation $\phi' \to \phi(\phi') = \phi' + c_2(\phi')^2 + \ldots$ is non singular. The equations of motion are nonetheless unchanged, since the extremums of L and L' coincide

$$\frac{\delta L(\phi)}{\delta \phi_i} = \frac{\delta L(\phi'(\phi))}{\delta \phi_i} = \sum_j \frac{\delta L'}{\delta \phi_j'} \left(\frac{\delta \phi_j'}{\delta \phi_i}\right). \tag{67}$$

The same ambiguity exists in the previously discussed approach to the classical string equations of motion. The β functions of a renormalizable theory, with couplings ϕ_i, are not unique. They depend on the definition of the coupling constant and the renormalization prescription. Thus if we redefine the couplings (the fields) $\phi \to \phi(\phi')$, the β functions, $\beta_i(\phi) = \Lambda \frac{\partial}{\partial \Lambda} \phi_i$, vary according to

$$\beta_i(\phi) = \Lambda \frac{\partial}{\partial \Lambda} \phi_i(\phi') = \beta_j'(\phi') \frac{\delta \phi_i}{\delta \phi_j'} . \tag{68}$$

However the zeroes of β_i, here identified with the equations of motion, are invariant under a nonsingular field redefinition. Since both sets of equations of motion, $\beta_i = 0$ and $\frac{\delta L}{\delta \phi_i} = 0$, must be satisfied, one would think that they coincide (at least order by order in perturbation theory) otherwise the fields would be over constrained. The above transformation properties in which $\frac{\delta L}{\delta \phi^i}$ transforms as a covariant vector and $\beta^i(\phi)$ as a contravariant vector in the field space suggests that the β functions of the nonlinear σ model are related to the effective Lagrangian by

$$\beta^i(\phi) = G^{ij}(\phi) \frac{\delta L}{\delta \phi^j} \tag{69}$$

where $G^{ij}(\phi)$ is a "metric," $G^{ij}(\phi'))\,\frac{\delta\phi'_\ell}{\delta\phi_i}\frac{\delta\phi'_k}{\delta\phi_j} = G_{\ell k}(\phi)$, in field space. A direct proof of this equivalence is lacking, although it is true to low orders where specific comparisons have been made.

Let us now derive the low order terms in the effective Lagrangian for the heterotic string, restricting attention to the massless bosonic fields. We start first with the spectrum of massless states which consists of the graviton, antisymmetric tensor, dilaton and the gauge bosons of $G=E_8\times E_8$ or $Spin(32)/Z_2$. Correspondingly we introduce a symmetric, traceless field $h_{\mu\nu}$, an antisymmetric field $B_{\mu\nu}$, a scalar field D and gauge fields A_μ^a in the adjoint representation of G. The knowledge of the existence of these massless particles alone, together with the fact that the S-matrix is Lorentz invariant, already allows us to deduce much more than the quadratic terms in these fields. It is a classic result[19] that the low energy couplings of massless spin two (spin one) bosons must be described by the Einstein (Yang-Mills) action in order to yield Lorentz invariant scattering amplitudes. Lorentz invariance and the existence of the massless particles is equivalent to gauge invariance. Thus we conclude that the effective Lagrangian must contain the Einstein, Kalb-Ramond, Yang-Mills actions, together with the kinetic energy term for the dilaton. Of course there may also be higher order terms, but we shall, following the above procedure, ignore these.

Finally there is an additional symmetry of the classical string action that we have not yet discussed. This symmetry is that of shifting the dilaton field by a constant. With the appropriate definition, the constant piece of this field couples to the Euler character of the world sheet on which the string propagates (see Eq. (66)) and thus plays the role of the string loop coupling constant or h. As such it appears as a multiplicative factor outside the classical Lagrangian. We are adopting a more conventional definition of the dilaton in which Newton's constant is independent of this field, which differs by a conformal transformation. With this definition we can determine how the constant piece of D(x) couples to any term in the Lagrangian.[20] The rule is simply that any term in the Lagrangian density (ignoring the factor of $\sqrt{-g}$) with conformal weight w couples to $\exp[(1+w)\kappa D/2]$ where $\kappa = \sqrt{8\pi G}$ is the ten dimensional gravitational coupling. The conformal weight of any term is calculated by adding the weights of the factors of $g_{\mu\nu}(+1)$ and $g^{\mu\nu}(-1)$ that appear. Thus $F^a_{\mu\nu}F^{\mu\nu a} = F^a_{\mu\nu}F^a_{\alpha\beta}g^{\mu\alpha}g^{\nu\beta}$ has conformal weight -2, and thus should be multiplied by $\exp\text{-}KD/\sqrt{2}$. Aside from these factors the action can only contain derivatives of D(x).

To this order we then have

$$\mathcal{L}_0 = \int d^{10}x \sqrt{-g}\,\{-\frac{1}{2\kappa^2}R - \frac{1}{4g^2}e^{-\kappa D/\sqrt{2}}F^a_{\mu\nu}F^{\mu\nu a} - \frac{3}{2}e^{-\kappa\sqrt{2}D}H_{\mu\nu\rho}H^{\mu\nu\rho}$$
$$- \frac{1}{2}\partial_\mu D\partial^\mu D\} \qquad (70)$$

where $H_{\mu\nu\rho} = \frac{1}{3}(\partial_\mu B_{\nu\rho} + \partial_\nu B_{\rho\mu} + \partial_\rho B_{\mu\nu})$.

Now we consider the vertices coupling any three massless particles. These we can calculate easily by evaluating the vector operators of Eqs. 33-36, between massless states. Of course some of the vertices are already contained in $\mathcal{L}_0$, being consequences of general coordinate and gauge invariance. However there will be new terms, higher order in powers of the momenta, that will require adding new terms, higher order in derivatives of the fields. Consider the matrix elements of the vertex for a graviton or an antisymmetric tensor or a dilaton of momentum k_2 and polarization $\rho_2^{\mu\nu}$, between states 1 and 3 with polarizations $\rho_1^{\alpha\beta}$, $\rho_3^{\alpha\beta}$ and momenta k_1, k_3 respectively. It yields the coupling

$$4g\,\rho_1^{\alpha\beta}\,\rho_2^{\mu\nu}\,\rho_3^{\lambda\rho}\,V_{\alpha\mu\lambda}(\tfrac{1}{2}k_i)\,[V_{\beta\nu\rho}(\tfrac{1}{2}k_i) + \tfrac{1}{4}\alpha' k^2_\beta k^3_\nu k^1_\rho], \qquad (71)$$

where $V_{\alpha\beta\gamma}$ is defined in Eq. (55). The two factors entering this amplitude correspond to the usual Yang-Mills factor arising from the right movers and the Yang-Mills result with an O(α') correction from the left movers. The O(α') corrections give new terms, the rest correspond to vertices already contained in $\mathcal{L}_0$.

From this term one can read off many new contributions to the effective action. For example if one takes $\rho_2^{\mu\nu}$ to be antisymmetric, corresponding to $B_{\mu\nu}$, and ρ_1, ρ_3 to be symmetric tracless, corresponding to $h^{\mu\nu} \equiv (g^{\mu\nu} - \eta^{\mu\nu})/2h$, it corresponds to a term in the effective Lagrangian of the form

$$- \frac{g}{8} (\partial_\alpha h_{\mu\beta} \partial^\beta \partial_\nu h^\alpha_\rho) (\partial^\mu B^{\nu\rho} + \partial^\rho B^{\mu\nu}). \tag{72}$$

Since on shell $k_i^2 = k_i \cdot k_j = 0$, one can rewrite this as

$$\frac{g}{32} \frac{1}{\kappa^2} W^{ab}_\mu R^{ab}_{\nu\rho} (\partial^\mu B^{\nu\rho} + \partial^\nu B^{\rho\mu} + \partial^\rho B^{\mu\nu}). \tag{73}$$

(Recall that the gauge coupling g is related to κ, by $g = 2\kappa$ $(\alpha'=1/2)$.) We see that this term corresponds to adding to $H_{\mu\nu\rho}$ the Lorentz-Chern-Simons term. A similar analysis of the $B_{\mu\nu}$, A_α, A_β vertex yields the gauge Chern-Simons coupling. Thus, not surprisingly, we are instructed to modify $H_{\mu\nu\rho}$ so that

$$H_{\mu\nu\rho} = \frac{1}{3} [\partial_\mu B_{\nu\rho} - \frac{1}{8g} \operatorname{tr} (A_\mu F_{\nu\rho} - \frac{2}{3} A_\mu A_\nu A_\rho) + \frac{1}{8g} \operatorname{tr} (\omega_\mu R_{\nu\rho} - \frac{2}{3} \omega_\mu \omega_\nu \omega_\rho)$$
$$+ \text{cyclic perm}]. \tag{74}$$

Taking $\rho_2^{\mu\nu}$ to be pure trace, corresponding to $D\eta^{\mu\nu}/\sqrt{8}$ and $\rho_1^{\mu\nu}$ and $\rho_3^{\mu\nu}$ to be traceless and symmetric, we get a new dilaton graviton coupling, which equals $- g/8\sqrt{8}\ Dh_{\alpha\beta,\lambda\rho} h^{\lambda\rho,\alpha\beta}$ in linearized approximation. The only term, generally covariant, that can yield this, consistent with the dilaton symmetries discussed above, is $e^{-\kappa D/\sqrt{2}}/8g^2 R_{\alpha\beta\lambda\rho} R^{\alpha\beta\lambda\rho}$. Finally, taking $\rho_2^{\mu\nu}$ to be symmetric and traceless and $\rho_1^{\mu\nu}$, $\rho_3^{\mu\nu}$ to be antisymmetric we find a new, $O(\alpha')$, coupling which can be written as $R_{\mu\nu\alpha\beta} H^{\lambda\mu\nu} H^{\alpha\beta}_\lambda$ or $\nabla_\alpha H_{\mu\nu\lambda} \nabla_\beta H^{\mu\nu\lambda}$.

To this order, the effective Lagrangian is given by

$$\mathcal{L}_1 = \int d^{10}x \sqrt{-g} \left[-\frac{1}{2\kappa^2} R - \frac{1}{8g^2} e^{-\kappa D/\sqrt{2}} (\mathrm{tr}\, F_{\mu\nu}F^{\mu\nu} - \mathrm{tr}\, R_{\mu\nu}R^{\mu\nu}) \right.$$

$$\left. - \frac{1}{2} \partial_\mu D \, \partial^\mu D - \frac{3}{2} e^{-\kappa D/\sqrt{2}} H_{\mu\nu\rho} H^{\mu\nu\rho} \right], \qquad (74)$$

with $H_{\mu\nu\rho}$ given in Eq. 74 (I have not included the RH^2 or $(\nabla H)^2$ terms). One might include in $\mathcal{L}_1$ terms, such as R^2 or $R_{\mu\nu}R^{\mu\nu}$, which vanish on shell to this order.[21] However these terms can always be removed by a field redefinition (for example if $g_{\mu\nu} \to g_{\mu\nu} + cR_{\mu\nu}$, then $\sqrt{-g}R \to \sqrt{-g}(R + c'R^2 + \ldots)$), and thus have no effect on the physical content of $\mathcal{L}_1$.

The equations of motion that follow from the above Lagrangian are satisfied by D=const, $H_{\mu\nu\rho}=0$, $R_{\mu\nu}=0$ if we embed the spin connection in the gauge group G. The last condition ensures that the dilaton equation of motion, which requires that $(\mathrm{tr}\, F_{\mu\nu}F^{\mu\nu} - \mathrm{tr}\, R_{\mu\nu}R^{\mu\nu}) = 0$, be statisfied. To this order in α' there is no requirement that the manifold be Kahler , only that it be Ricci flat.

So far the construction of $\mathcal{L}_1$ has been extremely easy, especially in comparison to the σ-model approach, where up to three loop calculations of the β functions would be required for a similar calculation. To go further we must consider the scattering amplitudes of the massless particles, subtract the Feynman diagrams that $\mathcal{L}_1$ produces, expand in powers of the external momenta and deduce the additional terms to be added to the effective Lagrangian that reproduce these interactions. Here I shall consider only a small part of the full story and describe the lowest-order quartic terms involving the gravitational field, that appear in both Type-II and heterotic closed string theories. (For a fuller account see ref. 22.)

Consider the four-point scattering amplitude of gravitons of momenta k^i and polarizations $\rho^i_{\mu\nu}(k^i)$ in the type II closed string theory. It is given by [23] (the heterotic amplitude for graviton scattering has an identical piece [22])

$$A_{GRAV}(1,2,3,4) = g^2\, t^{\alpha_1\beta_1\cdots\alpha_4\beta_4}\, t^{\gamma_1\delta_1\cdots\gamma_4\delta_4} \prod_{i=1}^{4} \rho^i_{\alpha_i\gamma_i}\, k^i_{\beta_i}\, k^i_{\delta_i} \cdot$$

$$\frac{\Gamma(-s/8)\,\Gamma(-t/8)\,\Gamma(-u/8)}{\Gamma(1+s/8)\,\Gamma(1+t/8)\,\Gamma(1+u/8)}\,, \qquad (75)$$

where $K(\rho^i, k^i) = t^{\alpha_1\beta_1\cdots\alpha_4\beta_4} \prod_{i=1} \rho^i_{\alpha i}\, k^i_{\beta i}$, K being the kinematical factor defined in(53). The tensor $t^{\alpha_1\cdots\beta_4}$ has the following properties[23]

$$t^{\alpha_1\beta_1\cdots} = -t^{\beta_1\alpha_1\cdots};\quad t^{\alpha_1\beta_1\alpha_2\beta_2\cdots} = t^{\alpha_2\beta_2\alpha_1\beta_1\cdots};$$

$$t^{\alpha_1\cdots\beta_4} = -\frac{1}{2}\,\varepsilon^{\alpha_1\cdots\beta_4} - \frac{1}{2}\,[\delta^{\alpha_1\alpha_2}\delta^{\beta_1\beta_2}\delta^{\alpha_3\alpha_4}\delta^{\beta_3\beta_4} + \ldots]$$

$$+ \frac{1}{2}\,[\delta^{\beta_4\alpha_1}\delta^{\beta_1\alpha_2}\delta^{\beta_2\alpha_3}\delta^{\beta_3\alpha_4} +], \qquad (76)$$

where the expressions within brackets are antisymmetric in each $(\alpha_i\beta_i)$ and symmetric under $(\alpha_i\beta_i) \leftrightarrow (\alpha_j\beta_j)$. It is most easily described as a Pfaffian

$$t^{\alpha_1\cdots\beta_4} = \mathrm{tr}\,(R_0^{\alpha_1\beta_1}\, R_0^{\alpha_2\beta_2}\, R_0^{\alpha_3\beta_4}\, R_0^{\alpha_4\beta_4}), \qquad (77)$$

where the trace is taken over the ground states $|a\rangle$, $|i\rangle$ which form a representation of the zero mode S_0^a oscillators and $R_0^{\alpha\beta} = \frac{1}{8}\,\bar{S}_0\gamma^{ij-}\, S_0$.

This amplitude contains poles at s,t,u=0, but these will cancel with massless exchange diagrams generated by $\mathcal{L}_1$. The product of the Γ-functions can be expanded about s=t=u=0 to yield $-\frac{512}{stu} + C(\alpha')^3 + \ldots$, where $C = [2\Gamma'(1)]^2$. The first correction to the pole terms is thus of order α'^3 and cannot, as is easily verified by power counting, arise as a four point contact term from $\mathcal{L}_1$. It therefore corresponds to a new interaction vertex, quartic in $h_{\mu\nu}$ and of order eight in powers of the momenta. General coordinate invariance requires that this be some product of four Rieman tensors. It is easy to establish, using (76), that it is of the form

$$\Delta \;=\; g^2 \cdot \exp(-3/2\,\kappa D)\; t^{\alpha_1\cdots\beta_4}\, t^{\gamma_1\cdots\delta_4} \prod_{i=1} R_{\alpha_i\beta_i\cdot\gamma_i\delta_i}\,. \qquad (78)$$

This addition changes the equations of motion for the graviton and the dilaton fields. If we had a solution to the equations of motion derived from $\mathcal{L}_1$, then in order for it to remain a solution to this order we require that Δ =0 (the dilaton equation) and that $\frac{\delta\Delta}{\delta g_{\mu\nu}}$ = 0. Witten and I have shown that these can be satisfied if the manifold is both Kahler and Ricci flat.[24] I shall illustrate the case of a four dimensional manifold (here Kahler is equivalent to hyper Kahler). In such a case the Rieman tension, $R_{\alpha\beta\gamma\delta}$, can be decomposed into a self dual and an anti-self dual pieces: $R = R^{+}+R^{-}$, where $R^{\pm}_{\alpha\beta\gamma\delta} = \pm 1/2\ \varepsilon_{\alpha\beta\alpha'\beta'}\ R^{\pm}_{\alpha'\beta'\gamma\delta}$. Using the form of $t^{\alpha_1 \ldots \beta_4}$, one can easily show that Δ is of the form $(R^{+}R^{-})^2$, where the precise form of the contractions of the indices of $R^{\pm}$ are irrelevant to the argument. The Kahler condition here is that either R^{+} or R^{-} vanish. Therefore it is obvious that Δ, and the variation of Δ, both vanish for a manifold of SU(2) holonomy. However it is clear that this will not be the case for a non-Kahler but Ricci flat manifold. Similar, but more intricate, arguments establish that manifolds with SU(N) holonomy satisfy the string equations of motion to this order.[24] The heterotic string theory produces additional terms, but these vanish if the spin connection is identified with the gauge connection.[22]

The fact that Ricci flat manifolds do not solve the string equations of motion means that we cannot embed in the string theory arbitrary solutions of Einstein's equation. In particular the Schwartzchild solution will have to be modified at short distances. It is tantalizing to conjecture that string theory, already at the classical level solves the problems of the ubiquitous singularities that appear in general relativity.

What has happened to all the massive modes of the string, have they been set to zero? The answer is that in constructing the effective Lagrangian by the above method we have, in fact, integrated out these modes, or solved for them as functions of the massless fields. This is how the higher order interaction terms are produced. Thus the massive fields are not zero; they are, for given values of $g_{\mu\nu}$, D, $B_{\mu\nu}$, A_μ, solutions of the massive equations of motion. Indeed if we were to construct $\mathcal{L}$, to all orders in α', we would find that it is non local since it must reproduce, for $\alpha' P^2 \approx 1$, the massive poles in the scattering amplitude. For that reason it is not clear how useful this approach is once one goes to high energies or small compactified manifolds where perturbation theory might break down.

REFERENCES

1. D.J. Gross, J. Harvey, E. Martinec and R. Rohm, Phys. Rev. Lett. 54 (1985) 502.
2. M. Green and J. Schwarz, Phys. Lett. 149B (1984) 117.
3. L. Alvarez-Gaumé and E. Witten, Nucl. Phys. B234 (1983) 269.
4. D.J. Gross, J. Harvey, E. Martinec and R. Rohm, Nucl. Phys. B256 (1985) 253.
5. P. Candelas, G. Horowitz, A. Strominger and E. Witten, Nucl. Phys. B258 (1985) 461.
6. M. Green and J. Schwarz, Phys. Lett. 136B (1984) 367; Nucl. Phys. B243 (1984) 285.
7. M. Henneaux and L. Mezincescu, Phys. Lett. 152B (1985) 340.
8. D.J. Gross, J. Harvey, E. Martinec, R. Rohm, Nucl. Phys. (to be published).
9. M. Green and J. Schwarz, Nucl. Phys. B243 (1984) 275; S. Mandelstam (to be published); M.B. Halpern, Phys. Rev. D12 (1975) 1689.
10. I.B. Frenkel and V.G. Kac, Inv. Math. 62 (1980) 23; G. Segal, Comm. Math. Phys. 80 (1982) 301.
11. A. Neveu and J. Scherk, Phys. Rev. D1 (1970) 2355; G. Frye and L. Susskind, Phys. Lett. 31B (1970) 537.
12. E. Witten, Phys. Lett. 149B (1984) 351.
13. J. Scherk, Rev. Mod. Phys. 47 (1975) 123; J. Shapiro, Phys. Rev. D11 (1975) 2937; M. Ademollo et al., Nucl. Phys. B94 (1975) 221.
14. M. Green and J. Schwarz, Phys. Lett. 151B (1985) 21.

15. E. Witten, in Symposium on Geometry, Anomalies and Topology (World Press, 1985).
16. D. Friedan, Phys. Rev. Lett. 45 (1980) 1057;
E. Fradkin and A. Tseytlin, Lebedev preprint (1984).
17. C. Callan, D. Friedan, E. Martinec and M. Perry, Princeton preprint (1985); A. Sen, Fermilab preprints.
18. L. Alvarez-Gaumé and P. Ginsparg, Harvard preprint (1985).
19. R. Feynman, Acta. Phys. Polon. 24, 697 (1963);
S. Weinberg, Phys. Rev. 138, 988 (1965);
S. Deser, Gen. Rel. Grav. 1, 9 (1970).
20. E. Witten, Princeton preprint (1985).
21. B. Zweibach, Berkeley preprint (1985).
22. D. Gross (to appear).
23. J. Schwarz, Phys. Reports, Vol. 89 No. 3 (1982).
24. D. Gross and E. Witten (to appear).

TOPOLOGICAL TOOLS IN TEN DIMENSIONAL PHYSICS

Edward Witten*

Joseph Henry Laboratories
Princeton University
Princeton, New Jersey 08544

I. INTRODUCTION

By now we have become accustomed to the idea that topological notions play some role in quantum field theory. Yet the aspects of quantum field theory which require topological ideas are somewhat special. Many recent works have made it clear, on the other hand, that in ten dimensional physics the role of topology is likely to be far more central.

Two basic notions in topology are homotopy groups and homology groups. The kth homotopy group of a space X, $\pi_k(X)$, measures the topological classes of maps from the k dimensional sphere S^k to X. The kth homology group of X, $H_k(X)$, measures roughly the independent topologically non-trivial k dimensional submanifolds of X. The homotopy groups are more familiar to physicists and are perhaps easier to define, but the mathematical theory of the homology groups is actually much simpler.

The non-trivial spaces with simplest homology are the spheres S^n, with $H_0(S^n) \simeq H_n(S^n) \simeq Z$, $H_k(S^n)=0$ for $k\neq 0,n$. What about homotopy? For $k\leq n$ the homotopy and homology groups are isomorphic, $\pi_k(S^n)\simeq H_k(S^n)$, $k\leq n$. For $k>n$ the story is completely different. Although the homology

*Research supported in part by NSF Grant PHY80-19754.

of spheres is trivial for k>n, the homotopy groups $\pi_k(S^n)$ are non-trivial. For instance, $\pi_4(S^3) \simeq \pi_5(S^3) \simeq Z_2$. In fact, the homotopy groups of spheres are extremely complicated and difficult to evaluate for k>n.

If the non-trivial spaces of simplest homology are the spheres, what are the non-trivial spaces of simplest homotopy? They are the Eilenberg-MacLane spaces K(G,n), G being an arbitrary abelian group. The space K(G,n) is defined to be a connected space with $\pi_n(K(G,n)) \simeq G$, $\pi_k(K(G,n))=0$ for $k\neq n$. It is not too hard to show that for each G and n, there is essentially one space K(G,n) with these properties. However, except for special choices of G and n, K(G,n) is a rather large space which can be simply characterized only by specifying its homotopy groups.

While the K(G,n) have simple homotopy, their homology is complicated. In fact, the homology of Eilenberg-MacLane spaces, like the homotopy of spheres, measures the mismatch between homotopy and homology. It is possible to make this connection precise. Moreover, homology is so much simpler than homotopy that the homology of Eilenberg-MacLane spaces is more easily computed than the homotopy of spheres; the relation between them is actually useful for calculating homotopy groups of spheres. This and other matters I have alluded to are explained in [1].

The reason I am mentioning these facts is that the first fourteen homotopy groups of the group E_8 are $\pi_k(E_8)=Z$ for k=3, $\pi_k(E_8)=0$ for $1\leq k\leq 14$, $k\neq 3$. If one considers phenomena on manifolds of dimension no more than fourteen, one does not encounter $\pi_k(E_8)$ for $k\geq 15$, and the fact that these groups are non-trivial is irrelevant. Thus, for certain purposes E_8 is equivalent to the Eilenberg-MacLane space K(Z,3). Among the finite dimensional Lie groups, E_8 has the simplest homotopy (in the range of dimensions of interest to us) and therefore the most complicated homology.

Is it more useful to have simple homotopy or simple homology? This depends on what one is interested in. One problem of importance is to classify the vector bundles on a manifold with given structure group G. This is a problem in homotopy; the answer depends on the homotopy groups of G. For instance, to classify SU(2) bundles on a ten manifold would be very complicated, because of the fact that SU(2) has many non-zero homotopy groups (like π_4 and π_5). By contrast, classifying the E_8 bundles on a ten manifold is rather simple, because the relevant homotopy groups of E_8 are so simple, and a general answer can be given. It was derived, in essence, in [2] (though there the emphasis was on E_6 bundles in eight dimensions rather than E_8 bundles in ten dimensions), and will be explained in section II. In section II this will be used to study various topological deflects--instantons, monopoles, lumps, vortex lines, and domain walls. Some of these have been studied before--the monopoles in [2], and the lumps in [3]. The study of domain walls in section III will lead us to some speculations about hyperbolic or Lorentzian algebras, whose possible role in string theory was suggested in [4]. Finally, in section IV we will apply the classification of E_8 bundles to study vacuum configurations and global anomalies.

II. CLASSIFICATION OF E_8 BUNDLES

Consider a charged field ψ propagating on a manifold M and interacting with a gauge field with gauge group G. At a given point x on M, the possible values of ψ form a vector space V_x (which furnishes a representation of G). As x varies, V_x varies smoothly. This structure--a family of vector spaces V_x, one for each $x \in M$, varying smoothly with x--forms a so-called vector bundle over M, with structure group G. We may call this vector bundle V. The vector space V_x is called the fiber of the vector bundle at x. Given two vector bundles V and W over the same space M, with generic fibers V_x and W_x, they are said to be isomorphic if it is possible to find a smoothly varying family of invertible maps $f_x: V_x \to W_x$. If two bundles are isomorphic there may be topologically inequivalent classes of isomorphisms between them. Given two isomorphisms $f: V \to W$ and $g: V \to W$ we compare them by studying $g^{-1}f: V \to V$ which is an isomorphism of V onto itself, or in other words a gauge transformation of V.

A particularly simple vector bundle U can be defined by picking a vector space U_0 and saying that the fiber U_x of U at each x is just U_0. This is called the trivial bundle. A vector bundle V is trivial if it is isomorphic to the trivial bundle. An isomorphism $f:V\to U$ is called a trivialization of V.

We will be interested in vector bundles V with E_8 structure group on manifolds M of dimension no more than fifteen. The simplest non-trivial case is that in which M is a sphere S^n. There are then two important statements:

(a) Vector bundles on a sphere S^n are made by taking trivial bundles on the northern and southern hemispheres, and gluing them together on the equator by a gauge transformation. This procedure is known to physicists from instanton theory. The equator is S^{n-1}, and gauge transformations on the equator are labeled topologically by $\pi_{n-1}(G)$. In the case of E_8, this means that E_8 bundles on S^n are trivial (for $n\leq 15$) unless n=4 (since $\pi_{n-1}(E_8)=0$ otherwise). But E_8 bundles on S^4 are labeled by an integer, since $\pi_3(E_8) \simeq Z$.

(b) For $n \neq 4$, $n \leq 15$, an E_8 bundle V on S^n is trivial. Let us now classify the possible trivializations $f:V\to U$. Given two trivializations $f,g:V\to U$, they differ by $gf^{-1}:U\to U$, which is a gauge transformaiton of the trivial bundle U. Such gauge transformtions are labeled by $\pi_n(E_8)$, so we conclude that for $n\leq 15$, $n\neq 3,4$, an E_8 bundle on S^n is trivial and can be trivialized in only one way. On the other hand, an E_8 bundle on S^3 is trivial but its possible trivializations are labeled by an arbitrary integer n.

Now let us consider not a sphere but some more general manifold M of dimension $n\leq 15$. We want to classify the E_8 bundles on M. The first step is to triangulate M. This means, in essence (figure 1), that we realize M as a collection of n-dimensional tetrahedra glued together on their boundaries. The vertices of the triangulation are called 0-simplices; an edge connecting two vertices is called a 1-simplex; a triangle bounded by three 1-simplices is called a 2-simplex; a tetrahedron bounded by four 2-simplices is called a 3-simplex; and so on. We will use the name $Q_{(k)}$ to denote a k-simplex. We pick an orientation for each $Q_{(k)}$. The union of all k simplices is called the k-skeleton. The notion of a triangulation is standard in topology; see for instance [5].

Given an E_8 bundle V, we would like to determine whether it is trivial or in other words whether it is isomorphic to the trivial bundle Q, with constant fiber T_o. In fact, let us try to prove that V is trivial. By finding where the proof can fail, we will discover a necessary condition for an E_8 bundle to be trivial; it will turn out to be easily shown that the necessary condition is sufficient.

The approach to proving that V is trivial is to first find a trivialization of V on the 0-skeleton, then on the 1-skeleton, then on the 2-skeleton, and so on until we finally reach the n-skeleton and obtain a trivialization of V on all of M. The starting point of the induction is the obvious fact that any E_8 bundle is trivial on the 0-skeleton, which is just a disjoint collection of points. Let us now analyze the inductive step. If V is trivial on the k skeleton $M^{(k)}$ of M, and a trivialization $\alpha: V^{(k)} \to T_o$ has been chosen, can α be extended to a trivialization of V on the k+1 skeleton $M^{(k+1)}$?

Given a particular k+1 simplex $Q_o^{(k+1)}$, its boundary is a k dimensional sphere S_o^k (figure 1). In the inductive reasoning, before

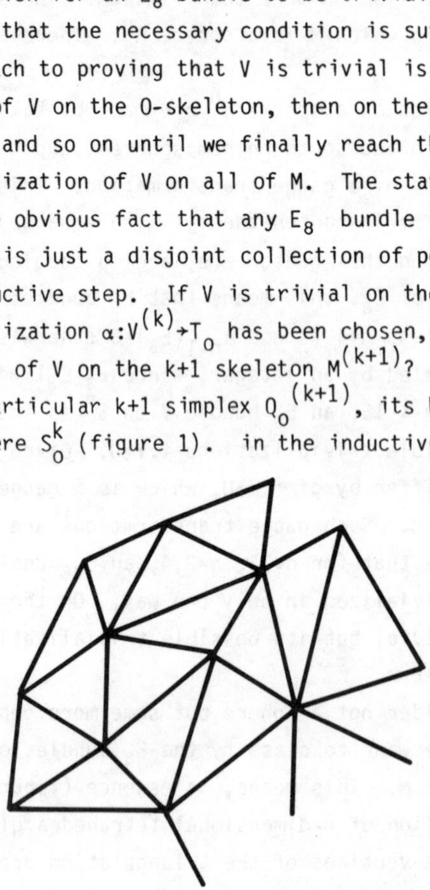

Fig. I

Sketched in the figure is the triangulation of a two dimensional surface

studying (k+1)-simplices we suppose that we are already given a choice of a trivialization α of V on k-simplices. A vector bundle V on $Q_o^{(k+1)}$ with a given trivialization α on the boundary of $Q_o^{(k+1)}$ is much the same thing as a vector bundle on the sphere obtained by identifying the boundary of $Q_o^{(k+1)}$ to a point. By collapsing the boundary of $Q_o^{(k+1)}$ to a point, we get (figure 2) a sphere S^{k+1}, with an E_8 bundle $\hat{V}$ which

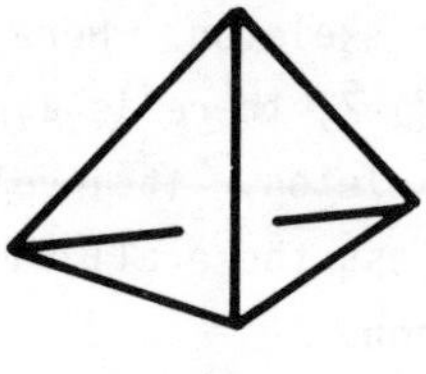

Fig. 2

Sketched here is a three simplex or solid tetrahedron $Q^{(3)}$ whose boundary is topologically a sphere S^2. If one identifies the boundary of $Q^{(3)}$ to a point, one gets topolgically S^3.

is "induced" from the bundle V on $Q^{(k+1)}$ and the trivialization α on the boundary of $Q_o^{(k+1)}$. α can be extended to a trivialization (which we will also call α) of V on all of $Q_o^{(k+1)}$ if and only if $\hat{V}$ is trivial. From our discussion of vector bundles on spheres, we know that $\hat{V}$ is always trivial if the gauge group G has $\pi_k(G)=0$.

Supposing that $\hat{V}$ is trivial on S^{k+1}, α can be extended from the k-skeleton over $Q_o^{(k+1)}$. However, the question arises of whether this extension is unique. In fact, we noted before that given a trivial bundle V on a sphere S^{k+1}, its possible trivializations correspond to the choice of an arbitrary element of $\pi_{k-1}(G)$. This same ambiguity shows up in extending α from the boundary of $Q^{(k+1)}$ over all of $Q^{(k+1)}$; if such an extension exists, the topological classes of possible extensions can be put in corrrespondence with the elements of $\pi_{k+1}(G)$.

Specializing to the case $G=E_8$, we know that the only non-zero homotopy group of E_8 in the range of dimensions of interest to us is $\pi_3(E_8)=Z$. It follows, then, that an E_8 bundle V on a manifold M is always trivial on the two skeleton, with a unique choice of trivialization. On the three skeleton, V is trivial, but there are many inequivalent trivializations, which depend on choices of arbitrary integers for three simplices. There is no particularly natural trivialization of V on the three skeleton, so let us pick an arbitrary trivialization α_0.

Now we come to the four skeleton. Here, we run into real trouble for the first time. As $\pi_3(E_8)=Z$, there is a possible obstruction to extending α_0 over the four skeleton. It involves an integer n_i for each four simplex $Q_i^{(4)}$. Unless these all vanish, α_0 cannot be extended over the four skeleton.

If we find that a particular trivialization α_0 on the three skeleton does not extend over the four skeleton, we should not give up; perhaps the choice of α_0 was unfelicitous. We must analyze how the n_i will transform if we use a different trivialization on the three skeleton rather than α_0.

Roughly speaking, the integer n_i is an "instanton number" on the four simplex $Q_i^{(4)}$. However, as $Q_i^{(4)}$ has a boundary, the "instanton number" of a configuration on $Q_i^{(4)}$ is not well-defined unless we adopt suitable boundary conditions on the boundary of $Q_i^{(4)}$. We have tied down the boundary conditions by means of the trivialization α_0; now we wish to discuss what will happen if we replace α_0 by some other trivialization $\tilde{\alpha}_0$.

In figure (3), we have sketched a four simplex $Q_i^{(4)}$, an adjoining four simplex $Q_j^{(4)}$, and the three simplex $Q_\alpha^{(3)}$ that connects them. Without changing anything outside $Q_i^{(4)}$ and $Q_j^{(4)}$ we can create an instanton-antiinstanton pair on $Q_i^{(4)}$ and let the antiinstanton migrate across $Q_\alpha^{(3)}$ from $Q_i^{(4)}$ to $Q_j^{(4)}$. In the process, the instanton number n_i of $Q_i^{(4)}$ changes by +1, and the instanton number n_j of $Q_j^{(4)}$ changes by -1. Also the configuration of $Q_\alpha^{(3)}$, though still trivial, differs by a topologically non-trivial gauge transformation from what it was before. The trivialization of V on $Q_\alpha^{(3)}$ has changed by one unit.

It is easy to pass from this to the general statement of what happens to the n_i under a change in the trivialization α_0 assumed on

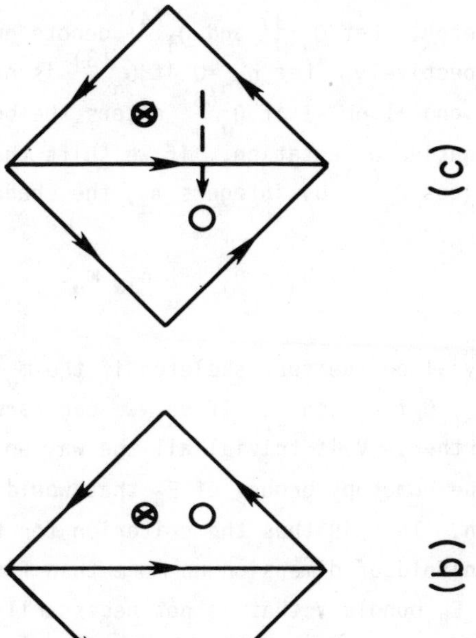

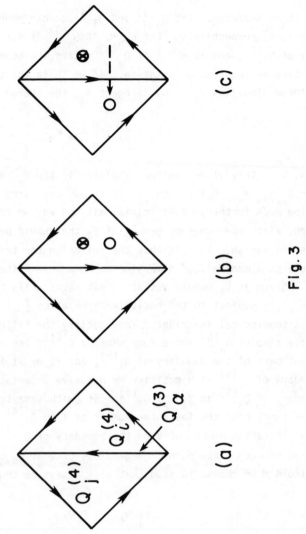

Fig. 3

Symbolically indicated by triangles are two four simplices Q(4); and $Q_j^{(4)}$; they are joined by a three simplex $Q_\alpha^{(3)}$ also drawn symbolically. In (b) an instanton-antiinstanton pair is created in $Q_i^{(4)}$. They are depicted as x and o respectively. In (c) the antiinstanton migrates across $Q_\alpha^{(3)}$ from $Q_i^{(4)}$ to $Q_j^{(4)}$. The trivialization of V on $Q_\alpha^{(3)}$ changes by one unit in the process.

the three skeleton. Let $Q_\alpha^{(3)}$ and $Q_i^{(4)}$ denote generical three or four simplices, respectively. Let $\eta_{i\alpha}=0$ if $Q_\alpha^{(3)}$ is not part of the boundary of $Q_i^{(4)}$, and +1 or -1 if $Q_\alpha^{(3)}$ enters the boundary of $Q_i^{(4)}$ with positive or negative orientation. If we shift the trivialization of V on three simplices $Q_\alpha^{(3)}$ by integers m_α, the change in the n_i is

$$n_i \to n_i + \sum_\alpha \eta_{i\alpha} m_\alpha. \tag{1}$$

Thus, V is trivial on the four skeleton if the m_α can be chosen so that $n_i + \Sigma_\alpha \eta_{i\alpha} m_\alpha = 0$ for each i. If so, we can carry out a similar reasoning much further. V is trivial all the way up to the fifteen skeleton, since the homotopy groups of E_8 that would be obstructions to this all vanish. This is thus the criterion for triviality of an E_8 bundle on a manifold of dimension no more than fifteen.

Given an E_8 bundle V that is not necessarily trivial, the integers $\{n_i\}$, subject to the equivalence relation $\{n_i\} \cong \{n_i + \Sigma_\alpha \eta_{i\alpha} m_\alpha\}$, are a topological invariant. They satisfy the following identity. For a five simplex $Q_x^{(5)}$ and a four simplex $Q_i^{(4)}$, let η_{xi} be zero if $Q_i^{(4)}$ is not part of the boundary of $Q_x^{(5)}$, and +1 or -1 if $Q_i^{(4)}$ enters the boundary of $Q_x^{(5)}$ with positive or negative orientation. Thus, the boundary of $Q_x^{(5)}$ is $\Sigma_i \eta_{xi} Q_i^{(4)}$. If (with some trivialization on three simplices) the instanton number of V on $Q_i^{(4)}$ is n_i, then the total instanton number of V on the boundary of $Q_x^{(5)}$ is $\Sigma \eta_{xi} n_i$. But now we come to a key point. Regardless of what E_8 bundle V on the manifold M we choose to study, it will always be true that for each x,

$$\sum_i \eta_{xi} n_i = 0. \tag{2}$$

Otherwise, V could not have an extension from the four skeleton over $Q_x^{(5)}$. But we know that V extends not just over $Q_x^{(5)}$ but indeed over all of M. After all, what we are studying is an E_8 bundle V over M.

Now, the fourth cohomology group of M with integer coefficients, denoted $H^4(M;Z)$, is defined as follows. An element of $H^4(M;Z)$ is a family of integers $\{\nu_i\}$, one for each four simplex $Q_i^{(4)}$, which obey $\Sigma n_{xi}\ \nu_i = 0$, for all x, and subject to the equivalence relation $\{\nu_i\} \cong \{\nu_i + \Sigma_\alpha n_{i\alpha}\ S_\alpha\}$, for any integers S_α. The condition $\Sigma n_{xi}\ \tilde{\nu}_i = 0$ is invariant under $\nu_i \to \nu_i + \Sigma_\alpha n_{i\alpha}\ S_\alpha$ because $\Sigma_i n_{xi} n_{i\alpha} = 0$, an identity that is usually described by saying that "the boundary of a boundary is zero." The group structure of $H^4(M;Z)$ is defined simply by saying that the sum of the two families $\{\nu_i\}$ and $\{\tilde{\nu}_i\}$ is the family $\{\nu_i + \tilde{\nu}_i\}$.

What we have learned in equations (1) and (2) is that for an E_8 bundle V the "instanton numbers" n_i define an element of $H^4(M;Z)$. This we will call $\lambda(V;Z)$ or simply $\lambda(V)$. What is more, we have learned that an E_8 bundle V on a maifold of dimension at most fifteen is trivial if and only if $\lambda(V;Z)=0$.

Now given two E_8 bundles V_1 and V_2 on a manifold M of dimension at most fifteen, when are they isomorphic? It is clearly necessary to have $\lambda(V_1;Z) = \lambda(V_2;Z)$ since the definition of λ makes clear that it is a topological invariant. It is only slightly harder to prove the converse. V_1 and V_2 are automatically trivial (and so isomorphic) on the three skeleton. If $\lambda(V_1;Z) = \lambda(V_2;Z)$ then (with proper choices of trivialization on the three skeleton) it is possible to find an isomorphism ϕ between V_1 and V_2 on the four skeleton. There is no obstruction to further extension of ϕ all the way up to the fifteen skeleton, since the relevant homotopy groups of E_8 are all zero.

The above argument shows that for M of dimension at most fifteen, there is at most one E_8 bundle V with given λ invariant. To complete the classification of E_8 bundles, we must show the converse. For any element γ of $H^4(M;V)$, we will show the <u>existence</u> of an E_8 bundle V with $\lambda(V;Z) = \gamma$.

On the three skeleton of M, V will necessarily be trivial. Defining V on the four skeleton means choosing an instanton number for each four simplex. Requiring that $\lambda(V;Z)=\gamma$ precisely tells us how to choose these instanton numbers. Having defined V on the four skeleton, we try to extend it over the five skeleton. There is a potential obstruction. A five simplex $Q_x^{(5)}$ over which we wish to extend V has a boundary which is a four sphere S^4 on which V has already been defined.

This S^4 is contractible in $Q^{(5)}$, so V can be extended from S^4 to $Q^{(5)}$ only if it is trivial on S^4 or in other words only if the total instanton number of V is zero on S^4. The fact that this is always true follows from the fact that the instanton numbers ν_i of V on four simplices $Q_i^{(4)}$ were chosen according to an element γ of $H^4(M;Z)$, and so obey $\Sigma\eta_{xi}\nu_i=0$ for any five simplex $Q_x^{(5)}$. Having thus extended V over the five skeleton, there is no obstruction to extending it further all the way up to the fifteen skeleton. This is so for the following reason. If V has been extended over the n skeleton, it can be further extended over an n+1 simplex $Q^{(n+1)}$ if it is trivial on the boundary of $Q^{(n+1)}$, which is an n sphere S^n. V will be trivial on S^n if $\pi_{n-1}(E_8) = 0$, and this is true for $5<n<15$. We have thus completed the classification of E_8 bundles on manifolds M of dimension at most fifteen. There is precisely one such bundle for every element of $H^4(M;Z)$.

The mathematical notions we have encountered can, of course, be generalized. First of all, one can define not just the fourth cohomology group $H^4(M;Z)$ but the general cohomology groups $H^k(M;Z)$ for arbitrary non-negative integers k. (They can be shown to be zero if k exceeds the dimension of the manifold, a property that is not shared by the homotopy groups.) We can even define cohomology groups $H^k(M;F)$ for any abelian group F.

For an n simplex $Q_x^{(n)}$ and an n-1 simplex $Q_y^{(n-1)}$, let n_{xy} be zero if $Q_y^{(n-1)}$ is not in the boundary of $Q_x^{(n)}$ and ± 1 otherwise (the sign depending on relative orientation of $Q_x^{(n)}$ and $Q_y^{(n-1)}$). To define the cohomology groups, we will use a slightly more abstract formulation than before. Let $Y = \Sigma m_x Q_x^{(n)}$ denote a formal linear combination of n simplices with integer coefficients m_x. Let $C^{(n)}$ be the space of all such formal linear combinations of n simplices. It is an abelian group with an obvious addition law. We define the "boundary" operator ∂ by saying that if $y = \Sigma m_x Q_x^{(n)}$ then $\partial y = \Sigma m_x n_{xy} Q_y^{(n-1)}$. Thus, ∂ is a map from $C^{(n)}$ to $C^{(n-1)}$. The statement that "the boundary of a boundary is zero" means that $\partial^2 = 0$.

Let F be an abelian group, and let $D^{(n)}$ be the space of all linear maps ϕ: $C^{(n)} \to F$. It is an abelian group with an obvious addition law. We define the "coboundary operator" $\delta:D^{(n)}\to D^{(n+1)}$ as follows. For any $\phi\varepsilon D^{(n)}$, we must define $\delta\phi$; to define $\delta\phi$ we must give its value $\delta\phi(y)$ for each $y \varepsilon D^{(n+1)}$. The map $y\to\delta\phi(y)$ must be an additive map $D^{(n+1)}\to F$. The required formula for defining $\delta\phi(y)$ is just $\delta\phi(y) = \phi(\partial y)$. Like the boundary operator, the coboundary operator δ obeys $\delta^2=0$, since $\delta^2\phi(y) = \delta(\delta\phi(y)) = \delta(\phi(\partial y)) = \phi(\partial(\partial y)) = \phi(0)=0$.

Let $X^{(n)}$ be the kernel of $\delta:D^{(n)}\to D^{(n+1)}$. Thus, $X^{(n)}$ consists of elements of $D^{(n)}$ annihilated by δ. And let $Y^{(n)}$ be the image of $\delta:D^{(n-1)}\to D^{(n)}$. Thus, $Y^{(n)}$ consists of elements y of $D^{(n)}$ that can be written $y=\delta z$ for some z in $D^{(n-1)}$. Both $X^{(n)}$ and $Y^{(n)}$ are abelian groups, the addition law being inherited from the $D^{(n)}$. Also, $Y^{(n)}$ is a subgroup of $X^{(n)}$ since $\delta^2=0$. It is possible to form the quotient group $X^{(n)}/Y^{(n)}$. An element of $X^{(n)}/Y^{(n)}$ is defined to be an equivalence class of elements of $X^{(n)}$ with two elements x and x' of $X^{(n)}$ considered equivalent if $x-x' \varepsilon Y^{(n)}$. The quotient group $X^{(n)}/Y^{(n)}$ is the object we want. It is denoted $H^n(M;F)$ and is called the nth cohomology group of M with coefficients in F. Our previous treatment amounted to the case n=4, F=Z.

The cohomology groups $H^n(M;Z)$ with coefficients in Z are in some sense the basic ones. A theorem called the "universal coefficients theorem" relates $H^n(M;F)$ for any F to $H^n(M;Z)$. An element z of $H^n(M;Z)$ is said to be a "torsion element" if nz=0 for some non-zero integer n. The torsion is by far the subtler part of the cohomology. It plays a key role in some of the physical applications we sketch later.

If desired, one can get rid of the torsion by picking the coefficient group F to be the real numbers R. The cohomology groups $H^n(M;R)$ are vector spaces over R whose dimensions (by definition) are the Betti numbers b_n.

Returning to the invariant λ that we defined for an E_8 bundle V, it may in general contain a torsion piece or a "free" (non-torsion) piece. The torsion piece of λ can only be understood by a more or less fancy treatment such as the treatment that we gave above. If one is only interested in the "free" part of λ, or in other words if one is content to define λ as an element of $H^4(M;R)$ rather than defining it as an element of $H^4(M;Z)$, a far more simple and perhaps more familiar treatment can be given.

In defining $\lambda(V;Z)$ with integer coefficients, we had to study four simplices; these are four manifolds with boundary. Almost all of the subtlety came from proper treatment of the effects of the boundary. If one is content to define only the much cruder object $\lambda(V;R)$ with real coefficients, this can be done just by studying closed four manifolds without boundary. $\lambda(V;R)$ is just a rule which to each closed four dimensional submanifold H of M assigns a number, the total instanton number of V on H. It can be defined as $\int_H \mathrm{Tr}\, F\wedge F$, where F is the field strength of any E_8 connection on V. The invariant we have called $\lambda(V;Z)$ is an example of a characteristic class--a cohomology class that can be naturally defined for every E_8 bundle just by virtue of the topology of E_8. Characteristic classes with real coefficients can be defined by means of differential forms such as $\mathrm{Tr}\, F\wedge F$ and have become more or less familiar to physicists. The much subtler characteristic classes with integer coefficients may be less familiar, but they are likely to play an important role in ten dimensional physics. They certainly enter in problems such as magnetic monopoles and global anomalies.

To conclude this section, let us note that while our main interest has been in E_8 bundles, some of the concepts carry over to a vector bundle W over a manifold M with arbitrary structure group or gauge group G. Let n be the lowest positive integer for which $\pi_n(G)\neq 0$, and let $F=\pi_n(G)$. Then a G bundle W is automatically trivial on the n skeleton of M (with a trivialization that is not unique but depends on an element of F for each n simplex). On the n+1 skeleton, W may not be trivial. Its non-triviality is measured by an element of F for each n+1 simplex. These elements are subject to conditions like those that appeared in equations (1) and (2) above, and they define an element of $H^{n+1}(M;F)$ which we may call $\gamma(W)$.

For instance, if G is SO(n), the first non-trivial homotopy group is $\pi_1(SO(n))=Z_2$. For a vector bundle W with SO(n) structure groups, the above procedure defines an element $\gamma(W)$ of $H^2(M;Z_2)$. It is usually called the second Stiefel-Whitney class $w_2(W)$. An important special case is that in which W is the tangent bundle of an oriented manifold M. In that case, $w_2(W)$ is usually called simply $w_2(M)$, the second Stiefel-Whitney class of M. It can be shown that M is a spin manifold (and spinor fields can be defined on M) if and only if $w_2(M)=0$. For an elucidation of these concepts see [6].

Now, suppose that G has structure group SU(n) or Spin(n) (Spin(n) is the simply connected double cover of SO(n)). In this case, the first non-zero homotopy groups is $\pi_3(SU(N)) = \pi_3(Spin(n)) = Z$. Hence the first obstruction to triviality of an SU(N) or Spin(n) bundle W is an element γ of $H^4(M;Z)$.

For G=SU(n), γ is known as the second Chern class of W, denoted $c_2(W)$. For G=Spin(n), γ seems to have no generally accepted name. However, 2γ is known as the first Pontryagin class $p_1(W)$. Of course, if γ were a real number, the map $\gamma \to 2\gamma = p_1$ would have an inverse $p_1/2=\gamma$. So γ and p_1 would contain the same information. In fact, γ and p_1 are not real numbers but elements of the abelian group $H^4(M;Z)$. In any abelian group, the operation $\gamma \to \gamma+\gamma=2\gamma$ makes sense, but the inverse operation $\gamma \to \gamma/2$ only makes sense under special conditions. Hence for a Spin(n) bundle, the basic invariant is γ rather than $p_1=2\gamma$. The reason that the Pontryagin class p_1 is defined is that for an SO(n) (as opposed to Spin(n)) bundle W, p_1 can be defined (by methods that go beyond what we have explained, since π_3 is not the first non-trivial homotopy group of SO(n)), but there is in general no way to make sense of $p_1/2$ for SO(n) bundles.

γ has no generally accepted name in the Spin(n) case (nor in the E_8 case we treated first). Because of the rather analogous role of Spin (10) and $E_8 \times E_8$ for heterotic superstrings, I will use the same name $\lambda(V)$ to denote the element of $H^4(M;Z)$ which is the first obstruction to triviality of an E_8 or Spin(n) bundle V.

One reason these ideas are significant for string theory is that as shown in [7,8], cancellation of various anomalies requires that the tangent bundle T of space time and the two E_8 bundles V_1 and V_2 should obey $\lambda(V_1) + \lambda(V_2) = \lambda(T)$. We will exploit this relation in the last section of these notes.

III. TOPOLOGICAL DEFECTS

In the last section we worked out a topological classification of E_8 gauge fields over a ten manifold. In this section we will apply what we have learned to study topological defects. Assuming compactification on $M^4 \times K$ for some K, we will consider in turn defects which in the four dimensional sense have space-time dimension zero, one, two, or three. These are the instantons, solitons, vortex lines, and domain walls respectively.

Apart from the topologial classification of E_8 bundles, the basic fact we will require is the Kunneth formula, which asserts that given two spaces N and K at least one of which has torsion-free cohomology, the cohomology of the product N×K is related to that of N and K by

$$H^q(N\times K) \cong \bigoplus_k H^{q-k}(N) \otimes H^k(K)$$

(The cohomology groups are taken with integer coefficients.) Taking N to be S^n, this reduces to

$$H^q(S^n\times K) \simeq H^0(S^n) \otimes H^q(K) \oplus H^n(S^n) \otimes H^{q-n}(K)$$

Since $H^0(S^n)$ and $H^n(S^n)$ are both Z, and for any abelian group G, $Z \otimes G \simeq G$, this actually can be written $H^q(S^n\times K) \simeq H^q(K) \oplus H^{q-n}(K)$, but this "simplified" form is sometimes confusing.

It should be noted that we will limit ourselves to the case in which it is only the topology of the gauge field, not the topology of space time, which is affected by the presence of a defect. Defects in the topology of space-time would be far more difficult to study.

III.1 Instantons

In this case instantons are gauge fields on $S^4\times K$ which reduce to the vacuum configuration on K. In view of the results of section II, they are classified by $H^4(S^4\times K) \simeq H^4(S^4) \otimes H^0(K) \oplus H^0(S^4) \otimes H^4(K) \simeq H^4(S^4) \oplus H^4(K)$. (We are using the fact that $H^0(K) \simeq Z$ for any connected space K.) The term involving $H^4(K)$ is irrelevant because it is uniquely determined by the vacuum configuration. The term involving $H^4(S^4)$ has nothing to do with K and picks out the instantons of four dimensional gauge theories. Therefore, the ten dimensional field theory has no gauge instantons that are not already known from four dimensional physics.

III.2 Particles

Now we turn our attention to particles that appear as topological defects. There are two cases to consider: magnetic monopoles, which have long range $E_8 \times E_8$ gauge fields; and "lumps," which do not (but which may have other long range fields, as we will discuss).

Monopoles were extensively treated in [2], and I will be brief. The vacuum state is described by an $E_8 \times E_8$ vector bundle V over K. Let S^2 be the "sphere at infinity" in uncompactified three space R^3. We can regard V as a bundle over $S^2 \times K$. (The fancy way to describe this is to say that we are "pulling back" V from K to $S^2 \times K$ via the projection $S^2 \times K \to K$.)

A monopole of given magnetic charge is specified by defining on $S^2 \times K$ a vector bundle W which differs from V by inclusion of a Dirac magnetic monopole gauge field. Now, in four dimensional unified gauge theories, we define the Dirac monopole on S^2 and try to extend it over R^3 without singularity; this can always be done. Here we must define the monopole on $S^2 \times K$ and try to extend it over $R^3 \times K$ without singularity. This <u>cannot</u> always be done; there is a restriction on the allowed values of magnetic charge.

If W, defined on $S^2 \times K$, extends to a vector bundle $\hat{W}$ defined on $R^3 \times K$, then--since $R^3 \times K$ is contractible onto K--$\hat{W}$ must be the pullback to $R^3 \times K$ of some $E_8 \times E_8$ vector bundle X defined on K. Then W, which is the restriction to $S^2 \times K$ of $\hat{W}$, must have the same property. We already know that W restricted to a copy of K is isomorphic to V. Hence, if W is the pullback of some X from K to $S^2 \times K$, X must be V itself. Thus, monopoles can exist only if W and V are isomorphic as $E_8 \times E_8$ bundles on $S^2 \times K$. Moreover, in that case monopoles certainly exist since V certainly extends from $S^2 \times K$ over $R^3 \times K$.

To determine whether W and V are isomorphic over $S^2 \times K$ is relatively straightforward. Their failure to be isomorphic would be measured by an element of $H^4(S^2 \times K) \simeq H^2(S^2) \otimes H^2(K) \oplus H^0(S^2) \otimes H^4(K)$. The second term is again irrelevant because it labels the vacuum, so the obstruction to existence of monopoles is in $H^2(S^2) \otimes H^2(K) \cong H^2(K)$. This obstruction was analyzed in detail in [2] for the case in which electromagnetism is part of a simple group (like O(10) or E_6) which is broken only by Wilson lines associated with $\pi_1(K)$. It was shown that if $\pi_1(K) \cong Z_n$, monopoles with a magnetic charge of n Dirac quanta always exist, but this is in general the lowest value.*

Now we turn to lumps, that is particles with no long-range $E_8 \times E_8$ gauge fields. In this case, at least for classifying the gauge fields, we can compactify three space from R^3 to S^3 and study gauge fields on $S^3 \times K$. The answer thus depends on $H^4(S^3 \times K) \simeq H^3(S^3) \otimes H^1(K) \oplus H^0(S^3) \otimes H^4(K)$. As usual, the relevant piece is $H^3(S^3) \otimes H^1(K) \simeq H^1(K)$. According to the universal coefficients theorem, there is no torsion part of $H^1(K)$; the free part is r copies of Z, r being $b_1(K)$, the first Betti number of K. In particular, $H^1(K)$ is zero if $\pi_1(K)$ is finite.

One may wonder if the lumps and monopoles considered here can be pair produced; otherwise they are scarcely of physical interest. Examples of topological defects which at least in the field theoretic limit cannot be pair produced were discussed in [9]. To address this question, consider first the lumps. The mere fact that lumps are classified by an element of the group $H^1(K)$ shows that each lump has an inverse, so that they can be pair produced. Turning now to monopoles, a monopole-antimonopole pair has no long range field, so it is a lump of some kind. Since lumps can be pair produced, the same is true for monopoles.

---

*The exposition here of the fact that the existence of monopoles is controlled by $H^2(K)$ made no assumption about K having SU(3) holonomy or any other property; in this respect it is an improvement on that in [2].

Incidentally, although the lumps we discussed have no long range $E_8 \times E_8$ gauge fields, some of them have long range fields of another variety. The point is that if $b_1(K) \neq 0$, then in the expansion of the two form $B_{\mu\nu}$ on $M^4 \times K$, one finds abelian gauge fields above and beyond those coming from $E_8 \times E_8$; some of the lumps discussed above behave as magnetic monopoles for these gauge fields [3].

III.3 Vortex Lines

Now we consider (figure 4) vortex lines, that is gauge fields that are localized in the x-y plane but extended in the z-direction. They were briefly discussed in [7]. Compactifying the x-y plane to a sphere S^2, vortex lines are classified by $H^4(S^2 \times K)$; as usual the relevant piece is $H^2(S^2) \otimes H^2(K) \cong H^2(K)$.

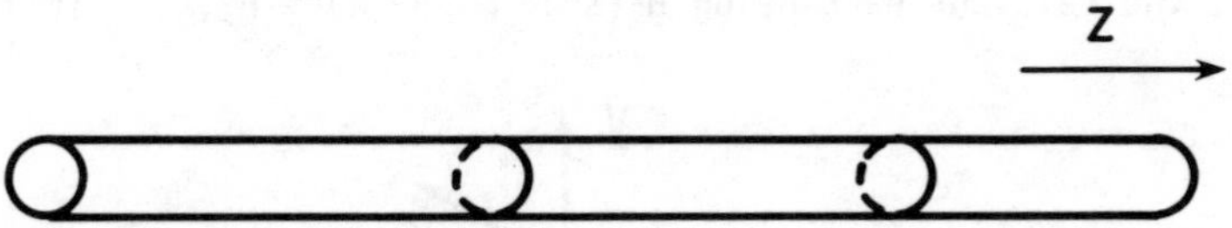

Fig. 4

A localized vortex line running in the z direction.

At first sight it might seem odd that the same $H^2(K)$ that appears in discussing monopoles also appears in discussing vortex lines. This, however, is no accident. A non-zero element of $H^2(K)$ may under suitable conditions prevent a monopole field on $S^2 \times K$ from extending over $R^3 \times K$; this must mean that passing through $S^2 \times K$ there is some kind of conserved flux--the stuff of a vortex line.

The above topological classification of vortex lines only describes vortex lines that are absolutely stable. We might have in addition vortex lines that are extremely long-lived but not rigorously stable. These can typically be produced by symmetry breaking well below the unification scale. They can be described and classified in four dimensional terms.

It may also be noted that the topological classification of vortex lines is not sensitive to energetics and so cannot predict whether a vortex line is truly localized in the x-y plane or is energetically capable of spreading. In practice, the latter situation is expected only if the "vortex line" can be made entirely from massless U(1) gauge bosons. Such "vortex lines" are just magnetic field lines and should be removed from the discussion. We will not pursue this further here.

III.4 Domain Walls

Now we come to a discussion of domain walls--by which we mean gauge fields that are localized in, say, the z direction but are independent of x and y (figure 5). Compactifying the z axis to a circle S^1, we are dealing with gauge fields on $S^1 \times K$. The classification is by $H^4(S^1 \times K)$, the relevant part being $H^1(S^1) \otimes H^3(K) \cong H^3(K)$. It is very

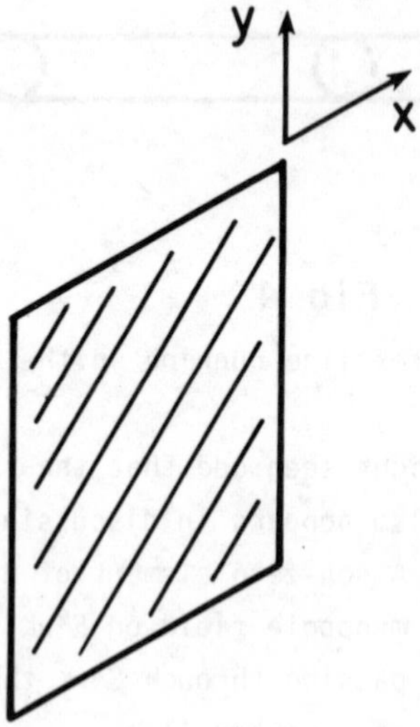

Fig. 5

A domain wall in the x-y plane. It poses a cosmological problem if the energy density is the same on each side.

difficult to make a phenomenologically viable model with $H^3(K)=0$. (For instance, on manifolds of SU(3) holonomy, $H^3(K)$ always contains at least $Z \oplus Z$, since the third Betti number $b_3(K)$ of such a manifold is at least two, receiving contributions from the holomorphic volume form and its dual.)

Domain walls are, of course, potentially a cosmological disaster. What can we do? We can hope that, for some reason, conditions in the early universe were such that the domain walls were not produced. Or we can try to solve the domain wall problem by means of microphysics. This is the possibility we will now discuss.

First we will discuss some relatively simple arguments that get rid of half of the domain walls. Let us recall the equation[10] $dH = tr_1 F^2 + tr_2 F^2 - trR^2$ of anomaly cancellation. Here F and R are the Yang-Mills and Riemann curvature two forms; tr_1 and tr_2 are traces in the first and second E_8's, and we will drop the trR^2 term in what follows, since it plays no role.

In $H^3(K)$ there is in general a "free" part (copies of Z) and a "torsion" part (finite groups such as Z_n). Let us first consider the free part. It is known that the free part of $H^3(K)$ is Z^n, n being the number of topologically non-trivial closed three surfaces in K. Let T be such a surface. It can be embedded in $M^4 \times K$ for any values of the coordinates x^λ of M^4. The domain walls of concern here are characterized by the two numbers $I_1 = \int_{S^1 \times T} tr_1 F^2$, $I_2 = \int_{S^1 T} tr_2 F^2$. Let $h(x^\lambda) = \int_T d\Sigma^{\alpha\beta\gamma} H_{\alpha\beta\gamma}$, x^λ being the four uncompactified coordinates and $\Sigma^{\alpha\beta\gamma}$ the volume element of T. In the field of a domain wall, it is reasonable to take h to be time independent and independent of x and y (parallel to the domain wall). Thus, h is a function of z only.

Let R be the z axis, which we now decompactify. The equation $dH = tr_1 F^2 + tr_2 F^2$, when integrated over $R \times T$, gives

$$\int_{-\infty}^{\infty} dz \frac{dh}{dz} = \int_{R\times T} (tr_1 F^2 + tr_2 F^2) = I_1 + I_2$$

or in other words $h(z=+\infty) - h(z=-\infty) = I_1 + I_2$. Thus, h jumps by a constant in passing through a domain wall with $I_1 + I_2 \neq 0$. Since the vacuum energy has an h^2 term, it is almost surely lower on one side of the domain wall than the other. This unequal energy causes the domain wall with $I_1 + I_2 \neq 0$ to pose no cosmological problem, since the region of higher energy will shrink and disappear.

Thus far we have been considering domain walls connected with the free part of $H^3(K)$. Now let us consider those connected with the torsion part. According to the universal coefficients theorem, the torsion part of $H^3(K)$ is the same as the torsion part of $H_2(K)$ (the second homology group of K with integer coefficients). In general, we can consider a domain wall with $E_8 \times E_8$ gauge fields that define two torsion elements α_1 and α_2 of $H^3(K)$ or $H_2(K)$.

I believe that domain walls with $\alpha_1 + \alpha_2 \neq 0$ are harmless for the same reason that those with $I_1 + I_2 \neq 0$ were harmless. I will not try to prove this but will only sketch the idea. The non-linear sigma model that describes the propagation of a first quantized string will like other field theories exhibit various topological effects. One of these is the appearance of instantons or θ angles, which are classified by $H_2(K)$. The free part of $H_2(K)$ determines ordinary, continuous θ angles, and the torsion part of $H_2(K)$ determines discrete θ angles. The vacuum energy will depend on the world sheet θ angles because of world sheet instantons; in the case of continuous θ angles this is discussed in [10]. I believe that a careful study of world sheet global anomalies [8] will show that the discrete θ angles jump when passing through a domain wall with $\alpha_1 + \alpha_2 \neq 0$. Hence the vacuum energy on one side of the domain wall will differ from that on the other side.

Assuming that this is true, it remains only to consider the domain walls with $I_1 + I_2 = 0$ and $\alpha_1 + \alpha_2 = 0$. How can we hope to eliminate them on microphysical grounds? The striking thing about them is that they have no long range fields. Also, naively adding together the contributions from the two E_8's, their total topological charges $I_1 + I_2$ and $\alpha_1 + \alpha_2$ are zero. Perhaps these "neutral" domain walls, although stable in field theory, are unstable in string theory. This would complete the solution of the domain wall problem.

A relatively pedestrian way for this to happen is that $E_8 \times E_8$ may be unified in a larger group G in such a way that only $I_1 + I_2$ and $\alpha_1 + \alpha_2$ (not the separate I_j and α_k) are true topological quantum numbers. In principle, G might simply be a finite dimensional group such as SO(496). (The plethora of massive states of the heterotic superstring[12] could well accommodate massive gauge bosons and Higgs bosons of spontaneously broken SO(496)!) However, this idea does not seem natural. Far more plausible is the idea that G is some "stringy" infinite dimensional group.

In the heterotic theory, $E_8 \times E_8$ is really a subgroup of $E_9 \times E_9$ (E_9 is the mathematical name for E_8 current algebra on a circle). We probably should think of unifying $E_9 \times E_9$, not $E_8 \times E_8$. What is a natural unified group that contains $E_9 \times E_9$? An interesting candidate was suggested by Goddard and Olive several years ago in the same paper in which they conjectured the use of E_9 in string theory [4]. They defined a hyperbolic algebra of rank 18, which I will call G, which contained both $E_9 \times E_9$ and $\widehat{SO(32)}$ ($\widehat{SO(32)}$ being the one dimensional current algebra of SO(32)).

One idea about the domain wall problem would thus be that the string theory really has G symmetry which is spontaneously broken to $E_9 \times E_9$ and thence to $E_8 \times E_8$. In G, the domain walls with $I_1 + I_2 = \alpha_1 + \alpha_2 = 0$ would be unstable.

This idea leads to a further speculation. Since G contains both $E_9 \times E_9$ and $\widehat{SO(32)}$, it might be that the same G invariant theory could break down to either $E_9 \times E_9$ or $\widehat{SO(32)}$--in which case the two heterotic superstring theories would be two vacuum states of the same theory.

The use of hyperbolic algebras such as G might be interesting for another reason. In some sense there must be symmetries associated with all the massive states of string theory. The required symmetry algebras would be enormous. Goddard and Olive showed in [4] that the roots of the hyperbolic algebras are more or less in correspondence with the states of the string. This certainly makes them intriguing candidates as part of the presently unknown symmetry structure of string theory.

Perhaps I should pause here for a word about what these hyperbolic or Lorentzian algebras are. By adding a point to the Dynkin diagram of a finite dimensional Lie algebra in a suitable way, one gets the Dynkin diagram of an affine Lie algebra--current algebra in one dimension. Historically, the theory of the single string had a big influence in the development of an understanding of the affine Lie algebras. Adding another point to the Dynkin diagram one gets the hyperbolic or Lorentzian Kac-Moody algebras which are to this day as mysterious as, say, the fundamentals of string theory. Among these algebras are the Lorentzian algebras in 10, 18, and 26 dimensions discussed in [4]; the 10 dimensional one is often called E_{10}. To me it seems logical that if the single string is related to affine Lie algebras, the string field theory should be related to--and perhaps should play a role in understanding--some larger algebras such as the hyperbolic algebras. Apart from very general facts surveyed in [13], and the results I have already mentioned, among the few concrete results known about hyperbolic algebras is Kac's proof that the Weyl character formula applies to the highest weight representations of these algebras (and indeed of any Kac-Moody algebras).

No speculation about string theory and hyperbolic algebras would be complete without mentioning one more fact. The highest weight representation of the hyperbolic algebras is more or less a Fock space of strings. This fact was exploited in [14]. The point is that E_{10} has a subgroup $E_9 \times U(1)$. A highest weight representation of E_{10} can be built in a "Fock space" of E_9 representations, with U(1) being the operator that counts the number of strings. But unfortunately these strings seem to obey neither bose nor fermi statistics, so that the utility of the approach is in doubt.

A concrete understanding of hyperbolic algebras would be a breakthrough for mathematics and perhaps for physics as well. See [15] for one of the few concrete approaches that have been suggested.

IV. THE VACUUM STATE AND GLOBAL ANOMALIES

We now turn our attention to the vacuum state. Thus, we wish to discuss the compact six manifold K and the $E_8 \times E_8$ bundle $V = V_1 \oplus V_2$ (V_1 and V_2 being the two E_8 bundles).

In section III we learned that an E_8 bundle X on a manifold M (such as $M^4 \times K$) of dimension no more than fifteen is uniquely characterized by an invariant $\lambda(X)$ which can be an arbitrary element of $H^4(M;Z)$. The tangent bundle T of the manifold M--assuming it is a spin manifold --has a similar invariant $\lambda(T) \varepsilon H^4(M)$ (though this does not characterize T completely). The elements $\lambda(V_1)$, $\lambda(V_2)$ and $\lambda(T)$ must for consistency be related by

$$\lambda(V_1) + \lambda(V_2) = \lambda(T).$$

At the level of differential forms this equation follows [7] from the equation $dH = tr_1F^2 + tr_2F^2 - trR^2$ of anomaly cancellation. The torsion part is required to avoid global world sheet anomalies [8].

Rewriting the above equation in the form $\lambda(V_1) = \lambda(T) - \lambda(V_2)$, we see that after choosing a manifold K, V_1 is determined if we are given V_2. For instance, it is amusing to consider the case $V_2 = 0$. By our general theory, there must be a unique E_8 bundle V_1 with $\lambda(V_1) = \lambda(T)$. What is it? The structure group of the tangent bundle of a manifold M of dimension n is SO(n). For $n < 16$, we can embed $SO(n) \subset SO(16) \subset E_8$ in the minimal way (so that the vector of 16 is a vector plus singlets of SO(n)). Via this embedding, we can construct from T an E_8 bundle V_1 which manifestly obeys $\lambda(V_1) = \lambda(T)$, and therefore by our general theory is the unique way of obeying this equation. What I have just described is the idea [17,18] of "embedding the spin connection in the gauge group," which is interesting for phenomenological reasons. We have seen that if one of the two E_8 bundles is known for some reason to be trivial, then at least topologically the second must be constructed by embedding the spin connection in the gauge group.*

---

*Here we are discussing global anomalies in space-time. The analogous question of global anomalies on a particle world line or a string world sheet has been treated in [16,8].

Now we turn our attention to global anomalies, which will occupy our attention for the remainder of these notes. In N=1 supergravity theory with $E_8 \times E_8$ gauge group, ordinary perturbative anomalies are known to cancel if certain counterterms are included in the effective action [10]. This discovery motivated the discovery of the heterotic superstring theory [12]. It means that the one loop effective action of the theory is invariant under diffeomorphisms and gauge transformations that can be reached continuously from the identity. It remains to consider the case of global anomalies [17]. Here we consider the group G of diffeomorphisms and gauge transformations of the manifold M and $E_8 \times E_8$ bundle $V_1 \oplus V_2$. In general G is not connected; its components form a discrete group G_o. We must ask whether the effective action is invariant under diffeomorphisms or gauge transformations that are not continuously connected to the identity or lie in other words in components of G corresponding to non-trivial elements of G_o.

Let M be a ten manifold with $E_8 \times E_8$ bundle $V_1 \oplus V_2$, and let f be a diffeomorphism and/or gauge transformation that is not continuously connected to the identity. Is the effective action of N=1 supergravity with gauge group $E_8 \times E_8$ invariant under f? In a lengthy analysis in [9], a general formula was given for the change of the effective action under f. The formula will be reviewed below. Using special properties of S^{10}, I evaluated this formula for the case $M=S^{10}$ and showed that the anomaly was zero in that case, so that the theory is consistent if formulated on S^{10}. It was not at all clear how the global anomaly formula could be evaluated for ten manifolds other than S^{10}. The first progress in this matter was made by S. Stoltz. He showed [19] that the global anomaly is always zero if the $E_8 \times E_8$ bundle is trivial. The proof used surgery theory. I then noticed that the treatment could be simplified and extended by using the classification of E_8 bundles explained in Section II. As I will sketch shortly, this makes it possible to reduce the whole question of global anomalies to a question in cobordism theory. The cobordism problem was then settled by R. Stong, whose results are described in an appendix to these notes. Stong showed that the relevant cobordism group is zero. (It was independently shown by T. Bahri that the group in question would have to be zero or Z_2.) As I will now sketch, the effect of all this is to show that the global space-time anomaly vanishes for every vacuum configuration that obeys the one condition $\lambda(V_1) + \lambda(V_2) = \lambda(T)$ that is known to be needed.

Given a diffeomorphism plus gauge transformation f, the first step toward understanding global anomalies is to construct the mapping cylinder $(M\times S^1)_f$. It is defined starting with the product $M\times I$ (I is the unit interval) by identifying (x,0) with (f(x),1). This gives a smooth manifold $(M\times S^1)_f$, which is a compact spin manifold if M is (as we assume). The action of f on the $E_8\times E_8$ bundle $V_1 \oplus V_2$ enables one to identify the fiber of $V_1 \oplus V_2$ at (f(x),1) with the fiber at (x,0), and so one can extend $V_1 \oplus V_2$ to an $E_8\times E_8$ bundle (which we will call $V_1 \oplus V_2$) over $(M\times S^1)_f$. In [9], it was shown that the global anomaly in the effective action of the supergravity theory on M is equal to a certain combination of eta invariants for wave operators on $(M\times S^1)_f$. This gave a formula for the global anomaly that is concrete but fairly untractable. However, if $(M\times S^1)_f$ is the boundary of a spin manifold B over which V_1 and V_2 can be extended, things simplify since the eta invariants can be expressed by the Atiyah-Patodi-Singer theorem as curvature integrals on B.

Before exploiting this fact, let us first discuss the question of whether a suitable B exists. This is a question in cobordism theory. Cobordism is a notion of equivalence between manifolds that is much coarser than topological classification. In (oriented) cobordism theory, two n dimensional compact oriented manifolds M and N are considered equivalent if there is an oriented n+1 dimensional manifold X whose boundary is M - N (that is X has two boundary components, isomorphic to M and N respectively; they enter with opposite orientation). To define the cobordism group, one considers formal linear combinations $\Sigma\, m_i M_i$ of oriented compact manifolds M_i with integer coefficients m_i, subject to the equivalence relation that $\Sigma\, n_i N_i$ is considered zero if it is the boundary of an oriented manifold. In particular, a manifold is considered trivial if it is a boundary. Subject to this equivalence relation, the formal linear combinations make up a group, the oriented cobordism group Ω_n.

Spin cobordism is defined similarly, but requiring that all manifolds considered be spin manifolds. Thus, a linear combination $\Sigma\, m_i\, M_i$ of n dimensional spin manifolds M_i is considered zero if it bounds a spin manifold X whose spin structure reduces to that of the M_i on each boundary component. The equivalence classes form the spin cobordism group Ω_n^{spin}.

It is an old result of Milnor's [19] that Ω_{11}^{spin} is zero, or in other words, that every eleven dimensional spin manifold is the boundary of a spin manifold. Thus, our mapping cylinder $(M\times S^1)_f$ certainly bounds a spin manifold B. However, for our applications we need to know whether B can be chosen so that the $E_8\times E_8$ bundle $V_1 \oplus V_2$ extends form $(M\times S^1)_f$ to an $E_8\times E_8$ bundle $W_1 \oplus W_2$ over B.

Considering first V_1, it is classified, in view of our results in Section II, in terms of the characteristic class $\lambda(V_1) \in H^4(M\times S^1)_f$. If there is an element $\alpha \in H^4(B)$ whose restriction to $(M\times S^1)_f$ is $\lambda(V_1)$, then by constructing (as in Section II) an E_8 bundle W_1 over B with $\lambda(W_1) = \alpha$, we get the desired E_8 bundle over B whose restriction to $(M\times S^1)_f$ is V_1. Conversely, if $\lambda(V_1)$ does not extend to a cohomology class $\alpha \in H^4(B)$, an extension W_1 of V_1 from $(M\times S^1)_f$ to B cannot possibly exist.

Thus, V_1 extends over B if and only if the cohomology class $\lambda(V_1)$ extends to a cohomology class in $H^4(B)$. The cobordism theory relevant for our purposes is hence a cobordism theory in which the basic object is a pair (M,β) consisting of an n dimensional spin manifold M and an element β of $H^4(M)$. The pair (M,β) is considered to vanish if M bounds a spin manifold X over which β can be extended. The cobordism group is called $\Omega_n^{spin}\,(K(Z,4))$. (Here $K(Z,4)$ is the Eilenberg-MacLane space discussed in the introduction. The name $\Omega_n^{spin}\,(K(Z,4))$ reflects the fact that elements β of $H^4(M)$ can be shown to be in one to one correspondence with maps $M\to K(Z,4)$.) In the appendix Strong shows that $\Omega_{11}^{spin}\,(K(Z,4))$ is zero. This means that B can always be chosen so that V_1 extends to an E_8 bundle W_1 over B.

Can V_2 be so extended at the same time? To answer this question, we use the fact that we are only interested in a situation in which

$\lambda(V_1) + \lambda(V_2) = \lambda(T)$. Here T is the tangent bundle of $(M\times S^1)_f$. (The requirement that $\lambda(V_1) + \lambda(V_2) = \lambda(T)$ follows from a two step reasoning. On M it must hold to avoid perturbative and world sheet anomalies [7,8]. On $(M\times S^1)_f$ it must hold because otherwise the transformation f should not be considered [8].) Let ε be the normal bundle of $(M\times S^1)_f$ in B; it is a trivial real line bundle. Then as ε is trivial, $\lambda(T) = \lambda(T \oplus \varepsilon)$. Here $T \oplus \varepsilon$ is the restriction to $(M\times S^1)_f$ of the tangent bundle $\hat{T}$ of B. So $\lambda(T) = \lambda(T \oplus \varepsilon)$ certainly extends to a cohomology class in $H^4(B)$, namely $\lambda(\hat{T})$. Also, Stong's results imply that B_0 can be chosen so that V_1 extends over B, and hence $\lambda(V_1)$ extends to $\lambda(W_1) \in H_4(B)$. Then $\beta = \lambda(\hat{T}) - \lambda(W_1)$ is an element of $H^4(B)$. Constructing as in Section II an E_8 bundle W_2 over B with $\lambda(W_2) = \beta$, we get the desired extension of V_2 from $(M\times S^1)_f$ to B. This argument shows not only that V_1 and V_2 can be assumed to both extend over B, but also that the extensions can be assumed to obey $\lambda(W_1) + \lambda(W_2) = \lambda(\hat{T})$. This fact will be useful shortly.

We can now readily show that the space time global anomaly vanishes for arbitrary choices of M, V_1 and V_2. In fact, given that $(M\times S^1)_f$ bounds a spin manifold B over which V_1 and V_2 can be extended, it was shown in [9] that the change in the effective action S under f is

$$\Delta S = 2\pi i \left[\int_B (\mathrm{tr}_1 F^2 + \mathrm{tr}_2 F^2 - \mathrm{tr} R^2) \wedge K - \int_{(M\times S^1)_f} H \wedge K \right] \quad (3)$$

Here K is a certain closed eight form built from the curvature and gauge field strength tensors R and F. And H is a three form which on $(M\times S^1)_f$ obeys $dH = \mathrm{tr}_1 F^2 + \mathrm{tr}_2 F^2 - \mathrm{tr} R^2$. Such an H exists because on $(M\times S^1)_f$ we have $\lambda(V_1) + \lambda(V_2) - \lambda(T) = 0$.*

To prove that the global anomaly is zero is now very simple. We have found that the various choices can be made so that on B, $\lambda(W_1) +$

---

*If K is non-zero in $H^8((M\times S^1)_f)$, (3) depends on the topological class of the solution H of $dH = \mathrm{tr}_1 F^2 - \mathrm{tr}_2 F^2 + \mathrm{tr} R^2$. This leads as in [9] to the quantization of torsion, a matter that is well understood from another viewpoint [3].

$\lambda(W_2) - \lambda(T) = 0$. This means that H can be defined to obey $dH = tr_1F^2 + tr_2F - trR^2$ not just on $(M\times S^1)_f$ but on B. With this choice (3) reduces to

$$\Delta S = 2\pi i \left[\int_B dH \wedge K - \int_{(M\times S^1)_f} H \wedge K\right]$$

$$= 2\pi i \left[\int_B d(H \wedge K) - \int_{(M\times S^1)_f} H \wedge K\right] = 0 \qquad (4)$$

Here we have used the fact that $dK = 0$, so $dH \wedge K = d(H \wedge K)$, and Stokes' law $\int_B d(H \wedge K) = \int_{(M\times S^1)_f} H \wedge K$. This (together with results in the appendix) concludes the proof that space-time global anomalies vanish for vacuum configurations that obey the one known condition $\lambda(V_1) + \lambda(V_2) = \lambda(T)$.

References

1. R. Bott and L. Tu, Differential Forms and Algebraic Topology (Springer-Verlag, 1982).
2. X.-G. Wen and E. Witten, "Electric and Magnetic Charges in Superstring Models," to appear in Nucl. Phys. B.
3. R. Rohm and E. Witten, to appear in Annals of Physics.
4. P. Goddard and D. Olive, in Vertex Operators in Mathematics and Physics, ed. J. Lepowsky et al. Springer-Verlag, 1985).
5. E.H. Spanier, Algebraic Topology (McGraw Hill, 1966), Chapter 3.
6. S.W. Hawking and C.N. Pope, Phys. Lett. 73B (1978) 42.
7. E. Witten, Phys. Lett. 149B (1984) 351.
8. E. Witten, in Anomalies, Geometry, and Topology, ed. A. White (World Scientific, 1985).
9. E. Witten, Comm. Math. Phys. 100 (1985) 197.
10. M.B. Green and J.H. Schwarz, Phys. Lett. 149B (1984) 117.
11. X.-G. Wen and E. Witten, Princeton preprint (1985), to appear in Phys. Lett.

12. D.J. Gross, J.A. Harvey, E. Martinec, and R. Rohm, Phys. Rev. Lett. 52 (1985) 502.
13. V. Kac, Infinite Dimensional Lie Algebras (Birkhauser, 1982).
14. A. Feingold and I. Frenkel, to appear.
15. B. Julia, in Applications of Group Theory in Physics and Mathematical Physics, ed. P. Sally et al. (American Mathematical Society, Providence, 1985), and references therein.
16. D. Friedan and P. Windey, Nucl. Phys. B235 [FS11] (1984) 395.
17. E. Witten, in Shelter Island II: Proceedings of the 1983 Shelter Island Conference on Quantum Field Theory and the Fundamental Problems of Physics, ed. N. Khuri et al. (MIT Press, 1985).
18. P. Candelas, G. Horowitz, A. Strominger, and E. Witten, Nucl. Phys. B258 (1985) 46.
19. S. Stoltz, unpublished.

Appendix: Calculation of $\Omega_{11}^{Spin}(K(Z,4))$

R. E. Stong

The objective of this note is to calculate the 11-dimensional Spin bordism group of the Eilenberg-MacLane space K(Z,4), and, in particular, to show that this group is zero. The techniques involved are largely standard, using the structure of the Spin bordism ring as described by Anderson, Brown, and Peterson [ABP], but the calculation is unpleasant.

Item 1. *$\Omega_{11}^{Spin}(K(Z,4))\otimes Q = 0$; i.e., the group is finite.*

Proof: $\Omega_*^{Spin}\otimes Q \cong Q[x_{4i}]$ is a polynomial ring on 4i-dimensional generators, and $H^*(K(Z,4);Q) = Q[i_4]$ is a polynomial ring on a 4-dimensional class. Thus $\Omega_*^{Spin}(K(Z,4))\otimes Q \cong \Omega_*^{Spin}\otimes H_*(K(Z,4);Q)$ is nonzero only in dimensions divisible by four. □

Item 2. *In low dimensions, Ω_*^{Spin} is given by*

n	0	1	2	3	4	5	6	7	8	9	10	11	12
Ω_n^{Spin}	Z	Z_2	Z_2	0	Z	0	0	0	2Z	$2Z_2$	$3Z_2$	0	3Z

Proof: This follows by reading [ABP]. Ω_*^{Spin} has no odd torsion and at the prime 2, MSpin is a product $\Pi \underset{\sim}{BO}(4n(I),\ldots) \times \Pi \underset{\sim}{BO}(4n(I)-2,\ldots) \times \Pi \underset{\sim}{K}(Z_2,n_i)$, and in this range one has $\underset{\sim}{BO} \times \underset{\sim}{BO}(8,\ldots) \times \underset{\sim}{BO}(10,\ldots)$. □

Item 3. *$\Omega_*^{Spin}(HP^\infty)$ is a free Ω_*^{Spin}-module on the classes given by the inclusions $HP^n \to HP^\infty$ of quaternionic projective spaces.*

Proof: The inclusions $HP^n \to HP^\infty$ realize homology generators by Spin bordism elements, and the Atiyah-Hirzebruch spectral sequence collapses to give a free module. □

<u>Item 4. The map $HP^\infty \to K(Z,4)$ by realizing the generator of $H^4(HP^\infty;Z) \cong Z$ gives an exact sequence</u>

$$\begin{array}{ccccccc} \tilde\Omega_{11}^{Spin}(HP^\infty) & \to & \tilde\Omega_{11}^{Spin}(K(Z,4)) & \to & \Omega_{11}^{Spin}(K(Z,4),HP^\infty) & \to & \tilde\Omega_{10}^{Spin}(HP^\infty). \\ \| & & \| & & & & \| \\ 0 & & \Omega_{11}^{Spin}(K(Z,4)) & & & & Z_2 \end{array}$$

One may now proceed to analyze the pair $(K(Z,4),HP^\infty)$ by studying each prime. One has $H^*(HP^\infty:Z_p) = Z_p[\sigma]$, where σ is 4-dimensional, and <u>for an odd prime</u> p, $H^*(K(Z,4);Z_p)$ is the free associative commutative algebra over Z_p on classes $St^I i_4$, where St^I is an admissible Steenrod operation. The generators occurring are i_4, $\mathcal{P}^1 i_4$ of dimension $4 + 2(p-1)$, $\beta\mathcal{P}^1 i_4$ of dimension $4 + 2(p-1) + 1$, and terms of higher dimension, with the map $HP^\infty \to K(Z,4)$ sending i_4 to σ. Thus, the first class in $H^*(K(Z,4),HP^\infty;Z_p)$ occurs in dimension $4 + 2(p-1)$. Hence one has

<u>Item 5. $\Omega_{11}^{Spin}(K(Z,4),HP^\infty)$ has no p-torsion if $4 + 2(p-1) > 11$; i.e., $2(p-1) > 11 - 4 = 7$ or $p > 3$.</u>

For the prime 3, the generators of $H^*(K(Z,4);Z_3)$ are

i_4,	$\mathcal{P}^1 i_4$,	$\beta\mathcal{P}^1 i_4$	$\mathcal{P}^3\mathcal{P}^1 i_4$
dim 4	dim 8	dim 9	dim 20

and terms of still higher dimension. The map $HP^\infty \to K(Z,4)$ sends i_4 to σ and $\mathcal{P}^1 i_4$ to $2\sigma^2$ (mapping $\mathbb{C}P^\infty \to HP^\infty$, σ goes to α^2, $\alpha \in H^2(\mathbb{C}P^\infty;Z_3)$, and $\mathcal{P}^1\sigma$ goes to $\mathcal{P}^1\alpha^2 = 2\alpha^4$) so that a basis for $H^*(K(Z,4),HP^\infty;Z_3)$ is given by

$\mathcal{P}^1 i_4 + i_4^2$,	$\beta\mathcal{P}^1 i_4$,	$i_4(\mathcal{P}^1 i_4 + i_4^2)$,	$i_4\beta\mathcal{P}^1 i_4$
dim 8	dim 9	dim 12	dim 13

and terms of higher degree. Because $\beta(\mathcal{P}^1 i_4 + i_4^2) = \beta\mathcal{P}^1 i_4$ and $\beta\{i_4(\mathcal{P}^1 i_4 + i_4^2)\} = i_4\beta\mathcal{P}^1 i_4$, the integral homology at the prime 3 consists of Z_3 in dimensions 8 and 12, together with terms of higher dimension. For the 3-primary component of the Atiyah-Hirzebruch spectral sequence one then has

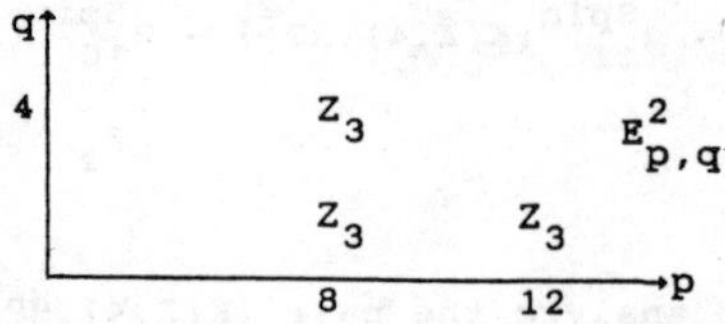

$$E^2_{p,q} = H_p(K(Z,4),HP^\infty;\Omega^{Spin}_q)_{(3)}$$

and all groups with $p + q = 11$ are zero. Thus one has

<u>Item 6</u>. <u>$\Omega^{Spin}_{11}(K(Z,4),HP^\infty)$ has no odd torsion; i.e., is a 2-group. Thus $\Omega^{Spin}_{11}(K(Z,4))$ is also a 2-group</u>.

<u>Note</u>: In dimensions less than 12, the only questionable behavior, except at the prime 2, is

$$\begin{array}{ccccccc} \tilde\Omega^{Spin}_8(HP^\infty) & \to & \tilde\Omega^{Spin}_8(K(Z,4)) & \to & \Omega^{Spin}_8(K(Z,4),HP^\infty) & \to & \tilde\Omega^{Spin}_7(HP^\infty). \\ \| & & & & \| & & \| \\ Z+Z & & & & Z_3+2 \text{ torsion} & & 0 \end{array}$$

The homomorphism $\Omega^{Spin}_8(K(Z,4)) \to Z_3$ sends M^{11} and $z \in H^*(M;Z)$ to $(\mathcal{P}^1 z + z^2)[M]$. According to Wu, $\mathcal{P}^1[M]$ is the mod 3 reduction of $p_1 z[M]$ where p_1 is the first Pontrjagin class. Thus the homomorphism lifts to Z, and so $\tilde\Omega^{Spin}_8(K(Z,4)) = Z + Z +$ 2 torsion.

In order to analyze the 2-primary structure, it is convenient to consider the map $f : BSpin \to K(Z,4)$ realizing the low dimensional homotopy. This provides a better approximation than does HP^∞.

One knows that $H^*(K(Z,4);Z_2)$ is the Z_2 polynomial ring on classes $Sq^I i_4 = Sq^{i_1} \ldots Sq^{i_r} i_4$ of dimension $4 + i_1 + \ldots + i_r$, where

$i_r > 1$, I is admissible ($i_{j-1} \geq 2i_j$) and the excess $(i_1-2i_2) + (i_2-2i_3) + \ldots + (i_{r-1}-2i_r) + i_r$ is less than 4. Also $H^*(BSpin;Z_2)$ is the quotient of $H^*(BSO;Z_2) = Z_2[w_i | i>1]$ by the ideal generated over the Steenrod algebra by w_2. One then has the homomorphism given by

dim 4	$f^*(i_4) = w_4$
dim 6	$f^*(Sq^2 i_4) = w_6$
dim 7	$f^*(Sq^3 i_4) = w_7$
dim 8	$f^*(i_4^2) = w_4^2$
dim 10	$f^*(i_4 Sq^2 i_4) = w_4 w_6$
	$f^*(Sq^4 Sq^2 i_4) = w_{10} + w_4 w_6$
dim 11	$f^*(i_4 Sq^3 i_4) = w_4 w_7$
	$f^*(Sq^5 Sq^2 i_4) = w_{11} + w_4 w_7$
dim 12	$f^*(i_4{}^3) = w_4^3$
	$f^*((Sq^2 i_4)^2) = w_6^2$

and the map f^* is a monomorphism in this range (actually it is monic in all dimensions) with cokernel given by

$$\underset{\text{dim } 8}{w_8}, \qquad \underbrace{w_{12}, \quad w_4 w_8}_{\text{dim } 12},$$

and higher dimensional terms. One has an exact sequence

$$H^*(K(Z,4), BSpin;Z_2) \xleftarrow{\ \delta\ } \cdots \rightarrow H^*(K(Z,4);Z_2) \rightarrow H^*(BSpin;Z_2)$$

which determines $H^*(K(Z,4), BSpin;Z_2)$ in the range of low dimensions. Specifically $H^9 = Z_2$ on δw_8 and $H^{13} = Z_2 + Z_2$ on δw_{12} and $\delta w_4 w_8$. At the prime 2 the integral homology is then $H_9(K(Z,4), BSpin;Z) = Z$ and $H_{13}(K(Z,4), BSpin;Z) = Z + Z_2$ ($Sq^1 \delta w_{12} = \delta w_{13}$ gives the Z_2). Considering the Atiyah-Hirzebruch spectral sequence with $E^2_{p,q} = H_p(K(Z,4), BSpin; \Omega^{Spin}_q)$ and converging to $\Omega^{Spin}_{p+q}(K(Z,4), BSpin)$, one then has

$$\Omega^{Spin}_j(K(Z,4), BSpin) = \begin{cases} Z & j=9 \\ Z_2 & j=10,\ 11 \\ 0 & j<9 \text{ or } j=12 \end{cases}$$

at the prime 2.

In order to determine $\tilde{\Omega}^{Spin}_*(BSpin)$, one recalls from [ABP] that the Thom space MSpin has

$$\tilde{H}^*(MSpin;Z_2) = (\mathcal{A}/\mathcal{A}Sq^1 + \mathcal{A}Sq^2)\ U + (\mathcal{A}/\mathcal{A}Sq^1 + \mathcal{A}Sq^2)\ w_4'^2\ U$$

plus higher terms, where $\mathcal{A}$ is the Steenrod algebra, U is the Thom class, and w_4' is the Stiefel-Whitney class from $H^*(BSpin;Z_2)$. This corresponds to the decomposition $MSpin = \tilde{BO} \times \tilde{BO}(8,\ldots) \times \ldots$ in low dimensions. In order to determine $\tilde{H}^*(MSpin \wedge BSpin;Z_2) = \tilde{H}^*(MSpin;Z_2) \otimes_{Z_2} \tilde{H}^*(BSpin;Z_2)$ as an $\mathcal{A}$ module, it then suffices to consider $\tilde{H}^*(BSpin;Z_2)$ as a module over $\mathcal{A}_1$, the subalgebra generated by Sq^1 and Sq^2. That is

1	Sq^1	Sq^2	Sq^3, Sq^2Sq^1	Sq^3Sq^1
dim 0	dim 1	dim 2	dim 3	dim 4
		$Sq^5+Sq^4Sq^1$	Sq^5Sq^1	
		dim 5	dim 6	

and one has

w_4 , $Sq^2w_4 = w_6$, $Sq^3w_4 = w_7$ $\quad (Sq^1w_4 = 0 = Sq^5w_4)$

w_4^2 $\quad (Sq^1w_4^2 = 0 = Sq^2w_4^2)$

w_8 $\quad Sq^2w_8 = w_{10}$ $\quad Sq^3w_8 = w_{11}$ $\quad (Sq^1w_8 = 0)$

w_4w_6 $\quad Sq^1w_4w_6 = w_4w_7$ $\quad Sq^2w_4w_6 = w_6^2$

$Sq^3w_4w_6 = 0$ $\quad Sq^2Sq^1w_4w_6 = w_6w_7.$

Thus

$$\tilde{H}^*(\mathrm{MSpin}\wedge\mathrm{BSpin};Z_2) \cong (\mathcal{A}/\mathcal{A}Sq^1+\mathcal{A}Sq^5)\, w_4\cdot U + (\mathcal{A}/\mathcal{A}Sq^1+\mathcal{A}Sq^2)\, w_4^2\cdot U + (\mathcal{A}/\mathcal{A}Sq^1)\, w_8\, U + (\mathcal{A}/\mathcal{A}Sq^3)\, w_4 w_6\, U$$

plus terms of degree at least 12.

Now the KO-theory Thom class gives a map MSpin → $\underset{\sim}{BO}$ realizing the Thom class U in Z_2-cohomology and the identity map (or first KO-theory Pontragin class) gives a map π_1 : BSpin → BO(4,...,∞). Squaring this gives π_1^2 : BSpin → BO(8,...,∞) (the filtration increases), and the products give

$$\pi_1 U \times \pi_1^2 U : \mathrm{MSpin}\wedge\mathrm{BSpin} \to \underset{\sim}{BO}(4,\ldots,\infty) \times \underset{\sim}{BO}(8,\ldots,\infty)$$

realizing the summands $(\mathcal{A}/\mathcal{A}Sq^1+\mathcal{A}Sq^5)\, U + (\mathcal{A}/\mathcal{A}Sq^1+\mathcal{A}Sq^2)\, w_4^2\cdot U$ in mod 2 cohomology. The class $w_4 w_6 U$ is realized by a map MSpin∧BSpin ⟶ $\underset{\sim}{K}(Z_2,10)$ and $w_8 U$ by a map MSpin∧BSpin ⟶ $\underset{\sim}{K}(Z,8)$ (U is integral and w_8 is integral; since $H_*(\mathrm{BSpin};Z)$ has all torsion of order 2 any class annihilated by Sq^1 is reduced integral.) The product

$$\pi_1 U\times\pi_1^2 U\times w_8 U\times w_4 w_6 U:\mathrm{MSpin}\wedge\mathrm{BSpin}\longrightarrow \underset{\sim}{BO}(4,\ldots)\times\underset{\sim}{BO}(8,\ldots)\times\underset{\sim}{K}(Z,8)\times\underset{\sim}{K}(Z_2,10)$$

induces a homomorphism in mod 2 cohomology which is an isomorphism in dimensions less than 12 and monic in dimension 12, and thus is a 2-primary isomorphism through dimension 11. Thus one has

$$\tilde{\Omega}_j^{\mathrm{Spin}}(\mathrm{BSpin}) = \begin{cases} Z & j=4 \\ Z+Z+Z & j=8 \\ Z_2+Z_2 & j=9 \\ Z_2+Z_2+Z_2 & j=10 \\ 0 & \text{all other } j\le 11 \end{cases}$$

ignoring odd torsion.

Of course, one then has an exact sequence

$\cdots \to \Omega_{j+1}^{Spin}(K(Z,4),BSpin) \xrightarrow{\partial} \tilde{\Omega}_j^{Spin}(BSpin) \to \tilde{\Omega}_j^{Spin}(K(Z,4)) \to \Omega_j^{Spin}(K(Z,4),BSpin) \to \cdots$ and in order to determine $\tilde{\Omega}_j^{Spin}(K(Z,4))$, one must analyze the homomorphism ∂. In order to do so, one considers $BO(8,\ldots,\infty) \xrightarrow{i} BSpin \xrightarrow{f} K(Z,4)$, where i is the inclusion of the fiber, the 7-fold connective cover. For any fibration $F \to E \to B$, one always has a map $\Sigma F \to (B,E)$ and a commutative diagram in any homology theory

$$\begin{array}{ccc} \tilde{h}_{j+1}(\Sigma F) & \longrightarrow & h_{j+1}(B,E) \\ \text{suspend} \uparrow \cong & & \downarrow \partial \\ \tilde{h}_j(F) & \longrightarrow & \tilde{h}_j(E). \end{array}$$

Now $\tilde{H}^*(BO(8,\ldots,\infty);Z_2)$ has a basis given by w_8 and w_{12} in dimensions 8 and 12 and one sees quite readily that

$$\tilde{\Omega}_j^{Spin}(BO(8,\ldots,\infty)) = \begin{cases} Z & j=8 \\ Z_2 & j=9,\ 10 \\ 0^2 & \text{all other } j \le 11. \end{cases}$$

For the map $BO(8,\ldots,\infty) \xrightarrow{i} BSpin \xrightarrow{\pi_1} BO(4,\ldots,\infty)$, one observes that the filtration of $i^*\pi_1$ in KO-theory increases. Thus the map

$$MSpin \wedge BO(8,\ldots,\infty) \to MSpin \wedge BSpin \to BO(4,\ldots)$$

actually lifts to $BO(8,\ldots)$. Thus, the characteristic class $\tilde{\pi}_1$ in $\widetilde{KO}(BSpin)$ gives nonzero values on $\tilde{\Omega}_j^{Spin}(BO(8,\ldots,\infty))$ for j=8, 9, and 10, and the homomorphism ∂ is monic in the corresponding dimensions. Thus, one has

<u>Proposition</u>. $$\tilde{\Omega}_j^{Spin}(K(Z,4)) = \begin{cases} Z & j=4 \\ Z+Z & j=8 \\ Z_2 & j=9 \\ Z_2+Z_2 & j=10, \\ 0^2 & \text{for all other } j \le 11. \end{cases}$$

and

<u>Corollary</u>. <u>$\Omega_{11}^{Spin}(K(Z,4)) = 0$</u>.

As a curious sideliight, notice that $\Omega_{10}^{Spin}(BSpin) \to \Omega_{10}^{Spin}(K(Z,4))$ is epic. Being given a closed Spin M^{10} and class $z \in H^*(M,Z)$, one then has $\rho z \cdot Sq^2 \rho z[M^{10}] = Sq^4 Sq^2 \rho z[M^{10}]$, where $\rho z \in H^*(M;Z_2)$ is the mod 2 reduction. To verify this, one need only check it when $\rho z = w_4(\xi)$ for some Spin bundle ξ, and $Sq^4 Sq^2 w_4(\xi)[M] = Sq^4 w_6(\xi)[M] = \{w_{10}(\xi) + w_4(\xi)w_6(\xi)\}[M] = \{Sq^2 w_8(\xi) + w_4(\xi)w_6(\xi)\}[M] = w_4(\xi)Sq^2 w_4(\xi)[M]$, since Sq^2 is zero into the top dimension for Spin manifolds.

Comparing with Lusztig, Milnor, and Peterson [LMP], one sees that for the symmetric bilinear form

$$[\ ,\ ] : H^*(M^{10},Z) \otimes H^*(M^{10},Z) \to Z_2 : [x,y] = \rho x \cdot Sq^2 \rho y[M^{10}]$$

defined for closed 10-dimensional Spin manifolds, one has $[x,x] = [v_4(M),x]$ where $v_4(M)$ is the 4-th Wu class (which is the reduction of an integral class). Thus one has

Corollary. For a closed 10-dimensional Spin manifold, the Stiefel-Whitney number $w_4 w_6[M^{10}] = v_4(M)Sq^2 v_4(M)[M] = [v_4(M), v_4(M)]$ is the rank mod 2 of the bilinear form $[\ ,\ ]$ on $H^*(M;Z)$.

References

[ABP] D. W. Anderson, E. H. Brown, Jr., and F. P. Peterson, The Structure of the Spin cobordism ring, Ann. of Math. 86 (1967), 271-298.

[LMP] G. Lusztig, J. Milnor, and F. P. Peterson, Semicharacteristics and cobordism, Topology 8 (1969), 357-359.

University of Virginia

UNIFICATION IN TEN DIMENSIONS

Edward Witten

Joseph Henry Laboratories
Princeton University
Princeton, New Jersey 08544

As an attempt to reconcile quantum mechanics with gravity, superstrings are not a new idea, but real phenomenology of superstrings became possible only upon the discovery of a generalized mechanism for cancellation of anomalies [1]. The discovery of a superstring theory with $E_8 \times E_8$ gauge group [2] has made the subject even more exciting, and by now there has been much exploration of the issues that are involved in attempting to make contact between superstrings and four dimensional physics [3–18] In attempting here a brief review of these matters, I will concentrate on a single theme. The theme will be to describe the similarities and differences between grand unification in four dimensions and unification in a higher dimensional Kaluza-Klein or superstring context. I will contrast four dimensional unification with higher dimensional unification in the following four areas: (i) the origin of flavor; (ii) the fine tuning problem and the nature of Higgs bosons; (iii) properties of the Yukawa couplings; (iv) magnetic monopoles and electric charges.

First we discuss the flavor problem, which is the modern version of Rabi's classic question 'Who ordered the muon?' The question is why observed fermions seem to fit into (at least) three generations which are identical as regards gauge quantum numbers. Why did nature choose to duplicate structure in this way?

To have any hope of understanding why there are three generations, we must first discuss what the really fundamental feature is of the generation structure that nature has chosen to duplicate. The most fundamental feature – unknown, of course, when Rabi originally posed the question – is that the quarks and leptons that make up a fermion generation transform in a so-called complex representation of the $SU(3) \times SU(2) \times U(1)$ gauge group. This means that the left-handed massless fermions transform in a representation V_L of the gauge group which is not equivalent to the representation V_R in which the right-handed fermions transform. While V_R and V_L are inequivalent, they are (by the CPT theorem) complex conjugates of each other, so the statement that V_L is not equivalent to V_R amounts to saying that V_L is not equivalent to its own complex conjugate or in other words that V_L is a complex representation of $SU(3) \times SU(2) \times U(1)$.

The fact that the fermion representation is complex means that the gauge interactions violate parity. It is often described by saying that there is a chiral asymmetry or a left-right asymmetry in the fermion quantum numbers. This left-right asymmetry is important not only because parity violation is interesting and important but also because the left-right asymmetry is the foundation of our understanding of why fermions that are light enough to be experimentally observeable exist at all.

As long as gauge symmetries are conserved, a left handed fermion can gain a mass only by pairing up with a right handed fermion of the same gauge quantum numbers (since a massive spin one half particle has two helicity states, which must have the same gauge charges). The fact that the left- and right-handed fermions transform differently under the gauge group means that they must remain massless as long as the gauge group is unbroken. What seems to happen in nature is that the observed quarks and leptons remain masssless down to a mass scale of a few hundred Gev at which the electroweak gauge group is broken down to a subgroup – electromagnetism. At that point, the fermions are in a real representation of the remaining gauge symmetries, and they can and do get masses, except possibly for the neutrinos. We do not understand why the scale of weak interaction symmetry breaking is so tiny compared to the Planck mass;

this is the gauge hierarchy problem. But we do at least understand the lightness of the quarks and leptons in terms of the lightness of the W and Z; it follows from the chiral asymmetry between left and right handed fermions.

Could it be that the left-right asymmetry in fermion quantum numbers is just an illusion? Perhaps there are mirror fermions at a Tev energy scale with $V + A$ couplings to the usual W bosons. (The mirror fermions could not be much heavier than a Tev, since their masses would violate $SU(3) \times SU(2) \times U(1)$, given that this is so for the usual fermions.) I think that the existence of mirror fermions is extremely unlikely. If they were discovered, we would lose our successful explanation of why the usual fermions are so light compared to the mass scale of grand unification or gravity; we would be faced with an embarrassing puzzle of why the usual fermions are significantly lighter than their mirror counterparts (much lighter, in the case of the first generation). Also, if mirrors exist, the intricate cancellation of $SU(3) \times SU(2) \times U(1)$ gauge anomalies among the fermions of a single generation is just an accident, since the mirrors would have canceled the anomalies anyway.

By continuously changing the parameters of a theory with unbroken gauge group $SU(3) \times SU(2) \times U(1)$, there is no way to disturb the left-right asymmetry in the quantum numbers of the massless fermions. It does not matter here whether the parameters that are being continuously varied are, say, unknown coupling constants in the underlying equations or artificial parameters that label theoretical assumptions. Because the left-right asymmetry only depends on qualitative facts about a theory, not on the details of how it is presumed to behave, it has been over the years a fruitful matter to think about in trying to understand grand unification both in four dimensions and in the context of higher dimensional theories. It is very likely that we will be able to predict the 'universality class' of the low energy world long before we can predict the details; since the left-right asymmetry depends only on the universality class of a theory with $SU(3) \times SU(2) \times U(1)$ gauge interactions, it will probably be determined on theoretical grounds long before we understand the Cabibbo angle or the mass of the electron.

There is another reason that the chiral asymmetry is important. Chiral asymmetry seems to be a delicate thing which is easily lost and less easily gained. Passing from a theory at one energy scale Λ with left-right asymmetry to an effective theory at a lower energy scale Λ', there are many ways that the chiral asymmetry of the original theory can be lost in going down to lower energies. For instance, chiral asymmetry is actually lost in nature in electroweak symmetry breaking at an energy of order one hundred Gev; it could easily have been lost in compactification from ten to four dimensions, if a certain slightly intricate chain of steps that I will mention later were not followed. While we know many ways to lose chiral asymmetry, we know of no ways to gain it in going from one energy scale to a lower one. To gain chiral asymmetry by a dynamical process starting with a non-chiral theory would require a mechanism for the binding of charged massless fermions, and I personally would consider such a mechanism strongly counter-intuitive. In fact, there are theorems which tend to show that spontaneous generation of chirality does not occur in the dynamics of non-chiral (vector-like) gauge theories.[19] If it is true that chiral asymmetry is easily lost but difficult or impossible to gain, then the chiral asymmetry that we observe in nature must be a trace of the most fundamental physical laws, whatever those may be. Indeed, along with general covariance and Yang-Mills gauge invariance, left-right asymmetry of fermion quantum numbers would appear to be one of the few really fundamental observations about nature.

Let us now survey some approaches to the problem of family replication, bearing in mind that the chiral asymmetry of the fermion families is one of the basic observations. Most work on the family problem in the context of conventional grand unification has been based on the idea that the unified gauge group G is larger than a minimal $SU(5), O(10)$, or E_6 unified group, large enough so that several generations of quarks and leptons fit into a single generation of G. The attempt to carry out this idea with G being a large $SU(N)$ group is a fascinating idea that runs into innumerable difficulties. An attempt which comes much closer to working is the idea[20] of taking G to be a large orthogonal group $O(4k+2)$ with fermions in the spinor representation of G. This idea of 'orthogonal family unification' beautifully fits several generations of fermions into an irreducible representation of G. The only thing which is really wrong with it is that it predicts at the same time an equal number of opposite chirality mirror generations, which cannot be superheavy since (given that the conventional fermions are light) their masses violate the gauge symmetries of the electroweak theory. Recently there have been heroic attempts to make a viable model in which the mirror fermions will be placed at the Tev mass region[21]. While it may turn out that that is how nature works, I consider it unlikely for reasons that I have already indicated. It is on the other hand a very striking fact that orthogonal family unification works just right in Kaluza-Klein theory without producing the mirror generations.

The basic idea behind this is the following. The Dirac equation in ten dimensions is $D\Psi = 0$, where D is the ten dimensional Dirac operator $D = \sum_{i=1...10} \Gamma^i D_i$, the $\Gamma^i, i = 1...10$ being the ten dimensional gamma matrices. Now, in fact, $D = D_4 + D_K$, where $D_4 = \sum_{i=1...4} \Gamma^i D_i$ is the four dimensional Dirac operator and $D_K = \sum_{i=5...10} \Gamma^i D_i$ is the Dirac operator of the compact Kaluza-Klein space. This means that the internal Dirac operator D_K is the 'mass' operator of the effective four dimensional theory. In fact, if $D_K\Psi = \lambda\Psi$, then the ten dimensional equation $D\Psi = 0$ reduces in four dimensions to $D_4\Psi + \lambda\Psi = 0$, so that Ψ will be observed in four dimensions as a fermion of mass λ. Now, the internal Dirac operator D_K acting on the compact space K will have a discrete spectrum, and the eigenvalues which are not zero will be of order $1/R$, R being the radius of K. Eigenvalues of order $1/R$ correspond to fermions of Planckian masses, which certainly would not have been observed to date. The observed quarks and leptons have masses that are essentially zero in Planck units, and they must correspond to zero eigenvalues of the internal Dirac operator.

Now, why would a Dirac operator have zero eigenvalues? Some of the basic facts about this question are known to physicists from instanton days. A Dirac operator can have zero eigenvalues for topological reasons, the number of zero eigenvalues being determined by the values of suitable topological invariants. Now, in the case at hand there is a separate Dirac equation for each choice of $SU(3) \times SU(2) \times U(1)$ quantum numbers, and the number of massless multiplets in a given $SU(3) \times SU(2) \times U(1)$ multiplet is simply the number of fermion zero modes for the corresponding Dirac operator on K. The family problem – the question of why there are **several** massless multiplets of given $SU(3) \times SU(2) \times U(1)$ quantum numbers even though the microscopic theory began with a single irreducible multiplet – is in this context simply the question of why a suitable Dirac operator has not one but several zero eigenvalues. But this is not an unusual behavior for Dirac operators; to cite a relatively familiar example, a color $SU(3)$ instanton acting on a single multiplet of fermions in the adjoint representation of $SU(3)$ would have six zero eigenvalues. We are thus led to the program[22–32] of relating the number of generations to topological invariants of K.

If we do manage to find several fermion generations, corresponding to several zero modes of the internal Dirac operator, what chirality will these have? To answer this, note that four dimensional chirality is measured by the product $\Gamma^{(4)} = \Gamma^1\Gamma^2...\Gamma^4$ of the usual four gamma matrices. On the other hand, chirality in the sense of the Dirac equation on K is measured by the internal chirality operator $\Gamma^{(K)} = \Gamma^5\Gamma^6...\Gamma^{10}$. And chirality in the ten dimensional sense (before compactification, so to speak) is measured by the product $\Gamma^{(10)} = \Gamma^1\Gamma^2...\Gamma^{10}$ of all ten gamma matrices. Now, it is just a fact of life of ten dimensional supergravity that in supergravity theories that have elementary gauge fields (otherwise, it is impossible to get chiral fermions [26]), the ten dimensional fermions have definite chirality, say $\Gamma^{10} = +1$. Since $\Gamma^{(10)} = \Gamma^{(4)} \cdot \Gamma^{(K)}$, the fact that $\Gamma^{(10)} = 1$ means that $\Gamma^{(4)} = \Gamma^{(K)}$. This wonderful equation says that chirality as measured by four dimensional experimentalists (who, in effect, measure $\Gamma^{(4)}$) coincides with chirality as understood by observers on K studying the internal Dirac operator D_K. A zero mode of D_K of $\Gamma^{(K)} = +1$ will give rise to a massless fermion in four dimensions of $\Gamma^{(4)} = +1$; a zero mode of $\Gamma^{(K)} = -1$ will give rise to a massless fermion of $\Gamma^{(4)} = -1$. To obtain left-right asymmetry in the effective four dimensional world requires that the Dirac operator on K has (for given $SU(3) \times SU(2) \times U(1)$ quantum numbers) more zero modes with one eigenvalue of $\Gamma^{(K)}$ than with the opposite eigenvalue. This is not an unlikely behavior at all; it amounts to saying that the Dirac operator on K has a non-zero character- valued index. These ideas in the Kaluza-Klein context were developed in [22,24,26].

There are many ways to implement these ideas. One approach which was originally proposed in an *ad hoc* way in [26] but recently proved [5] to be very natural and attractive in the context of $E_8 \times E_8$ superstrings is the following. Consider an $SO(16)$ gauge group in ten dimensions with fermions in the positive chirality spinor (or **128**) of $SO(16)$. After compactifying to four dimensions, we have on the compact six manifold K a spin connection which is a connection on the tangent bundle of K and so is a gauge field of $SO(6)$ (or perhaps a subgroup thereof). If we 'embed the spin connection in the gauge group' by setting the gauge fields of an $SO(6)$ subgroup of $SO(16)$ equal to the spin connection, then $SO(16)$ is broken down to a subgroup – which generically is $SO(10)$. Now, $SO(10)$ is essentially the only orthogonal group which is suitable for four dimensional grand unification because it has a complex representation, the **16**, which is just right for accomodating a standard generation. With other orthogonal groups one runs into the chirality problem I mentioned above, though one might try to deal with this problem along the lines of [21]. It was for this reason that in [26] an $SO(16)$ gauge group rather than some other orthogonal group was taken as the starting point. After breaking $SO(16)$ to $SO(10)$ by 'embedding the spin connection in the gauge group,' study of the Dirac equation shows that we indeed get chiral fermions in the **16** of $SO(10)$, the number of generations N_{gen} being related to one of the most basic topological invariants of K, namely its Euler characteristic. (In the superstring context, the precise relation turns out to be [5] $N_{gen} = (1/2) \cdot \chi(K)$, where $\chi(K)$ is the Euler characteristic of K, which in six dimensions can be any even number. In the model of [26] the number of generations was twice as large.) Choosing K to have a suitable Euler characteristic, one can well obtain several generations in this way, so that the flavor question *per se*, the question of how to get a multiplicity of generations starting with a unified underlying framework, is no mystery. What is unsatisfying is that several aspects of the construction are rather artificial. The gauge group and fermion representation that we started with and the choice of embedding the spin conection in the gauge group were all simply chosen to get the standard model in four dimensions after compactification. It is therefore very satisfying that [5] recent developments in superstring theory give a natural justification for these seemingly arbitrary choices. The group E_8 has a maximal $SO(16)$ subgroup, and the adjoint representation of E_8 contains a **128** of $SO(16)$, so these ingredients have been practically forced on us by recent developments about anomaly cancellation. Also, embedding the spin connection in the gauge group turns out in the superstring context to cancel various anomalies and to give (if K obeys certain conditions stated shortly) a solution of the equations of motion of the theory, without generating a cosmological constant in four dimensions. This is in fact the only presently known way to obtain compactified solutions of the superstring equations, and it is one of the few examples in any Kaluza-Klein theory of any sort in which a realistic candidate vacuum state does indeed obey the hoped-for equations. Actually, the metrics which seem to give solutions of the equations of the theory are Ricci flat Kahler metrics which are the so called metrics of $SU(3)$ holonomy. Their existence was conjectured by Calabi[33] and proved by Yau[34]. The choice of such a metric leads first of all to unbroken $N = 1$ supersymmetry in four dimensions, a property which may well be desireable, and second to a four dimensional gauge group which is E_6 rather than $SO(10)$. Just as $SO(10)$ is the one orthogonal group which is suitable for unification in four dimensions, E_6 is the one suitable exceptional group[35]. Compactification from ten to four dimensions on manifolds of $SU(3)$ holonomy turns out to lead in four dimensions to a theory with chiral fermions in the **27** of E_6 (which is, of course, the right representation [35]), the number of generations still being half the Euler characteristic. Actually, the logic I have followed here in sketching these matters was the reverse of that in [5]; there unbroken $N = 1$ supersymmetry was taken as the initial requirement, and the other features were deduced as consequences. Yet another possible line of development, sketched in the last section of [5] and pursued further in [36] begins with the requirement of a conformally invariant two dimensional sigma model as the starting point and deduces the other properties from this. It is very satisfying that similar conclusions can be reached from so many different starting points.[37].

This completes what I will say about the family problem. Now we turn to the question of Higgs bosons. If K is not simply connected there is a simple mechanism for grand unified symmetry breaking which involves no ingredients that aren't present anyway. (Some of the relevant issues were first considered in [38], the main difference being that when the fundamental group of K is finite, as in many cases of interest, one is led to topological questions, such as those sketched below, rather than to the dynamical issues considered in that paper.) Let γ be a non-contractible loop in K, beginning and ending at a point x. Let $U(x) = Pexp \oint_\gamma A \cdot dx$, A being the E_6 or $O(10)$ gauge field that is still unbroken after embedding the spin connection in the gauge group. For various reasons, such as a wish to keep unbroken $N = 1$ supersymmetry or a simple wish to obey the equations of motion, it is desireable for A to be a pure gauge locally, with zero field strength. If γ is non-contractible, this does not require $U = 1$. If $U \neq 1$, then in many ways $U(x)$ is a field like any other. In particular, if $U \neq 1$, then $O(10)$ (or E_6) is broken to the subgroup that commutes with U. It is possible in this way to get various more or less realistic gauge groups of rank five or rank six, such as $SU(3) \times SU(2) \times U(1) \times U(1)$ or $SU(3) \times SU(2) \times U(1) \times U(1) \times U(1)$ [7,12,13]. A rank four group cannot be obtained in this way, so there is a prediction of at least one new gauge interaction beyond the standard model. If K has $SU(3)$ holonomy, a rank five group can emerge [7] if and only if the fundamental group of K is non-abelian; otherwise there must be two new gauge interactions.

In many ways, U is similar to a Higgs boson ϕ in the adjoint representation of the gauge group, $U \sim expi\phi$. But there are essential differences. One difference arises if $\pi_1(K)$ is finite, which is so for many choices of K of current interest (for instance, it is true for all manifolds of $SU(3)$ holonomy). For instance, suppose $\pi_1(K) = Z_n$. Then n circuits of the curve γ considered before make a contractible loop, so it is necessarily so that $U^n = 1$. This means that the eigenvalues of U are quantized; they are n^{th} roots of unity.

To understand why this might be desireable, recall that one of the key mysteries in grand unified theories is the so-called fine tuning problem. This is the question of why the energy scale of weak interaction symmetry breaking is so much smaller than that of unification. The problem arises because eigenvalues of Higgs boson fields are continuously variable parameters, which depend continuously on the values of unknown coupling constants. A massless weak doublet can arise only if Higgs eigenvalues have special values, and conventionally this is artificial. But, as I have just explained, in the context under discussion here the relevant eigenvalues are naturally quantized and can take only discrete values. On this grounds alone it is not too surprising that one finds that there are possibilities for solving the fine tuning problem [7,12,13], though I will not review the details of this here.

Other special features of grand unified symmetry breaking by Wilson lines deserve mention. This procedure does not disturb the classical field equations, assuming that these were obeyed at $U = 1$, and does not induce a cosmological constant (at least not in the classical approximation) or break supersymmetry (if this is otherwise unbroken). There is no need to postulate Higgs bosons or a Higgs potential or any other ingredient that is not present automatically. Also, symmetry breaking by Wilson lines is more or less topological in nature (for instance, it involves discrete choices if the fundamental group of K is finite) so there is a chance that it can eventually be understood and predicted by general, qualitative arguments - which, when available, are almost always more convincing and satisfying than dynamical arguments. Many other topological approaches to grand unified symmetry breaking could be considered, but generally speaking the others would disturb the successful predictions concerning quark and lepton quantum numbers. (We have seen that unlike the situation in conventional grand unified theories, the choice of vacuum configuration in a higher dimensional theory determines the fermion quantum numbers; this was the key to getting the right chiral structure in the first place.) Symmetry breaking by Wilson lines is unusual as a topological approach to grand unified symmetry breaking that does not have harmful implications for the fermion quantum numbers.

Now we turn to a discussion of gauge and Yukawa couplings. Again we begin with some mathematical preliminaries. If $\pi_1(K) \neq 0$, then K has a 'covering space' K_0 with $\pi_1(K_0) = 0$. If the theory is formulated on K_0 the number of generations is $N_0 = (1/2) \cdot |\chi(K_0)|$. On K the number of generations is smaller, being in fact $N = (1/2) \cdot |\chi(K)|$. In passing from K_0 to K many or most of the quark and lepton states are 'lost.' To quantify the effect of this, let me describe symmetry breaking by Wilson lines in another way. On K_0 we have a discrete symmetry group G that acts freely; K is the quotient K_0/G. This means that for any $x \in K_0$ and $g \in G$, x and gx are considered equivalent as points in K. An ordinary scalar field ψ on K is the same thing as a field $\psi(x)$ on K_0 that obeys $\psi(gx) = \psi(x)$. This condition means that ψ 'lives' on K, not K_0. When we introduce symmetry breaking by Wilson lines, we must make a slight modification in this requirement. If U is the Wilson line corresponding to g (taken in whatever representation of the gauge group ψ is in) then the appropriate requirement is $\psi(gx) = U\psi(x)$. States that exist on K_0 but do not obey the condition just stated do not 'survive' when the theory is formulated on K rather than K_0.

The important point here is that, in general, the surviving states are not states that on K_0 were the $O(10)$ or E_6 partners of one another. If a given quark obeys $\psi(gx) = U\psi(x)$, its lepton partner will (typically) have different U and so will **not** obey this condition. This has the following consequence. In this theory as far as counting states is concerned the particles appear to form representations of a grand unified group. There are as many u quarks as d quarks or neutrinos or charged leptons. But they are not states that on K_0 were $O(10)$ or E_6 partners. This has an immediate beneficial consequence. Since the physical fermions are not $O(10)$ or E_6 partners of one another, there are no simple group theory relations among Yukawa couplings. This is a desireable state of affairs, since such relations (such as $m_d = m_e$, and its generalizations) are generally not in agreement with experiment. On the other hand, the observed gauge bosons of the four dimensional world all began as elements of a single $O(10)$ or E_6 representation (the adjoint representation) so their couplings are related to one another by the usual group theoretical formulas. As a consequence, the Georgi-Quinn-Weinberg relations among the gauge couplings will hold; these relations, of course, are quite successful. It should definitely be counted as a success of higher dimensional theories that – in this way – the group theory relations among gauge couplings are preserved but the (superficially analogous) group theory relations among Yukawa couplings are violated.

Another interesting point is the following. The usual X and Y bosons of grand unification – the $SU(5)$ partners of the photon – have $U \neq 1$ and do not exist. Of course, particles with the same quantum numbers and the same order of magnitude of mass and coupling of the X and Y bosons will exist and will mediate proton decay, but since the proton lifetime scales like the fourth power of the mass and the minus two power of the coupling of the heavy bosons, the relation between unification scale and proton lifetime might differ by a couple of orders of magnitude from what is conventionally calculated. One might almost say that minimal $SU(5)$ could be alive and well in ten dimensions!

I might also note that in discussions of proton decay, an important question is to determine the branching ratios. In particular, when the proton decays, is it more likely to emit an electron or a muon? This question is conventionally a question about X and Y boson couplings which in turn (since those are gauge bosons) amounts to the question 'which charged lepton is the $O(10)$ or E_6 partner of the d quark?' In higher dimensional unification, the latter question has no answer, as I have just explained, and the X and Y bosons do not exist. The question about branching ratios in proton decay is still meaningful, of course, but the ingredients in answering it will be somewhat different.

Turning now to our third topic, if Yukawa couplings do not obey simple relations of a group theoretical origin, what relations will they obey? Before stating the answer, I would like to note that it should come as no surprise that one has, potentially, a great deal of predictive power for the Yukawa couplings. In fact, the light fermions and Higgs bosons are all zero modes of suitable wave operators on the compact Kaluza-Klein space K. The Yukawa couplings are certain cubic terms that arise in expanding the exact ten dimensional theory in powers of the light fields, and they can be computed by taking a suitable product of wave functions and integrating it over K. Thus, if one knew everything about K (including its metric) one would simply determine the zero mode wave functions by solving the relevant wave equations on K (numerically, if need be) and then one could determine all the Yukawa couplings by evaluating a certain integral (which arises in expanding the ten dimensional action in powers of the light fields). While the procedure just stated is perfectly sound in principle, it presupposes an amount of information about K that will probably not be available at least for a very long time. The question really should be 'what simple relations among Yukawa couplings are there that can be deduced in a general way?' analogous to group theory relations in standard grand unification. The answer to this is that the Yukawa couplings obey relations of a topological origin. A preliminary discussion of this was made in [39] and stronger results were recently obtained in [40]. Topological relations emerge because the particular integrals that arise in evaluating Yukawa couplings have a topological interpretation, in terms of the so-called cohomology ring of K. While that description holds in limiting low energy field theory, it can be shown [37] that at least the original relations derived in [39] also hold in string theory, at least to all finite orders in sigma model perturbation theory.

What I find promising about this is the following. In the quark and lepton mass matrices, certain elements seem to be zero or very small. For instance, certain of the fermions are extremely light, and in the the Fritzsch form of the mass matrix, certain matrix elements are taken to be zero. The most straightforward approach to trying to get some matrix elements to vanish is to assume suitable global symmetries. However, attempts to explain the observed form of the fermion mass matrices via global symmetries have always led to difficulties. At least in simple approaches, symmetries that forbid unwanted elements of the mass matrices seem to also forbid elements of the mass matrices that are not zero or small in nature. Higher dimensional unification may ultimately shed a completely new light on this problem since the unwanted elements of the mass matrix could well vanish for topological reasons without unwanted consequences for other matrix elements. Although models with realistic mass matrices are probably still far way, it is already possible to say something about Yukawa couplings in toy models. For instance, in [7] it was determined which Yukawa couplings were zero in the four generation model introduced in [5]. (It was not necessary to use the full force of topological reasoning since a relatively simple tool which might be referred to as 'pseudosymmetries' sufficed in that case.) The result was that of the four generations, two obtained tree level masses and two were massless at tree level. Various Yukawa couplings vanished at tree level, and interestingly the vanishing Yukawa couplings were such that (like the Yukawa couplings in the real world) the pattern would be difficult to explain via global symmetries. Yukawa couplings in another toy model were recently studied in [10].

The last topic I wish to discuss concerns the allowed values of electric and magnetic charge in theories that are unified only in some higher dimension [41]. In any theory that includes electromagnetism, one can define at spatial infinity the $U(1)$ Dirac monopole gauge field. Monopoles exist if this can be extended throughout all space without encountering a singularity. This is impossible if the gauge group is $U(1)$ or more generally if it is any group such as $SU(3) \times SU(2) \times U(1)$ which has a $U(1)$ factor. However, many years ago 't Hooft and Polyakov showed that in four dimensional unified theories one can always 'unwrap' the monopole, obtaining in conventional grand unified theories states whose magnetic charge equals the Dirac quantum. But what if unification in $SU(5)$ or $O(10)$ or E_6 occurs not in four dimensions but only in ten dimensions? In this case we cannot unwrap the monopole in the four dimensional unified gauge group, because there is none. We must try to unwrap it in the **ten** dimensional gauge group. The topological problem is completely different, and there is no reason to expect the answer to be the same. To be precise, what must be done to determine whether the monopole exists is the following. We begin with the Dirac monopole gauge field not on S^2 but on $S^2 \times K$ (here S^2 is the 'sphere at infinity' and K is the Kaluza-Klein space) and we try to extend it over $R^3 \times K$ (R^3 being physical three space) without singularity. Can this be done? It turns out [41] that there is a non-trivial obstruction. When the dust settles, the picture that emerges is the following. Magnetic monopoles always exist, but the minimum allowed value of magnetic charge is typically larger than it would be in conventional grand unified theories. Generically, if for instance $\pi_1(K) = Z_n$, the minimum possible magnetic charge is not the Dirac quantum $(2\pi/e)$ but is rather n times larger.

This is purely a field theoretic result, which emerges upon solving the topological problem I just described. However, the result that the minimum allowed value of magnetic charge is not $(2\pi/e)$ but $n \cdot (2\pi/e)$ is surprising at first sight, and one may ask what is the 'physical' reason for this result. It has no simple 'physical' explanation in field theory, as far as I know, but upon coupling to string theory a simple 'physical' explanation emerges in a dramatic way. I have been tacitly assuming that grand unified symmetry breaking was carried out by Wilson lines (otherwise, in fact, the whole topological problem must be reexamined), so implicit in the whole discussion is the fact that the fundamental group of K is non-trivial; in fact, we took it to be Z_n. This being so, in a theory that only has closed strings, there are stable, superheavy states in which a closed string wraps around a non-contractible loop in K. If we are incredibly lucky, we might one day observe such particles in cosmic rays; but that is another story. In any case, quantization of string theory in the 'winding' sectors shows that these modes have electric charge e/n precisely when the minimum allowed value of magnetic charge is $n \cdot (2\pi/e)$. I should stress, perhaps, that the winding states of electric charge e/n are color singlet, unconfined states. Their existence 'explains' why the minimum magnetic charge would be n times larger than the Dirac quantum. The remarkable thing about these results is that the magnetic charge is determined by geometric-topological methods which seem to have nothing to do with string theory, but which somehow 'know' that one day this problem will be coupled to the theory of closed strings.

In summary, unification in higher dimensions preserves the standard successes of four dimensional unification. The light fermions fit in representations of $O(10)$ or E_6, and the Georgi-Quinn-Weinberg relations are valid. On the other hand, there are a few interesting differences from the standard framework. The Yukawa couplings do not obey $O(10)$ or E_6 relations but obey instead other relations of topological origin. The Higgs expectation values are quantized, giving perhaps the possibility of an insight into the hierarchy problem. There are particles (albeit superheavy) whose electric charges could not arise in any representation of the grand unified group; and conjugate to this some of the usual magnetic monopoles are missing. There is every reason to think that higher dimensional theories, and especially superstring theory, will give us the opportunity in coming years to rethink some of the issues which have been left open by the successes of conventional grand unification and which have fascinated and puzzled us so much.

REFERENCES

1. M. B. Green and J. H. Schwarz, Phys. Lett. **149B** (1984) 117.
2. D. J. Gross, J. A. Harvey, E. Martinec, and R. Rohm, Phys. Rev. Lett. **54** (1985) 502.
3. E. Witten, Phys. Lett. **149B** (1984) 351.
4. P. Frampton, H. van Dam, and R. Yanamoto,UNC preprint (1984).
5. P. Candelas, G. Horowitz, A. Strominger, and E. Witten, Nucl. Phys. **B258** (1985) 46. See also the articles by G. Horowitz and A. Strominger in this volume.
6. M. B. Green, J. H. Schwarz, and P. West, Cal Tech preprint, 1984.
7. E. Witten, Nucl. Phys. **B258** (1985) 75, Princeton preprints, 1985.
8. M. Dine, V. Kaplunovsky, M. Mangano, C. Nappi, and N. Seiberg, IAS preprint, 1985.
9. R. Nepomechie, Wu Yong-Shi, and A. Zee, Univ. of Washington preprint, 1985.
10. K. Pilch and J. Schellekens, Stony Brook preprints, 1985.
11. S. Cecotti, J.-P. Derendinger, S. Ferrara, L. Girardello, and M. Roncadelli, CERN preprint.
12. J. Breit, B. Ovrut, and G. Segre, Univ. of Pa. preprint (1985).
13. A. Sen, Fermilab preprints, 1985.
14. M. Dine, N. Seiberg, R. Rohm, and E. Witten, IAS preprint, 1985.
15. V. Kaplunovsky, Princeton preprint, 1985.
16. M. Dine and N. Seiberg, IAS preprint, 1985.
17. J. P. Derendinger, L. E. Ibañez, and H. P. Nilles, CERN preprint (July, 1985), to appear in Nucl. Phys. B.

18. I. Bars, USC preprint 85/015 (1985); I. Bars and M. Visser, USC preprint (1985); I. Bars, D. Nemeschansky, and S. Yankielowicz, SLAC preprint (August, 1985).

19. C. Vafa and E. Witten, Comm. Math. Phys. **95** (1984) 257.

20. M. Gell-Mann, P. Ramond, and R. Slansky, in **Supergravity**, ed. P. van Nieuwenhuysen et. al. (North-Holland Publishing Company, 1979); F. Wilczek and A. Zee, Princeton report, 1979 (unpublished), Phys. Rev. **D25** (1982) 553. A closely related idea can be attempted in the context of a four dimensional theory with E_8 gauge group, and has somewhat similar successes and difficulties; see I. Bars and M. Günaydin, Phys. Rev. Lett. **45** (1980) 859.

21. J. A. Bagger, S. Dimopoulos, E. Masso, and M. H. Reno, SLAC preprint (1984).

22. E. Witten, Nucl. Phys. **B186** (1981) 412.

23. G. Chapline and R. Slansky, Nucl. Phys. **B209** (1982) 461

24. C. Wetterich, Nucl. Phys. **B223** (1983) 109

25. S. Randjbar-Daemi, A. Salam, and S. Strathdee, Nucl. Phys. **B214** (1983) 491.

26. E. Witten, in the Proceedings of the 1983 Shelter Island Conference, ed. N. Khuri et. al. (MIT Press, 1985).

27. D. Olive and P. West, Nucl. Phys. B217 (1983) 248.

28. G. Chapline and B. Grossman, Phys. Lett. **125B** (1984) 109.

29. P. H. Frampton and K. Yamamoto, Phys. Rev. Lett. **52** (1984) 2016.

30. S. Weinberg, Phys. Lett. **138B** (1984) 47

31. P. H. Frampton and T. W. Kephardt, Phys. Rev. Lett. **53** (1984) 867

32. I. G. Koh and H. Nishino, Trieste preprint ICTP/84/129

33. E. Calabi, in **Algebraic Geometry and Topology: A Symposium in Honor of S. Lefschetz** (Princeton University Press, 1957) p. 58.

34. S.-T. Yau, Proc. Natl. Acad. Sci. **74** (1977) 1798.

35. F. Gursey, P. Ramond, and P. Sikivie, Phys. Lett. **60B** (1976) 177.

36. P. Candelas, G. Horowitz, A. Strominger, and E. Witten, to appear in the Proceedings of the Argonne-Chicago Symposium on Anomalies, Geometry, and Topology; C. G. Callan, Jr., D. Friedan, E. Martinec, and M. J. Perry, Princeton preprint, 1985. See also the articles of D. Friedan, E. Martinec, and S. Shenker in this volume.

37. See E. Witten, 'New Issues in Manifolds of $SU(3)$ Holonomy,' to appear in Nucl. Phys. B, for a discussion of more general vacuum states based on Calabi-Yau manifolds in which (while keeping unbroken supersymmetry) E_8 is broken to $SO(10)$ or $SU(5)$ rather than E_6. The construction involves the use of stable holomorphic vector bundles over the Calabi-Yau manifold other than the tangent bundle. A discussion of these more general vacuum states would take us too far afield.

38. Y. Hosotani, Phys. Lett. **129B** (1983) 193.

39. A. Strominger and E. Witten, to appear in Comm. Math. Phys.

40. A. Strominger, Santa Barbara preprint, 1985.

41. X. G. Wen and E. Witten, to appear in Nucl. Phys. B. See also my article, 'Topological Tools in Ten Dimensional Physics' in this volume.

SEMINARS

1. Two Dimensional Field Theory and Functional Techniques

CONFORMALLY INVARIANT FIELD THEORIES IN TWO DIMENSIONS CRITICAL SYSTEMS AND STRINGS

J.-L. GERVAIS
Physique Théorique, Ecole Normale Supérieure
24 rue Lhomond 75231 Paris cedex 05

At the present time it is hardly necessary to emphasize the fundamental importance of string models since super string theories are the most promising candidates for a completely unified theory of all interactions. An other key point is that string concepts have plaid an important role in the recent developments of theoretical physics and mathematics by suggesting many new important ideas, such as in particular supersymmetry (1), and have led to very interesting progress in the related critical models in two dimensions. In these notes I shall mostly concentrate on this latter aspect which is not, presently, so directly aimed at a unified theory of all interactions but is quite interesting in its own right.

The unifying feature of string theories and critical systems is that they are both associated with conformally invariant field theories. We shall first review the essential features of this connection considering only, for simplicity, bosonic strings. At the level of the present discussion, supersymmetric strings are not basically different. One essentially replaces the conformal group by its superconformal generalization.

The position of a string at time τ is specified by a field $X_\mu(\sigma,\tau)$ where τ distinguishes the various points along the line. Hence one has a two dimensionnal field theory in parameter space. When σ varies the string sweeps out a world sheet and one sets up the dynamics in such a way that it be invariant under reparametrization of the corresponding geometrical surface. One can rigourously show that it is always possible to choose the parametrization in such a way that the curves σ =cste and the curves τ =cste intersect at right angle. This choice is not unique since this orthogonality condition is left invariant by all conformal transformations of σ and τ. In string theories the conformal group is thus the residual symmetry of the system with an orthogonal choice of σ,τ parameters. Basically it is the group of all transformations of the form

$$\sigma'+\tau'=f(\sigma+\tau) \qquad \sigma'-\tau'=g(\sigma-\tau) \tag{1}$$

where f and g are two arbitrary real functions of one

variable. In the present dicussion we stick to the Lorentz covariant string quantization where conformal invariance is not explicitely broken.
Since a physical string has a finite length, σ varies over a finite range. It is always possible to redefine the parameters in such a way that the dynamics is periodic in σ with period 2π. It is often very convenient to go to Euclidean time by letting

$$\nu = i\tau \tag{2}$$

For real ν the conformal group can best be described as the group of analytic transformations of

$$z = e^{\nu + i\sigma} \tag{3}$$

Indeed with this variable, the strip $0 \leq \sigma \leq 2\pi$ is represented by the whole complex z plane. In this picture, a conformal transformation in given by

$$z' = F(z) \tag{4}$$

where F is an arbitrary complex function of one variable. In general, a quantity $\mathcal{O}(z, z^*)$ is called conformally covariant [2] if it transforms according to

$$\mathcal{O}'(z', z'^*) = \left(\frac{dF}{dz}\right)^{-\delta} \left(\frac{dF^*}{dz^*}\right)^{-\bar{\delta}} \mathcal{O}(z, z^*) \tag{5}$$

where δ and $\bar{\delta}$ are parameters depending on the quantity considered which are called conformal weights. This notion was rediscovered recently [3] and the corresponding fields were called primary. If we separate the real and imaginary parts according to

$$z = x_1 + i x_2 \tag{6}$$

the differential transforms as

$$dx'_\ell = \mu R_{\ell m}\, dx_m \tag{7}$$

where R is the rotation matrix with angle θ and where μ is a dilatation factor. μ and θ are given by the differential equations

$$\frac{\partial x'_1}{\partial x_1} = \frac{\partial x'_2}{\partial x_2} = \mu \cos\theta \qquad \frac{\partial x'_1}{\partial x_2} = -\frac{\partial x'_2}{\partial x_1} = \mu \sin\theta \tag{8}$$

Formula (5) becomes

$$\mathcal{O}'(x'_1, x'_2) = \mu^{-d} e^{-iJ\theta}\, \mathcal{O}(x_1, x_2) \tag{9}$$

Since μ and θ are the local dilatation and rotation parameters respectively, d is the dimension and J is the spin of the quantity considered.
For critical systems in two dimensions, x_1 and x_2 are the two coordinates. It is well known that a stastistical system at a point of transition of second order becomes

scale invariant. The corresponding rescaling of x_1 and x_2 is a particular case of (4). Polyakov has proposed that critical systems are invariant under the full conformal group (4).

One thus sees thas that both string theories and critical systems are based on conformally invariant field theories, with however different descriptions. σ and τ are string parameters while x_1, x_2 are the coordinates of the critical system.

In a conformally invariant field theory the improved energy momentum tensor is symmetric traceless and conserved. For two dimensional field theories in real σ, τ space this leads to

$$\left(\frac{\partial}{\partial\tau} \mp \frac{\partial}{\partial\sigma}\right)\left(T^0_0 \pm T^1_0\right) = 0 \tag{10}$$

Due to the periodicity in σ one can write

$$\begin{aligned} 2\pi\left(T^0_0 + T^1_0\right) &= \sum_m L_m z^{-m} \; ; \quad z = e^{i(\tau+\sigma)} \\ 2\pi\left(T^0_0 - T^1_0\right) &= \sum_m \bar{L}_m \bar{z}^{-m} \; ; \quad \bar{z} = e^{i(\tau-\sigma)} \end{aligned} \tag{11}$$

The operators L_m and $\bar{L}_m$ are the infinitesimal generators of conformal transformations. For a conformally covariant field which satisfies equation (5) one has

$$\begin{aligned} \left[L_m, \theta\right] &= z^m\left[z\frac{\partial}{\partial z} + (m+1)\delta\right]\theta \\ \left[\bar{L}_m, \theta\right] &= \bar{z}^m\left[\bar{z}\frac{\partial}{\partial \bar{z}} + (m+1)\bar{\delta}\right]\theta \end{aligned} \tag{12}$$

The operators L_m and $\bar{L}_m$ each satisfy the Virasoro algebra

$$\begin{aligned} \left[L_n, L_m\right] &= (n-m)L_{n+m} + \frac{c}{12}(n^3-n)\delta_{n,-m} \\ \left[\bar{L}_n, \bar{L}_m\right] &= (n-m)\bar{L}_{n+m} + \frac{c}{12}(n^3-n)\delta_{n,-m} \end{aligned} \tag{13}$$

where the central charge C depends on the model considered. Its actual value is a key point. We shall have more to say about this below. In general the field theories we are discussing are characterized by C and by the set of conformally covariant fields θ_α together with the set of conformal weights δ_α. Since the two Virasoro algebras (13) have the same properties we only consider explicitely the algebra of the L_m's most of the time.

As a first simple example, let us recall the essential

features of the standard bosonic (Veneziano) model. In this case one only considers massless two dimensional free fields X_μ. We shall denote by Λ_m the associated Virasoro generator. As it is well known they satisfy eq. (13) with

$$C = \mathcal{D} \tag{14}$$

the vertex for the emission of the lightest string state is simply given by

$$V_k(z) = :e^{ik^\mu X_\mu(z)}: \tag{15}$$

where k^μ is the energy momentum of the emitted particle. It is well known that V_k satisfies condition (12) with

$$\delta = \frac{k^2}{2} \tag{16}$$

The present discussion of string is in the covariant formalim where one has to make sure that the time like components of X_μ decouple from the physical S matrix. Such a ghost killing mechanism requires first of all that the vertex V_k have dimension 1. From eq. (16) this leads to, $k^2 = 2$ the emitted particle is a tachyon with mass $m^2 = -2$.

For critical models the δ_α and $\bar{\delta}_\alpha$ are critical exponents. Indeed it is easy to see that the global dilatations, rotations, and translations of x_1, x_2 are genrated by L_0 $\bar{L}_0$ L_{-1} and $\bar{L}_{-1}$. The vacuum state of the system must therefore be annihilated by these operators. As a result the two point function of any covariant operator can be computed up to a constant factor by means of equation (12). If $\delta = \bar{\delta}$ for instance, one finds

$$\langle 0 | \mathcal{O}(z,z^*), \mathcal{O}(z',z'^*) | 0 \rangle \sim \left(|z-z'|^2 \right)^{-2\delta} \tag{17}$$

Hence δ gives the power behaviour of the two point function at the critical point. It is quite obvious that δ must be positive for physical operators since the correlation functions must decrease when the separation increases.

As it is well known (2) (3), the derivative of a covariant operator is not covariant in general. An important exception is the case of an operator of vanishing weight. Its derivative with respect to has $\delta = 1$, $\bar{\delta} = 0$. Conversely, assume there exists an operator $I(z)$ with $\delta = 1$ Then it is obvious that

$$\left[L_n , \oint dz\, I(z) \right] = 0 \tag{18}$$

In stastistical mechanics such operators are called marginal. If they exist one has critical lines instead of

critical points since they can be added to the action with arbitrary coefficients without destroying conformal invariance. For the Veneziano model we have just recalled that $V_k(z)$ has weight 1. Indeed equation (18) for V_k is the basic ingredient for the decoupling of ghosts.
The notion of conformally covariant operator was introduced in string models [2] in order to dicuss posible generalizations of the Veneziano model. We now recall the essential points of this approach. Generalized string models involve other fields besides the free fields X_μ and the corresponding two dimensional field theory may have a non trivial interaction. Such is the case, for instance, if we have additional space components which are compactified. Quite generally we can consider that X_μ remains a free field which does not mix with the additional two dimensional fields. These will be characterized by the set of covariant operators O_α together with the set of weights δ_α. The ghost killing condition now requires the emission vertex to have conformal weight one under the action of the total Virasoro generator

$$J_m = \Lambda_m + L_m \tag{19}$$

where L_m is the Virasoro generator of the additional two dimensional dynamics. This is realized by choosing

$$\mathcal{V}_\alpha = V_k O_\alpha \qquad \frac{k^2}{2} + \delta_\alpha = 1 \tag{20}$$

We therefore see that the spectrum of lightest particles

$$m_\alpha^2 = 2(\delta_\alpha - 1) \tag{21}$$

will involve no tachyon provided that

$$\delta_\alpha \geqslant 1 \tag{22}$$

and this condition selects the conformally invariant field theories for which all covariant fields have weights larger than one. From the view point of critical systems this condition is unusual since it means that the corresponding two point function has a Fourier transform which has at most a logarithmic singularity at zero momentum. We shall come back to this later.
The last important general point about string theories is that they make sens only if the central charge of the total Virasoro algebra is equal to the critical value 26. This can be seen in many ways. The simplest one is to notice [4] that the central charge of the Faddev Popov ghosts is precisely -26 so that, with the above value, the central charge vanishes when all the fields are included and there is actually no breaking of conformal invariance. For the Veneziano model where only the X field enters this means that $\mathcal{D}$ =26. In the generalized models the total central charge is $C+\mathcal{D}$ where C is the central charge of the additional dynamics. Hence one must satisfy

$$\mathcal{D} = 26 - C \tag{23}$$

If C is an integer larger than one, this will effectively allow to lower the space time dimension.

At this point it is useful to recall some general properties of representations of the Virasoro algebra(13). From the group theory viewpoint it can be regarded as being in a Weyl Cartan basis, L_o being the only operator of the commuting subalgebra, and L_m with n 0 being step operators. Hence a highest weight vector will be such that

$$L_o|0,\mathcal{E}\rangle = \mathcal{E}|0,\mathcal{E}\rangle \; ; \quad L_m|0,\mathcal{E}\rangle = 0 \;, \; m>0 \tag{24}$$

An irreducible representation is characterized by the values of $\mathcal{E}$ and C. The corresponding vector space which is called a Verma module, is spanned by all vectors of the form

$$|\{m\},\mathcal{E}\rangle = \prod_{k>0} (L_{-k})^{m_k} |0,\mathcal{E}\rangle \tag{25}$$

where k, m_k are arbitrary positive integers. They are eigenstates of L_o:

$$L_o|\{m\},\mathcal{E}\rangle = (\mathcal{E}+N)|\{m\},\mathcal{E}\rangle \; ; \; N = \sum_k k\, m_k \tag{26}$$

All eigenvectors with the same eigenvalues are said to belong to the same level N. Kac [5] has considered the matrix of all inner products in a given module which is entirely determined from the Virasoro algebra together with the hermiticity condition

$$L_{-m} = L_m^{+} \tag{27}$$

Hence it is purely alebraic and only depends upon the values of $\mathcal{E}$ and C. It obviously factorizes into products of finite matrices at each level. Kac [5] has obtained a closed formula for each finite determinant. Define the quantity

$$\mathcal{E}(p,q) \equiv \frac{1}{48}\Big[(13-C)(p^2+q^2) - 24pq - 2(1-C) + (p^2-q^2)\sqrt{(C-1)(C-25)}\Big] \tag{28}$$

where p and q are arbitrary integers. If we consider a highest weight representation with $\mathcal{E} = \mathcal{E}(p,q)$ for some given p, q both larger than zero, the Kac determinant vanishes at the level N=pq. This vanishing shows that the metric of the Verma module need not be positive definite. The unitarity of the representation is thus in question. It is easy to show that the negative values of the highest weights are all excluded. For positive values we note that, for $C>1$, $\mathcal{E}(p,q)$ is always negative for $p>1$, $q>1$, i.e. when it corresponds to a zero of a Kac determinant. Hence, in this region the Kac determinants never change sign for positive $\mathcal{E}$ and one can show by explicit construction [6] that there exist a unitary representation for all $\mathcal{E}>0, C>1$. For $C<1$, on the contrary, there are Kac zeroes for positive $\mathcal{E}$ and the positivity of the metric is not assured. It has been shown that unitary representations only exist for [7]

$$C = 1 - 6/(r(r+1)) \quad , \; r \geq 3 \qquad (29)$$

$$\mathcal{E} = \mathcal{E}(p,q) \quad , \; 1 \leq p \leq q < r$$

where r, p, q are integers. Hence the allowed values of $\mathcal{E}$ precisely coincide with zeroes of Kac determinants.

Going back to conformally invariant field thories we recall that, given a covariant operator with weight δ, it is easy to see that the state

$$\lim_{z \to 0} O(z)|0\rangle \qquad (30)$$

is a highest weight vector with $\mathcal{E} = \delta$. The spectrum of highest weights coincides with the set of conformal weights which, in general, involves more than one values. The representation of the Virasoro algebra is thus reducible since the Hilbert space is the sum of the corresponding Verma modules. The covariant operators are intertwening operators between the different irreducible representations. For arbitrary $\mathcal{E}$ and C one can construct an infinite family of covariant operators with weights given by formula (28) for all p and q positive or negative as a natural byproduct of the exact quantum solution of quantum Liouville theory [8]. These operators are not all physical since formula (28) is not always positive. This shows nevertheless that the set of critical dimensions must coincide with Kac formula in general.

In view of formula (28), it is clear that one has to distinguish three regions for the possible values of C.

I-The region $C<1$

As we already pointed out, the Kac zeroes occur for positive highest weights. A systematic discussion has been given [3] which uses this fact, and shows the existence of special values of C

$$C = 1 + \frac{6(r+s)^2}{rs} \qquad (31)$$

where r and s are integers of opposite signs. The virtue of this formula is that then

$$(C-1)(C-25) = 36(r^2-s^2)/r^2s^2 \qquad (32)$$

is the ratio of squares of integers. In view of formula (29), unitarity is satisfied only for r+s=-1. This subseries remarkably reproduces a whole set of standard critical models[7]. In particular for r=3 and 5 one recovers the Ising (or 2 state Potts) model and the 3 state Potts model. A simple calculation shows that if we introduce

$$\sqrt{Q} = 2\cos\left[\frac{\pi}{12}(C-1)\left(1+\sqrt{\frac{(C-25)}{(C-1)}}\right)\right] \tag{33}$$

we obtain the correct number of spin components i.e. Q=2,3 for the Ising and the three state Potts model respectively. The Q state Potts model can be defined for continuous values of Q if we transform it into the random cluster model. For C<1 there exist various equivalent critical models. In particular the Q state critical Potts model is equivalent to a Coulomb gas model. For C<1 we have Q<4 and one is in a Coulomb phase. The point Q=4 corresponds to a point of transition of Kosterlitz-Thouless. Above Q=4 one enters into the plasma phase and the transition becomes first order. We shall come back to this below.

II-The region C>25

This region has some similarities with the region C<1 since in both cases the square root of formula (28) is real. A different approach is needed, however, since now the Kac determinants do not vanish for positive . The region C>1 is naturally covered by the quantum Liouville field theory since its central charge is given by[9]

$$C = 1 + 3/\hbar \tag{34}$$

where $\hbar$ is the Planck constant. The region C>25 corresponds to $\hbar$<1/8 i.e. to the weak coupling regime of Liouville theory which is connected to the semi classical limit $\hbar \sim 0$. In the exact quantum solution[9], special values of C were again found

$$C = 1 + 6(N+1)^2/N \tag{35}$$

They can be put under the form of equation (31) continued to r=N and s=1. The spectrum of weights is again given by formula (28) with

$$\begin{aligned} \mathcal{E} &= \mathcal{E}(1, 2m-N) \quad , \quad 0 \le m \le \nu \, , \quad N = 2\nu+1 \\ \mathcal{E} &= \mathcal{E}(1, 2m-1-N) \quad , \quad 1 \le m \le \nu \, , \quad N = 2\nu \end{aligned} \tag{36}$$

It is easily checked that all this values are larger than 1, and condition (22) is satisfied. The associated string model has no tachyon. However formula (23) shows that $\mathcal{D}$ <1! so that one has not gained much from this viewpoint. The existence of conformally invariant field theories for these values of C does however suggest that there are new critical models. These are models of a new type since all the known critical models have C<1, and a spectrum of conformal weights between 0 and 1. .The unusal feature of the new models is, as we already pointed out, that the two point functions are at most logarithmically divergent. The experimental feature of the transition are thus rather different from the standard ones.

III-The region $1<C<25$

In this case formula (28) gives complex values except when $p=\pm q$. The choice $p=q$ is unacceptable except for $p=q=1$, since it leads to negative ϵ. For three special values of C

$$C = 7 \ , \ 13 \ , \ 19 \tag{37}$$

local fields have been constructed[10] such that the spectrum of weights is given by formula (28) for

$$p=-q=1,2,3,\ldots \tag{38}$$

Condition (22) is again satisfied and the associated string theory has no tachyon. Condition (23) leads to

$$\mathcal{D} = 19 \ , \ 13 \ , \ 7 \tag{39}$$

and there exist new string models for these values of $\mathcal{D}$. From the view point of statistical model one therefore predicts [8] isolated points of second order phase transition for the above values of C. This may be a bit surprising since for $C>1$ one is outside of the Coulomb phase. One can directly see, however that these points must enjoy special properties. Formula (33) when continued for $1<C<25$ leads to Q complex in general. For the values (37) one obtains

$$\sqrt{Q} = 2\cos\left(\frac{\pi}{2} + i\pi\frac{\sqrt{3}}{2}\right) \tag{40}$$

$$\sqrt{Q} = 2\cos\left(\pi + i\pi\right)$$

$$\sqrt{Q} = 2\cos\left(3\frac{\pi}{2} + i\pi\frac{\sqrt{3}}{2}\right)$$

and one can verify that the three special values are the only ones for which Q is real even though the argument of the cosine is complex.

As a conclusion it is clear that the study of conformally invariant field theories from the double view point of string theories and cratical systems has unravelled an interesting structure. The string theories discussed here at not based on free field theories in two dimensions and, hence, the dual amplitudes are difficult to determine. The common feature of all the new conformally invariant field theories discussed here is the appearence of operators of dimension 1. For the associated string theories, it corresponds to the existence of a massless string state. This fact should play a key role in the complete understanding of these models. On the other hand, we must say that, at the present time, the relevance of these new models to particle physics is not yet clear. The supersymmmetric version of the present discussion has been worked out in all details.[11]

REFERENCES

(1) J.-L.GERVAIS,B.SAKITA Nucl.Phys.B34(1971)832
(2) J.-L.GERVAIS,B.SAKITA Nucl.Phys.B34(1971)477
(3) A.A.Belavin,A.M.Polyakov ,A.B.Zamolodchikov Nucl.Phys. B241(1980)333
(4) D.Friedan Les Houches Lectures Notes 1982
(5) V.KAC. Proceeding of the International Congress of Mathematicians Helsinsky 1978;Lecture Notes in Physics vol.94 p.441 Springer Verlag
(6) J.-L.GERVAIS,A.NEVEU COM.MATH.PHYS.100(1985)15
(7) D.FRIEDAN,Z.QIU,S.SHENKER in Vertex Operator in Mathematics and Physics ed.J.LEPOWSKY et al.Springer; Phys.Rev.lett.52(1984)1575
(8) J.-L.GERVAIS,A.NEVEU Nucl.Phys.B257 FS14(1985)59
(9) J.-L.GERVAIS,A.NEVEU NucL.Phys.B224(1983)329;B238(1984) 125
(10) J.-L.GERVAIS,A.NEVEU Phys.lett.151B(1985) 271
(11) J.-F.ARVIS Nucl.Phys.B212(1983) 151; B218(1983) 303. O.BABELON Nucl.Phys.B258(1985)680

Conformal Invariance and String Dynamics

Emil Martinec
Joseph Henry Laboratories, Princeton, NJ 08544

1. Introduction

Further progress in the understanding of string theory is likely to rest on the discovery and elucidation of its underlying invariances. Several directions are currently being pursued – string field theory on loop space[1], twistors[2], and two dimensional nonlinear models[3–5]. By the time our goal is achieved, we may find deep connections between these various approaches. In this talk I would like to discuss the last of these three, in particular results of work done in collaboration with C. Callan, D. Friedan, and M. Perry[4].

Strings extend our notions about the matter content of the world, its gauge symmetry, and even spacetime itself. Currently, the best studied part of string theory is its perturbation series, where all the invariance and geometry have been fixed. The path integral over surfaces embedded in spacetime is a first-quantized description of strings since the basic object is a string trajectory rather than a functional on string. The geometry of spacetime $\mathcal{M}$ is specified a priori, and strings are fluctuations about the background. Since string incorporates gravity, it is important to understand how the background arises from string dynamics; some framework should relate the two. A clue to the connection is provided by the study of the two-dimensional nonlinear model. The two-dimensional nonlinear model consists of the most general action principal for scalar fields X^μ, $\mu = 0, \ldots, D-1$,[3][6] parametrizing a spacetime $\mathcal{M}$

$$S = \int d^2\xi \sqrt{g}\Big[T(X) + G_{\mu\nu}(X) g^{ab}\partial_a X^\mu \partial_b X^\nu + B_{\mu\nu}(X)\frac{\epsilon^{ab}}{\sqrt{g}}\partial_a X^\mu \partial_b X^\nu \\ + \Phi(X) R^{(2)} + T_{\mu\nu\lambda\sigma}(X) g^{ab} g^{cd} \partial_a X^\mu \partial_b X^\nu \partial_c X^\lambda \partial_d X^\sigma + \ldots\Big] \tag{1.1}$$

where we have also coupled the theory to a two-dimensional metric g_{ab} in order to make the theory invariant under reparametrizations of the 2-d coordinates ξ^a, $a = 1,2$. We interpret X^μ as a string fluctuation about the background geometry specified by $T, G_{\mu\nu}, B_{\mu\nu}, \Phi$, *etc.* Clearly this reduces to the usual bosonic string theory[7,8] when we set $G_{\mu\nu} = \eta_{\mu\nu}$, $T = \Phi' = B = \cdots = 0$. These couplings are analogous to those in first-quantized particle theory; there, in the path representation for the propagator

$$\Delta(x,x') = \int \mathcal{D}x(t)\mathcal{D}g e^{i\int dt\sqrt{g}g^{-1}\dot{x}^\mu\dot{x}^\nu G_{\mu\nu}(x)dt} \tag{1.2}$$

The effect of $G_{\mu\nu}(x)$ is to reproduce the correct curved space propagator[9]. The sum over surfaces in the 2-d nonlinear model gives by analogy a first-quantized string propagator in a curved background. Reparametrization and gauge invariance of the background metric are point transformations of the nonlinear model, so the full nonlinear symmetry of the theory is manifest in the background. Transferring this to the fluctuations should teach us about the symmetries of string.

World-sheet reparametrization invariance is the key to consistency of the flat space string[8,7]. Surface reparametrization invariance guarantees that unphysical polarizations of string components decouple from physical processes, and accords with our geometrical picture of first-quantized string dynamics. In the first-quantized particle, reparametrization invariance on the world line means that the particle hamiltonian (which generates shifts in the world line parameter t) has vanishing expectation value. This yields the equation of motion

$$\nabla^2\phi(x) = 0$$

of the field whose propagator is being represented by (1.2). Similarly, coordinate transformations of 2-d quantum field theory are generated by the components of its stress-energy tensor T_{ab}. Reparametrization invariance implies $T_{ab} = 0$

in correlation functions, which is a severe constraint. The conformal group is part of the local diffeomorphism group of two dimensions, so preservation of the geometrical character of string requires that upon quantization (1.1) describes a conformally invariant theory*. Classically this would mean that only the $G_{\mu\nu}$ and $B_{\mu\nu}$ couplings would be allowed, but the necessary condition is conformal invariance in the full quantum theory, and so we shall see that other couplings are possible. Conformal invariance is equivalent to the vanishing of the trace of T_{ab} [11,12]. The trace has the form

$$T^a_a(\xi) = \beta^T\sqrt{g}+\beta^G_{\mu\nu}\sqrt{g}g^{ab}\partial_a X^\mu\partial_b X^\nu+\beta^B_{\mu\nu}\epsilon^{ab}\partial_a X^\mu\partial_b X^\nu+\beta^\Phi\sqrt{g}R^{(2)}+\dots \quad (1.3)$$

The β's are local functionals of the couplings $T, G_{\mu\nu}, B_{\mu\nu}, \Phi$,*etc.* The requirement $T^a_a = 0$ is a local equation on the world sheet, which we take to be a set of dynamical equations on the string background (note that there is one for each tensor coupling, hence the right number of equations). For the moment, let us concentrate on the massless particle couplings, which correspond to marginal operators of the 2-d quantum field theory; we will return later to the question of the relevant (tachyon) and irrelevant (massive tensor) couplings. We assume that the background is weakly coupled to the string; physically, this means that the typical size of a string fluctuation, $\sqrt{\alpha'}$, is small compared to the radius of the manifold, so that a plane-wave approximation is valid. To calculate, we choose the 2-d coordinate gauge $g_{ab} = e^{2\sigma}\delta_{ab}$ and work in complex coordinates $z, \bar{z}$ ($\delta_{z\bar{z}} = 1, \delta_{zz} = \delta_{\bar{z}\bar{z}} = 0$). The dimensionally continued action for massless backgrounds is

$$S = \frac{1}{2\pi\alpha'}\int d^2z\, e^{(d-2)\sigma}[\tfrac{1}{2}G_{\mu\nu}\partial X^\mu\bar{\partial}X^\nu + \tfrac{1}{2}B_{\mu\nu}\partial X^\mu\bar{\partial}X^\nu + \alpha'(-4\partial\bar{\partial}\sigma)\Phi] \quad (1.4)$$

The trace T^a_a is the variation of the effective action with respect to σ, and may

* Conformal invariance was first proposed in the early days of the dual resonance model as a criterion for model building [10]. At that time, different conformally invariant theories were considered to yield different hadronic models. Our proposal further asserts that all such theories (within a given class – bosonic, heterotic, type II, *etc.*) are solutions to the *same* underlying theory.

be calculated using standard background field perturbation theory[13], generating propagators and vertices from (1.4) and calculating the graphs with linear σ dependence. Apart from the dilaton, which gives a classical contribution to the trace, the terms which actually contribute to $T^a_a = t_{z\bar{z}}$ are the counterterms which must be added to the Lagrangian to cancel poles in $d-2$. Expansion of the volume element in powers of $d-2$ yields finite terms linear in σ. Terms with higher powers of σ are consequences of these together with the renormalization group pole equations[6,13]. The results are $(H_{\mu\nu\lambda} = 3\nabla_{[\mu}B_{\nu\lambda]})$

$$\begin{aligned}
\beta^{\Phi} &= \frac{1}{\alpha'}\frac{D-26}{48\pi^2} + \frac{1}{16\pi^2}[4(\nabla\Phi)^2 - 4\nabla^2\Phi - R + \tfrac{1}{12}H^2] + o(\alpha') \\
\beta^{G}_{\mu\nu} &= R_{\mu\nu} - \tfrac{1}{4}H_{\mu}{}^{\lambda\sigma}H_{\nu\lambda\sigma} + 2\nabla_{\mu}\nabla_{\nu}\Phi + o(\alpha') \\
\beta^{B}_{\mu\nu} &= \nabla_{\lambda}H^{\lambda}{}_{\mu\nu} - 2(\nabla_{\lambda}\Phi)H^{\lambda}{}_{\mu\nu} + o(\alpha')
\end{aligned} \tag{1.5}$$

What do these equations mean? When β^G and β^B vanish, the theory is globally scale invariant since $\sqrt{g}R^{(2)} = -4\partial^2\sigma$ is independent of global scale. Moreover, the combination of global scale invariance and locality of the 2-d field theory implies local conformal invariance[11]. The dependence of the theory on the background *two-dimensional* metric then resides solely in the trace anomaly β^{Φ}[7]. β^{Φ} is related to the Schwinger term c in the algebra of the traceless part of the stress tensor (the Virasoro algebra[14])

$$[T_{zz}, T_{ww}] = (T_{zz} + T_{ww})\delta'(z-w) + \frac{c}{12}\delta'''(z-w) \tag{1.6}$$

by the conservation of stress-energy[12]

$$\partial_z T_{z\bar{z}} + \partial_{\bar{z}} T_{zz} = 0 \tag{1.7}$$

The Jacobi identity implies that $c = const.$, but it is not obvious that β^{Φ} in (1.5) has this property. Fortunately, on applying the Bianchi identities to the set of equations (1.5), we find

$$\nabla^\mu \beta^G_{\mu\nu} = \nabla_\nu \beta^\Phi \tag{1.8}$$

so that when β^G and β^B vanish, β^Φ is indeed a constant in the background space. We expect this feature to continue to hold in higher orders of perturbation theory, and hope to report results soon[15].

Polyakov[7] considered the case $\beta^\Phi \neq 0$ for the special case of a flat background, and showed that the trace anomaly generates an effective dynamics for the scale part of the metric, $e^{2\sigma}$. In principle such a dynamics could restore the local scale invariance of the nonlinear model[12], and Fradkin and Tseytlin[8] have proposed that this could yield an off-shell dynamics for the nonlinear model. The question is whether this is the right way to go off-shell. In practice no acceptable quantization of the resulting Liouville-type theory has been found in the required range $1 < c < 25$, despite many attempts[16]. Moreover, since the scale factor acts as a longitudinal string coordinate, the graviton would in general be massive in such a theory, leading to a "spontaneous breakdown" of general coordinate invariance in the target space $\mathcal{M}$. Therefore we demand not just conformal invariance under $z \to f(z)$ but rather the stronger condition of Weyl invariance under $g_{ab} \to \Lambda(z,\bar{z}) g_{ab}$. This means we demand that $\beta^\Phi = 0$ as well. The meaning of (1.1) away from $T^a_a = 0$ is as yet unclear.

It turns out that the set of "beta-functions" (1.5) can be derived from the first variation of the target-spacetime action

$$\int d^D x \sqrt{G}\, e^{-2\Phi} [\tfrac{D-26}{3\alpha'} + R + 4(\nabla\Phi)^2 - \tfrac{1}{12} H^2] \tag{1.9}$$

It is interesting to note that to lowest order c is the cosmological constant. The overall factor of $e^{-2\Phi}$ is a reflection of the fact that the constant part of Φ multiplies the Euler character $\chi = \frac{1}{4\pi}\int d^2\xi \sqrt{g} R^{(2)}$ of the 2-d action, and is thus the coupling constant of string loop perturbation theory. We see that (1.9) is precisely the action for Einstein gravity coupled to an antisymmetric tensor and dilaton. Higher order contributions to the beta functions should produce higher derivative terms in this action. We find it quite remarkable that the beta functions of the 2-d nonlinear model are the gradients of a scalar function on coupling space. If true to all orders, it would indicate a deeper geometric structure than is first apparent in 2-d quantum field theory, for it is by no means guaranteed that the beta function (a vector field on the space of couplings) be the gradient of a scalar function.

The conformal invariance of the nonlinear model is not only the equation of motion for the background, but also that of the string fluctuations as well[17]. On-shell string vertex operators have non-vanishing scaling dimension. However there is no scale which a string one-point function could depend upon, so the first variation with respect to *string* fluctuations about a solution (which is simply a string vertex one-point function) must vanish also as a consequence of scale invariance. This argument relies on the scale invariance of the vacuum of the 2-d quantum field theory, which is only a property of the theory on the sphere (string tree level); on higher genus surfaces (string loops), dilations are not a symmetry and one usually doesn't expect the vanishing of tadpoles. Indeed, explicit calculations in the flat space bosonic string[18] confirm the presence of a divergence associated with dilaton emission into the vacuum. This just means that there are generically loop corrections to the string equations of motion, but it also means that the connection between conformal invariance and the equations of motion may only be valid at tree level unless one can find a way to go off-shell in the background of the nonlinear model. It then might be possible to, say, cancel the one-string-loop cosmological constant against the tree-level one in (1.9) by going away from $D = 26$. Conversely, a proof that the cosmological constant vanishes in superstring theory might amount to demonstrating that spacetime is ten-dimensional (non-perturbatively). The situation in supersymmetric string theories may be much simpler, where in backgrounds which admit a conserved supersymmetry charge it may be possible to prove that tadpoles vanish by supersymmetry[19]. Tree level solutions are then solutions to all orders of string loops.

2. The Heterotic String

Let us, then, turn to discuss one such theory, the heterotic string[20]. This nonlinear model has a (1,0)[21] or Majorana-Weyl supersmmetry on the world sheet: only the left-moving bosons have fermionic world sheet partners. We work in the older Neveu-Schwarz-Ramond fermionic string formalism[22], which when projected onto even world-sheet fermion number yields the superstring[23]. Right-moving fermions exist, but they are used to form a chiral $E_8 \otimes E_8$ or $SO(32)$ current algebra. The massless bosons of this string include the gauge bosons of the current algebra as well as the usual fields of (1.4). Backgrounds for these string modes can be written in chiral superspace $z, \bar{z}, \theta$; again the equation of motion is $T_{z\bar{z}} = 0$. The one-loop results for β^{Φ}, β^{G}, and β^{B} are unchanged; in addition, there is a beta function for A^a_μ

$$\beta^A = \nabla^\mu F^a_{\mu\nu} + \ldots \tag{2.1}$$

The general two-loop calculation of $T_{z\bar{z}}$ includes many terms; in particular one finds the expected $\mathrm{tr}\{F_{\mu\lambda}F^{\lambda}_{\ \nu}\}$ source for the metric dynamics β^G, which cancels against the higher-order curvature contribution $R_{\mu\lambda\sigma\tau}R_\nu^{\ \lambda\sigma\tau}$ when the spin connection is embedded in the gauge group (as for Calabi-Yau backgrounds[17]). One also discovers the Chern-Simons completions of the torsion

$$H \to \tilde{H} = dB + \frac{\alpha'}{8}[\mathrm{tr}(F \wedge A) - \mathrm{tr}(R \wedge \omega)] \tag{2.2}$$

as a consequence of the chiral anomalies of the world sheet Lorentz and gauge fermions. This feature of the Green-Schwarz anomaly cancellation mechanism[24] can already be seen at the lagrangian level[21] where the anomalies in the integration measures of ψ^μ and ψ^i result in contributions to the Wess-Zumino coupling

B of just the requisite variety:

$$\delta_\Lambda B = \mathrm{tr}\{\Lambda_{gauge}\, dA - \Lambda_{Lorentz}\, d\omega\} \tag{2.3}$$

Thus we find yet another geometrical structure manifested in the 2-d nonlinear model.

Little is known about the general solution of the beta function equations. One exception is the case $B = \Phi' = R_{\mu\nu} = 0$ where in addition the spin connection is embedded in the gauge group $\omega_\mu^{ij} = A_\mu^a T_a^{ij}$. Then the resulting nonlinear model has (1,1) supersymmetry ((2,2) if $\mathcal{M}$ is Kähler) and may be a solution to all orders[25]; explicit calculations have verified scale invariance through three loops[26]. More generally, when $B \neq 0$ the model is still left-right symmetric in its couplings when the torsion connection is equated with the gauge connection $\omega + H = A$. Solutions of this sort are currently under investigation[27]. Group manifolds are another class of solutions. Here the torsion H parallelizes the manifold, which has so much symmetry that the model is exactly soluble[28]. These solutions are of less phenomenological interest because they all break supersymmetry at the compactification scale[29]. From the equations of motion (1.5) it is not difficult to see that when $H \neq 0$, c always shifts away from the naive dimensionality of the target space to leading order. Unless this shift is corrected in higher orders, it is likely that supersymmetry would always break in such a case.

3. Vertex Operators

Suppose we have found an exact solution to the full 2-d quantum field theory: $T_a^a = 0$ to all orders. We then might like to catalogue the operators of the theory, their scaling dimensions, and operator products, in order to construct the string emission vertices and string scattering amplitudes in the presence of the background. In the flat space string theory we have

$$\begin{aligned} dim[e^{ik\cdot X}] &= k^2 \\ dim[\partial^n X^\mu] &= n \\ e^{ik\cdot X(z)} e^{ip\cdot X(w)} &\sim (z-w)^{p\cdot k}\, e^{i(k\cdot X(z)+p\cdot X(w))} \end{aligned} \tag{3.1}$$

and so on for more complicated operators. Using these building blocks, we can construct conformally invariant composite operators; these are the string vertices. In a purely metric background, the perturbative anomalous dimension operator is the Laplacian in the appropriate tensor space on the target manifold[6]

$$dim[T(X)] = \nabla^2 T(X) + o(\alpha') \tag{3.2}$$

and gains the appropriate generalization in arbitrary backgrounds. The traceless stress tensors T_{zz} and $T_{\bar{z}\bar{z}}$ may be expanded in modes

$$\begin{aligned} T_{zz} &= \sum_n L_n z^{n-2} \\ T_{\bar{z}\bar{z}} &= \sum_n \bar{L}_n \bar{z}^{n-2} \end{aligned} \tag{3.3}$$

The anomalous dimension operator (3.2) is of course the dilation operator $L_0+\bar{L}_0$. Those eigenvectors of L_0, $\bar{L}_0$ with eigenvalues $h = (1,1)$ which commute with the positive frequency Virasoro operators L_n, $\bar{L}_n$, $n > 0$, are string emission vertices just as in flat space[8]; their integrals over the world sheet are reparametrization invariant and generate string scattering amplitudes.

The nonlinear model could provide a means to study string backgrounds such as the Calabi-Yau compactifications[17] of the superstring. Let X^μ and Y^m parametrize 4-d Minkowski space and the 6-d Calabi-Yau manifold with metric G_{mn}, respectively. Conformal operators $V^{int}(Y)$ with $h_V = (1,1)$ of the internal space nonlinear model correspond to massless particles in four dimensions since $e^{ik\cdot X} V^{int}(Y)$, $k^2 = 0$, is an acceptable string emission vertex. For example, the radius of the internal space is undetermined in such a compactification, and therefore there is a massless dilaton field representing its fluctuations. This means that the Calabi-Yau nonlinear model has a corresponding $h = (1,1)$ operator. The massless dilaton vertex operator is

$$V_{dil} = e^{ik\cdot X} G_{mn}(Y) DY^m \bar{D} Y^n \tag{3.4}$$

Since these manifolds admit a complex structure J_{mn}, the dilaton has a massless axial partner

$$V_{axion} = e^{ik\cdot X} J_{mn}(Y) DY^m \bar{D} Y^n \tag{3.5}$$

arising from the antisymmetric tensor coupling B_{mn}. The coupling (3.5) is like the original dilaton coupling in (1.4) in that at zero momentum $k = 0$ it is the instanton density of this Kähler sigma model, and hence topological in character. Each of the massless particles in the low-energy effective theory has its corresponding vertex operator in the full string theory about this background. The existence of these massless particles are often consequences of index theorems (and supersymmetry) on the Calabi-Yau space[17]. The correlation functions of the massless particle vertex operators are of great phenomenological interest; they give the effective action of the compactified theory. The three-point functions are the operator product coefficients of the 2-d conformal quantum field theory[11]; they are also the Yukawa coupling constants of the tree level effective theory. Just as the number of each type of massless field is determined by an index, the operator product coefficients of the massless vertices may also contain topological information, being given by intersection numbers of differential forms[30]. Of particular interest as well are the dimension (1,0) fermion operators whose integrals are the conserved supersymmetry charges. Their study should reveal a great deal about the nature of supersymmetry in the string theory. It is an important problem to see how geometry is contained in the nonlinear model on these spaces.

Finally, I would like to discuss the more general couplings in (1.1). The metric is of indefinite signature, so eq.(3.2) has a whole manifold of solutions rather than the finite number characteristic of 2-d conformal field theories with positive metric Hilbert space. In a sense all vertex operators are marginal operators of the nonlinear model since they are by construction scale invariant when integrated over the world sheet. This is what gives us hope that the full nonlinear model (1.1) will describe the dynamics of all the string modes. Indeed, to lowest order, (3.2) gives the free equation for a massive tensor field $T_{\mu\nu\cdots}(X)\partial X^{\mu}\partial^{2}X^{\nu}\cdots$, the mass being the sum of the number of ∂'s in the vertex. By arranging for the anomalous dimension of the tensor T to cancel the canonical dimension of the derivatives, we have constructed a dimension (1,1) operator. If the operator is truly marginal, it is then a valid perturbation of the nonlinear model. The possibility that conformal invariance could be the dynamical principle of the full string theory, with $T^{a}_{a} = 0$ the equation that propagates string, is currently under investigation[15].

4. Conclusion

I hope that I have been able to convey a sense of the richness of structure in the 2-d nonlinear model, and how it parallels that of the string theory. The massless background fields incorporate a great deal of geometry – gauge invariance (and anomaly cancellation), index theory, topology – in an algebraic package, the conformal algebra of the nonlinear model. Algebraic geometry has already played a significant role in sorting out the various Calabi-Yau compactifications of the superstring. We would like to see how to find all this structure in the Virasoro algebra representations of the nonlinear model. Algebraic methods might then provide even more powerful tools for the analysis of string geometry; the unfolding of this structure should have profound consequences for string theory.

ACKNOWLEDGEMENTS

It is a pleasure to thank my collaborators C. Callan, D. Friedan, and M. Perry, as well as S. Shenker, A. Strominger, and E. Witten, for enlightening discussions.

REFERENCES

1. T. Banks and M. Peskin, talk at the Argonne Symposium on Anomalies, Geometry, and Topology (to appear in the proceedings), and SLAC-PUB-3740 (July 85);

   W. Siegel and B. Zwiebach, Berkeley preprint UCB-PTH-85/30 (July 85): K. Itoh, T. Kugo, H. Kunimoto, and H. Ooguri, Kyoto preprint KUNS-800-HE(TH) 85/04;

   D. Friedan, Chicago preprint EFI-85-27 (April 85);

   M. Kaku and J. Lykken, contribution to the Proceedings of the Argonne Symposium on Anomalies, *etc.*(March 85);

   A. Neveu and P. West, CERN-TH-4200/85 (June 85).
2. E. Witten, Princeton preprint (May 85).
3. D. Friedan and S. Shenker, Talk at the Aspen Summer Inst. (1984), unpublished; C. Lovelace, *Phys. Lett.* **135B** (1984), 75, and to appear; E. Fradkin and A. Tseytlin, Lebedev preprints (1984, 85).
4. C. Callan, D. Friedan, E. Martinec, and M. Perry, Princeton preprint (June 85).
5. A. Sen, Fermilab preprints (1985).
6. D. Friedan, PhD thesis (1980), *Ann. Phys.*, to be published.
7. A. M. Polyakov, *Phys. Lett.* **B103** (1981), 207, 211.

8. For reviews, see *e.g.* S. Mandelstam, *Phys. Rep.* **13C** (1974), 259; J. Scherk, *Rev. Mod. Phys.* **47** (1975), 123; and references therein.

9. B. DeWitt, *Rev. Mod. Phys.* **29** (1957), 377.

10. S. Fubini and G. Veneziano, *N. Cim.* **67A** (1970), 29, *Ann. Phys.* **63** (1971), 12;
E. Corrigan and C. Montonen, *Nucl. Phys.* **B36** (1972), 58;
J.-L. Gervais and B. Sakita, *Nucl. Phys.* **B34** (1971), 477.

11. Belavin, A. M. Polyakov, and A. B. Zamolodchikov, *Nucl. Phys.* **B241** (1984), 333;
D. Friedan, Z. Qiu, and S. Shenker, in *Proc. of a Conf. on Vertex Operators*, J. Lepowsky *et. al.*, eds. (Springer-Verlag 1984).

12. D. Friedan, in the 1982 Les Houches Summer School, *Recent Advances in Field Theory and Statistical Mechanics*, J. Zuber and R. Stora, eds. (North-Holland 1984).

13. L. Alvarez-Gaume, D. Z. Freedman, and S. Mukhi, *Ann. Phys.* **134** (1981), 85.

14. M. A. Virasoro, *Phys. Rev.* **D1** (1970), 2933.

15. C. Callan, D. Friedan, E. Martinec, and M. Perry, to appear.

16. J.-L. Gervais and A. Neveu, *Nucl. Phys.* **B199** (1982), 59; **B209** (1982), 125; **B224** (1983), 329; **B238** (1984), 125; **B238** (1984), 301;
R. Marnelius, *Nucl. Phys.* **B211** (1983), 14; **B221** (1983), 409;
E. D'Hoker and R. Jackiw, *Phys. Rev.* **D26** (1982), 3517;
E. D'Hoker, D. Z. Freedman, and R. Jackiw, *Phys. Rev.* **D28** (1983), 2583;
T. Yoneya, *Phys. Lett.* **148B** (1985), 111;
T. Curtright and C. Thorn, *Phys. Rev. Lett.* **48** (1982), 1309;
E. Braaten, T. Curtright, and C. Thorn, *Phys. Lett.* **B118** (1982), 115.

17. P. Candelas, G. Horowitz, A. Strominger, and E. Witten, *Nucl. Phys.* **B258** (1985), 46.

18. J. Shapiro, *Phys. Rev.* **D11** (1975), 2937;
M. Ademollo *et. al.*, *Nucl. Phys.* **B94** (1975), 221.

19. D. Friedan, E. Martinec, and S. Shenker, to appear.

20. D. Gross, J. Harvey, E. Martinec, and R.Rohm,
Nucl. Phys. **B256** (1985), 253.

21. C. Hull and E. Witten, Princeton preprint (June 1985).

22. A. Neveu and J. Schwarz, *Nucl. Phys.* **B31** (1971), 86;
P. Ramond, *Phys. Rev.* **D3** (1971), 2415.

23. F. Gliozzi, J. Scherk, and D. Olive, *Nucl. Phys.* **B122** (1977), 253.

24. M. Green and J. Schwarz, *Phys. Lett.* **B149** (1984), 117.

25. L. Alvarez-Gaume and P. Ginsparg, Harvard preprint HUTP 85/A030;
L. Alvarez-Gaume, S. Coleman, and P. Ginsparg, Harvard preprint HUTP 85/A037;
C. Hull, IAS preprints (May 1985).

26. L. Alvarez-Gaume, *Nucl. Phys.* **B184** (1981), 180.

27. C. Hull, MIT preprint (June 1985).

28. V. Knizhnik and A. B. Zamolodchikov, *Nucl. Phys.* **B247** (1984), 83.

29. D. Friedan and S. Shenker, unpublished.

30. A. Strominger and E. Witten, IAS preprint (1985).

FINITE N=2 SUPERSYMMETRIC σ-MODELS†

Paul Ginsparg
Lyman Laboratory of Physics
Harvard University
Cambridge, MA 02138

In this talk, I shall describe some recent work[1][2] on the finiteness of certain two dimensional supersymmetric non-linear σ-models. After introducing the models in question and some formalism, I will first show that N=4 supersymmetric models are on-shell ultraviolet finite to all orders in perturbation theory. This means that to all orders in perturbation theory, all ultraviolet divergences can be removed from all Greens functions by using only renormalization counterterms that arise from a (possibly non-linear) redefinition of the fields. I will then show that the same result holds for N=2 supersymmetric models, provided that the target manifold M on which the model is defined is Ricci flat. The relation between these models and superstring theories formulated on non-trivial background spacetimes is detailed extensively elsewhere in these proceedings (see, for example, the talks by D. Gross and E. Martinec), so I will have little to say in this regard.

Non-linear σ-models are the quantum field theories of maps from spacetime into a riemannian manifold M. Let (M_n, g) be an n-dimensional riemannian manifold with metric g_{ij}. For two dimensional spacetime, an N=1 supersymetric σ-model with scalar fields taking values on M_n is given by the lagrangian[3]

$$\begin{gathered}\mathcal{L} = \frac{1}{2}g_{ij}(\phi)\partial_\mu\phi^i\partial^\mu\phi^j + \frac{i}{2}g_{ij}\overline{\psi}^i\gamma^\mu D_\mu\psi^j + \frac{1}{12}R_{ijkl}\overline{\psi}^i\psi^k\overline{\psi}^j\psi^l \\ D_\mu\psi^i = \partial_\mu\psi^i + \Gamma^i_{jk}\partial_\mu\phi^j\psi^k, \quad \gamma^0 = \sigma_2,\ \gamma^1 = i\sigma_1,\end{gathered} \tag{1}$$

where Γ^i_{jk} is the Christoffel connection and R_{ijkl} is the Riemann curvature tensor. The ϕ's are bosonic fields which represent geometrically a coordinate system on M, and the fermions ψ^i are two-component Majorana fermions which behave like vectors under coordinate reparametrizations. The supersymmetry transformation rules for (1) are given by $\delta\phi^i = \overline{\epsilon}\psi^i$, $\quad\delta\psi^i = -i\partial\!\!\!/\phi^i\epsilon - \Gamma^i_{jk}(\overline{\epsilon}\psi^j)\psi^k$. The superspace form of the action (1), written in terms of a 2-dimensional real superfield $\Phi^i = \phi^i + \overline{\theta}\psi^i + \frac{1}{2}\overline{\theta}\theta F^i$ with θ a real two-component constant Majorana spinor, is the

† supported in part by NSF contract PHY-82-15249

naive extension

$$\mathcal{L} = \frac{1}{4i}\int d^2\theta\, g_{ij}(\Phi)\,\overline{D}\Phi^i D\Phi^j$$
$$D_\alpha\Phi^i = \left(\frac{\partial}{\partial\overline{\theta}^\alpha} - i(\gamma^\mu\theta)_\alpha\frac{\partial}{\partial x^\mu}\right)\Phi^i \tag{2}$$

of the purely bosonic σ-model. The superfield supersymmetry transformation rule is given by $\delta\Phi^i = \overline{\epsilon}_\alpha\left(\frac{\partial}{\partial\overline{\theta}_\alpha} + i(\partial\!\!\!/\theta)_\alpha\right)\Phi^i$.

A few years ago, it was shown[4] that the ultraviolet properties of supersymmetric non-linear σ-models can be studied using the geometrical properties of Kähler manifolds. Indeed the action (1),(2) admits a second supersymmetry[5][6] if and only if M is a Kähler manifold. (M_n, g) is said to be Kähler if there exists a tensor $f^i{}_j$ (the complex structure) which satisfies

$$f^i{}_k f^k{}_j = -\delta^i{}_j, \quad \nabla_i f^j{}_k = 0, \quad \text{and} \quad g_{ij}f^i{}_k f^j{}_l = g_{kl}. \tag{3}$$

(3) first of all implies that M is a complex manifold, i.e. that M can be covered patchwise with complex coordinates ϕ^α such that the transition functions in overlapping coordinate patches are holomorphic. In adapted complex coordinates ϕ^α and $\bar{\phi}^\alpha \equiv \phi^{\bar{\alpha}}$, where α and $\bar{\alpha}$ run from 1 to $n/2$, the complex structure $f^i{}_j$ becomes simply multiplication by i $(-i)$ on holomorphic (antiholomorphic) vectors (i.e. $f^\alpha{}_\beta = i\delta^\alpha{}_\beta$, $f^{\bar{\alpha}}{}_{\bar{\beta}} = -i\delta^{\bar{\alpha}}{}_{\bar{\beta}}$; $f^\alpha{}_{\bar{\beta}} = f^{\bar{\alpha}}{}_\beta = 0$). It then follows from the last condition in (3) that the only non-vanishing components of the metric in these coordinates are the mixed components $g_{\alpha\bar{\beta}}$. The conditions (3), giving $\nabla_l\left(g_{kj}f^k{}_i\right) = 0$, moreover require that the Kähler-form J, defined as

$$J = \frac{1}{2}g_{kj}f^k{}_i\, d\phi^i \wedge d\phi^j = ig_{\alpha\bar{\beta}}\, d\phi^\alpha \wedge d\bar{\phi}^\beta, \tag{4}$$

be closed: $dJ = 0$. This is equivalent to the curl-free conditions

$$\partial_\gamma g_{\alpha\bar{\beta}} = \partial_\alpha g_{\gamma\bar{\beta}} \qquad \partial_{\bar{\gamma}} g_{\alpha\bar{\beta}} = \partial_{\bar{\beta}} g_{\alpha\bar{\gamma}}, \tag{5}$$

which implies that there exists in any fixed coordinate patch on M a scalar function $K(\phi,\bar{\phi})$, called the Kähler potential, such that $g_{\alpha\bar{\beta}} = \partial_\alpha\partial_{\bar{\beta}}K$. In general, K is only defined patchwise and may vary from patch to patch by $K(\phi,\bar{\phi}) \to K(\phi,\bar{\phi}) + f(\phi) + \bar{f}(\bar{\phi})$, where $f(\phi)$ depends on the holomorphic transition functions between coordinates on the different patches.

On a Kähler manifold the standard formulæ of riemannian geometry are simplified. For example, the only non-vanishing components of Γ^i_{jk} are $\Gamma^\alpha_{\beta\gamma} =$

$g^{\alpha\bar{\rho}}\partial_{\beta}g_{\gamma\bar{\rho}}$ and $\Gamma^{\bar{\alpha}}_{\bar{\beta}\bar{\gamma}} = (\Gamma^{\alpha}_{\beta\gamma})^*$ so that the only non-vanishing components of the curvature tensor are $R^{\alpha}{}_{\beta\bar{\rho}\gamma} = \partial_{\bar{\rho}}\Gamma^{\alpha}_{\beta\gamma}$ (together with those related by the usual symmetries of the Riemann tensor). The cyclic and Bianchi identities reduce to

$$R_{\alpha\bar{\beta}\gamma\bar{\delta}} = R_{\gamma\bar{\beta}\alpha\bar{\delta}}, \quad R_{\alpha\bar{\beta}\gamma\bar{\delta}} = R_{\alpha\bar{\delta}\gamma\bar{\beta}} \tag{6a}$$

$$\nabla_{\lambda} R_{\alpha\bar{\beta}\gamma\bar{\delta}} = \nabla_{\alpha} R_{\lambda\bar{\beta}\gamma\bar{\delta}}, \quad \nabla_{\bar{\lambda}} R_{\alpha\bar{\beta}\gamma\bar{\delta}} = \nabla_{\bar{\beta}} R_{\alpha\bar{\lambda}\gamma\bar{\delta}}. \tag{6b}$$

Finally, the Ricci tensor on a Kähler manifold turns out to take the simple form

$$R_{\alpha\bar{\beta}} = g^{\mu\bar{\lambda}} R_{\bar{\lambda}\mu\alpha\bar{\beta}} = -\partial_{\alpha}\partial_{\bar{\beta}} \ln\det(g). \tag{7}$$

It can be shown[6] that each supersymmetry beyond N=2 requires an independent Kähler structure $f^{(a)i}{}_j$ satisfying

$$f^{(a)i}{}_k f^{(b)k}{}_j + f^{(b)i}{}_k f^{(a)k}{}_j = -2\delta^i{}_j \delta^{ab}. \tag{8}$$

It should be clear that N=3 supersymmetry automatically implies N=4 because if $f^{(1)}$ and $f^{(2)}$ are two Kähler structures which satisfy (3) and (8), so also does $f^{(3)i}{}_j = f^{(1)i}{}_k f^{(2)k}{}_j$. If M is an irreducible manifold (i.e. does not split into a product of lower dimensional manifolds), then N=4 is the largest number of supersymmetries that one can have. The reason is that if the manifold is irreducible, the tangent space provides a real irreducible representation of the holonomy group, but the $f^{(a)}$'s, since they are covariantly constant, commute with the action of the holonomy group. By Schur's lemma, a representation is real irreducible only if all the matrices which commute with the representation form a division algebra over the real numbers. The only non-trivial possibilities are the complex numbers (in which case the single complex structure plays the role of the imaginary unit) or the quaternions (in which case the $f^{(a)}$'s, a=1,2,3, represent the three imaginary units). The first case corresponds to N=2 supersymmetry in which the holonomy group of a manifold of $2n$ real dimensions is reduced from $SO(2n)$ to $U(n)$, and the manifold is Kähler. The second case is that of N=4 supersymmetry in which the holonomy group of a manifold of $4n$ real dimensions is further reduced to the $Sp(n)$ subgroup of $U(2n)$, and the manifold is then called hyperkähler.

We now briefly recall the methodology for analyzing the ultraviolet divergences induced by quantum corrections to the σ-model (2). Study of the ultraviolet structure of (2) is greatly facilitated by use of the superspace background field method (see [7] for details) which involves expanding Φ around a generic solution Φ_0 to equations of motion derived from (2), i.e. the superspace geodesic equations

$\overline{D}D\Phi_0^i + \Gamma^i_{jk}(\Phi_0)\overline{D}\Phi_0^j D\Phi_0^k = 0$. It is preferable to parametrize Φ^i in terms of normal coordinates centered upon Φ_0^i, giving Φ^i is in terms of the tangent vector ξ^i at Φ_0^i tangent to the unit speed geodesic joining Φ_0^i to Φ^i. The advantage of this procedure is that the functional Taylor expansion of the action around Φ_0 has manifestly covariant coefficients which can be written exclusively in terms of the curvature tensor and its covariant derivatives. For example, expanding to second order in ξ gives[7]

$$\mathcal{L} = \frac{1}{4i}\int d^2\theta \Big[g_{ij}(\Phi_0)\overline{D}\Phi_0^i D\Phi_0^j + \\ g_{ij}(\Phi_0)\overline{D}\xi^i D\xi^j + R_{iklj}\xi^k\xi^l\overline{D}\Phi_0^i D\Phi_0^j + O(\xi^3)\Big]. \tag{9}$$

It is easy then to verify that the one-loop divergences are proportional to $R_{ij}\overline{D}\Phi_0^i D\Phi_0^j$ and are thus generated by the Ricci tensor.

Given a regularization prescription which preserves supersymmetry, any divergences are related by supersymmetry and higher order counterterms can always be chosen in the form

$$\Delta\mathcal{L} = \frac{1}{4i}\int d^2\theta\, T_{ij}\,\overline{D}\Phi^i D\Phi^j, \tag{10}$$

i.e. simply duplicating the structure (2). The renormalization of the original lagrangian can thus always be expressed in terms of changes in the metric tensor, and the renormalization group equations[8] take the form

$$\mu\frac{d}{d\mu}g_{ij} = -\beta_{ij}(g), \qquad \beta_{ij}(g) = \frac{1}{2\pi}R_{ij} + \ldots .$$

It is important to note that in the normal coordinate expansion (9), the vertices of the theory contain explicitly neither the metric nor the Ricci tensor; they involve only the Riemann tensor and its covariant derivatives. Since loop computations involve only such vertices and their contractions with the metric g^{ij} (the propagator of the quantum field ξ^i is proportional to $g^{ij}(\Phi_0)$), the counterterms which can be generated in perturbation theory are somewhat constrained. There is no way, for example, to generate explicitly a counterterm of the form $g_{ij}S$, where S is a scalar function constructed in terms of curvatures and covariant derivatives (see [7] and references therein for more details). It is moreover possible to determine whether a given tensor may appear at a given order in the loop expansion by probing with a constant conformal rescaling of the metric: $g_{ij} \to \lambda^{-1}g_{ij}$. Since this would be equivalent to a rescaling of $\hbar$ (had it appeared) in the original action, it follows that λ can be used as a loop counting parameter. Thus if a tensor T_{ij} appears at ℓ-loop order, it must behave as $T_{ij} \rightarrow \lambda^{\ell-1} T_{ij}$ under the conformal rescaling. For example, at one-loop order, only tensor counterterms with zero conformal weight (ℓ=0) are allowed and there are only two such tensors: R_{ij} and $g_{ij}R$, where R is the Ricci scalar. But we have already argued that $g_{ij}R$ cannot appear in perturbation theory so we quickly conclude that the only possible tensor counterterm at the one-loop level is the Ricci tensor itself. Finally, a regularization procedure preserving N=2 supersymmetry will insure that counterterms can be chosen to preserve it as well. This means that the tensor T_{ij} appearing in (10) will be a so-called Kähler tensor, i.e. satisfying both the hermiticity condition $T_{ij}\, f^i{}_k f^j{}_l = T_{kl}$ (analogous to the last condition in (3), implying that only the mixed components $T_{\alpha\bar{\beta}}$ and $T_{\bar{\alpha}\beta}$ may be non-zero in complex coordinates), and the Kähler (curl-free) condition (5):

$$\partial_\lambda T_{\alpha\bar{\beta}} = \partial_\alpha T_{\lambda\bar{\beta}}, \quad \partial_{\bar{\lambda}} T_{\alpha\bar{\beta}} = \partial_{\bar{\beta}} T_{\alpha\bar{\lambda}}. \tag{11}$$

If the theory (2) is regulated in (2$-\epsilon$)-dimensions with supersymmetric dimensional regularization, the bare metric g^B is given in terms of the renormalized metric g^R by

$$g^B_{ij} = \mu^\epsilon \left[g^R_{ij} + \sum_{\nu=1}^{\infty} \frac{T^{(\nu)}_{ij}(g^R)}{\epsilon^\nu} \right], \tag{12}$$

representing the on-shell renormalization of the the model in terms of changes of the metric g_{ij}. The renormalization counterterms generated in perturbation theory are given by symmetric tensors $T^{(\nu)}_{ij}(g^R)$ constructed from the curvature tensor and its covariant derivatives, with universal coefficients having no further dependence on the underlying geometry of the σ-model manifold M. Since N=2 supersymmetry when M is taken to be Kähler requires that the only possible counterterms be Kähler tensors, the universality property of the coefficients of the background field expansion then tells us that the only tensors which may be generated on an *arbitrary* riemannian manifold M are those which when restricted to Kähler manifolds satisfy the same hermiticity and curl-free conditions.

Any tensor satisfying the condition (11) may be written, at least locally, in terms of a Kähler potential $T_{\alpha\bar{\beta}} = \partial_\alpha \partial_{\bar{\beta}} S(\phi, \bar{\phi}) = \nabla_\alpha \nabla_{\bar{\beta}}\, S(\phi, \bar{\phi})$. Kähler tensors thus come in two types: those for which S is not globally defined but instead can only be defined patchwise (changing from patch to patch by $S \rightarrow S + f(\phi) + \bar{f}(\bar{\phi})$, where $f(\phi)$ depends on the transition functions between patches), and those for which S is a globally defined scalar function, invariant from patch to patch. When may tensors of each of these types appear in perturbation theory? Counterterms generated in (ℓ+1)-loop order on a Kähler manifold can always be written using the Bianchi identities as $T_{\alpha\bar{\beta}} = \nabla_\alpha \nabla_{\bar{\beta}} S$, where S and $T_{\alpha\bar{\beta}}$ both have conformal weight ℓ and S is a polynomial constructed from curvatures and their covariant derivatives. If $T_{\alpha\bar{\beta}}$ generates a non-trivial cohomology class, then S must change from patch to patch according to $S \to S + f(\phi) + \bar{f}(\bar{\phi})$. It is not easy to see how a polynomial in the curvatures with all of its coordinate indices covariantly contracted can be non-invariant under a change of coordinates. In fact, the only way to construct such an S with well-defined conformal weight using only the local tensor constructions available in perturbation theory is in the form $S = \log\det M_{\alpha\bar{\beta}}$ (products of logarithms, for example, would not have well-defined conformal weight). Independent of the conformal weight of the tensor $M_{\alpha\bar{\beta}}$, however, $T_{\alpha\bar{\beta}} = \nabla_\alpha \nabla_{\bar{\beta}} S$ then necessarily has zero conformal weight and hence non-trivial generators of cohomology can only appear as tensor counterterms in one-loop order. But we already know that the only counterterm generated in one-loop order is the Ricci tensor R_{ij}, so it follows that the only two-index tensor of the first type which appears in σ-model perturbation theory to any order is the Ricci tensor $R_{\alpha\bar{\beta}}$ of (7). All counterterms generated in σ-model perturbation theory beyond one-loop order are of the second type, i.e. cohomologically trivial. (This result also follows immediately[9] from a manifest N=2 formulation of the theory, in which the corrections to the metric are calculated directly in terms of their coordinate invariant corrections to the Kähler potential.)

We now turn to discuss the relation between Ricci flatness and the holonomy group of a manifold. Let us begin with a manifold M of real dimension $4n$. The curvature 2-form

$$\Omega^i{}_j = \frac{1}{2} R^i{}_{jkl}\, d\phi^k \wedge d\phi^l \tag{13}$$

is valued in the Lie algebra of the holonomy group, generically $SO(4n)$. If the manifold is Kähler, then Ω is instead valued in the Lie algebra of $U(2n)$ and (13) becomes

$$\Omega^\alpha{}_\beta = R^\alpha{}_{\beta\delta\bar{\gamma}}\, d\phi^\delta \wedge d\bar{\phi}^\gamma. \tag{14}$$

The trace of the curvature 2-form picks out the $U(1)$ part of the holonomy group and, by (7), is related to the Ricci tensor by

$$\mathrm{tr}\,\Omega = \Omega^\alpha{}_\alpha = R^\alpha{}_{\alpha\delta\bar{\gamma}}\, d\phi^\delta \wedge d\bar{\phi}^\gamma = R_{\delta\bar{\gamma}}\, d\phi^\delta \wedge d\bar{\phi}^\gamma. \tag{15}$$

It follows that a Kähler manifold has no $U(1)$ holonomy if and only if it is Ricci flat. Now if M is hyperkähler, Ω is valued in the Lie algebra of $Sp(n)$. Since $Sp(n)$

lies entirely within the $SU(2n)$ subgroup of $U(2n)$, its generators are all traceless and we learn that hyperkähler manifolds are automatically Ricci flat.

We are now in a position to use a uniqueness result for the Ricci flat metric on a Kähler manifold originally due to Calabi[10] (for a discussion, see the talk by G. Horowitz in these proceedings). The result we shall use here is that for a given complex structure on the manifold, a Ricci flat metric is unique within a given topological class of the Kähler form. By definition, the topological class of the Kähler form J is not changed by adding a globally defined scalar S to the Kähler potential K. The complex structure is never changed by perturbation theory since preservation of supersymmetry means that perturbation theory acts only to renormalize the metric on the manifold, not its intrinsic complex structure[6]. We will give a simple perturbative demonstration of this uniqueness result after treating the N=2 case.

In the case of N=4 supersymmetric σ-models, we begin with a manifold M which is hyperkähler, and then the preservation of N=4 supersymmetry in perturbation theory requires that the metric plus induced counterterm T_{ij} preserve the hyperkählerity and consequent Ricci flatness. Thus, for hyperkähler manifolds it is guaranteed that

$$R_{ij}(g+T)=0. \tag{16}$$

The uniqueness property of the metric for Ricci flat manifolds can now be used to give an immediate proof of finiteness for hyperkähler manifolds (Other recent treatments of this result may be found in [11]). Substituting $T_{\alpha\bar{\beta}}=\partial_\alpha\partial_{\bar{\beta}}S$ in (16) implies that $R_{ij}(g+\partial\bar{\partial}S)=0$, and then the uniqueness of the Ricci flat metric within a given topological class requires that $\partial_\alpha\partial_{\bar{\beta}}S=0$ since the original metric g is already Ricci flat. Finiteness thus follows simply from the fact that counterterms from perturbation theory beyond one-loop cannot change the topological class of the metric, and hence uniqueness of the Ricci flat metric means the geometry is so constrained as to allow no counterterms at all. The N=4 condition is essential to this argument because otherwise the counterterms would not necessarily have to satisfy (16), i.e. in principle there might occur counterterms which alter the metric structure to have a riemannian connection with non-vanishing $U(1)$ holonomy and uniqueness arguments would no longer apply.

To extend this result to the more general case of Ricci flat N=2 models, then, we need to show that renormalization preserves Ricci flatness. In other words, if one starts with a Ricci flat M, then, to all orders in perturbation theory, all ultraviolet divergences can be removed from all Greens functions without introducing counterterms that spoil Ricci flatness. In complex coordinates adapted to ${f^i}_j$, (1) becomes

$$\mathcal{L} = g_{\alpha\bar{\beta}}\, \partial_\mu \phi^\alpha\, \partial^\mu \bar{\phi}^{\bar{\beta}} + i g_{\alpha\bar{\beta}}\, \overline{\psi}^\alpha \not{D} \psi^{\bar{\beta}} + \frac{1}{4} R_{\alpha\bar{\beta}\mu\bar{\lambda}}\, \overline{\psi}^\alpha \psi^\mu\, \overline{\psi}^{\bar{\beta}} \psi^{\bar{\lambda}}. \tag{17}$$

This has certain symmetries which become manifest if we write the fermions in Weyl form, as $\psi_\pm^\alpha$ and $\psi_\pm^{\bar{\alpha}}$, where $\pm$ denotes the eigenvalue of γ_5. The terms bilinear in the fermions take the form

$$\frac{i}{2} g_{\alpha\bar{\beta}}\, \psi_+^\alpha \overleftrightarrow{D}_+ \psi_+^{\bar{\beta}} + \frac{i}{2} g_{\alpha\bar{\beta}}\, \psi_-^\alpha \overleftrightarrow{D}_- \psi_-^{\bar{\beta}} \tag{18}$$

where

$$D_\pm \psi_\pm^\alpha = \partial_\pm \psi_\pm^\alpha + \Gamma^\alpha_{\gamma\beta}\, \partial_\pm \phi^\gamma\, \psi_\pm^\beta$$

$$\partial_\pm = \frac{\partial}{\partial x^\pm}, \quad x^\pm = \frac{1}{\sqrt{2}}(x^0 \pm x^1),$$

and the quartic fermion term is proportional to

$$R_{\alpha\bar{\beta}\gamma\bar{\delta}}\, \psi_+^\alpha \psi_+^{\bar{\beta}} \psi_-^\gamma \psi_-^{\bar{\delta}}. \tag{19}$$

All this is invariant under the $U(1)_+ \times U(1)_-$ transformations

$$\begin{aligned} \psi_+^\alpha &\to e^{i\eta}\, \psi_+^\alpha & \psi_-^\alpha &\to e^{i\eta'}\, \psi_-^\alpha \\ \psi_+^{\bar{\alpha}} &\to e^{-i\eta}\, \psi_+^{\bar{\alpha}} & \psi_-^{\bar{\alpha}} &\to e^{-i\eta'}\, \psi_-^{\bar{\alpha}} \end{aligned} \tag{20}$$

(with $\phi^\alpha \to \phi^\alpha$, $\bar{\phi}^\alpha \to \bar{\phi}^\alpha$). This symmetry group includes both vector transformations, $\eta = \eta'$, and axial transformations, $\eta = -\eta'$. While somewhat similar in classical field theory, the roles of these symmetries are very different in the quantum theory.

The vector transformations, together with N=1 supersymmetry, generate N=2 supersymmetry, or equivalently the full Kähler structure discussed earlier. In the quantum theory, there are supersymmetric regularizations of perturbation theory which do not break the vector symmetry (There is, for example, superspace regularization by dimensional reduction, reviewed in [12], and regularization by higher covariant derivative regulators in superspace, recently reviewed in [13].), so the Kähler structure of the theory is preserved by renormalization[4][6]. The net effect of renormalization is merely to replace the initial metric defining the theory (the "renormalized metric") by some other Kähler metric (the "bare metric"). There is nothing of course in our reasoning thus far that prevents the counterterms from being cutoff dependent, and the bare metric from being consequently divergent.

The situation is very different for the axial transformations, for here the symmetry can be spoiled by an anomaly. Computation of the anomaly is straightforward, because we can consider $\Gamma^{\alpha}_{\gamma\beta}\,\partial_\mu\phi^\gamma$ as a (composite) gauge field, and, in two dimensions, the anomaly is proportional to the trace, i.e. the $U(1)$ part of the field strength tensor ($\partial_\mu\, j_5^\mu \sim \mathrm{tr}F$ for a two dimensional gauge theory). The relevant field strength here is related to the curvature tensor, which, on a Kähler manifold, is just $R^\alpha{}_{\beta\bar\rho\gamma} = \partial_{\bar\rho}\Gamma^\alpha_{\beta\gamma}$. Its trace, picking out the $U(1)$ part of the holonomy group, turns out by the Kähler cyclic and Bianchi identities to be simply the Ricci tensor, $R^\gamma{}_{\gamma\alpha\bar\beta} = R_{\alpha\bar\beta}$. A one-loop calculation thus results in

$$\partial_\mu\, j_5^\mu = \frac{1}{\pi}\,\epsilon^{\mu\nu}\,R_{\alpha\bar\beta}\,\partial_\mu\phi^\alpha\,\partial_\nu\bar\phi^\beta, \tag{21}$$

where j_5^μ is the current associated with the axial symmetry. Notice that in two dimensions the anomaly equation can be non-polynomial in the scalar fields because they have zero canonical dimension. The statement that the manifold is Ricci flat is thus the statement that the axial current is conserved to one-loop. This observation is the key to our argument, but to exploit it to all orders of perturbation theory, we need an analog of the Adler-Bardeen theorem[14], the statement that (21) receives no corrections from higher orders of perturbation theory.

To prove the Adler-Bardeen theorem requires a regulator which maintains the axial symmetry while regulating all diagrams beyond one-loop. The most convenient regularization for supersymmetric calculations is superspace regularization by dimensional reduction[12]. This scheme is defined by keeping spacetime indices in two dimensions whereas momentum integrals are done in $(2-\epsilon)$-dimensions. It thus respects axial invariance at the classical level, since spinor algebra is done in two dimensions, but allows the existence of ϵ-dimensional operators that mix with the classical current. At the quantum level it is this mixing, which occurs only at the one-loop level, that breaks the conservation of the γ_5 current[15]. We do not wish to enter in detail into possible ambiguities in this procedure which may appear in high loop orders. All of our results are dependent in any event on the assumption that there exists *some* consistent supersymmetric regulator. We point out, though, that any possible difficulties in the two dimensional non-gauge theories we consider here are much less severe than those which may appear in four dimensional gauge theories.

We mention that we could also prove the Adler-Bardeen theorem by using a higher covariant derivative regulator[13]. Such a regulator regulates all diagrams beyond one-loop because of their negative superficial degree of divergence. Moreover on a Ricci flat manifold, the one-loop divergence vanishes and thus cannot enter as a subdivergence in higher order diagrams. A higher covariant derivative regulator can be introduced into the superspace form of the action (2) via the substitution $g_{ij}(\Phi)\,\overline{D}\Phi^i D\Phi^j \to g_{ij}(\Phi)\,\overline{D}\Phi^i_\Lambda D\Phi^j_\Lambda$. Here we have defined the bare superfield $\Phi^i_\Lambda = f(\hat{\overline{D}}\hat{D}/\Lambda)\Phi^i$, with f some regulator function satisfying $f(x/\Lambda) \to 1$ as $\Lambda \to \infty$. $\hat{D}$ is a supercovariant derivative defined to act on the superspace coordinate Φ^i by $\hat{D}\Phi^i = D\Phi^i$, and to act on a vector W^i by $\hat{D}W^i = DW^i + \Gamma^i_{jk}\,D\Phi^j\,W^k$. This definition ensures the manifest coordinate reparametrization invariance of the regulator. Although this regularization prescription is not practical for doing computations, it does show that if the anomaly does not appear at the one-loop level, it will not appear in any order of perturbation theory. Our ultimate results, of course, are independent of the choice of regulator.

Because our regulator preserves the vector symmetry, we know by standard arguments that the canonical vector current

$$j^\mu = g_{\alpha\bar{\beta}}\,\overline{\psi}^\alpha\,\gamma^\mu \psi^\beta \tag{22}$$

will have finite matrix elements which satisfy the vector Ward identities. The current is to be understood as built up as a sum in terms of the renormalized metric plus its counterterms. We shall denote by $g^{(n)}_{\alpha\bar{\beta}}$ and $j^{(n)}_\mu$ the metric and current up to and including the nth order counterterms. Now in two dimensions the canonical axial-vector current is given naively in terms of the canonical vector current by

$$j^\mu_5 = \epsilon^{\mu\nu}\,j_\nu. \tag{23}$$

This remains true in the presence of the supersymmetric dimensional regulator, so (23) allows us to determine the canonical axial-vector current from the vector current. Since (23) is a purely algebraic relation, it is trivial that the axial-current matrix elements are also finite. (This step would have been much more difficult if we had used higher covariant derivative regularization, for which (23) holds only up to terms which vanish in the limit as the regulator is removed.)

The Adler-Bardeen theorem states that

$$\partial_\mu\,j^{\mu(n)}_5 = \frac{1}{\pi}\,\epsilon^{\mu\nu}\,R^{(n-1)}_{\alpha\bar{\beta}}\,\partial_\mu\phi^\alpha\,\partial_\nu\bar{\phi}^\beta. \tag{24}$$

(This is of course a shorthand notation for expressing the effect of an operator insertion in arbitrary Green's functions.) The nth order current $j^{\mu(n)}_5$ is related to the $(n-1)^{\text{st}}$ order Ricci tensor $R^{(n-1)}_{\alpha\bar{\beta}}$ because of an implicit factor of the loop counting parameter $\hbar$ on the right hand side. We now wish to show that when the renormalized metric is Ricci flat, no counterterms to the metric need be added which induce a non-vanishing Ricci tensor. We shall proceed inductively, by assuming that to $(n-1)^{\mathrm{st}}$ order the theory has been renormalized with the $(n-1)^{\mathrm{st}}$ order metric, $g^{(n-1)}_{\alpha\bar{\beta}}$, Ricci flat, and show that this implies the same can be made true to n^{th} order. We know by renormalizability that the theory in n^{th} order can be made to have finite Greens functions by addition of a suitable counterterm $\delta g^{(n)}_{\alpha\bar{\beta}}$ to the metric. The question is whether the induced n^{th} order Ricci tensor will then continue to vanish. To assess this requires an additional step forward to $(n+1)^{\mathrm{st}}$ order. Renormalizability of the theory insures that it can to this order also be made to have only finite matrix elements (with the $(n+1)^{\mathrm{st}}$ order counterterms depending on the nth order ones). But it now follows from (24) that the divergence of the $(n+1)^{\mathrm{st}}$ order axial current is given by

$$\partial_\mu j_5^{\mu(n+1)} = \frac{1}{\pi}\,\epsilon^{\mu\nu}\,R^{(n)}_{\alpha\bar{\beta}}\,\partial_\mu\phi^\alpha\,\partial_\nu\bar{\phi}^\beta, \tag{25}$$

where $R^{(n)}_{\alpha\bar{\beta}}$ is the Ricci tensor calculated from the nth order metric. But since the axial current is finite to $(n+1)^{\mathrm{st}}$ order, the right hand side of (25) must be finite (Any divergences due to the compositeness of the operator on the right hand side involve additional bosonic loops and so only appear in higher order). This shows that any non-Ricci flat part of the metric is at worst a finite piece which can be removed from n^{th} order by a finite metric renormalization. (This is easily seen in perturbation theory, where the Ricci tensor $R^{(n)}_{\alpha\bar{\beta}} = -\partial_\alpha\partial_{\bar{\beta}}\ln\det g^{(n)}$ induced by the nth order metric $g^{(n)} = g^{(n-1)} + \delta g^{(n)}$ reduces, to linear order in $\delta g^{(n)}$, to $R^{(n)}_{\alpha\bar{\beta}} = -\partial_\alpha\partial_{\bar{\beta}}(g^{\mu\bar{\lambda}}_{(n-1)}\delta g^{(n)}_{\mu\bar{\lambda}})$ by assumption of Ricci flatness of $g^{(n-1)}$. $\delta g^{(n)}$ can then be decomposed into a part which maintains the vanishing of the Ricci tensor and a part which does not, but which is necessarily finite (as the regulator is removed) and so can be eliminated by a finite renormalization of the metric. This part of the argument is especially simple if we adopt minimal subtraction, fixing our counterterms to be sums only of negative powers of ϵ. In this case, the finiteness of the axial divergence to $(n+1)^{\mathrm{st}}$ order directly implies that there are no metric counterterms of the undesired form in n^{th} order, and there is no need to go back to make a further finite renormalization.) This completes our inductive step; we have shown that the theory to n^{th} order can be renormalized, possibly of course by divergent counterterms, while still retaining the Ricci flatness of the nth order metric.

The remainder of the proof of on-shell ultraviolet finiteness for $N=2$ supersymmetric σ-models defined on Ricci flat Kähler manifolds follows the previous discussion of $N=4$ theories. We give here for completeness a straightforward perturbative argument of the uniqueness result which suffices for our purposes (we do not need the far more difficultly proven existence result of [10] since by assumption we begin with a manifold admitting a Ricci flat metric). From our earlier discussion of the properties of counterterms generated in perturbation theory, we know that the nth order correction to the metric satisfies

$$\delta g^{(n)}_{\alpha\bar{\beta}} = \partial_\alpha \partial_{\bar{\beta}} S^{(n)}, \tag{26}$$

for some globally defined scalar field $S^{(n)}$, modulo terms arising from coordinate redefinitions. Because Ricci flatness is preserved, the nth order correction to the Ricci tensor (7) must vanish. We actually only need to compute the change in the scalar curvature,

$$0 = \delta R^{(n)} = -\tfrac{1}{2}\Box\Box S^{(n)}, \tag{27}$$

where $\Box$ is the usual Laplace operator $g^{ij}\,\nabla_i\nabla_j$ acting on scalar functions. If M is compact, this implies that $S^{(n)}$ is constant and thus that $\delta g^{(n)}_{\alpha\bar{\beta}}$ vanishes. (Proof: Multiply by $S^{(n)}$ and integrate by parts over the whole manifold. This shows that $\Box S^{(n)}$ vanishes. Now multiply again by $S^{(n)}$ and integrate once more, showing that $dS^{(n)}$ vanishes.) Thus, for compact M, there can be no divergent counterterms to any order of perturbation theory. Although our argument has been formulated for compact Ricci flat Kähler manifolds, we may now appeal to universality to extend the result to arbitrary Ricci flat Kähler manifolds. Because all the counterterms calculated in the background field normal coordinate expansion are polynomials in the curvature and its derivatives, they make no reference to the global structure of the target manifold; a divergent counterterm which would spoil on-shell finiteness for a noncompact manifold would spoil it as well for a compact one, where we know it does not. We point out that the finiteness of these theories up to four loops has also been verified recently by explicit superfield calculations[16].

Are $N=1$ theories defined on Ricci flat manifolds finite as well? Examination of the four-graviton amplitude in the tree approximation to superstring theory on background Ricci flat riemannian manifolds suggests that there is a 4-loop contribution to the β-function in the corresponding σ-model (see talk by D. Gross, these proceedings). This does not necessarily contradict the finiteness of Ricci flat $N=2$ models since there exist many tensors which are generally non-vanishing on generic riemannian manifolds but which vanish automatically due to the cyclic and Bianchi identities $(6a, b)$ on Kähler manifolds. The simplest such example is

$$R_{kl}\,\nabla_{(i}\nabla_{j)}\,R^{kl} - 2R_{kl}\,\nabla_{(i}\,\nabla^k R_{j)}{}^l, \tag{28}$$

a potential 3-loop counterterm in $N=1$ models (which would of course vanish for Ricci flat manifolds anyway). It is easy to imagine potential 4-loop counterterms constructed from four Riemann tensors having the necessary property of vanishing automatically on Kähler manifolds. It would be miraculous for the non-finiteness of $N=1$ models at four loops to emerge so easily from the string point of view.

References

[1] L. Alvarez-Gaumé and P. Ginsparg, Finiteness of Ricci flat supersymmetric non-linear σ-models, Harvard preprint HUTP-85/A016 (revised), to appear in Comm. Math. Phys.;
see also L. Alvarez-Gaumé and P. Ginsparg, A class of two-dimensional finite field theories, HUTP-85/A030, to appear in Proceedings of the Symposium on Anomalies, Geometry, and Topology (Argonne, March 1985).

[2] L. Alvarez-Gaumé, S. Coleman, and P. Ginsparg, Finiteness of Ricci flat $N=2$ σ-models, HUTP-85/A037, to appear in Comm. Math. Phys..

[3] D. Z. Freedman and P. K. Townsend, Antisymmetric tensor gauge theores and non-linear σ-models, Nucl. Phys. B177 (1981) 282.

[4] L. Alvarez-Gaumé and D. Z. Freedman, Kähler geometry and the renormalization of supersymmetric σ-models, Phys. Rev. D22 (1980) 846.

[5] B. Zumino, Supersymmetry and Kähler manifolds, Phys. Lett. 87B (1979) 203.

[6] L. Alvarez-Gaumé and D. Z. Freedman, Geometrical structure and ultraviolet finiteness in the supersymmetric σ-model, Comm. Math. Phys. 80 (1981) 443.

[7] L. Alvarez-Gaumé, D. Z. Freedman, and S. Mukhi, The background field method and the ultraviolet structure of the supersymmetric non-linear σ-model, Ann. Phys. 134 (1981) 85.

[8] D. Friedan, Non-linear σ-models in $2+\epsilon$ dimensions, Phys. Rev. Lett. 45 (1980) 1057;
D. Friedan, Ph.D. thesis, U. C. Berkeley (unpublished, 1980).

[9] P. Howe, K. Stelle, and P. Townsend, unpublished.

[10] E. Calabi, On Kähler manifolds with vanishing canonical class, in *Algebraic Geometry and Topology: a Symposium in Honor of S. Lefschetz* (Princeton

Univ. Press, 1957), p. 78;
S.-T. Yau, Calabi's conjecture and some new results in algebraic geometry, Proc. Nat. Acad. Sci. 74 (1977) 1798.

[11] C. M. Hull, Ultraviolet finiteness of supersymmetric non-linear σ-models, IAS preprint (1985), to appear in Nucl. Phys. B;
K. Y. Muck, Phys. Lett. 157B (1985) 263.

[12] J. Gates, M. Grisaru, M. Roček, and W. Siegel, *Superspace* (Benjamin, 1983).

[13] P. West, Higher Derivative Regulation of Supersymmetric Theories, Caltech preprint CALT-68-1226 (March 1985).

[14] S. L. Adler and W. A. Bardeen, Absence of higher order corrections in the anomalous axial-vector divergence equation, Phys. Rev. 182 (1969) 1517.

[15] M. T. Grisaru, B. Milewski, and D. Zanon, Supercurrents, anomalies, and the Adler-Bardeen theorem, Phys. Lett. 157B (1985) 174;
D. R. T. Jones, L. Mezincescu, and P. West, Anomalous dimensions, supersymmetry and the Adler-Bardeen theorem, Phys. Lett. 151B (1985) 219, and references therein.

[16] D. Zanon, Superspace loop calculations for N=2 non-linear supersymmetric σ-models in two dimensions, Harvard preprint HUTP-85/A064 (1985).

σ model approach to the heterotic string theory

ASHOKE SEN*

Stanford Linear Accelerator Center
Stanford University, Stanford, California, 94305

ABSTRACT

Relation between the equations of motion for the massless fields in the heterotic string theory, and the conformal invariance of the σ model describing the propagation of the heterotic string in arbitrary background massless fields is discussed. It is emphasized that this σ model contains complete information about the string theory. Finally we discuss the extension of the Hull-Witten proof of local gauge and Lorentz invariance of the σ-model to higher order in α', and the modification of the transformation laws of the antisymmetric tensor field under these symmetries. Presence of anomaly in the naive $N = \frac{1}{2}$ supersymmetry transformation is also pointed out in this context.

* Work supported by the Department of Energy, contract DE – AC03 – 76SF00515.

I shall begin my talk by discussing the relation between the fixed point equations of the σ model and the equations of motion for the various massless fields in the heterotic string theory.[1)–3)] I shall work in the light cone gauge and consider a background where the graviton field $g_{ij}(x)$, the antisymmetric tensor field $B_{ij}(x)$, and the gauge field $A_i^M(x)$ acquire vacuum expectation value (vev) only in the eight transverse directions and are independent of the longitudinal coordinates x^0 and x^9. The dilaton field ϕ is taken to be independent of all space-time coordinates, in which case it may be absorbed in various fields and coupling constants[4)] and never appear explicitly in our analysis.* The action for the first quantized heterotic string[5)] in such a background is given by,[1)–3),6)]

$$S = \frac{1}{4\pi\alpha'}\int d\tau \int_0^\pi d\sigma\Big(g_{ij}(X)\partial_\alpha X^i\partial^\alpha X^j + \varepsilon^{\alpha\beta}B_{ij}(X)\partial_\alpha X^i\partial_\beta X^j + ig_{ij}(X)[\bar\lambda^i\partial\!\!\!/\lambda^j$$
$$+ \bar\lambda^i(\Gamma^j_{k\ell}(X) + S^j_{k\ell}(X))\rho^\alpha\lambda^\ell\partial_\alpha X^k] + \bar\psi^s(i\partial\!\!\!/\delta_{st} + A_i^M(X)(T^M)_{st}\rho^\alpha\partial_\alpha X^i)\psi^t$$
$$+ \frac{i}{4}F_{ij}^M(X)\bar\psi^s\rho^\alpha(T^M)_{st}\psi^t\bar\lambda^i\rho_\alpha\lambda^j\Big). \tag{1}$$

where α' is the string tension, X^i are the eight bosonic fields, λ^i are the eight left-handed Majorana-Weyl spinors and ψ^s are the 32 right-handed Majorana-Weyl spinors respectively. We are working in the Neveu-Schwarz-Ramond representation, so that the λ^i's transform in the vector representation of $SO(8)$, whereas the ψ^s's transform in the 32 representation of $SO(32)$ or (16,1)+(1,16) representation of the $SO(16)\otimes SO(16)$ subgroup of $E_8\otimes E_8$. Also here,

$$S_{ijk} = \frac{1}{2}(\partial_i B_{jk} + \partial_j B_{ki} + \partial_k B_{ij}). \tag{2}$$

$$\Gamma_{ijk} = \frac{1}{2}(\partial_j g_{ik} + \partial_k g_{ij} - \partial_i g_{jk}). \tag{3}$$

and F_{ij}^M is the field strength associated with the vector potential A_i^M. The action

---

* These constraints on the background fields are needed to ensure that the 0 and 9 directions remain flat as a solution of the classical equations of motion, thus allowing us to choose the light-cone gauge.

(1) has an $N=\frac{1}{2}$ supersymmetry:†

$$\delta X^i = i\epsilon\lambda^i; \quad \delta\lambda^i = -(\partial_\tau - \partial_\sigma)X^i\epsilon.$$
$$\delta\psi^s = (-\epsilon\lambda^i A_i^M)(T^M)_{st}\psi^t \tag{4}$$

For a consistent formulation of the string theory in a given background, we require the sigma model described in Eq.(1) to be conformally invariant. This requires all the β-functions of the theory to vanish. Since there are three independent sets of dimension two operators in the theory, namely, $\partial_\alpha X^i \partial^\alpha X^j$, $\varepsilon^{\alpha\beta}\partial_\alpha X^i \partial_\beta X^j$, and $\bar{\psi}^s (T^M)_{st}\rho^\alpha \psi^t \partial_\alpha X^i$, that are not related to each other by supersymmetry transformation, we get three different sets of consistency conditions on the background fields. A fourth consistency condition comes from the requirement that the central charge of the Virasoro algebra in this conformally invariant field theory should be the same as in the corresponding free field theory. We have carried out a complete one loop calculation and part of the two loop calculation in this model. The consistency conditions turn out to be,

$$R_{i\ell} + S_{ikm}S_\ell^{km} + \frac{\alpha'}{4}(F_{ki}^M F_\ell^{Mk} - R_{imnp}R_\ell^{mnp}) + \cdots = 0, \tag{5}$$

$$D^i\left(S_{ijk} + \frac{\alpha'}{8}\left[(A_i^M F_{\ell k}^M + A_\ell^M F_{ki}^M + A_k^M F_{i\ell}^M) - \begin{pmatrix} A\to\omega \\ F\to R\end{pmatrix}\right]\right) + \ldots = 0 \tag{6}$$

$$D^k F_{k\ell}^M - f^{MNP}A^{Nk}F_{k\ell}^P - S_\ell^{ij}F_{ij}^M = 0 \tag{7}$$

$$R + \frac{1}{3}S^2 = 0 \tag{8}$$

where ω is the spin connection and f^{MNP} are the structure constants of the group. Here ...denotes terms from two loop contribution of order A^3 and ω^3,

† The transformation law of ψ given here was not needed in Ref. 1 to prove supersymmetry of the action (1), since we used the equations of motion of ψ in our proof. If we do not use the equations of motion of the ψ fields we need to use the explicit transformation laws of ψ given here.

as well as terms which vanish when we set the background antisymmetric tensor field and the Ricci tensor to zero. Eq.(7) contains only the complete one loop result, whereas Eq.(8) contains the complete two loop result. These equations turn out to be identical to the equations of motion for the massless fields derived from the Green-Schwarz, Gross-Witten modified Chapline-Manton action[7] :

$$S_{eff} = \int e^{\phi}\sqrt{g}[R + \frac{1}{3}H^2 + \frac{\alpha'}{8}(F^M_{ki}F^{Mki} - R_{imnp}R^{imnp})]d^{10}x \tag{9}$$

where,

$$H_{ijk} = S_{ijk} + \frac{\alpha'}{8}\Omega_3(A)_{ijk} - \frac{\alpha'}{8}\Omega_3(\omega)_{ijk} \tag{10}$$

$$\Omega_3(A)_{ijk} = \frac{1}{2}\Big[A^M_{[i}F^M_{jk]} - \frac{2i}{3}A^M_{[i}A^N_j A^P_{k]}Tr(T^M T^N T^P)\Big]. \tag{11}$$

and $\Omega_3(\omega)$ is obtained by replacing A^M_i by the spin connection ω in Eq.(11). From this we conjecture that there is an exact one to one correspondence between the consistency conditions for the propagation of a string in a given background, and the classical equations of motion for the massless fields derived from the ten dimensional effective action. The higher loop corrections in the σ model will correspond to the higher dimensional operators in the effective action for the massless fields.

I now want to emphasize the following points:

1) If we set the gauge connection to be equal to the spin connection, and set B_{ij} to zero, the action (1) reduces to that of an $N = 1$ supersymmetric non-linear σ model, plus the action for free fermions. Such models are known to have vanishing β- function if the background is Ricci flat and Kahler[8].

2) The exact one to one correspondence between the classical equations of motion and the equations for the vanishing of the β-function tells us that every solution of the classical equations of motion provide a consistent background for the formulation of the string theory. This includes not only the vacuum solution,

but also various topological and non-topological excitations around the vacuum, e.g. a classical monopole solution.

3) Action (1) is classically invariant under a gauge symmetry transformation,

$$A_i^M(X)T^M \to A_i^{M\prime}(X)T^M = U(X)A_i^M(X)T^M U^{-1}(X) + U(X)i\partial_i U^{-1}(X),$$
$$\psi \to \psi' = U(X)\psi, \tag{12}$$

where U(X) is any map from the manifold spanned by the coordinates X^i to the gauge group $E_8 \otimes E_8$ or $SO(32)$. Thus naively one would expect the equations for the vanishing of the β-functions to be invariant under the above symmetry transformation. Eq.(6), however, is not invariant under such symmetry, due to the presence of the gauge non-invariant terms of the form $A^M_{[i}F^M_{\ell k]}$. This is due to the fact that the symmetry (12) is anomalous due to the chiral nature of the fermions ψ^s. This is also responsible for the appearance of the Chern-Simons term in the effective action (9), which is not explicitly gauge invariant. Similar remarks hold also for the local Lorentz transformations.

4) Next I want to point out that the criterion that the fixed point equations of the σ-model are derivable from an action is a very strong constraint on the σ-model itself, and is probably true only for those σ-models which represent the propagation of strings in background fields. For example, in the σ-model approach, the $S^{ij}_\ell F^M_{ij}$ term in Eq.(7) appears from one loop fermion self-energy graphs, whereas the $D^i(A^M_{[i}F^M_{jk]})$ term in Eq.(6) appears from a two loop graph, one of whose internal loop is a fermion loop with anomalous contribution. When we derive these equations from the effective action (9), both these terms come from the variation of the $S_{ijk}\Omega_3(A)^{ijk}$ term in (9). Thus the criterion for the fixed point equations to be derivable from an effective action relates a non-anomalous one-loop contribution to the β-function to an anomalous two loop contribution. In particular, if we construct a new σ-model by adding 32 left-handed fermions which couple to the gauge field $A_i^M(X)$ in the same way as the 32 right-handed fermions ψ^s, the gauge symmetry (12) ceases to be anomalous,

and the $D^i(A^M_{[i}F^M_{jk]})$ term must disappear from Eq.(6). The $S^{ij}_\ell F^M_{ij}$ term, coming from the one-loop fermion self-energy contribution, knows nothing about the addition of the new fermions, and continues to be present in Eq.(7). Thus in the new σ-model constructed this way the fixed point equations are no longer derivable from an action.

5) The action (9) contains cubic as well as quartic and higher order terms in the massless fields, occuring due to the interchange of heavy intermediate states. This indicates that the calculation of the σ-model β-functions to all orders in the perturbation theory reproduces the full effective action for the massless fields, obtained by summing all the tree graphs with massless fields as external lines and massive fields as internal lines. (This is also equivalent to constructing the effective action by eliminating the massive fields by their classical equations of motion). In string theory, knowing this effective action we may calculate the scattering amplitude involving arbitrary external massless and massive states, by using the factorization properties of the amplitudes. Hence the σ-model described by the action (1) contains complete information about the heterotic string theory,* although we have not coupled the string to the massive fields explicitly.

Now I want to show how the result that the fixed point equations of the σ-model are identical to the classical equations of motion for the massless fields may be used to derive non-trivial information about the string effective action. We shall show, for example, that the Lorentz Chern-Simons term in the effective action must have as its argument the generalized spin connection which includes torsion.† Among other things, this will imply that the consistency condition[4] $\int F\wedge F - \int R\wedge R = 0$ must be replaced by $\int F\wedge F - \int \tilde{R}\wedge\tilde{R} = 0$ in the presence of torsion, where $\tilde{R}$ is the generalized curvature. The simplest way to see why

---

* This statement is not completely correct, since (1) does not contain the most general massless background fields. This difficulty may be avoided by working in the conformal gauge as in Ref.2.

† This investigation was inspired by a queation asked by A. Strominger during the talk.

this should be so is to write the part of the action (1) quadratic in λ as,

$$i\bar{\lambda}^a[\not{\partial}\delta_{ab} + \rho^\alpha(\omega_i^{ab} - S_i^{ab})\partial_\alpha X^i]\lambda^b \equiv i\bar{\lambda}^a[\not{\partial}\delta_{ab} + \rho^\alpha\bar{\omega}_i^{ab}\partial_\alpha X^i]\lambda^b \tag{13}$$

where a and b are tangent space indices, ω is the ordinary spin connection, and $\bar{\omega}$ is the generalized spin connection. [The only other term in the action involving λ is the four-fermion coupling, but we may ignore it completely in a two loop calculation of the β-function]. During the calculation of the β-function using the background field method,[9] we may treat $\bar{\omega}$ as a new parameter, independent of g_{ij} and B_{ij}. The connections $\bar{\omega}_i^{ab}$ and A_i^M then appear in the σ-model lagrangian exactly in the same way except for the fact that $\bar{\omega}$ couples to the left handed fermions λ, whereas A_i^M couples to the right handed fermions ψ^s. The presence of the Chern-Simons term involving A_i^M in the fixed point equations then automatically implies the presence of a similar Chern-Simons term with $\bar{\omega}$ as its argument.*

This result leads us naturally to ask whether the torsion S_i^{ab} appearing in the expression for $\bar{\omega}_i^{ab}$ should be replaced by the covariant torsion H_i^{ab} defined in Eq.(10) when we calculate the higher order terms in the β-function. The answer to this question is connected intimately as to how the local gauge and Lorentz invariance is restored in higher orders in the world sheet perturbation theory. Under local gauge and Lorentz transformation of the background fields, the one loop effective action involving the bosonic fields transforms as[10],

$$\delta S^{(1-loop)} = \frac{1}{8\pi}\int d\tau \int d\sigma\, \varepsilon^{\alpha\beta}(\partial_\alpha\theta^M A_i^M - \partial_\alpha\theta^{ab}\bar{\omega}_i^{ab})\partial_\beta X^i \tag{14}$$

where θ^M and θ^{ab} are the gauge and Lorentz transformation parameters respectively. As was pointed out by Hull and Witten[11], the anomalous variation of

* This can also be seen from the analysis of Ref. 11.

the effective action to one loop order may be cancelled by redefining the transformation laws of B_{ij} under local Lorentz and gauge transformations:

$$\delta B_{ij} = \frac{\alpha'}{4}(\theta^M \partial_{[i} A^M_{j]} - \theta^{ab}\partial_{[i}\bar{\omega}^{ab}_{j]}) \tag{15}$$

This anomalous variation of B_{ij}, however, induces an anomalous variation of S_{ijk} and hence also an anomalous variation of $\bar{\omega}$, which induces a further variation of the action in order α'^2. A simple way to get rid of this problem is to replace S^{ab}_i by H^{ab}_i in the original σ-model lagrangian, since H transforms covariantly under local gauge and Lorentz transformation. The new transformation law of B_{ij} is then given by Eq.(15) with $\bar{\omega}$ replaced by $\omega - H$, and the covariant torsion H_{ijk} is now determined from the equation,

$$H_{ijk} = \partial_{[i}B_{jk]} + \frac{\alpha'}{8}[\Omega_3(A) - \Omega_3(\omega - H)]_{ijk} \tag{16}$$

which can be solved iteratively for H.

In order to restore local gauge and Lorentz invariance in higher order in α', we must also take care of the fact that the presence of the four fermion coupling in (1) gives rise to new contribution to local Lorentz and gauge anomaly other than those discussed in Ref. 11. This may be analyzed by introducing auxiliary fields S^{ab}_α, R^{ab}_α, and replacing the four fermion coupling term in (1) by,

$$-\frac{1}{4\pi\alpha'}[S^{ab}_\alpha F^M_{ab}\bar{\psi}T^M\rho^\alpha\psi + iR^{ab}_\alpha\bar{\lambda}^a\rho^\alpha\lambda^b + 4S^{ab}_\alpha R^{ab\alpha}] \tag{17}$$

where S and R are defined to transform covariantly under the local gauge and Lorentz transformation. We may now construct an effective action involving the fields X^i, S^{ab}_α and R^{ab}_α by integrating out the ψ and λ fields. Since the connections coupling to ψ and λ fields contain new terms proportional to S and R respectively, the variation of this effective action under local gauge and Lorentz transformation now contains new terms given by,

$$-\frac{1}{8\pi}\int d\tau \int d\sigma\ \varepsilon^{\alpha\beta}(\partial_\alpha \theta^M F^M_{ab} S^{ab}_\beta - i\partial_\alpha \theta^{ab} R^{ab}_\beta) \tag{18}$$

besides those given in Eq. (14). This extra variation may be cancelled by adding new terms to the lagrangian given by,

$$\frac{1}{8\pi}\int d\tau \int d\sigma\ \varepsilon^{\alpha\beta}\partial_\alpha X^i (A^M_i F^M_{ab} S^{ab}_\beta - i\omega^{ab}_i R^{ab}_\beta) \tag{19}$$

Adding (19) to (17) and eliminating the auxiliary fields by their equations of motion we get the following extra terms in the action besides the four fermion coupling:

$$-\frac{i}{32\pi}F^M_{ab}[A^M_k \partial_\beta X^k \bar{\lambda}^a \rho_\alpha \lambda^b \varepsilon^{\alpha\beta} - \omega^{ab}_i \partial_\beta X^i \bar{\psi}\rho_\alpha T^M \psi \varepsilon^{\alpha\beta} + \frac{\alpha'}{4}\partial_\alpha X^i \partial^\alpha X^k A^M_k \omega^{ab}_i]. \tag{20}$$

The addition of these new terms, as well as the replacement of S by H in the σ-model action destroys the naive $N = \frac{1}{2}$ supersymmetry of the action. This symmetry, however, is anomalous,* since it involves field dependent phase transformation of the chiral fermions:

$$\begin{aligned} \delta\lambda^a &= e^a_{i,j}(i\epsilon\lambda^j)\lambda^i - e^a_i(\partial_\tau - \partial_\sigma)X^i\epsilon \\ \delta\psi^s &= (-\epsilon\lambda^i A^M_i)(T^M)_{st}\psi^t \end{aligned} \tag{21}$$

It is conceivable that the extra terms added to the lagrangian in order to restore the ten dimensional local gauge and Lorentz invariance will also restore the two dimensional $N = \frac{1}{2}$ supersymmetry.

I wish to thank J. Attick, W. Bardeen, S. Das, A. Dhar, E. Martinec, R. Nepomechie, M. Rubin, B. Sathiapalan, A. Strominger, T. Taylor, H. Tye, Y. S. Wu., S. Yankielowicz, and C. Zachos for useful discussions during various stages of this work.

---

* This has been noted by Attick, Dhar and Ratra in a different context[12].

REFERENCES

1. A. Sen, preprints FERMILAB-PUB-85/77-T, to appear in Phys. Rev. D; FERMILAB -PUB-85/81-T, to appear in Phys. Rev. Lett.
2. C. G. Callan, D. Friedan, E. J. Martinec and M. J. Perry, princeton preprint.
3. P. Candelas, G. T. Horowitz, A. Strominger and E. Witten, Nucl. Phys. *B258*, 46 (1985).
4. E. Witten, Princeton preprint(1984), M. Dine and N. Sieberg, Institute for advanced study preprint.
5. D. J. Gross, J. Harvey, E. J. Martinec and R. Rohm, Phys. Rev. Lett. 54, 502 (1985), Nucl. Phys. *B256*, 253 (1985), Princeton preprint.
6. E. S. Fradkin and A. A. Tseytlin, Phys. Lett. 158*B*, 316 (1985).
7. E. Bergshoeff, M. De Roo, B. De Witt and P. van Nieuwenhuizen, Nucl. Phys. *B195*, 97 (1982); G. Chapline and N. S. Manton, Phys. Lett. 120*B*, 105 (1983) M. B. Green and J. H. Schwarz, Phys. Lett. 149*B*, 117 (1984); D. J. Gross and E. Witten, as quoted in Ref.3, Proceedings of this workshop; R. I. Nepomechie, Univ. of Washington at Seattle report No. 40048-20 P5.
8. See for example, L. Alvarez-Gaume, S. Coleman and P. Ginsparg, preprint HUTP-85/A037, and references therein.
9. For a recent general treatment see S. Mukhi, preprint TIFR/TH/85/13.
10. P. Nelson and G. Moore, Phys. Rev. Lett. 53, 1519 (1984), preprint HUTP-84/A076; J. Bagger, D. Nemeschansky and S. Yankielowicz, preprint SLAC-PUB-3588, and references therein.
11. C. Hull and E. Witten, Princeton preprint; E. Witten, princeton preprint.
12. J. Attick, A. Dhar and B. Ratra, in preparation.

Aspects of Conformal Invariance and Current Algebra in String Theory

Spenta R. Wadia
Tata Institute of Fundamental Research
Homi Bhabha Road, Bombay 400 005, India

Abstract

We briefly review our work on the principle of conformal invariance in string theory. The ideas are illustrated in the soluble case of a string in (flat space time) x (compact lie group), which admits an algebraic formulation. The conformal anomaly cancellation formula for this case is derived and the massless spectrum is presented. We also present a preliminary discussion of current algebra vortices on the world sheet and gauge symmetry breaking.

The string model offers an attractive possibility for a unified description of all forms of matter and interaction since these particles arise as massless modes of a single entity: the string[(1)]. If we restrict ourselves to closed strings then all the interactions between these particles emerge from a single 3 string vertex. The string model not only offers a unification of physics below Planck length but can lead to a prediction of phenomena beyond Planck length. It is bound to shed more light on the phenomenon of blackholes.

Quantum mechanical consistency reflected in the principle of conformal anomaly cancellation fixes the dimension of the manifold in which the string lives. Therefore all but 4 dimensions should be curled up on a compact space of the dimensions of planck length. The study of string compactification is of importance. We briefly report here on our work in this direction[2]. The principle of conformal anomaly cancellation (both local and global) also fixes the gauge group to be either $E_8 \times E_8$ ot SO(32). The question of gauge symmetry breaking has been addressed in the framework of the low energy effective lagrangian. Here we briefly address ourselves to this problem and report some preliminary results[3].

The manifold on which the string compactifies gives the vaccuum gemetry. Determining it in the framework of the field theory of strings would involve computing the effective action of the massless modes and minimizing it. Such a program is important, however at the moment there does not exist a formulation of the field theory of strings which is independent of one or other assumed vaccuum geometry.

To circumvent this difficulty we adopt a self consistent approach in which we study the propagation of the string in an arbitrary background metric and coupled to the Kalb-Ramond antisymmetric tensor field . The principle of conformal invariance serves to restrict these background fields[4].

The string action is given by

$$S = \frac{1}{4\lambda^2} \int_{\sigma_2} d^2\xi \sqrt{g}\, g^{\mu\nu} \partial_\mu X_a \partial_\nu X_b G^{ab}(X) +$$

$$i \int_{\Sigma_2} dX_a \wedge dX_b A^{ab}(X) + \mu^2 \int \sqrt{g}\, d^2\xi \qquad (1)$$

$\xi=(\xi_1,\xi_2)$ parametrize the world sheet σ_2. $g_{\mu\nu}$ is a metric on the world sheet. $X_a(\xi)$ are string co-ordinates valued in a manifold M. $G_{ab}(X)$ is the metric on M. Σ_2 is the image of σ_2 in M. $A_{ab}(X)$ is the Kalb-Ramond field. $A_{ab}(X)$ may admit singularities. Hence the coupling of the string to this field is best written in terms of the gauge invariant field strength F = dA, by employing stokes theorem

$$\int_{\Sigma_2} dX_a \wedge dX_b A^{ab} = \int_{\Sigma_3} dX_a \wedge dX_b \wedge dX_c\, F^{abc}, \quad \partial\Sigma_3 = \Sigma_2. \qquad (2)$$

This is possible only if every image Σ_2 of the string world sheet σ_2, is a boundary of some 3-surface Σ_3 in M. This requires the second Betti number of M to vanish: $\beta_2(M) = 0$. (2) is the generalization of the Wess-Zumino term to string theory. Here since in general σ_2 is homeomorphic to a sphere with handles, the usual homotopy requirement $\pi_2(M) = 0$ is replaced by $\beta_2(M) = 0$. We further require that F, though closed, is not exact i.e. the 3rd Betti number is non-zero: $\beta_3(M) \neq 0$. Furthermore F is a integral, harmonic form.

An important property of (6) is that it is invariant under reparametrizations of co-ordinates on σ_2: $\xi_i \to f_i(\xi_1, \xi_2)$. But for the last term the action is also invariant under Weyl transformations: $g_{\mu\nu} \to e^{\alpha(x)} g_{\mu\nu}$. This transformation scales physical distances on the world sheet. A classical non-linear model defined in (6) can represent a string theory only if these symmetries are maintained in the 2-dim. quantum field theory. The 2-dimensional conformal algebra is infinite dimensional and may significantly constrain the environment in which the string propagates.

Finally the string field theory partition function is defined by

$$Z = \sum_{H=0}^{\infty} \left(\frac{1}{N}\right)^H Z_H$$

$$Z_H = \int Dg_{\mu\nu}\ DX_a\ e^{-S(X,g;\ G,A)} \tag{3}$$

H denotes the number of handles of σ_2. $\frac{1}{N}$ is the expansion parameter, which is related to the expection value of the dilation field: $\frac{1}{N} = e^{-2D_o^{(5)}}$. In (3), phenomena for which $Z \propto e^{-N}$ have not been accounted for.

CONFORMAL INVARIANCE AND THE STRING ON A GROUP MANIFOLD:

Reparametrizations of the world sheet σ_2, are generated by the 2-dim energy momentum tensor $T_{\mu\nu} = \frac{1}{\sqrt{g}} \frac{\delta S}{\delta g^{\mu\nu}}$. Its classical value is zero since it is the equation of motion for $g_{\mu\nu}$. In the quantum theory its expectation value in all physical states is equal to zero. From now on we present the discussion in the conformal gauge $g_{\mu\nu} = \delta_{\mu\nu} e^{\phi}$. In this gauge the reparametrizations are fixed upto conformal reparametrizations, which are analytic functions on the world sheet. Further assume that the manifold $M = R_d \times G$. R_d is d dimensional flat space time and G is a compact manifold. In this case we can choose the light cone gauge in R_d. Using standard procedures for string quantization, the string theory (1) in the conformal and light cone gauge is described as the following conformal invariant local field theory on the world sheet.

Introduce the complex co-ordinates $Z = \xi_1 + i\xi_2$ and $\bar{Z} = \xi_1 - i\xi_2$. The complex components of the stress tensor are $T_{Z\bar{Z}} = T_{11} + T_{22}$, $T_{ZZ} = T_{11} - T_{22} + 2iT_{12}$ and $T_{\bar{Z}\bar{Z}} = \bar{T}_{ZZ}$. Invariance under conformal reparametrizations and Weyl scalings requires the trace of the stress tensor to vanish: $T_{Z\bar{Z}} = 0$. Then the conservation law of the stress tensor reduces to the statement that $T_{ZZ} \equiv T(Z)$ is analytic and $T_{\bar{Z}\bar{Z}} \equiv \bar{T}(\bar{Z})$ is anti-analytic: $\partial_{\bar{Z}} T(Z,\bar{Z}) = \partial_Z \bar{T}(Z,\bar{Z}) = 0$. The string theory is conformal invariant provided the Virasoro algebra holds.

$$[T(Z), T(w)] = \frac{C}{12}\, \delta'''(Z-w) + T(Z)\, \delta'(Z-w)$$
$$[\bar{T}(\bar{Z}), \bar{T}(\bar{w})] = \frac{C}{12}\, \delta'''(\bar{Z}-\bar{w}) + \bar{T}(\bar{Z})\, \delta'(\bar{Z}-\bar{w}) \qquad (4)$$

C is the central charge and it is the sum of the conformal anomalies due to each degree of freedom in the string theory: $C = \sum_i C_i$. A necessary condition for a consistent string theory is

$$C = \sum_i C_i = 24. \qquad (5)$$

Note that in the r.h.s. 24 appears instead of the usual 26, because we are working in the light cone gauge. (5) is the principle of anomaly cancellation. It is a local statement on the string world sheet and hence it is independent of the number of handles of the string world sheet and hence true to all orders in the string field theory.

Returning to the model (1), where $M = R_d \times G$, let us denote the (d-2) transverse co-ordinates in R_d by $X_i(Z,\bar{Z})$ and those in the compact manifold G by $\theta_j(Z,\bar{Z})$. The metric on G can be written in terms of vierbiens: $G_{ab} = \sum_\alpha e_a^\alpha e_b^\alpha$. Then the stress tensor $T(Z,\bar{Z})$ has the following structure[6].

$$T(Z,\bar{Z}) \quad \frac{1}{K_1}\, \alpha_Z^i\, \alpha_Z^i + \frac{1}{K_2}\, J_Z^\alpha J_Z^\alpha \qquad (6)$$

where $\alpha_Z^i = \partial_Z X^i$ and $J_Z^{\alpha} = e_a^{\alpha}\, \partial_Z \theta^a$ are local currents. K_i are normalizations which are to be determined. K_i are in general not their classical values. It is of interest to investigate conformal invariant field theories emerging from a current algebra of the currents $J^a(Z,\bar{Z})$ on a general compact manifold G. Henceforth conformal invariant field theory will be denoted by CIFT.

However for our present purposes we restrict ourselves to G being a compact lie group. Its elements are denoted by $U(Z,\bar{Z}) = e^{i\, t^a \theta_a(Z,\bar{Z})}$. The anti-hermetian generators t_a satisfy the lie algebra $[t_a, t_b] = f^{abc} t_c$ and are normalized as $\mathrm{tr}\, t_a t_b = -2\delta ab$. In this case the action (1), (2) in the conformal gauge $g_{\mu\nu} = e^{\phi}\delta_{\mu\nu}$, reduces to

$$S = \frac{1}{2}\int_{\sigma_2} d^2\xi\, \partial_\mu X_i\, \partial_\mu X_i + \frac{1}{4\lambda^2}\int_{\sigma_2} d^2\xi\, \mathrm{tr}\, \partial_\mu U \partial_\mu U^{-1} +$$

$$i\,\frac{k}{N_G}\int_{\Sigma_3} d^3\, \mathrm{tr}\, \partial_\mu U U^{-1} \partial_\nu U U^{-1} \partial_\lambda U U^{-1} \varepsilon_{\mu\nu\lambda} \qquad (7)$$

which is (besides the first term) the Wess-Zumino model. k is an integer and N_G is a normalization factor depending on the group. The Wess-Zumino model is a CIFT only if $\alpha^2 = (\frac{6\lambda^2 k}{N_G})^2 = 1$ [7]. In fact a 1-loop background field calculation valid for small curvature ($\lambda \to 0$) gives

$$T_{Z\bar{Z}} = \frac{\delta}{\delta\phi} S_{eff} = \frac{C_v}{8\pi}(1 - \alpha^2)\, \mathrm{tr}\, \partial_\mu U_o \partial_\mu U_o^{-1} +$$

$$\frac{1}{48\pi}(24-(d-2)-\dim G)(\sqrt{g}(R^{(2)} + \mu^2) \qquad (8)$$

U_o is a background field and $R^{(2)}$ is the world sheet scalar curvature. C_v is the Cesimir of the adjoint representation defined by $f_{adc}\, f_{bdc} = C_v \delta_{ab}$. Clearly $T_{Z\bar{Z}} = 0$ only if

$\alpha^2=1$ and $(d-2) + \dim G = 24$. This latter equation is the conformal anomaly cancellation statment (5) for this model in the limit $\lambda \to 0$.

To go beyond this perturbative evaluation of the anomaly we proceed with the algebraic formulation. The crucial observation due to Knizhnik and Zamolodchikov[8] is that at the conformal invariant point $\alpha^2=1$, the action (7) is invariant under local gauge transformations $U(Z,\bar{Z}) \to \Omega(Z)U(Z,\bar{Z})\tilde{\Omega}(\bar{Z})$ and $X_i(Z,\bar{Z}) \to X_i(Z,\bar{Z}) + f(Z) + g(\bar{Z})$. This invariance is reflected in the conservation laws for the currents $J(Z,\bar{Z}) = \frac{12k}{N_G} \partial_Z U U^{-1}$, $\bar{J}(Z,E) = \frac{12k}{N_G} U^{-1}\partial_{\bar{Z}}U$, $\alpha^i_Z = \partial_Z X^i$ and $\alpha^i_{\bar{Z}} = \partial_{\bar{Z}}X^i$: $\partial_{\bar{Z}}J = \partial_Z\bar{J} = \partial_{\bar{Z}}\alpha^i_{\bar{Z}} = \partial_Z\alpha^i_{\bar{Z}} = 0$ and the gauge transformation laws for these currents expressed as a current algebra

$$[J^a(Z), J^b(w)] = f^{abc}\; J_c(w)\delta(z-w) + \frac{12k}{N_G}\,\delta_Z\delta(Z-w). \tag{9a}$$

$$[\alpha^i(Z), \alpha^j(w)] = i\,\partial_Z\delta(Z-w)\delta^{ij} \tag{9b}$$

Further since the currents are conformal tensors of dim 1, we have

$$[T(Z), J^a(w)] = \partial_Z(J^a(w)\delta(Z-w)) \tag{10a}$$

$$[T(Z), \alpha^i(w)] = \partial_Z(\alpha^i(w)\delta(Z-w)) \tag{10b}$$

Now given the definition (6) of the stress-tensor T(Z), (10) implies the normalizations to be $K_1 = -\frac{1}{2}$ and $K_2 = -(\frac{24\pi k}{N_G} + \frac{C_V}{2})$. The Virasoro algebra (4) is satisfied only if the central charge is given by

$$C = (d-2) + \frac{k\,\dim G}{k+C_V(\frac{N_G}{48\pi})}. \tag{11}$$

The normalization $N_G = 24\pi$ for orthogonal and exceptional groups. $N_G = 48\pi$ for SU(n). The conformal anomaly cancellation condition (5) now becomes

$$(d-2) + \frac{k \dim G}{k + C_v \frac{N_G}{48\pi}} = 24 \qquad (12)$$

It is crucial to realize that (12) is a consistency condition only on the local properties of the world sheet. For the same reason it does not receive any correction from string interactions which are characterized by global properties of the world sheet. In ref.2 (12) was derived only for the free string.

SPECTRUM:

It is easy to work out the spectrum of the above string model. We restrict ourselves to the free string which can be described in terms of oscillators defined using the equations of motion of the currents and laurent expansions:

$$\alpha^i_Z = \sum \frac{\alpha^i_n}{Z^{n+1}}, \quad \alpha^i_{\bar{Z}} = \sum \frac{\bar{\alpha}^i_n}{\bar{Z}^{n+1}}, \quad J^a(Z) = \sum \frac{J^a_n}{Z^{n+1}},$$

$$\bar{J}^a(\bar{Z}) = \sum \frac{\bar{J}^a_n}{\bar{Z}^{n+1}}.$$

In terms of these oscillators, the Virasoro oscillators $L_n = \oint \frac{dZ}{2\pi i} Z^{n+1} T(Z)$ and $\bar{L}_n = \oint \frac{d\bar{Z}}{2\pi i} \bar{Z}^{n+1} \bar{T}(\bar{Z})$, have expressions

$$L_n = \frac{1}{K_1} \sum_m \alpha^i_m \alpha^i_{n-m} + \frac{1}{K_2} \sum_m J^a_m J^a_{n-m}$$

$$\bar{L}_n = \frac{1}{K_1} \sum_m \bar{\alpha}^i_m \bar{\alpha}^i_{n-m} + \frac{1}{K_2} \sum_m \bar{J}^a_m \bar{J}^a_{n-m} \qquad (13)$$

In terms of these oscillators the local algebras (4),(9) and (10) take the following form:

$$[L_{n}, L_{m}] = (n-m) L_{n+m} + \frac{c}{12} (n^3 - n)\delta_{n+m,o} \quad (14)$$

$$[J^a_{n}, J^b_{m}] = f^{abc} J^c_{n+m} + n \frac{24\pi k}{N_G} \delta^{ab} \delta_{n+m,o} \quad (15)$$

$$[\alpha^i_{n}, \alpha^i_{m}] = -n\delta^{ij} \delta_{n+m,o} \quad (16)$$

$$[L_{n}, J^a_{m}] = -m\, J^a_{n+m} \quad (17)$$

$$[L_{n}, \alpha^i_{m}] = -m\, \alpha^i_{n+m} \quad (18)$$

The mass operator of the string is given by

$$\frac{\alpha'}{4} (\text{Mass})^2 = L_o + \bar{L}_o - 2 \quad (19)$$

and the physical states have zero conformal spin:

$$(L_o - \bar{L}_o)|\psi\rangle = 0 \quad (20)$$

The spectrum of massless states above the tachyon are the 'gravitational' multiplet $\alpha^i_{-1}\, \bar{\alpha}^j_{-1}|0\rangle$, the vector bosons $\bar{\alpha}^i_{-1}\, J^a_{-1}|0\rangle$, $\alpha^i_{-1} \bar{J}^a_{-1}|0\rangle$ and the charged scalars $J^a_{-1} J^b_{-1}|0\rangle$. The tachyon is defined by $J^a_n|0\rangle = \alpha^i_n|0\rangle = 0$, for $n > 0$.

Twisted Algebras and gauge symmetry breaking:

The problem of gauge symmetry breaking in string theory has been discussed at the level of the field theory of the massless modes by embedding the spin connection of the compact 6 dim. space time into the gauge group[10]. Thus at planck scale we have the gauge group $E_8 \times E_8$ breaking to $E_6 \times E_8$ by this mechanism. This identification of the gauge field with the spin connection has other important consequences also. Here we present some attemtps to realize these ideas at a stringy level[3].

It is clear from (15) that the zero mode oscillators J^a_o satisfy the Lie algebra of the group. If zero mode oscillators corresponding to some directions in the algebra are absent, then the symmetry is only partially realized. The simplest way to achieve this is to assume that $J^a(Z)$ has a branch cut at $Z = 0$, for some directions in the Lie algebra i.e. there is a vortex at $Z = 0$. If we make the conformal transformation $Z \to w(Z) = e^Z$, then the transformed currents are not single valued functions of the angle $\theta = \arg Z$. We say the currents are are twisted[11].

Now it turns out that it is not possible to satisfy the condition (20), i.e. $(L_o - \bar{L}_o)|\psi\rangle = 0$, unless we assume that some of the space-time currents α^i_Z are twisted. This tells us that for cowistency reasions the twisting of internal symmetry and spacetime currents is related.

A MODEL TWISTED STRING:

We explore these ideas in a simple model string theory where we hope to achieve SU(2) breaking to U(1). The string manifold is given by $M = R_{d_t} \times R_{d-d_t} \times [SU(2)_t]^{N_t} \times [SU(2)]^N$. Flat space time has d transverse dimensions of which d_t are twisted. $SU(2)_t$ stands for twisted SU(2). There are N_t copies of this and N copies of untwisted SU(2).

We assume the following natural twisting of SU(2): $J^{\pm}$ $J^{\pm}(\theta+2\pi) = -J^{\pm}(\theta)$ and $J^3(\theta+2\pi) = J^3(\theta)$, where $J^{\pm} = J^1 \pm J^2$. This means that the $J^{\pm}$ oscillators are half integral: $J^{\pm}(\theta) = \sum_n e^{i(n+\frac{1}{2})} J_{n+\frac{1}{2}}$. The J^3 oscillators are integrals before. Note that the stress tensor is still single valued.

Substituting these expansions into the local Kac-Moody algebra (9a), we get the twisted SU(2) Kac-Moody algebra (we assume k = 1),

$$[J^{+}_{n+\frac{1}{2}}, J^{-}_{m+\frac{1}{2}}] = 2J^{3}_{n+m+1} - \frac{1}{2}(n + \frac{1}{2})\,\delta_{n+m+1,0}, \tag{19}$$

(Note that there are no zero model oscillators corresponding to J).

$$[J^{+},J^{+}] = [J^{-},J^{-}] = 0$$

$$[J^{3}_{n}, J^{3}_{m}] = -n\delta_{n+m,0}$$

The statement (10a) of conformal transformation now becomes

$$[L_{n}, J^{\pm}_{m+\frac{1}{2}}] = -(m+\tfrac{1}{2})\, J^{\pm}_{n+\frac{1}{2}}\,, \quad [L_{n}, J^{3}_{m}] = -m\, J^{3}_{n+m} \tag{20}$$

For the twisted space-time oscillators, the algebra (16) and (18) is changed to

$$[\alpha^{i}_{n+\frac{1}{2}}, \alpha_{m+\frac{1}{2}}] = -(n+\tfrac{1}{2})\,\delta^{ij}\,\delta_{n+m+1,0}$$

$$[L_{n}, \alpha^{i}_{m+\frac{1}{2}}] = -(m+\tfrac{1}{2})\,\alpha^{i}_{n+m+\frac{1}{2}} \tag{21}$$

The twisting of space time is in a sence determined by the twisting of the current algebra by the consistency condition $(L_{o}-\bar{L}_{o})|\psi\rangle = 0$. The Hilbert space of the twisted oscillators is defined by $J^{\pm}_{n+\frac{1}{2}}|0\rangle = \alpha^{i}_{n+\frac{1}{2}}|0\rangle = 0$ for $n \geqslant 0$. The contribution of twisted oscillators to the Virasoro generators is given by

$$L_{n} = \frac{1}{K_{1}} \sum \alpha^{i}_{m+\frac{1}{2}}\alpha^{i}_{n-(m+\frac{1}{2})} + \frac{1}{K_{2}} \sum J^{a}_{m+\frac{1}{2}} J^{a}_{n-(m+\frac{1}{2})} + (L_{n})_{\text{untwisted}}$$

K_1 and K_2 which depend on the local properties of the world sheet as the same as before.

The crucial point is that zero point energy of the oscillators is now shifted from its untwisted value of -2. The formula for the mass of the ground state is

$$\alpha' M_G^2 = -\frac{1}{12} \sum_i \frac{1}{N_i}(1 + 6\eta_i^2 - 6\eta_i), \quad 0 \leqslant \eta_i < 1. \qquad (22)$$

where η_i is the twist mod(integer) for the ith oscillator. In our case $\eta_i = 0$ or $\frac{1}{2}$. N_i reflects the normalization of the stress tensor for that degree of freedom. For example for a space-time oscillator $N_i = 1$, for the Lie Algebra valued oscillators of the group $N_i = \frac{k}{k+C_v(\frac{N_G}{48\pi})}$. For SU(2), $N_i = \frac{1}{3}$ for k = 1 and so on. For the model under consideration where the string manifold is $M = R_{d_t} \times R_{d-d_t} \times [SU(2)_t]^{N_t} \times [SU(2)]^N$, (22) turns out to be

$$\frac{\alpha'}{4} M_G^2 = -\frac{1}{12} (N + d - \tfrac{3}{2}d_t) \qquad (23)$$

Note that there is no contribution to this from the twisted SU(2). Finally we have the conformal anomaly cancellation which reads

$$d + N_t + N = 24 \qquad (24)$$

The contribution to the anomaly from each SU(2) is simply the rank of SU(2), which is one, hence

$$\frac{\alpha'}{4}M_G^2 = -2 + \frac{1}{12} (N_t + \frac{3}{2} d_t). \qquad (25)$$

which says that, consistent with (24), there is always an upward shift in the mass.

Now we have a problem. If N_t and d_t are non-zero, it

is not possible to have massless gauge bosons in the untwisted space time dimensions, corresponding to the untwisted gauge generators. Not surpirsingly there are no gravitons either! Our tiwsted model cannot be a sensible model for gauge symmetry breaking from $SU(2) \rightarrow U(1)$. Nevertheless let us write down a few massive levels in the case when $M_G^2 = 0$:

$$\alpha^i_{-\frac{1}{2}} J^{\pm}_{-\frac{1}{2}}|0\rangle, \quad \alpha^i_{-\frac{1}{2}} \bar{\alpha}^j_{-\frac{1}{2}}|0\rangle, \quad \alpha^i_{-1} J^3_{-1}|0\rangle, \quad \alpha^i_{-1} \bar{\alpha}^j_{-1}|0\rangle, \quad \alpha^i_{-1} J^{\pm}_{-\frac{1}{2}} J^{\pm}_{-\frac{1}{2}}|0\rangle \text{ etc.}$$

The investigation of string theories, like the Heterotic String[12] in which the untwisted ground state is not a tachyon is desirable and may yield positive results. There it is natural to twist the 6 non-commuting generators of $SU(3) \subset E_8$ and the 6 dimensions of space time. This work is in progress.

Acknowledgement

I would like to thank Michael Green and David Gross for making it possible for me to attend this workshop. The Institute for Theoretical Physics at Santa Barbara is acknowledged for its hospitality and support. I also thank the TIFR Theoretical Physics Group for their support and encouragement. Discussions with Sanjay Jain and R. Shankar are also acknowledged.

References

1. For a review see J.H. Schwarz, Phys. Rep. 89, 223 (1982); M.B. Green, Surveys in High Energy Physics 3, 127 (1983).
2. S. Jain, R. Shankar and S. Wadia, 'Conformal Invariance and String Theory in Compact Space: Bosons', TIFR preprint, January 1985. Phys. Rev. D 32, 2713 (1985).
3. T.R. Govindarajan, T. Jayraman, A. Mukherjee and S. Wadia, in preparation.
4. The proposal that conformal invariance may serve to restrict the background fields to which a string is coupled was made in ref.2, where it was illustrated for a group manifold. It was independently made by E.S. Fradkin and A.A. Tseytlin, Lebedev Institute preprint and Phys. Lett. 158B, 316 (1985); D. Friedan and S. Shenker (unpublished); P. Candelas, et al in ref. (10).
5. E.S. Fradkin and A.A. Tseytlin, Lebedev preprint N 261 (1984).
6. H. Sugawara, Phys. Rev. 170, 1659 (1968).
7. E. Witten, Comm. Math. Phys. 92, 455 (1984).
8. V.G. Knizhnik and A.B. Zamolodchikov, Nuc. Phys. B247, 83 (1984).
9. This formula has been derived by a number of authors: S. Jain, R. Shankar and S. Wadia ref.2; D. Nemeschansky and S. Yankielowicz, Phys. Rev. Lett. 54, 620 (1985); P. Goddard and D. Olive, Nucl. Phys. B257, [FS 14] 83 (1985) and in ref. 8.
10. P. Candelas, G. Horowitz, A. Strominger and E. Witten, Nucl. Phys. B258, 46 (1985).
11. S.M. Roy and V. Singh, Tata Inst. Seminar January 1985 and Tata Institute Preprint 85-20. C. Vafa and E. Witten, Princeton preprint (1985); J. Lepowsky (private communication).
12. D.J. Gross, J. Harvey, E.J. Mortinec and R. Rohm, Phys. Rev. Lett. 54, 502 (1985); Nucl. Phys. B256, 253 (1985).

TORSION IN SUPERSTRINGS†

Itzhak Bars
Department of Physics, University of Southern California
Los Angeles, California 90089-0484
and
Dennis Nemeschansky and Shimon Yankielowicz‡
Stanford Linear Accelerator Center
Stanford University, Stanford, California 94305

In this talk we discuss string theories on a background manifold with torsion.[1] The talk contains two parts. In the first part we discuss candidate vacuum configurations for ten-dimensional superstrings. We compactify these on $M_4 \times K$, where M_4 is four-dimensional and K some compact six-dimensional manifold. In particular we are interested in investigating the existence of solutions with non-zero torsion on K. The compactification problem is approached both from the effective field theory point of view and directly using string considerations.

The second part of the talk is devoted to the construction of string theories in curved space with torsion. We discuss both the Neveu-Schwarz-Ramond[2] type string and the Green-Schwarz[3] type string. Particular emphasis is put on the resulting constraints on space-time supersymmetry in the Green-Schwarz approach.

We use two-dimensional non-linear sigma models to describe the propagation of strings in background geometries with torsion. The background field can be understood as arising from condensation of infinite number of strings. Torsion can be viewed as the field strength associated with the vacuum expectation value of the anti-symmetric tensor field B_{mn} which appears in the supergravity multiplet. We show that if the background fields only include the metric and torsion, a consistent string theory requires torsion to vanish. The possibility remains that

† Work supported by the Department of Energy, contract DE – AC03 – 76SF00515.
‡ On leave from Tel-Aviv University.

torsion is non-trivial when other background fields are included, e.g. gauge fields and dilaton.

The effective ten-dimensional field theory which appears in the zero-slope limit $\alpha' = 0$ of the superstring is $N = 1$ supergravity coupled to super Yang-Mills matter. The low-energy theory has the supergravity transformations of the Chapline-Manton action[4] modified by the appropriate Chern-Simons terms introduced by Green and Schwarz.[5] The effective Lagrangian contains, even at the classical level, operators of arbitrarily high dimensions. These arise from integrating out the massive modes. Recently problems associated with this approximation were raised by Dine and Seiberg[6] and Kaplunovsky[7] on the basis of phenomenological considerations. If for the purpose of discussing the vacuum this truncation is questionable, then our analysis would need modification.

To prove that the only viable compactifications of the ten-dimensional manifold are on manifolds without torsion we take advantage of the analysis of Candelas, Horowitz, Strominger and Witten.[8] They have analyzed in detail the conditions for $N = 1$ supersymmetry and found that space-time M_4 must be flat Minkowski space. Form the supergravity transformation of the fermionic fields it follows that the compact manifold K must admit a covariantly constant spinor ϵ with respect to the connection $\Omega_m = \omega_m - 4\beta H_m$

$$\nabla_m(\Omega)\epsilon \equiv (\nabla_m(\omega) - \beta H_m)\epsilon = 0 \qquad \beta = \frac{3\sqrt{2}}{8}\, e^{2\phi} \;, \tag{1}$$

where $\nabla_m(\omega)$ is the covariant derivative with spin connection ω. In Eq. (1) H_m is defined through $H_m = H_{mnp}\gamma^n\gamma^p$, where H_{mnp} is the field strength associated with the antisymmetric field B_{mn}. The indices m, n, p refer to the compact manifold K and the γ's are the $O(6)$ Dirac matrices. From eq. (1) one can read off the torsion of the new connection

$$T_{mnp} = 4\beta\, H_{mnp} \;. \tag{2}$$

Candelas, Horowitz, Strominger and Witten showed that

$$H\epsilon = 0 \tag{3}$$

where $H = H_m \gamma^m$. However, they only studied in detail the case $H_{mnp} = 0$. Equation (3) can be satisfied without H_{mnp} being equal to zero. This corresponds to manifolds with torsion. Following the analysis of Ref. [8] it follows that the compact manifold admits a complex structure $f^m{}_n$ which is covariantly constant

$$\nabla_p(\Omega)\, f^m{}_n = 0\,. \qquad (4)$$

The complex structure can be built from the covariantly constant spinor ϵ. Therefore in order to preserve supersymmetry the compact manifold must be a hermitean manifold.

In Ref. [8] the following relation between the scalar curvature $R(\omega)$ and torsion T_{mnp} was derived

$$R(\omega) = \frac{16}{3}\,\beta^2 H_{mnp} H^{mnp} = \frac{1}{3}\,T_{mnp} T^{mnp}\,. \qquad (5)$$

On manifolds with torsion it is straightforward to calculate the generalized Riemann tensor build from the connection Ω

$$R(\Omega)_{mnpq} = R(\omega)_{mnpq} + \nabla_p(\omega) T_{mnq} - \nabla_q(\omega) T_{mnp} + T_{rmp} T^r{}_{qn} - T_{rmq} T^r{}_{pn}\,. \quad (6)$$

For a totally antisymmetric torsion the generalized Ricci tensor is given by [1]

$$R(\Omega)_{mn} \equiv R(\Omega)^p{}_{mpn} = R_{mn}(\omega) - T^{pq}{}_m T_{pqn} + \nabla^p(\omega) T_{pmn}\,. \qquad (7)$$

The reparametrization invariance of the compactified string theory demands that the two-dimensional non-linear sigma model must be conformally invariant. This means that the β-function must vanish. In this talk we will only consider the case of identically vanishing β-function. However, one can imagine not having $\beta \equiv 0$ but just a theory at a non-trivial fixed point of the β-function. All the results presented here are based on $\beta \equiv 0$. In particular the one loop β-function should vanish. The β-function has been studied by several authors. [9],[10] They have shown at one loop that the β-functions vanishes, with metric and torsion as background fields, when the generalized Ricci tensor vanishes

$$R(\Omega)_{mn} = 0 \ . \tag{8}$$

Using Eqs. (7) and (8) one finds that the background manifold must satisfy

$$R(\omega) = T^{mnp} T_{mnp} \ . \tag{9}$$

This is in contradiction with Eq. (5) unless

$$T_{mnp} = H_{mnp} = 0 \ . \tag{10}$$

Hence if the Ricci tensor $R(\Omega)_{mn}$ is required to vanish the background manifold cannot have any torsion. If the background fields include the dilaton and/or gauge fields Eq. (10) may no longer hold, accordingly our conclusions based on Eq. (10) may be modified.

Next we would like to give another proof of the above result. Equations (1) and (3) are the important constraints that guarantee that space-time M_4 is a Minkowski space and that the four-dimensional theory has $N = 1$ supersymmetry at the compactification scale. These constraints together with other constraints obtained in Ref. [8] have been analyzed and solved [1],[11] in the presence of torsion with a holonomy group $SU(3)$. The hermitean metric g of the compact six-dimensional manifold K must then satisfy

$$\begin{aligned} &\partial_i g^{i\bar{j}} = \partial_{\bar{j}} g^{i\bar{j}} = 0 \qquad i,j = 1,2,3 \\ &\det g = 1 \ . \end{aligned} \tag{11}$$

In Eq. (11) we have introduced complex coordinates. The condition (11) can be rewritten in terms of the curl of the metric

$$\partial_i g^{i\bar{j}} = 2\epsilon^{ik\ell} \, \overline{\epsilon^{jnm}} \, g_{\overline{n}k} \partial_i g_{\overline{m}\ell} \ . \tag{12}$$

On a Kähler manifold the metric satisfies the condition of Eq. (11). However, Eq. (11) admits more general solutions that include torsion. In the complex basis the only non-zero components of the torsion are

$$T_{ij\bar{k}} = \tfrac{1}{2}\,\partial_{[i}g_{j]\bar{k}} \qquad T_{\bar{i}\bar{j}k} = (T_{ij\bar{k}})^* \; . \tag{13}$$

From Eq. (13) it is clear that on a Kähler manifold the torsion vanishes. The Ricci tensor has the following form

$$R(\Omega)_{i\bar{j}} = -g^{m\bar{s}}(\partial_m \partial_{[\bar{s}}g_{\bar{j}]i} - \partial_i \partial_{[\bar{s}}g_{\bar{j}]m}) \; . \tag{14}$$

The antisymmetrization makes the generalized Ricci tensor to depend on the curl of g. Hence the Ricci tensor vanishes only on Kähler manifolds. Therefore, on manifolds with torsion the generalized Ricci tensor does not vanish. This is the same result as obtained above.

Let us emphasize that the conclusion of a vanishing torsion is based on the requirement that the Ricci tensor vanishes. This condition followed form the sigma model analysis for models that include just the metric and torsion. For more general sigma models with more background fields this condition may be relaxed and torsion need not vanish.

So far we have given two proofs that torsion on the compact manifold must vanish when the Ricci tensor vanishes. Since not very much is known about string theories it is instructive to derive the same equations from different points of view. Next we show how some of the equations of the effective field theory approach can be derived directly from the string theory. In the previous analysis of the effective field theory with torsion Eqs. (5) and (9) played a crucial role. Equation (9) followed from the demand that the generalized Ricci tensor vanishes. On the other hand Eq. (5) had to be satisfied for the theory to be supersymmetric at the compactification scale. In the string theory this equation arises from the demand that the central charge in the Virasoro algebra must have the same value as in a ten-dimensional supersymmetric string theory in flat space. If the central charge is changed, then the critical dimension of the theory is changed. Friedan and Shenker[12] have shown that if the critical dimension is changed there are no zero mass fermions in the theory and therefore supersymmetry is broken.

In recent papers [13],[14] the critical dimension was computed for group manifolds with torsion. The torsion corresponds to the Wess-Zumino term which must be included in order to preserve conformal invariance of the theory. [15] For a purely bosonic string the critical dimension on $SU(N)$ is given by [13]

$$D = 26 - d_c = \frac{(N^2-1)k}{N+k} = d_G - \frac{N(N^2-1)}{k} + O\left(\frac{1}{k^2}\right) \tag{15}$$

where $d_G = N^2 - 1$ is the dimension of $SU(N)$ and k is the integer coefficient of the Wess-Zumino term. To make contact with the previous analysis one needs the relation between the string tension and the integer k [13]

$$\alpha' = \frac{2}{k} \; . \tag{16}$$

One can rewrite Eq. (15) in terms of curvature and torsion on the manifold

$$26 - (d_c + d_G) = \alpha'(-3R + T^2) + O(\alpha'^2) \; . \tag{17}$$

From Eq. (17) it follows that the critical dimension remains unchanged if $R = \frac{1}{3}T^2$. A similar analysis can be performed for the supersymmetric case using the computation of the critical dimensionality for this case. [13] Again one finds that $R = \frac{1}{3}T^2$ is needed to ensure that the critical dimension does not shift. To all orders in the string tension the critical dimension for group manifolds is given by

$$26 - (d_C + d_G) = -\frac{(3R - T^2)\alpha' d_G}{d_G + \alpha'(3R - T^2)} \; . \tag{18}$$

Therefore, if the dilaton is set to zero, the only way to preserve supersymmetry at the compactification scale is to have zero torsion. Recently Callan, Martinec, Perry and Friedan [16] have studied the non-linear sigma model with metric, torsion and dilaton background fields.

Next we would like to construct string theories on background manifolds with torsion. As we discussed earlier the requirement that the resulting four-dimensional effective field theory has $N = 1$ supersymmetry determines a lot of properties of the compact manifold. In particular the manifold must admit a covariantly constant spinor with respect to the connection with torsion Ω and consequently a covariantly constant complex structure. Furthermore, the metric must be hermitean. When the torsion vanishes the relevant manifolds are Ricci flat Kähler manifolds.[8] When torsion is included one should consider manifolds satisfying Eq. (11).

Recently two-dimensional supersymmetric non-linear sigma models with torsion have been analyzed.[17] It has been shown that the sigma model has an $N = 2$ supersymmetry if the manifold is hermitean and if the complex structure is covariantly constant relative to the connection that includes torsion.

The general structure of the action after elimination of the auxiliary fields is[17]

$$I(X,\lambda) = \frac{1}{2}\int d^2\sigma \Big[g_{mn}\partial_\mu X^m \partial^\mu X^n + \frac{3}{2} B_{mn}\epsilon^{\mu\nu}\partial_\mu X^m \partial_\nu X^n + i g_{mn}\overline{\lambda}^m_+ \not{D}^+ \lambda^n_+ + i g_{mn}\overline{\lambda}^m_- \not{D}^- \lambda^n_- + \frac{1}{4} R^+_{mnpq}(\overline{\lambda}^m_+ \rho_\mu \lambda^n_+)(\overline{\lambda}^p_- \rho^\mu \lambda^q_-)\Big] \tag{19}$$

where the ρ_μ's are the two-dimensional Dirac matrices and $\lambda_\pm$ are Majorana-Weyl spinors. In Eq. (19) $\pm$ refer to right-handed and left-handed fermions respectively. The antisymmetric tensor B_{mn} is the potential associated with the torsion $T_{mnp} = -B_{[mn,p]}$. Note that this definition of torsion does not include the Chern-Simons terms. It differs form the torsion that appears in the effective field theory $T = dB - \frac{1}{30}\omega_Y + \omega_L$. These extra terms involve a compensating dimensionful parameter. Such a parameter is the slope parameter $\alpha' \sim (\ell_{Planck})^{-2}$. In fact the Chern-Simons terms appear in the next order of the loop expansion in the sigma model with the coefficient α'.[16] This is a necessary consequence of Lorentz and gauge invariance as discussed by Green and Schwarz.[3]

The action (19) is invariant under the supersymmetry transformation

$$\delta X^n = \delta_+ X^n + \delta_- X^n = \bar{\epsilon}_+ \lambda^n_- + \bar{\epsilon}_- \lambda^n_+$$
$$\delta \lambda^n_\pm = -i \partial\!\!\!/ X^n \epsilon_\mp - \Gamma^n_{\pm mp} \lambda^m_\pm \delta_\pm X^p \ . \tag{20}$$

The sigma model has another supersymmetry

$$\delta X^n = \delta_+ X^n + \delta_- X^n = f^n_{-m} \bar{\epsilon}_+ \lambda^m_- + f^n_{+m} \bar{\epsilon}_- \lambda^m_+$$
$$\delta (f^n_{\pm m} \lambda^m_\pm) = -i \partial\!\!\!/ X^n \epsilon_\pm - \Gamma^n_{\pm mp} f^m_{\pm \ell} \lambda^\ell_\pm \delta_\pm X^p \tag{21}$$

provided the complex structure $f^n{}_m$ is covariantly constant.

The conformally invariant non-linear sigma model has another type of supersymmetry. This supersymmetry is the partner of the local Kac-Moody transformation and has the form

$$\delta X^i = 0 \qquad \delta \lambda^i_\pm = \delta^i_\pm \ . \tag{22}$$

Next we would like to elevate the two-dimensional supersymmetry to a space-time supersymmetry. We will work in ten dimensions. When the compact manifold is flat this amounts to going from the Neveu-Ramond-Schwarz version of the string theory to the Green-Schwarz superstring. For non-trivial curved background with torsion, we will use the light cone gauge to relate the Neveu-Ramond-Schwarz type of string theory to the Green-Schwarz superstring. The Green-Schwarz version of the action (19) is given by

$$I = \int d^2\sigma \left[\frac{1}{2} g_{ij} \partial_\alpha X^i \partial^\alpha X^j + \tfrac{1}{2} \epsilon^{\alpha\beta} B_{ij} \partial_\alpha X^i \partial_\beta X^j \right.$$
$$+ \frac{i}{4} \overline{S}_+ \gamma_+ \rho^\alpha D^+_\alpha S_+ + \frac{i}{4} \overline{S}_- \gamma_+ \rho^\alpha D^-_\alpha S_- \tag{23}$$
$$\left. + \frac{1}{4} R^+_{ijk\ell} \overline{S}_+ \gamma_+ \gamma^i \gamma^j \rho^\alpha S_+ \overline{S}_- \gamma_+ \gamma^k \gamma^\ell \rho_\alpha S_- \right]$$

where $S_+(S_-)$ is a right (left)-moving fermion. In flat space the action (23) has two eight component supersymmetries

$$\delta X^i = (p^+)^{-1/2}\sqrt{2}\,\bar{\epsilon}\gamma^i S \tag{24a}$$

$$\delta S^\gamma = i\frac{(p^+)^{-1/2}}{\sqrt{2}}\,(\gamma_- \gamma_M (\rho \cdot \partial x^M)\epsilon)^\gamma \tag{24b}$$

and

$$\delta X^i = 0 \qquad \delta S^\alpha = \delta^\alpha \tag{25}$$

where ϵ and δ are eight component real spinors of $O(8)$. In curved space the transformation (24b) gets modified by terms of the form $\Omega_i \gamma_- S\,\delta X^i$. The action (23) is invariant under the δ supersymmetry of Eq. (25) provided that δ is covariantly constant. In this case the quadratic term is automatically invariant and the quartic term is invariant since $R^{\pm}_{mnpq}\gamma^p\gamma^q\delta = 0$. The δ-supersymmetry is analogous to the transformation of Eq. (22). In curved space one cannot implement the full eight component ϵ supersymmetry. To see this let us study the relation between the action in Eq. (19) and that in Eq. (23) assuming $SU(3)$ holonomy. In the $SU(3)$ basis the fermions have the form

$$S^A = \begin{pmatrix} \psi^\alpha \\ \chi_1 \\ \psi_\alpha \\ \chi_2 \end{pmatrix} \qquad \begin{array}{l} A = 1,\ldots,8 \\ \\ \alpha = 1,2,3\,. \end{array} \tag{26}$$

This corresponds to the following decomposition of the spinor representation under $SU(8) \supset SO(6) \supset SU(3)$

$$8(\text{spinor}) \to 4 + \bar{4} \to 3 + 1 + \bar{3} + 1\,. \tag{27}$$

Under this decomposite the action in Eq. (23) takes the form of that in Eq. (19). For details see Ref. [1]. Since the action of Eq. (19) is a sigma model in two dimensions, it is clear that it cannot have the full ϵ-supersymmetry. To see what part survives let us focus on the compact part of the background manifold. The spinor ϵ has a similar decomposition under $SU(3)$ as S given in Eq. (26). The action for the superstring is invariant when the triplet ϵ^α and antitriplet ϵ_α vanish.

The supersymmetries of the curved space are associated with the singlet parameters $\epsilon_1 = \epsilon_2^*$. This means that the spinor ϵ must be covariantly constant. This is of no surprise if one studies the relationship of the supersymmetry transformations of the covariant action to the supersymmetry transformations of the light cone action of the Green-Schwarz string in flat space.[1] For the argument to be applicable to the present case one needs the covariant form of the Green-Schwarz action in curved space.[18]

REFERENCES

[1] Bars, I., Nemeschansky, D. and Yankielowicz, S., SLAC-PUB-3758 (1985).

[2] Neveu, A. and Schwarz, J., Nucl. Phys. B31, 86 (1971); Ramond, P., Phys. Rev. D3, 2415 (1971).

[3] Green, M. and Schwarz, J., Nucl. Phys. B181, 502 (1981).

[4] Chapline, G. and Manton, N., Phys. Lett. 120B, 105 (1983).

[5] Green, M. and Schwarz, J., Phys. Lett. 149B, 117 (1984).

[6] Dine, M. and Seiberg, N., IAS preprint, May and June (1985).

[7] Kaplunovsky, V., Princeton preprint (1985).

[8] Candelas, P., Horowitz, G., Strominger, A., and Witten, E., ITP preprint NSF-IT 84-170.

[9] Braaten, E., Curtright, T., and Zachos, C., Florida preprint UFTP85-01 (1985).

[10] Friedan, D., UC-Berkeley, Ph.D. Thesis (August 1980), LBL preprint LLB-11517.

[11] Bars, I., USC preprint 85/015 (1985).

[12] Friedan, D. and Shenker, S., unpublished.

[13] Nemeschansky, D. and Yankielowicz, S., Phys. Rev. Lett. 54, 620 (1985).

[14] Olive, D. and Goddard, P., Nucl. Phys. B257, 226 (1985); Krizhnik, V. and Zamolodnikov, Z., Nucl. Phys. B247, 83 (1984); Friedan, D. and Shenker, S., unpublished.

[15] Witten, E., Comm. Math. Phys. 92, 455 (1984).

[16] Callan, C., Martinec, E., Perry, M. and Friedan, D., Princeton preprint (1985).

[17] Howe, P. and Sierra, G., Phys. Lett. 148B, 451 (1984).

[18] Bagger, J., Nemeschansky, D. and Yankielowicz, S., in preparation.

AN INTRODUCTION TO THE MONSTER

I. B. Frenkel[1]

Department of Mathematics, Yale University
New Haven, CT 06520, USA

J. Lepowsky[2]

Department of Mathematics, Rutgers University
New Brunswick, NJ 08903, USA

A. Meurman[3]

Department of Mathematics, University of Stockholm
Stockholm, SWEDEN

ABSTRACT

In a brief exposition intended for string theorists, we discuss the Monster and its natural infinite-dimensional representation.

1. INTRODUCTION

The Fischer-Griess Monster, often denoted F_1, is a finite simple group with about 8×10^{53} elements. Predicted to exist in 1973 by B. Fischer and R. Griess, it was constructed by Griess [18]. We have found a natural infinite-dimensional Fock space representation of F_1, incorporating vertex operators and suggesting fundamental connections with string theory (refs. [9], [10], [11]). One could argue that the Monster is the most perfect structure in mathematics, and it appears that the miracles of string theory are very closely related to the miracles which allow the Monster to exist. Attempting to understand these miracles should lead to deeper interactions between mathematics and physics. Here we give a brief sketch of F_1 for string theorists, including a definition of F_1 based on vertex operators. This exposition complements an earlier one [10], which emphasized the infinite-

[1]Partially supported by a Sloan Foundation Fellowship.

[1,2]Partially supported by NSF Grant MCS83-01664.

[3]Partially supported by a grant from the Swedish Natural Sciences Research Council.

dimensional representation of F_1. See refs. [8], [10], and [19] for further discussion.

2. FINITE SIMPLE GROUPS

A group G is <u>simple</u> if it has no subgroups H other than $\{1\}$ and G such that $gHg^{-1} = H$ for all $g \in G$. The smallest nonabelian finite simple group is the group, say I, of rotations of the regular icosahedron. Compact Lie groups without center, such as the group E_8, are also simple, but are infinite. The nonabelian finite simple groups are the following:

(1) the alternating group A_n on n letters (the group of even permutations) for $n \geq 5$; the group A_5 is isomorphic to I

(2) the groups of "Lie type" over finite fields, for example, $SL(n)$ with entries in the 2-element field, for $n \geq 3$

(3) the sporadic groups, which have no apparent unifying pattern or general construction; F_1 is the largest of these, involving 20 or 21 sporadic groups as quotients of subgroups; string theorists will be amused that the number of sporadic groups is 26.

The precise description of all these groups is a long task. The classification theorem for finite simple groups - the assertion that this list is complete - is a landmark achievement of twentieth century mathematics. Its recently completed proof covers 10,000 to 15,000 journal pages and represents the work of over 100 mathematicians. See refs. [14], [15] for surveys. No one person has checked all the details of the proof, but outsiders to the classification project become more confident that the theorem is true every year that elapses without the discovery of a new finite simple group.

Both the search for sporadic groups and the classification effort made finite group theory a relatively isolated branch of mathematics for a long time. But suddenly the situation changed in 1978-79.

3. MONSTROUS MOONSHINE

Even before F_1 was proved to exist, it was strongly suspected to have a 196883-dimensional irreducible module (= representation), which would be the smallest possible nontrivial module.

Consider the modular function $j(e^{2\pi iz})$, which maps the quotient of the upper half plane by the standard action of $SL(2,\mathbb{Z})$, with the point at infinity adjoined, one-to-one onto the Riemann sphere. When this function is expanded in powers of $q = e^{2\pi iz}$, the coefficients are positive integers:

$$j(q) = q^{-1} + 744 + 196884q + 21493760q^2 + \cdots .$$

(cf. ref. [35]). The constant term 744 can be changed arbitrarily without destroying any of the fundamental properties of $j(q)$.

It was J. McKay who noticed the near coincidence $196884 = 196883 + 1$. Was there indeed a relationship between two traditionally such distant parts of mathematics? Interpreting the "1" as the dimension of the trivial F_1-module, J. Thompson extended the coincidence by checking that the first several coefficients of $j(q)$ (except for the constant term) are simple positive integral linear combinations of the conjectured dimensions of irreducible F_1-modules. These observations would be explained, he pointed out [37], if there were a natural infinite-dimensional $\mathbf{Z}$-graded F_1-module

$$V = \bigoplus_{n=-1}^{\infty} V_{-n}$$

such that the dimension of the F_1-module V_{-n} is the nth coefficient of $J(q) = j(q)-744$; in particular, $V_0 = 0$. (We have negated the subscripts because mathematicians conventionally consider modules with grading bounded above rather than below.) Defining the <u>character</u> of V to be

$$\mathrm{ch}\ V = \sum_{n\geq -1} (\dim V_{-n})q^n,$$

we require

$$\mathrm{ch}\ V = J(q). \tag{3.1}$$

By analyzing the expected action of nontrivial Monster elements on V, J. Conway and S. Norton dramatically amplified the conjectured relationship between finite group theory and modular function theory, calling it "Monstrous Moonshine" [5].

The numerology started people thinking that V might be some kind of analogue of the basic module of an affine Kac-Moody algebra (cf. refs. [22], [26]). This turned out to be true, but only in a very subtle sense.

It is interesting to note that modular invariance plays a crucial role in string theory. In particular, in the heterotic string [20], [21], this invariance limits the gauge group to either $E_8 \times E_8$ or SO(32), in agreement with the discovery of Green-Schwarz [17].

4. THE CONSTRUCTION OF F_1

In 1980, Griess [18] constructed F_1 as a group of automorphisms of a 196883-dimensional commutative nonassociative algebra, which we designate B_0. Very strange-looking, the Griess algebra B_0 was not destined to be stumbled upon by the usual axiomatic approach of non-associative algebraists, since it satisfies no low-degree identities besides the commutativity identity $xy-yx = 0$. For instance, B_0 is not a Jordan algebra. Nevertheless, B_0 looked to us a little like the Lie algebra E_8, in some as yet unknown presentation. This impression became the analogy explained below.

J. Tits has shown that F_1 is the full automorphism group of B_0 (see refs. [39], [40]). We shall define B_0 below, thereby giving the reader a precise definition of the Monster. Our definition is based on vertex operators, bringing another subject into the network of ideas.

5. AFFINE LIE ALGEBRAS

Let $\underline{g}$ be any Lie algebra and let $\langle\cdot,\cdot\rangle$ be a symmetric bilinear form on $\underline{g}$, invariant in the sense that

$$\langle[x,y],z\rangle = \langle x,[y,z]\rangle, \quad x,y,z \in \underline{g}. \tag{5.1}$$

(For instance, $\underline{g}$ might be a finite-dimensional semisimple Lie algebra and $\langle\cdot,\cdot\rangle$ a multiple of the Killing form. Or $\underline{g}$ might be an abelian Lie algebra and $\langle\cdot,\cdot\rangle$ an arbitrary symmetric form.) The corresponding (<u>untwisted</u>) <u>affine Lie algebra</u> is the infinite-dimensional Lie algebra

$$\hat{\underline{g}} = \underline{g} \otimes \mathbb{C}[t,t^{-1}] \oplus \mathbb{C}c$$

with brackets given by:

$$[c,\hat{\underline{g}}] = 0, \tag{5.2}$$

$$[x(m),y(n)] = [x,y](m+n) + \langle x,y\rangle m\delta_{m+n,0}c \tag{5.3}$$

for $x,y \in \underline{g}$ and $m,n \in \mathbb{Z}$. Here $\mathbb{C}[t,t^{-1}]$ designates the algebra of

Laurent polynomials in t, and $x(m) = x\otimes t^m$. If we write t as $e^{i\theta}$, then $\underline{g}\otimes\mathbb{C}[t,t^{-1}]$ becomes the space of maps $f: \mathbb{R} \to \underline{g}$ with periodicity $f(x+2\pi) = f(x)$ and finite Fourier expansion.

Suppose that ν is an automorphism of $\underline{g}$ such that $\langle\nu x,\nu y\rangle = \langle x,y\rangle$ for $x,y \in \underline{g}$ and $\nu^p = 1$ for some integer $p > 0$. For $n \in \mathbb{Z}$, let $\underline{g}_{(n)}$ be the $e^{2\pi in/p}$-eigenspace of ν in $\underline{g}$. The corresponding <u>ν-twisted affine algebra</u> is the Lie algebra

$$\hat{\underline{g}}[\nu] = \bigoplus_{n\in\mathbb{Z}} \underline{g}_{(n)}(n/p)\oplus\mathbb{C}c.$$

Here the space $\underline{g}_{(n)}(n/p)$ is spanned by the elements $x(n/p) = x\otimes t^{n/p}$ for $x \in \underline{g}_{(n)}$, and the brackets are again given by (5.2) and (5.3), this time for $m,n \in \mathbb{Z}/p$ and $x \in \underline{g}_{(pm)}$, $y \in \underline{g}_{(pn)}$. Then with $t = e^{i\theta}$, $\oplus\underline{g}_{(n)}(n/p)$ becomes the space of Fourier polynomial maps $f: \mathbb{R} \to \underline{g}$ with "twisted periodicity" $f(x+2\pi) = \nu f(x)$. For instance if $\underline{g}$ is abelian, $p = 2$ and $\nu = -1$, then f satisfies the antiperiodic boundary conditions $f(x+2\pi) = -f(x)$, and f involves only odd powers of $e^{i\theta/2} = t^{1/2}$.

For $\underline{g}$ semisimple, $\hat{\underline{g}}$ and $\hat{\underline{g}}[\nu]$ are examples of what mathematicians have termed Kac-Moody algebras, which are defined in terms of certain generators from a generalized Cartan matrix, and whose detailed study was begun by V. G. Kac, I. L. Kantor and R. L. Moody. The expression for the coefficient of c (the cocycle) in the bracket formula (5.3) for an untwisted affine Kac-Moody algebra seems not to have been recognized by mathematicians until around 1977. If ν is an inner automorphism of $\underline{g}$, then $\hat{\underline{g}}$ and $\hat{\underline{g}}[\nu]$ are isomorphic Lie algebras. See ref. [27] for further discussion.

6. VERTEX OPERATOR CONSTRUCTIONS

In 1977, one wanted to construct affine Kac-Moody algebras by means of concrete operators of some kind. The first result [30] was a Fock space realization of $\underline{s\ell}(2,\mathbb{C})\hat{}$, in a twisted form, based on an apparently new kind of differential operator which H. Garland recognized as similar to a vertex operator in string theory. This construction was generalized to a natural family of twisted affine algebras in ref. [24]. Subsequently, an analogous construction of the corresponding untwisted affine algebras $\hat{\underline{g}}$ was found (refs. [7], [34]), using the vertex operator of string theory. This is often called the homogeneous construction. See also refs. [42], [43].

We now describe still another vertex operator construction [8], corresponding to a different twisting. Let L be a lattice (the integral span of a basis) in a finite-dimensional Euclidean space with complexification $\underline{h}$. Denote the inner product by $<\cdot,\cdot>$ and assume that L is an even lattice, i.e., that $<\alpha,\alpha>$ is even for all $\alpha \in L$. View $\underline{h}$ as an abelian Lie algebra and let ν be the automorphism of $\underline{h}$ which multiplies each element by -1. Then the corresponding twisted affine algebra

$$\hat{\underline{h}}[-1] = \bigoplus_{n\in\mathbb{Z}+1/2} \underline{h}(n)\oplus\mathbb{C}c$$

can be realized as an algebra of half-integrally moded bosonic string oscillators acting by the canonical realization of the Heisenberg commutation relations on a Fock space S, with c acting as the identity operator. Here S is the space of polynomials on a basis of the space $\oplus_{n<0}\underline{h}(n)$, which we view as the space of creation operators. The space S has a natural nonpositive $\frac{1}{2}\mathbb{Z}$-grading which we denote as follows:

$$S = \oplus S_n \quad (n \in \tfrac{1}{2}\mathbb{Z},\ n \leq 0).$$

(If we want the character of S to have modular transformation properties, we can shift the degrees n by adding a suitable uniform constant.)

A lattice is unimodular or self-dual if it contains one point per unit volume. For $n \in \mathbb{Z}$, set

$$L_n = \{\alpha \in L \,|\, <\alpha,\alpha> = n\}.$$

We shall be especially interested in three lattices: $L = \mathbb{Z}\alpha_0$ where $<\alpha_0,\alpha_0> = 2$, the root lattice of $\underline{s\ell}(2,\mathbb{C})$ (which is not unimodular); $L = \Gamma$, the root lattice of E_8 - the unique (up to isometry) even unimodular lattice in 8 dimensions; and $L = \Lambda$, the Leech lattice - the unique even unimodular lattice in 24 dimensions such that Λ_2 is empty (see refs. [3], [25]; cf. refs. [29], [38]).

Set $\bar{L} = L/2L$, the lattice L with points differing by twice a lattice element identified. Then the set $\bar{L}$ has 2^ℓ elements, where $\ell = \dim \underline{h}$, and $\bar{L}$ can be viewed as a vector space over the 2-element field $\mathbb{Z}/2\mathbb{Z}$. Denote by $\bar{\alpha}$ the image of $\alpha \in L$ in $\bar{L}$. It is easy to construct a (not necessarily symmetric) bilinear map $\varepsilon_0: \bar{L}\times\bar{L} \to \mathbb{Z}/2\mathbb{Z}$ such that $\varepsilon_0(\bar{\alpha},\bar{\alpha}) = \langle\alpha,\alpha\rangle/2 \bmod 2$ for all $\alpha \in L$. Set $\varepsilon(\bar{\alpha},\bar{\beta}) = (-1)^{\varepsilon_0(\bar{\alpha},\bar{\beta})}$ for $\alpha,\beta \in L$ and define a multiplication on the set $F = \{\pm e_{\bar{\alpha}} | \bar{\alpha} \in \bar{L}\}$, the $e_{\bar{\alpha}}$ being a new set of symbols indexed by $\bar{L}$, by

$$e_{\bar{\alpha}}e_{\bar{\beta}} = \varepsilon(\bar{\alpha},\bar{\beta})e_{\overline{\alpha+\beta}} \quad \text{for } \alpha,\beta \in L.$$

Then F is a finite group which is a "finite Heisenberg group" if $L = \Gamma$ or Λ, and

$$e_{\bar{\alpha}}e_{\bar{\beta}} = (-1)^{\langle\alpha,\beta\rangle}e_{\bar{\beta}}e_{\bar{\alpha}} \quad \text{for } \alpha,\beta \in L.$$

Let T be an irreducible representation of F such that $-e_0 = -1$ in F acts as -1 on T. Then for $L = \mathbb{Z}\alpha_0$, Γ and Λ, we have $\dim T = 1$, 2^4 and 2^{12}, respectively.

Set $W = S\otimes T$ and give W the $\frac{1}{2}\mathbb{Z}$-grading

$$W = \oplus W_n \quad (n \in \tfrac{1}{2}\mathbb{Z},\ n \leq 0)$$

defined by: $W_n = S_n\otimes T$. The space W has a unique (up to constant multiple) symmetric bilinear form determined by the condition $h(n)^* = h(-n)$ $(h \in \underline{h},\ n \in \mathbb{Z}+\frac{1}{2})$ on S and by F-invariance on T. The restriction of this form to each W_n is nonsingular.

For $\alpha \in L$ define a "vertex operator" $X(\alpha,\zeta)$, ζ being a formal parameter, as follows:

$$X(\alpha,\zeta) = \exp(\textstyle\sum\alpha(-n)\zeta^{-n}/n)\exp(-\sum\alpha(n)\zeta^{n}/n)\otimes e_{\bar{\alpha}}, \tag{6.1}$$

where both sums range over $n \in \mathbb{Z}+\frac{1}{2}$, $n > 0$. Then the expansion coefficients $x_\alpha(n)$ defined by

$$X(\alpha,\zeta) = \textstyle\sum_{n\in(1/2)\mathbb{Z}} x_\alpha(n)\zeta^n$$

are operators on W. (It turns out that the operators (6.1), without the tensor factor $e_{\bar{\alpha}}$, had been written down in ref. [6] in connection with electromagnetic currents.)

The first main theorem about these operators [8] states: Suppose that L is the root lattice of a semisimple Lie algebra $\underline{g}$ with all root lengths equal and normalized to be $\sqrt{2}$, so that L is the lattice generated by L_2. Then under brackets, the operators $x_\alpha(n)$ for $\alpha \in L_2$ and $n \in \frac{1}{2}\mathbb{Z}$ generate a copy of the twisted affine Lie algebra $\hat{\underline{g}}[\nu]$, where ν is an automorphism of order 2 of $\underline{g}$ extending -1 on $\underline{h}$, which is identified with a Cartan subalgebra of $\underline{g}$. Moreover, $\hat{\underline{g}}[\nu]$ is spanned by the $x_\alpha(n)$ and the infinite-dimensional "Heisenberg algebra" $\hat{\underline{h}}[-1]$. In particular, the canonical realization of the

Heisenberg commutation relations on S can be extended naturally using (6.1) to a realization of $\hat{\underline{g}}[\nu]$ on $W = S \otimes T$.

This result generalizes the original construction [30] - the case $L = \mathbb{Z}\alpha_0$ - in a direction different from that of ref. [24].

Let $L = \Gamma$. Then the degree zero operators $x_\alpha(0)$, $\alpha \in \Gamma_2$, span a Lie algebra $\underline{k}$ isomorphic to $\underline{\underline{so}}(16)$, and the invariant bilinear form on $\underline{k}$ is determined from the vertex operator brackets (see the constant term in (5.2)). (Note that the E_8-symmetry of the untwisted construction (refs. [7], [34]) of $\hat{E}_8$ is broken down to $\underline{\underline{so}}(16)$. In addition to symmetry breaking, we also have a "dimensional reduction", corresponding to the fact that the number of zero modes in the harmonic oscillators of refs. [7], [34] has been reduced down to 0.) Now the degree $-\frac{1}{2}$ operators $x_\alpha(-\frac{1}{2})$ and $\underline{h}(-\frac{1}{2})$ span a $120 + 8 = 128$-dimensional space which transforms under $\underline{\underline{so}}(16)$ as one of the half-spin modules. On the other hand, the subspace $W_{-1/2} = S_{-1/2} \otimes T \simeq \underline{h} \otimes T$ of W is $8 \cdot 16 = 128$-dimensional, and is a copy of the <u>other</u> half-spin module of $\underline{\underline{so}}(16)$. In particular, the space $\underline{k} \oplus \underline{h} \otimes T$ carries a natural E_8-Lie algebra structure, in which the subalgebra $\underline{k}$ and its bracket action on its orthogonal complement $\underline{h} \otimes T$ are defined entirely by means of vertex operators. The brackets of pairs of elements of $\underline{h} \otimes T$ into $\underline{k}$ are then canonically determined from the symmetric bilinear form and the invariance condition (5.1). (The symmetric form on $\underline{k} \oplus \underline{h} \otimes T$ is a multiple of the Killing form of E_8.) To construct E_8, it does not matter which half-spin module (i.e., which chirality of spinors) one should adjoin to $\underline{\underline{so}}(16)$, although to pass between the two realizations of E_8 one would need to use a complicated outer automorphism of $\underline{\underline{so}}(16)$.

By contrast, the analogue of $\underline{k}$ for the Leech lattice (see below) admits a distinguished "chirality". Only the second, nonstandard, version of E_8, namely, $\underline{k} \oplus \underline{h} \otimes T$, leads by analogy to the Griess algebra, which is in a sense an even more unique structure than E_8.

7. THE CROSS BRACKET OPERATION

While the vertex operators (6.1) are defined for $L = \Lambda$ (the Leech lattice), there is no affine Lie algebra $\hat{\underline{g}}[\nu]$ because Λ_2 is empty. But it turns out in this case that the components $x_\alpha(n)$ of the vertex operators $X(\alpha,\zeta)$ for $\alpha \in \Lambda_4$ do generate something interesting under a new commutative nonassociative operation [8]: For $\alpha,\beta \in \Lambda_4$ and $m,n \in \frac{1}{2}\mathbb{Z}$, define

$$[x_\alpha(m)\times x_\beta(n)] = \frac{1}{2}([x_\alpha(m+1),x_\beta(n-1)]+[x_\beta(n+1),x_\alpha(m-1)]). \tag{7.1}$$

We call this the <u>cross-bracket</u> of $x_\alpha(m)$ and $x_\beta(n)$, because it is made up of two brackets which "cross"; strictly speaking it depends on $X(\alpha,\zeta)$ and $X(\beta,\zeta)$, not just on $x_\alpha(m)$ and $x_\beta(n)$. Of course, the definition (7.1) makes sense for any two sequences of operators.

The Leech lattice analogue [8] the above theorem on brackets of vertex operators says that under cross-brackets, the operators $x_\alpha(n)$ for $\alpha \in \Lambda_4$ and $n \in \frac{1}{2}\mathbb{Z}$ generate a cross-bracket algebra spanned by the identity operator and the components of $X(\alpha,\zeta)$ $(\alpha \in \Lambda_4)$, $\zeta\frac{d}{d\zeta}\alpha(\zeta)$ $(\alpha \in \underline{h})$ and $:\alpha(\zeta)\beta(\zeta):$ $(\alpha,\beta \in \underline{h})$, where $\alpha(\zeta) = \sum_{n\in\mathbb{Z}+1/2}\alpha(n)\zeta^n$ and $:\ :$ denotes normal ordering. In particular, the Λ_4 vertex operators generate a canonical copy of the Virasoro algebra among the operators quadratic in the half-integrally moded creation and annihilation operators. The degree zero operators in the cross-bracket algebra span a commutative nonassociative algebra $\underline{k}$, the quadratic operators $:\alpha(\zeta)\beta(\zeta):$ giving a Jordan subalgebra. The new product operation $\times$ on $\underline{k}$ and a symmetric bilinear form $\langle\cdot,\cdot\rangle$ on $\underline{k}$ are determined by the formula, analogous to (5.3),

$$[x(m)\times y(n)] = (x\times y)(m+n)+\langle x,y\rangle m^2\delta_{m+n,0} \tag{7.2}$$

for $x,y \in \underline{k}$ and $m,n \in \mathbb{Z}$. (Note the m^2 in (7.2).) In place of (5.1), we have the invariance condition

$$\langle x\times y,z\rangle = \langle x,y\times z\rangle \tag{7.3}$$

for $x,y,z \in \underline{k}$.

The operators $\underline{k}$ preserve the subspace $W_{-1/2} = S_{-1/2}\otimes T \simeq \underline{h}\otimes T$ of W. This provides the space

$$\underline{B} = \underline{k}\oplus\underline{h}\otimes T$$

with a canonical commutative nonassociative algebra structure, with an identity element equal to the suitably normalized degree zero Virasoro generator. (Just as in the E_8 case, the products of pairs of elements of $\underline{h}\otimes T$ into $\underline{k}$ are canonically determined by the invariance condition (7.3) on $\underline{B}$.) The dimension of $\underline{B}$ is 196884.

The algebra $\underline{B}$ is precisely the Griess algebra with a natural identity element adjoined: $\underline{B} = B_0 \oplus \mathbb{C}\cdot 1$ (see ref. [8]). In particular, the Monster may be defined as the automorphism group of $\underline{B}$. Note that the most distinguished sporadic group of all - the Monster - may be defined canonically using only the vertex operators (6.1) and the cross-bracket operation.

8. THE INFINITE-DIMENSIONAL REPRESENTATION OF F_1

The algebra $\underline{B}$ (or B_0) has many relatively easy automorphisms, preserving $\underline{k}$ and $\underline{h}\otimes T$ and forming the centralizer C of an involution (= automorphism of order 2) in F_1. The group C is related to the Conway group $\cdot 0$ of automorphisms of the Leech lattice [2]. The existence of the Monster hinges on the very subtle fact that $\underline{B}$ admits additional automorphisms [18].

In order to construct such automorphisms and hence F_1 in a conceptual way, we consider the involution θ of the space $W = S\otimes T$ (for the Leech lattice Λ) defined as follows: If p is a monomial of degree r in the generators $h(n)$ ($h \in \underline{h}$, $n \in \mathbb{Z}+\frac{1}{2}$, $n < 0$) of the polynomial algebra S, then θ multiplies $p\otimes t$ ($t \in T$) by $(-1)^r$. Let W^- be the -1-eigenspace of θ in W. Now consider the vertex operator construction of untwisted affine Lie algebras $\hat{\underline{g}}$ of refs. [7], [34], a construction incorporated in the heterotic string [20], [21] in the cases $\underline{g} = E_8\times E_8$ and $\underline{g} = \underline{so}(32)$. This construction is based on integrally-moded bosonic string oscillators in the same sense that W is based on half-integrally moded oscillators. We modify this construction by starting with the Leech lattice in place of a root lattice, and we define an involution analogous to θ on the resulting space, say U, analogous to W. Let U^+ be the +1-eigenspace of θ in U, and form the $\mathbb{Z}$-graded space

$$V^\natural = U^+ \oplus W^- = \bigoplus_{n=-\infty}^{1} V_n^\natural$$

($\natural$ = "natural"). There is a natural action of the group C on $V^\natural$. See refs. [9], [10] for the details. Cf. ref. [23], which describes an apparently similar space based on formulas in ref. [5]. This space seems, however, not to provide a natural F_1-module.

We construct a uniformly defined extra automorphism σ on each $V_n^{\natural}$ (refs. [9], [10]). The source of σ is a "triality" for $\underline{\underline{s\ell}}(2,\mathbb{C})^{\wedge}$ which intertwines the twisted and untwisted vertex operator constructions. In the E_8 "practice case" of our construction of $V^{\natural}$, the restriction of σ to a "bosonic" subspace V_1 of U^+ and a "fermionic" subspace V_2 of W^- gives a "supersymmetry" implementing and explaining the numerology pointed out in ref. [13] which led to the combination of the Neveu-Schwarz and Ramond models into superstrings. (See ref. [10].) The same phenomenon in the Leech lattice case underlies our construction of σ (see ref. [10]) and in particular gives, by analogy, a corresponding new kind of 24-dimensional "supersymmetry". See also ref. [1].

We obtain the action by cross bracket of a "commutative affinization"

$$\hat{\underline{B}} = \underline{B}\otimes\mathbb{C}[t,t^{-1}]\oplus\mathbb{C}c$$

of $\underline{B}$ with multiplication defined by (7.2), and a proof of the fact that σ defines an algebra automorphism of $\underline{B}$ (see refs. [9], [10]). In particular $V^{\natural}$ is an F_1-module. With respect to the canonical Virasoro algebra mentioned earlier, the "conformal fields" comprising $\hat{\underline{B}}$ have conformal weight 2. The module $V^{\natural}$ has character $J(q)$, as in (3.1). The finite-dimensional algebra $\underline{B}$ is better understood from the cross bracket operation than from its original finite-dimensional definition. See ref. [10] for a detailed exposition of $V^{\natural}$, including discussions of more analogies with E_8 and more links with string theory.

We remark that many of the properties of $V^{\natural}$ predicted in ref. [5] remain unproved, including a deeper connection with the theory of Riemann surfaces. Other recent (finite-dimensional) treatments of the construction of an extra automorphism of the algebra B_0 are given in refs. [4], [39].

Just as twisted affine Lie algebras can be constructed by means of twisted vertex operators, we anticipate that twistings of $\hat{\underline{B}}$ can be constructed using the general methods of ref. [28], leading to a subtle kind of dimensional reduction and symmetry breaking.

It would be very important to place all the elements of F_1 on an equal footing with one another. In the present formulation, which is analogous to the noncovariant light-cone quantization in string theory, the elements of C preserve a twisted and an untwisted sector of $V^\natural$, and the remaining elements, like σ, mix the two sectors. Such a unification would probably take place in some kind of "covariant" 26-dimensional formulation of the Monster and its natural module. It is interesting to note that the original 26-dimensional string model [41], which was later overshadowed by 10-dimensional models [13], [16], [31], [32], [33], has recently been revived (cf. refs. [12], [36]) in the heterotic string [20], [21].

The existence of the Monster, properly interpreted, might conceivably lead to a unique physically correct four-dimensional string theory.

REFERENCES

1. Chapline, G., Unification of gravity and elementary particle interactions in 26 dimensions?, Phys. Lett. 158B, 393-396 (1985).

2. Conway, J.H., A group of order 8,315,553,613,086,720,000, Bull. London Math. Soc. 1, 79-88 (1969).

3. Conway, J.H., A characterization of Leech's lattice, Inventiones Math. 7, 137-142 (1969).

4. Conway, J.H., A simple construction for the Fischer-Griess Monster group, Invent. Math. 79, 513-540 (1985).

5. Conway, J.H. and Norton, S.P., Monstrous moonshine, Bull. London Math. Soc. 11, 308-339 (1979).

6. Corrigan, E. and Fairlie, D.B., Nuc. Phys. B91, 527 (1975).

7. Frenkel, I.B. and Kac, V.G., Basic representations of affine Lie algebras and dual resonance models, Inventiones Math. 62, 23-66 (1980).

8. Frenkel, I.B., Lepowsky, J. and Meurman, A., An E_8-approach to F_1, in: Finite Groups - Coming of Age, Proc. 1982 Montreal Conference, ed. by J. McKay, Contemporary Math. 45, 99-120 (1985).

9. Frenkel, I.B., Lepowsky, J. and Meurman, A., A natural representation of the Fischer-Griess Monster with the modular function J as character, Proc. Nat. Acad. Sci. U.S.A. 81, 3256-3260 (1984).

10. Frenkel, I.B., Lepowsky, J. and Meurman, A., A Moonshine Module for the Monster, in: Vertex Operators in Mathematics and Physics-Proceedings of a Conference November 10-17, 1983, ed. by J. Lepowsky, S. Mandelstam and I.M. Singer, Publications of the Mathematical Sciences Research Institute #3, Springer-Verlag, New York, 231-273 (1985).

11. Frenkel, I.B., Lepowsky, J. and Meurman, A., Vertex operators and the Monster, to appear.

12. Freund, P.G.O., Phys. Lett. 151B, 387 (1985).

13. Gliozzi, F., Olive, D. and Scherk, J., Supersymmetry, supergravity theories and the dual spinor model, Nuclear Physics B122, 253-290 (1977).

14. Gorenstein, D., Finite simple groups, Plenum Press, New York, (1982).

15. Gorenstein, D., Classifying the finite simple groups, Anaheim Colloquium Lectures, January, 1985, Amer. Math. Soc., to appear.

16. Green, M.B. and Schwarz, J.H., Phys. Lett. 109B, 444 (1982).

17. Green, M.B. and Schwarz, J.H., Phys. Lett. 149B, 117 (1984).

18. Griess, R.L., Jr., The Friendly Giant, Invent. Math. 69, 1-102 (1982).

19. Griess, R.L., Jr., A brief introduction to the finite simple groups, in: Vertex Operators in Mathematics and Physics - Proceedings of a Conference November 10-17, 1983, ed. by J. Lepowsky, S. Mandelstam and I.M. Singer, Publications of the Mathematical Sciences Research Institute #3, Springer-Verlag, New York, 217-229 (1985).

20. Gross, D.J., Harvey, J.A., Martinec, E. and Rohm, R., Heterotic string theory (I). The free heterotic string, Nucl. Phys. B256, 253-284 (1985).

21. Gross, D.J., Harvey, J.A., Martinec, E. and Rohm, R., Heterotic string theory (II). The interacting heterotic string, to appear.

22. Kac, V.G., An elucidation of "Infinite-dimensional...and the very strange formula" $E_8^{(1)}$ and the cube root of the modular invariant j, Advances in Math. 35, 264-273 (1980).

23. Kac, V.G., A remark on the Conway-Norton conjecture about the "Monster" simple group, Proc. Natl. Acad. Sci. USA 77, 5048-5049 (1980).

24. Kac, V.G., Kazhdan, D.A., Lepowsky, J. and Wilson, R.L., Realization of the basic representations of the Euclidean Lie algebras, Advances in Math. 42, 83-112 (1981).

25. Leech, J., Notes on sphere packings, Canadian J. Math. 19, 252-267 (1967).

26. Lepowsky, J., Euclidean Lie algebras and the modular function j, Amer. Math. Soc. Proc. Symp. Pure Math., 37, 567-570 (1980).

27. Lepowsky, J., Introduction, in: Vertex Operators in Mathematics and Physics - Proceedings of a Conference November 10-17, 1983, ed. by J. Lepowsky, S. Mandelstam and I.M. Singer, Publications of the Mathematical Sciences Research Institute #3, Springer-Verlag, New York, 1-13 (1985).

28. Lepowsky, J., Calculus of twisted vertex operators, Proc. Nat. Acad. Sci. USA (1985), to appear.

29. Lepowsky, J. and Meurman, A., An E_8-approach to the Leech lattice and the Conway group, J. Algebra 77, 484-504 (1982).

30. Lepowsky, J. and Wilson, R.L., Construction of the affine Lie algebra $A_1^{(1)}$, Comm. Math. Phys. 62, 43-53 (1978).

31. Neveu, A. and Schwarz, J.H., Nucl. Phys. B31, 86 (1971).

32. Neveu, A. and Schwarz, J.H., Phys. Rev. D4, 1109 (1971).

33. Ramond, P., Phys. Rev. D3, 2415 (1971).

34. Segal, G., Unitary representations of some infinite-dimensional groups, Comm. Math. Phys. 80, 301-342 (1981).

35. Serre, J.-P., A course in arithmetic, Springer-Verlag, New York (1973).

36. Thierry-Mieg, J., Phys. Lett. 156B, 199 (1985).

37. Thompson, J., Some numerology between the Fischer-Griess Monster and the elliptic modular function, Bull. London Math. Soc. 11, 352-353 (1979).

38. Tits, J., Four presentations of Leech's lattice, in "Finite Simple Groups II", Proceedings of a London Math. Soc. Research Symposium, Durham, 1978, Academic Press, London/New York, 303-307 (1980).

39. Tits, J., Le monstre, Séminaire Bourbaki, 36e année, 1983/84, no. 620 (1983), Astérisque, to appear.

40. Tits, J., On R. Griess' "Friendly Giant", Invent. Math. 78, 491-499 (1984).

41. Veneziano, G., Nuovo Cim. 57A, 190 (1968).

42. Halpern, M.B., Quantum solitons which are SU(N) fermions, Phys. Rev. D12, 1684-1699 (1975).

43. Halpern, M.B., Prehistory of internal symmetry on the string, preprint (1985).

Curved Superspace σ-Models and Green Schwarz Covariant Actions*

Luca Mẹzincescu
Theory Group
Department of Physics
The University of Texas at Austin
Austin, TX 78712

1. INTRODUCTION

There exist two Lorentz covariant actions for the superstrings. They appear to be vastly different.

The first approach[1] has its origins in the Ramond[2] (R) and Neveau-Schwarz[3] (NS.) oscillators and it is related to the different possible local supersymmetries in two dimensions. The superstring is viewed as a truncation[4] of the R and NS sectors which are described by a superfield $x^m(\xi,\eta)$ (M = o, •• 9 spacetime index, ξ-parametrization of two dimensional world sheet, η-two dimensional spinors) with corresponding boundary values, propagating in an arbitrary two-dimensional supergravity background.

The second approach belongs to Green and Schwarz.[5] (G.S.) It exhibits a manifest 10-dimensional SuperPoincaré invariance along with a non manifest invariance under transformations whose parameters are ten dimensional local (on the world sheet) fermi fields. This action is also manifestly invariant under reparametrizations of the world sheet.

The GS action can be viewed as a superspace (10 dimensional) 2 dimensioal σ-model i.e., a σ-model whose target space is the

*Work supported by NSF grant number PHY 83-04629 and Robert A. Welch Foundation.

corresponding flat superspace, with a Wess-Zumino term, propagating in an arbitrary two-dimensional gravity background.[6]

While it may seem that the second approach is more suited for a quantum formulation of the superstring at the moment a Lorentz covariant quantization procedure exists only in the R-NS approach.[7]

A further step toward a comparison between these two actions will be to couple them to a nontrivial ten dimensional background.

In the R-NS appraoch this program[8] is described in the talk of E. Martinec. The upshot is that in order to preserve the 2 dimensional superconformal invariance at the quantum level one must satisfy (for the heterotic string) the Chapline-Manton supergravity equations.

In the G.S. approach the coupling to 10 dimensional supergravity background is equivalent to formulating σ-models in "curved superspace" with the additional requirement of the invariance under appropriate local fermionic transformations.

For the N=1 superstring this was accomplished by E. Witten[9], while for N=2 chiral superstring the coupling to supergravity is described in Ref. 10.

This talk will be devoted to a presentation of the curved superspace σ-models. In formulating a σ-model we deal with fields living in a certain manifold M, for the present case these supermanifolds will be solutions of superspace supergravity equations of motion. Section 2 will be devoted to a brief description of superspace supergravity. Section 3 will outline the construction of the superspace σ-models and their invariance under local fermionic transformations. We will end with brief conclusions.

2. SUPERSPACE GEOMETRY

The superspace formulation of 10-dimensional on shell supergravities is obtained by the tedious transcription in superspace of the component field formalism. This approach is devised after the 4-d formulation of supergravity in superspace by

Wess and Zumino.[11] In what follows we will summarize the N=1 and N=2 10-d chiral supergravities (the N=2 non chiral has not been yet formulated in superspace) following ref. 14 and 15.

The N=1 supergravity in 10 dimensions has the field content:

$$e_m{}^a, \ \psi_m, \ A_{mn}, \ \lambda, \ \phi \qquad (2.1)$$

where $e_m{}^a$ - 10 dimensional zehnbein ψ_m - the gravitino field which is a Majorana-Weyl spinor, A_{mn} antisymmetric two index gauge field, λ is Majorana Weyl spinor of opposite handedness with respect to the gravitino while ϕ is a real scalar field.

The N=2 chiral supergravity in 10 dimensions[13] has the field content

$$e_m{}^a, \ \psi_m, \ A_{mn}, \ b_{mnrs}, \ \lambda, \ a \qquad (2.2)$$

where $e_m{}^a$ is again the zehnbein, ψ_m - complex Weyl gravitino, A_{mn} - complex antisymmetric gauge field, b_{mnrs} - real antisymmetric gauge field (with a self dual field strength) λ-complex Weyl spinor of opposite handedness to the gravitino and a - complex scalar field.

The superspace transcriptions of N=1 d=10 supergravity was investigated in Ref. 14 and Ref. 9 while for N=2, d=10 chiral supergravity this was accomplished in Ref. 15. To this end on uses the formalism of Cartan and Weyl for differential geometry, generalized to superspace. In this approach the fundamental objects are the vielbein 1-forms and the connections 1-forms.

The vielbein forms: $E^A = dz^M \, E_M{}^A(z)$ (2.3)

accomplish the transition from the coordinate 1-forms ($dz^M = dx^M$, $d\theta$) to one of the preferred frames in superspace.

The connection 1-forms $\Omega_A{}^B$ are Lie algebra valued corresponding to the structure group in the tangent space which generates the equivalence class of the preferred frames. They do transform as gauge fields of the corresponding group.

The structure group in all supergravity theories is chosen to contain the corresponding orthogonal group in our case SO(9,1). In theories which do contain scalar fields however it is convenient to make a slight modification. These theories exhibit at the component level some additional global (unexpected) symmetries of the equations of motion beyond the unitary (global) symmetries related to the automorphism group of the supersymmetry algebra which by the construction of the superspace is induced upon the spinor coordinates.

The unexpected global symmetries are nonlinear transformation under a certain noncompact group G which are linear when restricted to a maximal compact subgroup H. The maximal subgroup H is the automorphism group of the supersymmetry algebra. That is no automorphism group for the N=1 d=10 supergravity and U(1) for N=2 chiral d=10 supergravity. The group G is obtained by adding to the generators of H noncompact generators whose number is equal to the number of scalar fields, in the model. For N=1 d=10 supergravity one gets the dilatations while for N=2 chiral d=10 supergravity this group becomes uniquely SU(1,1). SU(1,1) acts nonlinearly on the scalars (which live in SU(1,1)/U(1)) however it can be described in a linear form by adding an auxiliary scalar field and compensating for it with an additional local U(1) gauge symmetry. Correspondingly the slight modification in the structure group mentioned above is that one takes the connection form $\Omega_A{}^B$ to be Lie algebra valued in SO(1,9) x H. The charges under this H follow from the transformation rules of the θ-coordinates of the superspace under the automorphism group of supersymmetry algebra. The antisymmetric tensor A_{mn} in (2.2) transforms under SU(1,1) but is inert under the local U(1).

The gauge covariant quantities the torsion 2-form T^A and the curvature 2-form $R_A{}^B$ are defined through the structure equations

$$T^A = dE^A + E^B \Omega_B{}^A$$

$$R_A{}^B = d\Omega_A{}^B + \Omega_A{}^C \Omega_C{}^B \qquad (2.4)$$

They satisfy the Bianchi identities

$$DT^A - E^B R_B{}^A = 0$$

$$DR_A{}^B = 0 \qquad (2.5)$$

The components of $T^A = \frac{1}{2} E^C E^B T^A_{BC}$ and of $R_A{}^B = \frac{1}{2} E^D E^C R_{CD,A}{}^B$ get charge assignments under the group H which is useful in establishing the constraints. From Bianchi identities (2.5) one infers that one can express $R_A{}^B$ as a function of T^A_{BC} and its derivatives, which indicates that one can impose constraints only on the torsion components.

To proceed further one must locate the component field strengths (which follow from (2.1), (2.2)) in the corresponding θ-expansion of the superspace field strengths forms T^A and $R_A{}^B$ and component gauge fields (2.1) and (2.2) in the corresponding θ-expansion of the gauge forms E^A and $\Omega_A{}^B$. It turns out that the antisymmetric gauge fields contained in (2.1) or (2.2) do appear only through their corresponding field strengths. This suggest the introduction of the following gauge forms: a real two form C for (2.1) and a complex two form A and a real four form B for (2.2). Their field strengths are defined by analogy with the definition of the field strength in the component form. The system of forms T^A, $R_A{}^B$ and the field strengths corresponding to the forms just introduced do contain a multitude of superfluous degrees of freedom which are eliminated by imposing constraints on the components of T^A and on the field strengths corresponding to the antisymmetric tensors. This is a painful procedure which involves dimensional analysis and passing through the linearized form of the theory. Once the system of constraints is formulated one must verify their consistency by requiring that they do obey the Bianchi identities (2.5) together with the corresponding Bianchi identity for the field strengths of the antisymmetric tensors. The result of this is that the theory is on shell - i.e., the component fields which do appear in the θ expansions of various superfields satisfy the equations of motion of the full supergravity. Moreover all the field strength in the theory can be expressed as covariant derivatives of a single scalar superfield which is a group element of G. As mentioned before this is the dilatations group for N=1 d=10 supergravity and the group SU(1,1) for N=2 chiral supergravity (For the N=2 nonchiral supergravity this group should be the dilatations[10]). The next step should be to determine as much as possible of the gauge forms E^A, $\Omega_A{}^B$, ... It should be possible to do this by choosing an appropriate Wess-Zumino gauge. However for the applications we are going to describe in the next section an explicit knowledge of these forms is not required.

In the remainder of this section I will specify more elements of N=1 d=10 and N=2 chiral d=10 supergravity which will be used afterwards for the analysis of superspace σ-models.

In N=1 d=10 supergravity the superspace is parametrized by the coordinates $z = (x^m, \theta^n)$ where θ is a 10 dimensional Majorana-Weyl spincr.

The 3-form H is defined through

$$H = \frac{1}{3} E^B E^C E^A H_{ABC} = dC \tag{2.6}$$

where the two form C corresponds to the superspace description of the antisymmetric tensor in (2.1), H satisfies the Bianchi identity dH = 0.

Some of the constraints on the components of the torsion form T^A and 3-form H defined respectively in (2.4) and (2.6) are

$$T^a_{\alpha\beta} = -i(\Gamma^a)_{\alpha\beta},\ T^a_{\alpha b} = i\delta^a_b V_\alpha,\ T^a_{bc} = 0 \tag{2.7}$$

and

$$H_{\alpha\beta\gamma} = 0 \ , \ H_{a\alpha\beta} = i(\Gamma_a)_{\alpha\beta}$$

$$H_{ab\alpha} = i(\Gamma_{ab})_\alpha{}^\beta \lambda_\beta \tag{2.8}$$

All non vanishing components of $R_{Aa}{}^B$, T^A Ad(z). For example

$$T^a_{\alpha b} = \frac{1}{2}\, i\ \delta^a_b\ D_\alpha \phi \tag{2.9}$$

where $D_\alpha = E_\alpha{}^M \partial_M$, $\partial_M = (\partial_m, \partial_\mu)$. The $\theta=0$ component of $\phi(z)$ is the dilaton field

$$\phi(x) = \phi(z)|_{\theta=0} \tag{2.10}$$

The spinor field $\lambda_\alpha(x)$ is te $\theta=0$ component of $D_\alpha\phi(z)$ and so on for the rest of the degres of freedom.

The flat superspace corresponds to constant $\phi(z)$. Then all the components of T^A and H are zero except

$$T^a_{\alpha\beta} = -i(\Gamma^a)_{\alpha\beta} \text{ and } H_{a\alpha\beta} = i(\Gamma_a)_{\alpha\beta} \tag{2.11}$$

The flat space vielbeins are given by

$$E^a = dx^a - i\bar{\theta}\Gamma^a d\theta$$
$$E^\alpha = d\theta^\alpha \tag{2.12}$$

In N=2, d=10 chiral supergravity the superspace is parametrized by the coordinates $z=(x^m, \theta^\mu, \bar{\theta^\mu})$. θ^μ is a 16-component complex Weyl spinor.

The complex 3-form F and real 5-form G correspond to the field strengths of the antisymmetric tensors in (2.2). They are defined through

$$F = dA$$
$$G = dB + A\bar{F} - \bar{A}F \tag{2.13}$$

They satisfy the Bianchi identities

$$dF = 0$$
$$dG = 2F\bar{F} \tag{2.14}$$

As a consequence of the superspace constraints[15] on T^A, F and G all the components of $R_A{}^B$, T^A, F and G can be expressed through the scalar superfield V(z) which is an element of SU(1,1).

$$V = \begin{pmatrix} u & v \\ \bar{v} & \bar{u} \end{pmatrix} \qquad u\bar{u} - v\bar{v} = 1 \tag{2.15}$$

The SU(1,1) Lie algebra valued 1-form

$$\Omega = V^{-1}dV = \begin{pmatrix} 2iQ & P \\ \bar{P} & -2iQ \end{pmatrix} \tag{2.16}$$

has as matrix elements the 1-form Q which is to be identified with the U(1) part of the SO(1,9) x U(1) connection $\Omega_A{}^B$ and the 1-form $P = E^A P_A = E^a P_a + E^\alpha P_\alpha - E^{\bar\alpha} P_{\bar\alpha}$.

The superfields Λ_α

$$\Lambda_\alpha = -\frac{1}{2} P_\alpha = -\frac{1}{2}(\bar{u} D_\alpha v - v D_\alpha \bar{u}) \tag{2.17}$$

contain the physical spinor λ_α (2.2) as the $\theta = 0$ component. The flat superspace limit corresponds constants u and v.

3. SUPERSPACE σ-MODELS.

To construct σ-models one proceeds in the usual way. Given a manifold M, one needs the vielbeins, the structure group in the tangent space and the closed invariant forms on this manifold. We have seen that in formulating supergravity in superspace one is lead to introduce certain closed invariant forms. Correspondingly any solution of suprgravity equations for which these forms are nonvanishing will furnish possible Wess-Zumino terms.

The GS covariant superstring actions are such 2-dimensional σ-models in the limiting case when the superspace is flat. However this requirement is not sufficient to define the superstring. The superstring action must be invariant not only under reparametrizations of the target manifold but also under the reparametrizations of the world sheet - that is of the base manifold of the σ-model - this is achieved by making the σ-model to propagate in an arbitrary 2-dimensional gravity background.

To complete the construction of the action one must ask for a particular σ-model which admits an oscilator representation - i.e., it is a free two dimensional model. This last requirement selects a subset of the possible actions which possess a fermionic local (in the world sheet) symmetry. (For a more detailed discussion of GS action see 16 and 17).

In generalizing the superstring actions for other manifolds which are solutions of supergravity equations we will seek to generalize this local fermionic symmetry. We will see that supergravity superspace constraints (which do imply the equations of motion of suprgravity) are sufficient to guarantee the existence of such actions.

As an illustration of last paragraph I will indicate how Siegel's local symmetry[18] arises for the superparticle of Brink and Schwarz[19] in a supergravity background.[9]

The action for N=1 d=10 superparticle in a supergravity background is:

$$S = \frac{1}{2}\int d\tau V^{-1}(\dot{Z}^M E_M{}^a)(\dot{Z}^N E_N{}^b)\eta_{ab} = \frac{1}{2}\int d\tau V^{-1}\, E^a E^b \eta_{ab} \qquad (3.1)$$

where V^{-1} is the eihnbein, η_{ab} - is te 10-dimensional tangent space Minkowski metric.

This action is invariant under the following transformations[9]

$$\delta Z^a = \delta Z^M E_M{}^a = 0$$
$$\delta Z^\alpha = \delta Z^M E_M{}^\alpha = \dot{Z}^M E_M{}^a\,(\Gamma_a)^{\alpha\beta}\kappa_\beta \qquad (3.2)$$

and

$$\delta V^{-1} = -i\,\dot{Z}^M E_M{}^\alpha \kappa_\alpha + \delta Z^\alpha\, V_\alpha \qquad (3.3)$$

where $\kappa_\beta(\tau)$ is a ten dimensional Majorana Weyl spinor and V_α is defined by (2.7) and (2.9).

To establish this it is of great help the fact that $\delta Z^a = 0$.

The variation of the vielbein form E^a appearing in (2.1) is

$$\delta E^a = \delta(\dot{Z}^M E_M{}^a) = \dot{Z}^M \delta Z^P T^a_{PM} + D_\tau \delta Z^a - E^B \delta Z^P \Omega_{PB}{}^a \qquad (3.4)$$

Then the variation of S is

$$\delta S = \int d\tau V\, \{E^B \delta Z^C T^a_{CB} + D_\tau \delta Z^a - E^b\, \delta Z^P \Omega_{Pb}{}^a\}\dot{Z}^N E_N{}^b \eta_{ab} \qquad (3.5)$$

The last term in (3.5) vanishes by the antisymmetry of $\Omega_{pb}{}^a$. Then using the constraints (2.7) for the torsion components $T^a_{\alpha b}$ and $T^a_{\alpha\beta}$ which are the only ones which appear in (3.5) because $\delta Z^a = 0$, and finally explicitating δZ^α from (3.2) one gets

$$\delta S = \int d\tau V\{-i\, E^\beta\, (\not{E}\not{E})_\beta{}^\alpha \kappa_\alpha + i\,\delta Z^\alpha V_\alpha \eta_{ab} E^a E^b\} \qquad (3.6)$$

Then, because $\not{E}\not{E} = \eta_{ab}E^a E^b$, the variation of S under δZ^A can be compensated by the variation of the eihnbein (3.3).

The same procedure can be followed for the superstring.

As the space does not permit a detailed derivation I will state the results

For N=1 d=10 chiral superstring the invariant action is [9]

$$S = \int d^2\xi \frac{1}{2} E_i^a E_j^b \sqrt{-g}\, g^{ij} \eta_{ab} + \frac{1}{2} \varepsilon^{ij} E_i^B E_j^A C_{AB} \qquad (3.7)$$

where the g^{ij} – metric on the world sheet C_{AB} is defined through (2.6). The vielbein forms are $E_i^A = Z^M{}_{,i} E_M^A$ and ε^{ij} is the two dimensional antisymmetric tensor density.

The local gauge invariance is given by

$$\delta Z^a = 0 \quad , \qquad \delta Z^\alpha = (\not{E}_k)^{\alpha\beta} P_-^{k\ell} \kappa_{\beta\ell}$$

$$\delta\left(\sqrt{-g}\, g^{ij}\right) = i\sqrt{-g}\, g^{ij} P_-^{k\ell} V_\alpha (\not{E}_k)^{\alpha\beta} \kappa_{\beta\ell} \qquad (3.8)$$

$$+2i\varepsilon^{k(i} P_-^{j)\ell}\, V_\alpha (\not{E}_k)^{\alpha\beta} \kappa_{\beta\ell}$$

here $P_\pm^{ij} = \frac{1}{2}\left(g^{ij} \pm \varepsilon^{ij}/\sqrt{-g}\right)$ and $\kappa_{\beta\ell}(\xi)$ a local 10d spinor

It is to be mentioned that in proving these invariances one uses Nilsson[14] constraints (2.7), (2.8).

In order to generalize this to heterotic string (last reference[1]) one must also verify that the coupling to the two dimensional fermions is consistent with (3.8). (See[20] for the flat case).

For N=2 d=10 chiral superstring one obtains analogous results which are explained in ref. 10 we will only comment on the results. As explained in section 2 the three form which does appear in superspace formulation of N=2 d=10 chiral supergravity is a complex form. This form transforms under the unexpected SU(1.1). In order to construct a Lagrangian one needs however a real form. So one must take combinations $F \pm F^+$ which will break this symmetry. Moreover because the U(1.1) symmetry is broken one is also obliged to fix the U(1) fake gauge symmetry which is responsible for the elimination of the fake scalar degree of freedom.

4. CONCLUSIONS

I presented in this talk a new class of σ-models which may encompass the existing superstrings. One characteristic of these models is that in their definition one uses the equations of motion of supergravity at the classical level while the equations of motion of supergravity do appear in 8) only as a result of quantum corrections. It is in this respect imperative to develop a framework which will allow quantum computations in these models.

One such possibility is a variant of the light cone gauge where one should recover the models outlined in ref. 21 another possibility may be related to the new formulations of the superstrings given by Siegel.[22] However it is not clear to what extent these formulations are flexible enough as to accommodate other backgrounds than the flat ones.

Another problem is related to the coupling of the Yang-Mills field for the heterotic strings in nontrivial backgrounds which seems difficult to implement.[23]

Finally one should mention that there exist some suggestions about the possible compactifications of the chiral N=2 d=10 supergravity[24] which may be of interest as toy models which bypass the complications related to the embedding the gauge connection in the spin connection for the "realistic" strings.

References

1]. Deser, S and Zumino, B. Phys. Lett. 65B 369 (1976), Brink, L., DiVecchia, P. and Howe, P. Phys. Lett. 65B, 471 (1976). Gross, D., Harvey, J., Martinec, E. and Rohm, R. Nucl. Phys. B256, 253 (1985).

2]. Ramond, P. Phys. Rev. D3, 2415 (1975).

3]. Neveau, A. and Schwarz, J. Nucl. Phys. B31, 86 (1971).

4]. Gliozzi, F., Scherk, J. and Olive, D. Nucl. Phys. B122, 253 (1977).

5]. Green, M. and Schwarz, J. Phys. Lett. 136B, 367 (1984), Nucl. Phys. B243, 285 (1984).

6]. Henneaux, M. and Mezincescu, L. Phys. Lett. 152B, 340 (1985); Martinec, E. unpublished.

7]. Friedan, D., Shenker, S. and Martinec, E. preprint EFI 85-32 (1985); see also Neveau, A. and West, P. preprint CERN-TH.4233/85.

8]. Callan, C.G., Martinec, E., Perry, M. and Friedan, D. Princeton preprint (1985).

9]. Witten, E. Princeton preprint (1985).

10]. Grisaru, M., Howe, P., Mezincescu, L., Nilsson, B. and Townsend, P. preprint UTTG-15-85.

11]. Wess, J. and Zumino, B. Phys. Lett. 66B, 361 (1977).

12]. Chamsedine, A. Nucl. Phys. B185, 403 (1981).

13]. Green, M. and Schwarz, J. Phys. Lett. 109B, 444 (1982).

14]. Nilsson, B. Nucl. Phys. B188, 176 (1981).

15]. Howe, P. and West, P. Nucl. Phys. B238, 181 (1984).

16]. Green, M. Proceedings of Symposium on Anomalies Geometry Topology, W.A. Bardeen and A. White World Scientific Publ. Co. Singapore (1985).

17]. Mezincescu, L. talk 1985 Cambridge Workshop on Supersymmetry and Its Applications to be published.

18]. Siegel, W. Phys. Lett. 128B, 397 (1983).

19]. Brink, L. and Schwarz, J. Phys. Lett. 100B, 310 (1981).

20]. Curtright, T., Mezincescu, L. and Zachos, C. Univ. of Florida preprint 1985 to appear in Phys. Lett.

21]. Candelas, P., Horowitz, G., Strominger, A. and Witten, E. Nucl. Phys. B258, 46 (1985).

22]. Siegel, W. Preprints Berkeley (1985).

23]. Mezincescu, L. and Nilsson, B. unpublished.

24]. Candelas, P. Nucl. Phys. B256, 385 (1985).
Günaydin, M., Romans, L.J. and Warner, N.P. preprint Caltech CALT-68-1268 (1985).

COVARIANT SUPERSTRINGS[1]

Warren Siegel[2]
Department of Physics, University of California, Berkeley, CA 94720

ABSTRACT

We describe modification of the action of Green and Schwarz for the superstring to allow supersymmetric Lorentz covariant quantization. Insight is gained from the analogous treatment of the superparticle (with or without external fields), which leads to the usual superspace formalism. The first-quantization is treated more easily in a Hamiltonian approach, which leads to *spacetime*-supersymmetric generalizations of Kač-Moody and Virasoro algebras, describing the covariant derivatives and field equations of the superstring field theory.

1. INTRODUCTION

Elsewhere in these proceedings[1)] we have described an approach to covariant, interacting bosonic string field theory based on the Becchi-Rouet-Stora-Tyutin transformations of classical bosonic string mechanics. Here we summarize results of our recent paper "Classical Superstring Mechanics"[2)] (see also ref. 3)), in which we began the generalization of these results for the bosonic string to the superstring by describing the classical mechanics of the superstring. A covariant formulation of the classical superstring (which should be distinguished from the Neveu-Schwarz-Ramond style formulation, in which supersymmetry is far from manifest) has already been given by Green and Schwarz[4),5)], but it's insufficient to describe the

[1]This work supported in part by the National Science Foundation under Research Grant No. PHY81-18547.

[2]Address after September 1, 1985: *University of Maryland, College Park, MD 20742.*

auxiliary fields, and hence suitable only for light-cone quantization. The analogous problem for the superparticle has been solved by additional coordinates which weaken the constraints on the system and allow the existence of supersymmetric spinor derivatives[6]. We here apply a similar procedure to the superstring. We also describe interaction of the classical superstring with massless external superfields. Quantization is briefly discussed. We work mainly in a Hamiltonian formulation of the classical mechanics (see ref. 1)).

2. SUPERPARTICLE

The Brink-Schwarz Lagrangian for the superparticle[7],

$$L_{BS} = (\dot{x} - i\dot{\theta}\gamma\theta) \cdot p - g\tfrac{1}{2}p^2 \quad , \tag{2.1}$$

has the unfortunate feature of implying that the supersymmetry-covariant spinor derivative vanishes: Upon canonically determining the quantity conjugate to θ, we find

$$d = \frac{\partial}{\partial\theta} + \not{p}\theta = 0 \quad . \tag{2.2}$$

This condition kills all auxiliary fields and prevents the use of the standard superspace formalism. (All spinors here are Majorana in the sense that if a spinor's complex conjugate is linearly independent the two can be combined into a spinor twice as big which is linearly dependent on its own complex conjugate. The γ's are defined by $\gamma^a{}_{[\alpha\beta]} = \gamma^{a[\alpha\beta]} = 0$, $\gamma^{(a}{}_{\alpha\gamma}\gamma^{b)\beta\gamma} = 2\eta^{ab}\delta_\alpha{}^\beta$, where ( ) is symmetrization and [ ] antisymmetrization.)

The mechanics Lagrangian appropriate to Lorentz-covariant off-shell supersymmetry can be obtained from a consideration of the covariant derivatives and field equations which it should imply. For a massless scalar field, $p^2 = 0$ is the only equation of motion, but for a massless scalar superfield, the additional equation $\not{p}d = 0$ is neccesary and sufficient to impose that the superfield is an irreducible representation of (on-shell) supersymmetry[8]. (All other field equations, up to constants, can be derived from these.) The equations of motion $p^2 = \not{p}d = 0$ and the explicit forms of p and d in terms of the superspace coordinates x and θ are *all* the information contained in the classical mechanics of the superparticle. This can be obtained from a modification[6] of the Brink-Schwarz Lagrangian: By adding the coordinates d and ψ in two new terms $L = L_{BS} + i\dot{\theta}d - \psi i\not{p}d$, we get

$$L = \dot{z}^M e_M{}^A(z)\pi_A - \lambda^i G_i(\pi),$$

$$z^M = (x^m, \theta^\mu), \quad \pi_A = (p_a, id_\alpha),$$
$$\dot{z}^M e_M{}^A(z) = (\dot{x}^a - i\dot{\theta}^\alpha \gamma^a{}_{\alpha\beta}\theta^\beta, \dot{\theta}^\alpha),$$
$$\lambda^i = (g, \psi_\alpha), \quad G_i(\pi) = (\tfrac{1}{2}p^a p_a, i\gamma^{a\alpha\beta} p_a d_\beta). \tag{2.3}$$

The expressions π, λ, G, and $\dot{z}e$ (where e is the superspace vielbein) are each invariant under global spacetime supersymmetry. The first term in the Lagrangian serves only to determine the supersymmetry covariant derivatives π_A in terms of the coordinates $z^M(\tau)$ and their canonical conjugates $i\partial/\partial z^M$:

$$\pi_A = ie_A{}^M \partial_M = i(\partial_a, \partial_\alpha + i\gamma^a{}_{\alpha\beta}\theta^\beta\partial_a). \tag{2.4}$$

The second term is the products of Lagrange multipliers λ^i with constraints G_i. The λ^i not only imply the equations of motion $G_i = 0$, but are also gauge fields which can be gauged away (except at the boundaries of τ-space). (The covariant gauge is $g = 1$, $\psi_\alpha = 0$.) Therefore, the G_i are not only constraints, but the generators (currents) of the gauge symmetries:

$$\delta z^M = [-i\varsigma^i G_i, z^M], \quad \delta\pi_A = [-i\varsigma^i G_i, \pi_A],$$

$$\delta\lambda^i = \dot{\varsigma}^i + \varsigma^k \lambda^j f_{jk}{}^i \quad ([G_i, G_j] = if_{ij}{}^k G_k). \tag{2.5}$$

For $\varsigma^i = \varsigma\lambda^i$, these are τ reparametrizations (generated by $\lambda^i G_i$). L is invariant up to a total derivative: $\delta L = -(\varsigma^i G_i)^{\cdot}$. The "covariant derivative" under these transformations is the total τ-derivative $d/d\tau = \partial/\partial\tau - i\lambda^i G_i$, which would give the Schrödinger equation (for Hamiltonian $\lambda^i G_i$) in a nonrelativistic-style quantization. With G and e given by (2.3), we have explicitly (with $\varsigma^i = (\xi, \kappa_\alpha)$)[9]:

$$\delta x = \xi p + i\kappa(\gamma d + \not{p}\gamma\theta), \quad \delta\theta = \not{p}\kappa,$$
$$\delta p = 0, \quad \delta d = 2p^2\kappa,$$
$$\delta g = \dot{\xi} + 4i\kappa\not{p}\psi, \quad \delta\psi = \dot{\kappa}. \tag{2.6}$$

Thus the superparticle mechanics can be brought into full accord with the standard approaches to supersymmetry. However, for purposes of constructing field theories, the only relevant information is the explicit form of the covariant derivatives π and the equations of motion G (see ref. 1)).

3. SUPERPARTICLE IN EXTERNAL FIELD

Coupling to external super-Yang-Mills can be accomplished as usual by covariantizing $p_a \to \nabla_a = p_a + \Gamma_a$ and $d_\alpha \to \nabla_\alpha = d_\alpha + \Gamma_\alpha$ with vector and spinor gauge potentials. (In the abelian case, this corresponds to adding the term $\dot{z}^M e_M{}^A \Gamma_A$ to L, where the potential Γ_A transforms as $\delta\Gamma_A = -i[\pi_A, K]$ in terms of the Yang-Mills gauge parameter K, so $\delta L = \dot{K}$.) Without loss of generality, we can consider spacetime (x) and "antispacetime" (θ) dimensions such that the only physical fields are a vector and a spinor. ($D = 3, 4, 6$, and 10 for the former, $D' = 2(D-2)$ for the latter[10]. All other cases can be obtained by dimensional reduction.) The (on-shell) commutation relations of the covariant derivatives are then[11]:

$$
\begin{aligned}
\{\nabla_\alpha, \nabla_\beta\} &= 2\gamma^a{}_{\alpha\beta}\nabla_a, \\
[\nabla_\alpha, \nabla_a] &= 2\gamma_{a\alpha\beta}W^\beta, \\
[\nabla_a, \nabla_b] &= F_{ab} \quad (\gamma_{a(\alpha\beta}\gamma^a{}_{\gamma)\delta} = 0).
\end{aligned}
\tag{3.1}
$$

W^α is the field strength (the physical spinor at $\theta = 0$). The appropriate generalization of the field equations is then[12]:

$$
\begin{aligned}
\gamma^{a\alpha\beta} p_a d_\beta &\to \gamma^{a\alpha\beta}\nabla_a\nabla_\beta, \\
\tfrac{1}{2}p^2 &\to \tfrac{1}{2}\nabla^a\nabla_a + W^\alpha\nabla_\alpha.
\end{aligned}
\tag{3.2}
$$

However, the algebra of these operators doesn't close, and requires including part of the operator $\nabla_{[\alpha}\nabla_{\beta]}$ (in addition to implying all the constraints in (3.1).) In $D = 3$ this is also a field equation for a scalar superfield[13]. In $D = 4$ the same is true with appropriate ordering (which isn't determined by the classical mechanics). In $D = 6$ only the self-dual third-rank antisymmetric tensor piece of $\nabla_{[\alpha}\nabla_{\beta]}$ is required (the isovector vector piece drops out of the algebra), and this also is a field equation for the scalar field strengths of the scalar multiplet[14]. In fact, both the constraints (3.1) and the field equations of matter multiplets can be derived by this approach, which is therefore more powerful than approaches based on knowing matter multiplets[15]. (Similar results have been obtained based on a Brink-Schwarz-type action[16].) However, in $D = 10$ there is no matter multiplet to which to couple Yang-Mills, and Yang-Mills itself contains no physical scalars. The operators in (3.2) and $\nabla_{[\alpha}\nabla_{\beta]}$ must therefore be generalized to include spin operators which act on Lorentz indices (e.g., $\nabla_{[\alpha}\nabla_{\beta]}W^\gamma \sim \nabla_a W^\delta$ with an appropriate matrix factor).

Such equations can be derived for free theories by reducing to the dimension where the theory is (super)conformal ($D=4$ for Yang-Mills), applying superconformal transformations to generate all other field equations from $\frac{1}{2}p^2$ (see ref. 17) for the nonsupersymmetric case), and "oxidizing" back to the original dimension (but dropping field equations containing operators undefined in the nonconformal theory). It's sufficient to consider commutators with just the S-supersymmetry generator S^α, which gives

$$\tfrac{1}{2}p^2 \to \not{p}d \to \begin{cases} \tfrac{1}{2}\{p^b, J_{ab}\} + \tfrac{1}{2}\{p_a, \Delta\} = p^b M_{ab} + p_a(\aleph - \tfrac{D-2}{2}) \\ d_{[\alpha}d_{\beta]} + \cdots, \end{cases} \tag{3.3}$$

where M_{ab} is the spin part of the Lorentz generator J_{ab}, $\aleph$ is the canonical dimension of the field (the "spin" part of the dilatation generator Δ), and $d_{[\alpha}d_{\beta]} + \cdots$ denotes appropriate parts of that operator plus additional pieces proportional to p_a times internal symmetry and Lorentz generators. (The part $\gamma_{abc}{}^{\alpha\beta}d_\alpha d_\beta + \cdots$ is a generalization of the Pauli-Lubansky vector[18]). The equation $\frac{1}{2}\{p^b, J_{ab}\} + \cdots = 0$ is valid for all massless theories, and in the nonsupersymmetric case it and $\frac{1}{2}p^2$ give all field equations plus constraints, as well as determining $\aleph$.)

4. KAČ-MOODY FOR SUPERSTRINGS

We again construct the classical mechanics by generalizing the fundamental operators and the field equations quadratic in them. The fundamental operators are found by generalizing the algebra $[\hat{P}_a(\sigma_1), \hat{P}_b(\sigma_2)] = i\delta'(\sigma_2 - \sigma_1)\eta_{ab}$ of the bosonic string and $\{d_\alpha, d_\beta\} = 2\gamma^a{}_{\alpha\beta}p_a$ of the superparticle. (We here consider just a single set of modes, corresponding to the open string, either set of modes of the closed string, or the supersymmetric set of modes of the heterotic string.) The simplest result consistent with the Jacobi identities is[3]:

$$\begin{aligned}
\{D_\alpha(\sigma_1), D_\beta(\sigma_2)\} &= \delta(\sigma_2 - \sigma_1)2\gamma^a{}_{\alpha\beta}P_a(\sigma_1), \\
[D_\alpha(\sigma_1), P_a(\sigma_2)] &= \delta(\sigma_2 - \sigma_1)2\gamma_{a\alpha\beta}\Omega^\beta(\sigma_1), \\
\{D_\alpha(\sigma_1), \Omega^\beta(\sigma_2)\} &= i\delta'(\sigma_2 - \sigma_1)\delta_\alpha{}^\beta, \\
[P_a(\sigma_1), P_b(\sigma_2)] &= i\delta'(\sigma_2 - \sigma_1)\eta_{ab}, \\
[P, \Omega] = \{\Omega, \Omega\} &= 0 \quad (\gamma_{a(\alpha\beta}\gamma^a{}_{\gamma)\delta} = 0).
\end{aligned} \tag{4.1}$$

Note the analogy with the super-Yang-Mills algebra (3.1):

$$(D_\alpha, P_a, \Omega^\alpha) \leftrightarrow (\nabla_\alpha, \nabla_a, W^\alpha), \tag{4.2}$$

and also that the same constraint occurs on the γ-matrices, which implies $D = 3, 4, 6$, or 10[4),5)] when the maximal Lorentz invariance is assumed (i.e., all of $SO(D-1,1)$ for the D-vector P_a).

This algebra can be solved in terms of $\widehat{P}_a$ (which is itself expressible in terms of X), a spinor coordinate $\Theta^\alpha(\sigma)$, and its derivative $\delta/\delta\Theta^\alpha$:

$$\begin{aligned} D_\alpha &= \frac{\delta}{\delta\Theta^\alpha} + \gamma^a{}_{\alpha\beta}\widehat{P}_a\Theta^\beta + \tfrac{1}{2}i\gamma^a{}_{\alpha\beta}\gamma_{a\gamma\delta}\Theta^\beta\Theta^\gamma\Theta'^\delta, \\ P_a &= \widehat{P}_a + i\gamma_{a\alpha\beta}\Theta^\alpha\Theta'^\beta, \\ \Omega^\alpha &= i\Theta'^\alpha. \end{aligned} \tag{4.3}$$

Note that Ω has no zero mode (at least in this simplest solution). These are invariant under supersymmetry generated by

$$\begin{aligned} q_\alpha &= \int d\sigma \left(\frac{\delta}{\delta\Theta^\alpha} - \gamma^a{}_{\alpha\beta}\widehat{P}_a\Theta^\beta - \tfrac{1}{6}i\gamma^a{}_{\alpha\beta}\gamma_{a\gamma\delta}\Theta^\beta\Theta^\gamma\Theta'^\delta \right), \\ p_a &= \int d\sigma\, \widehat{P}_a, \end{aligned} \tag{4.4}$$

where $\{q_\alpha, q_\beta\} = -2\gamma^a{}_{\alpha\beta}p_a$. These supersymmetry generators are consistent with those of Green and Schwarz[4),5)] when the constraint $D_\alpha = 0$ is applied (as for the Brink-Schwarz case for the particle).

5. VIRASORO FOR SUPERSTRINGS

The smallest algebra which includes generalizations of the operators $\frac{1}{2}\widehat{P}^2$ of the bosonic string and $\frac{1}{2}p^2$ and $\not{p}d$ of the superparticle is generated by

$$\begin{aligned} \mathcal{A} &= \tfrac{1}{2}P^2 + \Omega^\alpha D_\alpha = \tfrac{1}{2}\widehat{P}^2 + i\Theta'^\alpha\frac{\delta}{\delta\Theta^\alpha}, \\ \mathcal{B}^\alpha &= \gamma^{a\alpha\beta}P_a D_\beta, \\ \mathcal{C}_{\alpha\beta} &= \tfrac{1}{2}D_{[\alpha}D_{\beta]}, \\ \mathcal{D}_a &= i\gamma_a{}^{\alpha\beta}D_\alpha D'_\beta. \end{aligned} \tag{5.1}$$

Note the similarity of $\mathcal{A}$ to both the gauge-fixed $\mathcal{A}$ of ref. 1) and (3.2). The algebra generated by these operators closes (see ref. 2) for the gorey details).

BRST quantization can again be performed (with some modification[19]), due to $[G_i, f_{jk}{}^l\} \neq 0$).

These results again can be described in a Lagrangian formulation.

6. CONCLUSIONS

Coupling to external super-Yang-Mills is analogous to the bosonic string and superparticle: Covariantize $D_\alpha \to D_\alpha + \delta(\sigma)\Gamma_\alpha$, $P_a \to P_a + \delta(\sigma)\Gamma_a$, $\Omega^\alpha \to \Omega^\alpha + \delta(\sigma)W^\alpha$. (The δ-function coupling makes the modification of the algebra essentially the same as in the case of the particle.) Assuming $\int d\sigma\, \mathcal{A}$ as kinetic operator (again ignoring ghosts), the vertex becomes

$$V = W^\alpha D_\alpha + \Gamma^a P_a - \Gamma_\alpha \Omega^\alpha \tag{6.1}$$

evaluated at $\sigma = 0$. Solving the constraints (3.1) in a Wess-Zumino gauge, we find

$$\begin{aligned} W^\alpha &\approx \lambda^\alpha, \\ \Gamma^a &\approx A^a + 2\gamma^a{}_{\alpha\beta}\Theta^\alpha\lambda^\beta, \\ \Gamma_\alpha &\approx \gamma^a{}_{\alpha\beta}\Theta^\beta A_a + \tfrac{4}{3}\gamma^a{}_{\alpha\beta}\gamma_{a\gamma\delta}\Theta^\beta\Theta^\gamma\lambda^\delta, \end{aligned} \tag{6.2}$$

evulated at $\sigma = 0$, where "$\approx$" means dropping terms involving x-derivatives of the physical fields A_a and λ^α. Plugging (4.3) and (6.2) into (6.1) gives

$$V \approx A^a \hat{P}_a + \lambda^\alpha\left(\frac{\delta}{\delta\Theta^\alpha} - \gamma^a{}_{\alpha\beta}\hat{P}_a\Theta^\beta - \tfrac{1}{6}i\gamma^a{}_{\alpha\beta}\gamma_{a\gamma\delta}\Theta^\beta\Theta^\gamma\Theta'^\delta\right). \tag{6.3}$$

Comparing with (4.4), we see that the vertices, in this approximation, are the same as the integrands of the supersymmetry generators, evaluated at $\sigma = 0$. (In the case of ordinary field theory, the vertices *are* just the supersymmetry generators p_a and q_α, to this order in θ.) This result corresponds to the vertices of Green and Schwarz[5], where their nonlinearities are simply the result of including all derivative terms in the expansion of the superfields Γ_α, Γ_a, and W^α in (6.1) to all orders in Θ. In practice, superfield techniques should be used even in the external field approach, so such explicit expansion (or even (6.2) and (6.3)) is unnecessary.

The classical description of the superstring given here is suitable for quantization. The steps are: (1) Use the commutation relations to derive the BRST operator Q, which will include terms higher than quadratic order in ghosts[19]. The condition $Q^2 = 0$ should reproduce the conditions $D = 10$, $\alpha_0 = 1$. This classical mechanics Lagrangian imposes weaker constraints than the Green-Schwarz one (which sets $D_\alpha = 0$ via Gupta-Bleuler), and thus should not impose stronger conditions. (2) Find the appropriate ground state (and corresponding string field) and BRST-invariant kinetic operator. Considering the remarks at the end of section 3, this may require modification of the generators (5.1) and BRST operator, perhaps to include Lorentz generators (acting on the ends of the string?) or separate contributions from the BRST transformations of Yang-Mills field theory. On the other hand, the ground state, rather than being Yang-Mills, might be purely gauge degrees of freedom, with Yang-Mills appearing at some excited level. In that case, modification might be unnecessary, and $\int d\sigma\, \mathcal{A}$ (plus ghost contributions) might serve as the analog of H of ref. 1), with (6.1) as external-Yang-Mills vertices. (3) Find the fully interacting BRST operator and string field theory. (4) Find the gauge-invariant string field theory action. Concerning the gauge invariance, it's interesting to note that, if we generalize D, P, and Ω to gauge-covariant derivatives $\nabla_\alpha = D_\alpha + \Gamma_\alpha$, $\nabla_a = D_a + \Gamma_a$, $\nabla^\alpha = \Omega^\alpha + W^\alpha$, with Γ_α; Γ_a, and W^α now functions of σ, describing the vector multiplets of all masses, then the fact that the only mode of the ∇'s missing is the zero-mode of Ω^α ($\int d\sigma\, \Omega^\alpha = 0$) directly corresponds to the fact that the only gauge-invariant mode of the connections is the zero-mode of W^α (the massless spinor, the massive spinors being Stueckelberg fields).

REFERENCES

1) W. Siegel, String field theory via BRST, and B. Zwiebach, contributions to these proceedings.
2) W. Siegel, Classical superstring mechanics, Berkeley preprint UCB-PTH-85/23 (May 1985), to appear in Nucl. Phys. B.
3) W. Siegel, Covariant approach to superstrings, Berkeley preprint UCB-PTH-85/16 (April, 1985), to appear in Anomalies, geometry and topology, ed. A. White (World Scientific, Singapore, 1985).
4) M. B. Green and J. H. Schwarz, Phys. Lett. **136B** (1984) 367.
5) M. B. Green and J. H. Schwarz, Nucl. Phys. **B243** (1984) 285.
6) W. Siegel, Class. Quantum Grav. **2** (1985) L95.
7) L. Brink and J. H. Schwarz, Phys. Lett. **100B** (1981) 310.
8) S. J. Gates, Jr., M. T. Grisaru, M. Roček, and W. Siegel, Superspace, *or* One thousand and one lessons in supersymmetry (Benjamin/Cummings, Reading, 1983) pp. 70, 88.
9) W. Siegel, Phys. Lett. **128B** (1983) 397.
10) L. Brink, J. H. Schwarz, and J. Scherk, Nucl. Phys. **B121** (1977) 77.
11) W. Siegel, Phys. Lett. **80B** (1979) 220.
12) M. T. Grisaru, W. Siegel, and M. Roček, Nucl. Phys. **B159** (1979) 429;

M. T. Grisaru and W. Siegel, Nucl. Phys. **B201** (1982) 292, **B206** (1982) 496;
S. J. Gates, Jr., M. T. Grisaru, M. Roček, and W. Siegel, Superspace, pp. 25, 382, 383.

13) W. Siegel, Nucl. Phys. **B156** (1979) 135.

14) J. Koller, Nucl. Phys. **B222** (1983) 319;
P. S. Howe, G. Sierra, and P. K. Townsend, Nucl. Phys. **B221** (1983) 331;
J. P. Yamron and W. Siegel, Unified description of the N = 2 scalar multiplet, Berkeley preprint UCB-PTH-85/15 (May, 1985), to appear in Nucl. Phys. B.

15) S. J. Gates, Jr., and W. Siegel, Nucl. Phys. **B163** (1980) 519;
S. J. Gates, Jr., K. S. Stelle, and P. C. West, Nucl. Phys. **B169** (1980) 347.

16) L. Lusanna and B. Milewski, Nucl. Phys. **B247** (1984) 396.

17) A. J. Bracken, Lett. Nuo. Cim. **2** (1971) 574;
A. J. Bracken and B. Jessup, J. Math. Phys. **23** (1982) 1925.

18) A. Salam and J. Strathdee, Nucl. Phys. **B76** (1974) 477;
S. J. Gates, Jr., M. T. Grisaru, M. Roček, and W. Siegel, Superspace, pp. 72-73;
R. Finkelstein and M. Villasante, Casimirs for the N-dimensional super-Poincaré algebra and the decomposition of the scalar superfield, UCLA preprint UCLA/84/TEP/13 (November, 1984).

19) E. S. Fradkin and T. E. Fradkina, Phys. Lett. **72B** (1978) 343.

An Invariant String Propagator

Andrew Cohen, Gregory Moore, Philip Nelson
Lyman Laboratory of Physics
Harvard University
Cambridge, MA 02138

Joseph Polchinski
Theory Group
Physics Department
University of Texas
Austin, TX 78712

Presented by J.P. at the Santa Barbara Workshop on Unified String Theories

ABSTRACT

We show that the Polyakov path integral can be used to define off-shell quantities in string theory.

The path integral of Polyakov gives an elegant description of strings and their interactions.[1] However, its use has been limited to obtaining the Koba-Nielsen expressions for S-matrix elements.[2] It is not yet clear what quantities make sense in string theory. We would like to show that the path integral can be used to define off-shell quantities as well. In particular it defines a natural n-point function in loop space as the sum of all world surfaces bounded by n specific spacetime curves. This work appears in refs. 3-6, to which the reader is referred for more detail, and was very much influenced by the work of O. Alvarez.[7]

The sum over surfaces is

$$\int \frac{dgdx}{V_{GC}V_W} \exp\left(-\frac{T}{2}\sqrt{g}\, g^{ab}\partial_a x^\mu \partial_b x^\mu + \lambda\sqrt{g}\, R\right) \tag{1}$$

where the integration is over embeddings x^μ of the world surface in d-dimensional spacetime and over an internal metric field g_{ab}

on the world sheet. In this paper we restrict attention to closed bosonic strings. The second term in (1) depends only on the world-sheet topology and determines the coupling of the string loop expansion. The classical action (1) has two local invariances:

$$\text{general coordinate:} \quad \delta\sigma^a = \xi^a(\sigma)$$
$$\text{Weyl} \quad : \quad \delta g_{ab} = \delta\phi(\sigma) g_{ab}(\sigma) \tag{2}$$

We are interested only in the critical dimension d=26, where the Weyl symmetry is a symmetry of the quantum theory.[1] In eq. (1), we have divided by the volume of the local symmetry group.

We first outline the general evaluation of (1).[3,4,7] We then discuss the additional features added by boundaries.[5,7] Locally, the three gauge freedoms ξ^a and $\delta\phi$ can be used to take $g_{ab}(\sigma)$ to the unit matrix. Globally, this is not quite possible. In general we can choose a family of fiducial metrics $\hat{g}_{ab}(\sigma,\tau)$, depending on a finite number of Teichmuller parameters τ, and every metric will be gauge equivalent to one of these. Infinitesimally, every variation of $g_{ab}(\sigma)$ can be decomposed as

$$\delta g_{ab} = \hat{g}_{ab}\delta\phi + \xi_{a;b} + \xi_{b;a} + \hat{g}_{ab,i}\delta\tau_i \tag{3}$$

Thus, we can change variables in the integration,

$$dg = (d\phi d\xi)' \, d\tau \, J(\phi,\tau) \tag{4}$$

The prime omits the conformal Killing vectors from the integration, which are those infinitesmal GC + Weyl transformations which leave g_{ab} invariant. In the language of ref. 7, the Teichmuller deformations $\hat{g}_{ab,i}$ are Ker $(P^\dagger)$, and the conformal Killing vectors are Ker (P). When $\hat{g}_{ab}$ can be chosen such that $(\hat{g}_{ab,i})^{;b} = 0$ (this is possible for the sphere, cylinder, and torus) $J(\phi,\tau)$ can be written

$$J = (\det{}' P^\dagger P)^{1/2} \, \frac{\det^{1/2} H(P^\dagger)}{\det^{1/2} H(P)} \tag{5}$$

where $P^\dagger P$ is the vector Laplacian

$$(P^\dagger P)_a{}^c = -\delta_a{}^c D^2 - D^c D_a + D_a D^c \tag{6}$$

and

$$H_{ij}(P^{\dagger}) = \int d^2\sigma \sqrt{\hat{g}}\, \hat{g}^{ac}\hat{g}^{bd} \chi_{iab}\chi_{jcd}$$

$$H_{rs}(P) = \int d^2\sigma \sqrt{\hat{g}}\, \hat{g}_{ab} V^a_r V^b_s \qquad (7)$$

where $\chi_{iab} = \hat{g}_{ab,i} - \frac{1}{2}\hat{g}_{ab}\hat{g}^{cd}\hat{g}_{cd,i}$

and V^a_r are a basis for the conformal Killing vectors.

The x-integration gives

$$\int dx\, e^{-S(x,\hat{g})} = V \left(\frac{\det' \nabla}{2\pi \int d^2\sigma \sqrt{\hat{g}}} \right)^{-13} \qquad (8)$$

where the factor of spacetime volume V is from the constant mode. The general coordinate times Weyl group is in general a connected piece times a discrete piece D. The volume is then

$$V_{GC} V_W = \int d\xi \int d\phi \text{ order } (D) \qquad (9)$$

The ϕ and ξ integrations cancel between the numerator and denominator. (It is important here that at d=26 the integrand of the numerator is in fact ϕ-independent.[1] It is shown in ref. 4 that there are no non-local sources of ϕ-dependence.) A factor of $V_{\tilde{G}}$ remains in the denominator, where $\tilde{G}$ is that subgroup of GC x Weyl which is not fixed by the gauge choice. $\tilde{G}$ in general has a continuous piece which is generated by the conformal Killing vectors and a discrete piece $\tilde{D} \subset D$. The final expression is

$$\text{sum} = V \int \frac{d\tau}{V_{\tilde{G}}} (\det' P^{\dagger}P)^{1/2} \frac{\det^{1/2} H(P^{\dagger})}{\det^{1/2} H(P)} \left(\frac{\det' \nabla}{2\pi \int d^2\sigma \sqrt{\hat{g}}} \right)^{-13} \qquad (10)$$

S-matrix elements are obtained by inserting vertex operators into the functional integral.[1] This procedure is Weyl invariant only on shell, and does not extend to define general off-shell quantities (though see ref. 8 for one application).

Let us mention some results of refs. 3,4 concerning eq. (10) without boundary loops. For genus zero, the sphere, $\tilde{G}$ = SL(2,C), the Möbius group. This is of complex dimension 3, non-compact, and has infinite volume; as a result, n-point functions vanish[4] for n < 3. In particular, the tree level cosmological constant and tadpole always vanish (related arguments appear in ref. 9). For genus zero with no external lines, the one loop cosmological constant, the result differs from what one gets by taking the spectrum of the free string and summing the corresponding field theoretic cosmological constants.[3] The difference arises due to the discrete component of $\tilde{G}$, the modular group. It is well known that duality means that a string theory differs combinatorically from a sum of field theories. This example, which involves no interactions, shows that even in the free theory, a string theory differs from a sum of field theories.

We now wish to sum over all surfaces bounded by specific spacetime loops.[5] This quantity is to be completely invariant under the local symmetries of the world sheet, as opposed to a light-cone or conformal gauge propagator. The boundary loops can be parameterized $x^\mu_L(\sigma)$, but the functional integral is to run over all surfaces bounded by any parameterization of this loop. Thus, the boundary condition on x is

$$x^\mu|_{\partial M} = x^\mu_L \circ \Sigma(\sigma) \tag{11}$$

where Σ any reparameterization of the boundary loops. A boundary condition on g_{ab} is also needed. Introduce an inward-pointing vector field n^a on the boundary. Then

$$n^a t^b g_{ab}|_{\partial M} = 0 \tag{12}$$

where t^a is the tangent to the boundary. The resulting sum is independent of the choice of n^a, so this represents no new information at the boundary, but is needed to get a self-adjoint Fadeev-Popov operator. It can be shown[5],[7] that no new anomaly in the Weyl symmetry arises at the boundary. The generalization of eq. (10) is now

$$\text{sum} = \int d\Sigma \int \frac{d\tau}{V_{\tilde{G}}} (\det{}'_{\text{mixed}} P^\dagger P)^{1/2} \frac{\det^{1/2} H(P^\dagger)}{\det^{1/2} H(P)}$$

$$\left(\frac{\det_{\text{Dirichlet}} \nabla}{2\pi \int d^2\sigma \sqrt{\hat{g}}} \right)^{-13} e^{-S(x_{c\ell}, \hat{g})} \tag{13}$$

The boundary condition on the x^μ determinant follows from (11). On the Fadeev-Popov determinant we have the mixed condition

$$n^a \xi_a = 0 \qquad t^a n^b \xi_{a;b} = 0 \tag{14}$$

The first condition expresses invariance of the coordinate region; the second results from (12). The classical solution $x^\mu_{c\ell}$ satisfies

$$\partial_a \sqrt{\hat{g}}\, \hat{g}^{ab} \partial_b x^\mu_{c\ell} = 0 \qquad x^\mu_{c\ell}|_{\partial M} = x^\mu_L \circ \sum \tag{15}$$

For the tree level two point function, the topology of the cylinder, it is convenient to take the coordinate region $0 \le \sigma^1 \le 1$, $0 \le \sigma^2 \le 1$ with σ^1 periodic and $\sigma^2 = 0,1$ the boundary. There is one real Teichmuller parameter λ and $\hat{g}_{ab} d\sigma^a d\sigma^b = (d\sigma^1)^2 + \lambda^2 (d\sigma^2)^2$. $\tilde{G}$ is the compact group of translation in the σ^1-direction. Then the propagator is

$$A = \int d\Sigma_i \int d\Sigma_f \int_0^\infty \frac{d\lambda}{\lambda^{13}}\, e^{4\pi\lambda}\, f(e^{-4\pi\lambda})^{-24}\, e^{-S(x_{c\ell},\hat{g})} \tag{16}$$

where i and f denote the initial and final loops and $f(x) = \prod_{n=1}^{\infty} (1 - x^n)$. It is also useful to give an operator expression for the propagator by introducing the space of functionals of parameterized loops, $\psi[x^\mu(\sigma)]$, where σ is periodic from 0 to 1. The Virasoro generators[2] act on this space. The propagator can then be written as

$$A = \sum_{m,n} (-1)^{m+n} \langle x_f | P_{diff} \frac{1}{H+6m^2+6n^2+2m+2n} P_{diff} | x_i \rangle \tag{17}$$

where $H = 2L_o + 2\tilde{L}_o - 4 = \frac{1}{8} p^2 + 2(N+\tilde{N}-4)$ and P_{diff} projects onto states annihilated by all the $R_n = L_n - \tilde{L}_{-n}$. The R_n generate the "diff" group, the reparameterizations in σ. P_{diff} is the operator representation of the integration in $\sum$.

Let us defer discussion of P_{diff}, and consider first several examples of diff-invariant states. The simplest are the "punctual" states [5], the pointlike loops $x^\mu_{i,f}(\sigma) \equiv x^\mu_{i,f}$. Then

$S_{c\ell} = (x_i - x_f)^2 T/2\lambda$. The resulting propagator has poles at the mass levels of the closed string. The residues are all positive and agree with the punctual propagator calculated in light-cone gauge. These pointlike states, which have been considered before,[10] are highly singular: the propagator diverges at a finite short distance, due to the large number of mass eigenstates which have an overlap with the pointlike state.[11]

Another set of diff-invariant states [6] is built by starting with any highest weight state $|T\rangle$:

$$L_n|T\rangle = \tilde{L}_n|T\rangle = 0 \text{ for } n>0$$

$$N|T\rangle = \tilde{N}|T\rangle \tag{18}$$

It can be shown that the following state is diff-invariant:

$$|\psi\rangle \equiv R|T\rangle = \sum_{I,J} L_{-I}\tilde{L}_{-J}\left(M^{-1}(L_o)\right)_{IJ}|T\rangle \tag{19}$$

Here L_{-I} is shorthand for $L_{-i_1}L_{-i_2}\cdots L_{-i_k}$, the sum runs over all sets $(i_1, \cdots, i_k)$ and $M(L_o)$ is the contravariant form.[12] The construction parallels those of ref. 13. Using external states $|\psi_i\rangle$, $|\psi_f\rangle$ in (17), the propagator becomes

$$A = \sum_m (-1)^m \langle T_f| \frac{1}{\frac{1}{2}p^2+2N+2\tilde{N}-4+6m^2+2m} |T_i\rangle \tag{20}$$

This is a simplification over (17), but there is still a large set of poles, some of negative residue. However, in a propagator we want the probes at the endpoints to be localized. The states $|\psi\rangle$ are not in general local because $M^{-1}(L_o)$ has poles in p^2 whenever $M(L_o)$ has a vanishing eigenvalue. (No string state is truly local; localized here implies gaussian falloff, while propagation is exponential falloff, in position space.) These are the points 13]

$$L_o = \frac{1}{8}p^2 + N = \frac{25-n^2}{24} \tag{21}$$

for n = 5,6,7, ••• .

In order that $|\psi\rangle$ be local, the state $|T\rangle$ must have zeroes in momentum space at the points (21). These zeroes coincide with all the poles in (20) except for the n=0 term. The final result is

$$A = \langle T_f| \frac{1}{\frac{1}{2}p^2+2N+2\tilde{N}-4} |T_i\rangle + \text{analytic} \tag{22}$$

These states do not have the short distance problems of the punctual states. The precise form of the analytic pieces in (22) depends on the choice of the states $|T\rangle$. One sees that the residue in (22) is positive: ghosts do not couple to these diff-invariant states. For example, let $|T_{f,i}\rangle$ be $e^{\mu\nu}_{f,i}(p)\, \alpha^{\mu}_{-1}\, \tilde{\alpha}^{\nu}_{-1}\, |0\rangle$, where transversality requires $p^{\mu}e^{\mu\nu}_{f,i}(p) = 0$. Then the 2-point function is

$$e^{\mu\nu}_{f}(p)\, e^{\mu\nu}_{i}(p)\, \frac{1}{p^2} + \text{analytic} \tag{23}$$

The residues at $p^2 = 0$ are all positive and correspond to intermediate scalar, graviton, and antisymmetric tensor states. As a consequence of the transversality of $e^{\mu\nu}_{f,i}$, the gauge of (23) is arbitrary: gauge dependent terms in the propagator would not contribute.

The operator P_{diff} is simply $RP_{tr}R^{\dagger}$, where R is defined in (19) and P_{tr} projects onto highest weight states. This satisfies $P_{diff}^{2} = CP_{diff}$. The constant C is infinite because the diff group is non-compact.

It is straightforward to extend the propagator to the case of 2 off shell + n on shell lines by inserting vertex operators into the cylinder. We have shown that the residues of the poles in this expression are in fact the Koba-Nielsen amplitudes.[6]

The amplitudes we have been studying were constructed to be invariant under the local symmetries of the path integral (1), that is, general coordinate and Weyl invariance. Of great interest is the appearance of the d-dimensional general coordinate invariance: what can we say about this? At the linearized level (the 2-point function), our expressions are d-dim GC invariant, because our diff-invariant states couple only to transverse gravitons: a satisfying result. However, this property cannot be maintained off shell at the interacting level, and when we factorize on internal graviton lines we find that they are in the Feynman gauge.[6] This is fascinating, since we have nowhere explicitly chosen such a gauge. Work is in progress on this question. The relation between our work and the approaches of ref. 14 is also an open question; in the path integral, the generators R_n play a central role, while in ref. 14, the positive frequency L_n and $\tilde{L}_n$ are the natural gauge generators. As an aside, we note that the spurious poles in (20) are all among the Stueckelberg poles of ref. 14, and that our expressions such as (17) can be simplified by the introduction of explicit Fadeev-Popov fields. Another extension under way is to the Ramond-Neveu-Schwarz string.

Acknowledgements

We would like to thank the organizers for a stimulating meeting, and many participants for helpful discussions. This work was supported in part by the National Science Foundation, PHY 82-15249 (A.C., G.M., P.N.) and PHY 83-04629 (J.P.), the Harvard Society of Fellows (G.M., P.N.), the Robert A. Welch Foundation (J.P.) and the Alfred P. Sloan Foundation (J.P.).

References

1]. A.M. Polyakov, Phys. Lett. 103B, 207 (1981).

2]. For a review, see S. Mandelstam, Phys. Rep. 13C, 261 (1974).

3]. J. Polchinski, "Evaluation of the One-Loop String Path Integral", Texas Preprint UTTG-13-85, to appear in Comm. Math. Phys. (1985).

4]. G. Moore and P. Nelson, "Absence of Nonlocal Anomalies in the Polyakov String", Harvard preprint HUTP-85/A057 (1985).

5]. A. Cohen, G. Moore, P. Nelson, and J. Polchinski, "An Off-Shell Propagator for String Theory", Harvard/Texas preprint HUTP-85/A058, UTTG-16-85 (1985).

6]. A. Cohen, G. Moore, P. Nelson, and J. Polchinski, "Another String Miracle", in preparation.

7]. O. Alvarez, Nucl. Phys. B216, 125 (1983).

8]. S. Weinberg, "Radiative Corrections in String Theories", Texas preprint UTTG-22-85 (1985).

9]. P. Candelas, G. Horwitz, A. Strominger, and E. Witten, Nucl. Phys. B258, 46 (1985). E. Witten, "Dimensional Reduction of Superstring Models", Princeton preprint (1985).

10]. E. Corrigan and D.B. Fairlie, Nucl. Phys. B91, 527 (1975); M. Green, Nucl. Phys. B124 (1977) 461; L. Susskind, unpublished.

11]. We thank T. Banks and D. Gross for this observation.

12]. R.C. Brower and C.B. Thorn, Nucl. Phys. B31, 163 (1971). V.G. Kac, Lecture Notes in Physics 94, 441 (1979).

13]. B.L. Feigin and D.B. Fuks, Functs. Anal. Prilozhen 17, 91 (1983). D. Friedan, "String Field Theory", Enrico Fermi preprint EFT 85-27 (1985).

14]. D. Friedan; W. Siegel and B. Zwiebach; T. Banks and M. Peskin; S. Raby; this Proceedings. A. Neveu and P.C. West, CERN preprint CERN-TH.4200/85; M. Kaku and J. Lykken, "Suprgauge Field Theory of Superstrings" CUNY preprint (1985).

INTERACTING-STRING PICTURE OF THE FERMIONIC STRING*

Stanley Mandelstam

Department of Physics
University of California
Berkeley, California 94720

In our main lectures we gave a review, hereinafter referred to as I, of the interacting-string picture of the Bose string. In the present lecture we shall outline a similar treatment of the Fermionic string. The quantization of the free Fermionic string was carried out by Iwasaki and Kikkawa [1]. In addition to the degrees of freedom x, representing the displacement of the string, one has Grassman degrees of freedom S distributed along the string. Aharonov, Casher and Susskind [2] have shown that one can picture the fermionic string as a string of dipoles. The general picture of the interaction of such strings by joining and splitting is the same as for the Bose string.

For reasons which will become evident in a subsequent section, we do not at present have what we believe to be the simplest formula for fermion string scattering amplitudes. We shall therefore give a less detailed treatment than we gave for the Bose string. We shall set up the functional-integration formalism, derive the analog mode, and indicate in general terms how the conformal transformation to the z-plane may be performed. We shall conclude by stating without proof the formula for the N-particle tree amplitude in the manifestly supersymmetric formalism.

There are two formulations of the fermionic string; the older

*This work supported by the National Science Foundations under Research Grant No. PHY81-18547

Neveu-Schwarz-Ramond formalism and the manifestly supersymmetric formalism of Green and Schwarz [3]. We shall present them in parallel, since both appear to have their strong points. The N.S.R. formalism possesses superconformal invariance (which we shall not discuss), but the final formulas are much simpler in the Green-Schwarz formalism.

Review of Free Fermionic Strings

Fermionic strings possess anti-commuting degrees of freedom S in addition to the commuting variables x. The critical dimension is ten, so that there are eight transverse dimensions. In the N.S.R. formalism the S's, like the x's, carry vector superscripts i, in the supersymmetric formalism they carry spinor superscripts α. When we wish to include both cases we shall use the notation S^a. The S's also carry subscripts 1 and 2, which refer to the Dirac components on the two-dimensional world sheet. The fermion part of the free Lagrangian (with imaginary time) is:

$$\mathscr{L}_F = -\frac{1}{2\pi}\ \{S_1(\frac{\partial}{\partial\tau}+i\frac{\partial}{\partial\sigma})S_1 + S_2\ (\frac{\partial}{\partial\tau}-i\frac{\partial}{\partial\sigma})\ S_2\ \}. \qquad (1)$$

In the supersymmetric formalism, the boundary condition at the ends of open strings is $S_1 = S_2$. In the N.S.R. formalism it is $S_1 = \pm S_2$, the signs being opposite at the two ends of a Bosonic string and the same at the two ends of a Fermionic string. The sign changes at a joining point in the N.S.R. formalism. With closed strings, the S's of the supersymmetric formalism and of the Fermions in the N.S.R. formalism are periodic, while the S's of the Bosons in the N.S.R. formalism are antiperiodic.

The easiest way of treating the open-string boundary conditions is to double the string, ie., to treat an open string as half a closed string as regards its S variables. The co-ordinate σ runs from $-\pi\alpha$ to $\pi\alpha$, and $S(\sigma)$ is defined as $S_1(\sigma)$ if $\sigma>0$ or as $S_2(-\sigma)$ if $\sigma<0$. Only the first term in the Lagrangian (1) will then be present.

Vertex Operator

The simple vertex operator, with δ-functions in the X and S variables, is not Lorentz invariant in Fermionic string models. A Lorentz-invariant vertex for the N.S.R. formalism is obtained by adding an extra factor $S_1^i \partial X^i/\partial\rho$ at the joining point [4,5]. More precisely, we take a point σ on the string near the joining point σ_α, and apply a factor

$$\lim_{\sigma\to\sigma_\alpha} \{\mp i(\sigma-\sigma_\alpha)\}^{3/4} S_1{}^i(\sigma) \{\partial X^i(\sigma)/\partial\rho\}. \tag{2}$$

The factor $\{\mp(\sigma-\sigma_\alpha)\}^{3/4}$ is required in order that the whole expression approach a finite limit as $\sigma\to\sigma_\alpha$. The sign is - for joining, + for splitting. We shall see presently that this factor cancels the factors for transforming to the z-plane. Owing to the boundary condition we could have replaced S_1 by S_2 or $\partial/\partial\rho$ by $\partial/\partial\rho^*$.

The equivalent factor takes a slightly different form in the supersymmetric formalism, as we cannot contruct a scalar operator out of SO(8) spinors and vectors; we also require conjugate spinors, which the model does not possess. We recall that any element of a supersymmetric string can be either a vector or a conjugate spinor, each having eight possible polarizations. The operator S^i in the N.S.R. vertex creates or destroys a vector element. We therefore make the ansatz that *the vertex operator between a state containing one string and a state containing two is non-zero only if the single string contains an extra vector element at the joining point. The vertex operator has a factor* $\sum_i |i\rangle\, \partial X^i/\partial\rho$, *where* $|i\rangle$ *represents an element with polarization* i. We shall see below how to rewrite this rule in terms of the operators S^α, and we shall also specify the $(\sigma-\sigma_\alpha)$ factors.

The single string of a closed-string vertex will have an extra tensor element at the joining point, and the factor will be $|ij\rangle(\partial X^i/\partial\rho)(\partial X^j/\partial\rho^*)$.

We have shown that the vertex operator is Lorentz invariant [6]. The proof works directly in σ-τ space.

Green and Schwarz [7] have given an alternative derivation of the vertex function. Working in mode space in the supersymmetric formalism, they apply necessary (but not sufficient) conditions for Lorentz invariance, and obtain a unique result.

Though the extra factor at the joining point does not prevent us in principle from repeating all the calculations we have given for the Bose string, it complicates the result considerably, since the co-ordinates $\tilde{z}$ of the joining point were completely eliminated from our final expressions. In the corresponding calculations for fermionic strings, the factors from the joining point would leave us with a result that depended explicitly on the $\tilde{z}$'s. Eq. (4.3b) of I, or the equivalent equation for loops, would then have to be used to obtain the $\tilde{z}$'s as a function of the Z's. The covariant approach appears to be free of this difficulty, but it is not manifestly supersymmetric and, even at the one-loop level, the manifestly supersymmetric results appear to be much simpler than those of the N.S.R. formalism.

Berkovits, working in the light-cone frame, has recently pointed out that no factor appears at the joining point if one performs the functional integral over supersheets instead of over world sheets; analogous methods had been used in the old dual-model operator formalism by Brink and Winnberg [8], and superconformal transformations have been studied further by Horsely [9]. The functional integration must be performed in the N.S.R. formalism, but we hope that it will be possible to translate the results into the supersymmetric formalism at the analog-model stage.

Ground State Wave Functions

In the Bose-string model the only variable on which the ground state depended was the momentum, but the fermionic ground states of the N.S.R. formalism, and all ground states of the supersymmetric formalism, and all ground states of the supersymmetric formalism, occur in multiplets. We shall arbitrarily regroup the zero-mode S's as follows:

$$\tilde{S}^a = \frac{1}{\sqrt{2}}(S^a + iS^{a+4}) \qquad \tilde{S}^{a\dagger} = \frac{1}{\sqrt{2}}(S^a - iS^{a+4}).\ 1 \leq a \leq 4. \tag{3}$$

Though we are using the same notation, we remind the reader that the a's are vector indices in the N.S.R. formalism, spinor indices in the supersymmetric formalism. We arbitrarily say that a state is "empty" in the a-mode ($1 \leq a \leq 4$) if $\tilde{S}^a\,|> = 0$, "full" if $\tilde{S}^{a\dagger}|> = 0$. The completely empty state will be denoted by the symbol $|0>$. The physical ground-state mulitplet, in which we shall be interested, will consist of the sixteen states obtained by applying zero, one, two, three or four $\tilde{S}^{\dagger}$'s to $|0>$. In the supersymmetric formalism we obtain the vector multiplet by applying an even number of $\tilde{S}^{\dagger}$'s to the state $|0>$, the (conjugate) spinor multiplet by applying an odd number.

The decomposition (3) entails loss of the manifest SO(8) invariance of the transverse space. We are left with SO(4) invariance in the N.S.R. formalism and with SO(6) x SO(2) invariance in the supersymmetric formalism. SO(8) invariance is never far below the surface, and the SO(8) invariance of all the formulas in readily seen.

With a suitable choice of axes, the vector states in the supersymmetric formalism are as follows:

$$(\sqrt{2})^{-1}\,\{|1>+i|2>\} = |0>, \tag{4a}$$

$$(\sqrt{2})^{-1}\,\{|1>-i|2>\} = \rho^i_{\alpha\beta}\,\tilde{S}^{1\dagger}\tilde{S}^{2\dagger}\tilde{S}^{3\dagger}\tilde{S}^{4\dagger}|0>, \tag{4b}$$

$$|i> = \lambda^i_{\alpha\beta}\,\tilde{S}^{\alpha\dagger}\tilde{S}^{\beta\dagger}\,|0>, \qquad 3 \leq i \leq 8 \tag{4c}$$

where the entry in the states on the left-hand side is the polarization, and $\lambda^i_{\alpha\beta}$ is the appropriate Clebsch-Gordan matrix for SO(6), and explicit form of which has been given by Brink, Scherk and Schwarz [10].

With outgoing particles we reverse our definition of "empty" and "full", so that "empty" states will again be those for which $\langle 0|\tilde{S}^a = 0$. We now define the "ground state" $|g\rangle$ of half the relevant particles to be the "empty" state and of the other half to be the "full" state; this definition is adopted so that "ground-state" scattering amplitudes are non-zero. The relevant particles are of course the fermions in the N.S.R. formalism, and all external particles and extra elements at the joining points in the supersymmetric formalism. The wave-functions of particles with "full" "ground" states will be of the form $\Pi S^a|g\rangle$, the actual combinations being the adjoints of (4).

The extra element of string at a joining point is also respresented by (4). If the "ground" state is defined to be "empty", the factor $\sum|i\rangle\partial X^i/\partial\rho$ can now be written:

$$\eta\, \{\frac{\partial X^L}{\partial\rho} + \tfrac{1}{2}\eta \sum_{i=3}^{8} \lambda^i_{\alpha\beta}\, \frac{\partial X^i}{\partial\rho}\, \tilde{S}^{\alpha\dagger}\tilde{S}^{\beta\dagger} + \eta^2\, \frac{\partial X^R}{\partial\rho}\tilde{S}^{1\dagger}\tilde{S}^{2\dagger}\tilde{S}^{3\dagger}\tilde{S}^{4\dagger}\}|0\rangle, \tag{5a}$$

where

$$X^R = (\sqrt{2})^{-1}(X^1 + iX^2) \qquad X^L = (\sqrt{2})^{-1}(X^1 - iX^2), \tag{5b}$$

$$\eta = \{\mp i(\sigma-\sigma_\alpha)\}^{\frac{1}{2}}. \tag{5c}$$

The corresponding formula with a "full" "ground" state is obvious. As before, the expression is to be evaluated at a point near the joining point, and the factor (5b) will cancel the factors necessary to transform to the z-plane.

Functional Integral and Analog Model

The functional integral over the X and S variables is performed in very much the same way as in the Bose-string model. The integrand now has factors of S^a and $\partial X^i/\partial\rho$ outside the exponential, both from the external-particle wave-functions and from the factor (2) or (5) at the joining point. It is straightforward to perform the integration, since the integral is still a Gaussian. The result will consist of a number of terms, one for each possible paring of *some* of the factors $\partial X^i/\partial\rho$ with one another and *all* the factors S^a with one another. The term corresponding to a given pairing will have the factor (3.9) of I, with $|\Delta|^{-(d-2)/2}$ replaced by $\{|\Delta_S|/|\Delta_X|\}^{(d-2)/2}$, together with:

i) A factor $i\,(\partial/\partial\rho)\sum_r P_r^i\, N(\rho,\rho_r)$ for each unpaired factor $(\partial X^i/\partial\rho)$,

ii) A factor $-S_{ij}\ (\partial^2/\partial\rho\partial\rho')N(\rho,\rho')$ for each paired $\partial X^i/\partial\rho$ and $\partial X^i/\partial\rho'$

iii) A factor $K_{11}^{ab}\ (\rho,\rho')$ for each paired S^a and S^b (open strings) or K_{mn}^{ab} for each paired S_m^a and S_n^b (closed strings), together with a factor ± 1 from interchange of the S's.

The function K is the Neumann function for the operators S in the Lagrangian (1). It satisfies the equation

$$2(\partial/\partial\rho^{(*)})K_{mn}^{ab}\ (\rho,\rho') = 2\pi\delta^{ab}\ \delta_{mn}\delta^2(\rho-\rho'), \tag{6}$$

where the star is present if m=n=1.

The conformal transformation to the z plane, the evaluation of the Neumann functions $N(\rho,\rho')$, and the evaluation of the factors of (3.9) of I in terms of the Z variables is exactly the same as for the Bose string. With regard to the Fermion Neumann functions, the first thing to notice is that they are conformal covariants rather than invariants. Thus:

$$K_{mn}(\rho,\rho') = \{\frac{\partial\rho}{\partial z}\,\frac{\partial\rho'}{\partial z'}\}^{(*)(-\frac{1}{2})}\ \bar{K}_{mn}(z,z'), \tag{7}$$

where $\bar{K}$ is the z-plane Neumann function. This is expressed by saying that the operators S have conformal weight $\frac{1}{2}$.

We may now discuss the factors of $\mp i(\sigma-\sigma_\alpha)$ in our operators at the joining point. When transforming to the z-plane the operators $\partial X^i/\partial\rho$ and S give rise to the factors $(\partial\rho/\partial z)^{-1}$ and $(\partial\rho/\partial z)^{-\frac{1}{2}}$ respectively. The factors $\mp i(\sigma-\sigma_\alpha)$ are therefore cancelled, and in their place we have factors $(2\frac{\partial^2\rho}{\partial z^2})^{-1}$.

Functions $\bar{K}$ which satisfy the correct equations could be defined as follows:

$$\bar{K}_{11} = 2(\partial/\partial z)N(z,z') \qquad \bar{K}_{22} = 2(\partial/\partial z^*)N(z^*,z'^*). \tag{8}$$

However, extra factors are required in order that the boundary conditions at the external particles and the joining points be satisfied.

The S's for Boson of the N.S.R. formalism satisfy antiperiodic boundary conditions. The ρ-plane Neumann function will therefore change sign when one of its arguments goes once round a meson string. This implies that the Neumann function changes sign when one of its arguments goes round a small circle surrounding a joining point. (Any three-particle interaction involves an odd number of mesons). There is no such sign change at a Fermion string, and there are none at all in the supersymmetric formalism. The factor $(\partial\rho/\partial z)^{-\frac{1}{2}}$ has a square-root singularity at all external particles and joining points, so that we require the z-plane Neumann function to have a square-root singularity at all external Fermions in the N.S.R. formalism, and at all external particles and joining points in the supersymmetric formalism. If we add an extra factor $(z-Z_r)^{\pm\frac{1}{2}}$ or $(z-\tilde{z}_\alpha)^{\pm\frac{1}{2}}$ to the Neumann function $\bar{K}(z,z')$ we must add factors $(z'-Z_r)^{\mp\frac{1}{2}}$ or $(z'-\tilde{z}_\alpha)^{\mp\frac{1}{2}}$ in order that the singularity at $z=z'$ be unaffected.

The extra factors will be simplest if we use the $\tilde{S}$'s instead of the S's; the new Neumann function will obviously be linear combinations of the old. We define $\bar{\bar{K}}_{mn}$ (z,z') as the factor obtained when an $\tilde{S}(z)$ is contracted with an $\tilde{S}^\dagger(z')$. Contracting two $\tilde{S}$'s or two $\tilde{S}^\dagger$'s obviously gives zero. The exclusion principle then determines whether the exponent is $\frac{1}{2}$ or $-\frac{1}{2}$. The formulas for the Fermionic z-plane Neumann functions are thus:

$$\tilde{\bar{K}}_{11}\ (z,z') = \prod_r \{(z-\bar{z}_{1r})^{\frac{1}{2}}(z-\bar{z}_{2r})^{-\frac{1}{2}}(z'-\bar{z}_{1r})^{-\frac{1}{2}}(z'-\bar{z}_{2r})^{\frac{1}{2}}\}2(\partial/\partial z)N(z,z'), \tag{9a}$$

$$\tilde{\bar{K}}_{22}\ (z,z') = K_{11}(z^*,z'^*), \tag{9b}$$

where the $\bar{z}_r$'s range over all external fermions in the N.S.R. formalism and all external particles and joining points in the supersymmetric formalism. The subscripts 1 and 2 denote particles where we have defined the "ground state" to be empty or full respectively. In loop amplitudes, the functions $z-\bar{z}_r$ are replaced by the appropriate elliptic or automorphic functions.

The evaluation of the Neumann functions is thus complete. We shall not discuss the evaluation of the determinant Δ_S. It is *not* the same as the determinant Δ_X, as the boundary conditions at the external strings and at the joining points are different. However, the partition-function factor $\prod_n (1-w^n)$ or $\prod_i (1-w)$ is the same, and this factor therefore cancels between the two determinants.

N-particle Tree Amplitude

We shall now state the manifestly supersymmetric form of the tree amplitudes without giving the derivation. The formula was obtained by using the supersymmetric functional-integration formalism to derive several general properties of the Z-integrand; analysis of the residues at the poles due to massless intermediate states then gave us a unique answer.

We introduce light-cone superspace in the form defined by us for ordinary supersymmetric models [11], and applied by Green and Schwarz [7] to superstrings. Corresponding to the external particle r we define a spinor co-ordinate θ_r^α. The state $\Pi\ \tilde{S}^{\alpha\dagger}|0\rangle$ has wave-function $\Pi\theta_r^\alpha$ (the distinction we previosuly made between "empty" and "full" ground states is now dropped). The Koba-Nielsen integrand will be a function of the θ_r's as well as of the Z_r's, and the θ-integration is performed according the the rules

$$\int d\theta_r^\alpha = 0 \qquad \int \theta_r^\alpha d\theta_r^\alpha = \alpha_r^{-1}. \tag{10}$$

The variables θ for different α's will be taken to commute; θ_r^α of course anti-commutes with θ_s^α.

Our result will be a product of the Koba-Nielsen integrand (4.7) of I and another factor. To write down the second factor we take a multiperipheral diagram(Fig. 1). The particles at the end of the chain are kept fixed, and the

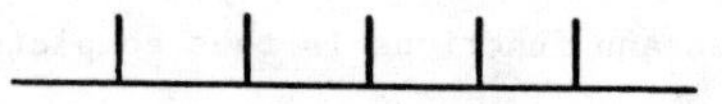

Fig. 1. Multiperipheral Diagram

result will be independent of the pair chosen. We sum over all possible positions of the other particles. We emphasize that the ordering of the particles in Fig. 1 has nothing to do with the original string diagram; the same factor multiplies the Koba-Nielsen integrand of all string diagrams.

The letters r and s will identify the external particles of Fig. 1, the letters n and m the verticles. The lines at the nth vertex will be denoted by in, jn, and kn in cyclic order. At any factor associated with a vertex, all P's and α's will be defined positively when directed inwards. The θ's associated with the internal lines are defined in terms of the external θ's by the equation

$$\alpha_{in}\,\theta_{in} + \alpha_{jn}\,\theta_{jn} + \alpha_{kn}\,\theta_{kn} = 0 \tag{11}$$

at each vertex. The θ's are regarded as directed quantities, so that the sign of the θ of a given internal line depends on the vertex with which it is associated.

Following Ref 7, we introduce the definitions:

$$P^{(n)} = \alpha_{in} P_{jn} - \alpha_{jn} P_{in}, \quad (12a)$$

$$\Theta_n = \alpha_{kn}^{-1} (\theta_{in} - \theta_{jn}), \quad (12b)$$

$$\gamma_n = \alpha_{in} \alpha_{jn} \alpha_{kn}, \quad (12c)$$

$$J_n = P^{(n)}L + \frac{\gamma_n}{2} \sum_{i=3}^{8} \lambda^i_{\alpha\beta} P^{(n)i} \Theta^\alpha_n \Theta^\beta_n + \gamma_n^2 P^{(n)} R_{\Theta^1_n \Theta^2_n \Theta^3_n \Theta^4_n}, \quad (12d)$$

$$K_{nm} = \tfrac{1}{4}\gamma_n \gamma_m \frac{1}{4!} \epsilon_{\alpha\beta\gamma\delta} \{\gamma_n \Theta^\alpha_n \Theta^\beta_n \Theta^\gamma_n \Theta^\delta_m - \gamma_m \Theta^\alpha_n \Theta^\beta_m \Theta^\gamma_m \Theta^\delta_m\} \times \sum_{i,j,k} \{-P^2_{in} \alpha^{-1}_{in} + P^2_{im} \alpha^{-1}_{im}\}. \quad (12e)$$

For each ordering of the external particles in Fig. 1, we pair some of the vertices with one another, leaving the rest unpaired. There is a term for each pairing; the factors multiplying the Koba-Nielsen integrand will be:

i) A factor $(Z_1 - Z_N)^{-1} \prod_{r=1}^{N-1} (Z_{r+1} - Z_r)^{-1}$.

ii) A factor J_n for each unpaired vextex,

iii) A factor K_{nm} for each paired vertices n and m.

iv) A factor $\delta^4(\sum_r \alpha_r \theta^\alpha_r)$.

The result generalizes the result obtained by Green and Schwarz [7] for the four-point amplitude, and its deviation does not require a great deal of algebra. The amplitude has the following properties:

a) It is manifestly supersymmetric, since the supersymmetry generators are $\sum \alpha_r \theta^\alpha_r$ and $\sum \partial/\partial\theta^\alpha_r$.

b) We may simply drop the first and last terms of (12d), and reinterpret all six-vector products as eight-vector products. In this sense the formula has manifest SO(8) invariance.

c) Since all momenta occur in the combinations $\alpha_r P_s - \alpha_s P_r$, we may now replace the eight-vector products by ten-vector products.

d) The amplitude has no poles as a function of an internal α at $\alpha=0$; it was constructed to have this property.

e) If all external particles are vectors, the only terms which depend on the α's are those containing factors $\varepsilon_r \cdot P_r$. where ε_r is the polarization. In conjunction with c), this implies that the amplitude is Lorentz invariant.

f) If any ε_r is replaced by P_r, a term corresponding to a given pairing of vertices will contain a factor α_r. Property e) then implies that the whole amplitude will vanish when ε_r is proportional to P_r.

The closed-string amplitude will involve superspace variables $\theta_1{}^\alpha$ and $\theta_2{}^\alpha$ corresponding to the operators $S_1{}^\alpha$ and $S_2{}^\alpha$. The Koba-Nielsen integrand will now be mulitplied by two factors similar to the open-string factor, one involving θ_1's and Z's, the other θ_2's and Z^*'s.

Finiteness of the Oriented Closed-Superstring Model

Green and Schwarz [12] have shown that the oriented closed-superstring models (type IIa and IIb) are finite at the one-loop level. Since we have not completed the loop calculations we cannot claim a rigorous result. However, we wish to outline reasoning that makes us fairly confident that the models will be finite in any order of perturbation theory.

Divergences or singularities may arise from configurations where the w's approach zero, or the Z's (including the invariant points of the T_r's) approach one another in any combination. Unless all the external Z's approach one another, the divergences or singularities are easily analyzed and are simply the expected unitarity effects.

The configuration where the Z's approach one another are those where one or more loops shrink to a point. Let us suppose inductively that we have already established finiteness at the n-1-loop level. Clearly the only configuration which might give divergences at the n-loop level is that where all loops together shrink to a point, with their relative dimensions remaining finite. We remarked in section 7 of I that such a configuration corresponded to that where all the invariant points of the T_n's approached one another; some of the w's might approach zero at the same time. By projective invariance this is equivalent to a configuration where the distances between the invariant points remain finite, but all Z's aproach one another. Now the behavior of the integrand when the Z's approach one another, keeping away from the invariant points, will not depend on the number of loops. We therefore expect finiteness at the n-loop level.

Another interesting question is whether the one-loop finiteness of the SO(32) open-string model persists in higher orders. As the finiteness in this case depends on a cancellation, we shall have to complete our calculations before we can answer this question.

REFERENCES

1. Y. Iwasaki and K. Kikkawa, Phys. Rev. D8, 440 (1973).

2. Y. Aharonov, A. Casher and L. Susskind, Phys. Rev. D5 (1971) 988.

3. M.B. Green, Surveys in High Energy Physics 3 (1983)127.
J.H. Schwarz, Phys. Reports 89, 223 (1982).

4. S. Mandelstam, Nucl Phys. B69, 77 (1974).

5. S. Mandelstam, Phys. Reports 13C, 260 (1974).

6. S. Mandelstam, Nucl. Phys. B83, 413 (1974).

7. M.B. Green and J.H. Schwarz, Nucl. Phys. B243, 479 (1984).

8. L. Brink and J.O. Winnberg, Nucl. Phys. B103, 445 (1976).

9. R. Horsley, Nucl. Phys. B138 493 (1978).

10. L. Brink, J. Scherk and J.H. Schwarz, Nucl. Phys. B121, 77 (1977).

11. S. Mandelstam, Nucl. Phys. B213, 149 (1983).

12. M.B. Green and J.H. Schwarz, Phys. Lett. 109B, 444 (1982).

SEMINARS

2. String Field Theory

STRING FIELD THEORY VIA BRST[1]

Warren Siegel[2]
Department of Physics, University of California, Berkeley, CA 94720

ABSTRACT

We describe the BRST approach to Lorentz covariant quantization of string field theory, as applied to the interacting bosonic string.

1. INTRODUCTION

The theory of relativistic strings[1)] is the only possibility presently known for a quantum theory of gravity (perhaps bringing unification as a byproduct), and may also be a useful approach to a direct description of hadrons. A Lorentz- and gauge-covariant field theoretic formulation of strings would provide the same advantages as in modern quantum theory of ordinary fields at the classical (model building), perturbative (graphs and finiteness), and nonperturbative (semiclassical solutions, topological properties, spontaneous symmetry breakdown) levels. However, the formulations of the string generally used are gauge-fixed ones suitable only for a limited study of perturbation theory: The covariant Neveu-Schwarz-Ramond formalism is not manifestly supersymmetric, and the light-cone formulation is not manifestly Lorentz covariant.

Previously[2)] we described an approach to covariant, interacting bosonic string field theory based on the Becchi-Rouet-Stora-Tyutin transformations of classical bosonic string mechanics[3),4)]. The quantized string field includes Faddeev-Popov ghosts and auxiliary fields missing from other covariant approaches (the traces of the physical tensor fields, such as the metric tensor), as well as allowing

[1]This work supported in part by the National Science Foundation under Research Grant No. PHY81-18547.

[2]Address after September 1, 1985: *University of Maryland, College Park, MD 20742.*

off-shell continuation of Green functions. The BRST transformations also contain the information for a straightforward construction of the gauge-invariant string field action. The path is: classical mechanics → mechanics BRST → field theory BRST → classical field theory.

The procedure for obtaining the BRST gauge-fixed interacting string action will be described here. In the following talk, Barton Zwiebach will describe our recent paper "Gauge String Fields"[5], in which we use this BRST-invariant formulation to derive the complete gauge-invariant action for the free string. (Recent attempts at this problem have been made using other methods[6]. Similar results to those of ref. 5) were obtained shortly afterward[7),8)].)

The BRST second-quantized formulation uses a real scalar string functional $\Phi[X(\sigma), C(\sigma), \tilde{C}(\sigma)]$, where $C(\sigma)$ and $\tilde{C}(\sigma)$ are a pair of real anticommuting ghost coordinates. If C and $\tilde{C}$ are set to zero, the functional Φ reduces to the one implicitly used in old covariant formulations. The latter functional contains enough degrees of freedom to describe the propagating modes of the string theory, but not enough to allow for a gauge-invariant local formulation[2]. In order to illustrate this point, consider the zero-mass sector of the closed oriented bosonic string. The physical fields are a graviton and a scalar. The old covariant formalism allows for a symmetric rank-two tensor for their description; the traceless piece is identified with the graviton, and the trace with the scalar. It is well known, however, that a general coordinate invariant description of Einstein's gravity requires a symmetric rank-two tensor with its trace included. This trace piece is therefore missing in the old covariant formalism, since it has been used to describe the physical scalar. In the BRST formulation of ref. 2) the required trace of the graviton field was found to lie in the ghost sector of the functional field; it appears as a field having a total ghost number (number of ghosts minus antighosts) equal to zero. Given that the gauge-fixed BRST-invariant action is local, the functional field must contain all the auxiliary fields necessary for the description of the higher-spin fields that appear in the string. The BRST formulation thus provides a convenient starting point for the construction of the gauge-invariant theory, the problem being reduced to the separation of the required auxiliary fields lying in the ghost sector from the ghosts, antighosts, and BRST auxiliary fields. The main points are that *the BRST-invariant action contains all the necessary auxiliary fields, and the BRST transformations contain all the information needed to separate the gauge-invariant action from the gauge-fixing and ghost terms.*

On the other hand, if we were to attempt to construct a general coordinate invariant action for gravity with just a traceless symmetric tensor, we would obtain the lagrangian given by Fradkin and Vilkovisky[9)], of the form

$$L \sim R - R\frac{1}{\frac{D-1}{D-2}\Box + R}R \quad .$$

This action has local Weyl scale invariance, and thus the trace of the metric can be gauged to zero without affecting general coordinate invariance. It results from introducing Weyl scale invariance into the Einstein action by rescaling the metric by a compensating scalar field, and then eliminating this compensator by its equation of motion. Not only is this action nonlocal, but the nonlocalities become more complicated when coupling to nonconformal matter (such as massive fields), in a way reminiscent of Coulomb terms or the nonlocalities in light-cone gauges. Thus, the construction of actions in such a formalism is not straightforward, and requires the use of Weyl invariance in a way analogous to the use of Lorentz invariance in light-cone gauges. Another alternative is to eliminate the trace of the metric from the Einstein action by a coordinate choice, but the remaining constrained (volume-preserving) coordinate invariance causes difficulties in quantization[10)].

2. BRST FOR YANG-MILLS

The BRST formalism[11)] is a more general procedure for quantization of gauge theories than the Faddeev-Popov approach, since it not only allows more general gauges, but the same BRST transformations which determine the action also give the conditions for unitarity, as well as determining the gauge-invariant part of the action and the physical states. The defining conditions for these transformations are: (1) The group (which is global) is given by a single (Abelian) anticommuting generator Q, and thus closure of the algebra requires nilpotency $Q^2 = 0$. (2) This generator's action on physical fields is given by their gauge transformations with the gauge parameter replaced by the corresponding (real) ghost. (3) The (real) antighost transforms into an auxiliary field.

In particular, in the case of Yang-Mills theory we have:

$$QA_a = i\nabla_a c$$

$$Qc = c^2$$

$$
\begin{aligned}
Q\tilde{c} &= \tilde{B} \\
Q\tilde{B} &= 0 \quad .
\end{aligned}
\tag{2.1}
$$

The Lagrangian then consists of three BRST-invariant terms:

$$
\begin{aligned}
L &= -\tfrac{1}{4}F^2 + \varsigma\tfrac{1}{2}\tilde{B}^2 + Q(\tilde{c}\partial \cdot A) \\
&= [-\tfrac{1}{4}F^2 - \frac{1}{\varsigma}\tfrac{1}{2}(\partial \cdot A)^2] + \varsigma\tfrac{1}{2}B^2 - i\tilde{c}\partial \cdot \nabla c \quad ,
\end{aligned}
\tag{2.2}
$$

where

$$
B = \tilde{B} + \frac{1}{\varsigma}\partial \cdot A \quad .
\tag{2.3}
$$

The three original terms are BRST invariant because of the vanishing of QF_{ab}, $Q\tilde{B}$, and Q^2, respectively.

In the BRST formalism for the string, the BRST auxiliary field which appears corresponds to one like B, not $\tilde{B}$. Therefore, in order to separate the gauge-invariant part of the string action, it is necessary to shift the BRST auxiliary fields by terms containing propagating fields in such a way that the redefined BRST auxiliary fields are BRST-invariant, like $\tilde{B}$.

3. PARTICLE

The classical mechanics of a free massless particle is described completely by its field equation $p^2 = 0$ in terms of $p_a = i\partial_a$ (conjugate to x^a). This follows from the action

$$
S_M = \int d\tau(p \cdot \dot{x} - g\tfrac{1}{2}p^2) \quad ,
\tag{3.1}
$$

where the metric g is a Lagrange multiplier as well as a gauge field for τ reparametrizations, generated by the corresponding current (Hamiltonian) $\frac{1}{2}p^2$. By the same method as for Yang-Mills, we find the BRST transformations

$$
\begin{aligned}
Qx &= i\hat{c}\dot{x} \\
Qp &= i\hat{c}\dot{p} \\
Qg &= i(\dot{\hat{c}g}) \\
Q\hat{c} &= i\hat{c}\dot{\hat{c}} \\
Q\tilde{\hat{c}} &= \tilde{B} \\
Q\tilde{B} &= 0 \quad ,
\end{aligned}
\tag{3.2a}
$$

and gauge-fix the Lagrangian by adding a BRST-invariant term:

$$L \quad \rightarrow \quad L + Q[\hat{\tilde{c}}(g-1)] \quad . \tag{3.2b}$$

(We work in a Landau gauge $\varsigma = 0$.)

We will work mainly in a Hamiltonian formulation of classical mechanics, which can be translated into the more familiar Lagrangian formulation. However, the mechanics Hamiltonian formulation will lead directly to the field theory Lagrangian formulation, and thus uses one less step than the mechanics Lagrangian approach: The kinetic operator in the field-theory Lagrangian is essentially just the first-quantized Hamiltonian. Furthermore, the Hamiltonian approach is somewhat simpler for calculations because it avoids the use of the proper time τ, which doesn't appear in the field theory: Unlike the nonrelativistic case, where $\mathcal{L} = \bar{\psi}(i\partial/\partial t - H)\psi$, in the relativistic case $\mathcal{L} = -\phi H \phi$, where $\phi = \bar{\phi}$ and $\partial\phi/\partial\tau = 0$. Also, the Lagrangian approach has Lagrange multipliers which are purely gauge degrees of freedom, which don't appear in the Hamiltonian approach. The corresponding local gauge algebra, however, remains in the Hamiltonian approach as the algebra of the equations of motion. (The mechanics Lagrangian approach may be advantageous in the interacting string picture, where τ reappears as the analog of the Schwinger parameters of ordinary field theory. This first-quantized path-integral method simplifies the evaluation of string Feynman graphs. However, it's applicable *only* to graphs, and is therefore only relevant to perturbation theory.)

In the Lagrangian approach one canonically quantizes ($p \rightarrow i\partial/\partial x$, $\hat{\tilde{c}} \rightarrow c$, $\tilde{c} \rightarrow \partial/\partial c$) and finds the operator (e.g., by a Noether procedure) which gives the above BRST transformations. In the Hamiltonian approach one skips the procedure of (3.2) and starts only with the gauge-group generators/field equations G_i, using the definition of Q directly in terms of them:

$$\begin{aligned}&[G_i, G_j\} = i f_{ij}{}^k G_k \\ &\rightarrow Q = c^i G_i + \tfrac{1}{2} i c^j c^i f_{ij}{}^k \frac{\partial}{\partial c^k} \quad , \quad Q^2 = 0 \quad .\end{aligned} \tag{3.3}$$

For example, in Yang-Mills $G_i = \vec{\nabla} \cdot \vec{E}(\vec{x})$, which are both generators of gauge transformations and field equations imposed by the Lagrange multiplier A_0. In the case of the particle, $G_i = \frac{1}{2}p^2$ is also the generator of τ-reparametrizations. Either way, we find

$$Q = \frac{\partial}{\partial c} \Box \quad . \tag{3.4}$$

External electromagnetism can be introduced via $L \to L - \dot{x}\cdot A$ or more directly by $p \to p + A$.

To go from first-quantization to second-quantization, we introduce a field $\Phi(x,c) = \psi(x) + c\phi(x)$, and first look for a free Lagrangian

$$L_{FT} = \tfrac{1}{2}\Phi K \Phi \tag{3.5}$$

invariant under the BRST transformations

$$\delta\Phi = \epsilon Q \Phi \quad , \quad \delta L = 0 \quad \to \quad \{Q, K\} = 0 \quad , \tag{3.6}$$

which can be satisfied by

$$K = [Q, \mathcal{O}] \quad , \tag{3.7}$$

($\{Q,K\} = 0$ as a consequence of the Jacobi identities) and we choose $\mathcal{O}$ of the form

$$\mathcal{O} = \tfrac{1}{2}\left[c, \frac{\partial}{\partial c}\right] \quad \to \quad K = \frac{\partial}{\partial c}\Box \quad \to \quad \int dc L = \frac{\partial}{\partial c}L = \phi \Box \phi \quad . \tag{3.8}$$

(The fact that K consists only of a $[Q,\ ]$ piece is related to the fact the classical gauge-unfixed Hamiltonian of the original mechanics Lagrangian consists only of a Lagrange-multiplier term.)

The BRST transformations

$$\delta\Phi = \epsilon\frac{\partial}{\partial c}\Box\Phi \quad \to \quad \delta\phi = 0 \quad , \quad \delta\psi = \epsilon\Box\phi \tag{3.9}$$

are then trivial because they are a special case of the gauge invariance

$$\delta\Phi = \frac{\partial}{\partial c}\Lambda \quad \to \quad \delta\phi = 0 \quad , \quad \delta\psi = \lambda \quad , \tag{3.10}$$

which simply says that ψ doesn't appear in the Lagrangian. In this trivial example, the interactions are simply those which preserve this gauge invariance:

$$L_{INT} = \Phi f\left(\frac{\partial}{\partial c}\Phi\right) \quad \to \quad \int dc L = \phi f(\phi) \quad . \tag{3.11}$$

4. FREE STRING

The generalization for the string is ($' \equiv \partial/\partial\sigma$):

$$x \to X(\sigma) \quad ,$$

$$p \to \hat{P}(\sigma) \equiv \frac{1}{\sqrt{2}}\left(i\frac{\delta}{\delta X} \pm X'\right)(\pm\sigma) \quad , \tag{4.1}$$

$$\tfrac{1}{2}p^2 \to \tfrac{1}{2}\hat{P}^2(\sigma) \quad ;$$

where $\hat{P}$ is a single operator for the open string (defined on $\sigma \in [-\pi, \pi]$ from $X(\sigma)$ defined on $[0, \pi]$), and two commuting operators for the closed string (for the choices $\pm$, where $X(\sigma)$ is already defined on $[-\pi, \pi]$). The boundary conditions on X are such that $\hat{P}$ is always periodic, and the expansion of $\frac{1}{2}\hat{P}^2$ in modes (powers of $e^{i\sigma}$) gives the Virasoro operators. The classical commutation relations (Poisson brackets) are

$$[\hat{P}_a(\sigma_1), \hat{P}_b(\sigma_2)] = i\delta'(\sigma_2 - \sigma_1)\eta_{ab} \quad , \tag{4.2a}$$

$$[\tfrac{1}{2}\hat{P}^2(\sigma_1), \tfrac{1}{2}\hat{P}^2(\sigma_2)] = i\delta'(\sigma_2 - \sigma_1)[\tfrac{1}{2}\hat{P}^2(\sigma_1) + \tfrac{1}{2}\hat{P}^2(\sigma_2)] \quad . \tag{4.2b}$$

Upon quantization, $\frac{1}{2}\hat{P}^2$ must be defined by appropiate normal ordering (the modes of $\hat{P}$ are creation and annihilation operators), and its commutation relation obtains an anomaly term.

Equations (4.1) and (4.2) are sufficient to describe classical string mechanics. Quantization can be performed[3),4)] by the Hamiltonian version[12)] of the BRST procedure (3.3). In the case of strings, $G_i = \frac{1}{2}\hat{P}^2(\sigma)$, and (4.2) gives

$$Q = -\sqrt{\frac{\pi}{\alpha'}} \int d\sigma\ \hat{C} \mathcal{A} \quad , \tag{4.3a}$$

$$\mathcal{A} = \tfrac{1}{2}\hat{P}^2 + i\hat{C}'\frac{\delta}{\delta \hat{C}} - \frac{1}{2\pi}\alpha_0 \quad , \tag{4.3b}$$

where $\hat{C} = (1/\sqrt{2})(\delta/\delta C \pm \tilde{C})$[13)] in analogy to $\hat{P}$. (This gives the usual, local reality condition for the string field, since Q is then of odd order in derivatives and thus real and antihermitian, generating unitary transformations.)

Now normal ordering is implicit in all definitions, and the constant α_0 has been introduced because of the resultant reordering ambiguity. The Virasoro-like operators $\mathcal{A}$ are not identical to those obtained by varying the first-quantized action with respect to the two-dimensional metric[4)], but they are more useful for second-quantization. Absence of anomalies in $Q^2 = 0$ implies spacetime dimension $D = 26$ and $\alpha_0 = 1$[3)]. Making all dependence on the zero modes c and $\tilde{c}$ of C and $\tilde{C}$ explicit (due to boundary conditions, $\tilde{C}$ has no zero mode for the open string), we find

$$\begin{cases} \textit{open}: & Q = \frac{\partial}{\partial c} H + cT_+ + \mathcal{Q}_+ \\ \\ \textit{closed}: & Q = \frac{\partial}{\partial c} H + cT_+ - \tilde{c}\Delta N + \frac{\partial}{\partial \tilde{c}}\frac{2}{\alpha'}\Delta T_+ + \mathcal{Q}_+ \quad , \end{cases} \tag{4.4}$$

where for the open string

$$H = -\frac{1}{\alpha'}\int d\sigma\ \mathcal{A} = \Box - \frac{1}{\alpha'}(N-1) \quad , \tag{4.5a}$$

$$T_+ = \int d\sigma \; \tfrac{1}{2} i \hat{C}' \hat{C} \quad , \tag{4.5b}$$

and for the closed string we sum over the open-string expressions for the $\pm$-modes (as from (4.1)):

$$Q = 2(Q_+ + Q_-) \quad , \quad H = 2(H_+ + H_-) = \Box - \frac{2}{\alpha'}(N-2) \quad , \tag{4.6a}$$

$$T_+ = T_{++} + T_{+-} \quad , \quad \Delta T_+ = T_{++} - T_{+-} \quad , \quad N = N_+ + N_- \quad , \tag{4.6b}$$

$$\Delta N = N_+ - N_- = i \int d\sigma \left(X' \cdot \frac{\delta}{\delta X} + C' \frac{\delta}{\delta C} + \tilde{C}' \frac{\delta}{\delta \tilde{C}} \right) \quad . \tag{4.6c}$$

Quantization is completed by defining a BRST invariant kinetic operator $K = -[\mathcal{O}, Q\}$ for some operator $\mathcal{O}$. For the string[2]:

$$\begin{cases} open: & \mathcal{O} = \frac{1}{2}[c, \frac{\partial}{\partial c}] \rightarrow K = \frac{\partial}{\partial c} H - cT_+ \\ closed: & \mathcal{O} = i \frac{\partial}{\partial \tilde{c}} \frac{1}{2}[c, \frac{\partial}{\partial c}] \rightarrow K = i \frac{\partial}{\partial \tilde{c}} \left(\frac{\partial}{\partial c} H - cT_+ \right) + i \frac{1}{2}[c, \frac{\partial}{\partial c}] \Delta N \quad . \end{cases} \tag{4.7}$$

Of the operators H, T_+, ΔN, only H contains dynamics (dependence on p, the zero-mode of $\hat{P}$). When expanding the string field in c (and $\tilde{c}$), H is the kinetic operator for the piece containing all physical and ghost fields. Explicitly,

$$\begin{cases} open: & \Phi = \psi + c\phi \\ closed: & \Phi = \hat{\phi} + ic\hat{\psi} + i\tilde{c}(\psi + c\phi) \quad , \end{cases} \tag{4.8}$$

when substituted into the lagrangian $L = \frac{1}{2} \Phi K \Phi$ and integrated over the ghost zero modes, gives

$$\begin{cases} open: & \frac{\partial}{\partial c} L = \frac{1}{2} \phi H \phi + \frac{1}{2} \psi T_+ \psi \\ closed: & i \frac{\partial}{\partial \tilde{c}} \frac{\partial}{\partial c} L = \frac{1}{2} \phi H \phi + \frac{1}{2} \psi T_+ \psi + \frac{1}{2} i \hat{\phi} \Delta N \phi - \frac{1}{2} \hat{\psi} \Delta N \psi \quad , \end{cases} \tag{4.9}$$

and in the BRST transformations $\delta \Phi = \epsilon Q \Phi$ gives

$$\begin{cases} open: & \delta\phi = \epsilon(\mathcal{Q}_+ \phi - T_+ \psi) \quad , \quad \delta\psi = \epsilon(\mathcal{Q}_+ \psi + H\phi) \\ closed: & \delta\phi = \epsilon(\mathcal{Q}_+ \phi - T_+ \psi + \Delta N \hat{\psi}) \quad , \quad \delta\psi = \epsilon(\mathcal{Q}_+ \psi + H\phi + i\Delta N \hat{\phi}) \quad , \\ & \delta\hat{\phi} = \epsilon(\mathcal{Q}_+ \hat{\phi} + iH\hat{\psi} + i\frac{2}{\alpha'} \Delta T_+ \psi) \; , \quad \delta\hat{\psi} = \epsilon(\mathcal{Q}_+ \hat{\psi} + iT_+ \hat{\phi} - \frac{2}{\alpha'} \Delta T_+ \phi) \; . \end{cases} \tag{4.10}$$

These results can also be obtained by the Lagrangian approach[2)] from the Lagrangian[14)]

$$L = \tfrac{1}{2} g_{mn} \hat{P}_a{}^m \hat{P}_b{}^n \eta^{ab} + \hat{P}_a{}^m \partial_m X^a \tag{4.11}$$

(in the action $S = \int d\tau d\sigma\ L$), where $\partial_m = (\partial/\partial\tau, \partial/\partial\sigma)$ and g_{mn} is the $det = -1$ piece of the two-dimensional (τ-σ) metric.

5. COMPONENT EXPANSIONS

Since H is a harmonic-oscillator type Hamiltonian, we expand the string field $\Phi[X, C, \tilde{C}]$, as a wave function in quantum mechanics, in terms of Hermite polynomials or creation and annihilation operators. We first expand all σ dependence in modes. For the open string the coordinates $X(\sigma)$, $C(\sigma)$, and $\tilde{C}(\sigma)$ and their respective conjugate momenta are expanded as follows:

$$\begin{aligned}
X(\sigma) &= \frac{1}{\sqrt{\pi}} \left(\frac{1}{\sqrt{2\alpha'}}\, x + \sum_{n\neq 0} i \frac{1}{n} \alpha_n \cos n\sigma \right) \quad ; \\
i\frac{\delta}{\delta X}(\sigma) &= \frac{1}{\sqrt{\pi}} \sum_{-\infty}^{\infty} \alpha_n \cos n\sigma \quad , \quad \alpha_0 = \sqrt{2\alpha'}\, p \quad ; \\
C(\sigma) &= \frac{1}{\sqrt{\pi}} \sum_{-\infty}^{\infty} \tilde{\beta}_n \cos n\sigma \quad , \quad \tilde{\beta}_0 = \tfrac{1}{2}\sqrt{2\alpha'}\, c \quad ; \\
\frac{\delta}{\delta C}(\sigma) &= \frac{1}{\sqrt{\pi}} \sum_{-\infty}^{\infty} \beta_n \cos n\sigma \quad , \quad \beta_0 = 2\frac{1}{\sqrt{2\alpha'}}\frac{\partial}{\partial c} \quad ; \\
\tilde{C}(\sigma) &= \frac{1}{\sqrt{\pi}} \sum_{-\infty}^{\infty} \beta_n (-i) \sin n\sigma \quad ; \\
\frac{\delta}{\delta \tilde{C}}(\sigma) &= \frac{1}{\sqrt{\pi}} \sum_{-\infty}^{\infty} \tilde{\beta}_n (-i) \sin n\sigma \quad ;
\end{aligned} \tag{5.1}$$

where

$$\begin{aligned}
\alpha_n = \alpha^\dagger{}_{-n} = -i\sqrt{n}\, a^\dagger{}_n \quad &, \quad [\alpha_m, \alpha_n] = n\delta_{m+n,0} \quad , \\
[p, x] = i \quad &, \quad [a_m, a^\dagger{}_n] = \delta_{mn} \quad , \\
\beta_n = \beta^\dagger{}_{-n} = i\frac{1}{\sqrt{n}} b^\dagger{}_n \quad &, \quad \tilde{\beta}_n = \tilde{\beta}^\dagger{}_{-n} = \sqrt{n}\, \tilde{b}^\dagger{}_n \quad , \\
\{\beta_m, \tilde{\beta}_n\} = \delta_{m+n,0} \quad &, \quad \{b_m, \tilde{b}^\dagger{}_n\} = -\{\tilde{b}_m, b^\dagger{}_n\} = i\delta_{mn} \ .
\end{aligned} \tag{5.2}$$

The operators a, b, and $\tilde{b}$ are real. (They yield real funtionals when acting on real bosonic functionals.) Using (5.1) one finds that the hatted operators (defined in

(4.1) and below (4.3b)) are:

$$\hat{P} = \frac{1}{\sqrt{2\pi}} \sum_{-\infty}^{\infty} \alpha_n e^{-in\sigma} \quad ,$$

$$\hat{C} = \frac{1}{\sqrt{2\pi}} \sum_{-\infty}^{\infty} \beta_n e^{-in\sigma} \quad ,$$

$$\frac{\delta}{\delta \hat{C}} = \frac{1}{\sqrt{2\pi}} \sum_{-\infty}^{\infty} \tilde{\beta}_n e^{-in\sigma} \quad ; \tag{5.3}$$

where $\delta/\delta\hat{C} = \frac{1}{\sqrt{2}}(C \pm \delta/\delta\tilde{C})$. The Virasoro-like operators (4.3b) are given by

$$\mathcal{A}(\sigma) = \frac{1}{2\pi} \sum_{-\infty}^{\infty} L_n e^{-in\sigma} \quad ,$$

$$L_n = \sum_m : (\tfrac{1}{2}\alpha_m \cdot \alpha_{n-m} + m\beta_m \tilde{\beta}_{n-m}) : \quad . \tag{5.4}$$

(More operators will be expanded by Barton in the following talk.) For the rest of this section we use units $\alpha' = 1$.

We now expand the field $\Phi(X, C, \tilde{C})$ in terms of the a, b, and $\tilde{b}$ operators. We find that up to the massless level the expansion of Φ is

$$\Phi = \{[B(x)\tilde{b}^\dagger{}_1] + c[\phi_0(x) + A^a(x)a^\dagger{}_{1a} + i\mathbf{C}(x)\tilde{b}^\dagger{}_1 + i\tilde{\mathbf{C}}(x)b^\dagger{}_1]\}|0> \quad , \tag{5.5}$$

where the harmonic oscillator vacuum $|0>$ is real in coordinate space. Here ϕ_0 is the tachyon, A^a is the massless vector, $\mathbf{C}$ is the ghost, $\tilde{\mathbf{C}}$ is the antighost, and B is the BRST auxiliary field. (We have dropped fields which can be gauged away by the analog of (3.10).) We then find

$$L = \tfrac{1}{2}\phi_0(\Box + 1)\phi_0 + (\tfrac{1}{2}A \cdot \Box A + i\mathbf{C}\Box\tilde{\mathbf{C}} + \tfrac{1}{2}B^2) \quad ;$$

$$\delta A = i\epsilon\partial\mathbf{C} \ , \ \delta\mathbf{C} = 0 \ , \ \delta\tilde{\mathbf{C}} = \epsilon(B - \partial \cdot A) \ , \ \delta B = i\epsilon\Box\mathbf{C} \quad . \tag{5.6}$$

The closed string is treated analogously, so we will consider only the massless level. The mode expansions of the coordinates and operators are:

$$X(\sigma) = \frac{1}{\sqrt{\pi}} \left(\frac{1}{\sqrt{2\alpha'}} x + \tfrac{1}{2} \sum_{\pm, n \neq 0} i \frac{1}{n} \alpha^{\pm}{}_n e^{\mp in\sigma} \right) \quad ;$$

$$i\frac{\delta}{\delta X}(\sigma) = \frac{1}{2\sqrt{\pi}} \sum_{\pm, n} \alpha^{\pm}{}_n e^{\mp in\sigma} \quad , \quad \alpha^{\pm}{}_0 = \sqrt{\frac{\alpha'}{2}}\, p \quad ;$$

$$C(\sigma) = \frac{1}{2\sqrt{\pi}} \sum_{\pm, n} \tilde{\beta}^{\pm}{}_n e^{\mp in\sigma} \quad , \quad \tilde{\beta}^{\pm}{}_0 = \tfrac{1}{2}\left(\sqrt{\frac{\alpha'}{2}}\, c \pm \sqrt{\frac{2}{\alpha'}} \frac{\partial}{\partial \tilde{c}} \right) \quad ;$$

$$\frac{\delta}{\delta C}(\sigma) = \frac{1}{2\sqrt{\pi}}\sum_{\pm,n}\beta^{\pm}{}_n e^{\mp in\sigma} \quad , \quad \beta^{\pm}{}_0 = \pm\left(\sqrt{\frac{\alpha'}{2}}\tilde{c} \pm \sqrt{\frac{2}{\alpha'}}\frac{\partial}{\partial c}\right) \quad ;$$
$$\tilde{C}(\sigma) = \frac{1}{2\sqrt{\pi}}\sum_{\pm,n}\pm\beta^{\pm}{}_n e^{\mp in\sigma} \quad ;$$
$$\frac{\delta}{\delta\tilde{C}}(\sigma) = \frac{1}{2\sqrt{\pi}}\sum_{\pm,n}\pm\tilde{\beta}^{\pm}{}_n e^{\mp in\sigma} \quad ; \tag{5.7}$$

where

$$\alpha^{\pm}{}_n = \alpha^{\pm\dagger}{}_{-n} = -i\sqrt{n}a^{\pm\dagger}{}_n \quad , \quad [\alpha^{\pm}{}_m, \alpha^{\pm}{}_n] = n\delta_{m+n,0} \quad ,$$
$$[p, x] = i \quad , \quad [a^{\pm}{}_m, a^{\pm\dagger}{}_n] = \delta_{mn} \quad ,$$
$$\beta^{\pm}{}_n = \beta^{\pm\dagger}{}_{-n} = i\frac{1}{\sqrt{n}}b^{\pm\dagger}{}_n \quad , \quad \tilde{\beta}^{\pm}{}_n = \tilde{\beta}^{\pm\dagger}{}_{-n} = \sqrt{n}\tilde{b}^{\pm\dagger}{}_n \quad ,$$
$$\{\beta^{\pm}{}_m, \tilde{\beta}^{\pm}{}_n\} = \delta_{m+n,0} \quad , \quad \{b^{\pm}{}_m, \tilde{b}^{\pm\dagger}{}_n\} = -\{\tilde{b}^{\pm}{}_m, b^{\pm\dagger}{}_n\} = i\delta_{mn} \; . \tag{5.8}$$

The complex conjugates of the +-oscillators are the −-oscillators. The hatted operators are found as before (but with $\pm$ referring to two sets of operators over the same range of σ):

$$\hat{P}^{\pm} = \frac{1}{\sqrt{2\pi}}\sum\alpha^{\pm}{}_n e^{-in\sigma} \quad ,$$
$$\hat{C}^{\pm} = \frac{1}{\sqrt{2\pi}}\sum\beta^{\pm}{}_n e^{-in\sigma} \quad ,$$
$$\frac{\delta}{\delta\hat{C}^{\pm}} = \frac{1}{\sqrt{2\pi}}\sum\tilde{\beta}^{\pm}{}_n e^{-in\sigma} \quad . \tag{5.9}$$

The expansion of the physical closed-string field at the second mass level is

$$\Phi = [c\tilde{c}(\tfrac{1}{2}h^{ab}a^{+\dagger}{}_{1a}a^{-\dagger}{}_{1b} + i\tfrac{1}{2}A^{ab}a^{+\dagger}{}_{1a}a^{-\dagger}{}_{1b} + \eta i b^{[+\dagger}{}_1\tilde{b}^{-]\dagger}{}_1) + \cdots]|0> \quad , \tag{5.10}$$

where h_{ab}, A_{ab}, and η describe the massless sector, consisting of the graviton, an antisymmetric tensor, and the dilaton. We have indicated only the physical fields explicitly. The important point is that both $h^a{}_a$ and η are needed to give the dilaton and the trace of the metric, and η appears multiplied by b-oscillators.

At the massive levels of the open and closed strings we also find Stueckelberg fields, which show the gauge invariance of the massive fields (which eat the Stueckelberg fields).

6. EXTERNAL FIELDS

External fields can be added by adding a potential term to $\frac{1}{2}\widehat{P}^2$, while preserving the algebra (4.2b):

$$\begin{aligned} G(\sigma) &= \tfrac{1}{2}\widehat{P}^2(\sigma) + V(\sigma) \quad , \\ [G(\sigma_1), G(\sigma_2)] &= i\delta'(\sigma_2 - \sigma_1)[G(\sigma_1) + G(\sigma_2)] \quad . \end{aligned} \tag{6.1}$$

For example, for the open string this is satisfied order-by-order in V for an interaction satisfying the usual conditions

$$\begin{aligned} [\tfrac{1}{2}\widehat{P}^2(\sigma_1), V(\sigma_2)] &= i\delta'(\sigma_2 - \sigma_1)V(\sigma_1) \quad , \\ [V(\sigma_1), V(\sigma_2)] &= 0 \quad . \end{aligned} \tag{6.2}$$

The former condition requires V to have "conformal weight" 1, and the latter requires that it be local: e.g., $V(\sigma) = \delta(\sigma)A(X(0)) \cdot \widehat{P}$ for an external vector field coupling to the end $\sigma = 0$, corresponding to the covariantization $\widehat{P} \to \widehat{P} + \delta(\sigma)A$, or to adding a term $\delta(\sigma)\dot{X} \cdot A$ to the Lagrangian. Upon quantization, ignoring ghosts, the vertex corresponding to the inverse properator $\int d\sigma\ \frac{1}{2}\widehat{P}^2 = p^2 + \cdots$ is $\int d\sigma\ V(\sigma) = A(X) \cdot \widehat{P}|_{\sigma=0}$.

7. INTERACTIONS

Interactions can be introduced in close analogy with the light-cone formalism. The fundamental part of the interaction is a δ-functional which equates the points on a string to the points on the two strings into which it splits. This delta functional can be written as

$$\delta\left[Z_2\left(\frac{\sigma}{\alpha}\right) + Z_3\left(\frac{\sigma - \pi\alpha}{1-\alpha}\right) - Z_1(\sigma)\right] \quad , \tag{7.1}$$

where $Z = X, C, \widetilde{C}$. The interacting BRST operator for the open string then becomes (to this order)

$$\begin{aligned} Q = &\int DZ\ (Q_0\Phi)\frac{\delta \cdot}{\delta\Phi} \\ &+ \int DZ_{1,2,3}\ \delta[\text{"}]\left(\frac{\delta}{\delta C_2(\pi)}\Phi[Z_2]\right)\left(\frac{\delta}{\delta C_3(0)}\Phi[Z_3]\right)\frac{\delta}{\delta\Phi[Z_1]} \quad . \end{aligned} \tag{7.2}$$

The anticommmuting derivatives are like those in (3.11), but act at the point where the string splits. Although α represents p_+ in the light-cone formalism, here it's a gauge-fixing parameter. (Q_0 is the first-quantized operator given in (4.3a).) The BRST-invariant action can then be written simply as

$$S = \int \Phi \mathcal{O} Q \Phi \quad , \tag{7.3}$$

with $\mathcal{O}$ as in (4.7). It can also be expressed as Q acting another functional[8], which in this case means

$$S = Q \int \tfrac{1}{2} \Phi \mathcal{O} \Phi \quad . \tag{7.4}$$

REFERENCES

1) Dual theory, ed. M. Jacob (North-Holland, Amsterdam, 1974);
J. Scherk, Rev. Mod. Phys. **47** (1975) 123;
J.H. Schwarz, Phys. Rep. **89** (1982) 223.
M.B. Green and J.H. Schwarz, Phys. Lett. **149B** (1984) 117;
D.J. Gross, J.A. Harvey, E. Martinec, and R. Rohm, Phys. Rev. Lett. **54** (1985) 502;
E. Witten, Phys. Lett. **149B** (1984) 351;
P. Candelas, G.T. Horowitz, A. Strominger, and E. Witten, Vacuum configurations for superstrings, Santa Barbara preprint (December 1984), to appear in Nucl. Phys. B.
2) W. Siegel, Phys. Lett. **149B** (1984) 157, 162; **151B** (1985) 391, 396.
3) M. Kato and K. Ogawa, Nucl. Phys. **B212** (1983) 443.
4) K. Fujikawa, Phys. Rev. D**25** (1982) 2584.
5) W. Siegel and B. Zwiebach, Gauge string fields, Berkeley preprint UCB-PTH-85/30 (July 1985).
6) T. Banks and M. Peskin, to appear in Anomalies, geometry and topology, ed. A. White (World Scientific, Singapore, 1985);
M. Kaku, *ibid.*;
D. Friedan, String field theory, Chicago preprint EFI 85-27 (April 1985);
A. Neveu and P.C. West, Gauge covariant local formulation of bosonic strings, CERN preprint CERN-TH. 4200/85 (June 1985).
7) T. Banks and M. E. Peskin, Gauge invariance of string fields, SLAC preprint SLAC-PUB-3740 (July 1985), and contribution to these proceedings.
8) K. Itoh, T. Kugo, H. Kunitomo, and H. Ooguri, Gauge invariant local action of string field from BRS formalism, Kyoto preprint KUNS 800 HE(TH) 85/04 (August 1985).
9) E.S. Fradkin and V.I. Vilkovisky, Phys. Lett. **73B** (1978) 209.
10) W. Siegel and S.J. Gates, Jr., Nucl. Phys. **B147** (1979) 77;
S.J. Gates, Jr., M.T. Grisaru, M. Roček, and W. Siegel, Superspace, *or* One thousand and one lessons in supersymmetry (Benjamin/Cummings, Reading, 1983) p. 242.
11) C. Becchi, A. Rouet, and R. Stora, Phys. Lett. **52B** (1974) 344;
I. V. Tyutin, Gauge invariance in field theory and in statistical physics in the operator formulation, Lebedev preprint FIAN No. 39 (1975), in Russian, unpublished;
T. Kugo and I. Ojima, Phys. Lett. **73B** (1978) 459;

I. A. Batalin and G. A. Vilkovisky, Phys. Lett. **69B** (1977) 309;
E. S. Fradkin and T. E. Fradkina, Phys. Lett. **72B** (1978) 343.

12) S. Hwang, Phys. Rev. **D28** (1983) 2614.

13) W. Siegel, Classical superstring mechanics, Berkeley preprint UCB-PTH-85/23 (May 1985), to appear in Nucl. Phys. B.

14) L. Brink, P. Di Vecchia, and P. Howe, Phys. Lett. **65B** (1976) 471.

GAUGE INVARIANT STRING ACTIONS*

Barton Zwiebach**

Department of Physics
University of California
Berkeley, California 94720 U.S.A.

ABSTRACT

The main ideas behind the recent construction of the gauge invariant string free actions starting from BRST formulations are reviewed and examined in detail.

1. INTRODUCTION

In this article I will describe recent work done together with Warren Siegel[1] in which we report the findings of gauge-invariant actions for the free bosonic strings. This subject matter is the continuation of the subject discussed by Siegel[2] earlier concerning the formulation of string theories using BRST invariance. He was able to use the BRST-first quantized string[3,4] to derive a gauge-fixed second-quantized field theory for bosonic strings.[5] For several reasons it is desirable to have a gauge invariant formulation. It will allow one to discover the underlying symmetry group of the string, it should facilitate finding classical solutions and, possibly, allow the study of semiclassical and nonperturbative phenomena. When the gauge-fixed theory was developed one year ago it was noted that at least for the first few levels of the string one could recover the gauge-invariant formulations in a straightforward way starting from the gauge-fixed actions.[5] This was done for the tachyon, vector and massive symmetric tensor of the open-string, and for the tachyon, graviton and massless scalar of the closed string. The problem to be discussed here is that of doing this reversal for the whole string field. This is not just expanding the component Lagrangian,

---

* This work was supported in part through funds provided by the U.S. Department of Energy (D.O.E.) under contract number DE-AC02-76ER03069 and NSF grant number PHY-8203424.

** Address after September 1, 1985: Center for Theoretical Physics, Laboratory for Nuclear Science and Department of Physics, Massachusetts Institute of Technology, Cambridge, Massachusetts 02139 U.S.A.

Typeset in T_EX by Roger L. Gilson

which at each level becomes more and more complicated, and finding for each level the field redefinitions required to go from gauge-fixed to gauge-invariant. Now we want to consider the *whole* string action, which is gauge-fixed, and on the *whole* string do the operations that are necessary to find the gauge-invariant formulation.

Siegel and the author have recently reported such construction[1] and thus found the gauge invariant actions for both the open and closed free bosonic string theories. Here, I will try to explain in simple terms and without too much technicalities the main ideas behind the construction of reference [1].

Other groups have been working on the same problem.[6–9]. The references in [6] are early attempts and discussion. References [7] and [8] have now reported the construction of the open string free action using an infinite number of string fields. Their answer is equivalent to our open string result. Reference [9] proposes other forms for free string actions.

2. FROM GAUGE-FIXED TO GAUGE-INVARIANT

Let us start by recalling BRST[10] formulation of a vector field. Take for simplicity a $U(1)$ theory and consider the gauge-fixed action obtained after adding a gauge fixing function proportional to $(\partial \cdot A)^2$ and Fadeev-Popov ghost c and antighost $\hat{c}$:

$$\mathcal{L} = \frac{1}{2} A_\mu \Box A_\mu + i\hat{c}\Box c \ . \tag{1a}$$

This Lagrangian has a global symmetry, BRST invariance under the following transformations:

$$\delta A_\mu = i\epsilon\partial_\mu c, \quad \delta c = 0, \quad \delta\hat{c} = \epsilon\partial \cdot A, \tag{1b}$$

where ϵ is an anticommuting constant. BRST transformations should be nilpotent ($\delta_{\epsilon_1}\delta_{\epsilon_2} = 0$), and this holds trivially when acting on A_μ or c. Acting on the antighost, $\delta\delta\hat{c} \sim \Box c$, and nilpotency holds only on-shell since $\Box c = 0$ is the equation of motion of the antighost. The gauge-invariant action that underlies the gauge-fixed action of eq. (1) cannot be found by any simple truncation of eq. (1) (say, throwing away the ghosts) and this is essentially due to the lack of off-shell nilpotency of the BRST transformations.

One can achieve off-shell nilpotency of BRST transformations using an auxiliary field[11] and this really makes a big difference. The gauge-fixed action now is:

$$\mathcal{L} = -\frac{1}{4} F_{\mu\nu}{}^2 + \hat{B}(\partial \cdot A) + \frac{1}{2}\hat{B}^2 + i\hat{c}\Box c \tag{2a}$$

where $\hat{B}$ is the auxiliary field whose elimination via its equation of motion would reproduce the Lagrangian (1a). The BRST transformations, now with off-shell nilpotency, are:

$$\delta A_\mu = i\epsilon\partial_\mu c, \quad \delta c = 0, \quad \delta\hat{B} = 0, \quad \delta\hat{c} = -\epsilon\hat{B} \tag{2b}$$

This is the form in which it is easiest to find the gauge-invariant action, and its gauge invariance. It is possible to set $\hat{B}$ and $\hat{c}$ to zero consistently as one can see in eq. (2b). This truncation in eq. (2a) gives us the gauge-invariant action $\frac{1}{4}F^2$, whose gauge invariance is found from the first eq. in (2b) replacing the ghost field $c(x)$ by a parameter $\lambda(x)$. It will be our goal to cast the gauge-fixed string action and its transformation laws in a way analogous to eqs. (2) so that the gauge-invariant form can be easily identified.

The gauge-fixed string action of [5] contains all the auxiliary fields analogous to $\hat{B}$ but in a somewhat different form. This form is illustrated by considering the corresponding vector field action:

$$\mathcal{L} = \frac{1}{2}A_\mu \Box A_\mu + \frac{1}{2}B^2 + i\hat{c}\Box c \ , \tag{3a}$$

and BRST transformations:

$$\delta A_\mu = i\epsilon\partial_\mu c, \quad \delta c = 0, \quad \delta B = i\epsilon\Box c, \quad \delta\hat{c} = -\epsilon(B - \partial\cdot A) \ . \tag{3b}$$

At first sight it seems that this action is an inconvenient as the one in (1a) but this is not true. Using the information in the BRST transformations we can easily turn (3a) into the form of (2a). First note that in (2b) $\delta\hat{B} = 0$ while in (3b) $\delta B \neq 0$. This suggests that we have to redefine the auxiliary field B such that its BRST transformation is zero. The necessary redefinition is $B' = B - \partial\cdot A$, and it follows that $\delta B' = 0$ since $(B - \partial\cdot A)$ is proportional to $\delta\hat{c}$ and the BRST transformations are nilpotent. This redefinition casts eqs. (3) into the form of eqs. (2). For the string field theory we will just do what was done above, namely, redefine the auxiliary fields by shifting them by pieces proportional to the propagating fields, such that they become BRST invariant. Redefine the action, truncate and find the gauge invariant form.

3. AUXILIARY FIELDS, STUECKELBERG FIELDS AND GAUGE INVARIANT FORMULATIONS OF MASSIVE FIELDS

Covariant formulations use a great number of variables to describe the propagation of physical fields. In four dimensions, for example, we use four variables A_μ to describe a vector field that carries only two propagating degrees of freedom. We know what are the particles we want to describe in the string, the question is whether we can find a string field that has a large enough number of variables to allow for the gauge covariant formulation of such particles.

For the bosonic string the naive candidate is the functional field $\Phi[X^\mu(\sigma)]$ with $\mu = 1, \ldots 26$. It was only realized very recently, however, that this functional field does not have enough fields for a gauge-covariant formulation.[5] This is most easily illustrated for the case of the closed, oriented bosonic string (Shapiro-Virasoro) whose ground state is a tachyon and at the zero mass level contains a graviton, an antisymmetric tensor and a scalar. Expanding the functional field, one finds:

$$\Phi\left[X(\sigma)\right] \to \phi_0 + h_{(\mu\nu)}a_1^{\dagger\mu}\tilde{a}_1^{\dagger\nu} + A_{[\mu\nu]}a_1^{\dagger\mu}\tilde{a}_1^{\dagger\nu} + \ldots, \tag{4}$$

here ϕ_0 can be used to describe the tachyon and $A_{\mu\nu}$ to describe the antisymmetric tensor, but the symmetric rank two tensor (with its trace included) $h_{\mu\nu}$ is not enough to describe gravity with general coordinate invariance *and* a massless scalar. This problem of missing fields only gets worse at higher mass levels. There are two obvious options; one could add more string fields (and this is related to the formulations[7,8] which use an infinite number of string fields) or one could add more oscillators, which could be arranged into new coordinates. We follow this second possibility and use the BRST-field of Siegel[5] which is a real anticommuting functional denoted by $\Phi\left[X(\sigma),\, C(\sigma),\, \tilde{C}(\sigma)\right]$, where $C(\sigma)$ and $\tilde{C}(\sigma)$ are real anticommuting ghost coordinates. One, therefore, has now real anticommuting oscillations b_n and $\tilde{b}_n$ associated with mode expansions that for the open string read:

$$C(\sigma) = \frac{1}{\sqrt{\pi}}\left[\frac{1}{\sqrt{2}}c + \sum_{n=1}^{\infty}\sqrt{n}\left(\tilde{b}_n + \tilde{b}_n^\dagger\right)\cos n\sigma\right] ,$$
$$\tilde{C}(\sigma) = \frac{1}{\sqrt{\pi}}\left[\sum_{n=1}^{\infty}\frac{1}{\sqrt{n}}\left(b_n + b_n^\dagger\right)\sin n\sigma\right] , \tag{5}$$

where c is a zero mode (for closed strings the oscillators carry $\pm$ labels and $\tilde{C}$ has also a zero mode).

For an open string there is also a lack of fields and this is related to the issue of gauge-invariant formulation of massive fields. The first few states in the open string are the tachyon, a massless vector (□ of $SO(24)$) and a massive second-rank symmetric tensor (□□ of $SO(25)$). We can describe this massive representation in a gauge-invariant way by adding fields, the so-called Stueckeberg fields, which can be gauged away using the gauge invariance. The case of massive vector fields is well-known; one adds a scalar field that can be gauged away. For the symmetric rank two tensor one used three fields ($h^{\mu\nu}$, B^μ, η) where the last two are Stueckelberg fields, and the gauge invariance is:

$$\delta h^{\mu\nu} \sim \partial^{(\mu}\epsilon^{\nu)}, \quad \delta B^\mu \sim \partial^\mu\epsilon + \epsilon^\mu, \quad \delta\eta \sim \epsilon \ . \tag{6}$$

One can use ϵ to gauge away η and then use ϵ^μ to gauge away B^μ being left only with the field $h^{\mu\nu}$ and no gauge invariance. The old-fashioned covariant field does not contain enough fields for such gauge-invariant formulation as we can easily see expanding out up to the second excited level

$$\Phi\left[X(\sigma)\right] = \phi(x) + A^\mu a_{1\mu}^\dagger + h^{\mu\nu}a_{1\mu}^\dagger a_{1\nu}^\dagger + B^\mu a_{2\mu}^\dagger \ . \tag{7}$$

Here one is missing the η field. This is not the case for the BRST field of reference [5]. For open strings, the BRST field is decomposed by making the dependence on the zero mode c of C explicit:

$$\Phi = \psi + c\phi \ . \tag{8}$$

The field ψ will be identified as BRST-auxiliary fields (such as the B field in eqs. (2) and (3)) and ϕ is the field that contains the physical fields, ghosts, antighosts and, as we will see, all the necessary Stueckelberg and auxiliary fields necessary for the second-quantized field theory. Let us expand the field ϕ up to second mass level. Making use not only of the bosonic oscillators, but of the anticommuting ones in (5) one sees that

$$\begin{aligned}\phi\left[X, C, \tilde{C}\right] = {} & \phi_0 + A^\mu a^\dagger_{1\mu} + i\tilde{C}b^\dagger_1 + iC\tilde{b}^\dagger_1 \\ & + \frac{1}{2}h^{\mu\nu}a^\dagger_{1\mu}a^\dagger_{1\nu} + B^\mu a^\dagger_{2\mu} \\ & + i\tilde{c}'b^\dagger_2 + ic'\tilde{b}^\dagger_2 + i\tilde{c}^\mu a^\dagger_{1\mu}b^\dagger_1 + ic^\mu a^\dagger_{1\mu}\tilde{b}^\dagger_1 \\ & + i\eta b^\dagger_1\tilde{b}^\dagger_1 \ .\end{aligned} \tag{9}$$

At the zero mass level, together with the vector, we find the corresponding Fadeev-Popov ghost and antighost C and $\tilde{C}$. At the next mass level we find $h^{\mu\nu}$ and B^μ, but also the ghosts and antighosts $(c', \tilde{c}')$ and $(c^\mu, \tilde{c}^\mu)$ corresponding to the invariances denoted by ϵ and ϵ^μ in eq. (6). Finally, we also find the Stueckelberg field η that was missing in the old fashioned functional field! It is a field with correct statistics since it appears accompanied by a bilinear in ghost oscillators.

One may at first think that all the fields appearing accompanied by an even number of ghost oscillators are either Stueckelberg or auxiliary fields that one is missing, but this is not quite correct. At higher mass levels there are antisymmetric tensors whose quantization require not only ghosts, but ghosts-for-ghosts, and so on.[12] Even-generation ghosts have correct statistics but they are not the Stueckelberg or auxiliary fields that one wants to identify. As we will see later, the proper criterion to find these consists of looking for $SU(1,1)$ singlets.[1] We shall have an $SU(1,1)$ symmetry acting on the ghost coordinates, such that the ghosts oscillators are assembled in real doublets $\begin{pmatrix} b^\dagger_n \\ \tilde{b}^\dagger_n \end{pmatrix}$ (like Majorana spinors of $SO(2,1) \approx SU(1,1)$ in three dimensions). The η field in eq. (9) is just the coefficient of the $SU(1,1)$ singlet $(b^\dagger_1\tilde{b}^\dagger_1)$. It is nice that such simple criterion allows one to find the fields that are necessary for the construction of the string action.

4. GAUGE-FIXED STRING ACTIONS AND $SU(1,1)$ GENERATORS

Let us recall the main facts about the gauge-fixed actions[5] and describe the action of the $SU(1,1)$ symmetry.[1] We will only consider here the open string (closed-string expressions and details of notation can be found in Ref. [1]). The gauge fixed action and BRST transformations are:[5]

$$L = \frac{1}{2}\phi H\phi + \frac{1}{2}\psi T_+\psi \ , \tag{10}$$

$$\delta\phi = \epsilon(Q_+\phi - T_+\psi) \tag{11}$$

$$\delta\psi = \epsilon(Q_+\psi + H\phi) \ , \tag{12}$$

where ψ and ϕ were defined in eq. (8) and H, T_+ and Q_+ are defined as coefficients of the zero mode expansion of the BRST-operator $Q_{\mathbf{BRST}}$:

$$Q_{\mathbf{BRST}} = -\sqrt{\pi}\int d\sigma\ \hat{C}\mathcal{A} = H\frac{\partial}{\partial c} + T_+c + Q_+ \ . \tag{13}$$

Here, $\mathcal{A}$ are Virasoro-like operators

$$\mathcal{A} = \frac{1}{2}\hat{P}^2 + i\hat{C}'\frac{\delta}{\delta\hat{C}} - \frac{\alpha_0}{2\pi} \ , \tag{14}$$

where:

$$\hat{P} = \frac{1}{\sqrt{2}}\left(i\frac{\delta}{\delta X} \pm X'\right)(\pm\sigma) \ ,$$
$$\hat{C} = \frac{1}{\sqrt{2}}\left(\frac{\delta}{\delta C} \pm \tilde{C}\right)(\pm\sigma) \ , \tag{15}$$

$\hat{P}$ and $\hat{C}$ being single operators defined on $\sigma\epsilon[-\pi,\pi]$ from $X(\sigma)$, $C(\sigma)$ and $\tilde{C}(\sigma)$ defined on $[0,\pi]$. Nilpotency of $Q_{\mathbf{BRST}}$ implies that

$$[H, T_+] = [H, Q_+] = [T_+, Q_+] = 0, \quad Q_+^2 = -HT_+ \ . \tag{16}$$

H is the kinetic operator for the physical and ghost fields (see (10)):

$$H = \Box - (N-1), \quad N = \sum_{N=1}^{\infty}\left(na_n^\dagger a_n - i\tilde{b}_n^\dagger b_n + ib_n^\dagger\tilde{b}_n\right) \ , \tag{17}$$

and operator T_+ is given by

$$T_+ = \sum_{1}^{\infty} b_n^\dagger b_n \ . \tag{18}$$

It is natural to define a ghost number operator T_3 that counts the difference between the number of ghost oscillators ($\tilde{b}$) and antighost oscillators (b):

$$T_3 = \frac{i}{2}\sum_{1}^{\infty}\left(b_n^\dagger\tilde{b}_n + \tilde{b}_n^\dagger b_n\right) \ . \tag{19}$$

One easily verifies, using $\{b_m, \tilde{b}_n^\dagger\} = i\delta_{mn}$, that

$$[T_3, T_+] = T_+ \ , \tag{20}$$

and this suggests that is could be useful to define a third operator T_- such that

$$[T_3, T_-] = -T_-, \quad [T_+, T_-] = 2T_3 \quad , \tag{21}$$

completing in this way $SU(2)$ commutation relations. The choice of T_- is suggested by eq. (18):

$$T_- = \sum_1^\infty \tilde{b}_n^\dagger \tilde{b}_n \quad , \tag{22}$$

and one can check that it works. (These operators have been written in $SU(2)$ form, rather than $SU(1,1)$ since we deal only with finite dimensional representations. As a consequence of this, the hermiticity properties are unusual, namely both T_+ and T_- are hermitian). One can verify that, as mentioned before $\begin{pmatrix} b_n \\ \tilde{b}_n \end{pmatrix}$ is an isospinor under the $SU(2)$. Furthermore, it is not too hard to check that Q_+ has definite ghost number:

$$[T_3, Q_+] = \frac{1}{2} Q_+ \quad , \tag{23}$$

and this relation together with $[T_+, Q_+] = 0$ (eq. (16)) implies that Q_+ is the $(+)$ component of an isospinor that we denote as $\begin{pmatrix} Q_+ \\ Q_- \end{pmatrix}$. One then sees that:

$$Q_- = i[T_-, Q_+], \quad \{Q_+, Q_-\} = 2iHT_3 \quad . \tag{24}$$

The above relations will be useful in constructing the gauge invariant action.

One last piece of information required refers to the definition of a suitable inverse operator for T_+. As we are familiar in $SU(2)$ representations, T_+ raises the value of T_3 by one unit, thus the inverse operator is that which decreases T_3 by a unit, namely T_-, up to a normalization constant. The precise expression is[1]

$$T_+^{-1} = (1 - \delta_{T_3, T}) \cdot \frac{1}{(T - T_3)(T + T_3 + 1)} T_-, \tag{25}$$

and this operator vanishes when acting on states with $T = -T_3$ on the right and states with $T = T_3$ on the left. It satisfies the following relations:

$$\begin{aligned} T_+ T_+^{-1} T_+ = T_+ \quad &, \quad T_+^{-1} T_+ T_+^{-1} = T_+^{-1} \quad , \\ T_+^{-1} T_+ = (1 - \delta_{T_3 T}) \quad &, \quad T_+ T_+^{-1} = (1 - \delta_{T_3, -T}) \quad . \end{aligned} \tag{26}$$

It is clear that T_+ does not have a strict inverse and this is so because T_+ vanishes when acting on some states (those with $T = T_3$). Fortunately, eqs. (26) is all one needs in order to find the gauge invariant action.

5. GAUGE-INVARIANT STRING ACTIONS

As discussed before, the gauge-fixed string action and transformations are analogous to those in eq. (3). The BRST auxiliary fields ψ appear quadratically in the action, and their BRST transformation is not zero. We therefore have to redefine the auxiliary field

$$\psi = \tilde{\psi} + A\phi \tag{27}$$

where A is an anticommuting operator. One may think that one needs to set $\delta\tilde{\psi} = 0$, but this is too strong. Not all of ψ appears in the action (10), in fact only the fields that are not annihilated by the hermitian operator T_+ do appear. The appropriate condition is therefore:

$$\delta(T_+\tilde{\psi}) = T_+\delta\tilde{\psi} = 0 \ \rightarrow T_+(\delta\psi - A\delta\phi) = 0 \ . \tag{28}$$

Using eqs. (11), (12) and (16) one finds two conditions:

$$(Q_+ - T_+A)T_+ = 0, \quad (Q_+ - T_+A)Q_+ = 0 \ , \tag{29}$$

and these are solved by $A = T_+^{-1}Q_+$, where T_+^{-1} was given in eq. (25) (use eq. (26) and (16)).

Having found the necessary shift, we now rewrite the action and transformation laws finding: (use eqs. (26)):

$$L = \frac{1}{2}\phi(H + Q_+T_+^{-1}Q_+)\phi + \phi Q_+(1 - \delta_{T_3T})\tilde{\psi} + \frac{1}{2}\tilde{\psi}T_+\tilde{\psi} \ , \tag{30}$$

and

$$\delta\phi = \epsilon\delta_{T_3,-T}Q_+\phi - \epsilon T_+\tilde{\psi} \ , \tag{31}$$

$$\delta\tilde{\psi} = \epsilon\delta_{T_3,T}(Q_+\tilde{\psi} + \tilde{H}\phi) \ . \tag{32}$$

Note that in the action eq. (30) only fields $\tilde{\psi}$ with $T \neq T_3$ appear and for those eq. (32) implies that their BRST transformation is zero: $\delta(1 - \delta_{T_3,T})\tilde{\psi} = 0$. Thus, the auxiliary fields are now in good shape and the only thing left to do is to separate the physical fields from the ghosts and antighosts. In order to do this we decompose eq. (31) into two subspaces:

$$\delta\,\delta_{T_3,-T}\phi = \epsilon\delta_{T_3,-T}Q_+\phi \ , \tag{33}$$

$$\delta(1 - \delta_{T_3,-T})\phi = -\epsilon T_+\tilde{\psi} \ . \tag{34}$$

Physical fields must have ghost number equal to zero: $T_3 = 0$. Moreover, physical fields should not transform into BRST auxiliary fields,[10] as it seems to happen in eq. (34). Therefore, physical fields must also have $T_3 = -T$. This together with $T_3 = 0$ implies that physical fields are singlets. We can now drop all the BRST auxiliary fields and the non-singlets of ϕ from the action finding

$$L = \frac{1}{2}\phi\delta_{T,0}(H + Q_+T_+^{-1}Q_+)\delta_{T,0}\phi \ , \tag{35}$$

where $\delta_{T,0}$ is a projector that kills non-singlets. T_+^{-1} (eq. (25)) is a complicated operator but it can be simplified greatly:

$$T_+^{-1}Q_+\delta_{T,0}\phi = i(1-\delta_{T_3,T})\cdot\frac{1}{(T-T_3)(T+T_3+1)}Q_-\delta_{T,0}\phi = iQ_-\delta_{T,0}\phi \ , \tag{36}$$

where we have used the fact that Q_- on a singlet is a state with $T_3 = -\frac{1}{2}$ and $T = \frac{1}{2}$. The action is therefore:

$$L = \frac{1}{2}\phi\delta_{T,0}(H - iQ_+Q_-)\delta_{T,0}\phi \ . \tag{37}$$

It is nice to use $SU(2)$ invariant notation

$$Q^2 = \frac{i}{2}Q_-Q_+ - \frac{i}{2}Q_+Q_- = -HT_3 - iQ_+Q_- \ , \tag{38}$$

and therefore due to the projector the term $(-iQ_+Q_-)$ can be replaced in the action by Q^2 giving us the final form[1]

$$L = \frac{1}{2}\phi(H + Q^2)\delta_{T,0}\phi \ . \tag{39}$$

This action is invariant under the following gauge transformations (implied by eq. (33)):

$$\delta\phi = Q_+\Lambda \ , \tag{40}$$

as one readily verifies:

$$\begin{aligned}
\delta L &= \phi\delta_{T,0}(H + Q^2)Q_+\Lambda \\
&= \phi\delta_{T,0}(HQ_+ - iQ_+Q_-Q_+)\Lambda \qquad &\text{(using (38))} \\
&= \phi\delta_{T,0}\left(HQ_+ - iQ_+(2iHT_3)\right)\Lambda \qquad &\text{(using (24))} \\
&= \phi\delta_{T,0}\left(HQ_+ + 2[Q_+, T_3]H\right)\Lambda = 0 \ . \qquad &\text{(using (23))}
\end{aligned} \tag{41}$$

Since the action in eq. (39) involves only the $SU(2)$ singlets of ϕ, we can add terms to the gauge transformations in order to gauge away the non-singlets. A term of the type $T_+\Sigma$ with Σ unconstrained contains no singlets, but no terms with $T_3 = -T$, similarly $T_-\chi$ with χ unconstrained contains no singlets and no terms with $T_3 = T$, thus the two terms together are enough (even redundant) to gauge away all nonsinglets:

$$\delta\phi = Q_+\Lambda + T_+\Sigma + T_-\chi \ . \tag{42}$$

This gauge transformation describes the enormous Abelian symmetry of the free open string.

For the closed string, the analogous results are:[1]

$$L = \frac{1}{2}\phi(H + Q^2)\delta_{T,0}\phi + \frac{i}{2}\hat{\phi}\Delta N \ \delta_{T,0}\phi \tag{43}$$

where both H and $Q_\pm$ are given by expressions analogous to those of the open string but with sums over $\pm$ modes. The field $\hat{\phi}$ is a Lagrange multiplier that imposes the constraint $\Delta N\phi = (N_+ - N_-)\phi = 0$, where N_+ and N_- are the number operators for the $+$ and $-$ modes, respectively. This action is invariant under:

$$\begin{aligned} \delta\phi &= Q_+\Lambda + T_+\Sigma + T_-\chi \\ \delta\hat{\phi} &= \Sigma_3 + T_+\hat{\Sigma} + T_-\hat{\Sigma} \end{aligned} \tag{44}$$

where

$$\Delta N\Lambda = \Delta N\Sigma_3 = 0 \ . \tag{45}$$

6. COMPONENT FIELDS, BOSONIZATION AND COMMENTS

Let us see briefly how to obtain component Lagrangians from the string action. Start with eq. (37) rewritten as:

$$L = \frac{1}{2}\phi(H + Q_+T_-Q_+)\delta_{T,0}\phi \ . \tag{46}$$

Expanding up to the second excited level the string field:

$$\delta_{T,0}\phi = \phi_0 + A^\mu a^\dagger_{1\mu} + \frac{1}{2}h^{\mu\nu}a^\dagger_{1\mu}a^\dagger_{1\nu} + B^\mu a^\dagger_{2\mu} + i\eta b^\dagger_1\tilde{b}^\dagger_1 \ , \tag{47}$$

and the operator Q_+:[1]

$$\begin{aligned} Q_+ = & \, i\partial_\mu \left(a^\dagger_{1\mu}b_1 + b^\dagger_1 a_{1\mu} + a^\dagger_{2\mu}b_2 + b^\dagger_2 a_{2\mu}\right) \\ & + \frac{i}{4}\left(b^\dagger_2 a_1\cdot a_1 - a^\dagger_1\cdot a^\dagger_1 b_2\right) \\ & + i\left(a^\dagger_2\cdot a_1 b_1 - b^\dagger_1 a^\dagger_1\cdot a_2\right) \\ & - \frac{3}{2}\left(b^\dagger_2\tilde{b}_1 b_1 + b^\dagger_1\tilde{b}^\dagger_1 b_2\right) \ , \end{aligned} \tag{48}$$

the action is now easily evaluated (use eqs. (17) and (22)):

$$L = L_{-1} + L_0 + L_1$$

$$L_{-1} = \frac{1}{2}\phi_0(\Box + 1)\phi_0 \tag{49a}$$

$$L_0 = \frac{1}{2}A\cdot\Box A + \frac{1}{2}(\partial\cdot A)^2 \tag{49b}$$

$$L_1 = \frac{1}{4}h^{\mu\nu}(\Box - 1)h_{\mu\nu} + \frac{1}{2}B^{\mu}(\Box - 1)B_{\mu} - \frac{1}{2}\eta(\Box - 1)\eta$$
$$+ \frac{1}{2}\left(\partial^{\nu}h_{\mu\nu} + \partial_{\mu}\eta - B_{\mu}\right)^2 + \frac{1}{2}\left(\frac{1}{4}h^{\mu}_{\mu} + \frac{3}{2}\eta + \partial\cdot B\right)^2 \ . \tag{49c}$$

The gauge invariances are found expanding eq. (40) where the nontrivial component transformations are those obtained setting Λ equal to the $-$ component of an isospinor:

$$\Lambda = \xi\tilde{b}_1^{\dagger} + \epsilon^{\mu}a_{1\mu}^{\dagger}\tilde{b}_1^{\dagger} + \epsilon\tilde{b}_2^{\dagger} \ . \tag{50}$$

We then find:

$$\delta A_{\mu} = \partial_{\mu}\xi \ , \tag{51}$$

and

$$\delta h_{\mu\nu} = \partial_{(\mu}\epsilon_{\nu)} - \frac{1}{2}\eta_{\mu\nu}\epsilon \ ,$$
$$\delta B_{\mu} = \partial_{\mu}\epsilon + \epsilon_{\mu} \ ,$$
$$\delta\eta = -\partial\cdot\epsilon + \frac{3}{2}\epsilon \ . \tag{52}$$

eqs. (49b) and (51) are the usual action and gauge invariance for a free photon. Eqs. (49c) and (52) are not in standard form for massive symmetric rank two. One has to let

$$h_{\mu\nu} = \hat{h}_{\mu\nu} + \frac{1}{10}\eta_{\mu\nu}\hat{\eta}, \quad \eta = -\frac{1}{2}\hat{h}^{\mu}_{\mu} - \frac{3}{10}\hat{\eta} \ , \tag{53}$$

obtaining

$$\delta\hat{h}_{\mu\nu} = \partial_{(\mu}\epsilon_{\nu)}, \quad \delta B_{\mu} = \partial_{\mu}\epsilon + \epsilon_{\mu}, \quad \delta\hat{\eta} = -5\epsilon \ . \tag{54}$$

It is now possible to gauge away the Stueckelberg fields B_{μ} and $\hat{\eta}$ and eq. (49c) becomes the usual Fierz-Pauli action for a massive symmetric rank two tensor.[13]

It was observed in Refs. [5, 1] that massive fields appear to be described in the string by action that arise via reduction over a circle from the corresponding massless field action formulated in one dimension higher. Rather than giving a general argument,[1] consider massive symmetric rank three tensor. The massless action requires a symmetric rank-three tensor $\phi_{\hat{\mu}\hat{\nu}\hat{\rho}}$ ($\hat{\mu}$, $\hat{\nu}$, $\hat{\rho}$ run over $D+1$ values and no traces have been taken away) with a gauge invariance $\delta\phi_{\hat{\mu}\hat{\nu}\hat{\rho}} = \partial_{(\hat{\mu}}\epsilon_{\hat{\nu}\hat{\rho})}$, $\epsilon^{\hat{\mu}}_{\hat{\mu}} = 0$.[14] Upon dimensional reduction one finds (μ, ν, p running over D values) the fields $\phi_{\mu\nu p}$, $\phi_{\mu\nu}$, ϕ_μ, ϕ. The string action uses these four fields. Indeed, at the third excited level the string states are the ☐☐☐ and ☐☐ (column) representations of $SO(25)$. The first one corresponds to the above field and the second one is a massive antisymmetric tensor requiring the fields $A_{[\mu\nu]}$ and A_μ (using this same prescription of dimensional reduction). The above six fields that are necessary at this mass level to indeed appear as $SU(2)$ singlets of the BRST-field:

$$\begin{aligned} S^{\mu\nu\rho}a^\dagger_{1\mu}a^\dagger_{1\nu}a^\dagger_{1\rho} + S^{(\mu\nu)}a^\dagger_{1\mu}a^\dagger_{2\nu} + A^{[\mu\nu]}a^\dagger_{1\mu}a^\dagger_{2\nu} + S^\mu a^\dagger_{3\mu} \\ + i\sigma\left(\tilde{b}^\dagger_1 b^\dagger_2 + \tilde{b}^\dagger_2 b^\dagger_1\right) + iV^\mu a^\dagger_{1\mu}\tilde{b}^\dagger_1 b^\dagger_1 \ . \end{aligned} \tag{55}$$

Denoting the general $SU(2)$ singlet made out of two isospinor operators by $(mn) = i\tilde{b}^\dagger_{(m}b^\dagger_{n)}$, we can tabulate the number of $SU(2)$ singlets that are made out of fermionic oscillators at the first few levels:

$$\begin{array}{ll} N=0 & - \\ 1 & - \\ 2 & (11) \\ 3 & (12) \\ 4 & (13),(22) \\ 5 & (14),(23) \\ 6 & (15),(24),(33),(11)(22) \end{array} \tag{56}$$

This counting can be reproduced using an extra bosonic coordinate that is effectively missing its zeroth and first mode $\{\hat{a}_2, \hat{a}_3, \hat{a}_4 \ldots\}$.[1] With such oscillators one verifies that it is not possible to have $N = 0$ or $N = 1$ singlets, that there is one singlet at $N = 2$ ($\hat{a}_2$), one at $N = 3$ ($\hat{a}_3$), two at $N = 4$ ($\hat{a}_2\,\hat{a}_2$, $\hat{a}_4$), two at $N = 5$ ($\hat{a}_2\,\hat{a}_3$, $\hat{a}_5$) and four at $N = 6$ ($\hat{a}_2\hat{a}_2\hat{a}_2$, $\hat{a}_3\hat{a}_3$, $\hat{a}_4\hat{a}_2$, $\hat{a}_6$) in agreement with the results listed in the table. As suggested by this counting, it is possible to bosonize the fermionic coordinates into a single bosonic one:[1]

$$\hat{C} = \frac{1}{\sqrt{2\pi}}e^{\sqrt{2\pi}\chi}, \quad \frac{\delta}{\delta\hat{C}} = \frac{1}{\sqrt{2\pi}}e^{-\sqrt{2\pi}\chi} \ . \tag{57}$$

Here, the exponentials are normal ordered and χ has the mode expansion:

$$\begin{aligned} \chi = \frac{1}{\sqrt{2\pi}}\left[(\hat{q}+\hat{p}\sigma) + \sum_{N=1}^{\infty}\frac{1}{\sqrt{n}}\left(\hat{a}_n e^{in\sigma} + \hat{a}^\dagger_n e^{-in\sigma}\right)\right] \ , \\ [\hat{p},\hat{q}] = -i, \quad [\hat{a}_m, \hat{a}^\dagger_n] = -\delta_{mn} \ , \end{aligned} \tag{58}$$

the above commutators indicating that this coordinate has timelike metric. It is therefore an extra time coordinate the one that gives rise to the additional fields required for a covariant formulation. This extra time coordinate is a Lorentz scalar and therefore there is no such thing as an $SO(25,2)$ symmetry. The constraints for $SU(2)$ singlets can be shown to imply[1] that in the bosonized language the operator $\hat{p}$ has a definite value when acting on fields and, that the operator $\hat{a}_1$ acting on fields is zero, this being equivalent to the statement that $\hat{a}_1^\dagger$ is missing (explaining the agreement between the bosonic and fermionic counting). The absence of these zeroth and first modes implies that this coordinate has peculiar geometrical properties (perhaps some sort of compactified time dimension?). From the first quantized viewpoint the extra coordinate corresponds to a wrong metric scalar field living in the two-dimensional world sheet with a coupling $R\chi$ to the two dimensional scalar curvature.[1]

It is possible that the bosonized form of the string action will help find some sort of geometrical picture, but further exploration is needed, perhaps along the lines of the first-quantized interpretation.

At any rate it seems likely that the full gauge-invariant interacting bosonic string theory will be constructed soon. Key aspects of the gauge-fixed interactions were given in [5] and it should be possible to construct the complete BRST invariant interactions. Some modification of the methods of Ref. [1] should enable one to find the underlying gauge invariant interactions.

REFERENCES

1. W. Siegel and B. Zwiebach, "Gauge String Fields", University of California at Berkeley preprint UCB-PTH-85/30, to appear in *Nucl. Phys. B.*
2. W. Siegel, see articles in these proceedings.
3. M. Kato and K. Ogawa, *Nucl. Phys. B* **212**, 443 (1983).
4. K. Fujikawa, *Phys. Rev.* **D25**, 2584 (1982).
5. W. Siegel, *Phys. Lett.* **149B**, 157, 162 (1984); **151B**, 391, 396 (1985).
6. T. Banks and M. Peskin to appear in "Anomalies Geometry and Topology", A. White, ed., (World Scientific, Singapore) (1985); M. Kaku, *ibid.*; D. Friedan, Chicago preprint EFI 85-27; C. B. Thorn, Lptens preprint 85/14; A. Neveu and P. C. West, CERN-TH-4200/85.
7. T. Banks and M. Peskin, SLAC preprint, SLAC-PUB-3740 (July 1985); see also the article in these proceedings.
8. K. Itoh, T. Kugo, H. Kunitomo and H. Ooguri, Kyoto University preprint, KUNS 800, HE(TH) 85/04, August 1984.
9. A. Neveu, H. Nicolai and P. C. West, CERN-TH. 4233/85; A. Neveu, J. H. Schwarz and P. C. West, CERN-TH. 4248/85.
10. C. Becchi, A. Rouet and R. Stora, *Phys. Lett.* **52B**, 344 (1974); I. V. Tyutin, "Gauge Invariance in Field Theory and in Statistical Physics in the

Operator Formulation", Lebedev preprint FIAN No. 39 (1975) in Russian, unpublished; T. Kugo and I. Ojima, *Phys. Lett.* **73B**, 459 (1978); I. A. Batalin and G. A. Vilkoviski, *Phys. Lett.* **69B**, 309 (1977); E. S. Fradkin and T. E. Fradkina, *Phys. Lett.* **72B**, 343 (1978).

11. N. Nakanishi, *Prog. Theor. Phys.* **35**, 1111 (1966); B. Lautrup, *K. Dan. Vidensk. Selsk. Mat. Fys. Medd.* **34**, No. 11, 1 (1967).
12. T. Curtright, unpublished; T. Kimura, *Prog. Theor. Phys.* **64**, 357 (1980); W. Siegel, *Phys. Lett.* **93B**, 170 (1980); J. Thierry-Mieg, Harvard University preprint HUTMP-79/B86 (June 1980), unpublished.
13. M. Fierz and W. Pauli, *Proc. Roy. Soc.* **A173**, 211 (1939).
14. C. Fronsdal, *Phys. Rev.* **D18**, 3624 (1978); T. Curtright, *Phys. Lett.* **85B**, 219 (1979); B. deWit and D. Z. Freedman, *Phys. Rev.* **D21**, 358 (1980).

GAUGE INVARIANCE OF STRING FIELDS*

Thomas Banks

Stanford Linear Accelerator Center
Stanford University, Stanford, California, 94305
and
Tel Aviv University, Ramat Aviv, Israel
and
Michael E. Peskin

Stanford Linear Accelerator Center
Stanford University, Stanford, California, 94305

1. STRING SYMMETRIES

This is a report on some work we have done[1] to understand the appearance of gauge bosons and gravitons in string theories. We have constructed an action for free (bosonic) string field theory which is invariant under an infinite set of gauge transformations which include Yang-Mills transformations and general coordinate transformations as special cases. Our work was motivated by some beautiful papers of Siegel[2] in which a covariant, gauge fixed free string action was constructed. At the end of this lecture I will show how to obtain Siegel's action from ours by a straightforward application of the Faddeev-Popov procedure. Two other groups[3] have independently arrived at the form for the gauge invariant string action that we will present here.

The Nambu action:

$$S = \int d^2\xi \sqrt{\epsilon_{ij}\epsilon_{kl}\frac{\partial x^\mu}{\partial \xi_i}\frac{\partial x_\mu}{\partial \xi_k}\frac{\partial x^\nu}{\partial \xi_j}\frac{\partial x_\nu}{\partial \xi_l}} \tag{1}$$

---

* Work supported by the Department of Energy, contract DE–AC03–76SF00515.

is not the action for string theory. Its function in string field theory is analogous to that of the relativistic particle action

$$\int \sqrt{\dot{x}^2} \tag{2}$$

in ordinary field theory. That is, the Feynman propagators and vertices in scalar field theory are operators in a Hilbert space of functions $\phi(x)$,which is determined by the quantum dynamics of the Lagrangian of equation (2). Since the Lagrangian is invariant under proper time reparametrizations its dynamics is determined by a Schrodinger-Wheeler-DeWitt (SWD)[4] equation:

$$(p^2 + m^2)\phi(x) = 0\,. \tag{3}$$

A classical field theory is defined by treating the wave function $\phi(x)$ as a classical field and finding a Lagrangian whose Euler Lagrange equation is the SWD equation.

In order to mimic this formalism in string field theory we introduce the canonical conjugate to $x^\mu(\sigma,\tau)$ in the action of Eq. (1). ($z_0 = \tau, z_1 = \sigma$)

$$P_\mu(\sigma,\tau) = \frac{\partial}{\partial \dot{x}^\mu}\sqrt{\det \frac{\partial x^\mu}{\partial z_i}\frac{\partial x^\mu}{\partial z_j}}\,. \tag{4}$$

The gauge invariant quantum dynamics of the string is constructed by making x^μ and P_μ into operators with canonical commutators:

$$[x^\mu(\sigma), P_\nu(\sigma')] = i\delta^\mu_\nu \delta(\sigma - \sigma')\,. \tag{5}$$

World sheet reparametrization invariance is imposed by constraining the wave functionals $\Phi[x(\sigma)]$ to satisfy:

$$(\pi^2 P^2 + x'^2)\Phi = 0 \tag{6}$$

$$x^{\mu\prime} P_\mu \phi = 0\,. \tag{7}$$

Since the system is invariant under proper time reparametrizations, these constraints include the SWD equation. Let us introduce some notation to rewrite these constraints in a more convenient form (we also specialize to open strings):

$$\begin{aligned} x^\mu(\sigma) &= x^\mu + \sum_{n>0} \frac{2}{\sqrt{n}} X_n^\mu \cos n\sigma \\ p^\mu(\sigma) &= \frac{1}{\pi}\Big\{p^\mu + \sum_{n>0} \sqrt{n} P_n^\mu \cos n\sigma\Big\}; \end{aligned} \tag{8}$$

$0 \le \sigma \le \pi$, and $[X_n, P_m] = i\delta_{nm}$. It is convenient to replace

$$X_n = \frac{i}{2\sqrt{n}}(\alpha_n - \alpha_{-n}) , \quad P_n = \frac{1}{\sqrt{n}}(\alpha_n + \alpha_{-n}) , \tag{9}$$

and to set $\alpha_0^\mu = p^\mu$; then the α_n have the commutation relations:

$$[\alpha_n^\mu, \alpha_m^\nu] = n\,\delta(n+m)\,\eta^{\mu\nu} . \tag{10}$$

$p(\sigma)$ and $x'(\sigma)$ are especially simple functions of the α_n:

$$\Big(\pi p \pm x'\Big) = \sum_{n=-\infty}^{\infty} \alpha_n e^{\mp in\sigma} . \tag{11}$$

The generators of reparametrizations of the string are the local Hamiltonian and momentum densities:

$$\mathcal{H}(\sigma) = \frac{1}{2\pi}\Big(\pi^2 p^2 + (x')^2\Big), \quad \mathcal{P}(\sigma) = p \cdot x' . \tag{12}$$

These quantities are summarized as:

$$\tfrac{1}{2}\Big(\pi p \pm x'\Big)^2 = \sum_{-\infty}^{\infty} L_n e^{\mp in\sigma} , \tag{13}$$

where the L_n are the Virasoro operators[5]

$$L_n = \tfrac{1}{2} \sum_{-\infty}^{\infty} : \alpha_{n+k}^\mu \alpha_{-k}^\mu : . \tag{14}$$

These operators satisfy the algebra:

$$[L_n, L_m] = (n-m)L_{n+m} + \frac{d}{12} n(n^2-1)\delta(n+m), \tag{15}$$

in which the central charge depends on d, the dimensionality of space.

The constraints are then equivalent (after a judicious choice of normal ordering constant in L_0) to:

$$(L_0 - 1)|\Psi >= 0 \tag{16}$$

$$L_n|\Psi >= 0 \ . \tag{17}$$

These equations are inconsistent because of the Schwinger term in the Virasoro algebra (15). We replace (17) by:

$$L_n|\Psi >= 0 \text{ for } n > 0 \tag{18}$$

which is sufficient to guarantee the constraint equations as equations for matrix elements between states satisfying (18).

If we were to follow the example of scalar field theory slavishly, we would now seek an action from which all of these constraint equations followed as equations of motion. However, we are searching for a gauge invariant theory, and, guided by hints from the old string literature[6] we will instead interpret only (16) as an equation of motion. Equations (18) will be gauge fixing conditions. Note that in making this artificial separation we are violating the manifest reparametrization invariance of the formalism. In the interacting theory, this will have the consequence that duality is manifest only after all diagrams of a given order are summed.

2. STRING FIELDS

It is now a simple matter to find a gauge invariant action from which (16) and (18) follow. Equation (18) is analogous to the Lorentz gauge condition in electrodynamics. The Maxwell Lagrangian may be written:

$$A_\mu \partial^2 P^{\mu\nu} A_\nu \tag{19}$$

where $P^{\mu\nu} = \eta^{\mu\nu} - \partial^\mu \partial^\nu / \partial^2$ is the projector on states satisfying the Lorentz condition. By analogy we write:

$$S = -\tfrac{1}{2}(\Phi \mid \mathcal{K}\Phi). \tag{20}$$

where:

$$\mathcal{K} = 2(L_0 - 1)P \ , \tag{21}$$

and P is the projector on states satisfying (18). This action has also been proposed by Kaku and Lykken.[7] It is invariant under the gauge transformations

$$\Delta\Phi = \sum L_{-n} C_n \ . \tag{22}$$

The projector P has been studied by Brower and Thorn and Feigin and Fuks[8] We found an infinite product representation of it which is useful for calculational purposes. We also showed that, when restricted to the massless levels of open and closed, bosonic and fermionic strings, our action reproduces the linearized actions and gauge transformations of Yang-Mills theory, general relativity, and supergravity. The spacetime fields arise as coefficients in the expansion of Φ in eigenmodes of L_0, as exemplified by the formula

$$\Phi[x(\sigma)] = \left\{\phi(x) - iA^{\mu}(x)\alpha^{\mu}_{-1} - \tfrac{1}{2}h^{\mu\nu}(x)\alpha^{\mu}_{-1}\alpha^{\nu}_{-1} - iv^{\mu}(x)\alpha^{\mu}_{-2} + \ldots\right\}\Phi^{(0)} \quad (23)$$

x^{μ} is the center of mass coordinate of the string. As we reported at the Argonne conference[9] the action (20) becomes non-local at the first positive mass squared level. In 26 dimensions a formula for $2(L_0-1)P$ which is valid through level $L_0 - p^2/2 = 3$ (in units where the Regge slope is one half) is

$$\begin{aligned}
\mathcal{K} = {} & 2(L_0-1) - L_{-1}L_1 - \tfrac{1}{2}L_{-2}L_2 \\
& + \tfrac{1}{2}(3L_{-2}+2L_{-1}^2)\frac{1}{8L_0+13}(3L_2+2L_1^2) \\
& - \frac{1}{3}L_{-3}L_3 \\
& -(3L_{-2}+2L_{-1}^2)L_{-1}\frac{12L_0+37}{48(3L_0+7)(L_0+4)(8L_0+21)}L_1(3L_2+2L_1^2) \\
& -(3L_{-2}+2L_{-1}^2)L_{-1}\frac{3}{48(3L_0+7)(L_0+4)}(8L_3+3L_1L_2) \ + h.c. \\
& +(8L_{-3}+3L_{-2}L_{-1})\frac{4L_0+7}{48(3L_0+7)(L_0+4)}(8L_3+3L_1L_2) \ + \ldots \,.
\end{aligned} \quad (24)$$

Although we have every reason to believe that string theories are indeed nonlocal when written as infinite component field theories in center of mass coordinates, the nonlocality of the *kinetic* term leads to incorrect counting of degrees of freedom at the quantum level. The obvious resolution of this problem is to introduce Stueckelberg[10] fields which make the action local. For example, nonlocality can be removed through the first positive mass squared level by introducing a single string field S with only a scalar component ($S = \int d^dx s(x)|0,x>$) and the action

$$\begin{aligned}
S = -\tfrac{1}{2}\Big\{ & (\Phi \mid 2(L_0-1)\,\Phi) \ - \ (S \mid 2(L_0+1)\,S) \\
& - (L_1\Phi + L_{-1}S \mid L_1\Phi + L_{-1}S) - \tfrac{1}{2}\,(L_2\Phi + 3S \mid L_2\Phi + 3S)\Big\},
\end{aligned} \quad (25)$$

The problem now is to generalize (25) to all mass levels. Such a generalized Stueckelberg action should reduce to (20) when the Stueckelberg fields are integrated out. This will be true if it is gauge invariant and if the Stueckelberg fields decouple from Φ when $L_n\Phi = 0$. This is not sufficient however. The action should also have the same number of degrees of freedom as the light cone gauge action of Kaku and Kikkawa.[11] A direct proof of this latter property for the formalism we will present below will be given in Ref. 12. In our original paper we showed instead that our action is equivalent to the gauge fixed covariant action of Siegel. In Ref. 2, Siegel argued that his formalism was equivalent to the light cone gauge.

Siegel's dynamical variable is a functional field $\Phi[x(\sigma), \theta(\sigma), \widehat{\theta}(\sigma)]$ which depends on two Grassmann variables in addition to $x(\sigma)$. Equivalently, it is an infinite collection of functional fields:

$$\Phi[x, \theta, \widehat{\theta}] = \sum \Phi^{m_1 \dots m_a}{}_{n_1 \dots n_b} \theta_{m_1} \dots \theta_{m_a} \widehat{\theta}_{n_1} \dots \widehat{\theta}_{n_b} \tag{26}$$

antisymmetric separately in upper and lower indices.

The first idea that we had for reproducing this zoo of fields was based on the concept of ghosts for ghosts.[13] The gauge transformations

$$\Delta\Phi = L_{-n} C^n$$

are redundant in the sense that C^n and

$$C^n + (L_{-m} G^{mn} + \frac{1}{2} V^n_{mk} G^{mk}) \tag{27}$$

are the same transformation of Φ, if G^{mn} satisfies

$$G^{mn} = -G^{nm} \tag{28}$$

and

$$[L_m, L_n] = V^k_{mn} L_k \qquad m, n, k > 0 \ . \tag{29}$$

Consequently, the Faddeev-Popov ghost action will be invariant under the transformations (29) (where C is now interpreted as the Faddeev-Popov ghost field), and we will need ghosts for ghosts. The transformations parametrized by G are themselves redundant and the process continues indefinitely.

The resemblance of all of this to the theory of a p form gauge field motivated us to develop a theory of differential forms in the space of strings. Remarkably, that formalism also leads to a natural solution of the Stueckelberg field problem. We will refer the reader to our Nuclear Physics paper[1] for the details of all of this and content ourselves with recording a few of the relevant equations.

3. DIFFERENTIAL FORMS IN THE SPACE OF STRINGS

It is useful to decompose the commutation relations of the Virasoro algebra as :

$$\begin{aligned}
[L_m, L_n] &= V_{mn}{}^p\, L_p \\
[L_{-m}, L_{-n}] &= -\, V_{mn}{}^p\, L_{-p} \\
[L_m, L_{-n}] &= W_{mn}{}^p\, L_p \;+\; W_{nm}{}^p\, L_{-p} \;+\; \eta_{mn}\mathbf{L}(m)\ ,
\end{aligned} \tag{30}$$

where m, n, p are positive integers. More explicitly,

$$\begin{aligned}
V_{mn}{}^p &= \delta(p-(m+n))(m-n) \\
W_{mn}{}^p &= \delta(m-(n+p))(m+n) \\
\mathbf{L}(m) &= 2L_0 + \frac{13}{6}(m^2-1)\ .
\end{aligned} \tag{31}$$

The last relations evaluated in $d = 26$. We now think of L_{-n} as covariant derivatives ∇_{z_n} on a complex manifold with coordinates $(z_n, \bar{z}_n)$ and L_n as the complex conjugate derivative. Formulas (30) are then statements about the torsion and curvature of the manifold.

An $\binom{a}{b}$-form on a complex manifold is a tensor field

$$\Phi^{m_1 \ldots m_a}{}_{n_1 \ldots n_b} \tag{32}$$

antisymmetric in upper (holomorphic) and lower (antiholomorphic) indices. By analogy with manifolds that have torsion, we can construct exterior derivatives d and $\delta^\star$ satisfying $d^2 = \delta^2 = 0$ from ∇_z and $\nabla_{\bar{z}}$. The formulas are:

$$\begin{aligned}
(dC)^{[m_1 \ldots m_a]}{}_{[n_1 \ldots n_{b+1}]} &= L_{[n_1}\, C^{[m_1 \ldots m_a]}{}_{n_2 \ldots n_{b+1}]} + a\, W_{p[n_1}{}^{[m_1}\, C^{p m_2 \ldots m_a]}{}_{n_2 \ldots n_{b+1}]} \\
&\quad - \tfrac{1}{2}\, b\, V_{[n_1 n_2}{}^p\, C^{[m_1 \ldots m_a]}{}_{p n_3 \ldots n_{b+1}]}, \\
(\delta C)^{[m_1 \ldots m_{a-1}]}{}_{[n_1 \ldots n_b]} &= L_{-p}\, C^{[p m_1 \ldots m_{a-1}]}{}_{[n_1 \ldots n_{b+1}]} + b\, W_{[n_1 p}{}^q\, C^{[p m_1 \ldots m_{a-1}]}{}_{q n_2 \ldots n_b]} \\
&\quad - \tfrac{1}{2}\,(a-1)\, V_{kl}{}^{[m_1}\, C^{k l m_2 \ldots m_{a-1}]}{}_{[n_1 \ldots n_b]}.
\end{aligned} \tag{33}$$

Here and henceforth, we make the convention that raised indices labeled as (m_i) are antisymmetrized together in the indicated order and lowered indices labeled as (n_i) are antisymmetrized together similarly.

⋆ Our definition of δ differs from the conventional one by a factor $(-1)^b$. Consequently a commutator appears in Eq. (36) instead of an anticommutator

d and δ are adjoints with respect to the scalar product between $\binom{a}{b}$-forms and $\binom{b}{a}$-forms, defined by combining the Hilbert space scalar product with contraction of holomorphic with antiholomorphic indices.

There are also two operations which change a holomorphic into an antiholomorphic index and vice versa:

$$\begin{aligned}\left(\Uparrow C\right)^{m_1\ldots m_{a+1}}{}_{n_1\ldots n_{b-1}} &= \eta^{m_1 q}\, C^{m_2\ldots m_{a+1}}{}_{q n_1\ldots n_{b-1}} \\ \left(\Downarrow C\right)^{m_1\ldots m_{a-1}}{}_{n_1\ldots n_{b+1}} &= \eta_{n_1 q}\, C^{q m_1\ldots m_{a-1}}{}_{n_2\ldots n_{b+1}}\end{aligned} \tag{34}$$

$$\eta_{mp} = p\delta_{mp} \qquad \eta^{mp}\eta_{pq} = \delta^m{}_q\,. \tag{35}$$

By our convention, the (m_i) are antisymmetrized together, and the (n_i) are antisymmetrized together. $\Uparrow$ and $\Downarrow$ are self adjoint with respect to the scalar product that we have defined.

All of this formalism works for any spacetime dimension (any coefficient of the Schwinger term in the Virasoro algebra). The special role of $d = 26$ becomes apparent when we compute the Laplacian

$$(d\delta - \delta d)C^{m_1\ldots m_{a-1}}{}_{n_1\ldots n_{b+1}} = \mathbf{K}\, \eta_{[n_1 p}\, C^{p m_1\ldots m_{a-1}}{}_{n_2\ldots n_{b+1}]}, \tag{36}$$

where

$$\mathbf{K} = 2(L_0 - 1 + (\text{sum of indices})) \tag{37}$$

This formula for K is valid only in 26 dimensions. It has the consequence that K commutes with all the other operators in the theory and may be treated as a c-number.

Finally we note that $\Uparrow$ and $\Downarrow$ allow us to symmetrize and antisymmetrize upper with lower indices. Under the symmetric group S_{a+b} of all the indices, an $\binom{a}{b}$-form can be in any of the representations in $a\left\{\begin{matrix}\square\\\square\\\square\end{matrix}\times\begin{matrix}\square\\\square\\\square\end{matrix}\right\}b$. An $\binom{a}{b}$-form in the representation $k\left\{\begin{matrix}\square\square\\\square\square\\\square\square\\\square\\\square\end{matrix}\right\}\ell$ of S_{a+b} is said to be (k,ℓ) symmetrized.

The $(\text{ghosts})^k$ of Φ can now be seen to be $\binom{k}{0}$-forms (the $(\text{antighosts})^k$ are $\binom{0}{k}$-forms) while Siegel's action contains $\binom{a}{b}$-forms for all a and b. It is natural to suppose that some of these are the Stueckelberg fields for which we have been searching. Indeed, if we further guess that the kinetic term will be given by K, then a $\binom{1}{1}$-form Φ^m_n will first appear at the first positive mass squared level, just where we first needed a Stueckelberg field.

It is possible to write down an action which contains only this single new string field, and which reproduces Eq. (20) when Φ_n^m is integrated out. This action was described in our paper and has recently been derived in work by Neveu, Schwarz and West.[14] Unfortunately, it does not contain the right number of quantum degrees of freedom, and does not reproduce Siegel's action upon Faddeev-Popov quantization. The problem is that it no longer has redundant gauge symmetries. We found that in order to preserve the redundant gauge structure, we had to introduce $\binom{k}{k}$-forms with (k,k) symmetry for arbitrary integer k. In addition we had to introduce an infinite set of new gauge transformations C_{2k+1} which are $\binom{k+1}{k}$-forms with $(k+1,k)$ symmetry.

The action for this infinite system of fields is $(\Phi = \Phi_0)$

$$S = -\tfrac{1}{2}\Big\{(-1)^k \left(\Phi_{2k} \mid \mathbf{K}\Phi_{2k}\right) - (-1)^k (k+1)^2 \left(d\Phi_{2k} + \delta\Phi_{2k+2} \mid\Uparrow (d\Phi_{2k} + \delta\Phi_{2k+2})\right)\Big\}. \tag{38}$$

The gauge transformations are:

$$\delta_C\, \Phi_{2k} \;=\; -d\, C_{2k-1} \;+\; \delta\, C_{2k+1}, \tag{39}$$

where C_{2k+1} is a $\binom{k+1}{k}$-form symmetrized according to $(k+1,k)$. Note that the right hand side must be explicitly (k,k) symmetrized. These gauge transformations are themselves invariant under the redundant transformations

$$\delta_{\mathcal{G}}\, C_{2k+1} \;=\; d\, \mathcal{G}_{2k} \;+\; \delta\, \mathcal{G}_{2k+2}, \tag{40}$$

where $\mathcal{G}_{2k+2}$ is a $\binom{k+2}{k}$-form symmetrized according to $(k+2,k)$. The $\mathcal{G}$ transformation law is in turn left invariant by transformations parametrized by $\binom{k+3}{k}$-forms C', symmetrized according to $(k+3,k)$, and so on. Each form that we have introduced has the total number of indices denoted by its subscript and cannot appear at a mass level lower than this number. Thus, at any given mass level, the proliferation of Stueckelberg fields and their successive gauge transformations eventually terminates. Indeed at any finite mass level, the infinite set of symmetrized Stueckelberg string fields, actually generates fewer spacetime fields than the single field Φ_2 without symmetrization. We believe that, in this sense, the Φ_{2k} are the minimal set of Stueckelberg fields for the string theory.

The action we have written has the form of a Feynman gauge kinetic term (each spacetime field has an action given by the appropriate Klein-Gordon operator) plus the square of a gauge fixing term:

$$\mathcal{F}_{2k-1} \;=\; \left(d\Phi_{2k-2} + \delta\Phi_{2k}\right)\,. \tag{41}$$

Thus we can define a gauge (which we call the Feynman-Siegel gauge) in which only the Klein-Gordon terms appear in the action. The Faddeev- Popov action in this gauge is:

$$S_C = (-1)^k \left\{ (\bar{C}_{2k-1} \mid \mathbf{K} C_{2k-1}) \right.$$
$$\left. - \frac{k(k+1)}{2} \left(\delta \bar{C}_{2k+1} - d\bar{C}_{2k-1} \,|\!\Uparrow (\delta C_{2k+1} - dC_{2k-1})\right) \right\}. \tag{42}$$

where C and $\bar{C}$ are the ghost and antighost fields. This action is invariant to the redundant gauge symmetries:

$$\delta_{\mathcal{G}}\, C_{2k+1} = d\,\mathcal{G}_{2k} + \delta\,\mathcal{G}_{2k+2}, \tag{43}$$

where $\mathcal{G}_{2k+2}$ is a $\binom{k+2}{k}$-form symmetrized according to $(k+2, k)$.

The ghost action has the Feynman-Siegel gauge form plus the square of a gauge fixing term for these redundant transformations. Thus we may fix the gauge and obtain ghosts of ghosts which are $\binom{k}{k+2}$-forms and $\binom{k+2}{k}$-forms with symmetry $(k+2, k)$. We must however be careful to remember the phenomenon of "hidden ghosts"[13], which occurs for any system with redundant gauge transformations. There are only three ghosts of ghosts, rather than the four one might have naively expected. Similarly, when we find that the ghost of ghost action has a gauge invariance, we only need four $(\text{ghosts})^3$ rather than eight. In general we need $p+1$ $(\text{ghosts})^p$.

The systematics of ghost counting is as follows. To each Stueckelberg field Φ_{2k} we associate the gauge transformation C_{2k+1} under which the field transforms as a divergence δ. The ghosts associated with this transformation and its redundancies are called $(\text{ghosts})^p$ of Φ_{2k}. There are $p+1$ $(\text{ghosts})^p$ and they are forms with the symmetry $(k+p, k)$. Table 1 shows how the ghost counting works up to forms with $a+b=4$. The pattern is clear: Φ_0, the Stueckelberg fields, and their descendant ghosts fill out all possible $\binom{a}{b}$-forms with general symmetry. This is precisely the field content of Siegel's action. The gauge fixed action is precisely of the Feynman-Siegel gauge form: each space time field will have a simple Klein-Gordon action.

The formalism we have presented can be simply generalized to closed strings and to open and closed superstrings. The correct treatment of spacetime supersymmetry is not at all straightforward, and is presently under study.[15]

TABLE 1

Number of Indices k	$\Phi_{(2k)}$ fields and their ghosts	Forms with k indices
0	Φ_0	$\begin{pmatrix}0\\0\end{pmatrix}$
1	$2\ Ghosts\ of\ \Phi_0 = 2\times\square$	$\begin{pmatrix}0\\1\end{pmatrix}(=\square)+\begin{pmatrix}1\\0\end{pmatrix}(=\square)$
2	$3(Ghosts)^2\ of\ \Phi_0 = 3\times\ydiagram{1,1}$ $\Phi_2 = \ydiagram{2}$	$\begin{pmatrix}0\\2\end{pmatrix}(=\ydiagram{1,1})+\begin{pmatrix}2\\0\end{pmatrix}(=\ydiagram{1,1})+\begin{pmatrix}1\\1\end{pmatrix}(=\ydiagram{1,1}+\ydiagram{2})$
3	$4(Ghost)^3\ of\ \Phi_0 = 4\times\ydiagram{1,1,1}$ $2\ Ghosts\ of\ \Phi_2 = 2\times\ydiagram{2,1}$	$\begin{pmatrix}0\\3\end{pmatrix}(=\ydiagram{1,1,1})+\begin{pmatrix}3\\0\end{pmatrix}(=\ydiagram{1,1,1})+\begin{pmatrix}2\\1\end{pmatrix}(=\ydiagram{1,1,1}+\ydiagram{2,1})$ $+\begin{pmatrix}1\\2\end{pmatrix}(=\ydiagram{1,1,1}+\ydiagram{2,1})$
4	$5(Ghosts)^4\ of\ \Phi_0 = 5\times\ydiagram{1,1,1,1}$ $3(Ghost)^2\ of\ \Phi_2 = 3\times\ydiagram{2,1,1}$ $\Phi_4 = \ydiagram{2,2}$	$\begin{pmatrix}0\\4\end{pmatrix}(=\ydiagram{1,1,1,1})+\begin{pmatrix}4\\0\end{pmatrix}(=\ydiagram{1,1,1,1})+\begin{pmatrix}3\\1\end{pmatrix}(=\ydiagram{1,1,1,1}+\ydiagram{2,1,1})$ $+\begin{pmatrix}1\\3\end{pmatrix}(=\ydiagram{1,1,1,1}+\ydiagram{2,1,1})+\begin{pmatrix}2\\2\end{pmatrix}(=\ydiagram{1,1,1,1}+\ydiagram{2,1,1}+\ydiagram{2,2})$

REFERENCES

1. Banks, T., and Peskin, M., SLAC-PUB-3740, July 1985.
2. Siegel, W., *Phys. Lett.*, 151B, 391; 396 (1985).
3. Siegel, W., and Zwiebach, B., Berkeley preprint UCB-PTH-85/ 30 (1985); Itoh, K., Kugo, T., Kunitomo, H. and Ooguri, H., Kyoto preprint KUNS–800 HE(TH) 85/04, August 1985.
4. DeWitt, B. S., *Phys. Rev.*, 160, 1113 (1967); C. W. Misner, Thorne, K. S., and Wheller, J. A., *Gravitation* (W. H. Freeman, San Francisco, 1973).
5. Virasoro, M., *Phys. Rev.*, D1, 2933 (1970).
6. Scherk, J., *Rev. Mod. Phys.*, 47, 123 (1975).
7. Kaku, M., presentation at the Symposium on Anomalies, Geometry, and Topology, Argonne, IL, 1985; Kaku, M. and Lykken, J., CUNY preprint (1985); Kaku, M., CUNY preprint (1985).
8. Brower, R. C., and Thorn, C. B., *Nucl. Phys.*, B31, 183 (1971); Fiegin, B. L., and Fuks, D. B., *Functs. Anal. Prilozhen.* 16, 47 (1982) [*Funct. Anal. and Applic.* 16, 114 (1982)].
9. Banks, T., presentation at the Symposium on Anomalies Geometry and Topology, Argonne, IL (1985).
10. Stueckelberg, E. C. G., *Helv. Phys. Acta..* 11, 225 (1938).
11. Kaku, M. and Kikkawa, K., *Phys. Rev.*, D10, 1110,1823 (1974).
12. Thorn, C. B., and Peskin, M. E., SLAC-PUB-3801.
13. Gates, S. J., Grisaru, M. T., Rocek, M., and Siegel, W., *Superspace.* (Benjamin/Cummings, Reading, MA, 1983).
14. Neveu, A., Schwarz, J., and West, P. C., CERN preprint, TH.4248/85; Neveu, A., and West, P. C., CERN preprint, TH.4200/85; Neveu, A., Nicolai, H., and West, P. C., CERN preprint, TH.4233/85
15. Banks, T., Friedan, D., Martinec, E., Peskin, M., and Preitschopf, C., in preparation.

SEMINARS

3. String Phenomenology

WHAT IS A CALABI-YAU SPACE?

Gary T. Horowitz
Physics Department
University of California
Santa Barbara, CA 93106

ABSTRACT

A pedagogical discussion is given of some of the mathematical tools needed for the investigation of complex manifolds with Ricci flat Kähler metrics.

Recent investigations of possible background configurations for superstrings have shown that the spacetime geometry is highly constrained.[1] If one looks for a configuration of the form $M^4 \times K$ where M^4 is four dimensional Minkowski space and K is a compact Riemannian six manifold, then the only known way of satisfying these constraints is for K to be what is called a Calabi-Yau space. This means:

Definition: A Calabi-Yau space is a compact, three dimensional complex manifold with a Ricci flat Kähler metric.

The mathematical machinery needed for the study of Calabi-Yau spaces include complex manifold theory and algebraic geometry. Since these fields are not yet familiar to many physicists, I will try to give an introduction to the basic concepts involved.[2] I will not go into the details of why these spaces are of interest for superstrings. The mathematics that will be used has obvious generalizations to higher dimensions. However to keep the discussion as concrete as possible, I will restrict myself to six real (or three complex) dimensional manifolds. <u>All manifolds will be assumed to be compact and have only one connected component.</u>

In order to minimize the confusion which might arise from introducing many new definitions, let me begin by mentioning the relation which exists between the concepts we are going to discuss.

Consider the set A of all six dimensional real manifolds. Contained within A lies the subset B of manifolds which admit a complex structure and hence can be viewed as complex three dimensional manifolds. There is a subset $C \subset B$ consisting of those complex manifolds which admit Kähler metrics, and finally a subset $D \subset C$ of manifolds admitting Ricci-flat Kähler metrics. All of these subsets are proper, in the sense that there are manifolds in A but not B, in B but not C, etc. The sets A, B, and C all contain an infinite number of manifolds. However D is believed to be just a finite set. Although the number of manifolds in D is not yet known, a reasonable guess[3] seems to be about 10,000. (Of course the number of phenomenologically interesting Calabi-Yau spaces is considerably smaller.)

Let me begin by reviewing some general properties of real manifolds, and then specialize to complex, Kähler, and finally Ricci flat Kähler spaces.

A. Real Manifolds

Let M be a real six dimensional manifold. We are interested in studying the cohomology of M. This is conveniently described in terms of differential forms. (This is known as de Rham cohomology.) Let x^μ be local coordinates on M and let ω be a p form:

$$\omega = \omega_{\mu\cdots\nu} dx^\mu \wedge \ldots \wedge dx^\nu \qquad (1)$$

where the coefficients $\omega_{\mu\cdots\nu}$ have p indices. The natural derivative operator on forms is the exterior derivative or curl:

$$d\omega = \frac{\partial\omega_{\mu\cdots\nu}}{\partial x^\sigma} dx^\sigma \wedge dx^\mu \wedge \ldots \wedge dx^\nu \qquad (2)$$

So d maps a p form into a p+1 form. Since partial derivatives commute, $d^2 = 0$. We now introduce the following definitions:

Definition: A p form ω is closed if $d\omega = 0$. A p form ω is exact if $\omega = d\alpha$ for some globally defined p-1 form α.

Notice that every exact form and every six form on M is automatically closed.

One can show that on $\mathbb{R}^n$ every closed form is exact. Hence, locally on M, one can express any closed form ω as $\omega = d\alpha$. But this is not in general true globally. The obstruction to doing so is the pth cohomology group*$H^p(M)$:

Definition: $H^p(M) \equiv$ {all closed p forms where two forms ω and ω' are considered equivalent if $\omega-\omega'$ is exact}

---

*This is cohomology over the field of real numbers. We will soon consider cohomology over the complex numbers. One can also define a more primitive notion of cohomology using only integer coefficients.

One often writes this as a quotient:

$$H^p(M) = \frac{\text{closed p forms}}{\text{exact p forms}}$$

For each p , $H^p(M)$ is a real vector space. The dimension of this space is called the pth Betti number b_p. H^0 is just the space of constant functions, so $b_0 = 1$. (This would no longer be true if we considered disconnected manifolds.) The Euler number χ of M is defined to be the alternating sum of the Betti numbers:

$$\chi(M) = \sum_{p=0}^{6} (-1)^p b_p \tag{3}$$

Given a Riemannian metric $g_{\mu\nu}$ on M, one obtains an inner product on the space of p forms:

$$\langle\omega|\tau\rangle = \int \omega^{\mu\cdots\nu}\tau_{\mu\cdots\nu}\sqrt{g}\,d^6x \tag{4}$$

where the indices of ω are raised with the metric. Using this inner product, one can define the adjoint $d^\dagger$ of d which maps p forms to p-1 forms. $d^\dagger$ is just the covariant divergence of the form. The Laplacian is defined to be $\Delta_d = dd^\dagger + d^\dagger d$ and solutions to $\Delta_d\omega = 0$ are called harmonic. Since

$$\langle\omega|\Delta_d\omega\rangle = \|d^\dagger\omega\|^2 + \|d\omega\|^2 \quad , \tag{5}$$

one has immediately that a form is harmonic if and only if it is both curl free and divergence free.

A fundamental theorem in this subject is the following

Hodge Theorem (Real Version)

Every p form ω has a unique decomposition:

$$\omega = \alpha + d\beta + d^\dagger\gamma \tag{6}$$

where α is harmonic, β is a p-1 form and γ is a p+1 form.

Notice that the three terms in the above decomposition are orthogonal with respect to the inner product (4) e.g. $\langle d\beta|d^\dagger\gamma\rangle = \langle d^2\beta|\gamma\rangle = 0$. If ω is closed then the last term must vanish (since $dd^\dagger\gamma = 0$ implies $\langle d^\dagger\gamma|d^\dagger\gamma\rangle = 0$). Thus the Hodge theorem implies that there is a unique harmonic representative for each equivalence class in $H^p(M)$. In other words, given any curl free p form ω, one can add to it the curl of some p-1 form so that the sum is divergence free. Notice that if you change the metric then the harmonic forms will change. However the total number of linearly independent harmonic p forms will not change and will always equal b_p.

If the manifold M with metric $g_{\mu\nu}$ is orientable, then there exists a covariantly constant volume form i.e. a 6 form $v_{\mu\cdots\nu}$ normalized so that $v_{\mu\cdots\nu}v^{\mu\cdots\nu} = 6!$ One can use this form to define the dual of a p form

$$^*\omega_{\mu\cdots\nu} = \frac{1}{p!}\,\omega^{\rho\cdots\sigma}v_{\rho\cdots\sigma\,\mu\cdots\nu} \tag{7}$$

If ω is harmonic, then the 6-p form $^*\omega$ is also harmonic. This implies Poincaré duality: $b_p = b_{6-p}$.

The last concept I wish to review about real manifolds is the holonomy group. Given a connection i.e. a derivative operator on M one defines the holonomy group as follows. Fix a point $p \in M$ and consider any closed curve γ containing p. Take any basis for the tangent space at p and parallel transport it along γ. The result will be a new basis at p which is related to the original one by an element of $GL(6,\mathbb{R})$. Repeating for all closed curves γ, one obtains a subgroup of $GL(6,\mathbb{R})$. This subgroup is independent of the original point p and is called the holonomy group of the connection. If the connection preserves a metric (and the manifold is orientable) then the holonomy group will be a subgroup of SO(6).

We now pause to consider a few examples. Since the Betti numbers are independent of the metric, one can compute them by picking a simple metric and counting the number of harmonic forms with respect to this metric.

1) $\underline{S^6}$ Consider the standard metric of constant curvature. It is not hard to show that with respect to this metric the only harmonic forms are the constant functions and multiples of the volume form. Thus $b_0 = b_6 = 1$, $b_i = 0 \quad 1 \le i \le 5$, and the Euler number is $\chi = 2$. For the connection defined by this metric, the holonomy group is SO(6), since there are no subspaces of the tangent space left invariant under parallel transport.

2) $\underline{S^3 \times S^3}$ Consider the standard metric on each S^3. In addition to $b_0 = b_6 = 1$, one now has $b_3 = 2$ since the volume form on each S^3 is harmonic. The remaining Betti numbers vanish and $\chi = 0$. The holonomy group is $SO(3) \times SO(3)$.

3) $\underline{S^2 \times S^2 \times S^2}$ This is similar to example 2. Consider the standard metric on each S^2. One now has $b_0 = b_6 = 1$, $b_2 = b_4 = 3$, $b_1 = b_3 = b_5 = 0$. Hence $\chi = 8$. The holonomy group is $SO(2) \times SO(2) \times SO(2)$.

4) $\underline{T^6}$ The six torus is not as trivial as one might think. With respect to a flat metric, all the constant forms are harmonic. Hence $b_0 = b_6 = 1$, $b_1 = b_5 = 6$, $b_2 = b_4 = 15$, $b_3 = 20$. This implies $\chi = 0$. The holonomy group of a flat connection is of course just the identity element.

These four examples were not picked completely at random. As we will see, they are simple examples of the four classes of manifolds mentioned earlier: S^6 cannot be viewed as a complex three manifold, $S^3 \times S^3$ is a complex manifold but does not admit any Kähler metric, $S^2 \times S^2 \times S^2$ admits Kähler metrics but not one which is Ricci flat, T^6 admits a Ricci flat Kähler metric.

B. Complex Manifolds

A complex structure on a real manifold M is a tensor field $J^\mu{}_\nu$ satisfying $J^\mu{}_\nu J^\nu{}_\sigma = -\delta^\mu{}_\sigma$ and a certain integrability condition. This integrability condition is precisely what is needed so that one can introduce local complex coordinates z^j on M so that the transition functions between different coordinate patches are holomorphic.* A complex manifold is simply a real manifold with a complex structure.

In order for a complex structure to exist M must clearly be even dimensional. It must also be orientable since the form $dz^1 \wedge d\bar{z}^1 \wedge dz^2 \wedge d\bar{z}^2 \wedge dz^3 \wedge d\bar{z}^3$ does not change sign under holomorphic change of coordinates and hence defines an orientation on M. But not every even dimensional orientable manifold admits a complex structure. In general it is a difficult mathematical problem to determine whether a given real manifold is also a complex manifold. The fact that S^6 does not admit a complex structure was realized relatively recently.[4] It is now known that the only sphere S^n which admits a complex structure is the two sphere S^2 which is the complex projective plane CP^1. One can show that the product of two odd dimensional spheres $S^p \times S^q$ always admits a complex structure. So $S^3 \times S^3$, $S^2 \times S^2 \times S^2$ and T^6 are all complex manifolds.

---

*Given a six dimensional real analytic manifold, one can always let the coordinates become independent complex variables and obtain a six complex dimensional manifold. This is not what is being considered here. Here one obtains a three complex dimensional manifold.

Two complex manifolds M and N are said to be equivalent if there exists a one-to-one, onto map $\varphi: M \to N$ such that when expressed in local complex coordinates, φ and φ^{-1} are holomorphic. It is important to keep in mind that a given real manifold may give rise to inequivalent complex manifolds. In other words, it may admit more than one complex structure. In fact, one can often continuously deform the complex structure. A simple example is given by the torus. Consider the complex plane $\mathbb{C}$ and pick a complex number $z = x+iy$ with $y > 0$. Now take the quotient of $\mathbb{C}$ by the lattice generated by $\vec{e}_1 = (1,0)$, $\vec{e}_2 = (x,y)$. The result is a complex one dimensional manifold T_z. For any z, the underlying real manifold is diffeomorphic to the two torus. However it is easy to show that T_z is equivalent to $T_{z'}$ only if $z' = (az+b)/(cz+d)$ with a,b,c,d integers and $ad-bc = 1$. More generally, on a two sphere with n handles ($n > 1$) there is a $3n-3$ (complex) dimensional space of complex structures.

We now want to repeat our discussion of forms and cohomology for complex manifolds. Consider a three dimensional complex manifold M with local coordinates (z^1, z^2, z^3). The tangent space to M is a <u>six</u> dimensional complex vector space. (This is the complexification of the six dimensional real tangent space of the underlying real manifold.) The cotangent space is spanned by dz^j, $dz^{\bar{j}}$ $j,\bar{j} = 1,2,3$. We define a (p,q) form to be a form which is p-fold linear in the dz^j's and q-fold linear in the $dz^{\bar{j}}$'s:

$$\omega = \omega_{j\ldots k\bar{j}\ldots\bar{k}}\, dz^j\wedge\ldots\wedge dz^k\wedge dz^{\bar{j}}\wedge\ldots\wedge dz^{\bar{k}} \qquad (8)$$

where $\omega_{j\ldots k\bar{j}\ldots\bar{k}}$ has p unbarred and q barred indices.

There are two analogs of the operator d for complex manifolds. The first maps (p,q) forms into (p+1,q) forms and is defined by:

$$\partial\omega = \frac{\partial\omega_{j\ldots k\bar{j}\ldots\bar{k}}}{\partial z^{\ell}}\, dz^{\ell}\wedge dz^{j}\wedge\ldots\wedge d\bar{z}^{\bar{k}} \tag{9}$$

The second maps (p,q) forms into (p,q+1) forms and is just the complex conjugate:

$$\bar{\partial}\omega = \frac{\partial\omega_{j\ldots k\bar{j}\ldots\bar{k}}}{\partial\bar{z}^{\bar{\ell}}}\, d\bar{z}^{\bar{\ell}}\wedge dz^{j}\wedge\ldots\wedge dz^{\bar{k}} \tag{10}$$

Clearly, on any three dimensional complex manifold, ∂ annihilates (3,p) forms and $\bar{\partial}$ annihilates (p,3) forms for any p. Since partial derivatives commute we have $\partial^2 = \bar{\partial}^2 = 0$, $\partial\bar{\partial} + \bar{\partial}\partial = 0$. It is easy to verify that

$$d = \frac{1}{2}(\partial + \bar{\partial}) \tag{11}$$

A form ω of type (p,0) is said to be <u>holomorphic</u> if $\bar{\partial}\omega = 0$. In other words, ω is holomorphic if the coefficient functions $\omega_{j\ldots k}$ are all holomorphic functions of the local coordinates.

The complex cohomology groups are defined in a similar manner to the real case:

<u>Definition</u>: The (Dolbeault) cohomology groups are

$$H^{p,q}_{\bar{\partial}}(M) \equiv \frac{\bar{\partial}\text{ closed }(p,q)\text{ forms}}{\bar{\partial}\text{ exact }(p,q)\text{ forms}}$$

One can show that on $\mathbb{C}^n$ these cohomology groups are trivial, so they again measure global properties of the complex manifold. For each (p,q), $H^{(p,q)}_{\bar{\partial}}$ is a complex vector space.

Given a Hermitian metric $g_{j\bar{k}}$ on M one obtains an inner product on (p,q) forms and can define the adjoints of the operators ∂ and $\bar{\partial}$. $\partial^{\dagger}$ takes the covariant divergence on unbarred indices and $\bar{\partial}^{\dagger}$ takes the covariant divergence on barred indices. One can now introduce two more Laplacians

$\Delta_{\partial} = \partial\partial^{\dagger}+\partial^{\dagger}\partial$ and $\Delta_{\bar{\partial}} = \bar{\partial}\bar{\partial}^{\dagger}+\bar{\partial}^{\dagger}\bar{\partial}$. The powerful Hodge theorem extends to the case of complex forms:

<u>Hodge Theorem (Complex Version)</u>

Every (p,q) form ω has a unique orthogonal decomposition

$$\omega = \alpha + \bar{\partial}\beta + \bar{\partial}^{\dagger}\gamma$$

where $\Delta_{\bar{\partial}}\alpha = 0$, β is a (p,q-1) form, and γ is a (p,q+1) form.

In particular, if $\bar{\partial}\omega = 0$ then the last term vanishes and we again have a unique representative α for each cohomology class $H^{p,q}_{\bar{\partial}}(M)$.

C. <u>Kähler Metrics</u>

For a general complex manifold with Hermitian metric there is no relation between the three Laplacians Δ_d, Δ_{∂}, and $\Delta_{\bar{\partial}}$. However there exist a special class of metrics for which all Laplacians agree. These are called Kähler metrics. To define these metrics, let $g_{j\bar{k}}$ be a Hermitian metric and consider the real (1,1) form:

$$J = i\, g_{j\bar{k}} dz^{j} \wedge dz^{\bar{k}} \qquad (12)$$

<u>Definition</u>: A <u>Kähler metric</u> is a Hermitian metric with $dJ = 0$.

If $g_{j\bar{k}}$ is Kähler, then the closed form J is called the <u>Kähler form.</u>

To develop some intuition for what this definition means, it is convenient to reexpress it in terms of real coordinates on M. Given a complex structure $J^{\mu}{}_{\nu}$, a Hermitian metric is just a Riemannian metric $g_{\mu\nu}$ satisfying $J^{\mu}{}_{\rho} J^{\nu}{}_{\sigma} g_{\mu\nu} = g_{\rho\sigma}$. <u>A Hermitian metric is Kähler if and only if $J^{\mu}{}_{\nu}$ is covariantly constant with respect to the connection</u>

defined by $g_{\mu\nu}$. Thus, Kähler metrics are the "nicest" class of metrics on a complex manifold in that the Riemannian structure is compatible with the complex structure.

A simple class of examples of Kähler metrics is furnished by one dimensional complex manifolds. Since dJ is a three form, it follows immediately that every Hermitian metric on a one dimensional complex manifold is Kähler. Of course in higher dimensions this is no longer the case. For any Hermitian metric on a three dimensional complex manifold the volume form V is related to the (1,1) form J Eq. (12) by:

$$V = \frac{1}{3!} J \wedge J \wedge J \tag{13}$$

For a Kähler metric J is closed and hence defines an element of $H^2(M)$. This cohomology class must be non-trivial, for if $J = d\alpha$, then $\int_M J \wedge J \wedge J = 0$ by Stokes' theorem. Similarly, $J \wedge J$ defines a non-trivial cohomology class in $H^4(M)$. Thus we learn that in order for a manifold to admit a Kähler metric, the even Betti numbers must satisfy $b_{2p} \geq 1$. This shows immediately that $S^3 \times S^3$ (which has $b_2 = 0$) does not admit any Kähler metric.

As mentioned earlier, one of the most important properties of Kähler metrics is the following:

Theorem: On a manifold with a Kähler metric

$$2\Delta_d = \Delta_\partial = \Delta_{\bar\partial} \tag{14}$$

This means that a form which is curl free and divergence free with respect to barred indices is also curl free and divergence free with respect to unbarred indices. In particular, consider a holomorphic p form $\bar\partial\omega = 0$. Since ω has no barred indices, $\bar\partial^+\omega = 0$ and hence $\Delta_{\bar\partial}\omega = 0$. By the above theorem, $\Delta_d\omega = 0$. Therefore we have the somewhat surprising result that holomorphic forms are automatically harmonic with respect to any Kähler metric. Conversely, $\Delta_d \omega = 0$ implies $\bar{\partial}\omega = 0$, so every harmonic (p,0) form is holomorphic.

The (complex) dimension of $H^{p,q}_{\bar{\partial}}(M)$ is called the <u>Hodge number</u>* $h^{p,q}$. There is a convenient way of summarizing the cohomology of a manifold which admits a Kähler metric in terms of the <u>Hodge diamond</u>:

$$
\begin{array}{ccccccc}
 & & & h^{3,3} & & & \\
 & & h^{3,2} & & h^{2,3} & & \\
 & h^{3,1} & & h^{2,2} & & h^{1,3} & \\
h^{3,0} & & h^{2,1} & & h^{1,2} & & h^{0,3} \\
 & h^{2,0} & & h^{1,1} & & h^{0,2} & \\
 & & h^{1,0} & & h^{0,1} & & \\
 & & & h^{0,0} & & &
\end{array}
$$

This diagram has the following properties:

(1) The sum of the Hodge numbers along the nth row is the nth Betti number:

$$b_n = \sum_{p+q=n} h^{p,q} \tag{15}$$

Notice that this is true even though b_p is the real dimension of $H^p(M)$ and $h^{p,q}$ is the complex dimension of $H^{p,q}_{\bar{\partial}}(M)$. (See property 2.)

---

* In some recent physics papers these numbers have been denoted $b_{p,q}$. Here we adopt the more standard mathematical notation.

(2) Complex conjugation maps (p,q) forms into (q,p) forms. Since $\Delta_{\partial} = \Delta_{\bar{\partial}}$ we know that ω is harmonic if and only if $\bar{\omega}$ is. Hence $h^{p,q} = h^{q,p}$ i.e. the diagram is symmetric about the verticle line.

(3) The volume form is a (3,3) form. Since the dual of a harmonic form is also harmonic, $h^{p,q} = h^{3-p,3-q}$. This implies that the diagram is symmetric about the center point.

Properties (1) and (2) give another topological restriction (in addition to $b_{2p} \geq 1$) on the existence of a Kähler metric: b_{2n+1} must be even. Properties (2) and (3) show that only 6 of the Hodge numbers are independent. We will see in the next section that the requirement of a Ricci flat Kähler metric reduces this number even further.

D) <u>Ricci Flat Kähler Metrics</u>

The curvature tensor of a Kähler metric takes a simple form. One can show that the only non-vanishing Christoffel symbols are $\Gamma^j{}_{k\ell}$ and their complex conjugates $\Gamma^{\bar{j}}{}_{\bar{k}\bar{\ell}}$ where

$$\Gamma^j{}_{k\ell} = g^{j\bar{m}} g_{k\bar{m},\ell} \qquad (16)$$

and the only non-vanishing components of the curvature tensor are

$$R^j{}_{k\ell\bar{m}} = -\Gamma^j{}_{k\ell,\bar{m}} \qquad (17)$$

and those related by symmetry and complex conjugation. Because of this, the Riemann tensor has an extra symmetry:

$$R^j{}_{k\ell\bar{m}} = R^j{}_{\ell k\bar{m}} \qquad (18)$$

It follows from (16) and (17) that the Ricci tensor of a Kähler metric can always be written locally as

$$R_{j\bar{k}} = \frac{-\partial^2(\ell n \det g)}{\partial z^j \partial \bar{z}^{\bar{k}}} \tag{19}$$

Therefore the Ricci form:

$$R \equiv iR_{j\bar{k}} dz^j \wedge d\bar{z}^{\bar{k}} \tag{20}$$

is closed $dR = 0$ and defines an element of $H^2(M)$. This cohomology class turns out to be independent of the Kähler metric you start with and is called the <u>first Chern class</u> <u>c_1</u>. (More generally, c_1 can be defined for any complex vector bundle. Here we are considering the tangent bundle of a complex manifold.) It is obvious that if $c_1 \neq 0$ i.e. the Ricci form is not exact, then there cannot exist a Ricci flat metric. The converse is far from obvious. However, Calabi has shown uniqueness[5] and Yau has proved existence[6] of a Ricci flat metric whenever $c_1 = 0$. (Hence the name Calabi-Yau spaces.*) More precisely, they proved:

<u>Theorem</u>: Given a complex manifold with $c_1 = 0$ and any Kähler metric $g_{j\bar{k}}$ with Kähler form J [Eq. (12)], then there exists a unique Ricci flat Kähler metric $\hat{g}_{j\bar{k}}$ whose Kähler form $\hat{J}$ is in the same cohomology class as J.

Notice that for most physical applications one should not think of the Ricci flat metric as unique. One can change this metric either by changing the cohomology class of J or by changing the original complex structure.

A useful property of manifolds with $c_1 = 0$ is the following: <u>$c_1 = 0$ if and only if there exists a non-vanishing holomorphic three form.</u> (For a simple proof, see Ref. [7].) This three form is in fact covariantly

---

*A complex manifold with $c_1 = 0$ is sometimes called a Calabi-Yau <u>manifold</u>.

constant with respect to a Ricci flat metric. Since $S^2 x S^2 x S^2$ has $b_3 = 0$, it has no harmonic three forms and hence no holomorphic three forms. It then follows that it cannot admit a Ricci flat Kähler metric.

If $c_1 = 0$, the Hodge numbers acquire an extra symmetry. By taking the dual of a harmonic (p,0) form using the covariantly constant three form, one obtains a harmonic (0,3-p) form. (Recall that lowering an index with a Hermitian metric changes its type.) Hence:

(4) If $c_1 = 0$, then $h^{p,0} = h^{0,3-p}$

Notice that if one tries to extend this to other harmonic forms, then one finds that the "dual" of a (p,q) form with $q \neq 0$ is symmetric in some of its indices and hence does not even define a form.

Property (4) reduces the six independent Hodge numbers to just four which we can take to be $h^{0,0}$, $h^{1,0}$, $h^{1,1}$, $h^{2,1}$. However $h^{0,0}$ is just the dimension of the space of constant functions and hence $h^{0,0} = 1$. $h^{1,0}$ usually is also trivial for the following reason. On any Riemannian manifold, one can express the Laplacian on p forms in terms of the covariant derivative ∇_μ and curvature of the metric. (This is called the Weitzenbock identity):

$$\Delta_d = -\nabla^2 + \text{curvature terms} \tag{21}$$

For a one form, the curvature term involves only the Ricci tensor. This shows that on a Ricci flat manifold, every harmonic one form must be covariantly constant and hence non-vanishing. On the other hand, a manifold with $\chi \neq 0$ does not admit any nowhere vanishing one forms. Therefore $\chi \neq 0$ implies $b_1 = 0$ which implies $h^{1,0} = 0$. One can also show that $\chi \neq 0$ implies $\pi_1(M)$ is finite. So there exists a compact simply connected covering space.

We have arrived at the following simple result: The cohomology of a Calabi-Yau space with $\chi \neq 0$ is characterized by two integers $h^{1,1}$ and $h^{2,1}$. In particular, the Euler number is $\chi = 2(h^{1,1}-h^{2,1})$. It turns out that $h^{2,1}$ gives the (complex) dimension of the space of complex structures that can be placed on M.* In terms of the compactification of the superstring discussed in [1], $h^{2,1}$ and $h^{1,1}$ give the number of families and anti-families of massless fermions.

Finally, we consider the holonomy group of a Hermitian metric on a complex manifold. If the metric is not Kähler, then the holonomy group will in general be SO(6). This is because the complex structure is not preserved under parallel transport. If the metric is Kähler, then the complex structure is preserved and the holonomy group is contained in U(3). If the metric is Kähler and Ricci flat, then the holonomy group is even further restricted. This can be seen as follows. The Lie algebra of the holonomy group is given by parallel transport around infinitesimal loops.** The change in a vector is then given by the curvature tensor. For a Kähler metric, the change is $R^j{}_{k\ell\bar{m}}v^\ell v^{\bar{m}}$ which are generators of U(3). If we take the trace of these generators and use (18) we obtain:

$$R^k{}_{k\ell\bar{m}} = R^k{}_{\ell k\bar{m}} = R_{\ell\bar{m}} \qquad (22)$$

Thus the holonomy group is contained in SU(3) if and only if the metric is Ricci flat and Kähler.

*In general, the Hodge numbers depend on the complex structure. However for compact manifolds admitting Kähler metrics, these numbers do not change under continuous deformations of the complex structure.

**This must be considered at each point of the manifold.

We conclude with a few more examples:

1) $\underline{\mathbb{C}P^3}$ The complex projective spaces admit Kähler metrics, and have the smallest possible Hodge numbers consistent with this fact: $h^{p,q} = \delta_{p,q}$. Since $h^{3,0} = 0$, there is no holomorphic three form and hence $c_1 \neq 0$. This shows that $\mathbb{C}P^3$ does not admit a Ricci-flat Kähler metric.

2) Consider the submanifold $Y_{4,5}$ of $\mathbb{C}P^4$ defined by

$$\sum_{i=1}^{5} z_i^5 = 0 \qquad (23)$$

where z_i are homogeneous coordinates. Let $y_i = z_i/z_5$ for $i = 1,\cdots,4$. Then the holomorphic three form

$$dy_1 \wedge dy_2 \wedge dy_3/y_4^4 \qquad (24)$$

is easily shown to be non-singular and non-vanishing everywhere on $Y_{4,5}$. Hence $Y_{4,5}$ has $c_1 = 0$ and admits a Ricci flat Kähler metric. One can show that $h^{2,1} = 101$ and $h^{1,1} = 1$, so $\chi = -200$. Since $h^{1,1} = 1$, there is a one dimensional space of cohomology classes of the Kahler form J.* In terms of the unique Ricci-flat metric guaranteed by Calabi and Yau's theorem, this corresponds to an overall constant rescaling of the metric. Since $h^{2,1} = 101$, there is a 101 (complex) dimensional space of complex structures on this manifold. This yields a 203 (real) dimensional space of Ricci flat metrics. It is unfortunate that not one of them is known explicitly.

---

*$h^{1,1} = 1$ means that there is a one <u>complex</u> dimensional space of harmonic (1,1) forms. However, since the Kähler form is real, one has only a one real dimensional space of possible cohomology classes for J.

The different complex structures on this manifold <u>are</u> known explicitly and can be realized as follows. Consider the general fifth order homogeneous polynomial in $\mathbb{C}^5$:

$$f(z^i) = A_{ijk\ell m} z^i z^j z^k z^\ell z^m . \qquad (25)$$

The equation $f = 0$ defines a smooth submanifold M_A of $\mathbb{C}P^4$ provided $df \neq 0$ whenever $f = 0$. One can show that two submanifolds M_A and $M_{A'}$ are always diffeomorphic but have different complex structures unless the tensors A and A' are related by a GL(5,$\mathbb{C}$) transformation. Since there is a 126 (complex) dimensional space of symmetric tensors A and dim GL(5,$\mathbb{C}$) = 25 we obtain 126-25 = 101 inequivalent complex structures.

Acknowledgments

This work was supported in part by NSF Grant PHY 85-06686 and by the Alfred P. Sloan foundation.

References

1) P. Candelas, G. Horowitz, A. Strominger, and E. Witten, Nucl. Phys. B258 (1985) 46; to appear in the proceedings of the Symposium on Anomalies, Geometry and Topology, Argonne IL. March (1985).

2) For a more detailed discussion of the material covered here see P. Griffiths and J. Harris, Principles of Algebraic Geometry (Wiley-Interscience, 1978) Chapter 0; S. Kobayashi and K. Nomizu, Foundations of Differential Geometry, Vol. II (Wiley, 1969) Chapter 9.

3) S.-T. Yau, private communication.

4) A. Adler, Amer. J. Math. 91 (1969) 657.

5) E. Calabi, in Algebraic Geometry and Topology: A Symposium in Honor of S. Lefschetz (Princeton University Press, 1957).

6) S.-T. Yau, Proc. Natl. Acad. Sci. 74 (1977) 1798.

7) A. Strominger and E. Witten, Comm. Math. Phys. to appear.

TOPOLOGY OF SUPERSTRING COMPACTIFICATION

Andrew Strominger

*The Institute for Advanced Study**
Princeton, NJ 08540

and

*Institute for Theoretical Physics***
University of California
Santa Barbara, CA 93106

ABSTRACT

Calabi-Yau manifolds obtained by dividing the torus by a discrete symmetry group are studied, and some new manifolds are constructed. The simplicity of these manifolds allows for a detailed understanding of topological aspects of superstring compactification. We explain in general and by way of explicit example how to compute the intersection matrix which determines the couplings of the low-energy field theory. It is found that generically most of the Yukawa couplings vanish for topological reasons.

Over the last year, there has been extensive investigation of superstring theories[1] compactified on Calabi-Yau manifolds,[a] So far, these are the only known compactifications consistent with the superstring equations of motion. The general tree level features of these compactifications are in remarkable agreement with observation.[2] Investigations of more detailed tree level predictions (which vary from manifold to manifold) such as particle masses, lifetimes and unification scales has uncovered a number of potentially severe phenomenological difficulties.[3] All of these quantities depend on the topology of the particular Calabi-Yau manifold used for compactification. The difficulties might be entirely absent for a Calabi-Yau manifold with appropriate topological characteristics. Progress in superstring phenomenology thus hinges on an increased understanding of these manifolds.

[†] Talk presented at The Santa Barbara Workshop on **Unified String Theories**, August 1985.
*Permanent address
**Present address
[a] *i.e.*, Kahler manifolds with vanishing first Chern class.

Of course it is also crucial that non-perturbative and quantum properties of these compactifications be understood. The associated problems of supersymmetry breaking and the cosmological constant are particularly vexing. At present, there is no good reason to believe that supersymmetry can be broken in a fashion that generates a realistic hierarchy,[4] or that the cosmological constant will be zero after supersymmetry breaking; but our understanding of these issues is far from complete. I will not discuss these issues in this talk. However it is worth noting that it is precisely the topological features of Calabi-Yau compactifications (discussed here) that one does expect will extend beyond the weak coupling regime (where they are derived) and be relevant in the full string theory — even if it is not weakly coupled. An example of this is the spectrum of massless particles. For large Calabi-Yau manifolds, (which means that the associated sigma model of the string theory is weakly coupled), the field theory approximation is valid. It is then easy to see that the spectrum of massless particles in four dimensions is given by the index of various differential operators on the Calabi-Yau manifold.[2] Masslessness even for a strongly coupled sigma model, where the field theory approximation is not valid, is still guaranteed by unbroken world sheet supersymmetry.[b] Thus topologically derived results have a tendency to retain validity outside the domain where they are easily derived.

In Ref. 5 we studied Calabi-Yau manifolds with negative Euler number constructed from algebraic varieties. This work was extended in Ref. 6, where it was shown that the entire low energy action can be computed topologically in the field theory limit. A general discussion of Calabi-Yau manifolds appears in Ref. 7, and a standard reference for the relevant mathematics (algebraic geometry) in Ref. 8.

In this talk, I will give a somewhat pedagogical discussion of the construction of a particularly interesting class of Calabi-Yau manifolds not previously investigated. These manifolds have positive Euler number (unlike those of Ref. 5) and are constructed by dividing the torus by discrete symmetries with fixed points and repairing the singularities. I will describe some explicit examples and compute the Yukawa couplings topologically. The interest in these manifolds lies in their simplicity. As we shall see, it is possible to obtain a physical picture of their topology. In a certain (orbifold or V-manifold) limit, they have a metric which is flat everywhere except at singularities of complex dimension zero or one. One can obtain an approximation to the Ricci-flat metric,[9] and to the full string theory,[10] by expanding around this limit.

[b]The effective potential for the massless fields associated with moduli is the logarithm of sigma model partition function as a function of those moduli. The Witten index, being non-zero for cases of interest, guarantees that world sheet supersymmetry is unbroken (even nonperturbatively). The partition function is therefore unity for any value of the moduli, the effective potential is flat, and the particles remain massless.

The basic idea in the construction of these manifolds can be understood by considering an example in one complex dimension. Consider the torus (T^2) obtained by identification of the complex line[c] (C^1):

$$z \simeq z + i$$

$$z \simeq z + 1$$

where z is a complex coordinate on C^1. This manifold has a discrete Z_2 reflection symmetry

$$R(z) = -z\,.$$

R has four fixed points, *i.e.*, points that are mapped into themselves under the action of R. These are $z = \{f_k\}$ where f_k is 0, $\frac{1}{2}$, $\frac{i}{2}$ or $\frac{1+i}{2}$ as illustrated in Fig. 1. The quotient space T^2/R therefore has four conical singularities and is not a smooth manifold.

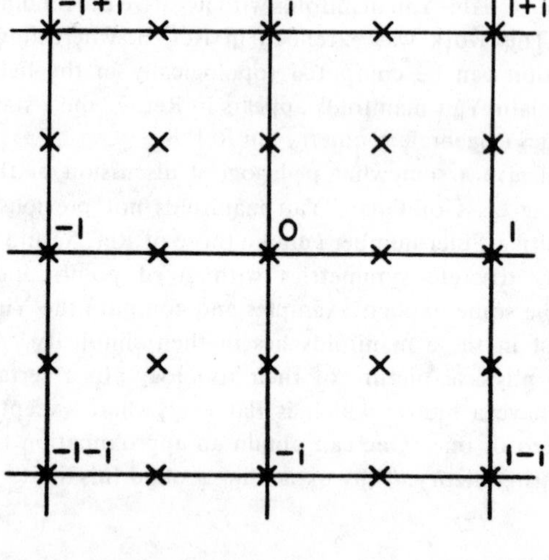

Figure 1: The two torus obtained by the identification $z \simeq z + 1$; $z \simeq z + i$ of the complex plane has four fixed points of the reflection $R(z) = -z$. They are located at $f_i = 0, \frac{1}{2}, \frac{i}{2}$ or $\frac{1+i}{2}$ and denoted by the x's.

[c]It is standard (and sometimes confusing) practice in the study of complex manifolds to refer to one complex dimensional objects as line or curves and two complex dimensional objects as surfaces.

Such objects are known as orbifolds or V-manifolds. Because R is order two, there is a deficit angle of π around each singularity. T^2/R can be visualized as two flat squares glued together at the edges.

A smooth manifold can be obtained by repairing the singularity. We simply cut out the singularity and glue in a smooth disc. The Euler number of the resulting manifold can be computed by an algorithm that generalizes to higher dimensions. Removing four discs from T^2 gives a manifold with Euler number $\chi(T^2 - \{f_k\}) = -4$ since a disc has Euler number one. $(T^2 - \{f_k\})/R$ then has half the Euler number since R acts freely (without fixed points) on $T^2 - \{f_k\}$. Gluing back in the four discs adds four to χ. The final result has $\chi = 2$ and is therefore a sphere. Thus a sphere can be obtained by identifying a torus under reflections and repairing the singularities.

In superstring theory we are interested in manifolds that admit Ricci flat metrics which, by Yau's proof[11] of the Calabi conjecture,[12] implies that c_1 (the first Chern class[d]) must vanish. T^2 has $c_1 = 0$ and admits a Ricci flat metric. On the two sphere (CP^1), c_1 is not zero and is a topological obstruction to obtaining a Ricci flat metric. Thus the above construction did not preserve c_1. When is c_1 preserved? A useful fact is that c_1 is zero, in n complex dimensions, if and only if there exists an everywhere non-vanishing, non singular holomorphic n form (denoted ω).[5,8] On T^{2n}, $\omega = dz_1 \wedge \ldots \wedge dz_n$. The reflection R maps dz_1 into $-dz_1$, so $\omega = dz_1$ does not live on the quotient space T^2/R and it does not have $c_1 = 0$.

In one dimension T^2 is the only manifold with $c_1 = 0$, so new Calabi-Yau manifolds cannot be obtained by dividing by discrete symmetries. This is not the case in higher dimensions. In two complex dimensions the singularities of any discrete action D that preserves ω (*i.e.*, acts as a discrete subgroup of $SU(2)$) can be "blown up" (*i.e.*, repaired) to yield a manifold with $c_1 = 0$. (The famous $K3$ surface is simply T^4 divided by reflections with blown up singularities.) Of course the topology of the object one must glue in at the singularity is more complicated than in one dimension. An analogous result is conjectured to hold in three complex dimensions, but it has only been shown for the case when D is an Abelian discrete subgroup of $SU(3)$.[13]

A classic three dimensional example is the Z manifold mentioned in Ref. 2. One begins with the three torus obtained by identifying

$$z_k \simeq z_k + 1$$

$$z_k \simeq z_k + \alpha \ , \qquad \alpha = e^{2\pi i/6}$$

where z_k are complex coordinates on C^3. This has the discrete symmetry

$$T(z_1, z_2, z_3) = (\alpha^2 z_1, \alpha^2 z_2, \alpha^2 z_3)$$

[d] A discussion for physics of the first Chern class is given in G. Horowitz (these proceedings) and Refs. 2, 5.

T preserves $\omega = dz_1 \wedge dz_2 \wedge dz_3$ since $\alpha^6 = 1$. T has 27 fixed points at

$$z = \frac{(1+\alpha)}{3}(n_1, n_2, n_3)$$

where $n_i = 0, 1, 2$, as illustrated in Fig. 2. Using the fact that the object used to repair a Z_3 singularity has Euler number 3, $\chi(T^6/T)$ can be computed by the same method that we computed $\chi(T^1/R)$. The answer is $\chi(T^6/T) = 72$.

It is interesting to note that Z has a Z_3 symmetry that acts freely[e]

$$Q(z_k) = \alpha^{2k} z_k + \frac{1}{3}(1+\alpha)$$

so one may obtain a new manifold with Euler number 24 and fundamental group Z_3. The 27 "bolts" (*i.e.*, regions where a singularity was repaired) of Z fall into 9 multiplets of 3 whose elements are cyclically permuted by the action of Q.

As derived in Ref. 2 and elaborated in Refs. 5, 14, superstring compactification on Z will lead to $\frac{\chi}{2} = 36$ families of quarks and leptons. These are in one to one correspondence with zero modes of the relevant Dirac operator on Z. These in turn are given by elements of H^2, the 36 cohomology classes of closed two forms.

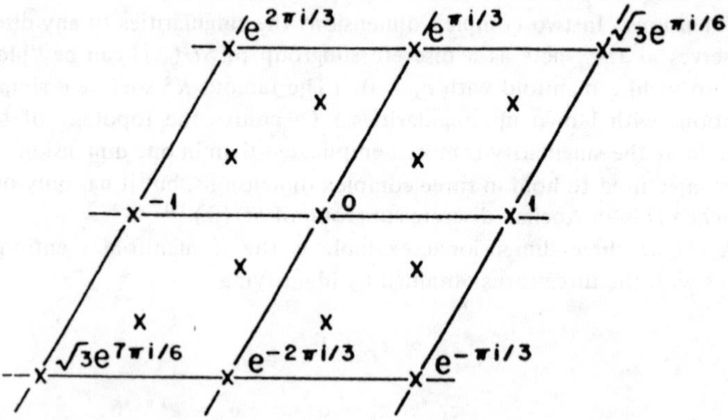

Figure 2: The two torus obtained by the identification $z \simeq z+1$; $z \simeq z+\alpha$, where $\alpha = e^{2\pi i/6}$, of the complex plane has three fixed points of the Z_3 action $T(Z) = \alpha^2 z$. They are located at 0, $\frac{1}{3}(1+\alpha)$ and $\frac{2}{3}(1+\alpha)$ and denoted by x's.

[e] This was found in collaboration with Don Page.

The Yukawa couplings between the ith, j^{th} and k^{th} families are given by[5]

$$g(F_i, F_j, F_k) = \int F_i \wedge F_j \wedge F_k$$

$$1 \leqslant i, j, k, \leqslant 36$$

where F_i is a closed two form in the cohomology class associated with the ith family. If the F_i are topologically normalized as generators of integral cohomology, this will be an integer. In such a basis the normalization matrix of family kinetic terms is not diagonal, but is given by[6]

$$N(F_i, F_j) = 3g(F_i, J, J)g(F_j, J, J)/8g(J, J, J) - g(F_i, F_j, J)/4$$

where J is the Kahler form. J can be expanded in terms of the F_i as

$$J = \sum_{i=1}^{36} \phi_i F_i$$

where ϕ_i are the 36 massless scalar fields associated with the 36 zero modes or "moduli", of the metric on Z.

Extracting the low energy tree level physics from superstring compactification on Z is this reduced to the topological problem of computing the "intersection matrix", $g(F_i, F_j, F_k)$ of Z. To evaluate this matrix, we will exploit the isomorphism between deRham cohomology (classes of closed forms) and homology (classes of non-contractible surfaces). This isomorphism is given by the inner product on H^2 (the group of closed two forms modulo exact forms) cross H_2 (the group of closed two surfaces modulo boundaries of three surfaces):

$$\langle S | \phi \rangle = \int_S \phi$$

where S is an element of H_2 and ϕ an element of H^2. This inner product depends only on the (co)homology class in $(H^2)H_2$. The deRham theorem states that the matrix $\langle S | \phi \rangle$ can be diagonalized, and one can thereby associate with every closed two form a linear combination of two surfaces. On orientable manifolds there is a further isomorphism, in n dimensions, between H_2 and H_{n-2} known as Poincare duality. This is given by the intersection number on $H_2 \otimes H_{n-2}$:

$$\langle S | S' \rangle = S \cap S'$$

where $S \in H_2$ and $S' \in H_{n-2}$, defined as follows. In n dimensions, an n surface and an $n - 2$ surface generically intersect in points. A sign can be associated with each intersection point by comparing the orientation defined by the wedge product of the volume forms of S and S' with the orientation of the n dimensional manifold. The sum of the intersection points, weighted by these signs, is a topological invariant known as the intersection number. In six dimensions, using both the inner product on $H^2 \otimes H_2$ and $H_2 \otimes H_4$, one thus has an isomorphism relating two forms to four surfaces.

Under this last isomorphism, the formula for the intersection matrix becomes the oriented number of points of common intersection of three non-contractible four surfaces[8]

$$g(F_i, F_j, F_k) = F_i \cap F_j \cap F_k$$

where F_i is now the four surface associated with the ith family. The self intersection of a surface is defined as the oriented intersection of two generic surfaces in the same homology class.

A convenient topological basis for the thirty-six four surfaces may be described as follows. In the process of repairing the singularities, the 27 fixed points are "blown up" to CP^2's. (Hence the name "blowing up.") This gives 27 non-contractible four surfaces (known as exceptional divisors) located at the 27 bolts $\frac{1}{3}(1 + \alpha)(n_1, n_2, n_3)$ for $n_i = 0$, 1 or 2. We denote these B_i. In addition, the original T^6 has 15 homology classes of four real dimensional tori. Of these, only nine have smooth, orientable, non-contractible images in Z. These are described by the nine equations, *e.g.*, $z_i + \alpha z_j = \frac{1}{2}$ and will be denoted T_{ij}.

The self-intersection of one of the tori T_{kj} is computed by taking two tori in the same homology class, *e.g.*, $z_1 + \alpha z_2 = \frac{1}{2}$ and $z_1 + \alpha z_2 = \frac{3}{8}$. These are in the same homology class because they bound the five manifold described by

$$z_1 + \alpha z_2 = \frac{3t}{8} + (1 - t)\,\frac{1}{2}\ ; \qquad 0 \leqslant t \leqslant 1\,.$$

Obviously the two equations have no simultaneous solution and the self intersection of the T_{kj}'s vanish. Non-zero intersection numbers involve three inequivalent T_{kj}'s, and is just the number of solutions of the three relevant linear differential equations. This is easily seen to be one for all cases. The T_{kj}'s do not go near any of the bolts so intersections involving a T_{kj} and a B_j vanish. Similarly as the bolts are separated there are no intersections involving inequivalent B_i.

Calculation of the triple self-intersection number of one of the B_i is more involved. Consider a general D-1 complex dimensiona holomorphic submanifold H embedded in a D dimensional complex manifold M. This is described by the holomorphic mapping $z^i(\rho_k)$ where z^i are coordinates on M and ρ_k are coordinates on H. The normal bundle of H, denoted $N(\rho)$, is the pullback to H of that part of the tangent bundle of M which is normal to H. Since the dimension of H is one less than that of M, N is a line bundle. A holomorphic deformation of H in M is a non-singular holomorphic section of the normal bundle N. The zero locus of this section will be the intersection of H with its holomorphic deformation, *i.e.*, the self-intersection of H in M. The zeros will always be D-2 dimensional holomorphic submanifolds, and one may have multiple zeros as well as simple zeros. A zero of order N on a D-2 dimensional submanifold I means that th self-intersection of H contains n submanifolds in the homology class of I. There remains an overall sign to be determined by comparing the orientation induced on I from H and its holomorphic deformation with the orientation induced from M. These are in fact equal because H, its deformation and I all carry a complex structure induced from M which determines the orientation.[8]

It is not always possible to deform a holomorphic surface in a holomorphic manner. The surfaces B_i in Z in fact cannot be deformed holomorphically. In this case there are only meromorphic sections of the normal bundle *i.e.*, sections with isolated singularities or dimension D-2. Such a section cannot be regarded as a small deformation of H. However it can be used to obtain a non-holomorphic deformation of H in M. A singularity S of a section of the normal bundle will be a pole of order n which can be written, in local coordinates near S

$$N(\rho) = \frac{1}{\rho^m} \cdot$$

To obtain a non-holomorphic deformation we alter the section near S letting

$$\begin{aligned} N(\rho) &= \frac{1}{\rho^m} \quad , \quad |\rho| \geqslant 1 \\ &= \rho^{*m} \quad , \quad |\rho| \leqslant 1 \,. \end{aligned}$$

This a holomorphic pole of order m may be replaced by an anti-holomorphic zero of order m. This means that the self-intersection of H contains m times S. The orientation induced by H and its deformation will be opposite to the orientation induced on S from M since the deformation of H is antiholomorphic in the neighborhood of the intersection. Thus the self-intersection contains $-mS$, with S regarded as an analytic submanifold of M.

In general a meromorphic section of N contains zero loci I_j with zeros of order n_j and singularities S_j with poles of order m_j. The self-intersection of H in M is then

$$H \cap H = \Sigma n_j I_j - \Sigma m_k S_k \,.$$

The right-hand side of this equation is called the *divisor* of the section of N. It is closely related to the first Chern class (c_1) of the normal bundle N — in physicist terms, the $U(1)$ field strength of the line bundle N. c_1 may be represented in cohomology by an element of H^2, or in homology as a surface in H_2. Poincaré duality then relates c_1, as an element of H_2, to an element of H_4. It is a basic fact of algebraic geometry that the linear combination of surfaces Poincaré dual to c_1 is precisely the above linear combination associated with the zeros and poles of meromorphic sections of N.[8]

To apply this to deformations of the B_i, we need to know the first Chern class $c_1(N)$ of the normal bundle of B embedded in the Z manifold. $c_1(T)$ of the tangent bundle of B_i is $3J$, where J is a CP^1 in B_i, since B_i is a CP^2. The total $c_1(z)$ is zero for a Calabi-Yau space. Since $c_1(z) = c_1(T) + c_1(N)$, $c_1(N)$ is $-3J$. Now $J \cap J = 1$ in CP^2, so J is Poincaré self dual in B_i. $c_1(N)$ thus equals the self-intersection of B_i in M,

$$B_i \cap B_i = -3J .$$

The triple self-intersection of B_i in M is then the self-intersection of the self-intersection $B_i \cap B_i$ in B_i

$$\begin{aligned} B_i \cap B_i \cap B_i &= (-3J) \cap (-3J) \quad (\text{in } B) \\ &= 9 \end{aligned}$$

since $J \cap J = 1$ in B_i.

In summary

$$g(T_{ik}, B_j, B_k) = 0$$

$$g(T_{ik}, T_{jl}, B_k) = 0$$

$$\begin{aligned} g(T_{il}, T_{jm}, T_{kn}) &= 1 \quad \text{for } il \neq jm \;, \quad jm \neq kn \;\text{ and }\; il \neq kn \\ &= 0 \quad \text{otherwise} \end{aligned}$$

$$\begin{aligned} g(B_i, B_j, B_k) &= 9 \quad \text{for } i = j = k \\ &= 0 \quad \text{otherwise} . \end{aligned}$$

Note that of 8,436 *a priori* independent Yukawa couplings all but 111 vanish! Thus it is not unreasonable to hope that couplings which must vanish for phenomenological reasons (such as proton decay), will vanish for topological reasons in some superstring compactification.

Given a choice of Kahler form on Z, one may now easily compute the normalization matrix. The answer is particularly simple in the orbifold limit. An orbifold corresponds to choosing a metric which is flat everywhere at the price of singularities near the bolts. For the Z manifold this is the metric induced from the flat metric on the original T^6. The Kahler form of the Z orbifold is also the Kahler form induced from T^6, namely

$$J = D \sum_{i=1}^{3} dz_i \wedge d\bar{z}_i$$

where D is a dilaton field. The associated linear combination of four surfaces is

$$J = D \sum_{i=1}^{3} T_{ii} .$$

The normalization matrix is then well behaved for the T_{kj}'s but vanishes for the B_i's. This means that in an orthonormal basis, the Yukawa couplings are divergent in the orbifold limit! Whether this is an artifact of the field theory approximation or if the full string theory on the Z orbifold is sick is not understood at present. (Indeed, the relation between physical picture obtained by direct analysis of string theory on orbifolds[10] and the picture discussed here is intriguing and ill-understood.) This difficulty occurs whenever the singularities are point-like. It does not occur in examples with singularities of complex dimension one, to which we now turn.

We begin with the complex torus obtained by the identifications $z_k \simeq z_k + 1$, $z_k \simeq z_k + i$ for each of three complex coordinates z_k. Now consider the Z_2 group generated by the reflection

$$A(z_1, z_2, z_3) = (z_1, -z_2, -z_3) .$$

This discrete symmetry preserves the holomorphic three form $dz_1 \wedge dz_2 \wedge dz_3$ which implies that there can only be fixed complex curves, as opposed to points or complex surfaces. This can be seen by locally decomposing the three form near the fixed point set as a wedge product of forms in the conormal (*i.e.*, those perpendicular to the fixed point set) and cotangent bundles of the fixed point set. The components in the cotangent bundle will be invariant under the Z_2 action, while those in the conormal bundle will acquire a minus (plus) sign if the complex dimension of the conormal bundle is odd (even).

A has 16 fixed complex curves in T^6 described by (z_1, f_i, f_j) where f_i is one of $0, \frac{1}{2}, \frac{i}{2}$ or $\frac{1+i}{2}$. One can construct a new Calabi-Yau manifold by first removing these curves, dividing the resulting manifold by A and then gluing back in the product of the fixed curve with the Eguchi-Hanson geometry,[15] *i.e.*, blowing up the singularity. The result is simply $K3 \times T^2$, and has Euler character $\chi(T^6/A) = 0$.

We now divide by a further Z_2

$$D(z_1, z_2, z_3) = (-z_1, z_3, z_2) .$$

This has 8 fixed curves in T^6/A described by (f_i, z, z) and $(f_i, z, -z)$. The topology of these curves can be determined as follows. Consider for example the curve $z_1 = 0$ and $z_2 = z_3$. This curve intersects the bolts of T^6/A (*i.e.*, the regions where the singularity was repaired) at four points with coordinates $(0, 0, 0)$, $(0, \frac{1}{2}, \frac{1}{2})$, $(0, \frac{i}{2}, \frac{i}{2})$ and $(0, \frac{(1+i)}{2}, \frac{(1+i)}{2})$. For ease of visualization, consider the case where the metric is nearly flat everywhere except very near the bolts. One can then define a deficit angle around each bolt in the curve $(0, z, z)$. Because of the Z_2 identification, each deficit angle will be π. The curve has a total deficit angle of 4π and is thus a CP^1. All 8 fixed curves are in fact CP^1's.

One must now determine whether or not these fixed curves intersect one another in T^6/A. Consider the two curves $(0, z, z)$ and $(0, z, -z)$ which appear to intersect at $(0, 0, 0)$. However, because the origin of an Eguchi-Hanson is not a point but a CP^1, $(0, 0, 0)$ is in fact a (non-contractible) CP^1. It can be parametrized by the coordinate $\rho = z_2/z_3$. $(0, z, z)$ intersects this CP^1 at $\rho = 1$ while $(0, z, -z)$ intersects at $\rho = -1$, so the two curves in fact do not intersect. It can similarly be shown that none of the 8 fixed curves intersect.

An alternate method (which serves as a useful check) of finding the Euler character of fixed point sets is provided by the Lefschetz fixed point theorem. A discrete symmetry D maps H^p into itself. One can thus define a $b_p \times b_p$ matrix D_p describing this map. The Lefschetz fixed point theorem says that the Euler character $\chi(D)$ of the fixed point set of D is

$$\chi(D) = \Sigma(-1)^p \operatorname{tr} D_p \cdot$$

For a product manifold such as $K3 \times T^2$, this Euler character will be the product of the Euler characters of the fixed point sets of D on $K3$ and T^2 separately. On T^2 $\operatorname{tr} D_0$ is one. H^1 consists of dz^1 and $d\bar{z}^1$ both of which go to minus themselves, so $\operatorname{tr} D_1 = -2$. H^2 consists of $dz^1 \wedge d\bar{z}^1$ so $\operatorname{tr} D_2 = 1$. Adding this all up, we find on T^2 that $\chi(D) = 4$. This of course was discussed earlier — it is just the four points $\{f_k\}$. On $K3$, $\operatorname{tr} D_0$ is again one. H^1 is empty, while H^2 consists of 16 bolts at (f_i, f_j) and the six two forms $dz_1 \wedge dz_2$, $d\bar{z}_1 \wedge d\bar{z}_2$, $dz_1 \wedge d\bar{z}_1$, $dz_2 \wedge d\bar{z}_2$, $dz_1 \wedge d\bar{z}_2$ and $d\bar{z}_1 \wedge dz_2$. All but four of the 16 bolts are mapped into one another under D and do not contribute to $\operatorname{tr} D_2$, while the remaining elements of H^2 give $\operatorname{tr} D_2 = 2$. H^3 is empty and $\operatorname{tr} D_4 = 1$. Adding this up we find $\chi(D) = 4$ on $K3$ and thus regain the result that $\chi(D) = 16$ on T^6/A.

Dividing T^6/A by D and blowing up the singularities, we obtain a new Calabi-Yau manifold. Using the fact that the Eguchi-Hanson space has $\chi = 2$, the Euler number can be computed using the previously discussed methods, and is 24. There remains a freely acting Z_2 symmetry

$$Q(z_1, z_2, z_3) = (-z_1 + \frac{1}{2}, -z_2 + \frac{1}{2}, z_3 + \frac{1}{2})$$

and $\chi(T^3/ADQ)$ is 12.

The intersection matrix can be computed by the methods described in this talk, and one again finds that most of the Yukawa couplings vanish. We will just compute enough of it to see that the couplings are not singular in the orbifold limit. Let T_k be the surface $z_k = \frac{3}{8}$ and let B be the surface $(z_1, 0, 0)$. Then $B \cap T_1$, is the CP^1 $(\frac{3}{8}, 0, 0)$. This is the minimal CP^1 of an Eguchi-Hanson geometry. As for the minimal CP^2's of the Z manifold, the self-intersection of this CP^1 is the Poincaré dual of the first Chern class of its normal bundle, namely minus two. It follows that

$$g(B, B, T_1) = -2 \cdot$$

Now let $J = D \sum_{k=1}^{3} T_k$, so that we are in the orbifold limit. Then

$$\langle B | B \rangle = \frac{D}{2}$$

is non-zero and finite. One may similarly show the entire normalization matrix is well behaved in this limit.

There are many other discrete symmetries of T^6 that preserves ω, but it is not easy to produce manifolds with small number of generations. Dividing by the group generated by

$$A(z_1, z_2, z_3) = (z_1, -z_2, -z_3)$$

$$B(z_1, z_2, z_3) = (-z_1, -z_2, z_3)$$

$$P(z_1, z_2, z_3) = (z_2, z_3, z_1)$$

$$F(z_1, z_2, z_3) = (z_1 + \frac{1}{2}, z_2 + \frac{1}{2}, z_3 + \frac{1}{2})$$

$$G(z_1, z_2, z_3) = (z_1 + \frac{i}{2}, z_2 + \frac{i}{2}, z_3 + \frac{i}{2})$$

appears to lead to a four generation model with a $Z_2 \times Z_2$ fundamental group, although there is some uncertainty because the resolution of non-Abelian singularities is not well understood. One need not begin with the "square" torus. The torus obtained from the root lattice of a Lie group, or any other torus with discrete symmetries that preserve the holomorphic three form, can be used.[16] So far none of the manifolds obtained in this way appear promising phenomenologically.

ACKNOWLEDGEMENTS

I am grateful to D. Gieseker, G. Horowitz, T. Killingback, R. Lazarsfield, D. N. Page, C. Vafa, E. Witten and S. T. Yau for valuable discussions. This research was supported in part by the National Science Foundation under Grant No. PHY82-17853, supplemented by funds from the National Aeronautics and Space Administration, at the University of California at Santa Barbara.

REFERENCES

1. M. B. Green and J. H. Schwarz, Nucl. Phys. **B181**, 502 (1982); **B198**, 252 (1982); **B198**, 441 (1982); Phys. Lett. **109**B, 444 (1982); M. B. Green, J. H. Schwarz, and L. Brink, Nucl. Phys. **B198**, 474 (1982).

2. P. Candelas, G. Horowitz, A. Strominger and E. Witten, Nucl. Phys. **B258**, 46 (1985) and Proc. of the *Argonne Symposium on Anomalies, Geometry and Topology* (1985).

3. M. Dine, V. Kaplunovsky, N. Mangano, C. Nappi, and N. Sieberg, IAS preprint (1985).

4. See for example M. Dine, R. Rohm, N. Seiberg and E. Witten, Phys. Lett. **156**B, 55 (1985); M. Dine and N. Sieberg, IAS preprints, or V. Kaplunovsky, Phys. Rev. Lett. **55**, 1036 (1985).

5. A. Strominger and E. Witten, to appear Comm. Math. Phys.

6. A. Strominger, ITP preprint (1985).

7. S. T. Yau, Proc. of the *Argonne Symposium on Anomalies, Geometry and Topology* and G. Horowitz, these proceedings.

8. P. Griffiths and J. Harris, **Principles of Algebraic Geometry**, (Wiley-Interscience, 1978).

9. D. Page, Phys. Lett. **80**B, 55 (1978).

10. L. Dixon, J. Harvey, C. Vafa and E. Witten, Princeton preprint (1985).

11. S. T. Yau, Proc. Natl. Acad. Sci. **74**, 1798 (1977).

12. E. Calabi, in **Algebraic Geometry and Topology**, a *Symposium in Honor of S. Lefschitz*, **78**, Princeton University Press (1957).

13. S. S. Roan and S. T. Yau, private communication.

14. E. Witten, Nucl. Phys. **B258**, 75 (1985).

15. E. Calabi, Ann. Sc. del'E.N.S. **12**, 266 (1979); T. Eguchi and A. S. Hanson, Phys. Lett. **74**B, 249 (1978).

16. L. Dixon, J. Harvey, A. Strominger, J. Thierry-Mieg, C. Vafa, E. Witten and various combinations thereof, unpublished.

SUPERSTRINGS AT LOW ENERGY

Chiara R. Nappi*

Joseph Henry Laboratories
Princeton University
Princeton, New Jersey 08544

The work I will describe in this talk has been done in collaboration with M. Dine, V. Kaplunovsky, M. Mangano and N. Seiberg.[1] We have investigated some aspects of the low-energy physics of the various models that can be obtained by compactifying string theories, with the aim of selecting those which might be phenomenologically viable.

Recently, after the discovery of the generalized anomaly cancellation,[2] an enormous amount of progress has been made in relating superstring theory to the real world.[3,4] In this respect, the most interesting superstring theory appears to be the d=10 superstring theory with $E_8 \times E'_8$ gauge group.[5] If one requires unbroken N=1 supersymmetry in four dimensions (a desirable feature if one hopes that supersymmetry might solve the gauge hierarchy problem), one needs to compactify these theories to $M_4 \times K$, where M_4 is d=4 Minkowski space and K is a six-dimensional Ricci-flat Kahler manifold, a so-called Calabi-Yau space. This compactification scheme breaks $E_8 \times E'_8$ down to $E_6 \times E'_8$ and puts the fermion families into 27 or $\overline{27}$ representations of E_6. The number of generations F [defined as the number of families (or 27) minus the number of antifamilies (or $\overline{27}$) of opposite chirality] is predicted in terms of the Euler characteristic of the compact manifold K.[4] To understand the particle content of a family, let us decompose E_6 under its maximal subgroup $SU(3)_C \times SU(3)_L \times SU(3)_R$ or $SU(3)^3$. Then the

* Research supported in part by NSF Grant PHY-80-19754

27 of E_6 goes into

$$27 \to (\bar{3},3,1) + (3,1,\bar{3}) + (1,\bar{3},3) \tag{1}$$

and known and unknown particles can fit as follows[6]

$$\begin{pmatrix} u \\ d \\ q \end{pmatrix} \quad (\bar{u}\ \bar{d}\ \bar{q}) \quad \begin{pmatrix} H_n & H^- & e^- \\ H^+ & H_d & \nu \\ e^+ & S_1 & S_2 \end{pmatrix} \tag{2}$$

Therefore each family of 27, aside from a generation of known quarks and leptons, contains an additional charge -1/3 quark we call g, extra leptons (H_n, H^+) and (H^-, H_d) which happen to have the quantum numbers of Weinberg-Salam Higgses and two singlets under $SU(3)_C \times SU(2)_L \times U(1)_Y$, namely S_1 and S_2, one of which, S_1, has the proper quantum numbers to represent the right-handed neutrino. The method of Wilson lines[4,7] can be employed to break E_6 down to some low-energy subgroup, if K has a nontrivial fundamental group. Wilson lines generate a discrete subgroup $\mathcal{G}$ of E_6 and the unbroken subgroup G of E_6 is the one whose generators commute with $\mathcal{G}$. If one requires that the unbroken subgroup of E_6 contains the standard model ($SU(3)_C \times SU(2)_L \times U(1)_Y$ must surely be contained in any acceptable low energy group), then there are 27 subgroups G that one can break E_6 to via the method of Wilson lines. It turns out[8] that E_6 cannot be broken directly to the standard model itself. The unbroken subgroup will always contain at least an extra U(1), so that extra neutral currents might survive down to low energy (incidentally, the existence of extra gauge bosons with masses in the Tev range seems to be compatible with the current data on neutral currents). Here is a classification of the various subgroups obtained from E_6 via Wilson lines:

1) Minimal extensions of the standard model, i.e. groups of the type

$$SU(3)_C \times SU(2)_L \times U(1) \times (U(1) \text{ or } U(1)^2) \tag{3}$$

$$SU(3)_C \times SU(2)_L \times SU(2)_R \times U(1) \times U(1) \tag{4}$$

2) Models with extended color, i.e. with color group larger than SU(3). For instance

$$SU(4)_C \times SU(2)_L \times U(1) \tag{5}$$

3) Models with extended flavor, i.e. where the flavor group is $SU(3)_L$. For instance

$$SU(3)_C \times SU(3)_L \times U(1) \times U(1) \tag{6}$$

4) Early unified models, i.e. models like SU(5) or O(10), where color and flavor are unified in the same group.

At some point between M_{GUT} and M_W, in order to recover standard phenomenology, we will need to break these groups down to $SU(3)_C \times SU(2)_L \times U(1)_Y$. In this theory the only particles around are the fermions in the families of $\underline{27}$ and their scalar partners. Between those, the only ones which we could use as Higgses to break G down to the standard model are the $SU(3)_C \times SU(2)_L \times U(1)$ singlets S_1 and S_2, since their Vev's will not jeopardize the standard model. However, their Vev's might not be enough to break G to $SU(3)_C \times SU(2)_L \times U(1)_Y$ or acceptable extensions of it. For instance, most of the early modified models at best can be broken down to SU(5), and therefore need to be discarded.

The question that immediately arises here is what is a natural scale for the expectation values $\langle S_{1,2} \rangle$. To answer it, we need to look at the scalar potential for $S_{1,2}$. We first recall that usually in supersymmetry phenomenology a desirable range for the effective scale of supersymmetry breaking is somewhere around M_W, let us say at 1 Tev. The way this might be achieved here is via the hidden sector scenario. Supersymmetry might be broken via gaugino condensation[9] in the hidden sector E'_8 at a much higher scale, but the effective breaking scale might well be around M_W, since the breaking is communicated to the observable sector via gravitational and related couplings. Whatever the mechanism of supersymmetry breaking might be, let us assume here that the scalar partners of fermions get masses of order M_W. For $S_{1,2}$ the mass^2 might well be negative (one needs a detailed renormalization group analysis to decide that[10]). Then the scalar potential for $S_{1,2}$ would look like

$$V(S) = - M_W^2 S^2 + \lambda S^4 + \ldots \tag{7}$$

The quartic term could come for instance from the D-term

$$\Sigma_a g_a^2 (\phi_i^+ T_a \phi_i - \bar{\phi}_i T_a \bar{\phi}_i^+)^2 \tag{8}$$

where the sum runs over the generators of G, ϕ_i is any scalar field, and λ (as well as g_a) is a dimensionless coupling of order one. By minimizing (7) one gets $\langle S \rangle \sim M_W$. Therefore nothing happens between M_{GUT} and M_W, but the various groups G stay unbroken down to low energy. This is the Grand Desert scenario.

We can perform a one-loop renormalization group calculation to see which models would allow reasonable values of $\sin^2\theta_W$ and M_{GUT}. As pointed out in [8], the Georgi-Quinn-Weinberg calculation is valid here since the gauge couplings do obey the standard E_6 relations after E_6 breaking. This is not a trivial statement, since Yukawa couplings, instead, do not need to be E_6 invariant, but only invariant under the unbroken subgroup G.[8] In order to perform the renormalization group analysis we need to know exactly the particle content of the theory. Therefore a remark is in order here: it can happen that some of the particles of a $\overline{27}$ (and necessarily the corresponding particle in a 27, since the number of generations is fixed) stay light after compactification, so that at the end one is left with a given number of generations plus extra light particles.[8,11,12] Actually, for the reasons explained in [8], a simple case where one can analyze the situation in detail is that in which there is one antifamily only. In this case one can analyze each single group G that one can break E_6 to and decide what extra particles outside of the generations are left light. For instance, in the case of $SU(3)_C \times SU(2)_L \times SU(2)_R \times U(1) \times U(1)$, one gets extra light H's and S_2 from the $\overline{27}$ and a 27.

The one-loop renormalization group equation is

$$\frac{1}{\alpha_N(\mu)} = \frac{1}{\alpha_{GUT}(M)} + \frac{1}{6\pi} b_N \ln \frac{M}{\mu} \tag{9}$$

where

$$\alpha_N(\mu) = \frac{4\pi}{g_N^2(\mu)} \tag{10}$$

and g_N is the gauge coupling of SU(N). The beta function b_N is given by

$$b_N = 3\left(3F + \frac{1}{2}n - 3N\right) \tag{11}$$

where n is the number of multiplets of extra light particles in the fundamental representation of SU(N). A first observation is that the Grand Desert scenario makes sense only for $F \leq 3$. Infact one can check that for F=4 both SU(2) and SU(3) have Landau poles, i.e. the couplings $\alpha_{2,3}$ diverge well below any reasonable unification mass. Infact if we start at $\mu = M_W$ with the values of the couplings usually adopted in model building, namely

$$\frac{1}{\alpha_{EM}(M_W)} = 128 \quad \frac{1}{\alpha_3(M_W)} = 9 \quad \frac{1}{\alpha_2(M_W)} = 26 \tag{12}$$

and we assume F=4, it turns out that the Landau mass at which the couplings diverge, i.e.

$$\mu_L = M_W \, e^{\frac{6\pi}{b_N \alpha_N(M_W)}} \tag{13}$$

is 10^{12} Gev for SU(3) and 10^{15} Gev for SU(2). Since models with less than three generations are not acceptable (a two generation topless model has been experimentally ruled out) Grand Desert scenario applies only to F=3. In this case, we have computed $\sin^2\theta_W$, M_{GUT} and α_{GUT} in the various models listed above. In principle we would consider a model acceptable if it satisfies the following bounds

$$0 < \alpha_{GUT} < \frac{1}{4} \qquad 0.19 < \sin^2\theta_W < 0.24 \qquad 10^{15}\text{ Gev} < M_{GUT} < 10^{19}\text{ Gev} \tag{14}$$

Actually, however, there is an extra constraint to take into account. It seems that M_{GUT} and M_{Planck} must be approximately equal to each other in any unified string theory with realistic couplings.[13] Taking all these constraints in mind, it appears that in the Grand Desert scenario the only models that can possibly work are minimal extensions of the standard model (3). They give infact.

$$\sin^2\theta_W = 0.206 \qquad M_{GUT} = 2.10^{17}\ \text{Gev} \qquad \alpha_{GUT} = 0.11 \tag{15}$$

Instead left right-symmetric models (4) do not work since they give $\sin^2\theta_W > 0.26$ and $M_{GUT} > M_{Planck}$. Extended flavor models (6) turn out to have $M_{GUT} > M_{Planck}$ and $\sin^2\theta_W < 0.15$. Extended color models (5) have M_{GUT} much too low to be acceptable ($M_{GUT} \sim 10^{14} \div 10^{15}$ Gev) and of course early unified models need to be discarded from the start.

There is however an interesting alternative to the above described scenario.[1,14] It is possible that $S_{1,2}$ acquire VEV's at some intermediate scale $M_I \sim 10^{10} \div 10^{11}$ Gev instead of the M_W as previously argued. In this case G would break down to the standard model at M_I, changing the game in the one-loop renormalization group analysis and rescuing some of the other models. It would also be possible in this case to have four generations. Infact since $S_{1,2}$ can couple to the H and g particles in the 27, they could give them a mass M_I and make them heavy. Below M_I the beta function would flatten down to

$$b_N = 3\,[2F + \frac{1}{2}n - 3N]\ . \tag{16}$$

Therefore, if M_I is high enough, perturbative couplings at M_{GUT} might still be compatible with the physical low energy values of the gauge couplings.

In order to see how it can happen that $S_{1,2}$ gets a Vev at a scale $M_I = 10^{10} \div 10^{11}$ Gev, let's look at the superpotential

$$W(\phi) = \lambda\phi\phi\phi + \bar{\lambda}\bar{\phi}\bar{\phi}\bar{\phi} + O(M_{GUT}^{-1})\ \phi\phi\bar{\phi}\bar{\phi} \tag{17}$$

(Since this superpotential needs to be invariant under G and G surely contains Z_3, the center of E_6, in (16) there is no term of the form $\bar{\phi}\phi\phi$.) Now the only way a quartic term $\lambda\ S^4$ in the scalar potential can be generated from the above superpotential is via a coupling $\phi_i S^2$ for some superfield ϕ_i in the 27. But such a coupling does not exist in E_6 and, although W does not need to be E_6 invariant, it still cannot contain couplings which would not be there in the first place if it were E_6 invariant.[8] Therefore $\lambda\ S^4$ can come only from the D-term (8). However if one has D-flat direction of $S_{1,2}$ Vev's, namely if

$$\langle S_{1,2}\rangle = \langle \bar{S}_{1,2}\rangle^{\dagger} \tag{18}$$

then the quartic term S^4 cannot be generated. From (17) the next term in the potential would be $M_{GUT}^{-2}\ S^6$ so that one needs now to minimize

$$V(S) = -M_W\ S^2 + M_{GUT}^{-2}\ S^6 \tag{19}$$

which indeed has a minimum at $\langle S\rangle = \sqrt{M_W\ M_{GUT}} \sim 10^{10} \div 10^{11}$ Gev. In order for this scheme to work, one needs Higgses $\bar{S}_{1,2}$ from the antifamilies. Actually D-flat directions are also needed in order for supersymmetry to stay unbroken. So if one insists that supersymmetry stays effectively unbroken down to M_W (and assumes that $S_{1,2}$ gets negative mass^2), then $\langle S_{1,2}\rangle$ is M_W or $\sqrt{M_W\ M_{GUT}}$ according to whether Higgses $\bar{S}_{1,2}$ from the antifamilies are available or not. This question can easily be answered for each group G, if one deals with Calabi-Yau with one antifamily only. For instance, it turns out that in the case of groups G of rank five no neutral singlets $\bar{S}_{1,2}$ can stay light, so no intermediate scale can be generated. Moreover, even if neutral singlets from the one antifamily are available to generate M_I, one can show that it is impossible to break any of the groups G to a subgroup of rank less than five. Hence it is not possible to get rid of all extra weak currents at the intermediate scale, but some will survive down to the weak scale. A word of warning is due here: this prediction of extra weak currents applies only to the case of Calabi-Yau with one $\overline{27}$ only. If there is more than one antifamily, one might get more $\bar{S}_{1,2}$ Higgses, and these could break all extra interactions, leaving us with the pure standard model. However, in the case of only one $\overline{27}$, we will end up either with $SU(3)_C\times SU(2)_L\times U(1)\times U(1)$ or $SU(3)_C\times SU(2)_L\times SU(2)_R\times U(1)$.

In order to compute the beta functions (11) and (16), let us assume that $S_{1,2}$ couple to g quarks and H's to give them a mass $\sim M_I$. Generically we assume tht only one pair of H doublets does not couple to $S_{1,2}$ and stays light to actually play the role of Weinberg-Salam Higgses (if however the model does not work we might try to see if leaving light more than one pair of Higgs doublets might help). Since the renormalization group equations for M_{GUT} and $\sin^2\theta_W$ are independent of the number of generations (they contain only differences of beta functions), the results for M_{GUT} and $\sin^2\theta_W$ are the same for three and four generations. Instead, α_{GUT} is small in the case of three families (generically $\alpha_{GUT} \sim 0.05$) and tends to be higher for four families, the exact value depending on the value of M_I [for each model we played around with different values of M_I in the attempt of satisfying the bounds (14)].

The models that appear to work with an intermediate scale are the following. First of all, in the class of minimal extensions of the standard model, the model in (3) with rank six group. This model, if we assume an intermediate mass $M_I \sim 10^{10}$ Gev, has $M_{GUT} \sim 10^{17}$ Gev and $\sin^2\theta_W \sim 0.22$. For F=4 the coupling at M_{GUT} is $\alpha_{GUT} \sim 0.17$. In the same class the left-right symmetric model (4) does well too, with $M_{GUT} \sim 10^{18}$ Gev, $\sin^2\theta_W \sim 0.23$ and $\alpha_{GUT} \sim 0.24$, a little bit too high (we have taken here $M_I \sim 10^{11}$ Gev). Finally, in the class of enlarged flavor model (6) gives (for $M_I \sim 10^{10}$ Gev and two pairs of Higgs doublets H) $M_{GUT} \sim 10^{17}$ Gev, $\sin^2\theta_W \sim 0.22$ and $\alpha_{GUT} \sim 0.18$ (for F=4).

In order to conclude that the models that survive the above analysis are actually phenomenologically acceptable, a lot of other issues need to be addressed, in particular proton decay and neutrino masses. Here I will restrict myself to a few general considerations on these issues.

Proton decay problems can come from baryon violating gauge interactions or Yukawa B-violating interactions. Models with conventional color, i.e. $SU(3)_C$, do not have any B-violating gauge interactions. Moreover, we can also assume only Yukawa couplings which conserve baryon number. Infact a symmetry group with conventional color will not force on us any B-violating Yukawa couplings, since it will not relate any B-violating coupling with any B-conserving one. Models with extended color, like $SU(4)_C$, could have problems due to gauge induced proton decay. Call X the generator of $SU(4)_C$ outside $SU(3)_C$. As any gauge generator, it maps something into something else in the same representation. Since g and $\bar{g}$ quarks are in the same representation of $SU(4)_C$ (namely the 6), the following graph is possible

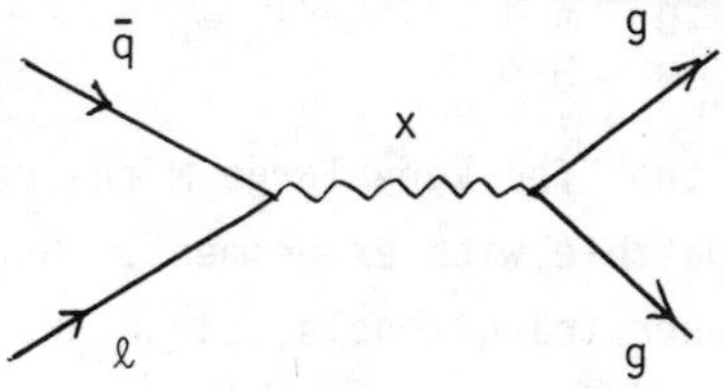

This process can conserve baryon number, if we allow unconventional assignment of baryon number for the g quarks (B = -1/6). So if there is no mixing between g and d quarks, conventional baryons that do not contain g quarks do not decay via gauge interactions. Actually the main problem in these models, as well as in early unified models, comes from Yukawa couplings. Typically, exchange of g quarks can generate baryon violating operators of dimension five. For instance, in these models Yukawa couplings that give masses to quarks and leptons are typically related by the symmetry group G to terms of the form

$$\lambda\ \bar{g}\ \bar{u}\ \bar{d} + \tilde{\lambda}\ g\ \bar{u}\ e^+ \qquad (20)$$

that give rise to an effective interaction of dimension five

$$O\left(\frac{\lambda\tilde{\lambda}}{M_g}\right)\ \bar{u}\ \bar{u}\ \bar{d}\ e^+. \qquad (21)$$

Unless $M_G \gtrsim 10^{15}$ Gev, such interactions would make proton lifetime much too short. To avoid this problem one can look for some discrete symmetry[8] that further restricts Yukawa couplings. A detailed analysis of each single model is therefore needed.[1]

The problem with neutrino masses is that one of the particles in the 27 of E_6 is the right-handed neutrino $\bar{\nu}$. To avoid neutrino masses in models that contain the right-handed neutrino one usually assumes that the right-handed neutrino has a large mass M,[15] gotten via coupling to some neutral Higgs. The 2×2 neutrino mass matrix is then

$$\begin{pmatrix} 0 & m \\ m & M \end{pmatrix}$$

with eigenvalues M and m^2/M, so that for very large M the neutrino mass would be small enough to be compatible with experiment. However this mechanism is not available in superstring models, since the only particles around are those in the 27. All the couplings which could generate a Majorana mass for the right-handed neutrino, for instance couplings like $\bar{\nu}^3$ or $\bar{\nu}^2\ S_2$ are not present in the superpotential, since there are not E_6 invariant, as already mentioned. (However the above mentioned see-saw mechanism might still be applicable in the global mass matrix for all neutral leptons.) Again, a more detailed analysis[1] is needed to assure that neutrinos can be uncoupled from all non-zero Vev's.

REFERENCES

[1] M. Dine, V. Kaplunovsky, M. Mangano, C. Nappi, N. Seinberg, Nucl. Phys. B, in press.

[2] M.B. Green and J.H. Schwarz, Phys. Lett. 149B (1984) 117.

[3] E. Witten, Phys. Lett. 149B (1984) 351.

[4] P. Candelas, G.T. Horowitz, A. Strominger and E. Witten, Nucl. Phys. B258 (1985) 46.

[5] D.J. Gross, J.A. Harvey, E. Martinec and R. Rohm, Phys. Rev. Lett. 55 (1985) 502, and Princeton preprints (1985).

[6] F. Gursey, P. Ramond, and P. Sikivie, Phys. Lett. 60B (1976) 177; Y. Achiman and B. Stech, Phys. Lett. 77B (1978) 389; Q. Shafi, ibid, 79B (1978) 301.

[7] Y. Hosotani, Phys. Lett. 129B (1984) 193.

[8] E. Witten, Nucl. Phys. B258 (1985) 75.

[9] M. Dine, R. Rohm, N. Seinberg and E. Witten, Phys. Lett. 156B (1985) 55. See also J.P. Derendinger, L.E. Ibanez, and H.P. Nillen, CERN-TH 4123/85.

[10] M. Mangano, Zeitschrift fur physik C., to appear.

[11] J. Breit, B. Ovrut and G. Segrê, Phys. Lett. B158 (1985) 33.

[12] A. Sen, Phys. Rev. Lett. 55 (1985) 33.

[13] V.S. Kaplunovsky, Phys. Rev. Lett. 55 (1985) 1036; M. Dine and N. Seiberg, Phys. Rev. Lett. 55 (1985) 366.

[14] See also S. Cecotti, J.-P. Derendinger, S. Ferrara, L. Girardello and M. Roncadelli, CERN preprint TH4103 (1985).

[15] M. Gell-Mann, P. Ramond and R. Slansky, "Supergravity," ed. D. Freedman (North Holland 1979); R.N. Mohapatra and G. Senjanovic, Phys. Rev. Lett. 44 (1980) 912.

IS THE SUPERSTRING SEMICLASSICAL? †

M. Dine
Institute for Advanced Study-Princeton, NJ 08540
and
City College of New York-New York, NY 10031
and
N. Seiberg
Institute for Advanced Study-Princeton, NJ 08540
and
Weizmann Institute of Science-Rehovot 76100

ABSTRACT

The various coupling constants and scales of the superstring are analyzed. Simple general considerations and an explicit calculation show that if the superstring is to describe our world, it is probably strongly coupled. Certain loopholes to the argument are discussed.

There has been enormous progress in superstring theory[1)] in the last year. The discovery of anomaly cancellations[2)], the construction of new, consistent theories[3)], and numerous technical developments have generated enormous excitement. The work on compactifications of the $E_8 \times E_8$ heterotic string on Calabi-Yau manifolds [4)] has provided real hope that these theories may possess four dimensional ground states which resemble our world.

Of course, interest in string theory ultimately stems from the hope that it can be a unified theory of all interactions. Here we discuss some issues which string theory must confront if it is to describe reality. This work began as an effort to identify the various couplings, scales, and expansion parameters of these theories. As Witten has stressed, superstring theory has no dimensionless parameters [5)]. Thus we might expect strong coupling. On the other hand,

---

† Talk given by N. Seiberg at the String Workshop, Santa Barbara 1985

there does exist a perturbation expansion; in fact, all we presently know of the theory is its perturbation expansion. We will explain here what the parameters of this expansion are. We will see that their actual values must be determined dynamically and that a weak coupling calculation, on very general grounds, cannot yield a ground state that looks at all like our world. Thus, if string theory is to be a unified theory of all interactions, it must be strongly coupled.

In the first part we will explain how the expectation value of the dilaton field serves as the expansion parameter of the full string theory. We will see that, upon compactification, the theory acquires new expansion parameters, corresponding to expectation values of fields which describe the size and shape of the internal manifold.

These expectation values label physically distinct state, all of which are perfectly good vacua, at the classical level, for string theory. If the internal manifold is large (small) in string units, the corresponding non-linear σ-model on the string would-sheet is weakly (strongly) coupled. Depending on the expectation value of the dilaton field, string diagrams with more and more complicated world-sheet topologies may or may not be neglected.

We expect the vev's of these various fields - and hence the various couplings - to be dynamically determined. To get some feeling for what may happen, we consider a field theory model in which the coupling constant is also the expectation value of a field - supersymmetric QCD coupled to an axion [6].

This theory is quite well understood, and is a good warmup for considering what happens in the case of compactification of string theories on Calabi-Yau spaces. There, we will see that gluino condensation in the second E_8 renders these vacua unstable [7,8] and causes a runaway to flat ten dimensional space and/or zero coupling. In fact, the problem which occurs for Calabi-Yau manifolds is more general [9,10]. The possible forms of the dilaton effective potential, at weak coupling, are very limited. As we will show, all of them are inconsistent with very basic facts of our world, such as the value of the cosmological constant and the principle of equivalence. Thus if the superstring does describe nature, it is probably strongly coupled. Finally, we will mention some possible loopholes in our arguments all of which if operative still require an understanding of string theory beyond perturbation theory.

That the dilaton field is the coupling constant of string theory can be seen in many ways. It was first noted by Shapiro and by Ademollo et al [11], examining string diagrams. Alternatively, it can be seen in Polyakov's approach [12]. In the presence of background fields, the two dimensional action contains a term (for closed strings)[13]

$$\frac{1}{4\pi}\int d^2\xi\sqrt{g^{(2)}(\xi)}R^{(2)}D(X) \tag{1}$$

where $g^{(2)}$ and $R^{(2)}$ are the two-dimensional metric and curvature, and D is the dilaton field. If D is a constant, the integral just gives 2D(# handles - 1), i.e. it gives a factor

$$\phi^2 = e^{-2D} \tag{2}$$

for every loop. Thus for closed strings, ϕ^2 is the loop expansion parameter.

Having in mind phenomenological applications, it is instructive to see how the role of the dilaton field as coupling emerges from a d=10 field theoretic description [8]. We can integrate out the heavy fields to obtain an effective Lagrangian for the massless modes of the string. The general supersymmetric Lagrangian with at most two derivatives was obtained in Ref.(14). The bosonic terms are

$$\frac{1}{e}\mathcal{L}^{(10)} = -\frac{1}{2\kappa^2}R - \frac{1}{4}\frac{\phi^{-1/2}}{g_{10}^2}F_{\mu\nu}^2 - \frac{3}{4}\frac{\kappa^2}{g_{10}^4}\phi^{-1}H_{\mu\nu\rho}^2 - \frac{1}{4\kappa^2}\left(\frac{\partial\phi}{\phi}\right)^2 + \dots \tag{3}$$

Here e is the vielbein and $\phi = e^{-D}$ as above.

This Lagrangian, classically, has many vacuum states labelled by $< \phi >$. These states are related by a classical symmetry of the full string theory [15]. Quantum mechanically, they are not equivalent; the gauge coupling, for example, depends on ϕ

This Lagrangian, of course, must be viewed as the Lagrangian for a cutoff theory, with cutoff $M_s \sim \alpha'^{-1/2}$. In addition to the terms above, it contains higher dimension operators obtained from integrating out higher modes of the string. In Eq. 3, the gravitational term is canonical, i.e. there are no factors of ϕ in front of R. With this choice, the Planck scale appears fundamental, while the cutoff, M_s, is ϕ-dependent. To see this, recall that for the heterotic string, $M_s \sim \alpha'^{-1/2} \sim g/\kappa$. Here g is the coupling at the gauge boson vertex, which we see, from $\mathcal{L}^{(10)}$, is $g_{10}\phi^{1/4}$. Thus $M_s \sim \phi^{1/4}$. For a smooth perturbative expansion for the heavy fields and for ease of calculations, it is more natural to make this ϕ dependence explicit is the Lagrangian rather than in the cutoff. This can be achieved by letting $g_{\mu\upsilon} \rightarrow \phi^{-1/2} g_{\mu\upsilon}$. Then the Lagrangian takes the form [8].

$$\frac{1}{e}\mathcal{L}^{(10)} = \phi^{-2}[-\frac{1}{2}R - \frac{1}{4}F^2 - \frac{3}{4}H^2 + 2(\frac{\partial\phi}{\phi})^2 + \ldots] \tag{4}$$

where we have set M_s=1 everywhere. Several features of this Lagrangian should be noted:

1. No factors of ϕ appear in the brackets. This is true, not only of the bosonic terms which we have indicated explicitly, but also of the fermionic terms (after rescaling of the fermions). With these rescalings, the supersymmetry transformation laws also do not contain factors of ϕ, which is quite suggestive.
2. A classical symmetry of string theory, described in Ref. (15), is particularly obvious here. Taking $\phi \rightarrow \lambda\phi$, the Lagrangian is rescaled by a constant, $\mathcal{L} \rightarrow \lambda^{-2}\mathcal{L}$. Witten's observation ensure that the higher dimension operators obtained by classically integrating out massive modes also take the form above (i.e. with ϕ^{-2} out front).
3. Weak coupling corresponds to $\phi << 1$, or $D = -ln\phi >> 1$. In reaching this conclusion, it is important that M_s is the only scale in this Lagrangian, and is also the cutoff. Also, it is clear that ϕ is a loop

counting parameter.

4. The three points above are not true for type I theory, due to the defferent rescaling required to make M_s ϕ-independent[8]. There, weak coupling corresponds to $\phi >> 1$, and ϕ does not count loops.
5. As an illustration, note that, for closed strings, the term ω_L in H introduced in Ref. (2), and the TrR^2 term found in Ref.(16), as tree effects, contain factors of ϕ^{-2} upon rescaling. On the other hand the contact terms added in Ref. (2) to cancel anomalies do not contain powers of ϕ (since they involve ϵ tensors), and are a one loop effect.

Upon compactification, additional coupling constants appear. These correspond, again, to expectation values of fields. Denoting the six compactified dimensions by y and the others by x, we write the internal metric as $g_{IJ} = X(x)g^{(0)}_{IJ}$, when $g^{(0)}_{IJ}$ is a fixed, reference metric satisfying $\int d^6y\sqrt{g^{(0)}} = M_s^{-6}$. $\sqrt{X}$ is thus the radius of the compact manifold measured in string unit; also $X^{1/2} = M_s/M_{GUT}$, where M_{GUT} is the unification scale. The string action now contains the term

$$\frac{1}{4\pi\alpha'}\int d^2\xi \; g_{IJ} \; \partial_\alpha x^I \partial^\alpha x^J \tag{5}$$

Thus X^{-1} is the coupling of the non-linear σ-model. Note that the metric, g_{IJ}, appearing here, is the rescaled metric of Eq. 4, not the canonical one [8].

Thus we see that string theory possesses two types of weak coupling expansions. ϕ measures the size of quantum fluctuations. X determines the importance of higher dimension operators in the d=10 effective field theory; it is also the coupling constant of the non-linear σ-model.

The four-dimensional Lagrangian takes classically the form

$$\frac{1}{e}\mathcal{L}^{(4)} = Y(-\frac{1}{2}R - \frac{1}{4}F^2_{\mu\nu} + ...) \tag{6}$$

where Y$= X^3\phi^{-2}$. Thus the four dimensional gauge coupling, $g^{(4)}$, and Planck mass, M_p, are given by

$$\frac{1}{g_4^2} = Y, \qquad M_p^2 = YM_s^2 \tag{7}$$

If both the full string theory and the non-linear σ-model are to be weakly coupled, we require the double hierarchy:

$$M_{GUT}^2 << M_s^2 << M_p^2 \tag{8}$$

Experimentally we know that the gauge coupling (Y) is of order one [(17)]. As pointed out in Ref. (8,18), if the string coupling is to be weak, $\phi^2 = X^3Y^{-1} \lesssim 1$, then $X \sim 1$, i.e.

$$M_{GUT} \sim M_s \sim M_p \tag{9}$$

This, of course, does not mean that the non-linear σ-model or the full string theory are strongly coupled. A criterion for strong coupling is not the size of the coupling but the relative size of the leading order vs. the next to leading order effects. We will see shortly, however, that quite general considerations imply that the string theory itself is truly strongly coupled.

Already in ordinary global supersymmetry, there are many theories in which coupling are determined by expectation values of fields, so it is instructive to review a well-studied case, supersymmetric QCD coupled to an axion multiplet[6)]. This is a theory with gluons, A_μ^a, gluinos χ_α^a, and a neutral chiral field, $Y = (Y, a, \psi)$. Before coupling the axion, the pure gauge theory has a discrete chiral Z_{2N} symmetry (for gauge group $SU(N)$) : $\chi \rightarrow e^{i\pi k/N}\chi$. Arguments based on $\mathrm{Tr}(-1)^F$ [(19)], effective Lagrangians [(20)], and instantons [(6,21)] show that in this theory supersymmetry is unbroken, while the discrete symmetry is broken by a gluino condensate, $< \bar{\chi}\chi >$. Coupling this theory to the axion multiplet Y leads to dynamical suppersymmetry breaking[6)]. The Lagrangian contains the term

$$\mathcal{L} = ... + \int d^2\theta \frac{1}{f_a} Y W_\alpha^2 \tag{10}$$

where f_a is the axion decay constant. In terms of component fields this is

$$\mathcal{L} = ... - \frac{Y}{f_a} F_{\mu\nu}^2 + \frac{a}{f_a} F\tilde{F} + \frac{1}{f_a^2}(\bar{\chi}\chi)^2, \tag{11}$$

where the last term comes from eliminating the auxiliary field of Y, F_Y, by its equation of motion, $F_Y \sim \frac{1}{f_a}\bar{\chi}\chi$. Note that $< Y >$ determines the gauge coupling, i.e.

$$\frac{1}{g_{eff}^2} = \frac{1}{g^2} + \frac{Y}{f_a} \tag{12}$$

Under a supersymmetry transformation,

$$\delta\psi = \frac{1}{fa}\bar{\chi}\chi\epsilon + ..., \tag{13}$$

so the gluino condensate breaks SUSY, and ψ is the corresponding Goldstone fermion. Also, if we naively replace $(\bar{\chi}\chi)^2$ by $|< \bar{\chi}\chi >|^2$, we see that a potential is generated for Y,

$$V_{eff}(Y) \sim \frac{1}{f_a^2} |< \bar{\chi}\chi >|^2 \sim e^{-\frac{3}{b_o g_{eff}^2}} \sim e^{\frac{-3}{b_o f_a} Y}, \tag{14}$$

where b_o is the one-loop β-function coefficient. That this procedure is in fact correct can be shown in a number of ways. The simplest is to note that the theory has a non-anomalous symmetry under which

$$Y(\theta) \to Y(e^{i\alpha}\theta) + i\frac{4b_o f_a}{3}\alpha \tag{15}$$

The unique suyperpotential invariant under this symmetry is

$$W_{eff} \sim e^{-\frac{3}{2b_o f_a} Y} \tag{16}$$

which yields the potential above. (Note that this represents a non-perturbative violation of the non-renormalization theorem). While classically there are an infinity of vacua labelled by Y, quantum mechanically a potential is generated for Y, which vanishes for large Y (zero coupling). Moreover, for larger and larger Y, the calculation which leads to this result becomes more and more reliable (since the coupling g^2_{eff} is weaker). For small Y, this calculation is not reliable and the behaviour of the theory is unknown. There is no particular reason to think, however, that there exist supersymmetric vacua in the strong-coupling domain [6)].

This analysis can be taken over directly to the case of the $E_8 \times E_8$, heterotic string[3)] compactified on Calabi-Yau manifolds [7)], if the second E_8 possesses an unbroken, non-Abelian subgroup. The effective Lagrangian below the compactification scale is

$$\frac{1}{e}\mathcal{L} = -YF^2_{\mu\upsilon} + aF\tilde{F} + \dots \tag{17}$$

Here the axion field, a, comes from $B_{\mu\upsilon}$, and its coupling from the Chern-Simons term in H. Now, gluino condensation in the second E_8 (if it has an unbroken non-abelian subgroup) generates a superpotential for the field Y and its superpartners [7)] (the axion and certain components of the 10-dimensional singlet fermion)

$$W_{eff} \sim e^{-\alpha Y}, \qquad \alpha = \frac{3}{2b_o} \tag{18}$$

Using a trivial dimensional reduction procedure [(7,15)], this yields

$$V_{eff} \sim \frac{1}{X^3} Y e^{-2\alpha Y} \tag{19}$$

This procedure does not correctly take into account all of the effect of compactification on Calabi-Yau spaces. These, however, will not alter the exponential falloff of the potential, nor the fact that it goes to zero for large X (fixed Y). For large Y or large X, the potential falls rapidly to zero. The system, thus rolls to weak coupling and/or flat d=10, where our approximations become more and more reliable. † One might try and save the situation by considering an

† Note that this analysis assumes that there are no larger non-perturbative effects ($\sim e^{-1/\phi^2}$) from string theory, but as we will see shortly, the conclusions are unchanged even if there are.

expectation value for H[7], $< H_{IJK} >= n\epsilon_{IJK}$. Here n is an integer [22], and ϵ is the covariantly constant three-form. Then there exists a stable ground state with zero cosmological constant in leading order [7]. For world-sheet conformal invariance, X must be of order one. If one closes ones eyes to the smallness of X and computes $m_{3/2}$ naively

$$m_{3/2} \sim \frac{n}{X^{3/2}Y^{1/2}} M_p \sim M_p \tag{20}$$

Correspondingly, the "soft breaking terms" in the effective low-energy Lagrangian will be of order M_p, and there is no hierarchy. This is hardly surprising, since SUSY was broken at tree level by terms of order one ($H \neq 0$).

This problem of runaway behaviour might also be avoided if we are in a region of small ϕ so that string perturbation theory is valid, but sufficiently small X that the leading order calculation in 1/X is not reliable [8]. In this case, the σ-model is strongly coupled (note that this is not the same as the statement that X$\sim$ 1!).

However, we would then expect a similar problem for the dilaton field, D ($D = -ln\phi$)[9,10]. In fact, for any compactification to four dimensions, for fixed radius, we can investigate the form of the dilaton effective potential, V_{eff}(D) at weak coupling. †

For D=∞, the theory is free and well-understood; for large D, the potential must tend smoothly to the D=∞ result. Thus there are three possible behaviors for the dilaton potential.

The potential may be identically zero, both perturbatively and non-perturbatively. Or it may be positive and tending to zero, so that the system rolls to weak coupling, or negative and tending to zero, so that the system rolls to strong coupling. The dilaton potential may be generated perturba-

† One might worry that V_{eff} is an off-shell quantity and hence not well-defined in string theory. However, at least at weak coupling, i.e. large D, the low energy theory should be describable in terms of an effective four dimensional field theory and V_{eff} is well-defined.

tively (as in the calculation of Ref. (23)), or non-perturbatively (as in the previous example of large Calabi-Yau vacua). If perturbative in L loop order, $V_{eff} \sim M_s^4 \phi^{2L-2}$. However, in order to interpret V_{eff} as a potential (cosmological constant) of the usual sort, we must restore Einstein's equations to their canonical form. This gives $V_{eff} \sim M_p^4 \phi^{2L+2}$.$V_{eff}$ goes to zero as $\phi \to 0$ for any fixed string configuration, whether or not it is a solution of the equations of motion. If non-perturbative, we expect, e.g., $V_{eff} \sim e^{-1/\phi^2}$, as in the Calabi-Yau case above.

None of these three possible behaviour for the potential in weak coupling is acceptable for the description of our world. If there is no dilaton potential, then not only is there a loss of predictive power (the vacuum must be chosen for us) but there is a massless particle with scalar couplings which mediates long range forces which compete with gravity. The couplings of this particle may be read off the Lagrangian of Eq. 3. In Planck units, they are of order one. Worse, the couplings of this particle do not respect the equivalence principle. Experimentally these couplings must be very small, less than $10^{-2} - 10^{-4}$(24), so this possibility is ruled out. Of course, it is hard to see how in any case, in a world with broken supersymmetry, one could avoid having a dilaton potential.

If there is a dilaton potential, one might hope that we live in the weak coupling regime, and are gradually rolling to weaker and weaker (or stronger and stronger) coupling. Again this possibility may be ruled out, on general grounds. The smallness of the cosmological constant implies the smallness of the potential, and thus its derivatives. The dilaton must, in fact, be so light that its Compton wavelenght is of order the size of the universe. So this situation is no better than that of the vanishing potential, and is again ruled out.

A third possibility, which cannot be ruled out by such simple arguments, is shown in the figure. Perhaps, the potential has an isolated zero at some

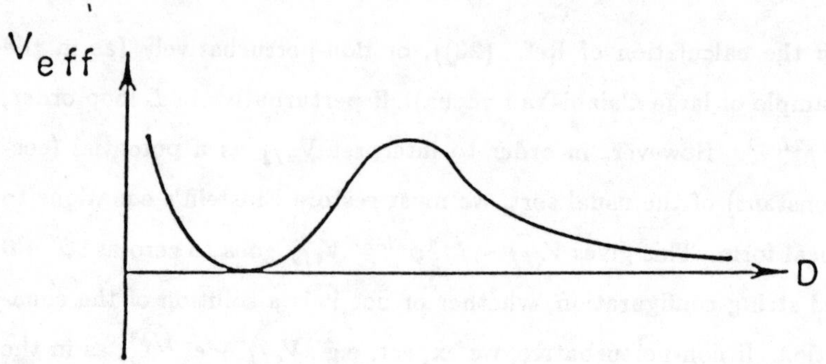

non-zero coupling, $\phi_o = e^{-D_o}$. Such a potential might have the form

$$V_{eff} \sim \phi^n[1 + O(\phi^2)]$$

or (21)

$$V_{eff} \sim e^{-1/\phi^2}[1 + O(\phi^2)]$$

At $D_0, V_{eff} = 0$, so the leading and next-to-leading order terms must be of the same size. By definition, this is strong coupling.

One can, of course, imagine loopholes of varying plausibility in this argument. Three possibilities which come to mind are:

1. The theory has a hidden, dimensionless parameter θ which is not the vev of any massless field (so that it does not adjust dynamically) and which can be arbitrarely small. Than, ϕ may be of order θ, e.g.

$$V_{eff}(\phi) = e^{-1/\phi^2}[\theta^2 + \phi^2 + O(\phi^4)] \tag{22}$$

The expansion in ϕ is then really an expansion in θ

2. Maybe the leading and next to leading order contributions are a good approximation to the potential, e.g.

$$V_{eff} \sim e^{-1/\phi^2}[1 + \phi^2 + \quad very \quad small \quad for \quad \phi \sim 1] \qquad (23)$$

This appears to require a miracle.

3. Perhaps $< \phi >$ and $V_{eff}(\phi)$ cannot be determined by weak coupling techniques, but some quantities (e.g. masses) can be. Again, this would appear to require great good fortune.

It seems much more likely that we will have to face the possibility that string theory, if it describes nature, is strongly coupled. Hopefully, some results from the weak coupling regime will carry over into the strong coupling regime. As stressed by Witten[9)] information of a topological character, such as the spectrum of light fermions and the unbroken gauge symmetry may have relevance even in strong coupling. It will be critical, of course, to find some principle which guarantees that at strong coupling there is a stable ground state with six compact dimensions, broken supersymmetry and vanishing cosmological constant.

Acknowledgements

It is a pleasure to thank E. Witten for discussions. The work of M.D. was supported in part by NSF grant PHY-82-17352 and that of N.S. by DOE contract DE-AC02-76ER02220.

References

1. For reviews see: J.H. Schwarz, Phys. Rep. 89 (1982) 223;M. B. Green, Surveys in H. E. Physics 3 (1982) 127.

2. M. B. Green and J. H. Schwarz, Phys. Lett. 149B (1984) 117.

3. D. J. Gross, J. A. Harvey, E. Martinec, and R. Rohm, Phys. Rev. Lett. 54 (1985) 502, Nucl. Phys. B256 (1985) 253 and Princeton preprint (1985).

4. P. Candelas, G. Horowitz, A. Strominger, and E. Witten, Nucl. Phys. B258 (1985) 46.

5. E. Witten, Phys. Lett. 149B (1984) 351.

6. I. Affleck, M. Dine and N. Seiberg, Nucl. Phys. B241 (1984) 493, B256 (1985) 557.

7. M. Dine, R. Rohm, N. Seiberg and E. Witten, Phys. Lett. 156B (1985) 55.

8. M. Dine and N. Seiberg, Phys. Rev. Lett. 55 (1985) 366.

9. E. Witten, unpublished.

10. M. Dine and N. Seiberg, IAS preprint (1985).

11. J. A. Shapiro, Phys. Rev. D11 (1975) 2937; M. Ademollo et.al., Nucl. Phys. B94 (1975) 221.

12. S. Weinberg, Texas preprint(1985).

13. E. S. Fradkin and A. A. Tseytlin, Phys. Lett. 158B (1985) 316. C. G. Callan, E. J. Martinec, M. J. Perry and D. Friedan, Princeton preprint (1985); M. Dine and N. Seiberg, unpublished.

14. A. H. Chamseddine, Nucl.Phys. B185 (1981) 403; E. Bergshoeff, M. de Roo, B. de Wit, and P. van Nieuwenhuysen, Nucl. Phys. B195 (1982) 97;G. F. Chapline and N. S. Manton, Phys. Lett. 120B (1983) 105. Another term needed for supersymmetry was added in Ref.(7).

15. E. Witten, Phys. Lett. 155B (1985) 151.

16. D. J. Gross and E. Witten, unpublished; last paper in Ref. (3).

17. M. Dine, V. Kaplunovsky, C. Nappi, M. Mangano and N. Seiberg, Princeton preprint(1985), to be published in Nucl. Phys.B.

18. V. Kaplunovsky, Phys. Rev. Lett. 55 (1985) 1033.

19. E. Witten, Nucl.Phys. B202 (1982) 253.

20. G. Veneziano and S. Yankielowicz, Phys. Lett. 113B (1982) 321.

21. V. A. Novikov, M. A. Shifman, A. I. Vainshtein, M. B. Voloshin and V. I. Zakharov, Nucl. Phys. B229 (1983) 394; V. A. Novikov, M. A. Shifman, A. I. Vainshtein and V. I. Zakharov, Nucl. Phys. B229 (1983) 381, 407.

22. R. Rohm and E. Witten, to appear.

23. R. Rohm, Nucl.Phys. B237 (1984) 553.

24. See e.g.:C. W. Misner, K. S. Thorne and J. A. Wheeler, Gravitation (W. H. Freeman, 1973).

E_6 SYMMETRY BREAKING IN THE SUPERSTRING THEORY†

Burt A. Ovrut*
The Rockefeller University
Department of Physics
New York, New York 10021

ABSTRACT

We derive two methods for determining the symmetry breaking of E_6 in the low energy superstring theory, and classify all breaking patterns. A method for calculating the effective vacuum expectation values is presented. We show that there are theories with naturally light SU_2^W Higgs doublets, and classify all theories in which this occurs.

1. Introduction

The theory of superstrings[1], which evolved from the string theories of the early 1970's[2], has recently undergone a great revival of interest, spurred by the work of Green and Schwarz[3]. An anomaly free, d=10 superstring theory is possible with gauge groups O(32)[3] or $E_8 \times E_8$[4]. The subsequent compactification of the ten dimensional space to $M_4 \times K$, with M_4 being Minkowski space and K a compact six-dimensional manifold, places further restrictions on the theory. In particular, the requirement that the compactification leave an unbroken N=1 local supersymmetry in d=4 implies that K has SU_3 holonomy[5]. The existence of spaces with SU_3 holonomy was conjectured by Calabi[6] and proved by Yau[7]. On such spaces one is naturally led to a four dimensional gauge theory with reduced gauge group $E_6 \times E_8$[5] Below the Planck scale the supergauge multiplet gives rise to an adjoint 496 of gluons and gluinos corresponding to E_8, and adjoint 78 of gluons and gluinos associated with E_6.

---

† Invited talk at the Workshop on Unified String Theory, I.T.P., Santa Barbara, California, 1985.

* On leave of absence from the University of Pennsylvania, Department of Physics, Philadelphia, Pennsylvania 19104.

In addition there are n^L 27's and n^R $\overline{27}$'s of E_6, each containing left handed fermions and their scalar superpartners. The number of generations, $N = n^L - n^R$, is restricted on topological grounds. For simply connected Calabi-Yau spaces K_o, N turned out to be hopelessly large, $N \geq 36$. This led the authors of Ref. 5 to consider multiply connected spaces $K = K_o/H$, where H is a discrete group which acts freely on K_o. For a specific choice of K_o, and $H = Z_5 \times Z_5$, the number of generations is reduced to N = 4. There is an additional benefit to having K be a multiply connected manifold. Define the Wilson loop

$$U = P\, e^{-i\int_\gamma T^a A^a_m \, dx^m} \qquad (1)$$

where A^a_m is the vacuum state E_6 gauge field, T^a, are the group generators and γ is a contour in K. Then, as pointed out in Ref. 5, one can have $U \neq 1$ even though the vacuum state gauge field strength ${F_{mn}}^a$ vanishes everywhere.[8] The reason for this is that, because of the "holes" in K, ${A_m}^a$ cannot necessarily be globally gauged to zero, even when ${F_{mn}}^a$ vanishes globally. Therefore, as long as contour γ is non-contractible, U is not necessarily unity. It follows from Eqn. (1) that U is an element of E_6. For a given vacuum configuration, ${A_m}^a$, there can be many inequivalent U matrices. Loosely speaking, there is one U for every "hole" in manifold K. If we define $\mathcal{H} = \{U\}$, then, with respect to matrix multiplication, $\mathcal{H}$ is a discrete subgroup of E_6. As an abstrct group $\mathcal{H} \subseteq H$. For example, if ${A_m}^a$ vanishes everywhere then $\mathcal{H} = \{1\}$. For non-trivial ${A_m}^a$, however, one can find that $\mathcal{H} = H$. Let $\mathcal{H}(=H)$ and $\mathcal{H}'(=H)$ correspond to vacua ${A_m}^a$ and ${A_m}^{a\prime}$ respectively. Then $\mathcal{H}$ and $\mathcal{H}'$ may be two inequivalent embeddings of H into E_6. The possible existence of non-trivial discrete group $\mathcal{H}$ has important ramifications. If ${A_m}^a$ is a fixed vacuum state and $\mathcal{H}$ the associated discrete subgroup, then denote by $\mathcal{G}$ the subgroup of E_6 that commutes with $\mathcal{H}$. Then, as discussed in Ref. 5, at energies below the Planck scale, E_6 will be spontaneously broken to $\mathcal{G}$.

This result is of fundamental importance in model building. For practical purposes the symmetry breaking is due to effective vacuum expectation values (VEV's)

$$\int_{\gamma} T^{a} A^{a}_{m} \, dx^{m} \qquad (2)$$

in the adjoint 78 representation of E_6. In this talk we derive methods for calculating patterns of E_6 symmetry breaking. For concreteness, we focus our discussion on the multiply connected manifold $K = K_o/H$ with $H = Z_5 \times Z_5$ and four generations. Many of our results, however, are valid for other manifolds and discrete groups.

2. E_6 Symmetry Breaking

We want to study the symmetry breaking patterns of E_6 on multiply connected Calabi-Yau manifolds $K = K_o/H$. In general, H can be any abelian or non-abelian discrete group that acts freely on K_o. The problem of E_6 breaking is simplified, however, if we restrict H to be an abelian group. For concreteness, we focus on the group $H = Z_5 \times Z_5$, (and manifold $K_o/(Z_5 \times Z_5)$ with four generations) although our discussion is valid for any abelian group. Let $A_m{}^a$ be a fixed vacuum configuration, $\mathcal{U}$ (= $\{U\}$) the discrete group of Wilson loops associated with it, and $\mathcal{G}$ the subgroup of E_6 that commutes with $\mathcal{U}$. Also, denote the $SU_3{}^c \times SU_2{}^W \times U_1$ gauge group of the standard model by $\tilde{\mathcal{G}}$. The following observations will be helpful in deriving E_6 symmetry breaking patterns. First, the two Wilson loop generators of $Z_5 \times Z_5$ commute and, hence, the associated effective VEV's can be simultaneously diagonalized. These VEV's can be extended to form a basis for the E_6 Cartan subalgebra. All the elements of this basis commute and, therefore, $\mathcal{G}$ must have the same rank as E_6, namely six. Second, $\mathcal{G}$ must be at least as large as $\tilde{\mathcal{G}}$ in order to successfully describe known phenomenology. Therefore, we must find the embeddings of $\tilde{\mathcal{G}}$ in E_6. This problem has already been studied extensively, in particular by Slansky[9], who has listed all the possible symmetry breaking patterns that preserve $\tilde{\mathcal{H}}$ and given the vector boson masses in terms of the symmetry breaking direction in weight space. Combined with our first observation it follows that $\mathcal{H}$ must be at least $SU_3{}^c \times SU_2{}^w \times U_1 \times U_1 \times U_1$. Third, note that if U is a Wilson loop in Z_5 then $U^5 = 1$. This follows from the fact that a non-contractable path repeated five times is contractable. Finally, note that if $\mathcal{H} = \{1\}$, E_6 remains unbroken. We now discuss our first method of determining E_6 breaking patterns.

A) Decomposition into maximal subgroups

E_6 has three maximal subgroups with rank 6: $SU_3{}^c \times SU_3{}^w \times SU_3$, $SU_2 \times SU_6$, and $SU_{10} \times U_1$. Let $\mathcal{H} = Z_5 \times Z_5$ and consider $SU_3{}^c \times SU_3{}^w \times SU_3$. If $U \in \mathcal{H}$ a possible form for U is

$$\begin{pmatrix} 1 & & \\ & 1 & \\ & & 1 \end{pmatrix} \times \begin{pmatrix} \alpha^j & & \\ & \alpha^j & \\ & & \alpha^{-2j} \end{pmatrix} \times \begin{pmatrix} \beta^k & & \\ & \beta^k & \\ & & \beta^{-2k} \end{pmatrix} \quad (3)$$

where j, k are integers, $\alpha^5 = \beta^5 = 1$, and, hence, $U^5 = 1$. Letting j and k vary from 0 to 4, we generate 25 distinct U's, which form a representation of the group $Z_5 \times Z_5$. The group $\mathcal{H}$ can be read off from the form of Eqn.(3). It is

$$\mathcal{H} = SU_3{}^c \times SU_2{}^w \times U_1 \times SU_2 \times U_1 \quad (4)$$

A second possible form for U is

$$\begin{pmatrix} 1 & & \\ & 1 & \\ & & 1 \end{pmatrix} \times \begin{pmatrix} \alpha^j & & \\ & \alpha^j & \\ & & \alpha^{-2j} \end{pmatrix} \times \begin{pmatrix} \beta^k & & \\ & \beta^{4k} & \\ & & 1 \end{pmatrix} \quad (5)$$

Again j, k are integers, $\alpha^5 = \beta^5 = 1$, and letting j and k vary from 0 to 4, we generate 25 distinct U's which form a representatoin of $Z_5 \times Z_5$. The group $\mathcal{H}$ can be read off from the form of Eqn. (5). It is

$$\mathcal{H} = SU_3^{\,c} \times SU_2^{\,w} \times U_1 \times U_1 \times U_1 \tag{6}$$

Eqs.(3) and (5) are the only possible embeddings of $\mathcal{K} = Z_5 \times Z_5$ into $SU_3^{\,c} \times SU_3^{\,w} \times SU_3$ that preserve $\tilde{\mathcal{H}}$. The embeddings of $\mathcal{K} = Z_5$ into $SU_3^{\,c} \times SU_3^{\,w} \times SU_3$ can be worked out in the same manner. The results are that $\mathcal{H}$ can be $SU_3^{\,c} \times SU_2^{\,w} \times SU_3^{\,c} \times SU_2$, and $SU_6 \times U_1$. The remaining breaking patterns can be found by embedding $Z_5 \times Z_5$ and Z_5 into maximal subgroups $SU_2 \times SU_6$ and $SO_{10} \times U_1$. It is clearly preferable to have a more general method, particularly one that allows a simple determination of the associated effective VEV's Such a method is most easily found by using the Dynkin formalism[10].

B) Method of Weyl weights

We can write the most general Wilson loop U as $\exp\{i\sum\lambda^i H^i\}$, where the H^i are the six generators of the Cartan subalgebra of E_6 and the λ^i are six real parameters. Arrange the λ^i into a vector $\lambda \equiv (a,b,c,d,e,f)$. Let α be a root vector of E_6. The mass of the vector boson corresponding to α is proportional to the inner product (λ,α). Since the vector bosons corresponding to $\tilde{\mathcal{H}}$ must be massless, $\lambda = (-c,c,a,b,c,0)$. We can then write U as a 27-dimensional diagonal matrix that depends on the three real parameters a,b, and c. The diagonal elements are given in Table 1, together with the transformation properties under $SU_3^{\,c} \times SU_2^{\,w}$, SO_{10}, and SU_5 of the elements of the 27-plet on which they act. Throughout this paper we label elements of the 27-plets that transform as an A under SO_{10} and a B under SU_5 by [A,B].

To find the group $\mathcal{H}$, we simply embed the discrete group in the above form for U. For $H = Z_5 \times Z_5$, we could let $e^{ia} = \alpha$, $e^{ib} = \beta$, and $e^{ic} = \alpha^j \beta^k$, where $\alpha^5 = \beta^5 = 1$ and j and k are integers.

$\mathcal{H} = Z_5 \times Z_5$ corresponds to two of the parameters a,b, and c being independent, while for $\mathcal{H} = Z_5$ there is only one independent parameter. Furthermore, by using Table 20 in Slansky's report[9], the relevant part of which is reproduced here as Table 2, we can find which $\mathcal{H}$ corresponds to a given $\mathcal{G}$.

As an example, let us find all the embeddings of $\mathcal{H} = Z_5 \times Z_5$ that break E_6 to $SU_4 \times SU_2^W \times U_1 \times U_1$ (Pati-Salam).[11] From Table 1 we see that if the (e,ν) and (u,d) are to lie in a 4 of SU_4 we must have $b=-c$. (We have used the fact that the (e,ν) corresponds to the weak doublet in the $[16,\overline{5}]$ and the (u,d) to the doublet in the $[16,10]$.) We can then either let a be independent or fix $a = 0, c, 2c, 3c$, or $4c$. (Because of the Z_5 symmetries $5c \equiv 0$, $6c \equiv c$, etc.)
For a independent we consult Table 2 and find that (λ,α) is zero only for $\alpha = \pm(000001)$, $\pm(0100\text{-}10)$, $\pm(0\text{-}10011)$, $\pm(10001\text{-}1)$, $\pm(100100)$, $\pm(\text{-}1101\text{-}1\text{-}1)$, $\pm(\text{-}10010\text{-}1)$, and the six roots (000000). So we have 20 massless vector bosons and these roots span a representation of $SU_4 \times SU_2^W \times U_1 \times U_1$. Similarly for $a=0$ we have 46 massless vector bosons and the gauge symmetry is $SO_{10} \times U_1$; for $a = c$ we have $SU_4 \times SU_2^W \times U_1 \times U_1$; for $a = 2c$, $SU_5 \times SU_2 \times U_1$; for $a = 3c$, $SU_5 \times SU_2 \times U_1$; and for $a = 4c$, $SU_4 \times SU_2^W \times U_1 \times U_1$. We break to the Pati-Salam group in three cases: i) a independent, $\mathcal{H} = Z_5 \times Z_5$, $e^{ic} = a^j$, $e^{ib} = \alpha^{-j}$, $e^{ia} = \beta^k$; ii) $a = c$, $\mathcal{H} = Z_5$, $e^{ic} = \alpha^j$, $e^{ib} = \alpha^{-j}$, $e^{ia} = a^j$; iii) $a = -c$, $\mathcal{H} = Z_5$, $e^{ic} = \alpha^j$, $e^{ib} = \alpha^{-j}$, $e^{ia} = \alpha^{-j}$.

By examining all the values for a,b, and c allowed by the discrete symmetry, one can exhaust all the symmetry breaking induced by $\mathcal{H}$. In particular we recover the embeddings found by the first method. In addition, we can find for which symmetries the effective VEV's are zero in some directions. We use this method in the next section to generate naturally light Higgs doublets.

3. Naturally Light Higgs Doublet Problem

We would like to have light Higgs doublets to set the electroweak scale and give masses through Yakawa couplings to the ordinary fermions. In particular, it is necessary that at least one of the light doublets be in the $[10,\overline{5}]$ or $[10,5]$ representations under $[SO_{10}, SU_5]$. At the same time, supersymmetric E_6 theories contain extra color triplets that can mediate nucleon decay via dimension 4 or 5 baryon number violating, $\not{R}$ invariant, operators. These triplets must be given very large masses. It is very difficult, even with fine tuning, to keep the doublet light while making the triplet heavy. Fortunately, the same mechanism that breaks the gauge symmetry gives us a natural method for splitting the doublets from the triplets. Although the method does not depend on which discrete symmetry we use, for simplicity we assume that $H = Z_5 \times Z_5$.

The nontrivial gauge fields that give rise to Wilson loops different from unity can also lead to a mass term through the coupling of the four-space part of the chiral superfields to the gauge fields in the compactified dimensions.[12] In other words, one of the 27's can couple to the $\overline{27}$ through a term involving an effective 78 VEV ($27 \times \overline{27} \times 78$ contains a singlet). These fields thus acquire a mass of order the inverse radius of the compactified dimensions, presumably the Planck mass, while the other four 27's remain massless. (Note that we cannot have a $27 \times \overline{27}$ bare mass term since, until E_6 is broken, the chiral superfields are all massless zero modes.) Let us call the 27 and $\overline{27}$ that pair off χ and $\overline{\chi}$, and the remaining 27's ψ. We then expect all the components of χ and $\overline{\chi}$ to gain huge masses and disappear from the spectrum. As we now show, however, it is possible for some of these components to remain naturally massless. This occurs when the diagonal entries of the U's that multiply these components are unity. That is, the associated effective VEV's are zero in these directions.

As an example, let us use Table 1 to find which components of χ and $\overline{\chi}$ are left massless for the breaking to $SU_4 \times SU_2^W \times U_1 \times U_1$

given in Section 2: i) for $b = -c$, a independent, none of the U^{diag} is one, the effective VEV then has no zeros, and all the components of χ and $\bar{\chi}$ are massive; ii) for $b = -c$, $a = c$, $U^{diag} = 1$ for the color triplet in the $[16,\bar{5}]$ and the singlet in the $[16,10]$, and hence those components of χ and $\bar{\chi}$ remain massless; iii) for $b = -c$, $a = -c$ the color triplet weak singlet in the $[16,10]$ and the $[16,1]$ remain massless.

Using Tables 1 and 2, we can find for which values of the parameters a,b, and c we obtain light doublets and what the resulting gauge symmetries are. The light doublets in χ can be used as Higgs fields to break $SU_2{}^W \times U_1$ and to give masses to quarks and leptons. The corresponding light doublets in $\bar{\chi}$ cannot couple to ordinary matter, and hence we ignore them. We list below all the cases in which at least one weak doublet in χ is light while the color triplets are all heavy:

i) $b = a + c$, a and c arbitrary
massless: doublet in $[10,5]$
gauge symmetry: $SU_3{}^c \times SU_2{}^W \times U_1 \times U_1 \times U_1$

ii) $b = 4c$, $a = 3c$
massless: doublet in $[10,5]$
gauge symmetry: $SU_5 \times SU_2 \times U_1$

iii) $b = 0$, $a = 4c$
massless: doublets in $[10,5]$ and $[16,\bar{5}]$, the singlet $[16,1]$
gauge symmetry: $SU_3{}^c \times SU_2{}^W \times U_1 \times U_1$

iv) $a = 2c$, b arbitrary
massless: doublet in $[10,\bar{5}]$
gauge symmetry: $SU_3{}^c \times SU_2{}^W \times U_1 \times U_1$

v) $a = 2c$, $b = 0$
massless: doublets in $[10,\bar{5}]$ and $[16,\bar{5}]$, singlet in $[16,10]$
gauge symmetry: $SU_3{}^c \times SU_2{}^W \times U_1 \times U_1$

vi) $a = 2c$, $b = 3c$
massless: doublets in $[10,5]$ and $[10,\bar{5}]$, the singlet $[1,1]$
gauge symmetry: $SU_3{}^c \times SU_2{}^W \times SU_2 \times U_1 \times U_1$

vii) $a = 2c$, $b = 4c$

massless: doublet in $[10,\overline{5}]$

gauge symmetry: $SU_5 \times SU_2 \times U_1$

viii) $b = 0$, a and c arbitrary

massless: doublet in $[16,\overline{5}]$

gauge symmetry: $SU_3{}^c \times SU_2{}^W \times U_1 \times U_1 \times U_1$

ix) $b = 0$, $a = 3c$

massless: doublet in $[16,\overline{5}]$

gauge symmetry: $SU_5 \times SU_2 \times U_1$

The embeddings of $\mathcal{H}$ that give rise to these values for a,b, and c can be readily found. For example, in case i) let $e^{ia} = \alpha^j$, $e^{ic} = \alpha^k$, and $e^{ib} = \alpha^{j+k}$, $\alpha^5 = 1$. This corresponds to $\mathcal{H} = Z_5 \times Z_5$. In case ii) let $e^{ic} = a^j$, $e^{ia} = \alpha^{3j}$, $e^{ib} = \alpha^{4j}$, which corresponds to $\mathcal{H} = Z_5$.

It is worth re-emphasizing that our light Higgs doublets were not obtained by fine tuning. Setting, for example, $b = 3c$, $a = 2c$ as in case vi) is a choice of parameters (optimistically a minimum for the vacuum configuration), but not a fine tuning, since e^{ia}, e^{ib}, and e^{ic} are restricted by the discrete symmetry to be fifth roots of unity. These Higgs doublets are totally massless until supersymmetry is broken spontaneously.

Acknowledgements

The results discussed in this talk were derived in collaboration with J. Breit and G. Segre. Similar results were also obtained independently by E. Witten and A. Sen.

Work supported in part by Department of Energy Grant Number DE-AC02-81ER40033B.

Rockefeller University Report Number RU/86/B/139.

References

1. J.H. Schwarz, Phys. Rep. 89 (1982) 223; M.B. Green "Surveys in High Energy Physics" 3 (1983) 127 are good reviews of the subject.

2. P. Ramond, Phys. Rev. D2 (1971) 2415; A. Neveu and J.H. Schwarz, Nucl. Phys. B31 (1971) 86, Phys. Rev. D4 (1971) 1109; L. Brink, D.I. Olive, C. Rebbi and J. Scherk, Phys. Lett. 45B (1973) 198 describe early models with fermions and bosons.

3. M. Green and J. Schwarz, Phys. Lett. 149B (1984) 117 for N=1 supersymmetry with open or closed strings. L. Alvarez-Gaume and E. Witten, Nucl. Phys. B234 (1983) 269 for N=2 supersymmetry with closed strings.

4. For recent discussions see also E. Witten, Phys. Lett. 149B (1984) 351; D.J. Gross, J. Harvey, E. Martinec and R. Rohm, Phys. Rev. Lett. 54 (1985) 46.

5. We are following here the discussion of P. Candelas, G.T. Horwitz, A. Strominger and E. Witten, "Vacuum Configuration for Superstrings", NSF-TTP-84-170 preprint (1984).

6. E. Calabi in Algebraic Geometry and Topology: A Symposium in Memory of S. Lefschetz (Princeton University Press, 1957), p. 58.

7. S.T. Yau, Proc. Nat. Acad. Sci. 74 (1977) 1798.

8. Similar mechanisms had been proposed for gauge symmetry breaking in Kaluza-Klein theories by Y. Hosotani, Phys. Lett. 129B (1984) 193.

9. R. Slansky, Phys. Rep. 79 (1981) 1.

10. E.B. Dynkin, Am. Math. Soc. Trans. Ser. 2 (1957) 111; 6 (1957) 245.

11. J.C. Pati and A. Salam, Phys. Rev. D10 (1974) 275.

12. Y. Hosotani, Phys. Lett. 126B (1983) 309; D.J. Gross, R. Pisarski and L. Yaffe, Rev. Mod. Phys. 53 (1981) 43.

Table 1. Diagonal elements of the Wilson loops U

U^{diag}	$SU_2{}^w \times SU_3{}^c$	SO_{10}	SU_5
exp{i(b-3c)}	(1,1)	1	1
exp{i(c+a-b)}	(2,1)	10	5
exp{2ic}	(1,3)		
exp{i(2c-a)}	(2,1)	10	$\bar{5}$
exp{i(c-b)}	(1,$\bar{3}$)		
exp{{-i(a+c)}	(1,1)	16	1
exp{ib}	(2,1)	16	$\bar{5}$
exp{i(a-c)}	(1,$\bar{3}$)		
exp{i(a-b-2c)}	(1,1)	16	10
exp{i(b-a)}	(1,$\bar{3}$)		
exp{-ic}	(2,3)		

Table 2. Nonzero E_6 Roots

Root α	(λ, α)	Root α	(λ, α)
(000001)	0	(0100-10)	0
(0-10011)	0	(10001-1)	0
(-11001-1)	3c	(-210000)	3c
(0-111-1-1)	a+b-2c	(00-12-10)	2b-a-c
(0-12-10-1)	2a-b-c	(1-11-110)	a-b-c
(101-10-1)	a-b-c	(1-11-11-1)	a-b-c
(0-11-101)	a-b-c	(001-1-10)	a-b-c
(0-11-100)	a-b-c	(00100-1)	a
(0-1101-1)	a	(00100-2)	a
(-1010-10)	a	(-1-11000)	a
(-1010-1-1)	a	(-100100)	b+c
(-1101-1-1)	b+c	(-10010-1)	b+c
(010-110)	2c-b	(000-120)	2c-b
(010-11-1)	2c-b	(-110-101)	2c-b
(-100-111)	2c-b	(-110-100)	2c-b
(-101-110)	a-b+2c	(-111-10-1)	a-b+2c
(-101-11-1)	a-b+2c	(-11-1011)	3c-a
(-12-1000)	3c-a	(-11-1010)	3c-a

TWISTING THE HETEROTIC STRING

J. A. Harvey*

Joseph Henry Laboratories
Princeton University
Princeton, New Jersey 08544

ABSTRACT

Gauge and Lorentz symmetry breaking are discussed in the context of the heterotic string. Symmetry breaking by Wilson lines and a singular limit of compactification on some Calabi-Yau spaces are shown to be equivalent to strings propagating on spaces with points identified under the action of discrete symmetries. It is also shown that certain twisted two dimensional heterotic string theories naturally incorporate the Frenkel-Lepowsky-Meurman construction of the monster goup F_1.

1. INTRODUCTION

String theories have the potential to give a truly unified description of all known interactions. Of course this unification comes only at the expense of an enormous increase in the symmetries of nature at a fundamental level. At low energies most of these symmetries must be hidden if string theories are to describe the real world. Ultimately we would like to determine whether or not various symmetries are broken by general topological or dynamical considerations. However at our present level of understanding there are many possible vacua for string theories and many possibilities for which symmetries are broken and which are unbroken. A more modest question that we can attempt to answer is simply whether or not we can find possible vacua that leave an unbroken gauge group which is close to the SU(3)×SU(2)×U(1) group we observe at low energies and with six of the ten spacetime dimensions

---

*Research supported in part by NSF Grant PHY80-19754.

compactified. Since string theories are more highly constrained than field theories we would like to formulate symmetry breaking directly in the string theory without having to resort to its low-energy field theoretic limit. Various aspects of symmetry breaking in string theories have been discussed in refs. 1, 5. Here I will discuss a method of symmetry breaking which was developed in ref. 2 and which has the advantage of being conceptually simple and reasonably tractable from a computational point of view. Since the $E_8 \times E_8$ heterotic string has the best phenomenological prospects I will discuss symmetry breaking exclusively in this theory.

2. TOROIDAL COMPACTIFICATION AND WILSON LINES

A simple but powerful way of breaking symmetries in string theory is to consider strings propagating on a space "modded out" by the action of a discrete group. For example if we want to break the ten-dimensional Lorentz group down to the four-dimensional Lorentz group by compactifying six of the coordinates the simplest consistent possibility is simply to take the six coordinates to parametrize a six-torus. In other words we mod out Euclidean six-space R^6 by the action of a Z^6 discrete group generated by six independent translations to form a six torus $T^6 = R^6/Z^6$. The six translations generate a lattice Λ and requiring that the theory be invariant under translation by a lattice vector requires that the center of mass momentum of the string lie on the dual lattice, Λ^*. In a closed string theory we must also allow for states with non-zero winding number on the torus. Viewed as strings on R^6 these are states that close only up to an element of the Z^6 group of translations.

This simple example has several interesting generalizations. In each case the procedure is to find a discrete symmetry G acting on the string theory, to project onto G-invariant states, and to include "twisted" string states that close only up to transformations by elements of G. From a path integral point of view the projection onto G-invariant states is a sum over all possible τ boundary conditions on the string. Since modular transformations mix σ and τ we must also include the twisted states with all possible σ boundary conditions in order to maintain modular invariance.

In order to describe gauge symmetry breaking we can take G to act also on the gauge degrees of freedom of the heterotic string. For example we can accompany each translation on R^6 in the previous example by a translation on the maximal torus of $E_8 \times E_8$. The group G is still Z^6 but each group element has the form $g = \alpha\beta$ where α acts on R^6 and β acts as a translation by $\pi\delta^I$ on the internal degrees of freedom, $\beta = e^{2\pi p^I \delta^I}$. Here p^I, I = 1..16 is the center of mass momentum in the bosonic formulation of $E_8 \times E_8$, the factor of π has been inserted for convenience and δ^I must not a lattice vector if the translation is to act non-trivially. Projecting onto G-invariant states correlates the momentum on T^6 with the internal quantum numbers or "internal momenta," leading in the field theory limit to what we would describe as symmetry breaking by Wilson lines. In string theory we must also include the twisted sectors. In this example these are string states that wind around T^6 and which have boundary conditions $X^I(\pi) = X^I(o) + \pi\delta^I$ for the $E_8 \times E_8$ coordinates. Using the normal mode expansion

$$X^I(\tau+\sigma) = x^I + p^I(\tau+\sigma) + \frac{i}{2} \sum_{n \neq o} \frac{\alpha_n^I}{n} e^{-2in(\tau+\sigma)} \tag{1}$$

we see that in this sector $p^I = \delta^I + L^I$ where L^I is a lattice vector. Since the weight vectors p^I label the U(1) charges of the Cartan sub-algebra of $E_8 \times E_8$ we see that in this sector the string states have fractional charges since δ^I is not a lattice vector if we have non-trivial symmetry breaking. In more interesting examples such as Calabi-Yau spaces the fundamental group will not be a product of Z factors but will be finite. For a manifold with fundamental group Z_n we can again introduce Wilson lines by associating translations around the non-contractible loops with group transformations. Now however n windings are equivalent to zero windings and thus $n\delta^I$ must be a lattice vector so the fractional charges are 1/n times the usual charges.[3]

In the case of a non-abelian fundamental group we cannot represent the action on $E_8{\times}E_8$ in such a simple way. Probably the simplest possibility is to use the fermionic formulation of $E_8{\times}E_8$ which has a linearly realized $O(16){\times}O(16)$ symmetry. As long as we can imbed the fundamental group in $O(16){\times}O(16)$ we can define its action on the fermions which transform as (16,1) + (1,16). There are certain cases in which we can still use the bosonic formulation however which are of interest since they bear a resemblance to compactification on orbifolds which will be discussed in the next section.

In the bosonic formulation of the heterotic string there are translational symmetries which represent the action of elements in the Cartan subalgebra and also discrete rotational symmetries which correspond to the automorphisms of the $E_8{\times}E_8$ lattice. Thus we could also represent certain group actions as rotations on the bosonic coordinates $X^I \to W^{IJ}X^J$ with W^{IJ} being a O(16) rotation which is an automorphism of the E_8 lattice, i.e. an element of the Weyl group of E_8. As a simple example consider a toy model where we represent the action on E_8 by the O(16) rotation $w = \mathrm{diag}\,(-1,-1,-1,\ldots -1)$. In terms of an automorphism of the E_8 Lie algebra w corresponds to the canonical involution $E_\alpha \to E_{-\alpha}$, $H^I \to -H^I$ in the Cartan-Weyl basis. This involution can be extended to the full E_8 affine Lie algebra.

In the untwisted sector the action of w on the previously massless states is given by

$$w|p^I\rangle = |-p^I\rangle\ , \quad w\,\alpha^I_{-1}|0\rangle = -\alpha^I_{-1}|0\rangle \tag{2}$$

so that the $\underset{\sim}{496}$ adjoint of E_8 splits into 120 states of eigenvalue +1 given by $1/\sqrt{2}\,(|p^I\rangle + |-p^I\rangle)$, $(p^I)^2 = 2$ and 128 states with eigenvalue -1 given by $\alpha^I_{-1}|0\rangle$, $1/\sqrt{2}\,(|p^I\rangle - |-p^I\rangle)$. In terms of Wilson lines the states of eigenvalue +1 can have zero momentum on the six-dimensional compact space and thus remain massless while the states of eigenvalue -1 acquire mass since they must have momenta on the six-dimensional space in order to be invariant. Thus we have broken E_8 down to O(16) with the $\underset{\sim}{120}$ adjoint of O(16) remaining massless and the spinor $\underset{\sim}{128}$ becoming massive.

In the twisted sector we have $X^I(\pi) = -X^I(o)$. Using the normal mode expansion (1) we see that this requires half integer mode numbers $n \varepsilon Z+1/2$, $p^I=0$ and $x^I=-x^I$. Since the center of mass coordinate is only defined modulo a lattice vector, $x^I=-x^I$ has non-zero solutions corresponding to the fixed points of the transformation w. w has 2^8 fixed points on the E_8 lattice but there is a subtlety due to the fact that the X^I have only left-moving modes. View the E_8 lattice Γ_8 as the abelian group generated by translations by the simple roots. Then in the heterotic string one must work not with Γ_8 but with a projective representation, $\bar{\Gamma}_8$. The two-cocycle in the vertex operators which determines this projective representation is necessary in order to ensure the proper commutation relations and is also required in order to have dual tree amplitudes. Thus rather than solving $x^I=-x^I$ mod Γ_8 or equivalently $2x^I \varepsilon \Gamma_8$ we should look for a faithful representation of the group $\overline{\Gamma_8/2\Gamma_8}$. The lowest dimensional faithful representation of this group has dimension $2^4=16$ so in effect we have to take the square root of the number of fixed points. This is also required in order to maintain modular invariance.

The half integer moding in the twisted sector also changes the zero-point energy. This change could be determined say by demanding Lorentz invariance in light-cone gauge but a quicker method is to simply regulate the zero point energy following Brink and Nielsen.[4] A bosonic oscillator with moding $n \varepsilon Z+a$, $o<a<1$, has a zero point energy

$$\sum_{n\varepsilon Z+a} \frac{n}{2} = \sum_{n\varepsilon Z} (n+a)^{-s} \Big|_{s=-1} = -\frac{1}{24} + \frac{1}{4} a (1-a). \qquad (3)$$

Thus the normal ordering term for the left-handed modes (including the 8 untwisted spacetime coordinates) is 16 (-1/24) + 8 (1/48) = -1/2 as compared to -1 in the untwisted sector. Note that the lowest 16-dimensional state and the states built on it are in a different SO(16) conjugacy class from the SO(16) states in the untwisted sector thus giving a non-abelian version of charge fractionalization.

Since w can be chosen to lie in a Cartan subalgebra of E_8 we would expect that twisting by w should be equivalent to a shift by some vector. It is not hard to check that this is indeed the case. In an orthonormal basis e_i, i=1..8 the roots of E_8 are given by

$$\pm e_i \pm e_j \qquad i \neq j$$

$$\frac{1}{2}(\pm e_1 \pm e_2 \ldots \pm e_8) \quad \text{even number of + signs} \tag{4}$$

and a twist by w is equivalent to a shift by $\delta = e_1$, which is half of a lattice vector. In fact for E_8 all automorphisms are inner, i.e. they can be represented as E_8 group elements so in this context any twist is equivalent to a Wilson line or in other words to accompanying a translation around a non-contractible loop by a gauge transformation. This is not quite true for $E_8 \times E_8$. The Weyl group of E_8 consists of the Weyl group for each E_8 plus an involution that interchanges the two E_8's. Modding out by this involution is not equivalent to a Wilson line and breaks $E_8 \times E_8$ down to E_8. It would be interesting if this could somehow be incorporated into realistic models thus getting rid of the "other" E_8 and its associated problems.

We have mentioned three ways of describing the action of a gauge transformation on the $E_8 \times E_8$ affine Lie algebra of the heterotic string; namely by a translation on the $E_8 \times E_8$ torus for an element in the Cartan subalgebra, by a discrete rotation for elements in the Weyl group, and by an O(16) transformation in the fermionic formulation for $E_8 \times E_8$ elements that lie in O(16)×O(16). While the latter description is certainly the most general the description in terms of twists by automorphisms of the $E_8 \times E_8$ lattice is rather similar to the construction of the Calabi-Yau Z-manifold discussed in Ref. 5. This suggests that we try to treat the compactification from ten to four dimensions also in terms of twisted strings.

3. COMPACTIFICATION ON ORBIFOLDS

Let us first recall the construction of the Z-manifold. One starts with a six torus T^6 which is a direct product of three two-tori, $T^6 = T_1 \times T_2 \times T_3$ with T_i defined by identifying its complex coordinate z_i by $z_i \sim z_i + 1$, $z_i \sim z_i + e^{i\pi/3}$. The lattice defining T^6 has various automorphisms, the one we will focus on is given by

$$\alpha: \quad z_i \rightarrow e^{2\pi i/3} z_i \tag{5}$$

which generates the cyclic group Z_3. One then considers the space T^6/Z_3 obtained by identifying points under the action of α. Since α has fixed points the resulting space is not a manifold but is what is called an orbifold or a V-manifold. As discussed in [5] it is possible to construct a Ricci-flat Kahler manifold from this orbifold by removing each one of the $3^3=27$ fixed points and gluing back in an appropriate manifold. This results in a Calabi-Yau manifold with Euler number 72 which leads to 36 generations of E_6 after embedding the SU(3) spin connection in one of the E_8 factors.

In analogy with twists acting on E_8 we might wonder whether it is possible to discuss string propagation on the Z-orbifold directly without worrying about repairing the singularities to form a manifold. Perhaps surprisingly the answer is yes. In fact string propagation on the orbifold captures much if not all of the topological structure of the resulting manifold in the cases where it is known how to repair the singularities. Let us review how this works for the simple example of the Z-orbifold discussed in [2].

Following our previous discussion we will have three sectors of string propagation, the untwisted sector, a sector twisted by α, and a sector twisted by $\alpha^2 = \alpha^{-1}$. If α acts only on space-time degrees of freedom as in eq. (5) extended appropriately to space-time fermions then the twisted sectors will not contain physical states in general. To see this recall that in a closed string thoery the mass operator must receive equal contributions from left- and right-moving modes or in other words $L_0 = \bar{L}_0$. This follows from invariance under rigid shifts in σ since $L_0-\bar{L}_0$ is the generator of infinitesimal shifts of σ. In the twisted sector the normal ordering terms in L_0 and $\bar{L}_0$ are shifted differently since the right movers consist of 8 bosonic and 8 fermionic space-time degrees of freedom while the left-movers consist of 8 bosonic space-time degrees of freedom and either 16 bosonic or 32 fermionic internal degrees of freedom. One way of ensuring that the normal ordering contributions match is to imbed the Z_3 group generated by α in an SU(3) subgroup of E_8 which amounts to a discrete version of imbedding the spin connection in the gauge group.

In analogy to the previous discussion of Wilson lines we consider the Z_3 group generated by elements of the form $\gamma=\alpha\beta$ where β acts on the $E_8\times E_8$ degrees of freedom. In the bosonic formulation we can take β to act as a translation $\beta=e^{2\pi i p^I \delta^I}$. In terms of the previous basis for the E_8 roots we can take the SU(3) roots to be $\pm(e_1+e_2)$, $\pm(e_1+e_3)$, and $\pm(e_2-e_3)$. An appropriate choice for δ is then

$$\delta = \frac{1}{3}(2e_1 + e_2 + e_3). \tag{6}$$

In the untwisted sector the right-moving ground state consists of a vector and spinor of SO(8). The decomposition of SO(8) in terms of SU(3) and the four dimensional helicity SO(2) gives

$$8_v = 3(0) + \bar{3}(0) + 1(1) + 1(-1)$$

$$8_s = 3(-1/2) + \bar{3}(1/2) + 1(1/2) + 1(-1/2) \tag{7}$$

where the helicity is listed in parantheses. Under the action of γ which lies in the center of SU(3) these states have eigenvalues determined by their triality, i.e. $e^{2\pi i/3}$ for the $\underset{\sim}{3}$, $e^{-2\pi i/3}$ for the $\underset{\sim}{\bar{3}}$, and 1 for $\underset{\sim}{1}$. γ invariant states are formed by combining these states with left-moving states with the complex conjugate eigenvalue. The left-moving states consist of the 8_v from the left-moving space-time coordinates and the adjoint of $E_8\times E_8$ given by

$$\alpha^I_{-1}|0\rangle_L\ ,\quad |p^I,\ (p^I)^2 = 2\rangle_L \tag{8}$$

It is not hard to check that these states decompose under $E_6\times SU(3)\times E_8$ as given below

γ eigenvalue	$E_6\times SU(3$
1	(1,1,248), (78,1,1), (1,8,1)
$e^{2\pi i/3}$	(27,3)
$e^{-2\pi i/3}$	$(\overline{27},\ \overline{3})$

(9)

The γ-invariant states arising from $8_v\times 8_v$ and $8_s\times 8_v$ consist of the supergravity multiplet with helicities (2,3/2,-3/2,-2) and ten super matter multiplets of helicity (1/2,0) and (-1/2,0). The γ-invariant states with gauge quantum numbers consists of the gauge supermultiplet for $E_6\times SU(3)\times E_8$ and chiral matter supermultiplets with helicity (1/2,0) transforming as (27,3) under $E_6\times SU(3)$ plus their antiparticles with helicity (-1/2,0) transforming as $(\overline{27},\overline{3})$.

Now consider the sector twisted by γ. If we form complex combinations of the compactified string coordinates, $Z_1=X_4+iX_5$, $Z_2=X_6+iX_7$, $Z_3=X_8+iX_9$, then these coordinates satisfy the boundary conditions

$$Z_i(\pi) = e^{2\pi i/3}\, Z_i(0). \tag{10}$$

Since the Z_i have both left- and right-moving modes this implies that the center of mass coordinate must be at one of the 27 fixed points of α. These twisted states are rather peculiar in that they sit at definite locations and have zero center of mass momentum. In particular this implies that they carry no knowledge of the size of the orbifold. The boundary condition on the Z_i's implies that the normal modes have $Z\pm 1/3$ quanta of oscillation for the compactified coordinates while the two uncompactified coordinates in light-cone gauge have integral quanta. Since γ respects space-time supersymmetry the space-time fermions have the same quanta of oscillation so that the normal ordering term is still zero.

To determine the massless modes in the sector twisted by γ we tensor the left-moving modes with right-moving modes to form γ-invariant states at each one of the 27 fixed points. The right-moving fermions have only two zero modes and quantization of these zero modes leads to two states with helicity (0,1/2) and γ eigenvalue 1. For the left-movers we have twisted boundary conditions for the spacetime bosons and also for the $E_8 \times E_8$ bosons. The latter satisfy

$$X^I(\pi) = X^I(o) + \delta^I \tag{11}$$

with δ^I given by Eq. (6). The center of mass momenta therefore lie on a "shifted" lattice $p^I = L^I + \delta^I$, $L^I \in \Gamma_8 \times \Gamma_8$. The twist (10) on the spacetime bosons implies a normal ordering term in the left-hand sector which is -2/3 using the formula (3). Massless states must therefore satisfy $N + \frac{1}{2}\Sigma(p^I)^2 - \frac{2}{3} = 0$, and will have eigenvalue $e^{\pi i p^I p^I}$ times the eigenvalue of the left-handed vacuum under the action of γ. It is easy to check that there are 27 states corresponding to p = δ+L with roots L having components along the e_i of the form

$$\begin{array}{ll} (-1,0,0,0..\pm1..0) & 10 \\ (0,-1,-1,(0)^5) & 1 \\ 1/2(-1,-1,-1,(\pm1)^5) & 16 \end{array} \tag{12}$$

corresponding to the decomposition of the $\underset{\sim}{27}$ of E_8 into SO(10) multiplets. One can use the chiral anomaly or the requirement of modular invariance to check that these states all have eigenvalue 1 under γ. The sector twisted by γ thus gives twenty seven (one for each fixed point) chiral matter multiplets in the $\underset{\sim}{27}$ of E_6. There are also nine E_6 singlet states with eigenvalue 1 arising from states created by

the fractionally moded space-time coordinates of the form $\alpha^{i}_{-1/3}$ $\alpha^{j}_{-1/3}|0\rangle_L$, $\bar{\alpha}^{i}_{-2/3}|0\rangle_L$, $i,j = 1,2,3$.
The sector twisted by $\gamma^2=\gamma^{-1}$ yields the antiparticles of these so altogether we find 36 chiral $\underset{\sim}{27}$'s of E_6.

This agrees with what one obtains by studying the manifold formed by repairing the singularities in the field theoretic limit. In that case the number of generations is one half of the Euler characteristic of the manifold. The Euler characteristic of the Z-manifold is easily calculated.[5] One starts with the six-torus T^6 with Euler characteristic zero and removes the 27 fixed points to form a space $\tilde{T}^6$ with Euler characteristic -27. Identifying points under the motion of the Z_3 group generated by α gives $\tilde{T}^6/Z_3$ with Euler characteristic -9. The Z-manifold is obtained by gluing in at the 27 fixed points a space X with Euler characteristic 3 to yield a total Euler characteristic of -9+3.27=72. In fact for the Z-manifold each generation is associated with a non-contractible four surface[6] with 9 of the four surfaces being those that come from T^6 that are invariant under the Z_3 symmetry and 27 coming from the fact that each X manifold used to repair the singularities has a non-contractible four surface. In the orbifold limit the size of the X manifold shrinks to zero and the wave functions of 27 of the chiral multiplets become concentrated on the fixed points of the Z_3 transformation corresponding to the twisted sectors of string propagation.

This analysis can easily be extended to more general models by starting with different six tori and modding out by more complicated crystallographic subgroups of SU(3). The various models that can be obtained this way are discussed in the second reference in [2]. One advantage of orbifolds is that they should make possible explicit calculations of Yukawa couplings as well as string loop effects. It is also possible that symmetry breaking effects may tend to drive Calabi-Yau spaces to the singular orbifold limit although this is just conjecture at present.

4. HETEROTIC MOONSHINE

One disappointing aspect of the previous discussion is the enormous choice of twistings or symmetry breakings that appear to lead to consistent string theories. We do not yet understand the principles that determine the correct (and hopefully unique) vacuum of string theory. There is a remarkable construction of the monster group F_1 in string theory where a particular twist is picked out from a mathematical point of view. This construction is reviewed in the talk by J. Lepowsky at this workshop. Although there is no clear connection between the monster and physics the elements of the construction of the monster in ref. 7 are so analogous to attempts to describe symmetry breaking in string theory that one feels that some elements of the construction may provide a clue as to the correct choice of twistings in realistic string theories.

The construction of the monster in ref. 7 can be naturally embedded in certain two-dimensional heterotic string theories. In generalizing the heterotic string one possibility is to take the left-moving modes to be those of the 26-dimensional bosonic string compactified on a 24-dimensional even self-dual lattice and the right-movers to be those of the two-dimensional N=2 superstring. However the N=2 theory has no transverse modes and is actually a field theory, not a string theory since there are no physical massive modes.

A more promising approach[8] is to again take the right-movers to be those of the ten-dimensional superstring but compactified on an eight-dimensional, even, self-dual lattice, i.e. Γ_8. There are 24 even self-dual lattices in 24 dimensions so there are 24 possible two-dimensional heterotic string theories of this type corresponding to which lattice is chosen for the left-moving modes. The most interesting of these is the one corresponding to the Leech lattice, Λ, which is the unique 24-dimensional even self-dual lattice without points of $(\text{length})^2=2$. In light cone gauge the physical degrees of freedom in this theory have the normal mode expansions

$$\chi^I(\tau+\sigma) = x^I + \tilde{p}^I(\tau+\sigma) + \frac{i}{2}\sum_{n\neq 0} \frac{\tilde{\alpha}_n^I}{n} e^{-2in(\tau+\sigma)} \qquad I=1..24, \quad \tilde{p}\epsilon\Lambda$$

$$\chi^i(\tau-\sigma) = x^i + p^i(\tau-\sigma) + \frac{i}{2}\sum_{n\neq 0} \frac{\tilde{\alpha}_n^i}{n} e^{-2in(\tau-\sigma)} \qquad i=1..8, \quad p\epsilon\Gamma_8 \quad (13)$$

$$S^a(\tau-\sigma) = \sum_n S_n^a e^{-2in(\tau-\sigma)} \qquad a=1..8$$

The mass operator is

$$\frac{1}{4}(\text{mass})^2 = N + \tilde{N} - 1 + \frac{1}{2}\sum_I (\tilde{p}^I)^2 + \frac{1}{2}\sum_i (p^i)^2 \qquad (14)$$

and invariance under shifts in σ gives the constraint

$$N + \frac{1}{2}\sum_i (p^i)^2 = \tilde{N} - 1 + \frac{1}{2}\sum_I (p^I)^2 . \qquad (15)$$

The right-handed ground state is again 16 dimensional and forms a representation of the S^a zero mode algebra. The massless states are formed by tensoring these states with left-moving states with $\tilde{N}=1$ or $\Sigma_I(p^I) = 2$. For the Leech lattice we thus get 24 massless supermultiplets of the form $|i \text{ or } a\rangle_R \times \tilde{\alpha}_{-1}^I |o\rangle_L$. At the next mass level for the left-moving states there are 24 states of the form $\tilde{\alpha}_{-2}^I |o\rangle_L$, 12·25 of the form $\tilde{\alpha}_{-1}^I \tilde{\alpha}_{-1}^I |o\rangle_L$ and 196,560 of the form $|p^I, (p^I)^2 = 4\rangle$. In general the number of states at each mass level is given by the partition function which for the left-movers is

$$\frac{\Lambda(q)}{q\prod_n (1-q^n)^{24}} \equiv \frac{\theta_\Lambda(q)}{(\eta(q))^{24}} = q^{-1} + 24 + 196884q + \ldots \qquad (16)$$

where $\theta_\Lambda(q) = \sum_{\ell\epsilon\Lambda} q^{\ell^2/2}$ is the theta function for the Leech lattice and $\eta(q)$ is Dedekind's η-function. Expect for the constant term this partition function is equal to the modular function $j(q) = q^{-1} + 744 + 196884q + \ldots$.

There are many mathematical connections between the monster group and modular functions which go under the general name of "monstrous moonshine." They started with the observation that each coefficient in $j(q)$ except for the constant term is a sum of dimensions of irreducible representations of the monster! In terms of string theory this means that at each mass level the states fall into monster multiplets except for the massless states. The question is then clearly whether one can actually construct the monster group as a symmetry group of the states of the string theory. For this to be possible the massless states must be removed but this can be done by using a twist that is the analog of the canonical involution discussed earlier in the context of breaking E_8 down to O(16). We thus consider the involution θ with $\theta\, \tilde{\alpha}^I_{-1}\, |o\rangle_L = -\tilde{\alpha}^I_{-1}|o\rangle_L$, $\theta|p^I\rangle_L = |-p\rangle_L$ etc. The projection onto θ invariant states removes the 24 massless states and the states at the next mass level of the form $\tilde{\alpha}^I_{-2}\,|o\rangle$, $1/\sqrt{2}\,(|p^I\rangle - |-p^I\rangle)$. However to preserve modular invariance the twisted sector must be added with boundary condition $X^I(\pi) = -X^I(o)$ and one must deal with a representation of the group $\overline{\Lambda/2\Lambda}$. The result of this is that the partition function for θ invariant states including untwisted and twisted sectors is as before but with the 24 removed. Furthermore Frenkel, Lepowsky and Meurman have shown that the monster group is in fact realized as a symmetry group of the theory. This is highly non-trivial since elements of F_1 must act between Hilbert spaces with different boundary conditions much as the supersymmetry operator does in the old superstring formolism.

There also appears to be moonshine associated with the superstring part of this construction.[7] The right-moving superstring partition function in this model is

$$\frac{16\theta_{\Gamma_8}(q)\prod_n (1+q^n)^8}{\prod_n (1-q^n)^8} = 2(8 + 2048q + 49152\, q^2 + \ldots) \qquad (17)$$

and the coefficients of the terms in parentheses are sums of irreducible representations of another sporadic group, .1, which is associated with automorphisms of the Leech lattice.

The connection between sporadic groups, modular functions and string theory is mysterious and probably quite deep. It may also have nothing to do with the real world. However it is a hint that there are very subtle mathematical structures in the heterotic string which we will probably have to understand better if we are to determine whether or not string theories describe the real world.

Acknowledgement

I would like to thank L. Dixon and C. Vafa for explanations of how one determines the charges of twisted string states and J. Lepowsky for discussions.

REFERENCES

1. E. Witten, Nucl. Phys. B258 (1985) 75;
J.D. Breit, B.A. Ovrut and G. Segrè, Phys. Lett. B158 (1985) 33;
A. Sen, Phys. Rev. Lett. 55 (1985) 33.
2. L. Dixon, J. Harvey, C. Vafa and E. Witten, Strings on Orbifolds I and II, Princeton preprints (1985).
3. X.-G. Wen and E. Witten, Electric and Magnetic Charges in Superstring Models, Princeton preprint (1985).
4. L. Brink and H.B. Nielsen, Phys. Lett. 45B (1973) 332.
5. P. Candelas, G. Horowitz, A. Strominger and E. Witten, Nucl. Phys. B258 (1985) 246.
6. A. Strominger and E. Witten, to appear in Comm. Math. Phys.;
A. Strominger, preprint NSF-ITP-85-109.
7. I.B. Frenkel, J. Lepowsky and A. Meurman, in "Vertex Operators in Mathematics and Physics, Publications of the Mathematical Sciences Research Institute No. 3 (Springer, Berlin, 1984).
8. J. Harvey and J. Minahan, in preparation.

QUANTIZED TORSION

Ryan Rohm
California Institute of Technology, Pasadena, CA 91125

I report on some properties of superstring compactifications with background antisymmetric tensor fields, $\langle H_{\mu\nu\rho}\rangle \neq 0$, in particular, the fact that generically the field must assume quantized values. This article will summarize a recent paper with E. Witten[1]. The word 'torsion' in the title refers to the fact that in the σ-model picture of the string moving in a background field, H couples as the torsion part of the affine connection. For our purposes, however, we will mean by 'torsion' the presence of nonzero $\langle H_{\mu\nu\rho}\rangle$.

Why should we consider torsion? In the limiting case in which the curved extra dimensions are large, it is sufficient to consider only the massless states of the string as background fields. The resulting classical field theory has Calabi–Yau manifolds as solution, with $\langle H\rangle = 0$, and the low energy field theory is phenomenologically promising. For the sake of completeness, we should consider as well the possibility that solutions exist with $\langle H\rangle \neq 0$. If we consider solutions in which the radius of the internal manifold is arbitrary, so that the corresponding σ-model action for the string has a β-function which vanishes identically, one can show that the closed part of H must be zero, so that H is determined from the requirement that $dH = tr(R \wedge R) - \frac{1}{30}\, Tr(F \wedge F)$[2]. If we consider instead the possibility of solutions in which the manifold has a fixed radius, involving a fixed point of the σ-model, less is known regarding possible solutions.

More importantly, there may be considerations outside classical field theory which lead us to consider solutions with torsion. The motivation for the current work grew out of work done by M. Dine, N. Seiberg, E. Witten and myself[3] on the low-energy phenomenology of Calabi-Yau compactifications of the heterotic string. The second E_8 couples to ordinary matter only through gravitational-strength interactions, and so we can treat it to a good approximation as an N = 1 supersymmetric Yang-Mills theory with no matter multiplets. Realistic four-dimensional phenomenology requires a gauge coupling constant which causes this Yang-Mills theory to become strongly interacting at a high energy scale. In this approximation, this does not lead to supersymmetry breaking. There is, however, in the low-energy limit of the string theory a gravitational-strength coupling to a scalar multiplet (consisting of an axion, a dilaton, and a Majorana fermion) whose auxiliary field acquires a vacuum expectation value when the gauge coupling becomes strong and gluino bilinears $\overline{\chi}^a\chi^a$ acquire vacuum expectation values. This gives rise to supersymmetry breaking and a vacuum energy of order $\langle\overline{\chi}^a\chi^a\rangle^2$. Surprisingly, though, the form of the lagrangian of the ten-dimensional effective field theory, with couplings of the form $(H_{ijk} - (\overline{\chi}^a\Gamma_{ijk}\chi^a))^2$ suggests that the vacuum energy vanishes if the antisymmetric tensor also assumes a nonzero vacuum expectation value, although supersymmetry is still broken. Thus the field-theory limit of the string suggests that there is a built-in supersymmetry-breaking mechanism in the $E_8 \times E_8$ string. One problem is to explain (or obtain) the presence of $\langle H_{ijk}\rangle$: if it is primordial, we lose the rationale for Calabi-Yau compactifications, whereas if we start with $H = 0$, $dH = 0$ seems to indicate that it cannot change. We will return to this question after considering the proper quantization condition.

Let us now discuss the quantization of H in string theory. We have a massless 2-index antisymmetric tensor field $B_{\mu\nu}(x)$ from which we form the field-strength $H_{\mu\nu\rho} = \partial_{[\mu}B_{\nu\rho]}$. This field has a natural coupling to the closed string via

$$I_H = \int d^2\xi \, \varepsilon_{\alpha\beta} \, B_{\mu\nu}\,(X(\xi))\, \partial^\alpha X^\mu(\xi)\partial^\beta X^\nu(\xi) \,. \tag{1}$$

This is the string analog of a charged point particle coupling to a Maxwell field:

$$I_M = \int A_\mu(X(\tau)) \left| \frac{\partial X^\mu}{\partial \tau} \right| d\tau \,. \tag{2}$$

The physically relevant object is H, not B, and so we must ask whether I is well-defined, as part of a first-quantized string action. In other words, is it possible to make consistent string quantum mechanics in the presence of H? Ambiguities are possible, because it may be that we cannot define B globally. Look at the integral of H over some closed three-surface V in our space-time manifold $M = M_4 \times K$. Define $S_V = \int_0 H = \int_0 H_{\mu\nu\rho}\, dx^\mu \wedge dx^\nu \wedge dx^\rho$. If V is the boundary of some four-manifold W, then $S_V = \int_{\partial W} H = \int_W dH = 0$, since H is closed. S_V is then a map from closed three-surfaces to the real numbers, which maps boundaries to zero; it is an element of the de Rham cohomology group $H^3(M,R)$ with real coefficients. If H is in a nontrivial cohomology class, so that S_V is nonzero for some V, then B cannot be globally defined, again by Stoke's theorem: if B were globally defined, we could use $H = dB$ to conclude that $S_V = 0$.

In this case, we must then prescribe how I is to be defined in the presence of the 'Dirac string singularities' of B. In general this is a rather subtle question; to begin with we will dodge the mathematical subtleties by assuming K to have a simple topology, so that $\pi_1(K) = \pi_2(K) = 0$. This being the case, any nontrivial three-surface in K defines a map of a three-sphere into K with the same homology class. (That is, any three-surface differs from a sphere by the boundary of a four-manifold.) We then can follow the classic argument: consider a map f of the two-sphere S^2 (string tree diagram) into K; then, since there is more than one way of contracting this map to a point, there are multiple ways of continuously defining the action assigned to a given map. We can define the action for a given map $f: S^2 \to K$ relative to the map f_0 which takes $S^2 \to$ point; we define $I_H[f_0] = 0$, and consider a set of maps f_u which interpolate between $f = f_1$ and f_0 (here we use our assumption that $\pi_2(K) = 0$). Then we define

$$I_H(f) - I_H(f_0) = \int_0^1 du \left[\frac{d}{du} I_H(f_u)\right]$$

$$= \int_0^1 du \frac{d}{du} \int_{f_u(S^2)} B$$

$$= \int_{B^3} H . \tag{3}$$

Because $dH = 0$, this definition is continuous under small variations of the contraction of the map f. However, the difference of two contractions g_1 and g_2 defines a map $S^3 \to K$, and then we have two ways of defining $I_H(f)$: $I_H(f) - \tilde{I}_H(f) = \int_{S^3} H = S_{S^3}$. As before, this H then defines an element of $H^3(M,R)$; the path-integral will be well-defined only if any ambiguity in the definition of I_H is a multiple of 2π. This implies that H determines an integral cohomology class, in $H^3(M,Z)$. We will discuss the implications of this condition after we deal with the case of more complicated manifolds.

Now we return to the realistic case of manifolds with more complicated topology. If $\pi_2(K) \neq 0$, then there are noncontractible maps of a two-sphere S^2, or other string diagram, into K. These mappings correspond to instantons in the σ-model action for the string, and give rise to certain interesting effects[5], but to us they are just a nuisance: they give an arbitrary additive parameter for each class of nontrivial mappings, corresponding to θ-angles (or values of certain axion-like pseudoscalar fields). They do not seem to cause any further difficulty in defining the string action.

There are other problems if $\pi_1(K) \neq 0$ as in realistic Calabi-Yau compactifications. We can repeat the above analysis, mapping a three-surface $\sum \times S^1 \to M$ to obtain the quantization condition; however, it is now possible that there are submanifolds of M which cannot be obtained this way. This analysis then does not determine whether the quantization condition continues to hold in this case. Here the analogy with Maxwell theory is helpful; although we cannot obtain general two-surfaces from mappings of $S^1 \times S^1$ (the 'path space' of the point particle), we know that in quantum mechanics the quantization condition is necessary to obtain a well-defined wavefunction. This can be shown from the point particle viewpoint by requiring that the action satisfy an additional condition, namely, that the action ascribed to a path which covers two loops in M be the same as the sum of the actions computed on each loop; we could call this requirement 'cluster decomposition.' The analogy for the string case would be to relate amplitudes with differing numbers of handles. We will then assume that the correct requirement for quantization of H is that all the closed three-surfaces give a quantization condition for H, so that as stated before, $H \in H^3(M,Z)$.

Now let us specify for which theories the quantization law holds. For bosonic theories, the result is valid for oriented but not for nonoriented strings, since in the latter case there is no corresponding field. For oriented closed superstrings (and heterotic strings) we again obtain the quantization condition; for type 1 superstrings, the quantization condition holds for a different reason, namely the existence of global anomalies.

For the type 1 and heterotic strings there are other considerations, reflecting the role of the antisymmetric tensor field in anomaly cancellation. Although we have shown that H is quantized, the quantization condition is not gauge-invariant, since $B_{\mu\nu}$ transforms under both gauge transformations and general coordinate transformations; instead, we should consider the gauge-invariant field-strength

$$\hat{H} = dB + \omega_{3L} - \frac{1}{30}\,\omega_{3YM}\,.$$

Then

$$\hat{S}_V = \int_V \hat{H} = n + \delta\,,$$

where δ represents the contributions to S from gauge and gravitational fields, as well as fermion determinants; because of the ambiguity in defining the ω's, it is only defined modulo an integer. δ is defined to be independent of B. In contrast to n, δ can change continuously, and we can increase it by one unit in such a way as to return all of the fields to their original values. So now $\langle H \rangle$ can change; since $d\hat{H} = tr(R \wedge R) - \frac{1}{30}\,Tr(F \wedge F)$, we've lost our conservation law $dH = 0$.

This does not vitiate our quantization law, however, since interpolating between states of differing values of n requires exciting gauge or gravitational fields on the internal manifold. More specifically, we can produce a change in $\int_W \hat{H}$ by one unit, for a given three- manifold W, by a time-dependent process which excites a gauge-field instanton. Consider first a process independent of x, y, z. Then $T \times W$ is a four-manifold, where T is the time axis; we construct on this four-manifold a Yang-Mills instanton with unit winding-number. We find

$$S_W(t_2) - S_W(t_1) = \int_W d\Sigma\,(\hat{H}(t_2) - \hat{H}(t_1))$$

$$= \int_{t_1}^{t_2} dt \left[\frac{d}{dt}\int_W \hat{H}(t)\right]$$

$$= \int_{W\times T} d\hat{H}$$

$$= \int_{W\times T} Tr(F \wedge F) = 1.$$

At t_2 the final gauge configuration is a gauge transformation of the original, and δ has changed by one unit; if we perform the gauge transformation at t_2, we exchange one unit of δ for one unit of n, so that $\langle H \rangle$ has changed by one.

We have illustrated a somewhat unphysical process independent of spatial coordinates; more realistically, we could construct a tunneling process by the 'bounce' method[6], replacing the time axis T by a 'radial' coordinate R. The end result is that in equilibrium, the field H is able to choose the quantum number which minimizes its energy.

Finally, we look at some of the consequences of this quantization condition for compactifications of superstrings. First, we see that if H assumes a nonzero vacuum expectation value parallel to the covariantly-constant tensor ε_{ijk}, as it would in order to cancel the vacuum energy arising from gluino condensation, the requirement that it also be an integral cohomology class implies a restriction on the complex structure of the manifold. Only for a certain discrete subset of the parameters determining the complex structure will this condition be satisfied, so that there are fewer massless scalars arising from deformations of the metric.

As for the phenomenology of the Calabi–Yau compactifications, this answers our question regarding the mechanism for cancelling the cosmological constant generated by gluino condensation; the condensation induces a nonzero potential for $\langle H \rangle$, $V(H) \simeq (H_{ijk} - (\overline{\chi}^a \Gamma_{ijk} \chi^a))^2$. H will then increase, one quantum at a time, by tunneling, until it matches the vacuum energy of the fermion condensate. The metric parameters and other massless scalar fields are able to make the final adjustments, so that H is both parallel to ε_{ijk} and an integral cohomology class.

Unfortunately there is an unpleasant consequence of this scenario: we are forced to have a manifold whose size is of order the Planck length by virtue of the quantization condition, and the requirement of reasonable 4-dimensional gauge couplings. N. Seiberg presented more details on this problem in his lecture.

This is all I have to say about the quantization of $\langle H \rangle$. I will now summarize some related work on the massless particle spectrum of strings in background fields with nonvanishing torsion. This involves an approximation technique that should prove useful in a general background field, although it has been used in detail only for more specialized situations closely related to classical solutions of the string theory. The approach is that of Born–Oppenheimer approximation to the string Hamiltonian: starting with the action for the string moving in a background field, we discard all of the modes except those responsible in the free-string theory for producing massless states. We then solve the resulting supersymmetric quantum-mechanics model, perhaps exactly, to obtain an approximation of the massless particle spectrum. The advantage of this approach over string perturbation theory is that it makes maximal use of the world-sheet supersymmetry of the action in suitable background fields.

We made a detailed analysis of only a particular class of background fields: we considered the $E_8 \times E_8$ heterotic string with a six-dimensional internal space which had the spin-connection embedded in the gauge group, with the additional assumption of a closed three-form H (torsion) which was also embedded in the gauge group. These assumptions on the background fields lead to an N = 1 supersymmetric σ-model action for the string. This supersymmetry is manifest if we use the Neveu–Schwarz–Ramond formulation of the superstring coordinates and the fermionic realization of the $E_8 \times E_8$ gauge degrees of freedom.

The results can be summarized as follows. Obtaining the correct set of states of the string requires summing over boundary conditions for the fermionic coordinates of the superstring and the gauge sector; since the antiperiodic boundary conditions break the world-sheet supersymmetry, we obtain four different sectors of states, with differing amounts of supersymmetry and different quantum-mechanical models for the set of massless states. The set of states with periodic boundary conditions in each direction give rise to chiral fermions in four dimensions, and also allow the most rigorous statements regarding the survival of massless states. The quantum-mechanical model which describes these states is a novel generalization of the de Rham complex; the supersymmetry operator is given by the exterior derivative operator d plus an operator H which corresponds to multiplication by the three form $H_{\mu\nu\rho}dx^{\mu}dx^{\nu}dx^{\rho}$. There are two sectors in which half of the fermions obey antiperiodic boundary conditions, yielding four-dimensional states which are either complex bosons or real fermions. They give rise to a quantum-mechanics model with N = 1/2 supersymmetry, with a single real supercharge which is a generalization of the Dirac or Rarita-Schwinger operator on the internal manifold. The bosonic states in the sector with antiperiodic boundary conditions on both sets of fermions have a quantum-mechanics model with no supersymmetry; in addition to a piece which generalizes the Laplacian, there are terms which explicitly mix the massless states with excited states of the string theory, so that for these states we have no improvement over ordinary perturbation theory.

REFERENCES

(For a more complete list of references, see [1].)

1. R. Rohm and E. Witten, "The Antisymmetric Tensor Field in Superstring Theory," Princeton preprint, October 1985.

2. P. Candelas, G. Horowitz, A. Strominger and E. Witten, Nucl. Phys. B258 (1985) p. 46.

3. M. Dine, R. Rohm, N. Seiberg, and E. Witten, Phys. Lett. 156B (1985) p. 55.

4. E. Witten, Nucl. Phys. B223 (1983) p. 422;
O. Alvarez, Berkeley preprint (1984).

5. X.-G. Wen and E. Witten, Princeton preprint, 1985.

6. S. Coleman, Phys. Rev. (1977) 2929;
S. Coleman and C. G. Callan, Jr., Phys. Rev. D16 (1977) 1762.

7. M. Dine and N. Seiberg, IAS preprints (1985)

APPENDIX

THE SUPER G-STRING[1]

V. Gates[2], Empty Kangaroo[3], M. Roachcock[4], and W. C. Gall[5]
Departure from Physics, University of Cauliflower, Broccoli, CA 94720

NOT TOO ABSTRACT

We describe a **new** string theory which gives all the phenomenology anybody could or will ever want (and more). It makes use of higher dimensions, higher derivatives, higher spin, higher twist, and hierarchy. It cures the problems of renormalizability of gravity, the cosmological constant, grand unification, supersymmetry breaking, and the common cold.

1. INTRODUCTION*

Actually, this paper doesn't need an introduction, since anyone who's the least bit competent in the topic of the paper he's reading doesn't need to be introduced to it, and otherwise why's he reading it in the first place? Therefore, this section is for the referee.

Various string theories have been proposed to solve the universe (or actually several universes, due to the use of higher dimensions)[1)]. Well, here's another one. Of course, this one's better because it solves problems the old ones didn't (or *really* solves problems the old ones only hand-waved away): (1) Proton decay is slowed by

[1]This work supported in parts by National Girdles and Foundations under Cary Grant No. PHYSICS85-12345.

[2]Address before September 1, 1985: *Massachusetts Institution for Technologists, Harvardbridge, MA 10101*

[3]Present address: *Brand-X University, Boson, Mass. 80800.*

[4]Mailing address: *Upto U., New York, NY 10036.*

[5]Address after September 1, 1985: *University of Merryland, Marcus Welby, MD 20742.*

*Complex conjugate.

the use of super-preservatives. As a result, the primary cause for its finite lifetime is cancer. (2) The hierarchy scale is found by renormalization group arguments to be of the order of $e^{4\pi D} \approx 10^{55}$, where D is the dimension of spicethyme (10). (3) The grand unification group is found to be $E(8) \otimes E(8) \otimes E(8) \otimes E(8)$, where the first two $E(8)$'s are from lattice compactification, the third $E(8)$ is from three-dimensional maximally extended supergravity, and the last $E(8)$ is for taxes. (4) Any particle we can't find is produced as a Skermion[2)].

Our string is a supersymmetric version of the G-string[3)], which is known to have maximal compactification[4)]. This is due to the appearance of generalizations of the Calliope-Yeow! metrics[5)]. Finiteness is proven to all orders. The masses of all hadrons can be predicted exactly. The no-content supergravity models[6)] can be obtained in the low-physics limit.

A preliminary version of these results was presented in[7)].

2. NOTATION

Before beginning, we introduce some notation (but not too much, because ambiguities are useful for hiding factors of $\sqrt{2}$ [8)] that we haven't checked yet). A $\wedge$ is used to indicate a wedge product of differential forms[9)] (for example, $dx^\mu \wedge dx^\nu \$_{\mu\nu}$ is a W2-form). Unless explicitly otherwise, we use index-free notation (i.e., we just leave all the indices off our equations). As a result, the Einstein summation convention is unnecessary (especially since nobody knows how to sum Einsteins anyway). Contravariant vectors are then distinguished from sandanistavariant vectors by context. "-1" is used to refer to the operator which produces 180° phase shifts (as in, e.g., the sublimation of ice). Before lattice compactification[10)], we work in 26 dimensions, with coordinates labeled as

$$a, b, c, d, e, f, g, h, i, j, k, l, m, n, o, p, q, r, s, t, u, v, w, x, y, z.$$

After lattice compactification, we work in 10 dimensions, with coordinates labeled as

$$0, 1, 2, 3, 4, 5, 6, 7, 8, 9.$$

Spinor coordinates are written as either Θ[11)] or θ[12)]. (Further superspace conventions are contained in[13)].) Letters indicating symbols that don't represent what you think they do are indicated by a " $\tilde{}$ " (as in, e.g., wall-$\tilde{\sigma}$). Greek letters are used to indicate culture, Gothic letters are used because they're pretty, Hebrew letters are used for religious reasons, and Cyrillic letters are avoided for political reasons.

3. CLASSICAL G-STRING

The action for the classical G-string is

$$\aleph = \frac{1}{\hbar}\oint \wp\Re\Im\top\bot\|\angle\forall\exists\neg\flat\natural\sharp\clubsuit\diamondsuit\heartsuit\spadesuit\coprod\sqcup\odot\uplus\hookleftarrow\rightleftharpoons\nearrow\nexists\smile\frown\bowtie\varpi\succ\models\Updownarrow]\llbracket . \qquad (\#)$$

(The inverse of this action has appeared in[14].) In component notation this becomes, unfortunately,

$$\begin{aligned}\aleph = \frac{1}{\hbar}\oint (& \wp\Re\Im\top\bot\|\angle\forall\exists\neg\flat\natural\sharp\clubsuit\diamondsuit\heartsuit\spadesuit\coprod\sqcup\odot\uplus\hookleftarrow\rightleftharpoons\nearrow\nexists\smile\frown\bowtie\varpi\succ\models\Updownarrow]\llbracket_1 + \\ & \wp\Re\Im\top\bot\|\angle\forall\exists\neg\flat\natural\sharp\clubsuit\diamondsuit\heartsuit\spadesuit\coprod\sqcup\odot\uplus\hookleftarrow\rightleftharpoons\nearrow\nexists\smile\frown\bowtie\varpi\succ\models\Updownarrow]\llbracket_2 + \\ & \wp\Re\Im\top\bot\|\angle\forall\exists\neg\flat\natural\sharp\clubsuit\diamondsuit\heartsuit\spadesuit\coprod\sqcup\odot\uplus\hookleftarrow\rightleftharpoons\nearrow\nexists\smile\frown\bowtie\varpi\succ\models\Updownarrow]\llbracket_3 + \\ & \wp\Re\Im\top\bot\|\angle\forall\exists\neg\flat\natural\sharp\clubsuit\diamondsuit\heartsuit\spadesuit\coprod\sqcup\odot\uplus\hookleftarrow\rightleftharpoons\nearrow\nexists\smile\frown\bowtie\varpi\succ\models\Updownarrow]\llbracket_4 + \\ & \wp\Re\Im\top\bot\|\angle\forall\exists\neg\flat\natural\sharp\clubsuit\diamondsuit\heartsuit\spadesuit\coprod\sqcup\odot\uplus\hookleftarrow\rightleftharpoons\nearrow\nexists\smile\frown\bowtie\varpi\succ\models\Updownarrow]\llbracket_5 + \\ & \wp\Re\Im\top\bot\|\angle\forall\exists\neg\flat\natural\sharp\clubsuit\diamondsuit\heartsuit\spadesuit\coprod\sqcup\odot\uplus\hookleftarrow\rightleftharpoons\nearrow\nexists\smile\frown\bowtie\varpi\succ\models\Updownarrow]\llbracket_6 + \\ & \wp\Re\Im\top\bot\|\angle\forall\exists\neg\flat\natural\sharp\clubsuit\diamondsuit\heartsuit\spadesuit\coprod\sqcup\odot\uplus\hookleftarrow\rightleftharpoons\nearrow\nexists\smile\frown\bowtie\varpi\succ\models\Updownarrow]\llbracket_7 + \\ & \wp\Re\Im\top\bot\|\angle\forall\exists\neg\flat\natural\sharp\clubsuit\diamondsuit\heartsuit\spadesuit\coprod\sqcup\odot\uplus\hookleftarrow\rightleftharpoons\nearrow\nexists\smile\frown\bowtie\varpi\succ\models\Updownarrow]\llbracket_8 + \\ & \wp\Re\Im\top\bot\|\angle\forall\exists\neg\flat\natural\sharp\clubsuit\diamondsuit\heartsuit\spadesuit\coprod\sqcup\odot\uplus\hookleftarrow\rightleftharpoons\nearrow\nexists\smile\frown\bowtie\varpi\succ\models\Updownarrow]\llbracket_9 + \\ & \wp\Re\Im\top\bot\|\angle\forall\exists\neg\flat\natural\sharp\clubsuit\diamondsuit\heartsuit\spadesuit\coprod\sqcup\odot\uplus\hookleftarrow\rightleftharpoons\nearrow\nexists\smile\frown\bowtie\varpi\succ\models\Updownarrow]\llbracket_{10} + \\ & \wp\Re\Im\top\bot\|\angle\forall\exists\neg\flat\natural\sharp\clubsuit\diamondsuit\heartsuit\spadesuit\coprod\sqcup\odot\uplus\hookleftarrow\rightleftharpoons\nearrow\nexists\smile\frown\bowtie\varpi\succ\models\Updownarrow]\llbracket_{11} + \\ & \wp\Re\Im\top\bot\|\angle\forall\exists\neg\flat\natural\sharp\clubsuit\diamondsuit\heartsuit\spadesuit\coprod\sqcup\odot\uplus\hookleftarrow\rightleftharpoons\nearrow\nexists\smile\frown\bowtie\varpi\succ\models\Updownarrow]\llbracket_{12} + \\ & \wp\Re\Im\top\bot\|\angle\forall\exists\neg\flat\natural\sharp\clubsuit\diamondsuit\heartsuit\spadesuit\coprod\sqcup\odot\uplus\hookleftarrow\rightleftharpoons\nearrow\nexists\smile\frown\bowtie\varpi\succ\models\Updownarrow]\llbracket_{13} + \\ & \wp\Re\Im\top\bot\|\angle\forall\exists\neg\flat\natural\sharp\clubsuit\diamondsuit\heartsuit\spadesuit\coprod\sqcup\odot\uplus\hookleftarrow\rightleftharpoons\nearrow\nexists\smile\frown\bowtie\varpi\succ\models\Updownarrow]\llbracket_{14} + \\ & \wp\Re\Im\top\bot\|\angle\forall\exists\neg\flat\natural\sharp\clubsuit\diamondsuit\heartsuit\spadesuit\coprod\sqcup\odot\uplus\hookleftarrow\rightleftharpoons\nearrow\nexists\smile\frown\bowtie\varpi\succ\models\Updownarrow]\llbracket_{15} + \\ & \wp\Re\Im\top\bot\|\angle\forall\exists\neg\flat\natural\sharp\clubsuit\diamondsuit\heartsuit\spadesuit\coprod\sqcup\odot\uplus\hookleftarrow\rightleftharpoons\nearrow\nexists\smile\frown\bowtie\varpi\succ\models\Updownarrow]\llbracket_{16}),\end{aligned}$$

$$\textit{where} \quad \theta = \vartheta = \Theta = \boldsymbol{\Theta} = \textcircled{H} = \smiley = \overset{\smiley}{\bowtie} \quad \textit{and} \quad \Omega^{-1} = \mho \quad ,$$

and we have used the Newton-Witten equation

$$F = ma \quad .$$

The G-string is unique in that it combines the properties of all known string theories. It has 26-dimensional modes propagating to the left, 10-dimensional modes propagating to the right, and 2-dimensional modes just sitting around wondering what the hell is going on. (These left- and right-footed modes only propagate on the surface of the string, because that's as far as you can get on one foot.) 4 dimensions then follows directly from the simple identity[15]

$$4^2 = 26 - 10.$$

In ten-dimensional (x) space the G-string has global supersymmetry, in two-dimensional ($\sigma - \tau$) space it has local supersymmetry, and in four-dimensional (honest-to-God) space it has no supersymmetry. Internal symmetry is introduced by applying Champagne factors: b, c, and d quarks[16] on one end of the string, and s, t, and u quarks on the other. Since the latter quarks are also the Mandelstam variables, we can introduce higher-derivative interactions through that end. (The t quark is also the tea quark of the MI tea-bag[13], so the latter model will be produced in the Regge limit where s and u go to infinity while fixing some tea. The string is reobtained in the inverse limit $\overset{\breve{\infty}}{\frown} \leftarrow \forall\exists\perp$.) The last term in the action is a Wess-Zumino term, which causes the coupling to be quantized (see below).

4. FIRST-QUANTIZED G-STRING

Since the coupling is quantized (see above), the action is finite to all orders. As a result, all higher-order corrections can be neglected, which is good, since nobody wants to calculate them anyway. (Similar remarks apply to anomalies.)

The most important property of the quantum G-string is that it provides more possibilities for compactification. This is accomplished by use of the coordinate

$$x^{\mu(\sigma)},$$

where the vector index is a function of the string coordinates. Effectively, this makes the spacetime dimension a function of σ. We can thus choose $D(\sigma) = 4$ *at the boundary* of the open string. As a result, all massless vector fields (photons, gluons, etc.), which couple only to the end of the string, couple only to four-dimensional spacetime, whereas gravity, which couples to the middle of the string, couples to all dimensions. The extra dimensions therefore act as "dark matter".

(More generally, we can choose D to be a nonlinear function of σ, thus naturally introducing nonlinear σ-models.)

The super G-string therefore allows for a much greater choice of effective theories. Thus, it not only produces QED[17] and QCD[18], but also QAD (quantum aerodynamics), QHD (quantum hydrodynamics), QUD (quantum uterine device), and QVD (quantum venereal disease).

This action is conformally invariant[19]. As a result, it describes particles of continuous mass[20]. Consequently, all masses of the known (and unknown) particles are predicted. However, since there are an infinite number of particles, lack of space prevents us from giving these results here. (Preliminary results appeared in[21].)

5. SECOND-QUANTIZED G-STRING

Due to the conformal symmetry of the super G-string, the second-quantized G-string is *the same* as the first-quantized one[22]. The only difference is that more parentheses are needed: e.g., $\Phi[X(\sigma)]$. Path[23] integrals are performed in terms of the sheets that the strings sweep out in spacetime. In the interacting case the nontrivial topology gives contour sheets, so we simplify the calculation by conformal transformations on the Green functions[24]. Loop integrals can be expressed in terms of Jacobi Theta functions[25], but since $\Theta^2 = 0$[26], these cancel against the Θ's of the anticommuting coordinates, giving another proof of finiteness. In performing explicit calculations, we use the interacting string picture, with all string fields expanded in terms of incoherent states. Amplitudes can then be expressed in terms of the two-dimensional Green function

$$G(\sigma,\tau) = \int d\nu \; I_\nu(\sigma) R(\sigma,\tau;\nu),$$

where $I = \Im J$ is the Imbessel function, R is the retarded potential, and ν is a dummy variable.

Since this formulation corresponds to field theory, it's useful to have the gauge invariance of the string manifest. This is much easier for the super G-string than other supersymmetric strings (Neveu-Schwarz, Green-Schwarz, or FAO-Schwarz[27]), since the Shoparound matrix is invertible on the Burma module. This produces Landau ghosts which exactly cancel the Faddeev-Popov ghosts (which is fortunate, since the Soviet government doesn't officially recognize the existence of ghosts[28]). As a result, the Verysorry algebra (which is such afine

algebra) can be nonlinearly realized on the interacting string field as a subgroup of the noncompact (via noncompactification) group SO(WHAT). Its grated extension O(4,CRYINGOUTLOUD) carries the entire super G-string as a (one-particle) irreducible representation. This result can be represented concisely in terms of the Stynkin diagram for averyffine SU(2)[29]:

and its corresponding Old toblow:

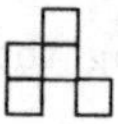

The gauge-invariant field-theoretic string action then follows directly by the usual group theory constructions[30], and is therefore too trivial to discuss further here. This result can also be obtained by the application of the twistor calculus to supercocycles, but if you've ever worked with those formalisms you know it's not worth the trouble[31].

6. THIRD-QUANTIZED G-STRING

Due to the conformal symmetry of the super G-string, the third-quantized G-string is *the same* as the second-quantized one. The only difference is that still more parentheses are needed: e.g., $\text{Œ}\{\Phi[X(\sigma)]\}$. Here σ is a coordinate, $X(\sigma)$ is a function, $\Phi[X]$ is a functional, and $\text{Œ}\{\Phi\}$ is a functionalal, describing the wave (particle) function of the universe. The universe begins as 26-dimensional, collapsing to 10-dimensional[32], with extra entropy coming from the phonons produced by the crystalization of the resulting 16-dimensional lattice. (No entropy comes from the 6 dimensions compactified into Cabala-Now spaces[33] because it gets Killed by the vectors of the leggoamy group RU(CRAZY).) Above the Hagedorn temperature the lattice undergoes a phase transition to an amorphous solid, explaining the homogeneity of the early universe.

The lattice also regularizes ultraviolet divergences (giving a *third* proof of finiteness, hence third-quantization[34]), and can be used to apply Monte Zuma calculational techniques[13]. (We also have a *fourth* proof of finiteness, but it requires use of the light-cone gauge[35], and is thus beneath the scope of this article[36].) Since higher-order corrections are negligible, quenching is an accurate approximation. However, these methods are not applicable for the early phase of the universe, where the amorphous solid has not yet become a lattice, corresponding to the fact that strong-coupling lattice methods are not accurate for this weak-coupling phase. Since the super G-string contains fermions, the string's latticization also solves the long-standing problem of putting fermions on a lattice. Finally, the lattice is furthermore useful for studying group theory, since it automatically gives representations of the Greasy-Fish Monster group. We thus obtain the celebrated result[37]:

$$e^{4\pi\cdot 10} \gg \text{any reasonable number you know.}$$

7. FOURTH-QUANTIZED G-STRING

There's no such thing as fourth quantization, but if there were, it would be *the same* as the third-quantized one, due to the conformal symmetry.

8. CONCLUSIONS

Our conclusions were already stated in the abstract and introduction, so go back and read them again. We could tell you what we're going to do in our next paper, but since we've already done everything in this paper, there won't be one (unless, of course, we find yet another string model that we like even better, in which case we'll write a paper telling you what's wrong with this one).

ACKNOWLEDGMENT

One of us (W.C.G.) would like to thank Ronald Reagan, but the rest of us (V.G., E.K., M.R.) won't let one of us because the rest of us hate his guts.

In fact, we don't really want to thank anybody, but if we don't, they'll get mad. On the other hand, if they don't read this paper, they won't know we didn't acknowledge them. Therefore, we would like to thank (WRITE YOUR NAME HERE)[38] for invaluable advice and encouragement.

NOTE ADDED IN PROOF

We have found a proof of Fermat's last theorem using the super G-string, but it's too small to fit in this margin.

After this work was completed, we received a preprint, but we don't know who wrote it because we were so afraid they might have produced some of our results that we didn't even open the envelope. Besides, we don't want to have to share our Nobel prize with anybody. However, we will acknowledge the work of Isaac Newton[39], because they don't award Nobel prizes posthumously. We have also heard that other people have done work along similar lines[40], but failed miserably.

NOTE ADDED IN PROOF OF NOTE ADDED IN PROOF

We decided to open the envelope after all, but it turned out to be just another paper by you-know-who[41], and we all know all his stuff is garbage, so we just threw it away.

REFERENCES

1) El Witten, What everybody's going to be working on as soon as word of this paper gets out, Princewiton preprint (Maybe 1985), in preparation.
2) Enrico Skermi, Phys. Mod. Rev. **199** (1960) 2222.
3) Georg Goubou, Phys. Rev. **1** (1906) 239.
4) Gomer Pyle, Private communication.
5) Definitely Gross, Just Horny, Amiable Martinet, and Roadsleadto Rome, The wet erotic T-shirt, Preprintston preprint (November 1984).
6) D.V. 10^{-15}-megalopolis, A new unlimited fuel source: NO-CONCERN preprints (Pick a month, 1984).
7) V. Gates, Empty Kangaroo, M. Roachcock, and W. C. Gall, private communication.
8) Pythagoras, private communication.
9) Orlando Florida, Topology, holonomy, homology, homotopy, homosapiens, cohomology, cohosalmon, and the mohorovic discontinuity, Colorado preprint OOO-000 (Tuesday 1981).
10) A. Chodos and J. Rabin, Monte Carlo simulation of a realistic unified gauge theory, Yale preprint YTP 83-41 (April 1983).
11) W. Siegel, Phys. Lett. **149B** (1984) 157, 162.
12) W. Siegel, Phys. Lett. **151B** (1985) 391, 396.
13) V. Gates, Empty Kangaroo, M. Roachcock, and W. C. Gall, Physica **15D** (1985) 289.
14) S.S. Schweber, An introduction to relativistic quantum field theory (Row, Peterson and Company, Evanston, 1961) pp. 163-164.
15) Simon Simple, Advanced applications of the Ugh equation $1 + 1 = 2$, Tòrino preprint (July 1984), to appear in Nuo. Cemento Litters.
16) Mury Gel-Man, private communication, # 137, p. 17, eq. (4.32).
17) I. Cain and U. Abel, Killing vectors in gauge groups, UC Santa Claus preprint UCSC-85/39 (January 1985).
18) Gluinos Ferrar, Supersymmetry breaking and electromagnetic boogie, Rutgers preprint BE4-5789 (May 1982).
19) Cloven 't Hooft, The Kosher-Riemann equation, Utrecht preprint, to disappear (March 1983).
20) Viet Nahm, Milk Cartan, and Vector Kac, Poles, cuts, and polecats, IMMT preprint (February 1972).

21) C.G. Wohl, R.N. Cahn, A. Rittenberg, T.G. Trippe, G.P. Yost, F.C. Porter, J.J. Hernandez, L. Montanet, R.E. Hendrick, R.L. Crawford, M. Roos, N.A. Törnqvist, G. Höhler, M. Aguilar-Benitez, T. Shimada, M.J. Losty, G.P. Gopal, Ch. Walck, R.E. Schrock, R. Frosch, L.D. Roper, W.P. Trower, and B. Armstrong, Rev. Mod. Phys. **56** (1984) S1.

22) Frie Danfrie, private communication, not to appear (it's private).

23) Frank Capra, The τ of physics (Bench Press, Venice, 1970) p. 99.

24) M. B. Green and J. H. Schwarz, Christawful transformations, Caltech preprint (October 1974).

25) Jacob E. Theta, Tables of lower meditational functions, Batman manuscripts (Wayne Publishing, Gotham City, 1939), p. 423.

26) C. Kent, Supersymmetry, Kryptonne National Lavatory preprint KAL-L/72 (September 1982).

27) F.A.O. Schwarz, A new toy model for superstring theory, Aspen preprint ASP-843 (August 1983).

28) A.I. Akhiezer and V.B. Berestetskii, Quantum electrodynamics (Interscience, New York, 1965), concluding remarks.

29) Twoloops Lautrec, Marker Off, and Alittlegusto Snotty, Cat-Snooty algebras and their Dinky little diagrams, Burkly preprint (June 1984).

30) John Iadfkgnsdfjbnd and Tom Hkjsdfbkjnsdjknvbkjnv, Another theorem on the dfvbdjhbvdh group in wkfjgndf of djfhbs rings and the Louisville transformation, *in* New results in 5-theory (Obscure Publishing, Louisville, 1842) p. 1596.

31) It's not worth the trouble to look up the references, either.

32) Parton Zwiebein, Fastidious physicists: remove unwanted dimensions, Daily Cal advertisement.

33) Phiphteen Candles, Gorey The Horrible, Andhe's Strummingher, and Endward Witty, Garbage compactifiers for superfluous dimensions, Kingston preprint (January 1985).

34) Stanley Manlystanley, as reported by Youknow Zumiknow, according to Chocolate-Bar Ḑąçķçķį, private rumor.

35) Kneel Mark, The covariant formalism is rubbish, private thought.

36) Covariant Formalism, Kneel Mark is rubbish, private thought.

37) We don't know who celebrated it because we weren't invited.

38) (WRITE YOUR NAME HERE), My favorite paper, My Favor. J. **1** (1980) 1.

39) I. Newton, Onway ethay opagationpray ofway ightlay underway ethay influenceway ofway avitygray, Cambridge preprint (December 1700), in Latin.

40) Private communication overheard in the mens' room at a recent conference, but we didn't see who it was because we didn't want to get up.

41) He never references us, so we're not gonna reference him.

LIST OF PARTICIPANTS

Andreas Albrecht
Inst. for Theoretical Physics
University of California
Santa Barbara, CA 93106
USA

Orlando Alvarez
Dept. of Physics
University of California
Berkeley, CA 94720
USA

Natan Andrei
Dept. of Physics
Rutgers University
Piscataway, NJ 08854
USA

Jonathan A. Bagger
SLAC, Theory Group
P.O. Box 4349 — Bin 81
Stanford, CA 94305
USA

Prof. Thomas Banks
SLAC, Theory Group
P.O. Box 4349 — Bin 81
Stanford, CA 94305
USA

Prof. I. Bars
Dept. of Physics
Univ. of S. California
Los Angeles, CA 90089
USA

A. Bengstsson
Dept. of Physics
Queen Mary College
Mile End Road
London E1 4NS
ENGLAND

Sid Bludman
Inst. for Theoretical Physics
University of California
Santa Barbara, CA 93106
USA

Robert Brandenberger
Inst. for Theoretical Physics
University of California
Santa Barbara, CA 93106
USA

Dr. Lars Brink
Inst. of Theoretical Physics
Chalmers Univ. of Technology
S-41296 Göteborg
SWEDEN

Daniel G. Caldi
Dept. of Physics, U-46
University of Connecticut
Stoirs, CT 06268
USA

Prof. P. Candelas
Dept. of Physics
University of Texas
Austin, TX 78712
USA

Martin Cederwall
Inst. of Theoretical Physics
Chalmers Univ. of Technology
S-41296 Göteborg
SWEDEN

Dr. G. Chapline
Physics Department
Lawrence Livermore Lab.
Livermore, CA 94550
USA

Utpal Chattopadhyay
Physics Department
University of Maryland
College Park, MD 20742
USA

L. Clavelli
Dept. of Physics & Astronomy
University of Alabama
University, AL 35486
USA

Joanne Cohn
Enrico Fermi Institute
University of Chicago
Chicago, IL 60637
USA

E. Corrigan
Dept. of Math Sciences
Science Laboratories
South Road
Durham DH1 3LE
ENGLAND

Prof. Carleton DeTar
Dept. of Physics, 201 JFB
University of Utah
Salt Lake City, UT 84112
USA

Prof. E. D'Hoker
Dept. of Physics
Columbia University
New York, NY 10027
USA

Dr. Michael Dine
Physics Department
City College of CUNY
New York, NY 10031
USA

Prof. P. Di Vecchia
Universität Gesamthochschule
Wuppertal, Postfach 100127
5600 Wuppertal 1
WEST GERMANY

Prof. Louise Dolan
Dept. of Physics
Rockefeller University
New York, NY 10021
USA

B. Durhuus
Kobenhavns Universitets
Matematiske Institut
Universitetsparken 5
2100 Kobenhavn O
DENMARK

Mark Evans
Dept. of Physics
The Rockefeller University
1230 York Avenue
New York, NY 10021-6399
USA

Prof. D. Fairlie
Dept. of Mathematical Sci.
University of Durham
South Road
Durham DH1 3LE
ENGLAND

Daniel Freedman
Dept. of Mathematics
MIT
Cambridge, MA 02139
USA

Prof. Daniel Friedan
Enrico Fermi Institute
5630 S. Ellis
Chicago, IL 60637
USA

Prof. Murray Gell-Mann
Dept. of Physics, 452-48
Caltech
Pasadena, CA 91125
USA

Prof. J.-L. Gervais
Lab. de Physique Theorique
Ecole Normale Superieure
24 Rue Lhomond
F-75231 Paris Cedex 05
FRANCE

Prof. Paul Ginsparg
Lyman Lab of Physics
Harvard University
Cambridge, MA 02138
USA

Prof. P. Goddard
University of Cambridge
Dept. of Appl. Math.
and Theoretical Physics
Silver Street
Cambridge CB3 9EW
ENGLAND

Jeffrey Goldstone
Dept. of Physics, 6-313
MIT
Cambridge, MA 02139
USA

Dr. Michael B. Green
Dept. of Physics
Queen Mary College
Mile End Road
London E1 4NS
ENGLAND

Prof. David Gross
Dept. of Physics
Princeton University
P.O. Box 708
Princeton, NJ 08544
USA

Bernard Grossman
Rockefeller University
1230 York Avenue
New York, NY 10021
USA

Dr. M. Günaydin
Lawrence Livermore Lab.
L454
Livermore, CA 94550
USA

Jim Hartle
Dept. of Physics
University of California
Santa Barbara, CA 93106
USA

Prof. Jeffrey Harvey
Dept. of Physics
Princeton University
Princeton, NJ 08544
USA

Prof. G. Horowitz
Dept. of Physics
University of California
Santa Barbara, CA 93106
USA

Prof. Antal Jevicki
Dept. of Physics
Brown University
Providence, RI 02912
USA

Adrian Kent
DAMTP, Silver Street
Cambridge CB3 9EW
ENGLAND

T. Killingback
Physics Department
Jadwin Hall
Princeton University
Princeton, NJ 08544
USA

In-Gyu Koh
Physics Department
Korea Advanced Institute
of Science & Technology
P.O. Box 150
Cheongryang, Seoul
KOREA

Prof. J. Lepowsky
Dept. of Mathematics
Rutgers University
New Brunswick, NJ 08903
USA

Prof. Stanley Mandelstam
Dept. of Physics
University of California
Berkeley, CA 94720
USA

Michelangelo Mangano
Physics Department
Jadwin Hall
Princeton University
Princeton, NJ 08544
USA

Neil Marcus
Lawrence Berkeley Lab.
50A - 3115
University of California
Berkeley, CA 94720
USA

Emil Martinee
Jadwin Hall
Physics Department
Princeton, NJ 08544
USA

Pawel Mazur
Inst. for Theoretical Physics
University of California
Santa Barbara, CA 93106
USA

L. Mezincescu
Dept. of Physics
University of Texas
Austin, TX 78712
USA

Prof. Gregory Moore
Dept. of Physics
Harvard University
Cambridge, MA 02138
USA

Emil Mottola
Inst. for Theoretical Physics
University of California
Santa Barbara, CA 93106
USA

Ivan Muzinich
Inst. for Theoretical Physics
University of California
Santa Barbara, CA 93106
USA

Prof. Chiara Nappi
Dept. of Physics
Princeton University
Princeton, NJ 08544
USA

Dr. Philip Nelson
Lyman Lab. of Physics
Harvard University
Cambridge, MA 02138
USA

Prof. D. Nemeshansky
Stanford Linear Accelerator Centre
P.O. Box 4349
Stanford, CA 94305
USA

B. Nilsson
TH—CERN
CH-1211 Genéve 23
SWITZERLAND

Dr. David Olive
Theoretical Physics Group
Blackett Laboratory
Imperial College
Prince Consort Road
London SW7 2BZ
ENGLAND

Dr. Burt Ovrut
Dept. of Physics E1
University of Pennsylvania
Philadelphia, PA 19104
USA

Dr. Duong H. Phong
Dept. of Mathematics
Columbia University
New York, NY 10027
USA

Prof. J. Polchinski
Dept. of Physics
University of Texas
Austin, TX 78712
USA

Z. Qiu
Enrico Fermi Institute
5640 S. Ellis Avenue
Chicago, IL 60637
USA

Dr. Ryan Rohm
Dept. of Physics 452-48
Calif. Inst. of Technology
Pasadena, CA 91125
USA

Larry Romans
Inst. for Theoretical Physics
University of California
Santa Barbara, CA 93106
USA

Augusto Sagnotti
Dept. of Physics
University of California
Berkeley, CA 94720
USA

Bunji Sakita
City College of CUNY
Dept. of Physics
New York, NY 10031
USA

Dr. Richard Scalettar
Dept. of Physics
California State Univ.
Long Beach, CA 90840
USA

Dr. N. Seiberg
Inst. for Advanced Studies
Princeton, NJ 08540
USA

Ashoke Sen
Theoretical Physics
Fermilab, MS 106
P.O. Box 500
Batavia, IL 60510
USA

Prof. J. Shapiro
Dept. of Physics
Rutgers University
P.O. Box 849
Piscataway, NJ 08854
USA

Prof. Stephen Shenker
Dept. of Physics
Enrico Fermi Institute
University of Chicago
Chicago, IL 60637
USA

Prof. W. Siegel
Dept. of Physics & Astronomy
University of Maryland
College Park, MD 20742
USA

Mark Srednicki
Dept. of Physics
University of California
Santa Barbara, CA 93106
USA

Kelly Stelle
Imperial College of Science and Technology
The Blackett Laboratory
Prince Consort Road
London SW7 2BZ
ENGLAND

Prof. A. Strominger
Inst. for Advanced Studies
Princeton, NJ 08540
USA

Prof. Robert Sugar
Dept. of Physics
University of California
Santa Barbara, CA 93106
USA

Prof. Charles Thorn
Physics Department
215 Williamson Hall
University of Florida
Gainesville, FL 32611
USA

Jennie Traschen
Inst. for Theoretical Physics
University of California
Santa Barbara, CA 93106
USA

C. Vafa
Dept. of Physics
Harvard University
Cambridge, MA 02138
USA

Luc Vinet
Lab. de Phys. Nucléaire
Université de Montréal
P.O. Box 6128
Montreal, Quebéc
H3C 3J7
CANADA

Dr. Spenta Wadia
Tata Inst. of Fundamental Research
Homi Bhabha Road
Bombay 400 005
INDIA

Nick Warner
CERN-Theory Division
1211 Geneva 23
SWITZERLAND

Frank Wilczek
Inst. for Theoretical Physics
University of California
Santa Barbara, CA 93106
USA

P. Windey
LBL - Theory Group
Bd 50A, Rm 3115
Berkeley, CA 94720
USA

Prof. Ed Witten
Dept. of Physics
Princeton University
Princeton, NJ 08540
USA

Yong-Shi Wu
Dept. of Physics
University of Utah
201 JFB
Salt Lake City, UT 84112
USA

Prof. Larry Yaffe
Dept. of Physics
Princeton University
P.O. Box 708
Princeton, NJ 08544
USA

Prof. S. Yankielowicz
Physics Department
Tel-Aviv University
Ramat-Aviv, Tel-Aviv
ISRAEL

Barton Zwiebach
Center for Theoretical Physics
MIT
Cambridge, MA 02139
USA